Size Effect in Concrete Materials and Structures

Xiuli Du · Liu Jin

Size Effect in Concrete Materials and Structures

Science Press
Beijing

Springer

Xiuli Du
Key Laboratory of Urban Security
and Disaster Engineering of Ministry
of Education, College of Architecture
and Civil Engineering
Beijing University of Technology
Beijing, China

Liu Jin
Key Laboratory of Urban Security
and Disaster Engineering of Ministry
of Education, College of Architecture
and Civil Engineering
Beijing University of Technology
Beijing, China

ISBN 978-981-33-4942-1 ISBN 978-981-33-4943-8 (eBook)
https://doi.org/10.1007/978-981-33-4943-8

Jointly published with Science Press, China
The print edition is not for sale in China Mainland. Customers from China Mainland please order the print book from: Science Press.

This Springer imprint is published by the registered company Springer Nature Singapore Pte Ltd.
The registered company address is: 152 Beach Road, #21-01/04 Gateway East, Singapore 189721, Singapore

Preface

Scaling is the most important aspect of every physical theory. In solid mechanics, the scaling problem of main interest is the effect of structure size on its nominal strength of quasi-brittle materials, e.g. concrete, rocks, sea ice and ceramics, etc. The question of the size effect has been investigated and discussed for quite a long time.

The focus of this monograph is on the size effect in concrete since it is the most widely used quasi-brittle material. Better understanding of the effects of structure size on the load capacity will improve the design of concrete structures. Concrete has highly heterogeneous and complex microstructures, and it consists of hydrated cement paste, aggregate and the interfacial transition zone between the cement paste and aggregate. The heterogeneity makes concrete materials and structures present obvious size effect in their nominal strengths.

In the last century, several classic size effect laws suitable for concrete-like quasi-brittle materials under static load were proposed. However, almost no efforts were conducted to study the size effect in concrete under dynamic load, and systematics investigations on the size effect in concrete components involving beams, columns and beam-to-column joints are also lacking.

In this monograph, the authors present their theoretical, experimental and numerical investigations on the size effect in concrete materials and structures in the last decades.

Specifically, the questions the authors intend to address in this book include:

1. What is the size effect in concrete materials under dynamic load? How the strain rate would affect the size effect?
2. What is the correlation between the component-level size effect and the material-level size effect?
3. How to build the size effect laws for concrete components?

The main contents of this book consist of following two parts.
Part I: Material Size Effect. The meso-or-microstructures of concrete (Chap. 2); introduction of the meso-scale simulation approaches (Chap. 3); the analytical and numerical modelling of static size effect of concrete (Chap. 4); the modelling of dynamic size effect of concrete (Chap. 5).
Part II: Component Size Effect. Size effect in the shear and flexure failure of concrete beams (Chap. 6); size effect in the compressive failure of concrete columns (Chap. 7); seismic performance and size effect in compressive shear failure of concrete columns (Chap. 8); size effect in shear failure of concrete beam-to-column joints (Chap. 9).

The works included in this monograph have been granted by the National Natural Science Foundation of China (51421005, 51822801) and National Key Research and Development Program of China (2018YFC1504302, 2019YFC1511003). Moreover, the publication of the book is supported by the Double First-Class Construction Fund of China. We appreciate these financial supports, and we also would like to express our gratitude and thanks to all our colleagues and students for their contributions to these works.

In addition, the second author wishes to express his thanks to his supervisor, Xiuli Du, for having excited his interest in structural mechanics and engineering. The monograph is also dedicated to Prof. Xiuli Du on the occasion of his 60th birthday.

Beijing, China

Xiuli Du
Liu Jin

Contents

Chapter 1
Introduction

Scaling is the most important aspect of every physical theory. In solid mechanics, the scaling problem of main interest is the effect of structure size on its nominal strength of quasi-brittle materials, e.g. concrete, rocks, sea ice and ceramics, etc. The question of size effect has been investigated and discussed for quite a long time. As is known, better understanding of the effects of structure size on the load capacity will improve the design of concrete structures. This chapter describes the concept of size effect, source of size effect, the available size effect laws and the scope of this book.

1.1 Concept of Size Effect

Usually, one would prefer the concept that all intrinsic properties of materials would be material constants and can be exactly measured in laboratory tests, regardless of the size of tested samples. However, later, researchers found that some results obtained experimentally show sample size dependence, for instance the strength of some materials may decrease with the increase of sample size. This is the so-called size effect that refers to the phenomenon that the mechanical performance and parameters represented by nominal strength vary with the change of the structural size of materials. It is an inherent characteristic of quasi-brittle materials, including rock, ice, concrete, various composite materials, and toughened ceramics, etc.

As an artificial quasi-brittle material, concrete is composed of aggregate particles, mortar matrix and various additives; and the nonlinearity of its global mechanical properties is mainly caused by the internal heterogeneity. As shown in Fig. 1.1, the mechanical behavior represented by nominal strength determined by tests are no longer completely repeating properties of concrete material, but parameters that depends on the geometric dimensions (or structural size).

The investigations on the size effect of materials have a long history. Speculations about the size effect can be traced back to Leonardo da Vinci (1500 s) in Renaissance period and he found there was a certain correlation between strength and length of

X. Du and L. Jin, *Size Effect in Concrete Materials and Structures*,
https://doi.org/10.1007/978-981-33-4943-8_1

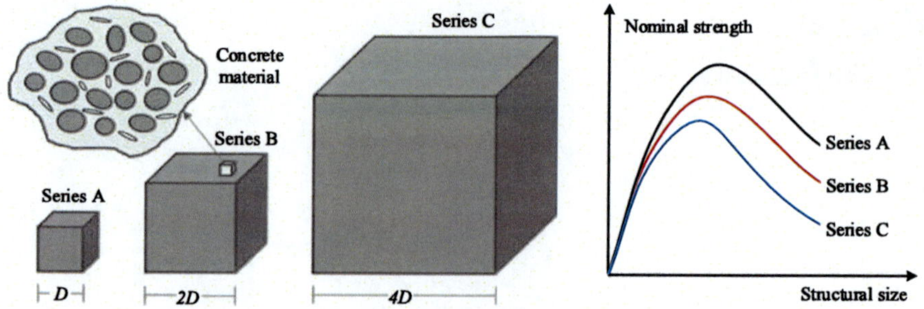

Fig. 1.1 Size effect on mechanical behavior of concrete

rope. From then on, some scholars have discussed the phenomenon that the strength of the material changes with the size (Galileo Galilei 1638; Mariotte 1686). Not much progress was achieved until Griffith (1921) introduced the idea of fracture mechanics to the study of material size effect. However, in Griffith's view, the flaws or cracks at the moment of failure were still only microscopic, and their random distribution controlled the macroscopic strength of the material but did not invalidate the concept of strength. Thus, Griffith discovered the physical basis of statistical idea but not a new kind of size effect (Bažant 1999). The statistical theory of size effect began to emerge after Peirce (1926), formulated the weakest-link model for a chain and introduced the extreme value statistics. Weibull (1939, 1951) proposed the famous Weibull distribution based on the weakest-link model, and basically completed the basic framework of statistical size effect theory. Subsequently, a large quantity of researches conducted by many scholars (Mihashi and Izumi 1977; Zech and Wittmann 1978; Carpinteri 1989; Kittl and Diaz 1988, 1990; etc.) mainly focused on the validation, modification, and application of Weibull theory. According to the characteristics of quasi-brittle materials, Kaplan (1961, 1963) first applied fracture mechanics theory to the study of concrete and measured the fracture toughness of concrete using a three-point bending beams. Kesler et al. (1972) and Walsh (1972, 1976) found that the classic linear elastic fracture mechanics (LEFM) (Knott 1973) was not suitable to describe the fracture of quasi-brittle materials such as concrete. Therefore, many nonlinear fracture mechanics theories for concrete materials have emerged, and the most representative ones are the equivalent elastic crack model (Jenq and Shah 1985; Karihaloo and Nallathambi 1989) and the cohesive crack model (Hillerborg et al. 1976; Hillerborg 1985a, 1985b). Fictitious crack model proposed by Hillerborg et al. (1976) is one of the famous cohesive crack models that can well describe the fracture behavior of concrete materials, which brought the fracture mechanics of concrete into a new stage of development. In the 1980s, Bažant (1984) developed a crack band model and proposed a size effect law (SEL) for quasi-brittle materials based on nonlinear fracture mechanics. This SEL describes the size effect behavior of the nominal strength of quasi-brittle materials after stable crack propagation, and a size effect theory based on the energy release criterion is preliminarily established. Carpinteri (1994a, 1994b) studied the fractal characteristics of cracks in different materials, and proposed the multi-fractal scaling law (Carpinteri and Ferro 1994), which provides a new idea for the study of size effect of concrete. In the twenty-first century, Duan et al. (2006) established the boundary size effect theory based on the Hu-Duan boundary effect model (Hu and Wittmann 2000), considering the effect of cracks of different sizes on the nominal strength of fracture failure concrete. Hoover and Bažant (2014) have developed the universal-size effect law for concrete in a relatively large size range through the experimental investigation on three-point notched bending beams. Recently, with the improvement of experimental conditions and the development of high-performance computers, the investigations on the size effect of concrete is no longer limited to the level of concrete materials, but to continue to explore the level of reinforced concrete members (such as reinforced concrete members, steel fiber reinforced concrete members FRP concrete members, etc.).

1.2 Source of Size Effect

At present, the researches on the size effect of concrete are mainly attributed to two important levels, namely the size effect of concrete materials and the size effect of concrete members. As for the material level of size effect, this is a complex material science issue, which is mainly affected by aggregate size, aggregate distribution, mix proportion, concrete strength grade, initial defects within concrete, etc. The analysis on the source of size effect of concrete materials should focus on the performance of material internal composition.

There are several different factors that may cause the size effect on the nominal strength of concrete materials (Bažant and Planas 1997): (1) Boundary layer effect (or the wall effect). Due to the influence of concrete pouring, large aggregates tend to be distributed in the central area of the specimen while small aggregates tend to be distributed in the boundary layer of the specimen, resulting in a boundary layer independent of the specimen size. The thickness of the boundary layer depends on The particle size of the largest aggregate. Therefore, the surface layer, whose thickness is independent of the structure size and is of the same order of magnitude as the maximum aggregate size, has different properties. The size effect is due to the fact that in a smaller specimen, the boundary layer occupies a large portion of the cross-section, while in a large specimen, it occupies a small part of the cross-section. In most cases, the size effect formed by this factor does not seem to be obvious. The second type of boundary layer effect is due to the difference in elastic properties between the boundary layer and the central area of the specimen. The normal stress parallel to the boundary leads to transverse stresses in the interior, and there is no such stress on the boundary of the specimen. A third type of boundary layer size effect arises from the Poisson effect. The boundary of the specimen may be in a plane stress state, while the interior of the specimen may be in a plane strain state. (2) Diffusion phenomena, such as heat conduction or pore water transfer. This size effect is due to the fact that the diffusion half-times (i.e., half-times of cooling, heating, drying, etc.) are proportional to the square of the structural size. At the same time, the diffusion process changes the material properties and produces residual stresses which in turn produce inelastic strains and cracking. Because of the difference in diffusion time between large and small specimens, the degrees of cracking in specimens with different sizes are different, causing this type of size effect. (3) Hydration heat or other phenomena associated with chemical reactions. This effect is related to the previous one in that the half-time of dissipation of the hydration heat produced in a concrete specimen is proportional to the square or the thickness (size) of the specimen. Therefore, the larger the concrete specimen size, the higher the internal temperature. The uneven temperature distribution inside the specimen will cause cracking and change the material properties, which will lead to this type of size effect.

In practical tests, these above three types of explanations on the source of size effect could be eliminated if the specimens with different structural sizes are geometrically similar. The first source of size effect can be negligible for sufficiently thick specimens. The second source of size effect can be also negligible if the specimens are sealed and is at constant temperature. The third source of size effect is significant only for very massive structures. In addition to these three types of explanations on the source of size effect, there are some other explanations on the source of size effect by many scholars based on different theoretical frameworks, forming different size effect laws, such as statistical size effect law originated by Weibull (1939), fracture mechanics size effect law proposed by Bažant (1984), and multi-fractal scaling law established by Carpinteri (1994a). These classical size effect laws would be briefly introduced in the next section.

As for the size effect of concrete members, actually, it is a structural mechanics issue, which is mainly affected by two aspects: (1) heterogeneity and mechanical nonlinearity of concrete materials; and (2) the interaction between different materials. Among these two aspects, the interaction between different materials is more complex and it has various types, including the interaction between steel bar and concrete, the interaction between FRP and concrete, the interaction between steel tube and concrete, etc. According to different research objects, these two aspects also have different contributions to the influence of size effect behavior of concrete member. However, compared with the studies of material size effect, the current investigations on the size effect of concrete members are still incomplete, and the size effect on structural strength of concrete members still needs to be investigated and discussed in-depth.

1.3 Size Effect Laws

Based on the above two important levels on the size effect of concrete, combined with the type of load suffered by concrete, the size effect on nominal strength can be studied separately from the following three parts: static size effect of concrete material, dynamic size effect of concrete material and size effect of concrete members.

1.3.1 Static Size Effect of Concrete Materials

The size effect on nominal strength of concrete materials under static load is the basis issue of size effect of concrete. Since Gonnerman (1925) noticed the size effect on concrete strength, a large quantity of researchers carried out physical experimental tests on concrete material specimens for trying to explore the relationship between nominal strength and structural size of concrete materials under static load, including three-point bending beam tests, notched specimen tensile tests, Brazilian disc splitting-tensile tests, cubic specimen uniaxial compressive tests, etc. According to the experimental results, some classic size effect theories based on different theoretical frameworks have been proposed.

Herein, we can define the nominal strength of concrete materials under static load as $\sigma_{\text{material}}^{\text{static}}$, and it can be expressed as

$$\sigma_{\text{material}}^{\text{static}} = \sigma_{\text{con}}(D, \ldots) \tag{1.1}$$

where σ_{con} is the functions depended on the structural size D of concrete materials and it is closely related to the internal characteristics (or fracture properties) of concrete. In reality, Eq. (1.1) is the fundamental framework of size effect laws of concrete materials under static load.

According to the time sequence of establishment and development, Fig. 1.2 lists several size effect laws widely accepted in static size effect of concrete materials, including Weibull statistical size effect law (Weibull 1939, 1951), fracture mechanics size effect law (Bažant 1984), multi-fractal scaling law (Carpinteri and Ferro 1994), boundary size effect law (Duan et al. 2006), universal size effect law (Hoover and Bažant 2014), etc. Here in the following, these five kinds of size effect theories suitable for quasi-brittle materials under static load would be briefly introduced.

1. *Weibull statistical size effect law*

Weibull may be one of the pioneers to explore the statistical essence of size effect (Weibull 1951). According to the weakest link model, the theory based on the randomness of material strength was originally established by Weibull (1939, 1951). It is assumed that the nominal strength of concrete was determined by the material element with the weakest strength. As the structural size gradually increases, the probability of encountering material elements with weak-strength also gradually increases, and the strength of concrete decreases accordingly. The statistical size effect law of Weibull can be expressed as:

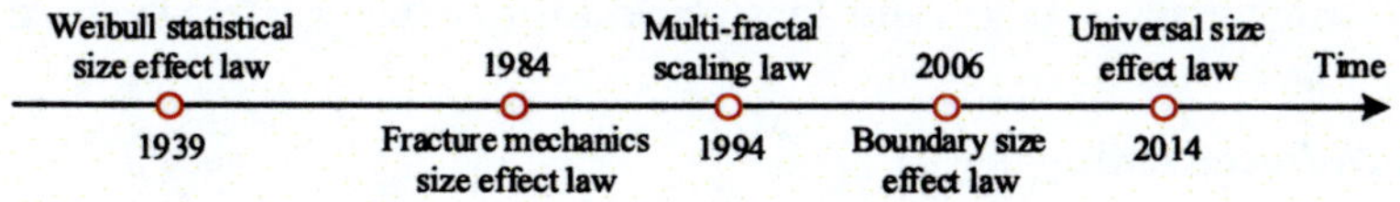

Fig. 1.2 Several typical size effect laws for concrete material

$$\bar{\sigma} = \int_{-\infty}^{\infty} \sigma d P_f(\sigma) \tag{1.2}$$

$$P_f(\sigma) = 1 - \exp[-c(\sigma)V] \tag{1.3}$$

$$c(\sigma) = \frac{1}{V_0} \frac{\sigma - \sigma_u}{\sigma_0}^m \tag{1.4}$$

where $\bar{\sigma}$ is the mean nominal strength before failure; σ is the strength before failure of material element; $P_f(\sigma)$ is the failure probability of a structure that fails; V is the volume of structure; $c(\sigma)$ is a function giving the spatial concentration of failure probability of the material; V_0 is the reference volume understood as the volume of specimens on which $c(\sigma)$ was measured; σ_u is the strength threshold, which may usually be taken as zero; m is the Weibull modulus, usually between 5 and 50; and σ_0 is a scale parameter.

Provided Weibull modulus tends to infinity, $m \to \infty$, namely the material tends to be homogeneous, the strength $\bar{\sigma}$ would tend to be a constant. Or inversely, the more heterogeneous (smaller m) the material is, the material would demonstrate more pronounced size effect.

Weibull statistical size effect law has three basic assumptions: (1) when an element of the material reaches its ultimate strength, the material fails; (2) failure of each element is independent; and (3) the ultimate strength of the elements are randomly distributed.

The simple relation of size effect based on Weibull proposition of the weakest link is justified in some tests, for instance the Brazilian disk tests of granodiorite and limestone. However, as far as quasi-brittle structures are concerned, applications of the classical Weibull theory face some objections (Bažant 1999). One is that every structure is mathematically equivalent to a uniaxial stressed bar, which means that no information on the structural geometry and failure mechanism is taken into account. Secondly, the statistical size effect law does not take into account the difference in load type (or force mode), but uses a single volume effect to summarize the size effect under various force modes, which is obviously unreasonable. In addition, the energy release due to stress redistributions caused by macroscopic failure process zone, or stable crack growth before ultimate strength, gives rise to a deterministic size effect which is ignored. Therefore, the Weibull theory is valid only if the structure fails as soon as a microscopic crack becomes macroscopic. In view of these limitations, among concrete structures Weibull theory appears applicable to some extremely thick plain (unreinforced) structures, e.g. the flexure of an arch dam acting as a horizontal beam (but not for vertical bending or arch dams nor gravity dams, because large vertical compressive stresses cause long cracks to grow stably before the maximum load).

2. *Multi-fractal scaling law*

Fractal theory was established by Mandelbrot (1982), Mandelbrot et al. (1984) and Schwarzenberger and Falconer (1990) in the 1980s, and it has been widely applied in many advanced technical fields. Unlike the characteristic that the physical dimension is an integer, the fractal dimension of an object is a non-integer in an abstract mathematical domain (fractal space). Concrete has a certain level of similar characteristics in structural composition and its internal crack propagation has obvious fractal characteristics. Therefore, the fractal method can be utilized to quantitatively describe the fracture failure behavior of concrete.

Carpinteri (1994a, b) pointed out that the difference in crack fractal characteristics under different observation scales is the source of the size effect. That is to say, the reason for the size effect of materials is that the arrangement structure of the same micro-crack has different fractal dimensions d_f at different observation scales. When the observation scale is small, the fractal dimension d_f is obviously less than 1.0. When the observation scale is large, the fractal dimension d_f is very close to 1.0. In the failure analysis of relatively small structures, small-scale fractal structures and fractal dimensions play a controlling role. Conversely, when a larger structure fails, it is the fractal structure and fractal dimension obtained from the large scale that determine it. The effect of heterogeneity on the mechanical properties of materials depends on the ratio of the size of the largest heterogeneous phase (such as aggregate particle) to the macro structural size of specimens. For large structures (relative to the microstructure size), the microstructure can be regarded as homogeneous. Thus, when the size tends to infinity, the size effect disappears and the strength tends to a certain value. On the other hand, for small specimens, the influence of microstructure heterogeneity is more prominent, resulting in an increase in strength with the decreasing size. In an ideal state, the size tends to zero and the strength tends to be infinite. Based on the above analysis, Carpinteri established the mathematical relationship between the fracture characteristics of materials and the fractal dimension and proposed the multi-fractal scaling law (Carpinteri and Ferro 1994) based on fractal theory. The multi-fractal scaling law of nominal strength can be expressed as

$$\sigma_{\mathrm{Nu}} = f_{\mathrm{t}}\sqrt{1 + \frac{l_{\mathrm{ch}}}{D}} = \sqrt{A_1 + \frac{A_2}{D}} \tag{1.5}$$

where σ_{Nu} is the nominal strength of materials; f_{t} is minimal nominal strength of materials; l_{ch} is characteristic length of materials; D is structural size; and A_1 and A_2 are constants dependent on the structural size. It should be noted that the multi-fractal scaling law is effective only when the fractal dimension can be considered as a constant. The bi-logarithmic curve of the multi-fractal scale law for nominal strength is given in Fig. 1.3. In addition, the multi-fractal scaling law is based on the following basic assumptions: (1) in a sufficiently small size range, the failure is caused by a single fractal crack propagation; and (2) fractal fracture energy is a material constant that can define energy dissipation.

3. *Fracture mechanics size effect law*

The fracture mechanics size effect law based on the release of fracture energy was proposed by Bažant (1984) according to the crack band model (Bažant and Oh 1983).

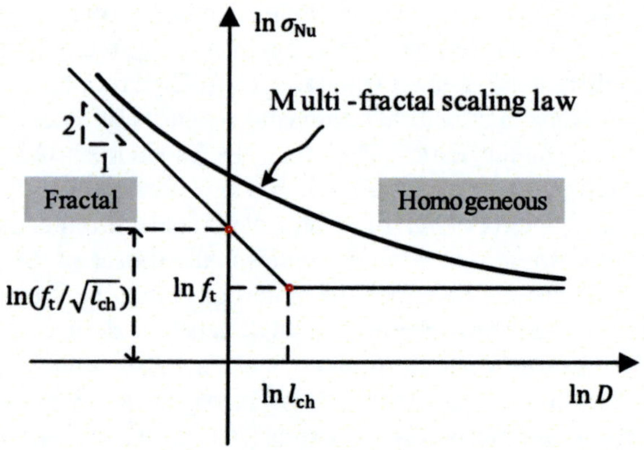

Fig. 1.3 Bi-logarithmic curve of the multi-fractal scale law for nominal strength

For the size effect of the so-called quasi-brittle materials and structures, Bažant and Chen (1997) have given an excellent comprehensive review with 377 references. They discussed three main types of size effect: the statistical size effect due to randomness of strength, the energy release size effect, and the possible size effect due to fractality of fracture or microcracks.

Bažant (1984) pointed out that the size effect of concrete materials is caused by the steady growth of a large crack or a large fracture process area containing micro cracks before concrete materials reaching the maximum load, especially due to the development of large crack or micro crack areas caused by stress redistribution and stored energy release. In other words, the size effect of concrete materials is caused by the dissipation of strain energy during macro crack propagation. Specifically, Bažant pointed out that energy balance can be achieved only if the nominal strength is considered to decrease with structural size (see Bažant 2004a, b; Bažant and Chen 1997)

It should be noted that the fracture mechanics size effect law has several basic assumptions:

(1) The propagation of crack band requires that the energy provided per unit area of the fracture plane is approximately constant;
(2) The energy released due to the propagation of crack or crack band is a function that can be described by fracture length or the size of fracture process zone;
(3) Failure modes of geometrically similar specimens with different structural sizes are the same;
(4) The specimen does not fail at the beginning of the crack.

Based on the above basic assumptions, the size effect law of concrete materials can be expressed as

$$\sigma_{Nu} = \frac{Bf_t'}{\sqrt{1 + D/D_0}} \tag{1.6}$$

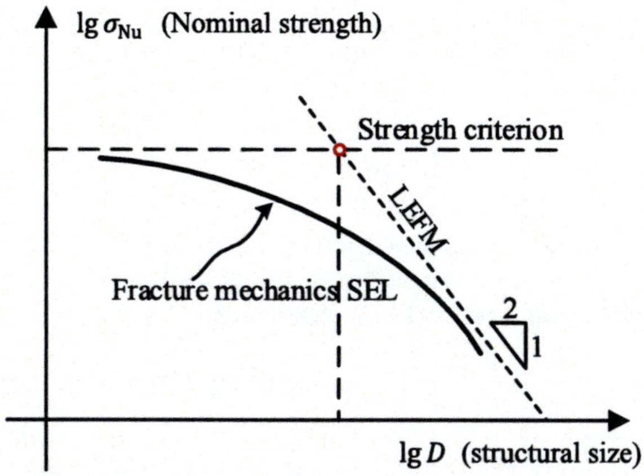

Fig. 1.4 Bi-logarithmic curve of the fracture mechanics size effect law for nominal strength

where σ_{Nu} is the nominal strength of materials; f_t' is the tensile strength of specimens with standard size; D is the structural size of specimens; B is the dimensionless geometry-dependent parameter; and D_0 is the size dependent parameter called transitional size. The parameter β can be defined as the ratio of D to D_0, which is called the brittleness number. The case $\beta \to \infty$ represents a perfectly brittle behavior, and the case $\beta \to 0$ represents a perfectly non-brittle (plastic, ductile) behavior.

The bi-logarithmic curve of fracture mechanics size effect law for nominal strength is also given in Fig. 1.4. It can be seen that Eq. (1.6) is actually an asymptotic curve between the plasticity theory (the horizontal line, strength criterion) and the theory of linear elastic fracture mechanics (LEFM, the straight line of slope −1/2), which characterizes the change of nominal strength of concrete materials with the structural size.

This SEL is most widely utilized due to its simplicity and versatility, e.g. data from specimens under tensile and bending (or even compression for pure numerical fitting) can all be fitted, and the two empirical parameters do not have to be linked specifically to the tensile strength or fracture toughness.

According to Bažant, "Smaller structures fail in a ductile manner, which usually involves distributed cracking with strain-softening leading to an energetic size effect, i.e., the decrease of the nominal strength of structures with increasing structure size. A structure far larger than the fracture process zone (FPZ) fails in an almost perfectly brittle manner and, if the failure occurs right at the crack initiation, the failure load is governed by the statistically weakest point in the structure, which gives size to the statistical size effect".

4. *Boundary size effect law*

The boundary effect model originally proposed by Hu and Wittmann (2000) is to deal with quasi-brittle fracture (or non-LEFM fracture) outside the LEFM area specified by the ASTM standards. Subsequently, Duan et al. (2006) established boundary size effect law based on the boundary effect model. The boundary size effect law actually refers to the influence of cracks with different sizes on the nominal strength σ_{Nu} of fractured concrete specimens. It can be expressed as

$$\sigma_{\mathrm{Nu}} = \frac{f_t'}{\sqrt{1 + a_0/a_{\mathrm{FPZ}}}} \tag{1.7}$$

where f_t' is tensile strength of a_0 is the size of initial crack or initial notch. a_{FPZ} is a material constant independent of specimen geometry, and it is proportional to Hillerborg et al.'s (1976) characteristic length l_{ch}

$$a_{\mathrm{FPZ}} = l_{\mathrm{ch}}/(1.12^2\pi) \approx 0.25 l_{\mathrm{ch}} \tag{1.8}$$

$$l_{\mathrm{ch}} = E \cdot G_F/(f_t')^2 \tag{1.9}$$

where G_F is the specific fracture energy based on the fictitious crack model of Hillerborg et al. (1976) and E is the Young's modulus.

Figure 1.5 shows the asymptotic curve of boundary size effect law for nominal strength. It can be seen that Eq. (1.7) is also the asymptotic curve between the horizontal straight line with height of $\lg f_{\mathrm{t}}$ and the theory of LEFM (the straight line of slope −1/2). That is to say, the theoretical curve of boundary size effect law is similar to the theoretical curve determined by the fracture mechanics size effect law of Bažant (1984). As the crack size $a_0 \to 0$, the maximum tensile stress trends to the constant strength of f_t'. As the crack size a_0 is large enough, the nominal strength approaches the descending straight line determined by LEFM.

Subsequently, Hu and Duan (2007, 2008, 2010) modified Eq. (1.7) so that it can simultaneously consider the influence of crack size and structural size on the

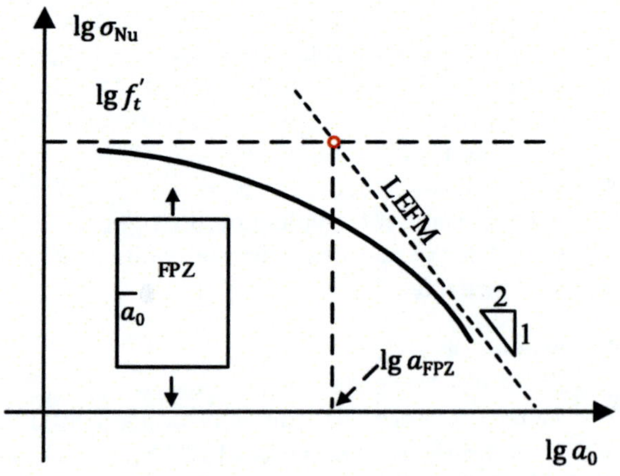

Fig. 1.5 Bi-logarithmic curve of the boundary size effect law

nominal strength of geometrically similar concrete specimens. Actually, the difference between the boundary size effect law and the fracture mechanics size effect law lies in the definition of fracture energy. The fracture energy utilized in the fracture mechanics size effect law is initial fracture energy, which is the area under the initial tangent of the cohesive softening stress-separation curve; while the fracture energy utilized in the boundary size effect law is total fracture energy, which is the total area under the stress-separation curve.

5. *Universal size effect law*

Concrete is a typical quasi-brittle material, and the size effect theories based on the nonlinear fracture model of concrete are divided into two types: Type I and Type II size effect. The Type I size effect occurs in structures that have no notches nor pre-existing large cracks and fail immediately after the initiation of a macro-crack from a smooth surface. In this type the macro-crack at maximum is approximately of the size of the FPZ. The Type II size effect occurs when either a deep notch or a long stress-free crack has grown in a stable manner before reaching to the maximum load. In this type the notch or pre-existing macro-crack is much larger than the FPZ. In practice, there are also structures where the crack at maximum load is neither larger than the FPZ nor negligible. Therefore, the transition of the nominal strength between Type I and Type II size effects should be described analytically. For this issue, based on the previous theoretical researches (Bažant 1996; Bažant and Li 1996; Bažant and Yu 2009), Hoover et al. conducted comprehensive fracture tests for three-point bend concrete beams (Hoover et al. 2013; Hoover and Bažant 2013) and proposed the universal size effect law (Hoover and Bažant 2014). Considering the statistical size effect, the universal size effect law has two forms: the non-statistical and statistical universal size effect laws. They can be stated respectively as follows

$$\sigma_{\mathrm{Nu}} = \left[\frac{E'G_f}{g_0 D + (1-\lambda)c_f g_0' + \lambda E' G_f / f_{r\infty}^2}\right]^{1/2} \left(1 + \frac{r\lambda D_b}{\bar{D} + l_p}\right)^{1/r} \tag{1.10a}$$

$$\sigma_{\mathrm{Nu}} = \left[\frac{E'G_f}{g_0 D + (1-\lambda)c_f g_0' + \lambda E' G_f / f_{r\infty}^2}\right]^{1/2} \left[\left(\frac{r l_s}{D + l_s}\right)^{rn/m} + \frac{r\lambda D_b}{\bar{D} + l_p}\right]^{1/r} \tag{1.10b}$$

where G_f is initial fracture energy; c_f is characteristic length; λ is the transition parameter; E' is Young's modulus; $f_{r\infty}$ is nominal strength for very large structures, assuming no statistical size effect; r is empirical constant; D_b is depth of the boundary layer of cracking (roughly equal to the FPZ size); l_p is material characteristic length, which is related to the maximum aggregate size; l_s is statistical characteristic length; m is Weibull modulus; n is number of spatial dimensions of fracture scaling; other parameters and the detailed introduction can refer to the literature (Hoover and Bažant 2014).

The universal size effect law describes the dependence of nominal strength of specimen or structure on both its size and the crack (or notch) length, over the entire range of interest, and exhibits the correct small-size and large-size asymptotic properties as required by the cohesive crack model (or crack band model).

1.3.2 Dynamic Size Effect of Concrete Materials

Concrete, as a kind of rate-dependent material, the apparent strength and the deformation characteristics of concrete under dynamic loads are significantly different from the ones under static loads. As loading strain rate changes, the macro mechanical properties and the failure behavior of concrete vary significantly.

As presented in Chapter 1.3.1, lots of efforts have been focused on size effect of concrete under static loading. However, less efforts have been conducted to study the size effect of concrete under dynamic load. Accordingly, what is the size effect in concrete materials under dynamic load, and how the strain rate would affect the size effect turns to be a challenging topic.

Herein, we can also define the nominal strength of concrete materials under dynamic loads as $\sigma_{\text{material}}^{\text{dynamic}}$. The relationship between $\sigma_{\text{material}}^{\text{dynamic}}$ and $\sigma_{\text{material}}^{\text{static}}$ is expressed as

$$\sigma_{\text{material}}^{\text{dynamic}} = \sigma_{\text{material}}^{\text{static}} \cdot \mathrm{K}_1 \cdot \mathrm{K}_2 \cdot \ldots \cdot \mathrm{K}_n \tag{1.11}$$

where the $\mathrm{K}_1, \mathrm{K}_2, \ldots, \mathrm{K}_n$ are coefficients or functions related to dynamic loads (i.e. strain rate). Actually, under dynamic load, strain rate presents significant influences on enhancement of strength and the trend of size effect. The systematic investigations on dynamic size effect on nominal strength of concrete materials will be presented and discussed in detail in Chap. 5 of this book.

1.3.3 Size Effect of Concrete Members

It is known that for concrete members (including beams, columns, and beam-to-column joints) the mechanical properties and failure mechanisms are much more complicated than those of concrete materials. In addition to the basic contribution of the concrete part, the nominal strength of the concrete members also includes the contribution of other materials, such as steel rebar, steel fiber, steel tube and FRP, etc. For instance, the nominal shear strength of concrete beams with stirrups is composed by two parts, i.e. concrete part and the contribution by stirrups.

The physical characteristics and distribution forms of these composed materials have an important influence on the mechanical performance and the corresponding size effect of concrete members. Moreover, there are many forms of external loads that concrete members may be subjected. The axial compression ratio, shear span ratio and loading eccentricity are also important factors that cannot be ignored. However, the available typical size effect laws suitable for concrete materials are often utilized to describe the size effect of concrete members. They can only present the unique factor (i.e., structural size) on the nominal strength, and they cannot describe the influence of other important factors. How to consider the other important influencing factors into the size effect laws for concrete members is also a challenging topic of this book.

Similarly, the size effect law on nominal strength of concrete members can be also developed considering different influencing factors based on the size effect laws for concrete materials. Herein, we can define the nominal strength of concrete members as σ_{member}, the relationship between σ_{member} and $\sigma_{material}$ can be expressed as:

$$\sigma_{member} = \sigma_{material} \cdot \Omega_1 \cdot \Omega_2 \cdot \ldots \cdot \Omega_n \tag{1.12}$$

where the $\Omega_1, \Omega_2, \ldots, \Omega_n$ are coefficients or functions related to different influencing factors, including longitudinal reinforcement ratio, stirrup ratio, steel ratio, fiber ratio, FRP ratio, axial compression ratio, shear span ratio and loading eccentricity, etc.

Equation (1.12) provides the theory framework of size effect on nominal strength of concrete members. The first part of Eq. (1.12) (i.e. $\sigma_{material}$) can reflect the behavior that the strength of concrete members changes with the structural size, which is similar with the static size effect of concrete materials. The second part of Eq. (1.12) (i.e. $\Omega_1, \Omega_2, \ldots, \Omega_n$) can reflect the influence of related factors on the nominal strength and the corresponding size effect of concrete members. However, this is not a simple linear combination and different related factors also affect each other. Therefore, the systematic investigations on the size effect law on nominal strength of concrete members involving beams, columns, and beam-to-column joints will be detailed discussed in Chaps. 6~9 of this book.

1.4 Scope

In this book, the authors present systematically their experimental, numerical and theoretical researches on the size effect in concrete materials and structures in recent years. This book is organized into nine chapters and the main contents are introduced briefly as follows:

Part I: Material size effect

Chapter 2 presents the meso-structure of concrete materials and gives a brief introduction of three components of concrete, namely aggregate particle, mortar matrix and the interfacial transitional zone (ITZ) between the aggregate particle and mortar matrix. The knowledge of the structure of the individual components of concrete and their relationship to each other is useful for exercising some investigations on the properties of concrete.

Chapter 3 introduces one of the main methodologies adopted in this book to study the size effect at both material and component levels, namely, the meso-scale simulation approach. After the review of the available mesoscopic simulation models, the geometric model, material model including strain rate effect and interaction model in the meso-scale modelling approach are described. Furthermore, the reasonability and efficiency of the meso-scale approach in simulating the behavior of concrete materials and RC structural components (including beams, columns and joints, etc.) are validated.

Chapter 4 mainly focuses on static size effect on nominal strength of concrete materials. The mechanism analysis of size effect on nominal tensile strength is conducted from theoretical perspective firstly. The static size effect on tensile strength, flexural strength and compressive strength of concrete materials is also investigated from both experimental and numerical perspectives. Moreover, the influences of material parameters on size effect of concrete materials are studied and a modified size effect law considering the maximum aggregate size is presented from theoretical perspective.

Chapter 5 conducts investigations on dynamic size effect on nominal strength of concrete materials. Applying the meso-scale simulation approach, the size effect on spitting-tensile/tensile/compressive strength of concrete materials under dynamic loads is simulated and analyzed. The influences of meso-structure and initial loads on the dynamic size effect are also discussed. In addition, the static-dynamic unified size effect law (SD-SEL) considering the contribution of strain rate for predicting the nominal strength of concrete materials is introduced.

Part II: Component size effect

Chapter 6 shows the shear failure and the corresponding size effect of concrete beams without stirrups, beams with stirrups, CFRP-wrapped concrete beams and some lightweight aggregate concrete beams from experimental and numerical perspectives. The size effect law reflecting various influencing factors (shear-span ratio, fiber ratio, CFRP ratios, etc.) for evaluating the shear strength of concrete beams is introduced, which can also describe the difference between component size effect and material size effect. In addition, the size effect on flexural failure of RC beams is also provided.

Chapter 7 mainly focuses on the compressive failure and size effect behavior of ordinary and high strength RC columns, stirrups-confined concrete columns, FRP-confined concrete columns, and normal/lightweight aggregate concrete filled steel tube (CFST) columns. Based on the classical size effect law of Bažant, the size effect law considering various parameters (stirrup ratio, constraint ratio, etc.) for describing the compressive strength of columns is presented.

Chapter 8 is devoted to explore the seismic performances and the corresponding size effect of RC columns, CFST columns and FRP-confined concrete columns. In addition, the corresponding size effect law considering various influencing factors (shear-span ratio, constraint coefficient, etc.) for describing the shear strength of columns is also provided.

Chapter 9 discusses the shear failure behavior of RC beam-to-column joints with different stirrup ratios and axial compression ratios from experimental and numerical perspectives. Based on the results of experiments and simulations, a semi-empirical and semi-theoretical formula is established, and it can quantitatively reflect the influence of stirrup ratio and axial compression ratio on size effect of shear strength of RC joints.

References

Bažant ZP (1984) Size effect in blunt fracture: concrete, rock. Metal ASCE J Eng Mech 110(4):518–535

Bažant ZP (1996) Size effect aspects of measurement of fracture characteristics of quasibrittle material. Adv Cem Mater 4(3–4):128–137

Bažant ZP, Chen EP (1997) Scaling of structural failure. Appl Mech Reviews 50(10):593–627

Bažant ZP (1999) Size effect on structural strength: a review. Arch Appl Mech 69(9):703–725

Bažant ZP, Li ZZ (1996) Zero-brittleness size-effect method for one-size fracture test of concrete. ASCE J Eng Mech 122(5):458–468

Bažant ZP (2004a) Scaling theory for quasibrittle structural failure. Proc Natl Acad Sci USA 101(37):13400–13407

Bažant ZP (2004b) Probability distribution of energetic-statistical size effect in quasibrittle fracture. Probabilistic Eng Mech 19(4):307–319

Bažant ZP, Oh BH (1983) Crack band theory for fracture of concrete. Mater Struct 16(3):155–177

Bažant Z, Planas JP (1997) Fracture and size effect in concrete and other quasibrittle materials. CRC Press, USA

Bažant ZP, Yu Q (2009) Universal size effect law and effect of crack depth on quasi-brittle structure strength. ASCE J Eng Mech 135(2):78–84

Carpinteri A (1989) Decrease of apparent tensile and bending strength with specimen size: two different explanations based on fracture mechanics. Int J Solids Struct 25(4):407–429

Carpinteri A (1994a) Fractal nature of material microstructure and size effects on apparent mechanical properties. Mech Mater 18(2):89–101

Carpinteri A (1994b) Scaling laws and renormalization groups for strength and toughness of disordered materials. Int J Solids Struct 31(3):291–302

Carpinteri A, Ferro G (1994) Size effects on tensile fracture properties: a unified explanation based on disorder and fractality of concrete microstructure. Mater Struct 27(10):563–571

Duan K, Hu X, Wittmann FH (2006) Scaling of quasi-brittle fracture: Boundary and size effect. Mech Mater 38(s1–2):128–141

Galileo Galilei L (1638) Discorsi i Demostrazioni Matematiche intono a due Nuove Scienze, Elsevirii, Leiden; English transl by Weston T, London 178–181

Gonnermann HF (1925) Effect of size and shape of test specimen on compressive strength of concrete. Proc ASTM 25:237–250

Griffith AA (1921) The phenomena of rupture and flow in solids. Phil Trans 221A:179–180

Hillerborg A (1985a) Results of three comparative test series for determining the fracture energy G_F of concrete. Mater Struct 18(5):407–413

Hillerborg A (1985b) The theoretical basis of a method to determine the fracture energy G_F of concrete. Mater Struct 18(4):291–296

Hillerborg A, Modéer M, Petersson PE (1976) Analysis of crack formation and crack growth in concrete by means of fracture mechanics and finite elements. Cem Concr Res 6(6):773–781

Hoover CG, Bažant ZP (2013) Comprehensive concrete fracture tests: Size effects of types 1 & 2, crack length effect and postpeak. Eng Fract Mech 110:281–289

Hoover CG, Bažant ZP (2014) Universal size-shape effect law based on comprehensive concrete fracture tests. ASCE J Eng Mech 140(3):473–479

Hoover CG, Bažant ZP, Vorel J, Wendnera R, Hublera MH (2013) Comprehensive concrete fracture tests: Description and results. Eng Fract Mech 114:92–103
Hu X, Duan K (2008) Size effect and quasi-brittle fracture: the role of FPZ. Inter J Fract 154(1):3–14
Hu X, Duan K (2007) Size effect: influence of proximity of fracture process zone to specimen boundary. Eng Fract Mech 74(7):1093–1100
Hu X, Duan K (2010) Mechanism behind the size effect phenomenon. J Eng Mech 136(1):60–68
Hu X, Wittmann F (2000) Size effect on toughness induced by crack close to free surface. Eng Fract Mech 65(2–3):209–221
Jenq Y, Shah SP (1985) Two parameter fracture model for concrete. J Eng Mech 111(10):1227–1241
Kaplan MF (1961) Crack propagation and the fracture of concrete. ACI J Proc 58(11):591–610
Kaplan MF (1963) Strains and stresses of concrete at initiation of cracking and near failure. ACI J Proc 60(7):853–880
Karihaloo BL, Nallathambi P (1989) Fracture toughness of plain concrete from three-point bend specimens. Mater Struct 22(3):185–193
Kesler CE, Naus DJ, Lott JL (1972) Fracture mechanics: its applicability to concrete. Proceedings of the Society of Materials Science Conference on the Mechanical Behavior of Materials, pp 113–124
Kittl P, Diaz G (1990) Size effect on fracture strength in the probabilistic strength of materials. Reliab Eng Syst Safety 28(1):9–21
Kittl P, Diaz G (1988) Weibull's fracture statistics, or probabilistic strength of materials: state of the art. Res Mech 24:99–207
Knott JF (1973) In fundamentals of fracture mechanics. Butterworths, London
Léonardo da Vinci (1500 s) Notebooks of Leonardo da Vinci. Edward McCurdy, London, 546; and Les Manuscrits de Léonardo de Vinci, transl in French by Ravaisson-Mollien C, Inst de France, vol 3
Mandelbrot BB (1982) The fractal geometry of nature. W.H. Freemann and Company, New York
Mandelbrot BB, Passoja DE, Paullay AJ (1984) Fractal character of fracture surfaces of metals. Nature 308:721–722
Mariotte E (1686) Traité du movement des eaux, (posthumously), de la Hire M (ed); English transl by Desvaguliers JT, London, 249; also Mariotte's collected works, 2nd edn, The Hague
Mihashi H, Izumi M (1977) Stochastic theory for concrete fracture. Cem Concrete Res 7:411–421
Peirce FT (1926) Weakest link theorems on the strength of long and composite specimens. J Textile Inst Trans 17(7):355–368
Schwarzenberger R, Falconer K (1990) Fractal geometry: mathematical foundations and applications. The Mathematical Gazette 74(469):316
Walsh PF (1976) Crack initiation in plain concrete. Mag Concrete Res 28(94):37–41
Walsh PF (1972) Fracture of plain concrete. Indian Concrete J 46(11):1–3
Weibull W (1951) A statistical distribution function of wide applicability. ASME J Appl Mech 18(2):293–297
Weibull W (1939) The phenomenon of rupture in solids. Proc Royal Sweden Inst Eng Res 153:1–55
Zech B, Wittmann FH (1978) A complex study on the reliability assessment of the Containment of a PWR, Part II: probabilistic Approach to describe the behavior of materials. Nucl Eng Des 48:575–584

Chapter 2
Concrete on the Meso-level

Concrete, known as the most widely used construction material in civil engineering, generally consists of water, cement powder, fine aggregate (sand), coarse aggregate (gravel), admixture and additives. When cement is mixed with water, calcium silicate minerals contained in cement enter into chemical reactions with water, generating chemical bonds between its molecules and forming cement past via a hydration reaction, which will, in turn, produce a large number of hydrate in a wide range of sizes. In the concrete material system, cement paste serves as an adhesive material between aggregates. Moreover, if the aggregates are added into the material system during the mixing procedure of cement paste, a thin porous layer between aggregates and cement pastes formed, namely the Interfacial Transition Zones (ITZs), which generally has lower stiffness than other elementary components (aggregate and mortar, etc.), thus, weakening the concrete structure.

Indeed, like many other heterogeneous material, particles in the concrete varies from several tens of nanometers (automatic scale) to thousands of meters (structural scale), concrete exhibits a multi-scale nature. First, we give a broad overview on the multiscale material system of concrete. From Fig. 2.1, we see that the lowest material scale considered here for a concrete structure is the atomic or molecular level, i.e. 10^{-8}–10^{-4} m, namely the micro-scale, at which level the analyses on material properties of the microstructural phases are often conducted based on statistical mechanics and Scanning Electron Microscopy (SEM) techniques. The middle level refers generally to the elementary phase level, i.e. 10^{-4} and 10^{-1} m, namely the meso-scale. At this level, concrete can be considered as a three-phase material comprised of coarse aggregate, mortar matrix and the ITZs between them. Determining the mechanical behavior of concrete is actually a complex procedure since it depends not only on the meso- and on microstructural information of concrete including size, shape, spatial distribution, physical properties of aggregate, mortar, ITZs and the interactions between them, but also all undergoing physical and chemical reactions, etc. The highest material scale shown in Fig. 2.1, namely macro-scale, refers to a structural size that is at least several times bigger than the biggest aggregate size. Wittmann (1983) is probably the first researcher who made a direct linkage between

X. Du and L. Jin, *Size Effect in Concrete Materials and Structures*,
https://doi.org/10.1007/978-981-33-4943-8_2

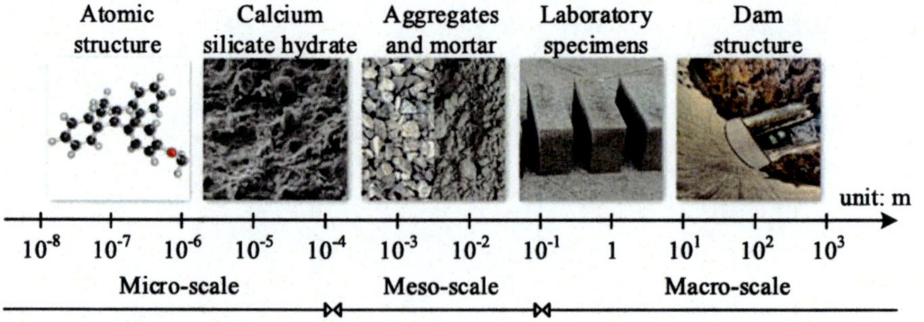

Fig. 2.1 Multi-scale material system of concrete

the mechanical properties of concrete and its structural characteristics at a lower level. We have to mention that research efforts on this topic has been extended to nano-level, a scale level even lower than the micro-level, however, it is beyond of current scope. Considering the high structural complexity of concrete, constituting exact models for reliably predicting the material behavior becomes substantially difficult however very necessary. Indeed, the current awareness about the information transfer-feedback system of a heterogeneous material is far from enough to clearly state the accurate linkage between the microstructural (or, meso-structural) information and the physical properties of the whole, more research effort on this topic is therefore urgently required by the development of advanced materials and the enlarging engineering needs.

Accounting all possible factors that influence the mechanical properties of concrete at one time appears technically changeling and raises practical difficulties, therefore, it is reasonable to adopt several restrictions and avoid the no-necessary complications. In the following text, we restrict ourselves to the mesoscale at which concrete can be viewed as a 3-phase material comprised of coarse aggregates randomly dispersed in a mortar matrix with ITZs between them. Figure 2.2

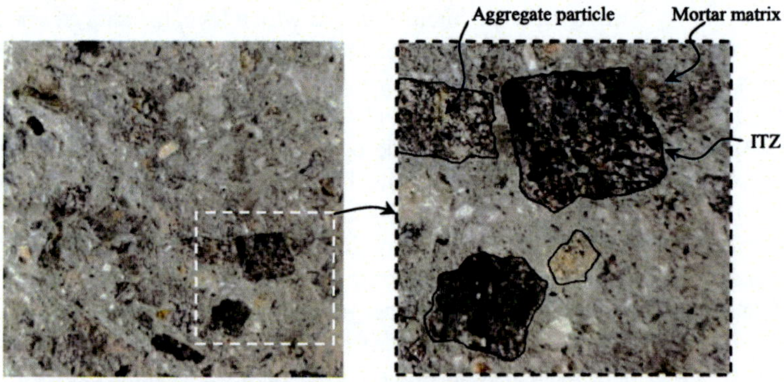

Fig. 2.2 Cross section of a concrete cubic specimen

illustrates a typical cross section of a concrete cubic specimen with ITZs marked by black solid lines.

2.1 Coarse Aggregate Particles

Rock-forming mineral-aggregate particles are added to the cement paste for strengthening the whole. Coarse aggregate refers to a group of particles having sizes greater than 4.75 mm, which are dug or dredged directly from a pit, river, lake or seabed, or alternatively, produced by crushing rocks (limestone, dolomite, etc.) and boulders. An otherwise group of particles having sizes in the range of 0.075 and 4.75 mm are generally known as sands. Figure 2.3 shows two commonly used coarse aggregates (Gravels and Cobble). Regarding the particle size, the concept of the maximum aggregate size (MAS), which signifies the upper limit of nominal particle size in coarse aggregates, is frequently referred by researches. For instance, for reinforced concrete, the MAS of coarse aggregate is expected to be in the range between ¾ of the minimum spacing of reinforcement and ¼ of the minimum size of concrete structure section. Additionally, it is worth to mentioning also a crucially important features of the size distribution of the coarse aggregates, namely gradation, which has a significantly effect on the strength and durability of a concrete material. According to the particle-size distribution rules, gradation can be furthermore divided to two groups: continuous gradation and discontinuous gradation. If the particle-size distribution follows a continuous function, that is all intermediate-size of coarse aggregates are presented in the concrete, the gradation is continuous, otherwise, it is discontinuous.

Besides, concrete properties are closely related to the characteristics of the aggregate used in the fabrication procedure including its type, shape, surface texture, and its physical properties, volume fraction etc. To provide further information, aggregate can be classified into several broad types such as normal weight aggregate, lightweight/heavyweight aggregate and recycled aggregate, according to the aggregate type. As examples, Fig. 2.4a shows a group of widely used lightweight aggre-

Fig. 2.3 Typical aggregate particles in normal concrete materials: **a** natural gravel; and **b** cobble stone

Fig. 2.4 Different types of aggregate particles: **a** lightweight aggregate; and **b** recycled aggregates

gates with bulk densities no greater than 1120 kg/m^3, which are comprised of pumice, diatomite, scoria, volcanic cinders, expanded shale or clay or perlite or vermiculite, etc. It is worth mentioning that, in spite of its low bulk density, lightweight aggregate can produce similar structural performance to that of the normal weight aggregate, meeting the high-performance requirements of civil engineering. Figure 2.4b shows the recycled aggregates made from recycled concrete from demolished building, concrete sidewalks, bridges, and pavements. The waste concrete materials are firstly thoroughly screened to remove any metal, scrap, or other impurities, and then crushed down to a smaller aggregate size so that it can be reused for other construction and landscaping purposes. After removal of contaminants, crushed concrete can be used in many new constructions such as new concrete for pavements, shoulders, sidewalks, and bituminous concrete (Surahyo 2019).

Otherwise, aggregates can be grouped mainly in four headings according to the shape: rounded, irregular, flaky/elongated, and angular; in six headings according to the surface texture: glassy, smooth, granular, rough or pitted, crystalline, and honeycombed or porous. Both shape and surface texture can possibly affect the density and strength of the concrete material, a rounder shape and a smoother surface may increase aggregate performance and reliability. Contrarily, a flaky or elongated shape may have a detrimental effect, thus more highly sanded mixture and more amount of cement and water are suggested to ensure the concrete performance (Surahyo 2019). A sharp angular (crushed) aggregate does not help to improve the material performance, but angular particles interlock better and can result in higher compressive strength due to the good bonding characteristic of the aggregate-to-mortar bond. Due to the same bonding effect, a rough and porous textured aggregate result in a better compressive and a flexural strength of concrete than a smooth textured aggregate. Moreover, aggregates are expected to be resistant to the material degradation once subjected to the cycles of freezing and thawing, wetting and drying, which depends furthermore on the aggregate lithology, strength, absorption and porosity. To be considered adequate in strength, the aggregate is expected to develop a full strength of the cement material.

2.2 Mortar Matrix

An otherwise group of rock-forming mineral-aggregate particles having sizes in the range of 0.075 and 4.75 mm and generally known as sands and serves as a main component of mortar matrix. Figure 2.5 shows, as an example, the river sand and manufactured sand that is crushed from coarse aggregates through crushing machines, respectively. Regarding the fineness of sand, we mention the concept currently known as the Fineness Modulus (FM) that indicates the influences of sands on cement dosage and shrinkage, the greater the FM is, the coarser the sand is. More precisely, the sand can be divided in 4 groups according to the specific series of sieve: FM is 3.7–3.1 for coarse sand ($d < 0.5$ mm); 3.0–2.3 for middle sand (0.35 mm $< d < 0.5$ mm); 2.2–1.6 for fin sand (0.25 mm $< d <$ 0.35 mm) and 1.5–0.7 for extra fin sand ($d < 0.25$ mm).

Besides, the fabrication procedure of the other main component of mortar matrix, cement paste, is very simple. It is enough to mix cement powder, water and additives in a room temperature then add sands into the cement paste to obtain a mortar, during the mixing process for fabricating cement paste or after. Letting the mixture of sand and cement paste stand for hours, the mortar is hardened. However, all physical–chemical reactions associated to this hardening process are not clearly understood and more research effort on this topic is highly required.

In this subsection, we focus mainly on a set of physical–chemical transformations following the general thermodynamics and kinetics laws, which is generally known as hydration process, on the example of the most common used Portland cement. The Portland cement is actually a mixture of clinker and calcium sulfate and is probably first fabricated in the nineteenth century by Romains. Briefly speaking, the clinker is made by limestone and clay (or, marl) mix finely grinded and homogenized in a ratio of 80–20%, then baked in a cement kiln preheated to 1450 °C. After a fast cooling process, the hardened nodules are crushed and mixed with calcium sulfate to produce the Portland cement. Surahyo (2019) investigated such hydration and mentioned that the Portland cement sets and hardens by reacting chemically with water and then becomes stiffer with time and sets. Indeed, the Portland cement in its

Fig. 2.5 River sand and manufactured sand

anhydrous state consists mainly of angular particles typically in the size range from 1 to 50 μm. It is produced by pulverizing a clinker, which is a heterogeneous mixture of several compounds formed by the high-temperature reactions between the calcium oxide and silica, aluminium oxide, and iron oxide during the clinkering process. The main chemical composition of a typical clinker compound includes C_3S (45–60%), C_2S (15–30%), C_3A (5–15%), and C_4AF (5–10%) (see Singh 2017) with C = CaO, S = SiO_2, A = Al_2O_3 and F = Fe_2O_3. Moreover, when Portland cement is mixed with water, the calcium sulfate and the high-temperature compounds of calcium tend to dissolve in the solution, and the liquid phase gets rapidly saturated with various ionic species. By mixing calcium, sulfate, aluminate, and hydroxyl ions and stand shortly, i.e. a few minute, the needle-shaped crystals of ettringite (calcium sulfoaluminate hydrate) firstly appear. Figure 2.6 shows a scanning electron micrograph of cement paste from which the ettringite can be observed (see, for instance, Fig. 2.6a). Next, a few hours later, large prismatic crystals of calcium hydroxide and very small fibrous crystals of calcium silicate hydrates, namely C–S–H, begin to fill the pores that are formerly occupied by water and the dissolving cement particles (see, for instance, Fig. 2.6b). Since the stability of the ettringite depends on the alumina-to-sulfate ratio of the Portland cement, the ettringite would be decomposed in order to form the monosulfate hydrate if the cement past stands for a few days. The monosulfate hydrate has hexagonal-plate morphology, so as calcium aluminate hydrates, and the latter are generally formed in the hydrated pastes of either undersulfated or high-C_3A Portland cements (Mehta and Monteiro 2006).

Additionally, numerous admixtures including air-entraining admixtures, accelerating admixtures, superplasticizers and retarding admixtures, etc. can be added to the concrete batch during mixing or just prior to placement to enhance the setting rate, to accelerate hydration reaction and to improve the strength of the mortar (Setareh and Darvas 2017). Apart from that, many other factors such as cement type, tamping degree, porosity, water-cement ratio and curing environmental conditions (such as temperature and humidity) may affect physical and mechanical properties of mortar. Among all, we highlight the pronounceable role of the water-cement ratio, which

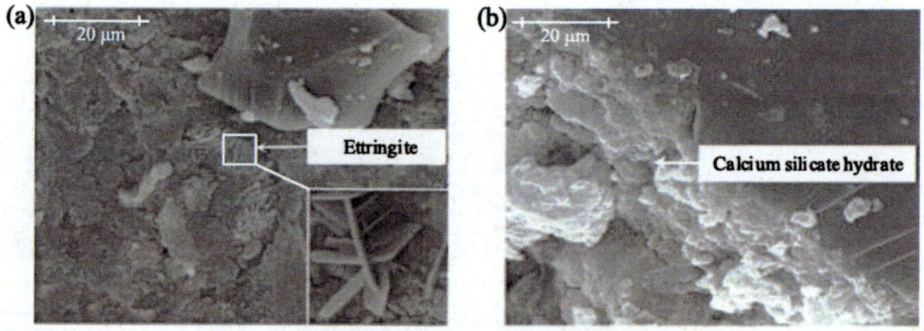

Fig. 2.6 The scanning electron micrograph of cement paste: **a** Ettringite; and **b** Calcium silicate hydrates

generally varies between 0.35 and 0.40 in a concrete material. Effect of water-cement ratio on the mechanical properties of mortar matrix are investigated by many researchers (among all, see, for instance, the work of Setareh and Darvas 2017).

2.3 Interfacial Transitional Zone (ITZ)

Using scanning electron micrograph, Fig. 2.7 illustrates the interfacial transitional zone (ITZ) formed between a coarse aggregate and the hydrated cement paste, the thickness of ITZ in a concrete structure usually varies from 10–50 μm (Tasong et al. 1999; Xiao et al. 2013). However, the ITZ is not a definite zone but, in a more appropriate term, a region of transition, it forms by Calcium Hydroxide (C–H) crystals with a chemical formulas of $Ca(OH)_2$ lined up at the aggregate and mortar interface. Its effective thickness depends on the microstructural feature being studied and the associating courses of hydration.

From a mechanical point of view, ITZ remains as a "weak point" of the concrete structure since C–H crystals does not provide any extra strength, its effect on the mechanical or transport properties of the concrete remains no-straightforward, however, important. Maso (1980) investigated on ITZ in a concrete structure and provided the following information. In a freshly compacted concrete, C–H crystals gather around the large aggregate particles, leading to a high water-ratio ratio at the aggregate surface. Ettringite and C–H crystals form through chemical transformations that re-range, for instance, calcium, sulfate, hydroxyl and aluminate ions resulting from the dissolution process of calcium sulfate and calcium aluminate compounds. Nevertheless, the particle sizes of ettringite and C–H crystals are relatively larger, leading to a high-porosity ITZ and thus, create a "week point" of the concrete. In addition, the plate like calcium hydroxide crystals tend to form in oriented

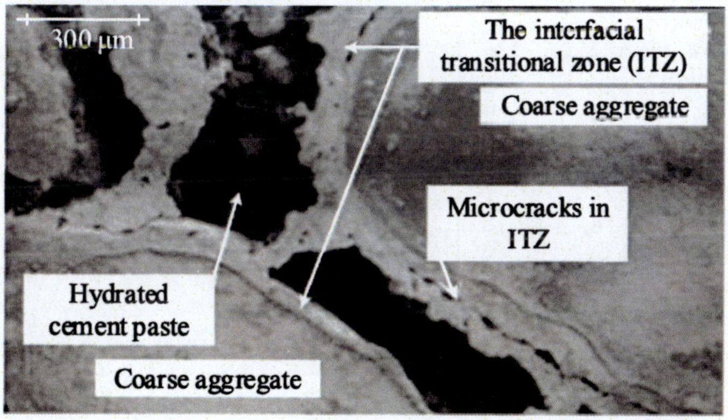

Fig. 2.7 The ITZ in scanning electron micrograph of microscopic concrete material (Singh 2017)

layers with the c-axis perpendicular to the aggregate surface. Then, with the hydration transformation evolves over time, the poorly crystalline C–S–H and the ettringite and C–H crystals of smaller sizes fill progressively the porous space between large Ettringite and C–H crystals, the latter procedure enhance the density and the strength of ITZ (Mehta and Monteiro 2006). Simultaneously, it appears possible that microcracks manifest themselves in the ITZ, deteriorating the ITZ structure (see Fig. 2.7). If the structure is subjected to a remotely load, the microcracks develop progressively in the ITZ and build cross-bridges form between them, then eventually form important microcracks running around aggregate surface with a size of a few microns (see Fig. 2.8).

Figure 2.9 demonstrates the nominal compressive strength evolution with respect to strain for different materials: coarse aggregate, mortar matrix and concrete material containing ITZ, the decisive effect of ITZ on the material degradation is underlined by comparing the compressive strength of the concrete material with that of coarse aggregate and mortar matrix. (Singh 2017).

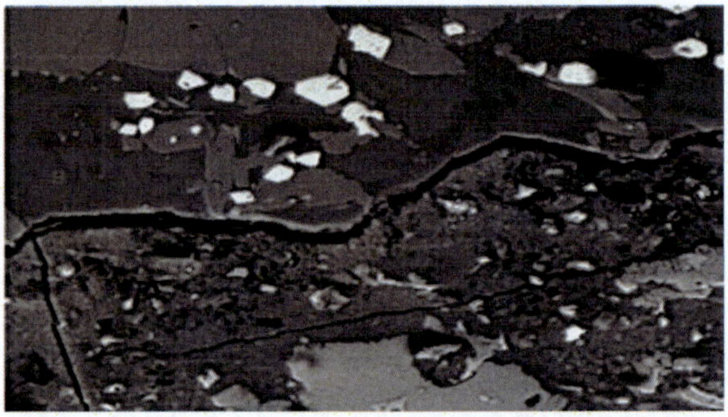

Fig. 2.8 Micro-cracks running around aggregate surface using scanning electron micrograph (Grzeszczyk and Starzyk 2008)

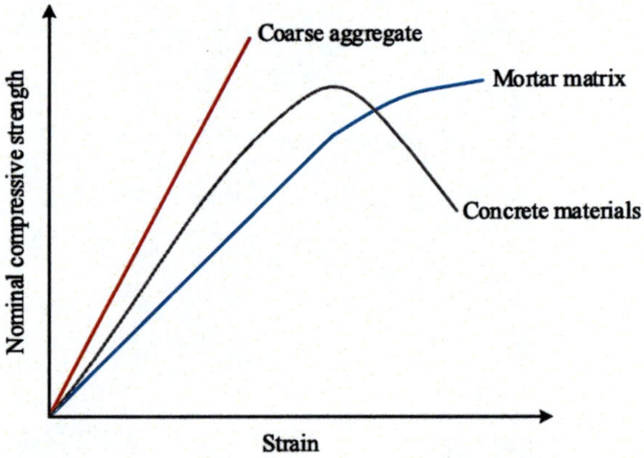

Fig. 2.9 Strength-strain responses of concrete materials, mortar matrix and coarse aggregate

References

Grzeszczyk St, Starzyk K (2008) Importance of cement type in the microstructure of interfacial transition zone in HPC. Cement Wapno Beton 75:89–98

Maso JC (1980) Proceedings of the seventh international congress on the chemistry of cements. Septima, Paris

Mehta PK, Monteiro PJ (2006) Concrete microstructure, properties and materials. McGraw-Hill, USA

Setareh M, Darvas R (2017) Concrete structures, 2nd edn. Springer, Switzerland

Singh H (2017) Steel fiber reinforced concrete: behavior, modelling and design. Springer, Singapore

Surahyo A (2019) Concrete construction: practical problems and solutions. Springer, Switzerland

Tasong WA, Lynsdale CJ, Cripps JC (1999) Aggregate-cement paste interface: part I. Influence of aggregate geochemistry. Cement Concr Res 29:1019–1025

Wittmann FH (1983) Structure of concrete with respect to crack formation. Fract Mech Concr 43(5):6

Xiao J, Li W, Corr DJ, Shah SP (2013) Effects of interfacial transition zones on the stress-strain behavior of modeled recycled aggregate concrete. Cement Concr Res 52:82–99

Chapter 3
Methodology: Meso-Scale Simulation Approach

Concrete can be regarded as a composite material consisting of aggregate particles, mortar matrix, and ITZ between them at the meso-scale level considering the heterogeneous characteristics of meso-structure. Starting from this idea, a new branch of mechanics, the meso-mechanics has been gradually formed and developed. By utilizing the meso-mechanical analysis and simulation approaches, not only the macroscopic deformation behavior but also the localized deformation and the fracture process as well as the damage mechanism of concrete subjected to external loads can be reflected and revealed. Moreover, the size effect of concrete related to its internal structure can be better described by the mesoscopic models, compared with those macroscopic ones. Accordingly, the mesoscopic numerical simulation was often adopted as one of the main methodologies to study the size effect of concrete materials and components (Chapters 4–9). After a brief introduction of the available meso-mechanical simulation methods, the details of the mesoscopic model introduced in this book was presented in this chapter, including the geometric model, constitutive relationship, interface model, and its validation at both material and component levels.

3.1 Mesoscopic Numerical Methods

The basic idea of mesoscopic numerical analysis method is to discretize the structure according to its inhomogeneous components and then calculate the macroscopic stress-strain relationship. First, the stress-strain field needs to be solved, and then the macroscopic stress-strain relationship is obtained by homogenization means. Of course, the damage and destruction process as well as plastic yielding of composite materials can be further studied according to the mesoscopic field (Zheng et al. 2010).

The mesoscopic numerical tests of concrete can not only directly reflect the whole process of the mesoscopic damage and destruction within the concrete, but also help understanding the influence of aggregate shape, gradation, distribution pattern

X. Du and L. Jin, *Size Effect in Concrete Materials and Structures*,
https://doi.org/10.1007/978-981-33-4943-8_3

and ITZ on the macroscopic mechanical properties of concrete. If the calculation model is reasonable and the mechanical parameters of each phase are accurate, the mesoscopic numerical tests can partially substitute the physical tests to avoid the limitation of test conditions and the influence of artificial operation error on the results. At present, with the development of computer technology and the maturity of finite element numerical simulation, the research on the macroscopic mechanical properties of concrete and its damage as well as failure process at the meso-scale has become a hot topic. Since Roelfstra et al. (1985) first proposed the concept of "numerical concrete", according to the understanding of the concrete mesoscopic structures, domestic and foreign researchers have developed and put forward many mesoscopic mechanical models, such as the lattice model (Schlangen and Garbociz 1996; Chiaia et al. 1997; Schlangen and Garbociz 1997; van Mier and van Vliet 1999, van Mier et al. 1999; Vervuurt et al. 1996), stochastic mechanical behavior model (Zhu and Tang 2002, 2004; Tang and Zhu 2003; Zhu et al. 2005), random particle model (Cundall and Strack 1979), rigid body-spring model (Kawai 1977), and random aggregate model (Liu and Wang 1996; Wang et al. 1999; Kwan et al. 1999), etc. In these meso-mechanical models, the nonlinearity of concrete material is reflected by the heterogeneity of spatial distribution of meso-components. The following is a brief introduction of these classic models.

3.1.1 Lattice Model

The lattice model, developed on the basis of physics 50 years ago, is a grid model that discretizes the continuum into a lattice system connected by elastic truss or beam elements at meso-scale, as shown in Fig. 3.1.

Originally, it was mainly used to solve classical elastic problems. The shape of grid is usually regular triangular, quadrilateral or random irregular. The grid is composed of truss or beam elements, representing a small part of the bulk material (such as the solid matrix of rock or concrete). The simple constitutive relationship and failure criterion are adopted for each element while the random distribution of aggregate

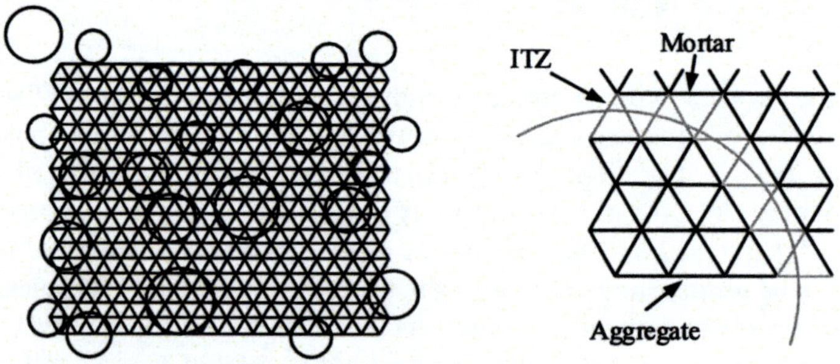

Fig. 3.1 Schematic diagram of lattice model for concrete

and its mechanical parameter is considered. The truss elements can only transmit axial force, while the beam elements can transmit not only axial force, but also shear force and bending moment, thus can simulate more complex stress state. During calculation, linear elastic analysis is carried out on the whole system under external loadings to obtain the local stress of each element in the lattice structure. If the failure criterion is exceeded, the element is removed from the system. The failure process of the element is irreversible. At early stages, due to limited numerical calculation capacity, the lattice model was remained at the theoretical level. Since the late 1980s, many scholars have used this model to simulate the failure process of heterogeneous materials. Schlangen and Garbociz (1996, 1997), Chiaia et al. (1997), van Mier and van Vliet 1999, van Mier et al. 1999 and Vervuurt et al. (1996) firstly applied the lattice model to study the fracture of concrete. In their model, concrete was assumed as a three-phase composite material composed of coarse and fine aggregate, mortar matrix and ITZ at the mesoscopic level.

The framework of lattice model is to generate firstly the three-phase composite structure of concrete according to the particle size distribution of aggregate; and then, project the regular or irregular triangular mesh onto the generated composite structure. The elements belonging to aggregate, mortar matrix and ITZ part are endowed with corresponding mechanical property parameters to reflect the heterogeneity of concrete materials. After the failure of element, the load is redistributed to calculated continuously until the next failure element is obtained. The calculation is repeated until the entire nonlinear system is completely destroyed.

In 2003, Lilliu and van Mier (2003) extended the two-dimensional lattice model to three-dimensional and carried out uniaxial tensile numerical simulation on the concrete. Figure 3.2a is a lattice model with regular triangular mesh. Man and van Mier (2008) used a three-dimensional lattice model to study the size effect on the flexural behavior of concrete beams. Typical damage and fracture behaviors of concrete beams of different sizes are shown in Fig. 3.2b.

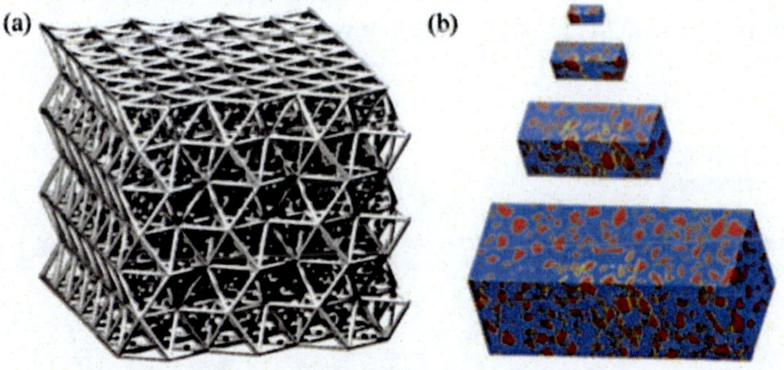

Fig. 3.2 Three-dimensional lattice model and its application: **a** Lattice model with triangular mesh; and **b** Damage and fracture behavior of concrete beams of different sizes

3.1.2 Stochastic Mechanical Property Model

The stochastic mechanical property model was proposed by Tang (Zhu and Tang 2002; Tang and Zhu 2003; Zhu et al. 2004), as shown in Fig. 3.3. From the viewpoint of meso-scale, concrete is also assumed as tri-phase composites consisting of aggregate, mortar matrix and ITZ. By means of meso-mechanics, statistics and numerical simulation, the model was established for the damage and fracture of concrete. To captures the nonlinear essential heterogeneous characteristic, the material properties of each component (aggregate, mortar matrix and ITZ) are assigned according to a given Weibull distribution. By this mean, an inhomogeneous numerical model for concrete is obtained, which is an abstract meso-mechanical model. In order to reflect the dispersion of the internal structure within each component, it is assumed that the mechanical properties of the material satisfy Weibull distribution. The probability density function of Weibull distribution is expressed as

$$f(u) = \frac{m}{u_0}\left(\frac{u}{u_0}\right)^{m-1} \exp\left(-\frac{u}{u_0}\right)^{m} \tag{3.1}$$

where, u represents the value that satisfies the distribution parameters (such as strength, elastic modulus, etc.); u_0 is the parameter related to the average of all the parameters of element; m defines the shape of Weibull distribution density function. The damage evolution of the meso element is described by the brittle damaged elastic constitutive relationship (Mazars and Pijaudier-Cabot 1987; Yu 1993). The criterion of maximum tensile stress and the criterion of Mohr-Coulomb are taken as the failure criteria of concrete elements, in which the tensile criterion has priority. Here, it is believed that the nonlinearity of concrete is caused by the damage related to the initiation, propagation and convergence of micro-cracks instead of the plastic deformation. Based on the stochastic mechanical property model, Zhu et al. (2002) simulated the fracture process of concrete samples under uniaxial and biaxial static loadings, and obtained the biaxial strength envelope.

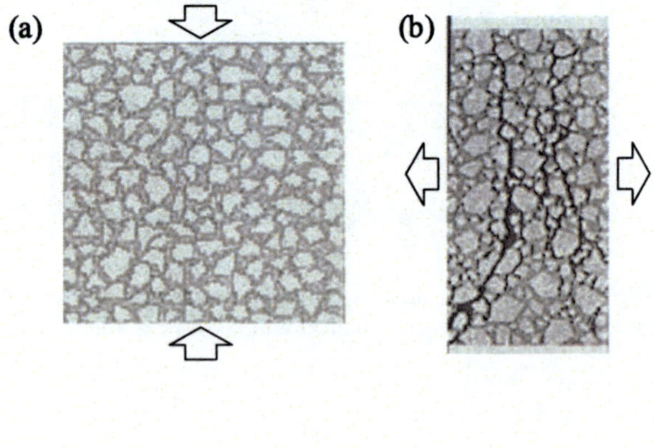

Fig. 3.3 Failure patterns of concrete simulated by stochastic mechanical property model under **a** compressive loads; and **b** tensile loads

3.1.3 Random Particle Model

The random particle model was proposed by Professor Cundall and Strack (1979) and later developed into the discrete element method (Cundall and Hart 1993; Yen and Chaki 1992; Fu and Dekelbab 2003). In the random particle model, concrete is a two-phase composite consisted of aggregate and mortar matrix. Firstly, according to the actual particle size and amount of aggregate, the aggregate is randomly distributed in the mortar matrix. The aggregates are characterized by circular or spherical particles. Then the aggregate and mortar matrix are divided into triangular elements and corresponding mechanical parameters are assigned to them. It is worth noting that the aggregate is regarded as rigid body in the original model proposed by Cundall and Strack (1979). Namely, the cracks will not pass through the aggregate but only occur in the mortar matrix. Because the shape of the circular or spherical element is regular, the contact retrieval is simple and easy while the blocks in other shapes can be bound together by multiple spheres through bonding. Therefore, the shape of the element in the discrete element for particles is often circular or spherical. To be applied for different problems, researchers have developed polygonal element (Camborde et al. 2000), ellipsoidal element (Tang 2004) and other non-smooth non-spherical elements (Matterw 2003).

In the 1990s, Bažant and Tabbara (1990) improved the rigid particle model of Cundall, assuming that aggregate particles are elastic and can be deformed. The circular particles are also randomly distributed, and the influence of the interface between aggregate and mortar matrix is considered. The interface layer is designed to have the property of strain softening and to retain the original stiffness when the element is unloaded. Moreover, the interface element is assumed to transfer only the axial stress of the particles, ignoring the shear capacity of interface. Namely the systems are equivalent as a connection of truss elements while the crack of bulk concrete is modelled by the tensile failure of elements. When the strain of bond belt reaches a certain value, its stress-strain curve is characterized by linear strain softening curve. Therefore, the fracture energy of matrix phase is considered as an important material parameter.

The meso-mechanical analysis model proposed by Zhong and Chang (1999) is also based on the assumption of the random particle model. However, the difference is that the matrix itself is considered to contain micro-cracks or microdefects, which further expand and connect under stress. The criterion in linear elastic fracture mechanics is used to judge the crack propagation. Of course, the simulation of this model requires not only the elastic mechanical parameters and geometric parameters of the aggregate, but also the parameters of aggregate and mortar matrix. The mechanical parameters such as cohesion, friction angle, Type-I fracture and Type-II fracture toughness should be given. However, it is difficult to select due to the lack of experimental data. Furthermore, the aggregate is considered elastic and does not break, which is different from the actual experimental phenomenon, or at least not suitable for studying concrete containing soft aggregate.

3.1.4 Rigid Body-Spring Model

Rigid body-spring model (RBSM) was established by Kawai (1977), which is a numerical calculation method applicable to limit analysis and crack behavior. The basic idea of this method is to take the centroid displacement of discrete block elements as the basic unknown variable. The real structural displacement field is approximated by the fragmented rigid body displacement. The elastic deformation inside the structure is reflected by the connection spring between the elements, and the stress inside the structure is characterized by the spring stress. According to the principle of computational mechanics, the governing equation of displacement is established. Because the displacement mode of the method is rigid body displacement mode, it is also known as rigid body finite element method. As shown in Fig. 3.4, two triangular (or polygonal) rigid body elements 1 and 2 were connected by two types of springs regardless of size. Each element has two translational motions and one rotational degree of freedom at its center of gravity, and the growth of micro-cracks is along the edges of the element. At present, the theoretical study and engineering application of this method mainly focus on the field of geotechnical engineering and the microscopic simulation of concrete materials.

The rigid-body spring method not only has the computational accuracy of finite element method, but also inherits the advantages of discontinuous numerical method in the simulation of discontinuous medium problem. Therefore, it has obvious advantages in solving the continuous-discontinuous media problem (Bolander and Saito 1998). However, in the rigid body spring method, the crack propagation can only be carried out along the edges of the block, resulting in significant influence of mesh shape on the computation results (Yao et al. 2015; 2016).

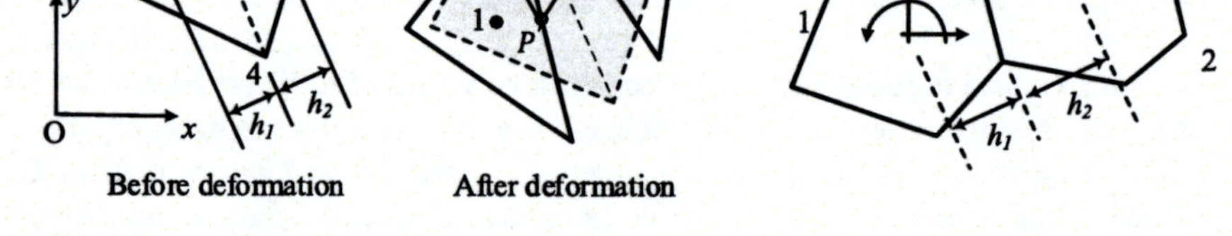

Fig. 3.4 Element of Rigid Body Spring Element Mode: **a** Triangular; and **b** Arbitrary polygonal

3.1.5 Random Aggregate Model

Random aggregate structure (RAS) was proposed by Liu and Wang (1996). In the model, concrete is regarded as a three-phase composite composed of aggregate, mortar matrix and the interface between them. Firstly, the number of aggregate particles was determined according to the Walraven and Reinhardt (1981) formula that transforms Fuller aggregate grading curve into two-dimensional aggregate grading curve. Then, the aggregate particles are distributed randomly in the mesoscopic model by the Monte Carlo method, projecting the finite element mesh onto the generated structure, as shown in Fig. 3.5a. Optionally, the coarse aggregate and cement mortar within the section of specimen are meshed directly by finite elements, such as plane Delaunay triangular element shown in Fig. 3.5b (De and Taerwe 1993). Moreover, based on the location, different properties are determined and given to the elements to characterize the heterogeneous structure of the concrete. Different from the abstract stochastic mechanical property model, the random aggregate model is a typical phenomenological model, which can represent the spatial random distribution of aggregate particles in concrete.

Due to the limitation of computation power, most of the early efforts were conducted based on two-dimensional plane model. With the development of high speed and large capacity computers, a large number of researchers have tried to extend the plane analysis model to the three-dimensional model to better simulate the process of damage, fracture and failure as well as the macroscopic mechanical properties of concrete, as shown in Fig. 3.6 (Zhou et al. 2017). The number of nodes in the 3D model is huge, which requires huge computer capacity for calculation. If the serial program is used to solve the problem, it will take a long time. Therefore, it is necessary to build a parallel computing platform to solve such super-large degree of freedom equations.

Fig. 3.5 Two methods to mesh random aggregate model: **a** Projection method; and **b** Direct method

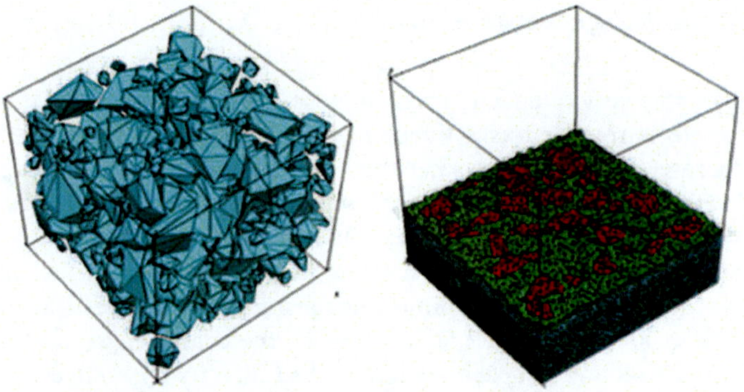

Fig. 3.6 Three-dimensional random aggregate structure and meshing

3.1.6 Mesoscopic Element Equivalence Method

In order to extend the application scope of meso-mechanics methods, the computational efficiency should be improved while the bottleneck problem of computation cost should be overcome. Based on the random aggregate model of concrete, Du et al. (2013a, b) and Du and Jin (2012a) proposed a new meso-mechanics method with high computational efficiency, i.e. Meso-Element Equivalence Method (MEEM). This method starts from the description of the essential characteristic of the meso-scale heterogeneity of concrete materials, and then meshes the typical meso-random aggregate model of concrete materials into a series of units with the same size. In the obtained units, aggregate, mortar matrix, ITZs and pore occupy different volume fraction as shown in Fig. 3.7a. Furthermore, each unit can be treated as isotropic homogeneous medium with mechanical parameters obtained by the equivalence methods for composite materials, as sketched in Fig. 3.7b.

Thus, a nonlinear finite element model with uniform and isotropic properties in each element while different properties between elements is formed, as shown in Fig. 3.7c. In Fig. 3.7c, different elements have different colors, indicating different mechanical properties, such as the effective elastic modulus, Poisson's ratio and the effective strength as well as other mechanical parameters. Based on this method,

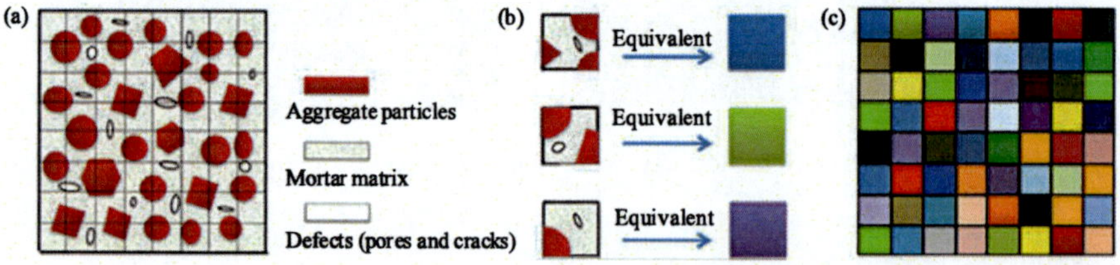

Fig. 3.7 The basic idea of mesoscopic element equivalence method: **a** Mesoscopic structure of concrete; **b** Equivalence of characteristics; and **c** Meso-element equivalent model

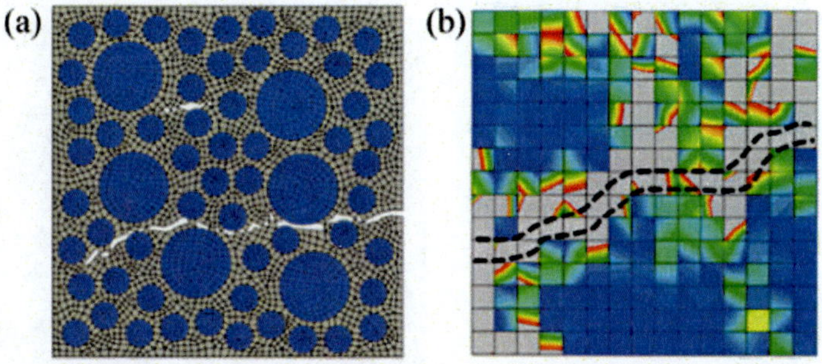

Fig. 3.8 The crack failure pattern of concrete specimen under uniaxial tensile loading obtained by **a** Random aggregate model; and **b** Mesoscopic element equivalence method

Du et al. (2013a, b) and Du and Jin (2012a, b, c) simulated the compressive, tensile and flexural behavior of concrete and the mechanical behavior of reinforced concrete members, and discussed and analyzed the influence of defects such as ITZs and pores. Figure 3.8 shows the crack destruction morphology of specimens under uniaxial tensile loadings obtained by the random aggregate structure and the meso-element equivalence method.

The numerical cases show that the degree of freedom of the system decreases because the number of mesoscopic elements is greatly reduced compared with other mesoscopic mechanic methods. It not only guarantees certain accuracy but also greatly improves the computational efficiency of the model, especially in the three-dimensional cases.

3.1.7 Other Numerical Methods

1. *M-H method*

Mohamed and Hansen (1999a, b, c) proposed a micro-mechanical model (In fact, it is more accurate to call it the mesoscopic structure) based on the in-depth study of concrete meso-structure and failure mechanism, namely, the M-H model, as shown in Fig. 3.9. This model also starts from the concrete mesoscopic structure, and assumes that concrete is a three-phase composite material model composed of aggregate, mortar matrix and ITZs between them at the mesoscopic level. The randomness of aggregate distribution and component properties are considered in this model. On this base, the concept of fracture energy of concrete is introduced to give the constitutive relation of strain softening under uniaxial tensile failure. Then the smeared crack model is used to describe the constitutive relationship of element under tensile loadings. The finite element method is used to implement the model. In addition, it is believed that the main mechanism of failure is tensile cracking. Thence, it is

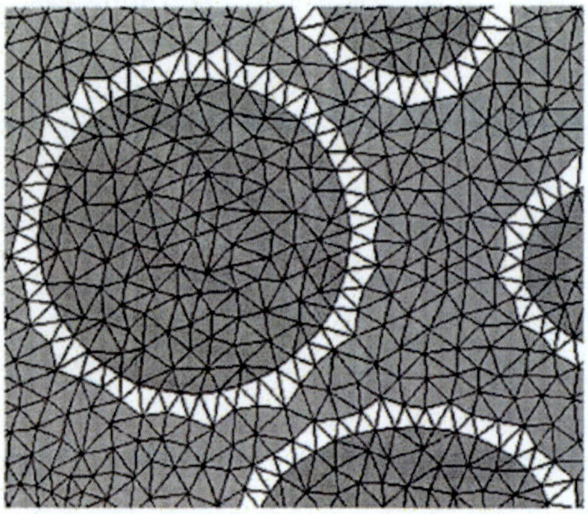

Fig. 3.9 M-H mesoscopic analysis model

assumed that the element only suffers from tensile failure whilst there is no shear or compressive destruction.

A typical crack propagation pattern of concrete specimens under uniaxial tensile condition obtained from M-H model is shown in Fig. 3.10. The M-H model is successful in simulating mechanical tests of concrete subjected to tensile failure, such as uniaxial tensile test, uniaxial compression test and four-point shear test. However, there is no literature report on the macroscopic response of concrete in multi-axial state.

2. *The interface element method*

Interface element method is a numerical method to solve the static and dynamic problems of inhomogeneous and discontinuous media. It is proposed by Zhuo and Zhang (2000) and Zhang et al. (1997) based on the rigid body spring element method (Kawai 1977) to solve homogeneous and linear statics problems. This method absorbs the basic idea of rigid body spring element model, and improves and expands it. Based on the principle of virtual work, replacing the spring element with the interface element

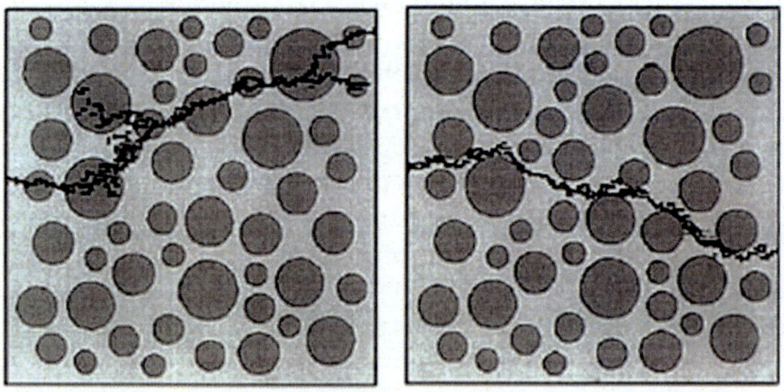

Fig. 3.10 Crack propagation pattern of concrete

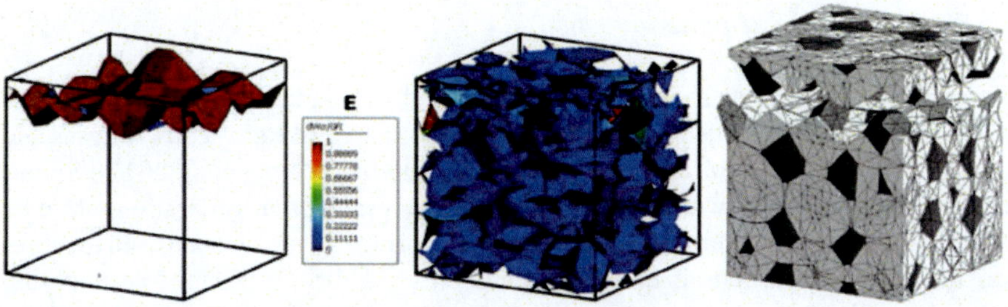

Fig. 3.11 Failure mode of concrete under uniaxial tensile loading obtained by interface element method

to reflect the elastic, viscos and plastic deformation characteristics, the deformation of the element is equivalent to that on the interface. Finally, the governing equations with 6 displacement components at the centroid of each element as basic variable are established, and a relatively complete theoretical system is formed. With the three-dimensional interface element method, Caballero et al. (2006) obtained the failure pattern of concrete specimens subjected to uniaxial tensile loadings, as shown in Fig. 3.11.

3.2 Geometric Model

Within the scope of this book, both the size effect of concrete materials and components was involved. For a structural component, such as beam, column, or beam-to-column joint, there is always reinforcement in its composition except for concrete material, including steel bars within components, or even fiber reinforced polymer (FRP) sheets and steel tube enfolding concrete, etc. In this section, the geometric models for plain concrete (i.e. random aggregate model), steel rebar, FRP sheet and steel tube in the meso-scale simulation were introduced.

3.2.1 Random Aggregate Model of Concrete

Among the available mesoscopic numerical methods for plain concrete, the random aggregate model is one of the most widely applied methods at present, which is also adopted here to simulate the mechanical failure behavior of concrete and the corresponding size effect. The details of the random aggregate model, including the shape generation of aggregate particles, methods of obtain optimum particle size distribution, aggregate placing methods, and meshing techniques are described as follows.

1. *Shape of aggregate particle*

Different shapes such as spherical, ellipsoidal and polyhedral have been applied to represent the aggregates in 3D models while planer shapes like polygons, ellipses and circles are applied to represent aggregates in 2D models (Zhang et al. 2017). Spherical (or circle) shape is widely used due to the convenience in generation while its rationality and acceptability has been verified. Ellipsoids have also been used to represent the aggregate particles in mesoscale concrete modelling. In 3D models, applying ellipsoids and spheres to represent uncrushed aggregate such as pea-gravel particles with smooth surfaces is realistic. However, spherical and ellipsoidal particle shape may not realistically represent the stress concentrations of the angular corners of the crushed aggregates with rough surfaces, flaky shape and elongated surfaces (Wang et al. 1999). Polyhedrons have been used to represent the actual aggregate shape of the crushed aggregate particles (Zhang et al. 2019). Due to the angular nature of aggregate particles, local stress concentrations would occur, thus leading to the crack initiation (Häfner et al. 2006), which can be realistically represented by models with polyhedral particles. Typically, generated aggregate particles with different shapes inside cylinder specimens by Zhou et al. (2017) are shown in Fig. 3.12.

2. *Aggregate particle size distribution*

Aggregate particle size distribution is an important part in continuum meso-structure generation, which influences the mechanical properties and fracture behavior of concrete (Meddah et al. 2010). The typical Fuller's curve (Fuller and Thompson

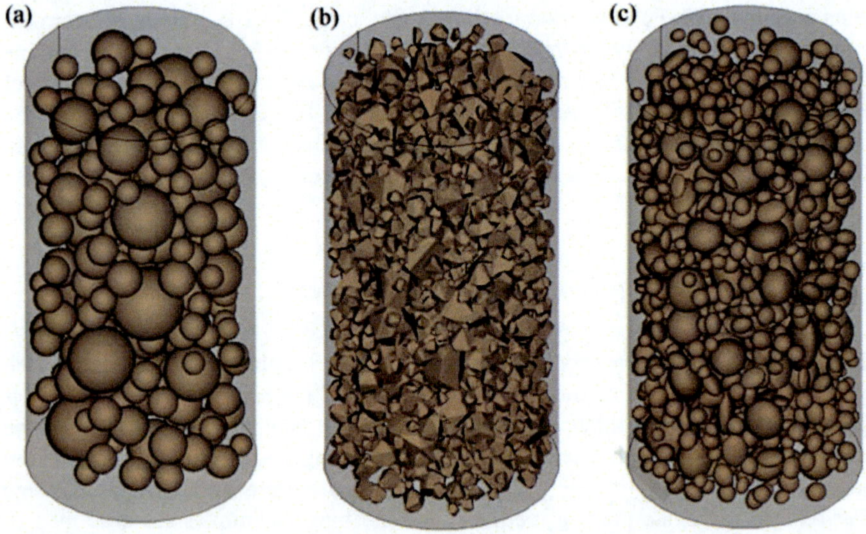

Fig. 3.12 Mesoscopic cylinder specimens of concrete with **a** spherical particles; **b** polyhedral particles; and **c** ellipsoidal particles

1907) has been widely used in most meso-scale modelling studies to obtain optimum particle size distribution of concrete. It can give the optimum compaction, density and strength in concrete and also a good workability and segregation resistance (Wriggers and Moftah 2006). In the Fuller's curve, aggregate percentage by weight passing through a sieve diameter d can be given by following equation:

$$p(d) = 100\left(\frac{d}{d_{\max}}\right)^n \tag{3.2}$$

where $p(d)$ is the percentage of aggregate by weight passing through a sieve with diameter d; $d_{\max}$ is the maximum diameter of aggregate particles; and n is the exponent of the equation, whose value range is $n = 0.45 \sim 0.70$ (usually $n = 0.50$).

As for spherical or circle aggregates, according to Walraven's equation (Walraven and Reinhardt 1981) which makes it possible to analyze meso-scale concrete materials based on Fuller's grading curve (Fuller and Thompson 1907), the percentage P_i of cumulative area of aggregates with diameter under d_i is

$$\begin{aligned} P_i(d < d_i) = P_k \Bigg[& 1.065\left(\frac{d_i}{d_{\max}}\right)^{\frac{1}{2}} - 0.053\left(\frac{d_i}{d_{\max}}\right)^4 \\ & - 0.012\left(\frac{d_i}{d_{\max}}\right)^6 - 0.0045\left(\frac{d_i}{d_{\max}}\right)^8 - 0.0025\left(\frac{d_i}{d_{\max}}\right)^{10} \Bigg] \end{aligned} \tag{3.3}$$

where d_i is the diameter of the ith group of aggregate; P_k is the aggregate volume that defined as the ratio between the total volume of (fine and coarse) aggregates and that of macro structure, for concrete, usually $P_k = 0.75$. In Eq. (3.3), for $d_{\max} = d_i$, then $P_i = 0.993P_k$ for a 2D specimen (certainly, $P_i = P_k$ for a 3D specimen according to the Fuller grading curve (Fuller and Thompson 1907)). Based on Eq. (3.3), the number of aggregate particles with different sizes can be generated for certain concrete specimens.

3. *Aggregate placing method*

After determining the number of aggregate particles, aggregate placement should be done inside the volume of concrete specimen. Generally, three conditions need to be satisfied when placing the aggregate particles based on Monte Carlo method: (1) aggregates should be completely inside the domain of specimen; (2) there should not be any overlaps between aggregate particles; and (3) there should be a minimum distance between two aggregate particles because there should be a coating of mortar in between aggregate particles (Thilakarathna et al. 2020). Aggregate placement in concrete specimen can be done by adopting various methods. Among them, most widely used method is the "take-and-place" method (Wang et al. 1999; Schlangen and van Mier 1992; Bazant and Tabbara 1990). In this method, aggregate particles are placed in the volume from largest particle to smallest particle until the specified volume fraction is achieved. Placing aggregate particles in the volume starting from the largest particle rather than from the smallest particle drastically improves the efficiency of the particle placing algorithm (Häfner et al. 2005). Particles are placed using the coordinates generated by a random number generator and the overlaps and gaps between the aggregates are checked. This method is a simple and convenient. Placement of the final few particles will take a long time because the probability of overlapping particles at this stage is higher. This method can be recommended as the best method to generate and distribute spherical aggregate particles due to its simplicity. With the current computer capacities, the generation time is insignificant (Thilakarathna et al. 2020).

4. *Meshing technique*

There are many techniques to mesh the geometry domain of the mesoscopic specimen. One type of these methods is direct mesh, by which the material interfaces would coincide with finite element boundaries (Zohdi and Wriggers 2001). Therefore, a clear boundary between the matrix and particles would be generated. Typically, spherical, ellipsoidal and polyhedral aggregates with aligned meshing conducted by Hypermesh (Computing 1997) are given in Fig. 3.13. Another type of method to project the regular mesh onto the generated heterogeneous structure where material interfaces do not coincide with the finite element boundaries. In this case, material within a given finite element may be discontinuous.

Cubic meshes (or voxels) are applied to discretize the concrete volume by projecting a regular mesh into the random aggregate structure, which are shown in Fig. 3.14 (Montero-Chacón et al. 2014). Zohdi and Wriggers (2001) utilized an unaligned cubic mesh in their meso-scale models. When the concrete cylinder

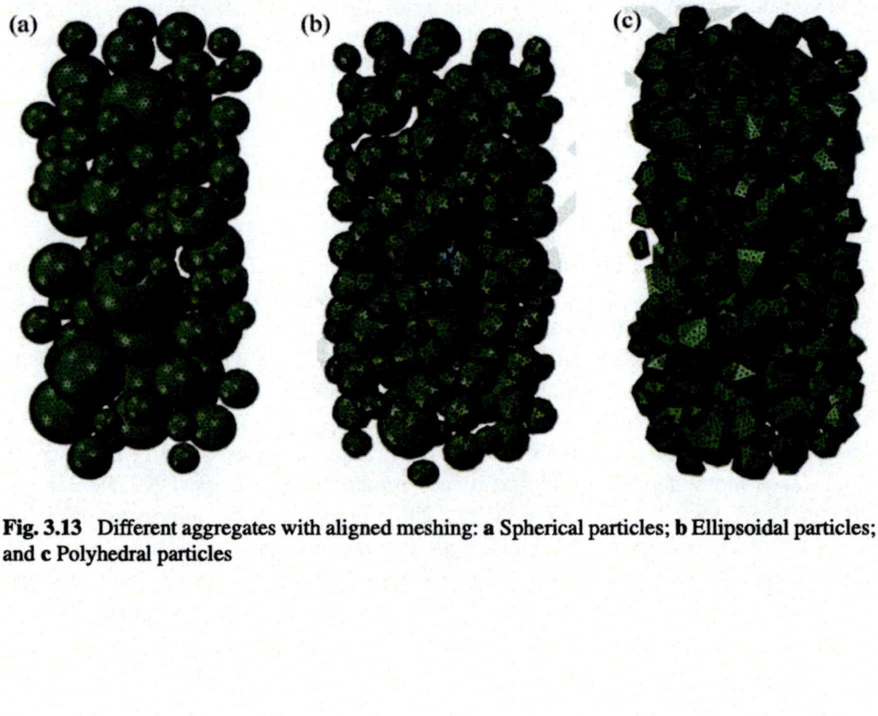

Fig. 3.13 Different aggregates with aligned meshing: **a** Spherical particles; **b** Ellipsoidal particles; and **c** Polyhedral particles

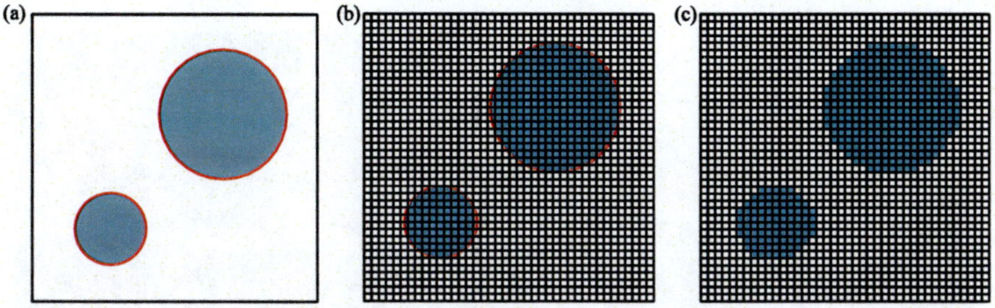

Fig. 3.14 Meso-scale model with voxel meshes: **a** Particles arrangement, **b** Mapping of the distribution, and **c** Resulted pixel-based discretization

model is meshed with cubic meshes, the real shape of the concrete cylinder cannot be obtained because of the curved shape of cylinder model (Dupray et al. 2009). Spherical aggregate particles also have the same issue and they cannot be meshed with a smooth particle shape by cubic meshes. Similarly, the circle aggregate particles cannot be meshed to the smooth shape by square or triangle meshes in 2D models while only an approximation of the geometry can be obtained. In the case with smaller mesh sizes, the geometry can get closer to the actual one. Voxel unaligned meshing is mainly used when the meso-structure of the concrete developed using digital image-based approach. It is convenient to represent pixels of the scanned model using a voxel mesh (Nagai et al. 2000).

In addition, Voxel representation of the meso-scale mesh is convenient due to its simplicity. However, due to fracture and damage investigations of concrete, the voxel mesh fails to realistically describe the post-peak behavior and the residual strength compared with tetrahedral mesh (Montero-Chacón et al. 2014). Figure 3.15 presents the typical random aggregate models with different aggregate grades.

3.2.2 Steel Rebar

For traditional reinforced concrete (RC) structural components, there are usually two types of modelling methods, namely, the integral and separated models. In the integral models, steel rebars are not explicitly modelled. Alternatively, they are smeared into the concrete elements where they are located according to their amounts, without considering the interaction between rebar and concrete. While, steel rebars are explicitly simulated in the separated models, taking the bond-slip behavior at the rebar-concrete interface. In general, compared with the integral model, the separated model needs higher calculation cost. However, more accurate results can be obtained by separated model. Thence, it is adopted in the following simulations.

After the generation of the random aggregate model for concrete, steel rebars are set up individually and simply inserted into concrete according to the reinforcement details, as shown in Fig. 3.16a. The bond between concrete and the steel bars plays a

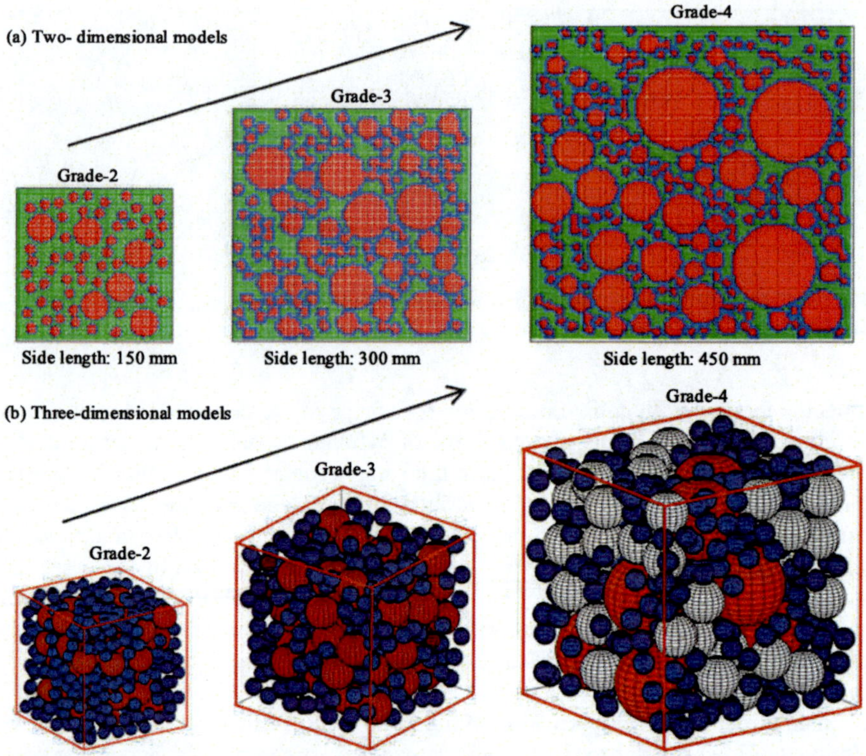

Fig. 3.15 Random aggregate models with different aggregate grades: **a** Two-dimensional models; and **b** Three-dimensional models

crucial role for the structural behavior. It contains three major mechanisms: chemical adhesion, friction and mechanical interlocking. As illustrated in Fig. 3.16b, the spring element connecting rebar nodes and the surrounding concrete nodes is employed to implement the nonlinear behavior between them. The mechanical behavior of spring is defined based on the bond stress-slip relationships reported in the available literatures.

3.2.3 *FRP Sheet*

FRP has been considered to be an improved material for the strengthening of existing RC structures due to their high efficiency, convenience, simple construction, wide application, high strength and corrosion resistance (Hosen et al. 2015; Nayak et al. 2018). In practice, strengthening with FRP is an effective method to enhance the bearing capacity of concrete members in the world (Cholostiakow et al. 2018; Davood

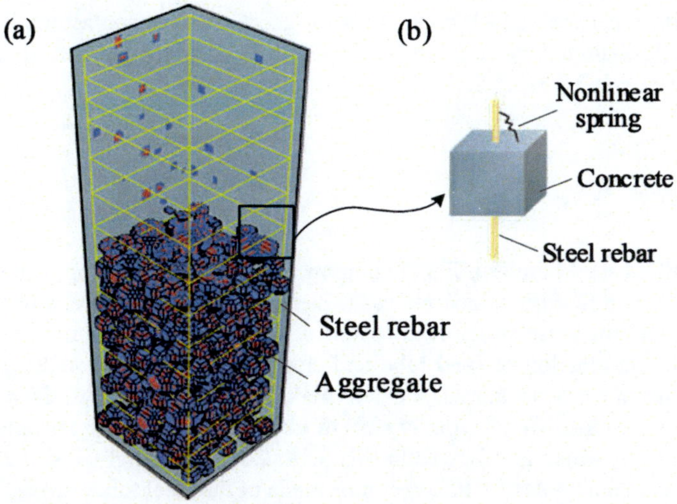

Fig. 3.16 Meso-scale model of the reinforced concrete components

and Amirhomayoon 2013). The shear strength and behavior of RC members, particularly members reinforced with FRP sheets, is an interesting area of study and concern of researches.

According to its chemical properties, FRP can be divided into basalt fiber reinforced polymer (BFRP), glass fiber reinforced polymer (GFRP), carbon fiber reinforced polymer (CFRP) and aramid fiber reinforced polymer (AFRP) (Kalfat et al. 2013), as shown in Fig. 3.17. It is noted that CFRP has more advantages than other materials due to its outstanding high strength and light weight performance, and it is recognized that it can reasonably replace steel materials (Sayed et al. 2014).

The main role of CFRP sheet bonding in concrete surface is to limit the propagation of cracks. Furthermore, CFRP is tensioned in the member plane whilst the stiffness in other directions is very small. Therefore, with reference to the work of Supaviriyakit et al. (2004), CFRP sheet is meshed with shell element. Certainly, CFRP sheet can also be modeled with a solid element. To describe the interaction

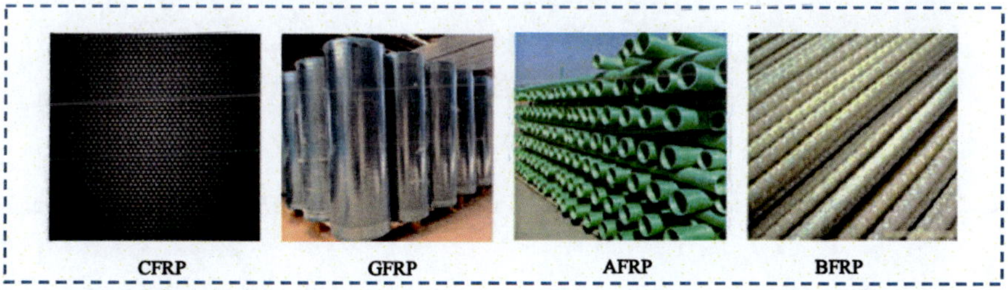

Fig. 3.17 Different types of FRP

at the interface between CFRP-concrete, a surface-to-surface contact model is often employed (Mohammad et al. 2018; Aljaafreh 2016; Supaviriyakit et al. 2004), as introduced later in Sect. 3.5.2.

3.2.4 Steel Tube

Concrete-filled steel tubular (CFST) components refers to the composite structural members formed by filling concrete in the steel tube while they are widely used in large span and high-rise structures because of their high performances (Han et al. 2014). The combination of steel tube and concrete can not only make up for the shortcomings of the two materials, but also make full use of their material properties. The common cross-sectional types of steel tubes include circular, square and special-shaped sections, which are showed in Fig. 3.18. Correspondingly, Fig. 3.19 depicts the cross-sectional types of CFST columns made of the steel tubes in Fig. 3.18.

Similar to FRP sheet, steel tube in the mesoscopic numerical model can be simulated by shell or solid element. In addition, the interaction between steel tube and the inside concrete can also be simulated by the surface-to-surface contact model.

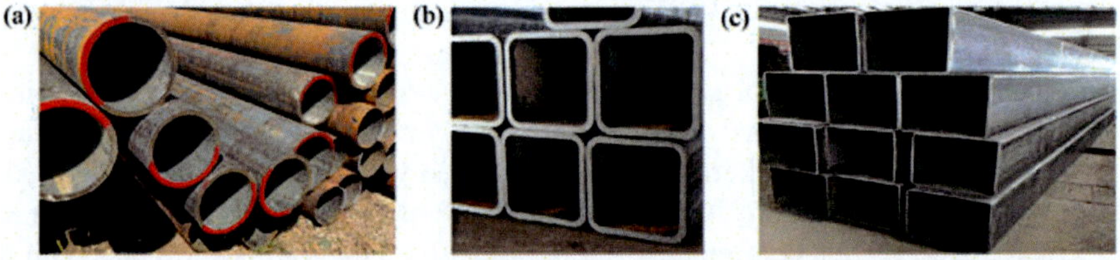

Fig. 3.18 Steel tubes with different cross-sectional shapes: **a** Circular; **b** Square; and **c** Rectangular

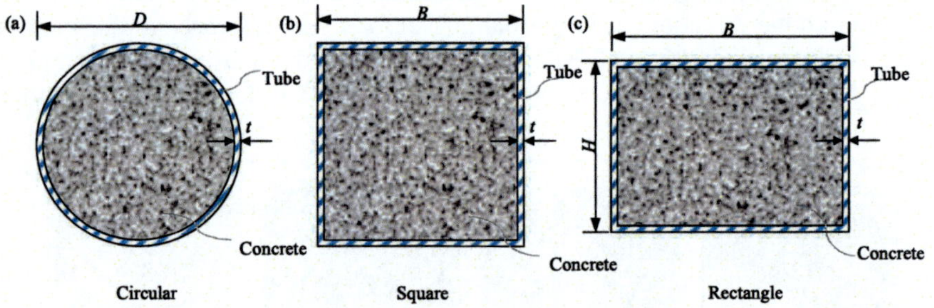

Fig. 3.19 Common cross-sectional types of CFST columns: **a** Circular; **b** Square; and **c** Rectangular

3.3 Material Model

In numerical simulations, reasonable material models are needed to describe their behavior under different kinds of loads. Corresponding to the geometry models described in Sect. 3.2, three types of material models are introduced here, namely, damaged plasticity model for the meso-components of plain concrete, elastoplastic model for both steel rebars and tubes, and elastic-brittle model for FRP materials.

3.3.1 Damaged Plasticity Model

As mentioned above, concrete is often treated as tri-phase composites consisting of aggregate, mortar matrix and ITZ in the meso-scale numerical models. Aggregate is less possible to be damaged or fractured due to its higher strength than mortar matrix and ITZs. Thence, similar to that in the work of Pedersen et al. (2013) and Chen et al. (2018), the mechanical behavior of aggregate in the present study was set as elastic. As for mortar and ITZs, it has been proved by Grote et al. (2001) that the performance of mortar is similar to that of concrete. On the other hand, the ITZ is a thin layer of mortar around aggregate particles with higher porosity. There is no essential difference between the apparent property of mortar matrix and ITZs (Grondin and Matallah 2014; Scrivener et al. 2004). Therefore, both the mechanical behavior of mortar matrix and ITZs were described by damage-plasticity model proposed by Lubliner et al. (1989) and improved by Lee and Fenves (1998). Based on the assumption of scalar (isotropic) damage, this model is designed to describe the behavior of concrete-like materials subjected to arbitrary loading conditions, including cyclic loading. The model considers the degradation of the elastic stiffness induced by plastic straining both in tension and compression. It also accounts for stiffness recovery effects under cyclic loading.

The model offers a general capability for modeling progressive material damage, in which a scalar damage variable d $(0 < d < 1)$ is introduced. The stress-strain relation can be expressed as

$$\sigma = (1 - d)\boldsymbol{D}_0^{\mathrm{el}} : \left(\varepsilon - \varepsilon^{\mathrm{pl}}\right) \tag{3.4}$$

in which $\boldsymbol{D}_0^{\mathrm{el}}$ is the initial elasticity matrix, ε is the strain tensor, d means the damage variable and $\varepsilon^{\mathrm{pl}}$ is the plastic strain tensor.

The damaged plasticity model is a continuum, plastic based model and it assumes that the main failure mechanisms of concrete are cracking in tension and crushing in compression. For the plastic-damage constitutive model, the tensile stress σ_{t} and compressive stress σ_{c} have the form as

$$\sigma_{\mathrm{t}} = \sigma_{\mathrm{t}}(\tilde{\varepsilon}_{\mathrm{t}}^{\mathrm{pl}}) \tag{3.5a}$$

$$\sigma_c = \sigma_c(\tilde{\varepsilon}_c^{pl}) \tag{3.5b}$$

in which, $\tilde{\varepsilon}_t^{pl}$ and $\tilde{\varepsilon}_c^{pl}$ denote the equivalent plastic strain generated by tension and compression respectively. The evolution equations of the hardening variables, $\tilde{\varepsilon}_t^{pl}$ and $\tilde{\varepsilon}_c^{pl}$, are conveniently formulated by considering uniaxial loading conditions first and then extended to multi-axial conditions.

When the specimen is unloaded from any point on the strain softening branch of the stress–strain curves, the unloading response is weakened: the elastic stiffness of the material appears to be damaged (or degraded). The degradation of the elastic stiffness is characterized by two damage variables tensile damage factor d_t and compressive damage factor d_c, which are assumed to be functions of the plastic strains, temperature, and field variables:

$$d_t = d_t(\tilde{\varepsilon}_t^{pl}) \quad 0 \le d_t \le 1 \tag{3.6a}$$

$$d_c = d_c(\tilde{\varepsilon}_c^{pl}) \quad 0 \le d_c \le 1 \tag{3.6b}$$

It is noted that the damage variables can take values from zero, representing the undamaged material, to unity, which represents total loss of strength. The stress-strain relations under uniaxial tension and compression loadings were illustrated in Fig. 3.20.

It is assumed that the stress-strain curve is linear before reaching the peak stress (i.e. strength), and when the peak stress is exceeded, the softened stress-strain or stress-displacement curve can be utilized to characterize the stiffness degradation behavior of the material (Chi et al. 2017).

If E_0 is the initial (undamaged) elastic stiffness of the material, the stress-strain relations under tension and compression loading are expressed respectively as

$$\sigma_t = (1 - d_t)E_0\left(\varepsilon_t - \varepsilon_t^{pl}\right) \tag{3.7a}$$

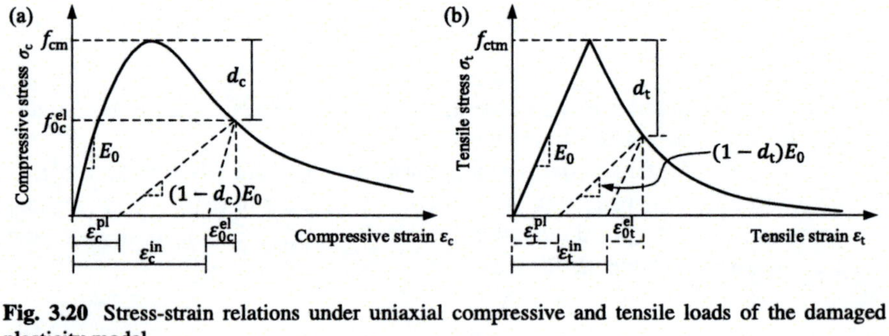

Fig. 3.20 Stress-strain relations under uniaxial compressive and tensile loads of the damaged plasticity model

$$\sigma_c = (1 - d_c)E_0\left(\varepsilon_c - \varepsilon_c^{pl}\right) \tag{3.7b}$$

in which, ε_t and ε_c respectively denote the total tensile and compressive strains. Herein, the damage evolution law recommended by the Chinese code (Ministry of Construction of People's Republic of China 2011) for determining the damage factor d_c and d_t was utilized. The damage factor d_c and d_t has the form as

$$d_c = \begin{cases} 1 - \frac{\rho_c n}{n-1+x^n}, x \le 1 \\ 1 - \frac{\rho_c}{\alpha_c(x-1)^2+x}, x > 1 \end{cases} \tag{3.8a}$$

$$\rho_c = \frac{f_{c,r}}{E_c \varepsilon_{c,r}} \tag{3.8b}$$

$$n = \frac{E_c \varepsilon_{c,r}}{E_c \varepsilon_{c,r} - f_{c,r}} \tag{3.8c}$$

$$x = \frac{\varepsilon}{\varepsilon_{c,r}} \tag{3.8d}$$

$$d_t = \begin{cases} 1 - \rho_t\left[1.2 - 0.2x^5\right], x \le 1 \\ 1 - \frac{\rho_t}{\alpha_t(x-1)^{1.7}+x}, x > 1 \end{cases} \tag{3.9a}$$

$$x = \frac{\varepsilon}{\varepsilon_{t,r}} \tag{3.9b}$$

$$\rho_t = \frac{f_{t,r}}{E_c \varepsilon_{t,r}} \tag{3.9c}$$

in which, $\alpha_c(\alpha_t)$ presents parameter value of the falling section of the stress-strain curve of concrete under uniaxial compression (tension), which is closely associated with concrete strength grade (Ministry of Construction of People's Republic of China 2011); $f_{c,r}\left(f_{t,r}\right)$ is the uniaxial compressive (tensile) strength of concrete; and $\varepsilon_{0,r}$ $(\varepsilon_{t,r})$ is the corresponding compressive (tensile) strain corresponding to the peak compressive (tensile) stress.

Moreover, in the work of Santos et al. (2020), the damage factor d_c and d_t was given respectively by

$$d_c = 1 - \frac{\sigma_c}{f_{c,r}} \tag{3.10a}$$

$$d_t = 1 - \frac{\sigma_t}{f_{t,r}} \tag{3.10b}$$

Zhang and Li (2012) introduced calibration techniques for concrete, under uniform and non-uniform confinement, based on Lubliner's theory through 3D simulation.

Grassl et al. (2013) used a combination of plasticity and damage mechanics, to create a constitutive model that was employed to study the failure of concrete structures. The model was designed to analyze the properties of the failure process in concrete structures experiencing multi-axial loading.

3.3.2 Elastoplastic Model

The mechanical behavior of steel materials presents obvious elastic phase followed by plastic hardening phase, and is always described by the elastoplastic model in the numerical analysis. The conventional elastoplastic model (Drucker 1988), premised upon the idea that the interior of a yield surface is an elastic region, has contributed to the prediction of the elastoplastic deformation behavior of solids such as metals, geo-materials, and concretes. However, it cannot describe cyclic loading behavior, thus limited to describe the behavior in the case of monotonic loading.

Various elastoplastic models aimed at describing cyclic loading behavior have been proposed to date. They require a pertinent description of the plastic strain rate induced by the rate of stress inside the yield surface, being termed the unconventional plasticity model by Drucker (1988) and also designated as the cyclic plasticity model (Hashiguchi and Ozaki 2008). They are classified into two types, e.g., the subloading surface model (Hashigucht 2000) based on the concept of expansion/contraction, and the multi surface model (Iwan 1967), the two surfaces model (Dafalias and Popov 1975; Krieg 1975), and the nonlinear kinematic hardening model (Chaboche 1991; Frederick and Armstrong 2007) based on the concepts of kinematic hardening. The models based on the concept of the kinematic hardening assume the yield surface enclosing the purely elastic domain so that the tangent modulus decreases discontinuously at the yield point or the contact point of assumed surfaces, violating the smoothness condition. While the subloading surface model fulfills not only the continuity condition but also the smoothness condition (Hashigucht 2000), leading to the description of smooth transition from the elastic to the plastic state. The subloading surface model has been applied to describe the behavior of soils (Hashigucht 2000, Mase and Hashiguchi 2009). Hashiguchi et al. (2012) modified that model and make it fit for the prediction of real deformation behavior of metals and verified the model by comparisons with various cyclic loading test data.

In the case of uniaxial loading, the ideal elastoplastic model can be illustrated by Fig. 3.21a while considering the kinematic hardening, a widely utilized bilinear elastoplastic model is shown in Fig. 3.21b. In Fig. 3.21b, "o-a" is the elastic stage, and "a–b" is the hardening stage and the magnitude of the hardening modulus (i.e. the tangent modulus) was set as 1% of Young's modulus (Xiao 2015).

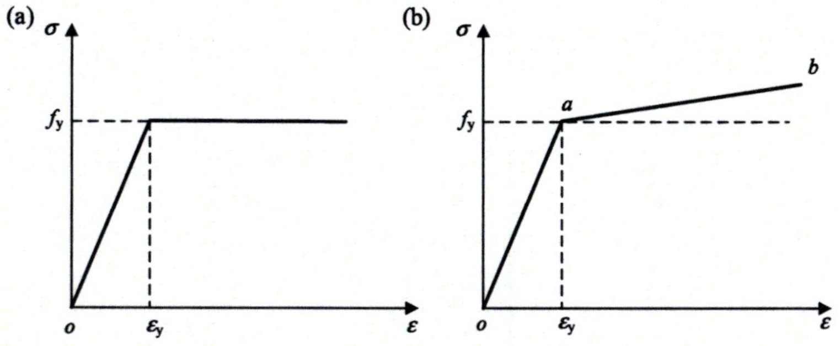

Fig. 3.21 Typical elastoplastic models: **a** Ideal elastoplastic model; and **b** Bilinear elastoplastic model with kinematic hardening

3.3.3 Elastic-Brittle Model

FRP always have the advantages in the high tensile strength, light weight, and great corrosion resistance. However, the fracture of the FRP composites shows obvious brittleness. Thus, the elastic-brittle model is always adopted to describe the behavior of FRP composites.

Damage plays an important role in many fibrous composite materials with non-ductile matrices. Their elastic-brittle behavior is characterized by the formation and evolution of micro-cracks (surface discontinuities) and cavities (volume discontinuities). Pronounced irreversibility of these defects is a consequence. These defects cause primarily stiffness degradation and only small permanent deformations remain in the stress-free body after unloading as long as the material is not close to complete deterioration. Matzenmiller et al. (1995) proposed a damage model for FRP composites exhibiting elastic-brittle behavior. The reduction of elastic modulus in the model is controlled by five damage variables. The damage surface is defined in terms of Hashin's criteria (1980) expressed in the effective stress space. The evolution of a damage variable controlling a failure mode is governed by effective stress components acting on the failure plane corresponding to this failure mode.

The CFRP is taken as the example of applying elastic-brittle model in the finite element analysis of FRP composite materials (Barbato 2009; Quantrill et al. 1996; Jumaa and Yousif 2019; Moller et al. 2005). The tensile strength of CFRP is much higher than the compressive strength. When the longitudinal tensile strain reaches the ultimate tensile strain, the CFRP will be broken. Furthermore, the stress in the other direction will be much less than the stress in the tensile direction. The anisotropy of the CFRP sheet is not considered. Herein, the constitutive model of CFRP sheet is shown in Fig. 3.22, as follows:

$$f_{\mathrm{fs}} = \varepsilon_{\mathrm{f}} E_{\mathrm{f}} \leq f_{\mathrm{fu}} \tag{3.11}$$

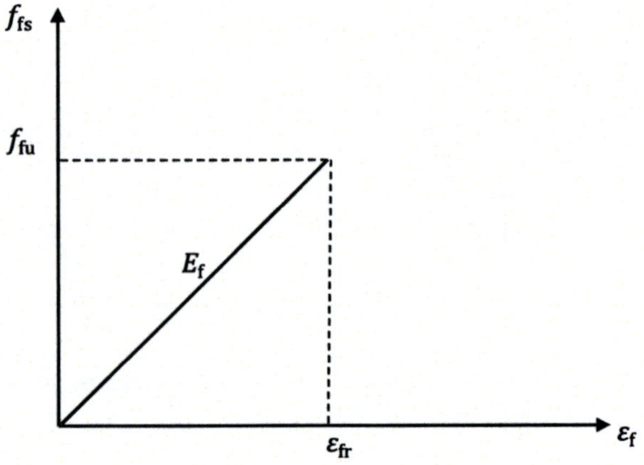

Fig. 3.22 The elastic-brittle model of CFRP

where f_{fs} is the stress of CFRP sheet; ε_f is the corresponding strain; f_{fu} is the ultimate stress; and E_f is the elastic modulus of CFRP.

3.4 Strain Rate Effect

Concrete is a kind of rate-dependent material. Under dynamic loads, the nominal strength and the deformation characteristics of concrete are significantly different from those of concrete under static loads (Li and Meng 2003), and vary with respect to loading rate. This is the so-called strain rate effect. The interpretations of the strain rate effect on material strength clearly involves three main aspects: (1) Lateral confinement effect (Hao et al. 2013a, b, 2010): the lateral inertial force from Poisson's effect and end friction can restrict the lateral deformation of the specimens; (2) Evolution of the cracks (Hao et al. 2016): cracks can form and propagate through coarse aggregate particles under impact loading instead of simply initiating and propagating along the weaker zones (e.g. the ITZ and the mortar matrix); and (3) Viscosity effect (Hao et al. 2016): the movement of free water in micro-defects within the concrete results in the resistance to crack propagation under dynamic loads. Generally, the split Hopkinson pressure bar (SHPB) test, drop weight test and other test methods can be utilized to obtain the dynamic increasing behavior (strain rate effect behavior) of concrete strength. Based on a large amount of test data, Fig. 3.23 lists the relationship between dynamic increasing behavior of nominal strength and strain rate. Herein, the DIF (Dynamic Increasing Factor) is the ratio between the dynamic and static strength under uniaxial loads. CDIF and TDIF represent the DIF for compressive and tensile strengths, respectively.

It has been proved that Young's modulus, Poisson's ratio, energy-absorption capacity, ultimate strain, peak strain, are rate-sensitive quantities, but at a much lower intensity than the compressive strength and tensile strength (Dilger et al. 1984; Bischoff and Perry 1991). Therefore, only the strength enhancement behavior

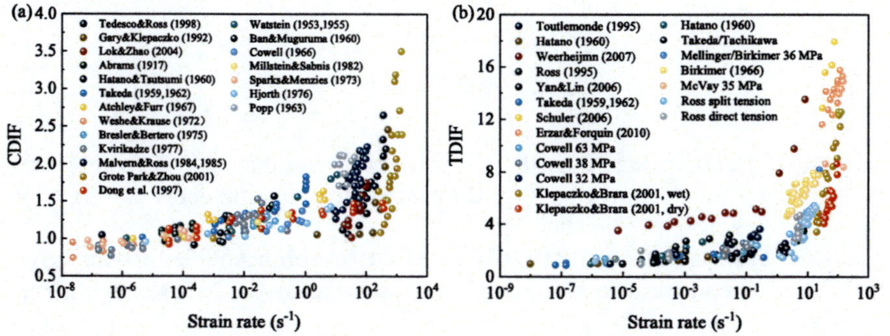

Fig. 3.23 Relationship between DIF and strain rate for **a** Compressive strength; and **b** Tensile strength

is considered in many simulations of concrete materials (Hao et al. 2013b; Zhou and Hao 2008b; Cusatis 2011; Du et al. 2014). For meso-components of concrete materials, the strain rate effect on compressive and tensile strengths can be incorporated into the above plasticity damaged model. So far, many empirical and theoretical relations have been developed to describe the strain rate effect on the properties of concrete materials. Among them, the relationship between DIF and strain rate recommended by the Comite Euro-international du Beton code (CEB 1993) on the basis of the available test results is one of the most classic empirical relations, which is widely applied to simulate the dynamic mechanical behavior of concrete. Nevertheless, concrete is regarded as a homogeneous material in this model, without considering the mesoscopic heterogeneity. Therefore, Hao and Hao (2011) proposed relationships to describe the strain rate effect of the strength for the meso-components of concrete. These two types of models are briefly intruded as follows.

3.4.1 Code Recommendations

1. *Compressive strength*

In CEB (1993), the DIF for the compressive strength is valid for monotonically increasing compressive stresses or strains at a range of approximately $30 \times 10^{-6}\ \text{s}^{-1} < |\dot{\varepsilon}_{\text{d}}| < 30 \times 10^{2}\ \text{s}^{-1}$ (here $\dot{\varepsilon}_{\text{d}}$ is the applied strain rate). For a given strain rate the compressive strength may be estimated from following equation:

$$\text{CDIF} = f_{\text{cd}}/f_{\text{cs}} = (\dot{\varepsilon}_{\text{d}}/\dot{\varepsilon}_{\text{cs}})^{1.026\alpha_{\text{s}}}, \text{ for } |\dot{\varepsilon}_{\text{d}}| \leq 30\text{s}^{-1} \tag{3.12a}$$

$$\text{CDIF} = f_{\text{cd}}/f_{\text{cs}} = \gamma_{\text{s}}(\dot{\varepsilon}_{\text{d}}/\dot{\varepsilon}_{\text{cs}})^{1/3}, \text{ for } |\dot{\varepsilon}_{\text{d}}| > 30\text{s}^{-1} \tag{3.12b}$$

with

$$\lg \gamma_s = 6.156\alpha_s - 2 \tag{3.13}$$

where $\dot{\varepsilon}_{cs}$ is the strain rate 30×10^{-6} s^{-1} under compression; f_{cd} is the compressive strength under the strain rate $\dot{\varepsilon}_d$; f_{cs} is the compressive strength under the strain rate $\dot{\varepsilon}_{cs}$; α_s and γ_s are the coefficients.

In CEB-FIB Model Code 2010 (2010), the CDIF has been modified and for a given strain rate the compressive strength can be estimated by the following simplified equations:

$$\text{CDIF} = f_{cd}/f_{cs} = (\dot{\varepsilon}_d/\dot{\varepsilon}_s)^{0.014}, \text{ for } |\dot{\varepsilon}_d| \le 30\text{s}^{-1} \tag{3.14a}$$

$$\text{CDIF} = f_{cd}/f_{cs} = 0.012(\dot{\varepsilon}_d/\dot{\varepsilon}_s)^{1/3}, \text{ for } |\dot{\varepsilon}_d| > 30\text{s}^{-1} \tag{3.14b}$$

2. *Tensile strength*

In CEB code (1993), the DIF for the tensile strength is valid for monotonically increasing tensile stresses or strains at a range of approximately 3×10^{-6} $s^{-1} <$ $|\dot{\varepsilon}_d| < 3 \times 10^2$ s^{-1}. For a given strain rate the tensile strength may be estimated from following equation:

$$\text{TDIF} = f_{td}/f_{ts} = (\dot{\varepsilon}_d/\dot{\varepsilon}_{ts})^{1.016\delta_s}, \text{ for } |\dot{\varepsilon}_d| \le 30\text{s}^{-1} \tag{3.15a}$$

$$\text{TDIF} = f_{td}/f_{ts} = \beta_s(\dot{\varepsilon}_d/\dot{\varepsilon}_{ts})^{1/3}, \text{ for } |\dot{\varepsilon}_d| > 30\text{s}^{-1} \tag{3.15b}$$

with

$$\lg \beta_s = 7.112\delta_s - 2.33 \tag{3.16a}$$

$$\delta_s = 1/\left(1 + 8f_c'/f_{c0}'\right) \tag{3.16b}$$

where $\dot{\varepsilon}_{ts}$ is the strain rate 3×10^{-6} s^{-1} under tension; f_{td} is the tensile strength under the strain rate $\dot{\varepsilon}_d$; f_{ts} is the tensile strength under the strain rate $\dot{\varepsilon}_{ts}$; β_s and δ_s are the coefficients; f_c' is the quasi-static compressive strength of concrete; and $f_{c0}' = 10\,\text{MPa}$.

However, comparison between the available experimental results with the CEB empirical equations indicates that the TDIF of CEB substantially underestimates the TDIF of concrete (Malvar and Ross 1998; Zhou and Hao 2008a). In the study of Malvar and Ross (1998), a modified empirical formula was proposed based on test data and it can be written as follows:

$$\text{TDIF} = f_{\text{td}}/f_{\text{ts}} = (\dot{\varepsilon}_d/\dot{\varepsilon}_{\text{ts}})^{\delta_s}, \text{for } |\dot{\varepsilon}_{\text{d}}| \leq 1\text{s}^{-1} \tag{3.17a}$$

$$\text{TDIF} = f_{\text{td}}/f_{\text{ts}} = \beta_s(\dot{\varepsilon}_d/\dot{\varepsilon}_{\text{ts}})^{1/3}, \text{for } |\dot{\varepsilon}_{\text{d}}| > 1\text{s}^{-1} \tag{3.17b}$$

In CEB-FIB Model Code 2010, the relationship between TDIF of concrete and strain rate can be expressed as:

$$\text{TDIF} = f_{\text{td}}/f_{\text{ts}} = (\dot{\varepsilon}_{\text{d}}/\dot{\varepsilon}_{\text{ts}})^{0.018}, \text{for } |\dot{\varepsilon}_{\text{d}}| \leq 10\text{s}^{-1} \tag{3.18a}$$

$$\text{TDIF} = f_{\text{td}}/f_{\text{ts}} = 0.0062(\dot{\varepsilon}_{\text{d}}/\dot{\varepsilon}_{\text{ts}})^{1/3}, \text{for } |\dot{\varepsilon}_{\text{d}}| > 1\,\text{s}^{-1} \tag{3.18b}$$

3.4.2 Hao Hong's Model

Some available researches (Zhou and Hao 2008a; Grote et al. 2001) have discovered that the dynamic mechanical behavior of mortar matrix is similar to that of concrete and the DIF models for concrete are also appliable for mortar matrix. Due to the lack of relative tests, it is really difficult to obtain the dynamic mechanical behavior of ITZ. In many studies (Zhou and Hao 2008a, b; Du et al. 2014), the ITZ could also be assumed to have the similar DIF with mortar matrix, which have got some appropriate numerical results. As for the DIF of aggregate particles, researchers (Wang et al. 2006; Li et al. 2004; Zhao 2000) conducted experimental investigation and simulation on the dynamic mechanical behavior (strain rate effect) of rock (aggregate). The limited research results demonstrated that the DIF for rock (aggregate) is in the similar range to that for concrete. Therefore, in some mesoscopic simulations, the assumption that the DIF model for the meso-components of concrete is the same is acceptable and reasonable.

However, it needs to be clear that the strain rate effects of meso-scale components of concrete are different from each other in real concrete and many efforts need to be done to get a closer modeling of the real situation in simulations. To achieve this goal, some scholars employ different DIF models based on experimental and numerical results to describe the dynamic mechanical behavior of meso-scale components of concrete, especially for mortar matrix and aggregate particles. Among these models, the DIF model recommended by Hao et al. (2013a, 2017) is another classic model and it is also widely used. In addition, the DIF obtained in dynamic tests includes structural effect and material property, which could overestimate the material strain rate effect. The structural effect (the inertia confinement effect) should be removed from the experimental data to get the pure material DIF (material strain rate effect) for simulations. Considering these factors, Hao and Hao (2011) have summarized and verified the DIF of mortar matrix based on their experimental and numerical results (Ye et al. 2019; Hao et al. 2017; Chen et al. 2015; Hao and Hao 2014; Hao et al. 2013a). In Hao et al.'s model, the DIF of mortar matrix for the compressive strength ($CDIF_m$) is expressed as:

$$CDIF_m = 0.0419\lg\dot{\varepsilon}_d + 1.2165, \text{ for } \dot{\varepsilon}_d \le 30/s \quad (3.19a)$$

$$CDIF_m = 0.8988(\lg\dot{\varepsilon}_d)^2 - 2.8255\lg\dot{\varepsilon}_d + 3.4907, \quad \text{for } \dot{\varepsilon}_d > 30/s \quad (3.19b)$$

The DIF of mortar matrix for the tensile strength ($TDIF_m$) is expressed as:

$$TDIF_m = 0.26\lg\dot{\varepsilon}_d + 2.06, \quad \text{for } \dot{\varepsilon}_d \le 1/s \quad (3.20a)$$

$$TDIF_m = 2\lg\dot{\varepsilon}_d + 2.06, \quad \text{for } 1/s < \dot{\varepsilon}_d \le 2/s \quad (3.20b)$$

$$TDIF_m = 1.4433\lg\dot{\varepsilon}_d + 2.2276, \quad \text{for } 2/s < \dot{\varepsilon}_d \le 150/s \quad (3.20c)$$

As for the DIF model of aggregate particles, Hao and Hao (2013) also applied their proposed method to remove the lateral inertia confinement effect, the CDIF relations of three rock types (namely granite, limestone and tuff) are derived and verified in simulations of SHPB tests.

The $CDIF_G$ relations of the granite are derived as:

$$CDIF_G = 0.018668\lg\dot{\varepsilon}_d + 1.291887, \quad \text{for } 1/s < \dot{\varepsilon}_d \le 220/s \quad (3.21a)$$

$$CDIF_G = 1.8547(\lg\dot{\varepsilon}_d)^2 - 7.9014\lg\dot{\varepsilon}_d + 9.6674, \quad \text{for } 220/s < \dot{\varepsilon}_d \le 1000/s \quad (3.21b)$$

The $CDIF_L$ relations of the limestone are derived as:

$$CDIF_L = 0.0262\lg\dot{\varepsilon}_d + 1.2827, \quad \text{for } 1/s < \dot{\varepsilon}_d \le 220/s \quad (3.22a)$$

$$CDIF_L = 1.898704(\lg\dot{\varepsilon}_d)^2 - 8.46954\lg\dot{\varepsilon}_d + 10.76523, \quad \text{for } 220/s < \dot{\varepsilon}_d \le 1000/s \quad (3.22b)$$

The CDIF relations of the tuff are derived as:

$$CDIF_T = 0.0237\lg\dot{\varepsilon}_d + 1.2791, \quad \text{for } 1/s < \dot{\varepsilon}_d \le 220/s \quad (3.23a)$$

$$CDIF_T = 1.297645(\lg\dot{\varepsilon}_d)^2 - 5.89126\lg\dot{\varepsilon}_d + 8.01433, \quad \text{for } 220/s < \dot{\varepsilon}_d \le 1000/s \quad (3.23b)$$

Figure 3.24 depicts the best-fitted curves of the above CDIF relations of three rock types under high strain rates.

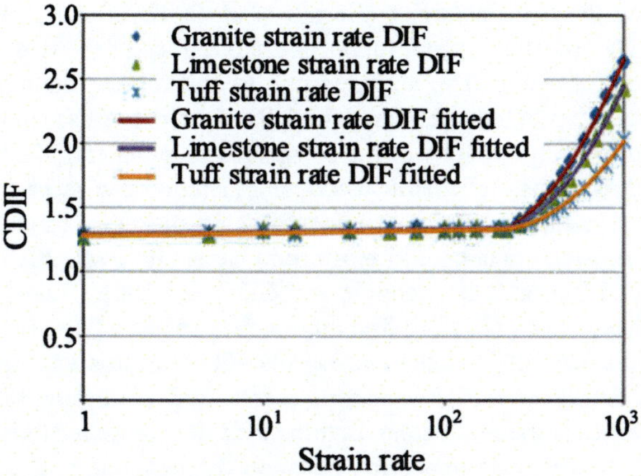

Fig. 3.24 CDIF relations of three rock types under high strain rates

The TDIF relations of aggregate particles can be uniformly expressed as:

$$\text{TDIF} = 0.0598\lg\dot{\varepsilon}_{\mathrm{d}} + 1.3588, \quad \text{for}\, \dot{\varepsilon}_{\mathrm{d}} \leq 0.1/\mathrm{s} \tag{3.24a}$$

$$\text{TDIF} = 0.5605(\lg\dot{\varepsilon}_{d})^2 + 1.3871\lg\dot{\varepsilon}_{\mathrm{d}} + 2.1256, \quad \text{for}\, 0.1/\mathrm{s} < \dot{\varepsilon}_{\mathrm{d}} \leq 50/\mathrm{s} \tag{3.24b}$$

$$\text{TDIF} = 6.100, \quad \text{for}\, \dot{\varepsilon}_{\mathrm{d}} > 50/\mathrm{s} \tag{3.24c}$$

3.5 Interaction Model

The structural components involved in this book, namely, beams, columns, and beam-to-column joints, are always composed of more than one type of material. The advantages of different materials are fully utilized through good interaction between them. Accordingly, to simulate the mechanical behavior of components more realistically, the interaction should be taken into account in the meso-scale numerical model. Here, the involved interaction can be divided into two categories, namely, the node-to-node model for rebar-concrete interaction and the surface-to-surface contact model for concrete-to-steel tube or FRP sheet.

3.5.1 Node-to-Node Interaction Model

The mechanical properties of RC structures, such as cracking, deflections, loading capacity and hysteretic behavior under seismic excitations, are highly dependent on the bond behavior (Harajli et al. 2002, Desnerck et al. 2015, Liu et al. 2019). The transfer of axial forces at the interface of the two materials requires good bond performance. The slippage due to insufficient bond strength not only causes the capacity of the steel cannot be fully exerted, but also causes cracking and the reduction of bearing capacity of the components. In general, the bond strength is consisting of three parts, i.e. chemical adhesion, friction and mechanical interlocking.

In a number of previous research projects for modelling the bond characteristics between concrete and reinforcing steel (Jin et al. 2019, Kwak and Kim 2001, Cruz and Barros 2004, Jendele and Cervenka 2006, Huang 2010), two types of models have been used. The first approach is to use a "bond-zone connect" element (Kwak and Kim 2001, Cruz and Barros 2004). In this approach, the characteristics of the contact surface between the concrete and reinforcing steel is represented by the properties of the bond zone. In this model it is assumed that the contact element provides a continuous connection between the concrete and reinforcing steel. Hence, finer mesh is needed within the bond zone area in order to achieve reasonable accuracy. The second one is to use nonlinear-spring element to connect the concrete and the reinforcing steel elements at the nodes (Jin et al. 2019, Jendele and Cervenka 2006, Huang 2010). In comparison, for modelling the global structural behavior of reinforced concrete structures the bond-link element provides a reasonable compromise between accuracy and computational efficiency.

As shown in Fig. 3.25, the nonlinear-spring model in studies of Huang (2010) and Jin et al. (2019) was defined as two-node element, having three translational and three rotational degrees of freedom. For the sake of simplification, it was assumed that the slippage between concrete and rebar occurs only along the length of the steel rebar.

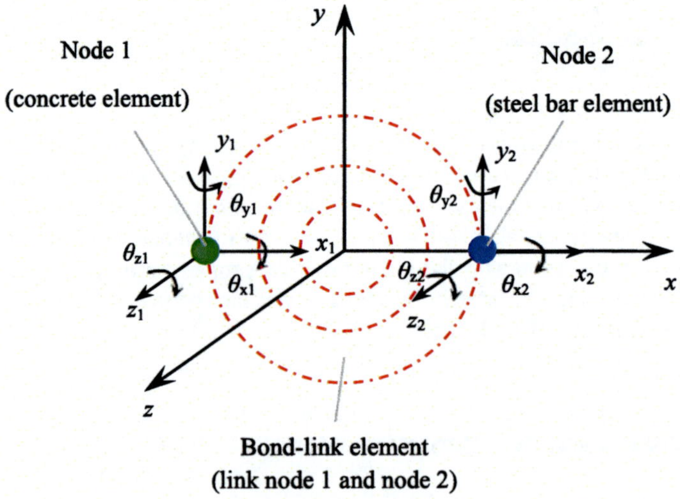

Fig. 3.25 Nonlinear-spring model between steel and concrete (Jin et al. 2019, Huang 2010)

3.5.2 Surface-to-Surface Contact Model

Surface-to-surface contact model was always used in mechanical simulation and nonlinear stress analysis (Wu et al. 2007, Mashaly et al. 2011, Wang et al. 2017), allowing to create pairs of surfaces that may come into contact with each other during the analysis, thereby connecting the nodes on the surfaces together. The processor will determine the distance between the nodes on these surfaces at each time step of the analysis. When the nodes are sufficiently close to each other, a force will be applied to prevent penetration. The contact pairs can consist of any two arbitrary surfaces. Surface-to-surface contact will connect the nodes on one surface to the faces of the other surface (and optionally vice versa). However, in order to speed up the contact search, the user should only specify the contact pairs that will definitely interact within the given event duration.

The interaction between contacting surfaces consists of two components: normal interaction due to loads perpendicular to surfaces in contact, and tangential one, referring to possible relative sliding between two adjacent surfaces.

Different contact formulations can be used to describe the normal behavior, i.e. hard and soft contact models as shown in Fig. 3.26a. The clearance was defined as the normal separation between two surfaces. In hard contact formulation, the contact constraint is applied when the clearance becomes zero. There is no limit on the magnitude of contact pressure that can be transferred by the contact surfaces. The surfaces start to separate when the contact pressure becomes zero or negative, in which case the contact constraint is removed. Thus, a dramatic change in contact pressure may occur when the contact condition changes from "open" (a positive clearance) to "close" (zero clearance). On the other hand, the normal stress can be exponential function of the clearance in the soft contact formulation (Wang et al. 2017), which is suitable to describe the contact behavior of soft, thin layer of grouting materials between the contraction joint surfaces. The adopted exponential function is given as follows:

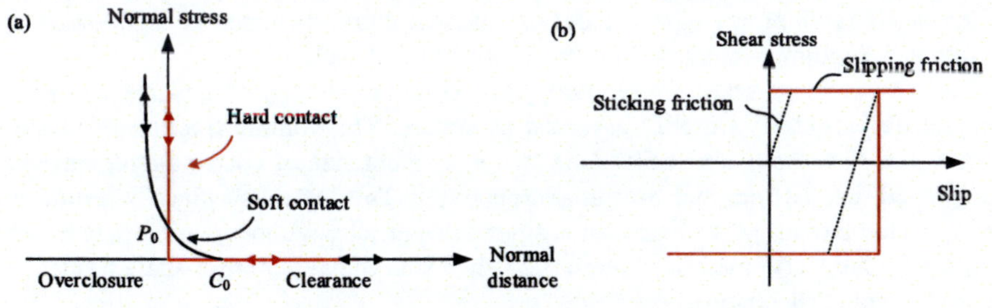

Fig. 3.26 Surface-to-surface contact model: **a** Pressure-overclosure relations in the normal direction; and **b** Friction behavior in the tangential direction

$$p = \begin{cases} 0, c \geq c_0 \\ \frac{P_0}{\exp(1)-1}\left[\left(1 - \frac{c}{c_0}\right)\left(\exp\left(1 - \frac{c}{c_0}\right) - 1\right)\right], & c < c_0 \end{cases} \quad (3.25)$$

where c_0 is the initial contact distance, p is the normal contact pressure, and P_0 is the pressure value at zero opening.

In the tangential direction of the surfaces, the Coulomb friction was used and simulated by a penalty method, similar to that in the study of Tang et al. (2019). The tangential slipping occurs when the shear stress on the contact surface is greater than the maximal friction between the surfaces, as illustrated in Fig. 3.26b. The friction coefficient μ between non-corroded reinforcement and concrete can be set as 0.3 (Wang and Liu 2004).

3.6 Validation of Simulation Method

After introducing the details of the meso-scale numerical model, the reasonability and efficiency of the model is presented through comparison between the simulation results with the available experimental observations at both material and component levels.

3.6.1 Material

The mechanical properties of concrete materials show the nonlinearity due to the heterogeneity of meso-components. Accordingly, the meso characteristics of concrete materials should be definitely described in meso-scale numerical simulation. Therefore, concrete is usually regarded as a tri-phase composite material consisting of coarse aggregates, mortar matrix and ITZs. Here in this part, a 3D meso-scale random aggregate model of concrete cubic specimen is established to explore the splitting-tensile behavior of concrete materials.

In the 3D meso-scale model, two-grade concrete is adopted and the shape of aggregate particle is simplify assumed as sphere. The volume fraction of coarse aggregate particles is about 50% and two equivalent size of coarse aggregates are employed, i.e. 12 mm and 30 mm, respectively. The "take-and-place" method is adopted to randomly distribute coarse aggregate particles without overlapping based on the Monte Carlo method. Figure 3.27 shows a typical 3D meso-scale model of concrete cubic specimen.

The loading method and boundary conditions for the 3D meso-scale model of concrete cubic specimen are introduced as follows: (1) The vertical motion of the bottom surface of concrete models is constraint; (2) In order to prevent the local crush, the rigid plate (whose width is a, thickness is $0.1a$, and length is equal to the specimen length b. $a / b = 0.06$) is set on the middle area of the upper surface of

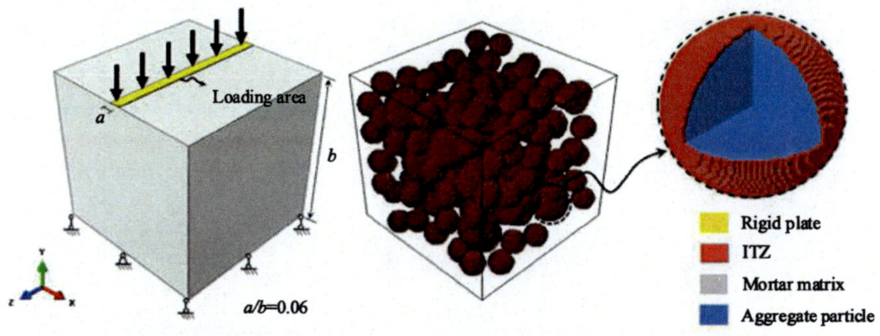

Fig. 3.27 3D meso-scale modeling model of concrete cubic specimen for splitting-tensile loads

concrete specimen. The contact surface between the rigid plate and concrete cubic specimen is set as tie constraint; (3) The velocity v is applied on the upper surface of the rigid plate. Thus, the nominal strain rate $\dot{\varepsilon}$ can be calculated as $\dot{\varepsilon} = v/b$, in which b is the side length. Moreover, the quasi-static strain rate is defined as $\dot{\varepsilon}_0 = 10^{-5}/\text{s}$; (4) The other surfaces of models are left free.

The ITZ in reality is a layered structure with 10–50 μm thickness. However, an expensive computational cost will be demanded if the real ITZ thickness is applied in the meso-scale models, especially for 3D models (Zhou and Hao 2008a). Therefore, the ITZ thickness is set as 0.5–2 mm in most meso-scale numerical studies (Zhou and Hao 2008a; Pedersen et al. 2013; Zhang et al. 2019; Zhou et al. 2017). Kim and Abu Al-Rub (2011) set ITZ thickness as 0.1–0.8 mm and found that the change of ITZ thickness has slight influence on the post-peak behavior of concrete. Song and Lu (2012) conduct a convergence analysis (the ITZ thickness, along with the nominal grid size, is varied in a range of 0.5–2 mm). They concluded that the limited variation of the equivalent ITZ thickness only affects the descending branch, while the stress increase phase is almost identical. In order to investigate the effect of ITZ thickness on the meso-scale simulation results, the splitting-tensile behavior of cubic specimens (side length = 150 mm, ITZ thickness = 0.5 mm, 1.0 mm and 2.0 mm) under a low strain-rate ($\dot{\varepsilon} = 10^{-3}/\text{s}$) and a high strain-rate ($\dot{\varepsilon} = 10/\text{s}$) were simulated based on the meso-scale numerical method. The numerical results of load- displacement curves with different ITZ thicknesses are plotted in Fig. 3.28. Under the low strain rates (such as $\dot{\varepsilon} = 10^{-3}/\text{s}$), one can find that the ITZ thickness (0.5 mm, 1.0 mm and 2.0 mm) has a slight influence on the load-displacement curve, especially on the post-peak behavior of concrete materials. As for the peak load, although there are some slight differences in models with different ITZ thicknesses, the disparity could be acceptable. Under the high strain rates (such as $\dot{\varepsilon} = 10/\text{s}$), the load-displacement curves are basically the same and the ITZ thickness has an ignorable effect on the peak load. Therefore, taking into account the computation efficiency, the appropriate thickness is utilized as 2 mm without much loss of accuracy in 3D simulations.

The damaged plasticity model is widely employed in describing the static and dynamic mechanical behavior of concrete materials. The experimental observations

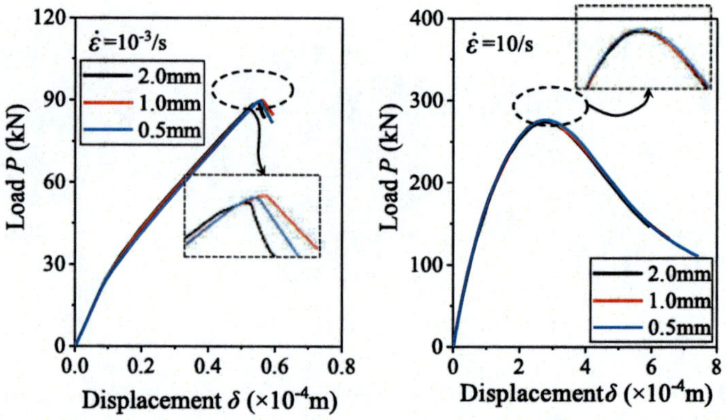

Fig. 3.28 Effect of ITZ thickness on the splitting tensile behavior of concrete under low and high strain-rates

in the study of Grote et al. (2001) have indicated that the dynamic mechanical behavior of mortar matrix is closely similar with that of concrete materials. Due to the lack of test results, the properties of ITZ are still not well understood. It is well accepted that the mechanical properties of ITZ are similar to those of mortar matrix and they could be described by the weakened mechanical parameters of mortar matrix (Kim and Abu Al-Rub 2011). As for the aggregate phase, Zhou and Hao (2008b) and Snozzi et al. (2011) conducted tensile and compressive tests on concrete specimens and found that no damage generates in the aggregate phase, indicating that tensile and compressive strengths of aggregate particles are much larger than those of mortar matrix and ITZ. Pedersen et al. (2013) assumed the aggregate as linear elastic body in numerical simulations. Actually, this assumption may not be reasonable sometimes, because cracks may penetrate into some aggregate particles under loads, especially in the case of high strain rates. Therefore, similar to the treatment on ITZ, the mechanical properties of aggregate particles can be treated as a kind of mortar matrix with stronger mechanical parameters. Therefore, the mechanical properties of aggregate, mortar matrix and ITZ are characterized by the damaged plasticity model in the meso-scale simulations.

For the simulations of failure of concrete materials under dynamic load, the strain rate effect of materials should not be ignorable. Dilger et al. (1984) and Bischoff and Perry (1991) carried out the tests on dynamic mechanical properties of concrete and indicated that, comparing with tensile and compressive strengths, the other mechanical parameters like modulus of elasticity, fracture energy and Poisson's ratio are less sensitive to strain rate. Therefore, only the strength enhancement behavior of the meso-components, which is characterized by DIF, were considered in the simulations. The details have been introduced in Sect. 3.4.

The 8-nodded hexahedral elements (type C3D8R: Continuum, 3D, 8-node, reduced integration) are adopted and the sharing node is applied to connect the

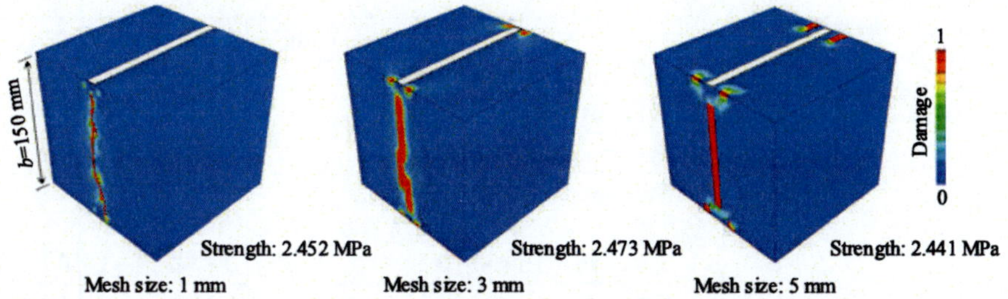

Fig. 3.29 Numerical results of concrete cubic specimens having different mesh sizes under static load

aggregates, mortar matrix and ITZs. To discuss the influences of mesh size on numerical models, the splitting-tensile behavior of concrete cubic specimens meshed with average sizes of 1, 3 and 5 mm were simulated under static load (quasi-static load $\dot{\varepsilon} = 10^{-5}/\text{s}$) and the typical numerical results are presented in Fig. 3.29.

It can be seen from Fig. 3.29 that the mesh size of meso-scale model basically has no obvious influence on the apparent strength of specimen and models meshed with small sizes exhibit meticulous failure patterns. Therefore, the approximate mesh size of meso-scale numerical models is adopted as 2 mm. It should be noted that for the purpose of preserving the well-posedness of the boundary value problem and including a characteristic length of microstructure for simulations of a deterministic size effect, a nonlocal theory should be employed as a regularization technique. The tensile post-failure behavior is given in terms of a fracture energy cracking criterion by specifying a stress-displacement curve instead of a stress-strain curve to avoid or relax unreasonable mesh sensitive results (Du et al. 2013a; Huang et al. 2016). The use of a fracture energy approach provides mesh-independent results for load-displacement diagrams only.

The size effect tests on the splitting-tensile behavior of concrete cubic specimens under static load conducted by Du et al. (2017) are selected for the validation of the above 3D meso-scale numerical model. The related parameters (e.g. Young's modulus, compressive strength, tensile strength, etc.) for the meso-components applied in meso-scale simulations refer to the tests of Du et al. (2017), which are listed in Table 3.1.

Actually, it is difficult to obtain the mechanical parameters of ITZ and they can be determined by an inversion method, which is a repeated trial algorithm. It means that a series of mechanical parameters for ITZ phase are involved to simulate the macroscopic behavior of concrete and the numerical result is compared to that in tests. Then the parameters by which the most appropriate numerical results could be obtained are selected. After large quantity of repeated numerical trials, the mechanical parameters shown in Table 3.1 are adopted in simulation. On this base, the failure patterns and nominal splitting-tensile strength of concrete cubic specimens with different structural sizes under static load (quasi-static load $\dot{\varepsilon} = 10^{-5}/\text{s}$) are obtained. Typically, the comparison of splitting-tensile behavior between numerical

Table 3.1 Mechanical parameters of concrete meso-components for simulation of splitting-tensile behavior

	Aggregate particle	Mortar matrix	ITZ
Young's modulus, E (GPa)	50[a]	30[a]	26[b]
Poisson ratio, ν	0.16	0.20	0.22
Dilatant angle, ψ (°)	30	30	30
Eccentricity, η (%)	0.1	0.1	0.1
Stress ratio, f_{b0}/f_{c0}	1.16	1.16	1.16
Fracture energy G_c (J/m^2)	60	50	30
Compressive strength, σ_c (MPa)	130[a]	30[a]	28.4[c]
Tensile strength, σ_t (MPa)	10[a]	3[a]	2.8[c]

Note [a]Data were obtained from the tests (Du et al. 2017); [b]Data were got from the literatures (Du et al. 2013); and [c]Data were obtained by the inversion method

and test results of concrete cubic specimens are presented in Fig. 3.30. It can be seen that the main crack appears in the middle line of cubic specimen in the simulations, which is in good agreement with the test results. As shown in Fig. 3.30b, on the failure surface some aggregate particles are split by cracks, which is also similar with that in the tests. In addition, the splitting-tensile strength of numerical models with different structural sizes are approaching those of test specimens, as described in Fig. 3.30c. These comparisons indicate that the 3D meso-scale numerical model with the applied mechanical parameters could yield relatively acceptable and reliable predictions of the mechanical failure behavior of concrete materials.

3.6.2 Beam

Different from the concrete materials, the mechanical failure behavior and corresponding size effect of RC members are more complex. The size effect behavior of RC members is mainly affected by two aspects: (1) the heterogeneity and mechanical nonlinearity of concrete materials; (2) the complex interaction between reinforcement and concrete. Therefore, the above two aspects should be taken into account in order to analyze the size effect behavior of RC members more truly and accurately through numerical simulation. Firstly, the validation of simulation method on RC beams was conducted in this section.

As introduced in Sect. 3.2, considering the heterogeneity and mechanical nonlinearity of concrete material itself, concrete is regarded as a three-phase composite material composed of coarse aggregate particles, mortar matrix and ITZ. After

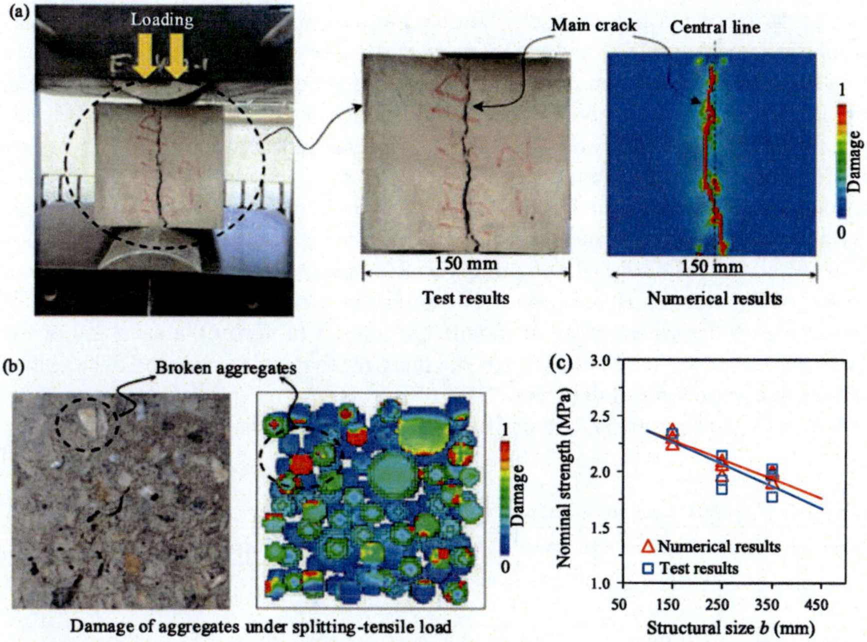

Fig. 3.30 Comparison of splitting-tensile behavior between numerical results and test results: **a** Failure patterns of specimen; **b** Failure patterns of aggregates; and **c** Nominal strength

inserting the reinforcement cage, the 3D meso-scale mechanical analysis model of RC cantilever beam can be generated. The loading is controlled by displacement, as presented in Fig. 3.31.

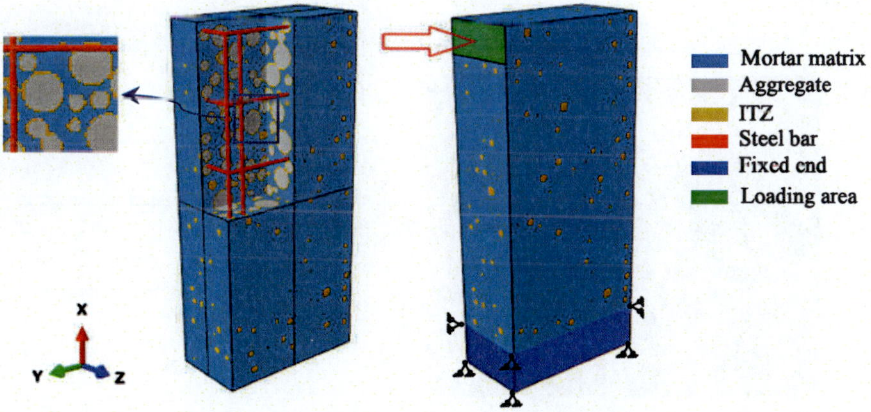

Fig. 3.31 3D meso-scale mechanical analysis model of RC cantilever beam

The damaged plasticity model is adopted to describe the mechanical properties of mortar matrix and ITZ, and the aggregate particles are set as elastic body due to their high strength. Reinforcing steel is generally regarded as homogeneous material and its mechanical behaviors are described by the ideal elastic-plastic model. The non-linear spring is adopted to simulate the interaction between the concrete and reinforcing bars while the bond-slip constitutive model recommended in the Chinese code (Ministry of Construction of People's Republic of China 2011) is adopted to characterize the behavior springs.

Jin et al. (2016) designed five groups of RC cantilever beams to study the shear failure of RC beams. Herein, the group of beam specimens with section size of 160 mm × 400 mm are taken as simulation objects to verify the accuracy of the meso-numerical model. Table 3.2 lists the main mechanical parameters of steel bars and meso-components of concrete.

Figure 3.32 presents the comparison between the test results and the simulation

Table 3.2 Mechanical parameters of concrete and steel bars

Mechanical parameters	Aggregate	Mortar matrix	ITZ	Longitudinal bar	Stirrups
Compressive strength σ_c (MPa)		39.9[a]	31.9[b]		
Tensile strength σ_t (MPa)		3.9[a]	3.2[b]		
Fracture energy G_c (J/m^2)		50	30		
Elastic modulus E (GPa)	70	42.8[a]	38.7[b]	196[a]	210[a]
Poisson ratio ν	0.2	0.2	0.2	0.3	0.3
Yield strength f_y (MPa)				392.1[a]	297.9[a]
Reinforcement ratio ρ (%)				1.14[a]	0.14[a]

Note [a]Data were obtained from the tests (Jin et al. 2016); and [b]Data were obtained by the inversion method

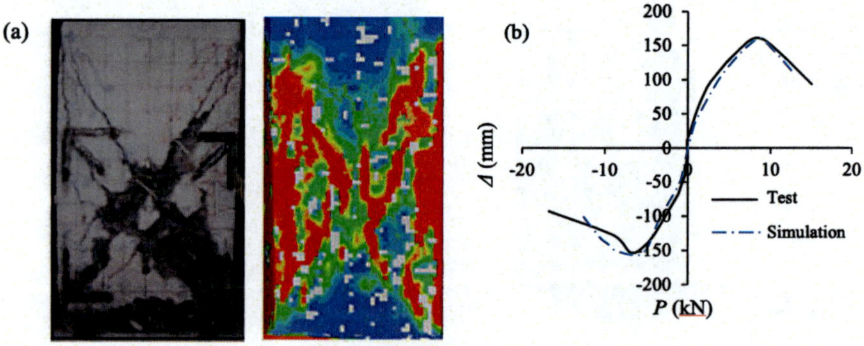

Fig. 3.32 Verification of the meso-scale numerical model for RC beams: **a** Comparison of failure patterns; and **b** Comparison of skeleton curves

results. As presented in Fig. 3.32 a, the failure patterns of the cantilever beam obtained from simulation is similar to the test results, and X-type critical cracks are produced under cyclic load. In addition, as shown in Fig. 3.32b, the peak load and skeleton curve in simulations are also similar to the test results. Therefore, the rationality and accuracy of the 3D meso-scale numerical analysis model are verified.

3.6.3 Column

To validate the simulation method for RC columns, a mesoscopic numerical model of RC columns has been established according to the experimental program in Jin et al. (2018b), as shown in Fig. 3.33. The steel reinforcement cage is set up individually after the generation of the concrete partition based on the reinforcement detail in Jin et al. (2018b), and then it is simply inserted into the concrete partition. The interaction between concrete and steel bar was considered according to the method in Sect. 3.5.1. The boundary and loading conditions are described in Fig. 3.33b. Considering the friction provided by the test setup, the base of the model is set to be pinned-end to simulate those in the experiment. A constant load is applied on the top surface while a displacement-controlled horizontal cyclic load is applied at the upper area of the side surface. Other surfaces of the model are left free.

The mechanical behavior of steel bar is described by an ideal elastic-plastic model. The aggregate is set as elastic, while the mortar matrix and ITZ are modeled by the damaged plasticity model. The major mechanical parameters of the three meso-components, as well as the steel rebar, used in the simulations are provided in Table 3.3. It is to be noted that, the Young's modulus, Poisson's ratio, compressive strength and tensile strength for the aggregate particles and the mortar matrix are obtained from Jin et al. (2018b), while the mechanical properties for the ITZs

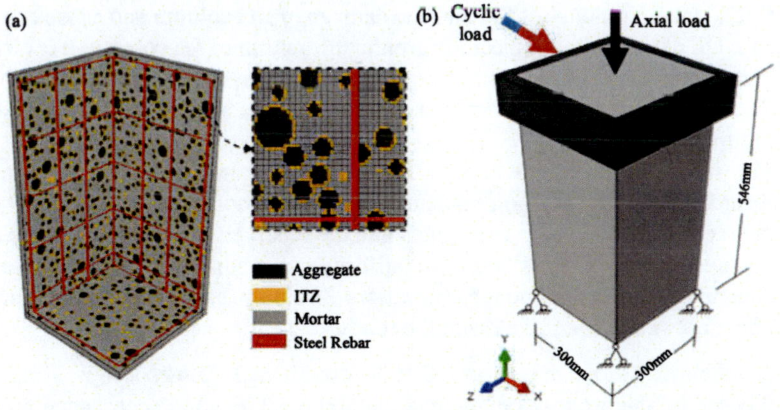

Fig. 3.33 Numerical model of RC columns: **a** Mesoscale components; and **b** Boundary conditions

Table 3.3 Mechanical parameters of the meso components and steel rebar

	Mortar matrix	ITZ	Aggregate particle	Steel rebar HRB400/HPB300
Young's modulus E (GPa)	32.5[a]	26[b]	70[a]	210[a]
Poisson ratio ν	0.20[a]	0.22[b]	0.16[a]	0.3[a]
Dilatant angle ψ (°)	18	15	–	–
Eccentricity η (%)	0.1	0.1	–	–
Stress ratio σ_{b0}/σ_{c0}	1.16	1.16	–	–
Compressive strength f_c (MPa)	31.6[a]	23.9[b]	80[a]	–
Tensile strength f_t (MPa)	2.92[a]	2.4[b]	8[a]	–
Yield strength f_y (MPa)	–	–	–	400/300[a]

Note [a]Data are obtained from the tests (Jin et al. 2018b); [b]Data are given with reasonable assumption; and the other parameters are default values or determined by referring to Marzec et al. (2007) and Syroka-Korol et al. (2014)

are obtained using an inversion method. The other parameters are default values or determined by referring to Marzec et al. (2007) and Syroka-Korol et al. (2014).

Figure 3.34 presents the comparison of the final failure pattern and the load-deflection relationship between the real column and the numerical model. It can be clearly seen from Fig. 3.34 that both the numerical model and the real column exhibits a compression-shear failure mode. The main crack is a diagonal splitting crack because the high compression ratio of 0.6. It can be also noticed that there is a wider plastic band from the center to the left corner of bottom. The possible explanations on the difference between the numerical and experimental results are interpreted as:

(1) Cyclic loading increases the internal damage of RC column and cause concrete strength deterioration, especially around the bottom of the column. It is obvious that in a real specimen, the spalling of the concrete cover is frequent and vast. The real plastic zone may not be as large as the one be seen from the surface of the specimen.
(2) The steel reinforcement cage is simply inserted into the concrete partition regardless of the meso-components in the mesoscopic model. Specifically, the interactions between the aggregates and the rebar are not taken into consideration during the numerical tests, but uniformly set a nonlinear spring between the concrete partition and the rebar partition. This may lead to an uncertain mutual effect between the concrete and the reinforcement.

Nevertheless, it still can be drawn that the established 3D mesoscopic simulation model based on the proposed numerical method exhibits good accordance with the experimental results.

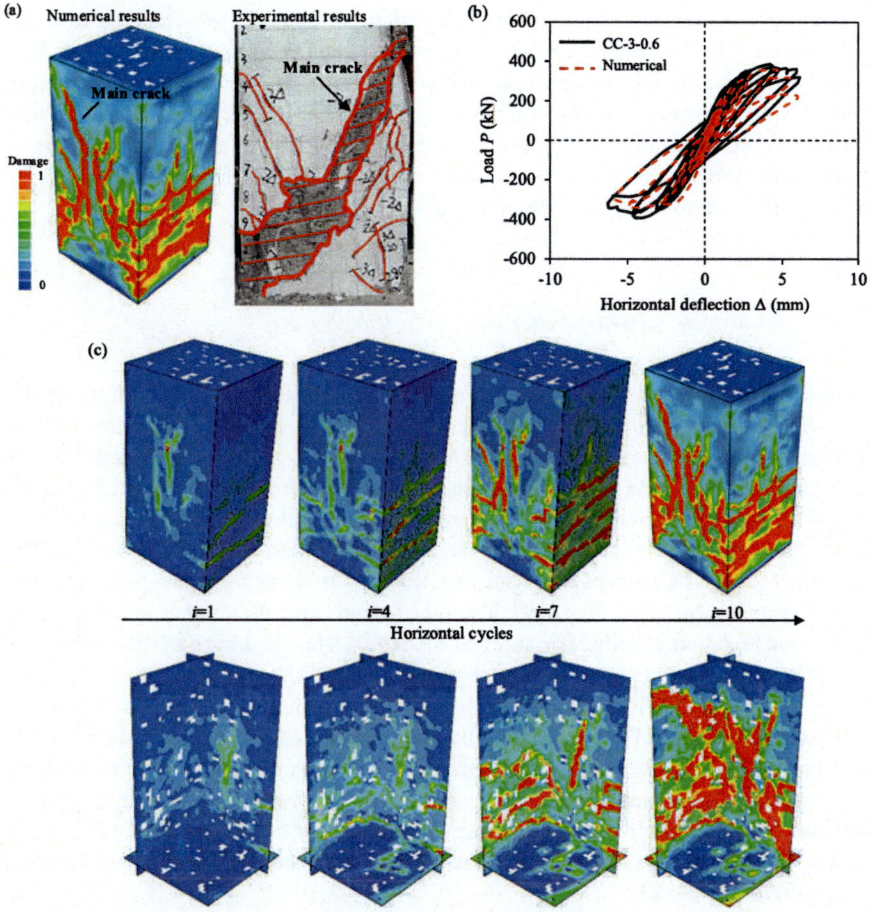

Fig. 3.34 Simulation results of the RC column sized of 300 mm × 300 mm × 546 mm: **a** Final failure pattern; **b** Load-deflection relationship; and **c** Failure process

One of the most important advantages of mesoscale method compared to macro-scale method is that, the mesoscopic mechanical model could provide a detailed failure process inside the specimen vividly, such as how cracks initiated and propagated among aggregates, mortar matrix, ITZs and rebar. Figure 3.34c presents the failure process of the cyclically loaded RC column at the meso-scale. It is well known that fracture and damage are multi-level phenomena, where micro-cracks may coalesce into macro-cracks. Generally, difference between the elastic moduli and the pronounced differential shrinkage at early stages increase the stress concentrations in the ITZs, which result in the weakness of the ITZs. More initial defects must be expected in the ITZs than that in mortar, and consequently the micro-cracks are initiated in the ITZs. With the increasing load, the micro-cracks remain stable up to about 30% of the ultimate strength and then they start to grow in length, width and number. The propagation of micro-cracks within the ITZs corresponds to 50% of the ultimate strength. Larger cracks are resulting from coalescence of interface cracks in areas where neighboring micro-cracks get to closer. The coalescence of cracks and their rapid propagation corresponds to 70–90% of the ultimate strength. With the increasing deformation of the specimen and the yielding of the rebar, the RC column is finally corrupted by a diagonal splitting crack.

3.6.4 Beam-to-Column Joint

Similarly, for validation of the simulation method on the beam-to-column joints, the concrete of RC joints at the meso-level is generally considered as a three-phase heterogeneous composite material, including aggregate, mortar matrix and ITZ (Jin et al. 2018a). Aggregate particles are idealized into spheres, and the spatial position of aggregate (minimum and maximum equivalent particle sizes: 12 mm and 30 mm, respectively; volume fraction: 35%) is determined randomly. The thickness of ITZs was set as 2 mm. Moreover, steel bars are inserted to form a 3D meso-scale model of RC joints as shown in Fig. 3.35. The meso-structures of concrete are divided by eight-node hexahedral reduction integral elements. The steel bars are discretized by beam elements, and the average size of grid elements is 2 mm.

The damaged plasticity model is adopted to describe the mechanical properties of mortar matrix and ITZs, and the aggregate particles are set as elastic body due to their high strength. The ideal elastic-plastic constitutive model is used to describe the mechanical behavior of steel bar, and von Mises yield criterion was employed. The model recommended in the Chinese code (Ministry of Construction of People's Republic of China 2011) is adopted to characterize the bond-slip relationship between the concrete and steel bars.

The mechanical parameters of the meso-components of concrete and steel bar are shown in Table 3.4. The mechanical parameters of aggregate and mortar matrix are measured. The mechanical parameters of ITZs are determined by

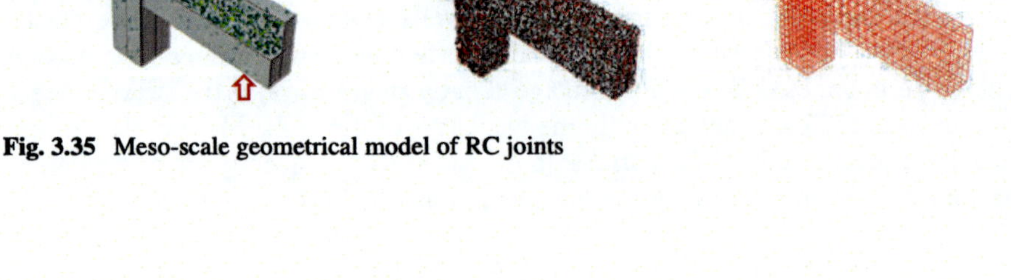

Fig. 3.35 Meso-scale geometrical model of RC joints

Table 3.4 Mechanical parameters of the meso-components of concrete and steel bars

Meso-scale composition	Aggregate	Mortar	ITZ	Steel bar
Compressive strength f_c (MPa)	–	36.2[a]	29.0[b]	–
Tensile strength f_t (MPa)	–	3.4[a]	2.7[b]	–
Fracture energy G_c (J/m^2)		45.0[a]	27.0[b]	
Young's modulus E (GPa)	70.0[a]	32.1[a]	25.7[b]	203.0[a]
Poisson ratio ν	0.2[a]	0.2[a]	0.2[b]	0.3[a]
Yield strength f_y (MPa)	–	–	–	496[a]

Note [a]Data are obtained from the tests (Jin et al. 2018a); and [b]Data are given with reasonable assumption

repeated trial calculation. As shown in Table 3.4, the uniaxial compressive/splitting-tensile strength of concrete obtained by simulation based on this set of parameters is 34.20 MPa/3.10 MPa, which is in good agreement with the test value of 34.55 MPa/3.06 MPa. Therefore, it can be considered that the mechanical parameters of ITZs utilized are reasonable.

The numerical simulation results of RC joints were compared with the physical tests conducted by Jin et al. (2018a). The specimen whose stirrup ratio ρ_{sv} is 0.84% and structure size is 300 mm × 300 mm was taken as an example to conduct model verification.

The shear failure of RC joints is simulated using 3D numerical model with average mesh sizes, i.e. 1 mm, 2 mm, 5 mm. The failure patterns of tested and simulated specimens are shown in Fig. 3.36. As shown in Fig. 3.36, the failure patterns obtained by simulation (mesh size is 2 mm) and tests are basically the same. With diagonal cracks appearing in core area of joints and cracks extending to the beam end, the core area of joints shows typical shear failure.

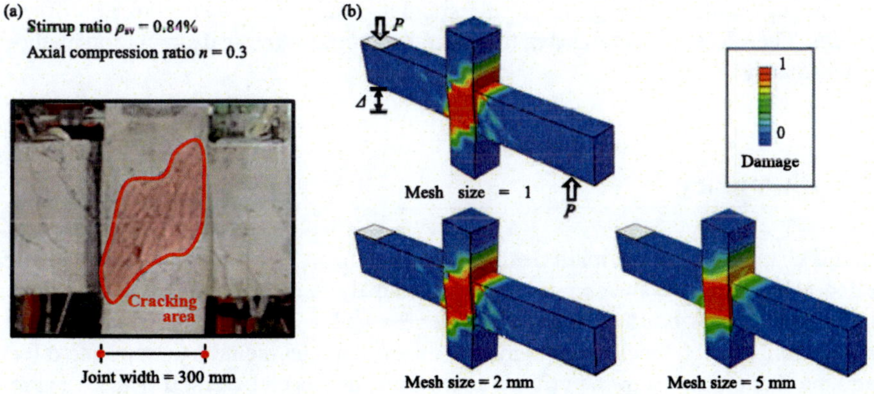

Fig. 3.36 Shear failure patterns of the RC beam-to-column joints: **a** Test observations; and **b** Simulation results with different average mesh sizes

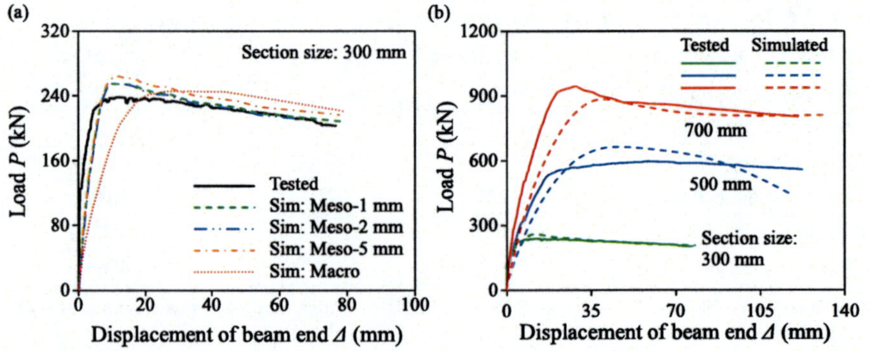

Fig. 3.37 Load-displacement curves of joints with axial compression ratio n = 0.3 and stirrup ratio $\rho_{sv} = 0.84\%$: **a** Determination of simulation method; and **b** P-Δ curves of test and simulation (Mesh size: 2 mm)

Figure 3.37 shows the relation curves of load (P)-the beam end displacement (Δ) of RC joints with different section sizes. It also gives the comparison between simulation and test results obtained from different numerical methods (mesoscopic and macroscopic) and different mesh sizes (1 mm, 2 mm, 5 mm).

As can be seen from Fig. 3.37a, the simulation results of the mesoscopic analysis model are more consistent with the experimental results than those of the macroscopic model. It has been pointed out by Du et al. (2013a) that the macroscopic model fails to reflect the mesoscopic heterogeneity of concrete, and thus the size effect of RC members is not able to be fully reflected when macroscopic simulation methods are utilized. Therefore, mesoscopic simulation is selected as the numerical method to study shear failure and size effect of joints in the following numerical study. Considering the calculation amount of numerical simulation and the accuracy of simulation results, the mesh size adopted in the meso-scale simulation is set as 2 mm. As observed in Fig. 3.37b, the simulated data agree well with experiment results. Therefore, it can be drawn that the mesoscopic numerical simulation method is reasonable.

3.7 Summary

In this chapter, one of the main methodologies adopted in this book to study the size effect at both material and component levels, namely, the meso-scale numerical simulation approach is briefly introduced. After the review of the available mesoscopic simulation models, the geometric model, material model including strain rate effect and interaction model in the utilized meso-scale approach based on random aggregate model are described. Finally, the reasonability and efficiency of the meso-scale approach in simulating the failure behavior of concrete material and RC structural components (including beams, columns and joints, etc.) are presented.

References

Aljaafreh T (2016) Strengthening of lightweight reinforced concrete beams using carbon fiber reinforced polymers (CFRP) [D]. The University of Texas, USA Arlington

Barbato M (2009) Efficient finite element modelling of reinforced concrete beams retrofitted with fibre reinforced polymers [J]. Comput Struct 87(3/4):167–176

Bažant ZP, Tabbara MR (1990) Random particle models for fracture of aggregate or fiber composites [J]. ASCE J Eng Mech 116(8):1686–1705

Bischoff PH, Perry SH (1991) Compressive behaviour of concrete at high strain rates [J]. Mater Struct 24(6):425–450

Bolander JE, Saito S (1998) Fracture analyses using spring networks with random geometry [J]. Eng Fract Mech 61:569–591

Caballero A, Lopez CM, Carol I (2006) 3D meso-structural analysis of concrete specimens under uniaxial tension [J]. Comput Methods Appl Mech Eng 195(52):7182–7195

Camborde F, Mariotti C, Donze FV (2000) Numerical study of rock and concrete behavior by discrete element modeling [J]. Comput Geotech 27(4):225–247

Chaboche JL (1991) On some modifications of kinematic hardening to improve the description of ratchetting effects [J]. Int J Plasticity 7(7):661–678

Chen G, Hao Y, Hao H (2015) 3D meso-scale modelling of concrete material in spall tests [J]. Mater Struct 48(6):1887–1899

Chen H, Xu B, Mo YL, Zhou T (2018) Behavior of meso-scale heterogeneous concrete under uniaxial tensile and compressive loadings [J]. Construct Build Mater 178:418–431

Chi Y, Yu M, Huang L, Xu L (2017) Finite element modeling of steel-polypropylene hybrid fiber reinforced concrete using modified concrete damaged plasticity [J]. Eng Struct 148:23–35

Chiaia B, Vervuurt A, VanMier JGM (1997) Lattice model evaluation of progressive failure in disordered particle composites [J]. Eng Fract Mech 57(2/3):301–313

Cholostiakow S, Di BM, Pilakoutas K (2018) Effect of beam depth on shear behavior of FRP RC beams [J]. ASCE J Compos Construct 23(1):04018075

Comité euro-international du béton, Fédération Internationale de la Précontrainte (1993) CEB-FIP model code 1990: Design code [M]. Thomas Telford Publishing, Lausanne, Switzerland

Computing A (1997) HyperMesh reference manual—version 2.1 [J]. Altair Computing, Incorporated, 1090:1091

Cruz JS, Barros J (2004) Modeling of bond between near-surface mounted CFRP laminate strips and concrete [J]. Comput Struct 82(17):1513–1521

Cundall A, Hart RD (1993) Numerical modeling of discontinua [M]. Pergamon, Analysis and design methods

Cundall PA, Strack ODL (1979) A discrete numerical model for granular assemblies [J]. Geotechnique 29(1):47–65

Cusatis G (2011) Strain rate effects on concrete behavior [J]. Int J Impact Eng 38(4):162–170

Dafalias YF, Popov EP (1975) A model of nonlinearly hardening materials for complex loading [J]. Acta Mech 21(3):173–192

Davood M, Amirhomayoon TK (2013) Experimental study on effect of EBR and EBROG methods on debonding of FRP sheets used for shear strengthening of RC beams [J]. Compos Part B: Eng 45(1):1704–1713

De Schutter G, Taerwe L (1993) Random particle model for concrete based on delaunay triangulation [J]. Mater Struct 26(156):67–73

Desnerck P, Lees JM, Morley CT (2015) Bond behaviour of reinforcing bars in cracked concrete [J]. Construct Build Mater 94:126–136
Dilger WH, Koch R, Kowalczyk R (1984) Ductility of plain and confined concrete under different strain rates [C]. J Proc 81(1):73–81
Drucker DC (1988) Conventional and unconventional plastic response and representation [J]. Appl Mech Rev 41(4):151–167
Du X, Jin L (2012a) Meso-element equivalence model for analysis of macroscopic mechanical properties of concrete materials [J]. Chinese J Comput Mech 29(5):654–661 (in Chinese)
Du X, Jin L (2012b) Research on macroscopic mechanical properties of concrete considering the effects of pores and microcracks [J]. Eng Mech 29(8):101–107 (in Chinese)
Du X, Jin L (2012c) Research on the macroscopic mechanical properties of concrete considering the influence of interface in transition zone [J]. Eng Mech 29(12):72–79 (in Chinese)
Du X, Jin L, Ma G (2013a) A meso-scale analysis method for the simulation of nonlinear damage and failure behavior of reinforced concrete members [J]. Int J Damage Mech 22(6):878–904
Du X, Jin L, Ma G (2013b) Meso-element equivalent method for the simulation of macro mechanical properties of concrete [J]. Int J Damage Mech 22(5):617–642
Du X, Jin L, Ma G (2014) Numerical simulation of dynamic tensile-failure of concrete at meso-scale [J]. Int J Impact Eng 66:5–17
Du M, Jin L, Li D, Du X (2017) Mesoscopic simulation study of the influence of aggregate size on mechanical properties and specimen size effect of concrete subjected to splitting tensile loading [J]. Eng Mech 34(09):54–63 (in Chinese)
Dupray F, Malecot Y, Daudeville L, Buzaud E (2009) A mesoscopic model for the behaviour of concrete under high confinement [J]. Int J Numer Anal Methods Geomech 33:1407–1423
Frederick CO, Armstrong PJ (2007) A mathematical representation of the multiaxial Bauschinger effect [J]. Mater High Temp 24(1):1–26
Fu G, Dekelbab W (2003) 3-D random packing of polydisperse particles and concrete aggregate grading [J]. Powder Technol 133(1–3):147–155
Fuller WB, Thompson SE (1907) The laws of proportioning concrete [J]. Trans Am Soc Civil Eng 59:67–143
Grassl P, Xenos D, Nyström U, Rempling R, Gylltoft K (2013) CDPM2: a damage-plasticity approach to modelling the failure of concrete [J]. Int J Solids Struct 50(24):3805–3816
Grondin F, Matallah M (2014) How to consider the Interfacial Transition Zones in the finite element modelling of concrete? [J]. Cem Concr Res 58:67–75
Grote DL, Park SW, Zhou M (2001) Dynamic behavior of concrete at high strain rates and pressures: I. Experimental characterization [J]. Int J Impact Eng 25(9):869–886
Häfner S, Eckardt S, Könke C (2005) A geometrical inclusion-matrix model for the finite element analysis of concrete at multiple scales [M]. Professur Informatik im Bauwesen
Häfner S, Eckardt S, Luther T, Könke C (2006) Mesoscale modeling of concrete: geometry and numerics [J]. Comput Struct 84:450–461
Han LH, Li W, Bjorhovde R (2014) Developments and advanced applications of concrete-filled steel tubular (CFST) structures: Members [J]. J Construct Steel Res 100:211–228
Hao Y, Hao H (2011) Numerical evaluation of the influence of aggregates on concrete compressive strength at high strain rate [J]. Int J Protective Struct 2(2):177–206
Hao Y, Hao H (2013) Numerical investigation of the dynamic compressive behaviour of rock materials at high strain rate [J]. Rock Mech Rock Eng 46(2):373–388
Hao Y, Hao H (2014) Influence of the concrete DIF model on the numerical predictions of RC wall responses to blast loadings [J]. Eng Struct 73:24–38
Hao Y, Hao H, Li ZX (2010) Numerical analysis of lateral inertial confinement effects on impact test of concrete compressive material properties [J]. Int J Protective Struct 1(1):145–167
Hao Y, Hao H, Jiang GP, Zhou Y (2013a) Experimental confirmation of some factors influencing dynamic concrete compressive strengths in high-speed impact tests [J]. Cem Concr Res 52:63–70
Hao Y, Hao H, Li ZX (2013b) Influence of end friction confinement on impact tests of concrete material at high strain rate [J]. Int J Impact Eng 60:82–106

Hao H, Hao Y, Li J, Chen W (2016) Review of the current practices in blast-resistant analysis and design of concrete structures [J]. Adv Struct Eng 19(8):1193–1223
Hao Y, Huang X, Hao H (2017) Mesoscale modelling of concrete reinforced with spiral steel fibres under dynamic splitting tension [J]. Adv Struct Eng 21(8):1197–1210
Harajli M, Hamad B, Karam K (2002) Bond-slip response of reinforcing bars embedded in plain and fiber concrete. ASCE J Mater Civ Eng 14(6):503–511
Hashiguchi K, Ozaki S (2008) Constitutive equation for friction with transition from static to kinetic friction and recovery of static friction [J]. Int J Plast 24(11):2102–2124
Hashiguchi K, Ueno M, Ozaki T (2012) Elastoplastic model of metals with smooth elastic–plastic transition [J]. Acta Mech 223(5):985–1013
Hashiguchi K (2000) Fundamentals in constitutive equation: continuity and smoothness conditions and loading criterion [J]. Soils Found 40(4):155–161
Hashin Z (1980) Failure criteria for unidirectional fiber composites [J]. J Appl Mech 47(2):329–334
Hau-Kit Man, van Mier JGM (2008) Size effect on strength and fracture energy for numerical concrete with realistic aggregate shapes [J]. Int J Fract 154:61–72
Hosen MA, Jumaat MZ, Islam ABMS (2015) Side Near Surface Mounted (SNSM) technique for flexural enhancement of RC beams [J]. Mater Des 83:587–597
Huang Z (2010) Modelling the bond between concrete and reinforcing steel in a fire [J]. Eng Struct 32(11):3660–3669
Huang YJ, Yang ZJ, Chen XW, Liu GH (2016) Monte Carlo simulations of meso-scale dynamic compressive behavior of concrete based on X-ray computed tomography images [J]. Int J Impact Eng 97:102–115
Iwan WD (1967) On a class of models for the yielding behavior of continuous and composite systems [J]. J Appl Mech 34(3):612–617
Jendele L, Cervenka J (2006) Finite element modelling of reinforcement with bond [J]. Comput Struct 84(28):1780–1791
Jin L, Du XL, Li D, Su X (2016) Seismic behavior of RC cantilever beams under low cyclic loading and size effect on shear strength: an experimental characterization [J]. Eng Struct 122:93–107
Jin L, Miao LY, Han J, Du X, Wei N, Li D (2018a) Size effect tests on shear failure of interior RC beam-to-column joints under monotonic and cyclic loadings [J]. Eng Struct 175:591–604
Jin L, Zhang S, Li D, Xu H, Du X, Li Z (2018b) A combined experimental and numerical analysis on the seismic behavior of short reinforced concrete columns with different structural sizes and axial compression ratios [J]. Int J Damage Mech 27(9):1416–1447
Jin L, Lan Y, Zhang R, Du X (2019) Impact performances of RC beams at/after elevated temperature: a meso-scale study [J]. Eng Fail Anal 105:196–214
Jumaa GB, Yousif AR (2019) Numerical modeling of size effect in shear strength of FRP-reinforced concrete beams [C]. Structures 20:237–254 Elsevier
Kalfat R, Al-Mahaidi R, Smith ST (2013) Anchorage devices used to improve the performance of reinforced concrete beams retrofitted with FRP composites: state-of-the art review [J]. J Compos Constr 17(1):14–33
Kawai T (1977) New element models in discrete structural analysis [J]. J Soc Naval Architects Japan 1977(141):174–180
Kim S, Abu Al-Rub RK (2011) Meso-scale computational modeling of the plastic-damage response of cementitious composites [J]. Cem Concr Res 41(3):339–358
Krieg RD (1975) A practical two surface plasticity theory [J]. J Appl Mech 42(3):641–646
Kwak H, Kim S (2001) Bond-slip behavior under monotonic uniaxial loads [J]. Eng Struct 23(3):298–309
Kwan AKH, Wang ZM, Chan HC (1999) Mesoscopic study of concrete II: nonlinear finite element analysis [J]. Comput Struct 70(5):545–556
Lee J, Fenves GL (1998) Plastic-Damage model for cyclic loading of concrete structures [J]. J Eng Mech 124(8):892–900
Li QM, Meng H (2003) About the dynamic strength enhancement of concrete-like materials in a split Hopkinson pressure bar test [J]. Int J Solids Struct 40(2):343–360

Li HB, Zhao J, Li JR, Liu YQ, Zhou QC (2004) Experimental studies on the strength of different rock types under dynamic compression [J]. Int J Rock Mech Min Sci 41(S1):68–73
Lilliu G, Van Mier JGM (2003) 3D lattice type fracture model for concrete [J]. Eng Fract Mech 70(7–8):927–941
Liu G, Wang Z (1996) Numerical simulation of fracture of concrete materials using random aggregate model. [J]. J Tsinghua Univ 36(1):84–89. (in Chinese)
Liu B, Bai G, Xu Z, Ma J, Han Y (2019) Experimental study and finite element modeling of bond behavior between recycled aggregate concrete and the shaped steel [J]. Eng Struct 201:109840
Lubliner J, Oliver J, Oller S, Onate E (1989) A plastic-damage model for concrete [J]. Int J Solids Struct 25(3):299–326
Malvar LJ, Ross CA (1998) Review of strain rate effects for concrete in tension [J]. ACI Mater J 95(6):735–739
Marzec I, Bobiñski J, Tejchman J (2007) Simulations of spacing of localized zones in reinforced concrete beams using elasto-plasticity and damage mechanics with non-local softening [J]. Comput Concr 4(5):377–402
Mase T, Hashiguchi K (2009) Numerical analysis of footing settlement problem by subloading surface model [J]. Soils Found 49(2):207–220
Mashaly E, El-Heweity M, Abou-Elfath H, Osman M (2011) Finite element analysis of beam-to-column joints in steel frames under cyclic loading [J]. Alexandria Eng J 50(1):91–104
Mattew RW (2003) Smooth convex three-dimensional particle for the discrete element method [J]. ASCE J Eng Mech 129(5):539–547
Matzenmille A, Lubliner J, Taylor RL (1995) A constitutive model for anisotropic damage in fiber-composites [J]. Mech Mater 20(2):125–152
Mazars J, Pijaudier-Cabot G (1987) Continuum damage theory—application to concrete [J]. ASCE J Eng Mech 115(2):345–365
Meddah MS, Zitouni S, Belâabes S (2010) Effect of content and particle size distribution of coarse aggregate on the compressive strength of concrete [J]. Constr Build Mater 24:505–512
Ministry of construction of People's Republic of China (2011) Code for design of concrete structures GB 50010-2010 [S]. China Architecture & Building Press, Beijing (in Chinese)
Mohamed AR, Hansen W (1999a) Micromechanical modeling of concrete response under static loading: Part I: Model development and validation [J]. ACI Mater J 96(2):196–203
Mohamed AR, Hansen W (1999b) Micromechanical modeling of concrete response under static loading: Part II: model prediction for shear and compressive loading [J]. ACI Mater J 96(3):354–358
Mohamed AR, Hansen W (1999c) Micromechanical modeling of crack-aggregate interaction in concrete materials [J]. Cem Concr Compos 21(5–6):349–359
Mohammad A, Sarsam K, Al-Bayati N (2018) Shear strength enhancement of lightweight aggregate reinforced concrete deep beams by using CFRP strips [C] MATEC Web of Conferences. EDP Sci 162:04011
Moller B, Graf W, Hoffmann A, Steinigen F (2005) Numerical simulation of RC structures with textile reinforcement. Comput Struct 83:1659–1688
Montero-Chacón F, Marín-Montín J, Medina F (2014) Mesomechanical characterization of porosity in cementitious composites by means of a voxel-based finite element model [J]. Comput Mater Sci 90:157–170
Nagai G, Yamada T, Wada A (2000) Three-dimensional nonlinear finite element analysis of the macroscopic compressive failure of concrete materials based on real digital image [J]. Comput Civ Build Eng pp 449–456
Nayak AN, Kumari A, Swain RB (2018) Strengthening of RC beams using externally bonded fibre reinforced polymer composites. Struct 14(3):137–152
Pedersen RR, Simone A, Sluys LJ (2013) Mesoscopic modeling and simulation of the dynamic tensile behavior of concrete [J]. Cem Concr Res 50:74–87
Quantrill RJ, Hollaway LC, Thorne AM (1996) Experimental and analytical investigation of FRP strengthened beam response: Part I [J]. Mag Concr Res 48(177):331–342

Roelfstra PE, Sadouki H, Wittmann FH (1985) The numerical concrete [J]. Mater Struct 18(5):327–335

Santos LRD, Cardoso HDS, Caldas RB, Grilo LF (2020) Finite element model for bolted shear connectors in concrete-filled steel tubular columns [J]. Eng Struct 203:109863

Sayed AM, Wang X, Wu Z (2014) Finite element modeling of the shear capacity of RC beams strengthened with FRP sheets by considering different failure modes [J]. Construct Build Mater 59(2):169–179

Schlangen E, Garbociz EJ (1996) New method for simulating fracture using an elastically uniform random geometry lattice [J]. Int J Eng Sci 34(10):1131–1144

Schlangen E, Garboczi EJ (1997) Fracture simulations of concrete using lattice model computational aspects [J]. Eng Fract Mech 57(2–3):319–332

Schlangen E, van Mier JGM (1992) Simple lattice model for numerical simulation of fracture of concrete materials and structures [J]. Mater Struct 25:534–542

Scrivener KL, Crumbie AK, Laugesen P (2004) The interfacial transition zone (ITZ) between cement paste and aggregate in concrete [J]. Interface Sci 12(4):411–421

Snozzi L, Caballero A, Molinari JF (2011) Influence of the meso-structure in dynamic fracture simulation of concrete under tensile loading [J]. Cem Concr Res 41(11):1130–1142

Song Z, Lu Y (2012) Mesoscopic analysis of concrete under excessively high strain rate compression and implications on interpretation of test data [J]. Int J Impact Eng 46:41–55

Supaviriyakit T, Pornpongsaroj P, Pimanmas A (2004) Finite element analysis of FRP-strengthened RC beams [J]. Songklanakarin J Sci Technol 26(4):497–507

Syroka-Korol E, Tejchman J, Mróz Z (2014) FE analysis of size effects in reinforced concrete beams without shear reinforcement based on stochastic elasto-plasticity with non-local softening [J]. Finite Elem Anal Des 88(10):25–41

Tang TN (2004) Triaxial test simulation with discrete element method and hydrostatic boundaries [J]. J Eng Mech 130(10):1188–1194

Tang C, Zhu W (2003) Damage and fracture of concrete—numerical test [M]. Science Press, Beijing (in Chinese)

Tang B, Chen S, Li X, Xiong L, Chen H, Feng Q (2019) Seismic performance of RC frames with EPSC latticed concrete infill walls [J]. Eng Struct 197:109437

Thilakarathna PSM, Kristombu Baduge KS, Mendis P, Vimonsatit V, Lee H (2020) Mesoscale modelling of concrete—a review of geometry generation, placing algorithms, constitutive relations and applications [J]. Eng Fract Mech p 106974

van Mier JGM, van Vliet MRA (1999) Experimentation, numerical simulation and the role of engineering judgment in the fracture mechanics of concrete and concrete structures [J]. Construct Build Mater 13(1):3–14

van Mier JGM, Vervuurt A, van Vliet MRA (1999) Materials engineering of cement-based composites using lattice models [J]. Comput Fract Mech Concr Technol. Southampton, WIT Press/Computational Mechanics Publications, pp 1–32

Vervuurt A, Schlangen E, van Mier JGM (1996) Tensile cracking in concrete and sandstone: Part 1-basic instruments [J]. Mater Struct 29(185):9–18

Walraven JC, Reinhardt HW (1981) Theory and experiments on the mechanical behavior of cracks in plain and reinforced concrete subjected to shear loading [J]. HERON 26(1A):26–35

Wang X, Liu X (2004) Modeling bond strength of corroded reinforcement without stirrups [J]. Cem Concr Res 34(8):1331–1339

Wang ZM, Kwan AKH, Chan HC (1999) Mesoscopic study of concrete I: generation of random aggregate structure and finite element mesh [J]. Comput Struct 70(5):533–544

Wang QZ, Li W, Song XL (2006) A method for testing dynamic tensile strength and elastic modulus of rock material using SHPB [J]. Pure appl Geophys 163(5–6):1091–1100

Wang G, Wang Y, Lu W, Yu M, Wang C (2017) Deterministic 3D seismic damage analysis of Guandi concrete gravity dam: a case study [J]. Eng Struct 148:263–276

Wriggers P, Moftah SO (2006) Mesoscale models for concrete: homogenisation and damage behavior [J]. Finite Elem Anal Des 42:623–636

Wu JS, Chang J, Hung J (2007) The effect of contact interface on dynamic characteristics of composite structures [J]. Math Comput Simul 74(6):454–467

Xiao S (2015) Numerical study of dynamic behaviour of RC beams under cyclic loading with different loading rates [J]. Mag Concr Res 67(7):325–334

Yao C, Jiang QH, Shao JF (2015) Numerical simulation of damage and failure in brittle rocks using a modified rigid block spring method [J]. Comput Geotech 64:48–60

Yao C, Jiang QH, Shao JF, Zhou CB (2016) A discrete approach for modeling damage and failure in anisotropic cohesive brittle materials [J]. Eng Fract Mech 1555:102–118

Yen KZY, Chaki TK (1992) A dynamic simulation of particle rearrangement in powder packing with realistic interactions [J]. J Appl Phys 71(7):3164–3173

Yu T (1993) Damage theory and its application [M]. National Defense Industry Press, Beijing (in Chinese)

Zhang J, Li J (2012) Investigation into lubliner yield criterion of concrete for 3D simulation [J]. Eng Struct 44:122–127

Zhang J, Fan J, Hu D (1997) Theory and application of rigid body spring element [M]. Chengdu University Science Technology Press, China Chengdu (in Chinese)

Zhang Z, Song X, Liu Y, Wu D, Song C (2017) Three-dimensional mesoscale modelling of concrete composites by using random walking algorithm [J]. Compos Sci Technol 149:235–245

Zhang H, Sheng P, Zhang J, Ji Z (2019a) Realistic 3D modeling of concrete composites with randomly distributed aggregates by using aggregate expansion method [J]. Construct Build Mater 225:927–940

Zhang Y, Wang Z, Zhang J, Zhou F, Wang Z, Li Z (2019b) Validation and investigation on the mechanical behavior of concrete using a novel 3d mesoscale method [J]. Materials 12(16):2647

Zhao J (2000) Applicability of Mohr-Coulomb and Hoek-Brown strength criteria to the dynamic strength of brittle rock [J]. Int J Rock Mech Min Sci 37(7):1115–1121

Zheng X, Zheng X, Lin H (2010) Research progress of multi-scale method in mechanical analysis of composite materials [J]. Adv Mech 40(1):41–56 (in Chinese)

Zhong XX, Chang CS (1999) Micromechanical modeling for behavior of cementitious granular materials [J]. ASCE J Eng Mech 125(11):1280–1285

Zhou XQ, Hao H (2008b) Modelling of compressive behaviour of concrete-like materials at high strain rate [J]. Int J Solids Struct 45(17):4648–4661

Zhou XQ, Kuznetsov VA, Hao H, Waschl J (2008) Numerical prediction of concrete slab response to blast loading [J]. Int J Impact Eng 35(10):1186–1200

Zhou R, Song Z, Lu Y (2017) 3D mesoscale finite element modelling of concrete [J]. Comput Struct 192:96–113

Zhou XQ, Hao H (2008a) Mesoscale modelling of concrete tensile failure mechanism at high strain rates [J]. Comput Struct 86(21/22):2013–2026

Zhu WC, Tang CA (2002) Numerical simulation on shear fracture process of concrete using mesoscopic mechanical model [J]. Construct Build Mater 16(8):453–463

Zhu W, Tang C, Zhao W, Teng J (2002) Numerical simulation of fracture process of concrete samples under static load [J]. Eng Mech 19(6):148–153 (in Chinese)

Zhu W, Tang C, Teng J, Zhao W (2004) Numerical experiment on the influence of micromechanical properties of concrete on macroscopic fracture process [J]. J Three Gorges University 26(1):22–26 (in Chinese)

Zhu WC, Tang CA, Wang SY (2005) Numerical study on the influence of meso-mechanical properties on macroscopic fracture of concrete [J]. Struct Eng Mech 19(5):519–533

Zhuo J, Zhang Q (2000) Interface element method for mechanics of discontinuous media [M]. Science Press, Beijing (in Chinese)

Zohdi TI, Wriggers P (2001) Aspects of the computational testing of the mechanical properties of micro heterogeneous material samples [J]. Int J Numer Methods Eng 50:2573–2599

Chapter 4
Static Size Effect in Concrete Materials

As known, concrete materials with different sizes have different internal components and show significant size effect. The investigation on size effect of concrete structures has fallen into two major directions (Du et al. 2017a, b): (1) material size effect, and (2) component size effect. In reality, the material size effect is a complex materials science problem, and it is mainly affected by aggregate size, aggregate distribution, mix proportion, concrete strength grade and initial defects within concrete, etc. A number of laboratory experimental tests, numerical simulations and theoretical studies have been performed to explore the size effect on nominal strengths of concrete materials. These studies have greatly promoted the understanding in the size effect on nominal strengths of concrete materials. Nevertheless, there are still many issues that need to be addressed further. First, there are some conflicts and arguments in the size effect on nominal strengths of concrete materials due to discreteness and differences in test conditions. The in-depth systematic mechanism analysis should be required. The mechanical behavior of large-sized concrete specimens needs to be further investigated (by tests or simulations) to describe the size effect in a full-size range. Second, in spite of these valuable theoretical and experimental studies on the size effect on nominal strengths of concrete materials, the researches on effects of material specific parameters (e.g. maximum aggregate size, aggregate content, aggregate type, water cement ratio, etc.) on nominal strengths of concrete materials are relatively less (Kadlеček and Modrý. 2002; Qian et al. 2004). However, many experimental and simulation researches (Ince et al. 2015; Khaloo et al. 2009; Sim et al. 2013) have revealed that the failure behavior of concrete materials is essentially influenced by these material specific parameters and the influence of these parameters may not be independent of one another. Therefore, it is necessary to make more efforts focusing on the influential mechanism of these material parameters (e.g. maximum aggregate size, aggregate type, water cement ratio) on size effect on nominal strengths of concrete materials.

Here in this chapter, the size effect analysis on nominal tensile strength is shown from theoretical perspective. Then, the static size effect on tensile, flexural and

X. Du and L. Jin, *Size Effect in Concrete Materials and Structures*,
https://doi.org/10.1007/978-981-33-4943-8_4

compressive strengths of concrete materials are provided from experimental perspective and numerical perspective. The material parameters (including maximum aggregate size (MAS), aggregate type and concrete strength gradation, etc.) on size effect of concrete materials are also discussed. In addition, the size effect law considering the influence of maximum aggregate size on nominal strength is presented from theoretical perspective (Jin et al. 2020a, b, c; Du 2016; Du et al. 2016).

4.1 Tensile Strength of Concrete Materials

The mechanism of size effect in concrete or other quasi-brittle materials should be theorized with physical soundness first, then the size effect theory can be adopted by specifications for various materials and all kinds of structures. The size effect on nominal strength σ_n should be related to the structure size independent material constants of f_t. Since all laboratory test results are penetrated by structural size effect phenomenon, the aforementioned material constants are hardly to be obtained by experimental techniques. In addition, the so-called constant material parameters are significantly influenced by the mix variables within the material, especially for concrete-like cementitious materials. Surely, the evaluation of size effect should always be described based on the same batch of concrete mix. However, this leads to an inconvenient situation of testing the material parameters prior to size effect analysis in each case. Determination of material parameters on a basis of theoretical modelling can provide convenience for size effect analysis. This section will present the theoretical modelling of concrete tensile strength based on the morphological fracture features from an energetic point of view.

4.1.1 Morphological Material Model for Concrete

The theoretical modelling is based on Fuller grading curve (Fuller and Thompson 1907) and Walraven's equation (Walraven and Reinhardt 1981) from a universal perspective. Walraven's equation (Walraven and Reinhardt 1981) makes it possible to analyze mesoscale concrete-like cementitious materials within a 2-D specimen based on Fuller grading curve, which has the form as

$$P_i(D > D_i) = P_k\left[1.065\left(\frac{D_i}{D_m}\right)^{\frac{1}{2}} - 0.053\left(\frac{D_i}{D_m}\right)^4 - 0.012\left(\frac{D_i}{D_m}\right)^6 - 0.0045\left(\frac{D_i}{D_m}\right)^8 - 0.0025\left(\frac{D_i}{D_m}\right)^{10}\right] \tag{4.1}$$

where P_i is the percentage of cumulative area of aggregates with diameter under D_i in a 2-D specimen; D_i is the aggregate size that been focused on and i represents the value of diameter of the aggregate; D_m is the maximum aggregate size within the full size range of utilized aggregates and m represents the value of maximum diameter of the utilized aggregates; P_k is the aggregate volume that defined as the ratio between the total volume of (fine and coarse) aggregates and the macro structure, for concrete, normally $P_k = 0.75$ (Walraven and Reinhardt 1981).

As depicted in Fig. 4.1, consider a 2-D specimen ($b = 1$), the cross-section length is l_{macro} and the height is l_r (which is denoted as the reference length for further derivations), made with multi-grade concrete-like cementitious materials under direct tension. The shape of aggregate is assumed to be round for the convenience of theoretical derivation, and the strength of aggregate is assumed to be strong enough that cannot be damaged during the failure process of the material. Then, the total work for the fracture of the macro structure can be calculated by the following conceptual equation considering of the meso structures

$$W_{total} = W_i + W_m \tag{4.2}$$

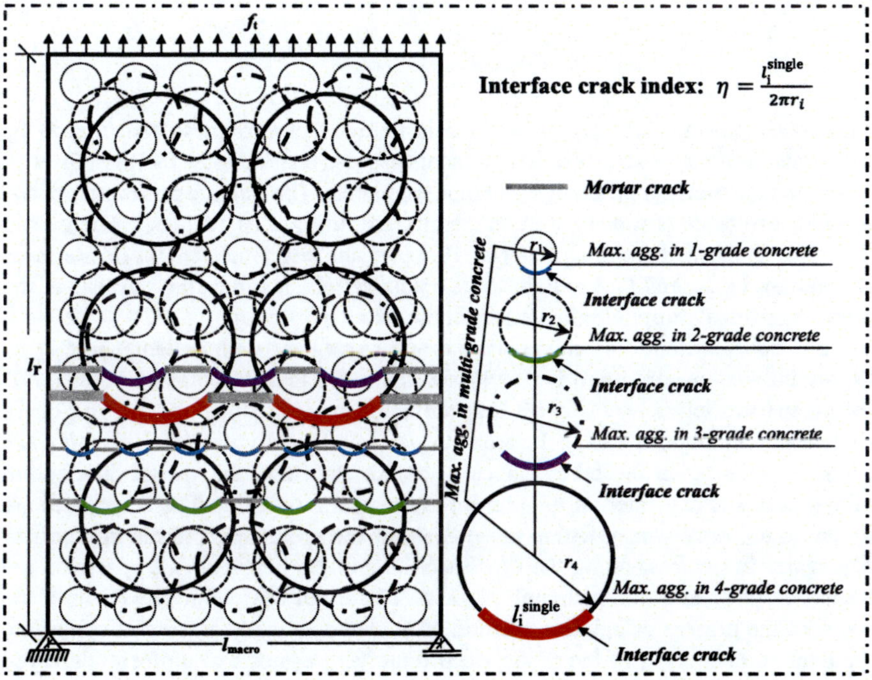

Fig. 4.1 Morphological analysis model for concrete-like cementitious materials

in which, $W_{\rm i}$ and $W_{\rm m}$ are respectively the work for the fracture of interfaces and mortar. Considering the differences between research scales, the total work for the fracture of the macro structure should be calculated from the macroscopic perspective by

$$W_{\rm total} = G_{\rm f} b l_{\rm macro} \tag{4.3}$$

While, the work for the fracture of interfaces and mortar should be determined from the mesoscopic perspective as

$$W_{\rm i} = G_{\rm f,i} b l_{\rm i} \tag{4.4}$$

$$W_{\rm m} = G_{\rm f,m} b l_{\rm m} \tag{4.5}$$

in which, $l_{\rm i}$ and $l_{\rm m}$ are respectively the total crack length in interfaces and mortar along the torturous crack path. Substituting Eqs. (4.3)–(4.5) into Eq. (4.2), one can obtain the fracture energy $G_{\rm f}$ from a mesoscopic of view as

$$G_{\rm f} = \frac{l_{\rm i}}{l_{\rm macro}} G_{\rm f,i} + \frac{l_{\rm m}}{l_{\rm macro}} G_{\rm f,m} \tag{4.6}$$

where $(l_{\rm i}/l_{\rm macro})$ and $(l_{\rm m}/l_{\rm macro})$ act as the contribution factors of meso structures to the performance of macro structure. This concept has also been adopted by Rosselló et al. (2006) in their mesoscopic analyses of model concrete. The fracture parameters of $G_{\rm f,i}$ and $G_{\rm f,m}$ can be determined based on the mixtures utilized as the meso structures in the concrete-like cementitious material. However, the total crack lengths in interfaces and mortar, i.e., $l_{\rm i}$ and $l_{\rm m}$, are hard to be obtained, which is the core purpose of the present universal morphological material model.

First, consider that only mono-sized round aggregates with a diameter of $D_i = 2r_i$ are utilized, according to Eq. (4.1), for $D_m = D_i$, then $P_i = 0.993P_{\rm k}$ for a 2-D specimen (certainly, $P_i = P_{\rm k}$ for a 3-D specimen according to the Fuller grading curve (Fuller and Thompson 1907). In order to deterministically evaluate the influence of aggregate size on the performance of macro structure, the spatial distribution of aggregates is neglected in the present theoretical modelling. The aggregates are forced to uniformly array within the domain of the specimen, and the numbers of aggregates in one row along the cross-section and the height of the specimen are respectively N_x and N_y. Although N_x and N_y are not always integers within the scope of the present calculation method, it is mathematically accurate to calculate the total number $N_{\rm total}$ of the mono-sized round aggregates that uniformly arrayed within the specimen by

$$N_{\rm total} = N_x N_y \tag{4.7}$$

In addition, based on the mix design of the material, the total number N_{total} of aggregates utilized within the specimen can be calculated by

$$N_{\text{total}} = \frac{S_{\text{a}}^{\text{total}}}{S_{\text{a}}^{\text{single}}} \tag{4.8}$$

in which, ${S_{\text{a}}}^{\text{total}}$ is the area of total aggregates utilized within the specimen that can be calculated by ${S_{\text{a}}}^{\text{total}} = l_{\text{r}} l_{\text{macro}} P_i$, and ${S_{\text{a}}}^{\text{single}}$ is the area of a single aggregate that can be calculated by ${S_{\text{a}}}^{\text{single}} = \pi {r_i}^2$.

Within the scope of morphological analysis model adopted for concrete-like cementitious materials in present study (Fig. 4.1), the fracture surface should always appear along the cross-section under direct tension. Considering of the mesoscopic tensile crack (i.e., the combination of interface cracks and mortar cracks) that finally fracture the specimen, we should pay more attention on N_x. The ratio between N_x and N_y can be calculated by $\frac{N_x}{N_y} = \frac{l_{\text{macro}}}{l_{\text{r}}}$, then one can replace N_y in Eq. (4.7) by $N_y = \frac{l_{\text{r}}}{l_{\text{macro}}} N_x$. Combine Eqs. (4.7) and (4.8), the aggregate number N_x along the cross-section can be obtained

$$N_x = \frac{l_{\text{macro}}}{r_i} \sqrt{\frac{P_i}{\pi}} \tag{4.9}$$

For 2-D specimens using mono-sized round aggregates, the total crack lengths in interfaces can be calculated by the product of aggregate number (N_x) along the cross-section with the interface crack length around a single aggregate (${l_{\text{i}}}^{\text{single}}$). Herein, define interface crack index (η) as the proportion of interface crack length around a single aggregate (${l_{\text{i}}}^{\text{single}}$) in the circumference of the round aggregate ($2\pi r_i$), which has the form as $\eta = \frac{l_{\text{i}}^{\text{single}}}{2\pi r_i}$. Then, the interface crack length around a single aggregate can be given as

$$l_{\text{i}}^{\text{single}} = 2\pi r_i \eta \tag{4.10}$$

The value of interface crack index η should be varied in [0, 0.5], and it can be calculated by the following equation

$$\eta = \frac{1}{\pi} \arccos\left(\frac{\omega_{\text{i}}}{\omega_{\text{m}}}\right) \tag{4.11}$$

in which, ω_{i} and ω_{m} are respectively the strain energy density of interfaces and mortar. Moreover, since the reference length ${l_{\text{r}}}^{\text{meso}}$ for the meso structures is selected to be the same, the ratio between fracture energies is same with that between strain energy densities, then Eq. (4.11) can be expressed as $\eta = \frac{1}{\pi} \arccos\left(\frac{G_{\text{f,i}}}{G_{\text{f,m}}}\right)$.

By combining Eqs. (4.9) and (4.10), the total crack length in interfaces along the torturous crack path can be calculated by

$$l_{\mathrm{i}}^{\mathrm{mono-sized}} = N_x l_{\mathrm{i}}^{\mathrm{single}} = 2\eta l_{\mathrm{macro}} \sqrt{\pi P_i} \tag{4.12}$$

Based on the above derivations for mono-sized aggregates, it is easy to extrapolate that, for 2-D specimens using continuously-graded round aggregates like that depicted in Fig. 4.1, the total crack length in interfaces along the torturous crack path can be calculated by

$$l_{\mathrm{i}} = \alpha l_{\mathrm{macro}} \tag{4.13}$$

and

$$\alpha = 2\eta \sum_{0}^{m-\Delta D} \sqrt{\pi (P_{i+\Delta D} - P_i)} \tag{4.14}$$

where α is defined as the partition coefficient for the determination of the total crack length in interfaces. ΔD is the increment for the calculation within the full size range of aggregates utilized in the specimen.

As can be seen from the morphological analysis model depicted in Fig. 4.1, for 2-D specimens using continuously-graded round aggregates, the total crack length in mortar should be calculated by the following equation

$$l_{\mathrm{m}} = l_{\mathrm{macro}} - N_x l_{\mathrm{i}}^{*\mathrm{single}} = (1 - \beta) l_{\mathrm{macro}} \tag{4.15}$$

in which, $l_{\mathrm{i}}^{*\mathrm{single}} = 2r_i \sin(\pi\eta)$ is the length of arc chord corresponding to the interface crack around a single aggregate, and

$$\beta = 2\sin(\pi\eta) \sum_{0}^{m-\Delta D} \sqrt{\frac{P_{i+\Delta D} - P_i}{\pi}} \tag{4.16}$$

where β is defined as the partition coefficient for the determination of the total crack length in mortar.

Substituting Eqs. (4.13) and (4.15) into Eq. (4.16), one can obtain the fracture energy G_{f} of concrete-like cementitious materials as

$$G_{\mathrm{f}} = \alpha G_{\mathrm{f,i}} + (1 - \beta) G_{\mathrm{f,m}} \tag{4.17}$$

One can clearly note from Eq. (4.17) that, the meso structures individually contribute to the fracture properties of concrete-like cementitious materials. The properties of interfaces and mortar are represented by $G_{\mathrm{f,i}}$ and $G_{\mathrm{f,m}}$, and the scales of meso structures are hidden inside the partition coefficients α and β.

Moreover, since $G_{\mathrm{f,i}} = \omega_{\mathrm{i}} l_{\mathrm{r}}^{\mathrm{meso}}$ and $G_{\mathrm{f,m}} = \omega_{\mathrm{m}} l_{\mathrm{r}}^{\mathrm{meso}}$, then Eq. (4.17) can be expressed in a more compact way as

$$G_f = \omega^{meso} l_r^{meso} \tag{4.18}$$

in which, l_r^{meso} is the reference length for meso structures; ω^{meso} is the mesoscopic strain energy density, which has the form as

$$\omega^{meso} = \alpha\omega_i + (1 - \beta)\omega_m \tag{4.19}$$

Considering the formulation of fracture energy from the macroscopic of view, i.e., $G_f = \omega l_r$, and combining it with Eq. (4.18), the macroscopic strain energy density can be derived by

$$\omega = \rho_r \omega^{meso} \tag{4.20}$$

where $\rho_r = \left(\frac{l_r^{meso}}{l_r}\right)$ is a scale factor defined as the reference length ratio between the utilized mesoscale and macroscale. Equation (4.20) is convenient for the derivation of other material parameters from a continuum mechanics point of view.

4.1.2 Multi-grade Analysis Method for Cementitious Systems

The building technology of massive structures enables large diameter aggregates to be used. In concrete mix design, gradation requirement results in a mixture of many different sizes of aggregates, which can be represented by the maximum aggregate size or just by gradation number. It must be clarified that, the maximum aggregate size cannot simply be replaced by gradation number in concrete-like cementitious materials, but these two meso-level design parameters are closely related. The way of obtaining a good quality concrete is to combine fine and coarse aggregates to produce optimum grading (Neville 1995). Aggregate gradation is one of the key design parameters affecting the workability and strength properties of concrete mixtures. Shi et al. (2009) presented a meso-analysis of concrete behavior due to effect of aggregate gradation based on discrete element method. They found that, for a given aggregate content and aggregate size range, the strength of concrete varied with different grading curves. Meddah et al. (2010)'s experimental results showed that, for the same coarse aggregate content, a continuous granular system seems to provide the highest compressive strength. As for the case that a fixed grading curve (herein, the Fuller grading curve (Fuller and Thompson 1907) is adopted during the mix design of concrete-like cementitious materials, the maximum aggregate size in each gradation can be replaced by the gradation number.

Figure 4.2 depicts the normalized cumulative distribution curve (P_i/P_k versus D_i/D_m) based on Walraven's equation (Walraven and Reinhardt 1981). Firstly, it is assumed that the full size range (i.e., order $i = m$ in Eq. (4.1)) of utilized aggregates could be separated into (n + 1) size groups (where n = 0, 1, 2, ...) with a fixed

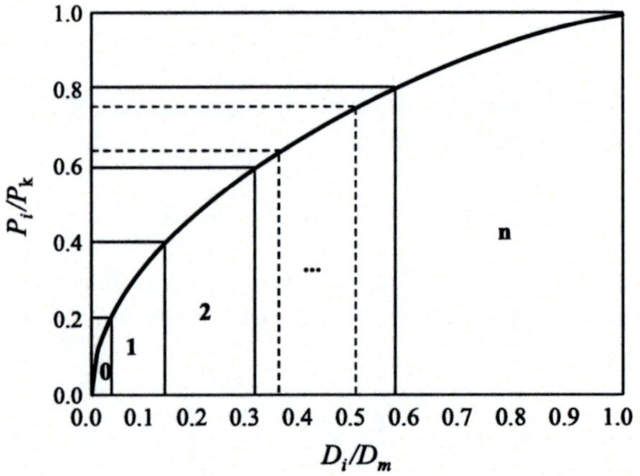

Fig. 4.2 Normalized cumulative distribution curve (P_i/P_k versus D_i/D_m) of the Walraven's equation (Walraven and Reinhardt 1981) in conjunction with the assumed $(n + 1)$ aggregate size groups

increment $\Delta P_n/P_k = (P_m/P_k)/(n + 1) = 0.993/(n + 1)$. One can note from Fig. 4.2 that, with fixed increment ($\Delta P_n/P_k$) in the (P_i/P_k) axis, the increment ($\Delta D_n/D_m$) in the (D_i/D_m) axis for each size group is enlarged with increasing n. The aggregates in the size group-0 could always be treated as fine aggregates in a multi-grade cementitious system, for they are proportionally too small compared to the maximum aggregate size D_m and of which the amount is too large for modelling. Then, from a unified research scale point of view, it is convenient to define any cementitious system as a n-grade material (in the sense of coarse aggregates) for the study of heterogeneity generated by the utilized maximum aggregate size in each gradation.

According to the above assumptions, a multi-grade analysis method for cementitious systems can be established based on the proposed morphological material model. Due to the intrinsic drawback of overlap calculation of the adopted theoretical modelling technique, the partition coefficients α and β must be calibrated prior to its utilization in multi-grade concrete-like cementitious materials (Li et al. 2018). Then, based on Eqs. (4.19) and (4.20), the macroscopic strain energy density ω can be expressed as

$$\omega = \rho_{\mathrm{r}}\left[\alpha'\omega_{\mathrm{i}} + \left(1 - \beta'\right)\omega_{\mathrm{m}}\right] \tag{4.21}$$

in which, α' and β' are respectively the calibrated partition coefficients. Equation (4.21) can be served as the general form of material-level energetic size effect model, which is convenient for the derivation of material parameters from a continuum mechanics point of view. Reference to the derivation process for multi-grade concrete in (Li et al. 2018), α' (based on Eq. (4.14)) and β' (based on Eq. (4.16)) can be expressed as.

Table 4.1 Fraction parameters for coarse aggregates within the scope of present study

Gradation g_j (j=1, 2, …, n)	Fraction parameter a_j (j=1, 2, …, n)				
	a_1	a_2	a_3	..	a_n
1	1/1	0	0	0	0
2	1/2	1/2	0	0	0
3	1/3	1/3	1/3	0	0
⋮	⋮	⋮	⋮	⋮	⋮
n	1/n	1/n	1/n	1/n	1/n

$$\alpha' = 2\eta \sum_{1}^{n} \sqrt{\pi \gamma a_j} + X_j = 2\eta \sqrt{g_j \pi \gamma} + X_j \quad j = 1, 2, 3 \ldots, n \tag{4.22}$$

$$\beta' = 2\sin(\pi\eta) \sum_{1}^{n} \sqrt{\frac{\gamma a_j}{\pi}} + Y_j = 2\sin(\pi\eta)\sqrt{\frac{g_j \gamma}{\pi}} + Y_j \quad j = 1, 2, 3 \ldots, n \tag{4.23}$$

in which, $\gamma = (0.993nP_k)/(n+1)$ is the area percentage of coarse aggregates; a_j, as presented in Table 4.1, is the fraction parameter for coarse aggregates within the scope of present study; g_j represents the gradation; j represents the number of gradation; X_j and Y_j are the adjustment coefficients respectively for different gradations.

In the study of Li et al. (2018), the adjustment coefficients X_j and Y_j are empirically determined by linear regression for multi-grade concrete with a specific case of $\gamma = 0.45$ based on meso-level numerical tests. It is undisputed that, the aggregate volume P_k, which is in conjunction with n controls γ, should have significant influences on the global performance of concrete-like cementitious materials (Grassl et al. 2010; Beygi et al. 2014). Herein, considering of the variations of aggregate volume P_k in a cementitious system, the adjustment coefficients X_j and Y_j will be directly determined through theoretical derivations. Based on Eqs. (4.21)–(4.23), the fracture energy G_f can be written in detail as

$$G_f = \left\{ \left(2\eta\sqrt{g_j \pi \gamma} + X_j\right)\omega_i + \left[1 - 2\sin(\pi\eta)\sqrt{\frac{g_j\gamma}{\pi}} - Y_j\right]\omega_m \right\} \times l_r^{\mathrm{meso}} \tag{4.24}$$

Herein, based on Eq. (4.24), two critical limits for fracture energy G_f are adopted for the determination of adjustment coefficients X_j and Y_j as follows.

Limit 1: Consider a limit situation with interface crack index $\eta = 0.5$, namely, the interfaces are initially debonded with the aggregates, i.e., $\omega_i = 0$:

When the aggregate volume $P_k = 0$ (corresponding to $\gamma = 0$), the material becomes plain mortar, then the fracture energy G_f for different gradations should all be equal to that of mortar (i.e., $G_{f,m} = \omega_m l_r^{\mathrm{meso}}$), which can be expressed as

$$G_f|_{\eta=0.5, P_k=0} = \left[X_j \times 0 + \left(1 - Y_j\right) \times \omega_m\right] \times l_r^{\mathrm{meso}} = \omega_m \times l_r^{\mathrm{meso}} \tag{4.25}$$

which yields

$$Y_j|_{\eta=0.5,P_k=0} = 0 \tag{4.26}$$

The densest aggregate volume $P_k = 1$ (corresponding to $\gamma = (0.993n)/(n + 1)$), although it is not possible in practice, can be utilized as the limitation of theoretical derivation. In this case, there would be no mortar, but only the aggregates with zero or thin thickness interfaces. In other words, the fracture surface under direct tension would only be composed by interfaces. Consequently, the fracture energy G_f for different gradations should all be equal to zero, which can be expressed as

$$\begin{aligned} G_f|_{\eta=0.5,P_k=1} = & \left\{\left(n\sqrt{\frac{0.993n\pi g_j}{n+1}} + X_j\right) \times 0 \right. \\ & \left. + \left[1 - 2\sin\left(\frac{\pi}{2}\right)\sqrt{\frac{0.993ng_j}{\pi(n+1)}} - Y_j\right] \times \omega_m\right\} \\ & \times l_r^{meso} = 0 \end{aligned} \tag{4.27}$$

which yields

$$Y_j|_{\eta=0.5,P_k=1} = 1 - 2\sin\left(\frac{\pi}{2}\right)\sqrt{\frac{0.993ng_j}{\pi(n+1)}} \tag{4.28}$$

Herein, we preliminarily assume that the adjustment coefficient Y_j has linear relation with the aggregate volume P_k, then based on Eqs. (4.26) and (4.28), one can deduce the equation for Y_j as

$$Y_j = \left[1 - \sqrt{\frac{3.972ng_j}{\pi(n+1)}}\right]P_k \tag{4.29}$$

Limit 2: Consider the other limit situation with interface crack index $\eta = 0$, namely, the mechanical properties of interfaces are identical to that of mortar, i.e., $\omega_i = \omega_m$. At this circumstance, the mesoscopic crack should only be able to propagate in the mortar. The existence of aggregates would possibly make the crack path torturous compared to that in plain mortar (which is comparatively straight). Nevertheless, from the practical perspective, e.g. testing a RVE specimen, considering of the boundary layer effect, the protective cover should only be composed by mortar, then if the mechanical properties of interfaces are ideally identical to that of mortar, the lowest energy and the shortest crack path for the fracture of specimen would be located within the boundary layer. As from the theoretical perspective, for this kind of extreme situation (i.e., $\omega_i = \omega_m$), the possible lowest energy must be dissipated in the straight crack path (i.e., shortest) within mortar. Therefore, the fracture energy G_f of the material should be equal to that of mortar (i.e., $G_{f,m} = \omega_m l_r^{meso}$) with $\eta = 0$. This concept has also been established and discussed in Rosselló et al. (2006) as: since the crack path is always through the mortar or follows the weaker mortar-aggregate interface, the *actual* fracture energy will always be equal to or lower than the average mortar fracture energy. Lilliu and van Mier (2003) presented a 3D lattice type fracture model for concrete and they conducted a series of numerical tests by varying the aggregate volume. The results showed that, when the interfaces were assigned to equal to the mortar, debonding (i.e., interface failure) and mortar failure cannot be distinguished anymore, consequently the peak-load and ductility of the material were hardly affected by aggregate volume. Therefore, the fracture energy G_f (with any aggregate volume P_k) for different gradations can be expressed as

$$G_f|_{\eta=0} = \left[X_j \times \omega_m + \left(1 - Y_j\right) \times \omega_m\right] \times l_r^{meso} = \omega_m \times l_r^{meso} \tag{4.30}$$

Then, based on Eq. (4.29), one can get the equation for X_j as

$$X_j = \left[1 - \sqrt{\frac{3.972ng_j}{\pi(n+1)}}\right] P_k \tag{4.31}$$

Substituting Eqs. (4.29) and (4.31) into Eq. (4.24), the facture energy G_f can be calculated based on the properties and scales of meso structures.

Hitherto, the formulation of the multi-grade analysis method for cementitious systems based on the proposed morphological material model is completed. A general material-level energetic size effect model has been shown in the form of strain energy density as Eq. (4.21), which can derive other material parameters from a continuum mechanics point of view. Then, the material-level size effect behavior of concrete-like cementitious materials can be revealed by the variation of the mix variables including ω_i, ω_m, η, g_j, γ, P_k and ρ_r.

4.1.3 Validation and Analysis

Rosselló et al. (2006) designed and tested a series of simple model concrete with spherical aggregates and different interfaces to investigate the interface properties on the mechanical behavior of concrete. The difference and advantage of the simple model concrete is that, although it is not applicable to graded concrete, the test results of which can be profitably used as a challenge and benchmark for checking theoretical and numerical models, seeking to predict the fracture and mechanical behavior of concrete.

The coarse aggregates were of commercial spheres of mullite with average dimeter of 7.4 mm, i.e., $n = 1$. The volume percentage of coarse aggregates was 25.8%, i.e., $\gamma = 0.258$. Correspondingly, within the scope of present study, the total aggregate volume can be deduced as $P_k = 0.52$. Then, based on Eqs. (4.29) and (4.31), the

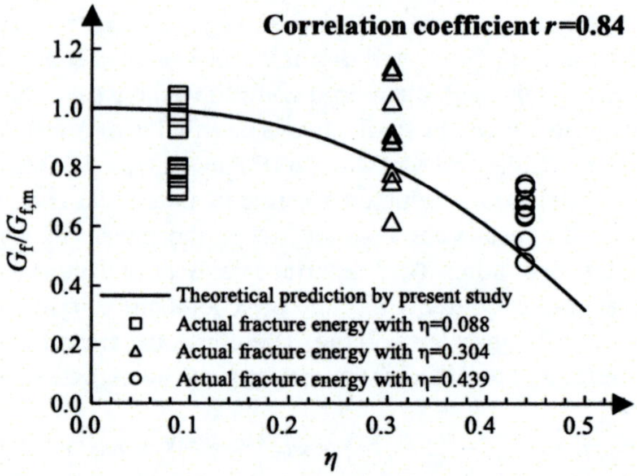

Fig. 4.3 Normalized fracture energy ($G_f/G_{f,m}$) versus interface crack index η, comparison of Rosselló et al. (2006)'s test data with theoretical prediction by present study

adjustment coefficients $X_1 = Y_1 = 0.11$. The aggregates were selected to be strong enough with an average of 16 MPa that cannot be broken during the fracture tests. The mortar was designed as the combination of cement and sand (i.e., fine aggregates) with a fracture energy of 52 J/m^2. The kinds of interface were devised into three kinds and the estimated fracture energy was around 50 J/m^2, 30 J/m^2 and 10 J/m^2, respectively. Then, based on Eq. (4.11), the interface crack index η for each kind of interface can be calculated as 0.088, 0.304 and 0.439, respectively. Their test results showed that, the actual fracture energy of concrete should always be equal to or lower than the average mortar fracture energy. The test data of normalized fracture energy ($G_f/G_{f,m}$) vs. interface crack index η are shown in Fig. 4.3.

Based on the proposed theoretical approach in present study, the fracture energy G_f in Eq. (4.24) can be transformed into a normalized equation as

$$\frac{G_f}{G_{f,m}} = \left(2\eta\sqrt{g_j \pi \gamma} + X_j\right)\cos(\pi\eta) + \left[1 - 2\sin(\pi\eta)\sqrt{\frac{g_j\gamma}{\pi}} - Y_j\right] \tag{4.32}$$

Substituting the test parameters of n, γ, P_k, g_j, X_j and Y_j ($j = 1$) into Eq. (4.32), one can draw the theoretical prediction of normalized fracture energy ($G_f/G_{f,m}$) with continuously varied interface crack index η, as shown in Fig. 4.3. One can clearly note that, the theoretical prediction curve coincides with the trend of test data with a correlation coefficient $r = 0.84$.

According to Walraven and Reinhardt (1981), the aggregate volume for concrete is normally $P_k = 0.75$. Based on the proposed multi-grade analysis method, and reference to the Chinese code for mix design of hydraulic concrete (DL/T 5330–2015 (Ministry of construction of People's Republic of China 2015)), set $n = 4$ and $m = 150$, namely, the cementitious system is distinguished by four gradations within the full aggregate size range of 0–150 mm. Then, the area percentage of coarse aggregates $\gamma = (0.993 \times 4 \times 0.75)/(4 + 1) = 0.5958$ for the four gradations.

Table 4.2 Assumed ΔP_n–ΔD_n relationship based on the Walraven's Eq. (1981)

Size group n	0	1	2	3	4	Sum
ΔP_n	0.14895	0.14895	0.14895	0.14895	0.14895	0.74475
Approximate ΔD_n (mm) (Approximate size range (mm))	5 (0–5)	16 (5–21)	26 (21–47)	38 (47–85)	65 (85–150)	150 (0–150)

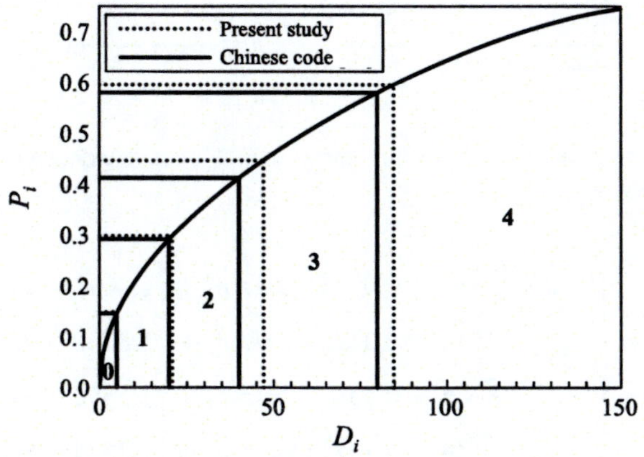

Fig. 4.4 Comparison of size groups for the full size range of utilized aggregates with $D_m = 150$ mm determined by present study with that by DL/T 5330–2015 (2015)

Based on Eq. (4.1), the increment $\Delta P_n = (0.993 \times 0.75)/(4 + 1) = 0.14895$, and the corresponding approximate increment ΔD_n and size range for each size group is listed in Table 4.2. Figure 4.4 depicts the comparison of the size groups for the full size range of utilized aggregates with $D_m = 150$ mm determined by present study with that by the Chinese code DL/T 5330–2015 (Ministry of construction of People's Republic of China 2015), which shows relatively good consistency. The discrepancies around 2-grade and 3-grade are due to the utilization of integer boundaries in the size groups defined by the Chinese code DL/T 5330–2015 (Ministry of construction of People's Republic of China 2015), which is for the convenience in practice. Based on Eqs. (4.29) and (4.31), Table 4.3 summarizes the adjustment coefficients X_j and Y_j for the four gradations of concrete. Then, based on Eq. (4.32), the normalized fracture energy ($G_f/G_{f,m}$) for the four gradations of concrete can be respectively calculated by Eq. (4.33), and Fig. 4.5 depicts the variation of normalized fracture energy ($G_f/G_{f,m}$) with interface crack index η and gradation number g_j.

Table 4.3 Adjustment coefficients X_j and Y_j for four gradations of concrete

Gradation g_j ($j = 1, 2, 3$ and 4)	1	2	3	4
X_j	−0.0075	−0.3150	−0.5550	−0.7575
Y_j	−0.0075	−0.3150	−0.5550	−0.7575

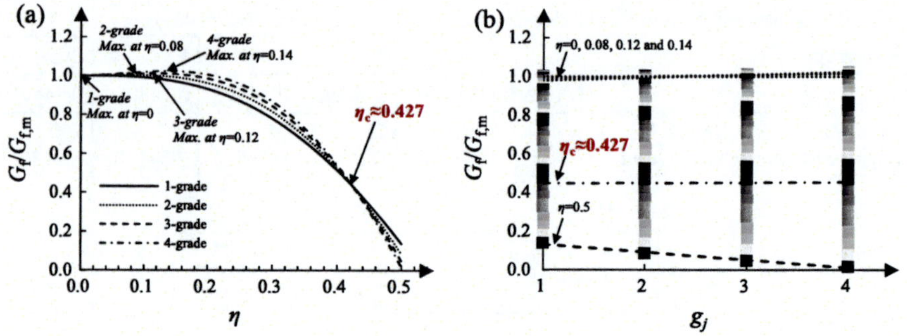

Fig. 4.5 Variation of normalized fracture energy ($G_f/G_{f,m}$): **a** Interface crack index η; and **b** Gradation number g_j

$$\begin{cases} \frac{G_f}{G_{f,mo}}|^{1-\text{grade}} = \left[2\eta\sqrt{0.5958\pi} - 0.0075\right]cos(\pi\eta) + \left[1.0075 - 2\sin(\pi\eta)\sqrt{\frac{0.5958}{\pi}}\right] \\ \frac{G_f}{G_{f,mo}}|^{2-\text{grade}} = \left[2\eta\sqrt{1.1916\pi} - 0.3150\right]cos(\pi\eta) + \left[1.3150 - 2\sin(\pi\eta)\sqrt{\frac{1.1916}{\pi}}\right] \\ \frac{G_f}{G_{f,mo}}|^{3-\text{grade}} = \left[2\eta\sqrt{1.7874\pi} - 0.5550\right]cos(\pi\eta) + \left[1.5550 - 2\sin(\pi\eta)\sqrt{\frac{1.7874}{\pi}}\right] \\ \frac{G_f}{G_{f,mo}}|^{4-\text{grade}} = \left[2\eta\sqrt{2.3832\pi} - 0.7575\right]cos(\pi\eta) + \left[1.7575 - 2\sin(\pi\eta)\sqrt{\frac{2.3832}{\pi}}\right] \end{cases} \tag{4.33}$$

It can be concluded from Fig. 4.5a that, when the mechanical properties of interfaces are weak (i.e., relatively large interface crack index η), the fracture energy generally decreases with decreasing mechanical properties of interfaces (i.e., increasing interface crack index η). The theoretical analysis result for 1-grade (i.e., mono-sized aggregates) shows that, the maximum value of ($G_f/G_{f,m}$) = 1 is obtained at $\eta = 0$, which is coincide with Rosselló et al. (2006)'s meso-level test results. Nevertheless, for 2–4 gradations, when the mechanical properties of interfaces are strong (i.e., relatively small interface crack index η), the fracture energy slightly increases with decreasing mechanical properties of interfaces (i.e., increasing interface crack index η) first and then it turns to decrease, which results in a maximum value of fracture energy. Specifically, the maximum value of ($G_f/G_{f,m}$) = 1.003 is obtained at $\eta = 0.08$ for 2-grade, the maximum value of ($G_f/G_{f,m}$) = 1.012 is obtained at $\eta = 0.12$ for 3-grade, and the maximum value of ($G_f/G_{f,m}$) = 1.024 is obtained at $\eta = 0.14$ for 4-grade. This is because that, the degree of tortuosity of the mesoscopic crack increases with decreasing mechanical properties of interfaces dramatically. For weak interfaces, the interface cracks possess most of the mesoscopic crack, no doubt that the weaker the interface is (i.e., the more percentage of interface cracks in the mesoscopic crack), the weaker the mechanical properties of concrete are. While, for strong interfaces, although the mortar cracks possess most of the mesoscopic crack, compare to the increasing interface crack index η (i.e., decreasing mechanical properties of interfaces), the total length of mesoscopic crack grows faster, more energy would be consumed in the curved interface crack comparing to that consumed in the straight mortar crack in plain mortar, results in larger fracture energy. Consequently, there is a trend that the fracture energy increases with decreasing mechanical properties of relatively strong interfaces. This phenomenon has also been concluded by Shah and Chandra (1968) that, a reduction of interface strength leads to significant increase of fracture energy. Moreover, one can note from Fig. 4.5a that, the rate of growth in fracture energy with relatively strong interfaces increases with the gradation number, which partially demonstrates that continuous granular systems provide higher quality of cementitious systems.

Figure 4.5b presents the variation of fracture energy with maximum aggregate size of 21, 47, 85 and 150 mm (represented by the gradation number of 1, 2, 3 and 4). It can be inferred from the figure that, when the interfaces are relatively strong (i.e., small interface crack index η), the mechanical properties of concrete increase with increasing maximum aggregate size; while the trend is opposite when the interfaces are relatively weak. Therefore, a critical interface crack index η_c should be existed to make the fracture energy stable with the variation of maximum aggregate size, and corresponding to the intersect point in Fig. 4.5a, $\eta_c \approx 0.427$. Akçaoğlu et al. (2004) concluded that with increasing the maximum aggregate size, the properties of interfaces become critical. When the interface crack index η is larger than the critical value of 0.427, the fracture energy decreases with increasing maximum aggregate size; while as the interface crack index η is smaller than the critical value of 0.427, the fracture energy generally increases with increasing maximum aggregate size. This is indicating that, by controlling the mix variables, the established multi-grade analysis method should have a potential to design concrete materials based on required performance.

4.2 Splitting-Tensile Strength of Concrete Materials

The tensile strength is one of the main parameters in the design of concrete structures and it mainly influences the cracking properties and durability of concrete materials. The researches on the size effect on tensile strength are relatively insufficient compared with the ones on compressive strength of concrete. It is important to conduct efforts on the size effect of tensile strength in concrete materials. Due to limitations of test equipment, it is hard to carry out the direct tensile tests and lots of researchers measured the splitting-tensile behavior of concrete materials instead in laboratory experiments (Malhotra 1976; Carmona 2009; Wang et al. 2014; Rocco et al. 1999; Zhou et al. 1998; Ince et al. 2015; etc.). On the whole, the above studies have greatly promoted the understanding in the size effect on splitting-tensile strength of concrete materials. Nevertheless, there are still many issues that need to be addressed further. As mentioned previously, it is necessary to make more efforts focusing on the influential mechanism of these material parameters (e.g. aggregate size, water cement ratio) on size effect on splitting-tensile strength of concrete materials. Here in this part, the size effect on splitting-tensile strength of concrete materials is provided from experimental perspective and numerical perspective, and the influences of maximum aggregate size (MAS) and concrete strength grade (CSG, i.e. water cement ratio *w/c*) are also presented (Jin et al. 2020b).

4.2.1 Experimental Analysis

First, four groups of MAS (d_{max} = 10, 20, 30 and 40 mm) and three groups of CSG (C20, characteristic compression strength of cubic concrete between 20.0 and 23.0 MPa; C40, characteristic compression strength of cubic concrete between 40.0 and 46.0 MPa; and C60, characteristic compression strength of cubic concrete between 60.0 and 69.0 MPa, according to the Chinese code GB/T 50010–2010 ((Ministry of construction of People's Republic of China 2011))) were considered. A total of 84 concrete cubic specimens with different side lengths (b = 150–350 mm) for splitting-tensile tests were designed. The details about the production and introduction of tested specimens can refer to the work (Jin et al. 2020b). All specimens were tested in the hydraulic universal testing machine (see Fig. 4.6) having a load capacity of 2000 kN. Between the pressure plates and cubic specimens, the steel arc pad (the radius is 75 mm and length is not less than side length of specimen, see Fig. 4.6) was set. In order to prevent the loading area from being crushed, the plywood pad (the width is 20 mm, thickness is 3–4 mm, and length is not less than side length of specimen) was applied in all specimens prior to tests. Specimens for splitting-tensile tests should be continuously and evenly loaded without any impact, and the loading rate was controlled by applying vertical forces of 0.04 MPa/s, according to the Chinese code GB/T 50081–2002 (Ministry of construction of People's Republic of China 2003).

Fig. 4.6 Test setup of concrete under splitting-tensile loads

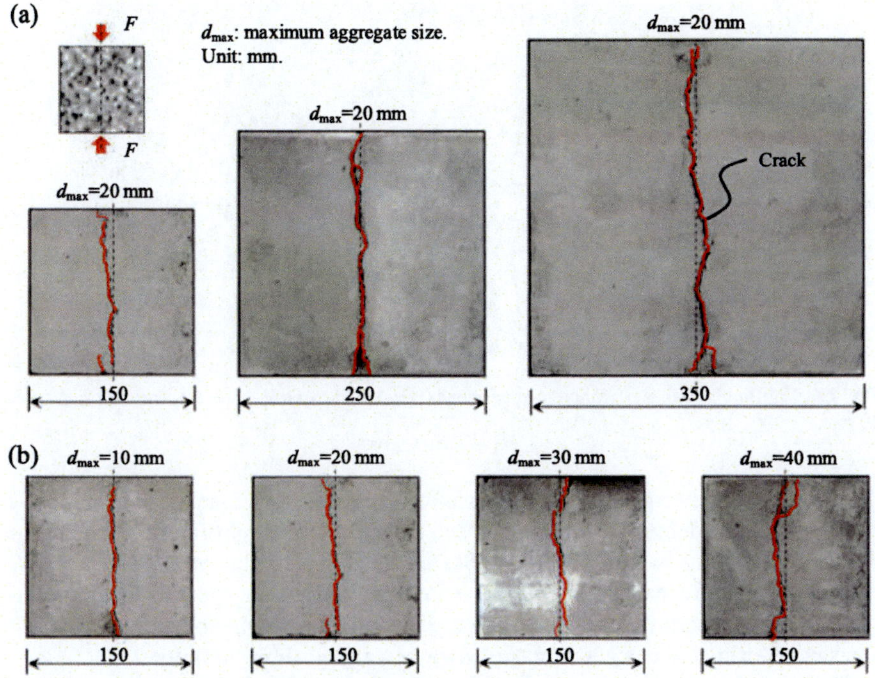

Fig. 4.7 Typical failure patterns of concrete cubic specimens in tests: **a** With same MAS and different structural sizes; and **b** With different MAS and same structural sizes

1. *Failure patterns in test*

Figure 4.7a presents splitting-tensile failure patterns of concrete cubic specimens having different side lengths with the same MAS (d_{max} = 20 mm). As visible in Fig. 4.7a, the final failure patterns of specimens with different side lengths are basically similar. The splitting cracks take place along the central line and the specimen is split into approximately equal halves. Figure 4.7b presents typically the final failure patterns of concrete specimens (side length b = 150 mm) having different MAS. As visible in Fig. 4.7b, the MAS has an obvious influence on the failure patterns of concrete. As the MAS improves, the tortuosity of crack path becomes greater, which may be because that the fracture cracks need to bypass or split the aggregate particles having larger sizes.

The influence of MAS on the splitting-tensile failure patterns also can be found on the fracture surfaces, as depicted in Fig. 4.8. One could find that the splitting damage mainly occur first in the ITZ phase. In term of the fracture surfaces of concrete specimens with d_{max} = 40 mm, splitting cracks occur around coarse aggregate particles, and only some aggregate particles are penetrated by cracks. This causes that the fracture surfaces of specimens with d_{max} = 40 mm are rougher. As for the fracture surfaces of concrete specimens with d_{max} = 10 mm, more small aggregate particles are more likely to be bypassed by splitting cracks. The fracture surfaces become smoother and the splitting failure pattern becomes more brittle.

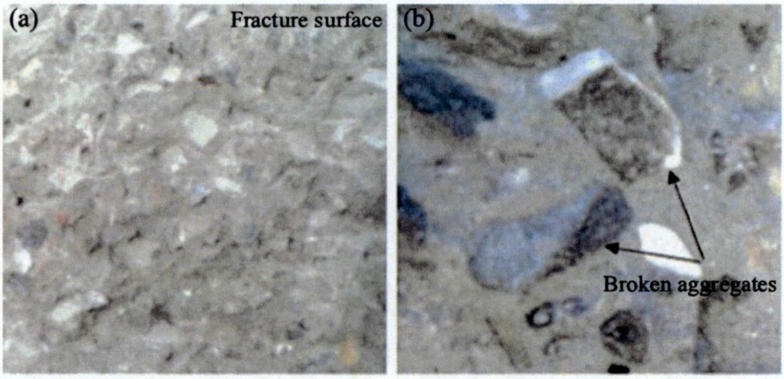

Fig. 4.8 Splitting surfaces of concrete specimens ($b = 150$ mm): **a** $d_{\max} = 10$ mm; and **b** $d_{\max} =$ 40 mm

Figure 4.9 shows the failure patterns of concrete specimens (side length by 150 mm) with different CSG. One can see clearly that the splitting failure patterns of cubic specimens having different CSG are basically similar and splitting cracks take place along the central line. However, cracks in concrete specimens with high-strength grade expand quickly once they appear, and even the phenomenon of bursting appears when specimens get damaged, showing a typical brittle failure.

The typical fracture surfaces of concrete specimens with different CSG are given in Fig. 4.10. The splitting fracture surfaces are quite different between concrete specimens having the strength grade C60 and C20. The splitting cracks of concrete specimens with C20 surround the aggregate particles and the damage mainly occurs in the mortar matrix and ITZs. The fracture surfaces of concrete specimens with C20 are relatively rougher. The cracks in concrete specimens having high-strength grade of C60 not only occur in the mortar matrix and ITZs, but also penetrate many aggregate particles. That is to say, more aggregate particles are pulled off by splitting cracks, making the specimen exhibit a typical brittle failure.

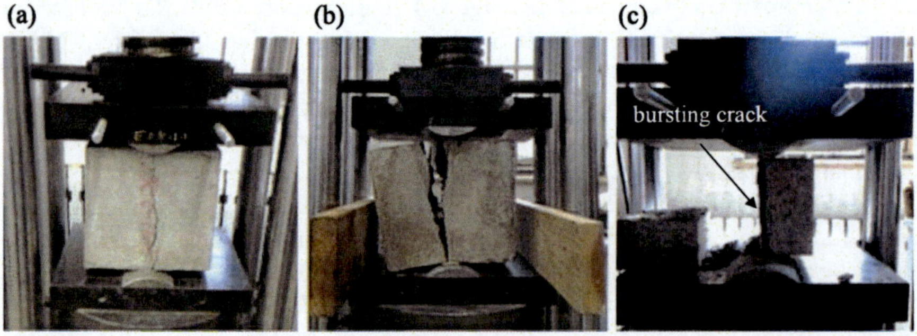

Fig. 4.9 Failure patterns of concrete specimens ($b = 150$ mm) with different CSG: **a** C20; **b** C40; and **c** C60

Fig. 4.10 Splitting surfaces of concrete specimens ($b = 150$ mm) with different CSG

2. *Splitting-tensile strength in test*

According to the Chinese code GB/50010–2010 (Ministry of construction of People's Republic of China 2011), the measured load of loading area can be converted to the splitting-tensile stress f_{st} (nominal strength) as

$$f_{st} = \frac{2P_u}{\pi A} \quad (4.34)$$

where f_{st} = the nominal splitting-tensile strength; P_u = the applied peak load (N); and A = the area of splitting failure surface (mm^2).

Figure 4.11 presents splitting-tensile strength in concrete cubic specimens decreases basically with increasing side length under different MAS, showing an obvious size effect, especially in cubic specimens with small(er) side lengths. The downtrend of splitting-tensile strength is slowing down with increasing side length and the size effect is gradually weakened in larger-sized specimens. These findings are consistent with the efforts of Zhou et al. (2010). In addition, the MAS has an

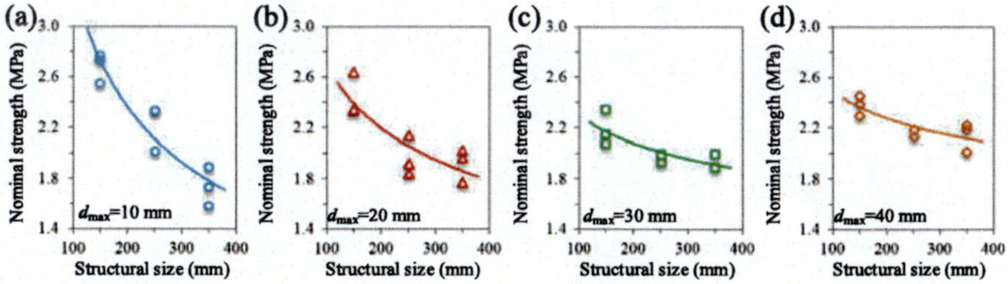

Fig. 4.11 Size effect on splitting-tensile strength of concrete having different MAS: **a** MAS: 10 mm; **b** MAS: 20 mm; **c** MAS: 30 mm; and **d** MAS: 40 mm

important influence on the size effect on splitting-tensile strength: the downtrend of nominal strength in specimens with smaller MAS is more obvious and the increase in MAS can reduce the sensitivity to the size effect on splitting-tensile strength. This may be attributed to the fact as follows.

On the one hand, the inherent brittleness is the crucial reason for the size effect. The micro-structure of ITZs is highly damaged by pre-existence micro-voids and micro-cracks, and the ITZs often turn out to be weak-links within concrete. Specimens with smaller MAS have more aggregate particles, making that there are more weak-links and defects with poor mechanical properties. The weak links and defects have a significant influence on the size effect on fracture behavior of concrete materials according to the theory of Weibull (1939). Many micro-cracks (or damage) are initiated at the weakest link of concrete and prefer to propagate along the weak interfacial zone rapidly for forming main splitting cracks through the whole specimen, resulting that the failure pattern of specimens (with small MAS) having more weak-links exhibits more brittle. On the other hand, when meeting aggregate particles, the splitting cracks are forced either to propagate through tougher aggregate particles or deflect and travel around weak interfacial zone. In specimens with smaller aggregates, the advancing crack is prone to bypass the aggregate particles and few aggregate particles are pulled off, causing that the tortuosity of cracking path is smaller. It is observed that the fracture surface of specimens with small aggregates is smooth while the surface becomes rougher and some coarse aggregates are snapped in specimens with larger MAS. These factors result in that the failure patterns exhibit more brittle and the size effect on splitting-tensile strength is more obvious in concrete specimens with smaller MAS.

As observed in Fig. 4.12, splitting-tensile strength of concrete specimens with different CSG decreases as the side length increases. However, the downtrend of splitting-tensile strength is also different. For high(er) strength grade (i.e., C60), splitting-tensile strength is more sensitive to the increasing side length, showing that the size effect is more obvious. This may be because that under high(er) strength grade, the homogeneity and workability of concrete specimens are better. The discrepancy of mechanical properties between the aggregate particles and mortar

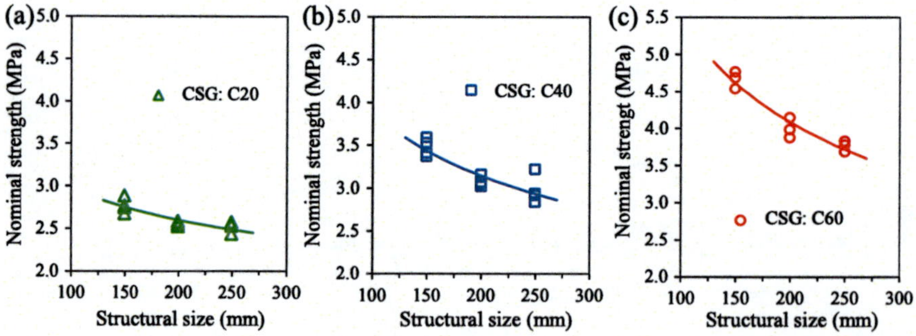

Fig. 4.12 Size effect on splitting-tensile strength for different CSG: **a** C20; **b** C40; and **c** C60

matrix becomes smaller, and splitting cracks have more possibilities to penetrate more aggregate particles when they meet aggregate particles. The main cracks extend in a short time once they show up, and fracture surfaces are smooth and less tortuous in the cubic specimens having high-strength grade. Therefore, the failure pattern exhibits more brittle and the size effect on splitting-tensile strength becomes more obvious in specimens with high(er) strength grade.

4.2.2 Numerical Analysis

Similarly, for investigating the influence of MAS and CSG on the mechanical behavior and the corresponding size effect from numerical perspective, especially in large-sized concrete specimens, the splitting-tensile failure behavior of concrete cubic models having different side lengths (side length b = 150–600 mm) under different MAS (d_{max} = 10, 20, 30 and 40 mm) and different CSG (C20, C40 and C60) were simulated by Jin et al. (2020b) using the meso-scale numerical method described in Chap. 3. In each case, three concrete cubic specimens (only the aggregate distribution is different) were conducted and 114 concrete cubic models in total were simulated. The 3D cubic concrete model showed in Fig. 3.27 was applied and the modeling details (including loading method and boundary condition) are no longer given here (mentioned in Sect. 3.6.1 of Chap. 3). The numerical model was meshed with an 8-nodded hexahedral element (type C3D8R: Continuum, 3-D, 8-node, reduce integration) and the average element size is around 1 mm. In addition, the damaged plasticity model described in Sect. 3.3.1 of Chap. 3 has been adopted to simulate the mechanical failure behavior of concrete three-phase components (aggregate, mortar matrix and ITZ) and the mechanical parameters of three-phase components used in meso-scale simulation can refer to Table 3.1. Detailed modeling introduction can refer to this literature (Jin et al. 2020b). Figure 4.13 depicts 3D meso-scale numerical models of concrete cubic specimens having different side lengths and MAS, which is utilized in the following simulations.

1. *Failure pattern*

Figure 4.14 depicts the typical stress distribution in the splitting surface of concrete cubic specimens obtained from simulation. One can see that the compressive stress is mainly concentrated in the area near the top and bottom surface while the tensile stress is mainly distributed in the core zone of specimens. Therefore, the area (such as Point A and D) near top and bottom surface is in the state of tri-axial compression while the area (such as Point B and C) in the core zone is in the state of tension–compression. This causes that splitting damage first appears in the core zone of cubic specimen. Typical failure patterns of concrete cubic models with different MAS obtained from simulation are described in Fig. 4.15. It can be seen that with the increase in MAS the path of splitting cracks becomes more tortuous and complex. On the splitting surface of cubic specimens with aggregates of small sizes, the damage area is larger and

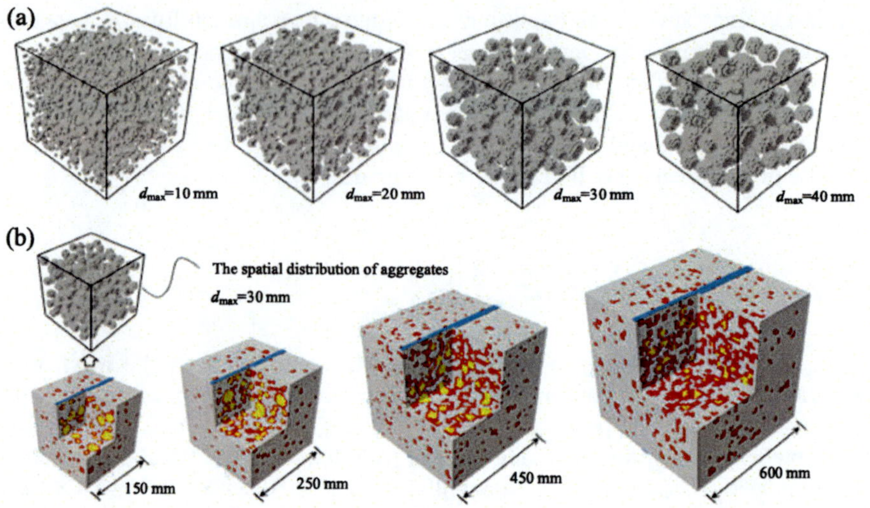

Fig. 4.13 3D meso-scale numerical models of concrete cubic specimens for splitting-tensile simulation: **a** With 150 mm size length for different MAS; and **b** With $d_{\max} = 30$ mm for different structural sizes

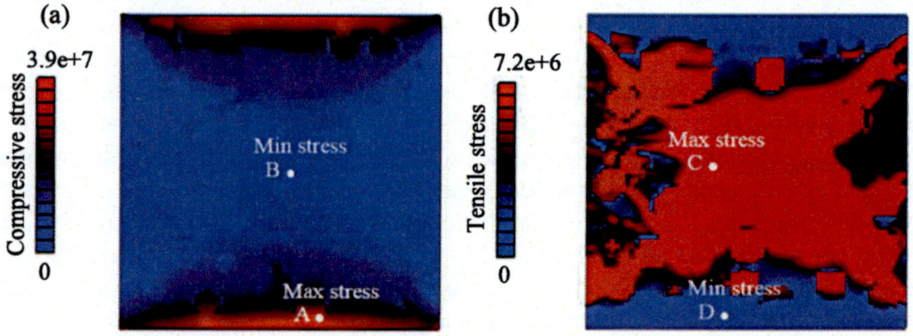

Fig. 4.14 Typical stress distribution in the splitting surface of concrete cubic specimens obtained from simulation: **a** Compressive stress; and **b** Tensile stress

more uniform (see Fig. 4.15b), and more small aggregate particles get damaged by splitting cracks (see Fig. 4.15c. These indicate that the roughness of fracture surfaces is smaller and the splitting failure pattern is more brittle, which are consistent with the test results in Figs. 4.7 and 4.8.

Typical failure patterns of concrete cubic models with different CSG obtained from simulation are shown in Fig. 4.16. One can see that there are more damage areas and the damage distribution is more uniform on the splitting surface of specimens with high strength grades. In addition, more splitting damage appears in aggregate

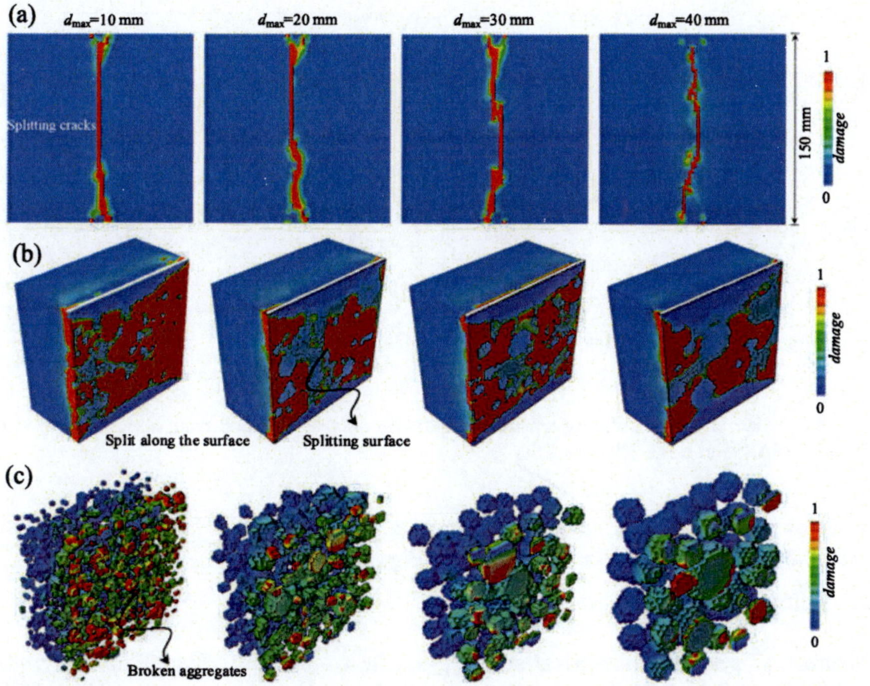

Fig. 4.15 Typical failure patterns of concrete cubic specimens having different MAS obtained from simulation: **a** The cracks in lateral view; **b** Damage in splitting surface; and **c** Damage of aggregate particles

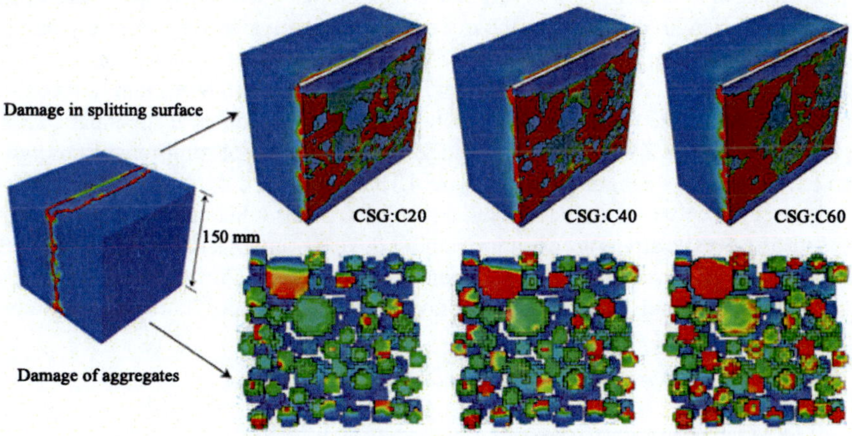

Fig. 4.16 Numerical failure patterns of concrete cubic specimens having different CSG

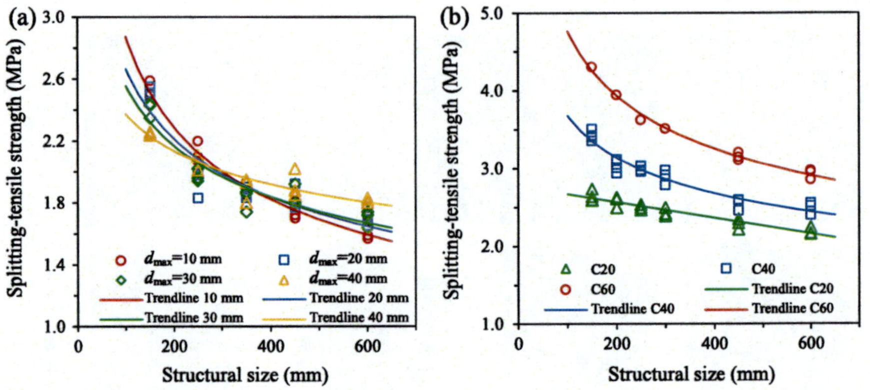

Fig. 4.17 Size effect on splitting-tensile strength of concrete obtained from simulation: **a** For different MAS; and **b** For different CSG

particles on the splitting surface and the roughness of fracture surfaces is smaller, which are also similar to the test results in Fig. 4.10.

2. *Splitting-tensile strength*

Figure 4.17 gives the numerical splitting-tensile strength of concrete models with different MAS and CSG. One can notice that splitting-tensile strength decreases with the increase in side length, showing an obvious size effect. However, the downtrend of splitting-tensile strength slows down with the increasing side length and the size effect is weakened gradually. In other words, the size effect of splitting-tensile strength is not obvious in large-sized specimens (size >450 mm), and even the nominal strength could approach a size independent value for extremely large-sized specimens. Moreover, MAS and CSG have a significant influence on splitting tensile strength and corresponding size effect. The increase in MAS can reduce the sensitivity to the size effect on splitting tensile strength. Concrete specimens with high strength grades are more sensitive to the size effect on splitting-tensile strength than ones with low strength grades. Figure 4.18 depicts the comparison of splitting-tensile strength between simulation and test results. One could note that most numerical splitting-tensile strengths are approaching the experimental data, indicating that the simulation results are in good agreement with test results. Furthermore, the meso-scale numerical models can predict more nominal strength data of concrete specimen with larger structural sizes.

3. *Compare with Bažant's size effect law*

The typical size effect law (SEL) proposed by Bažant (Bažant and Planas 1997; Bažant 1984) was selected to analyze the size effect on nominal strength of concrete materials. Through a large quantity of experimental and theoretical approaches based on non-linear fracture mechanics, *Type-2* SEL is applicable to brittle failure of heterogeneous materials including concrete material and it can be written as

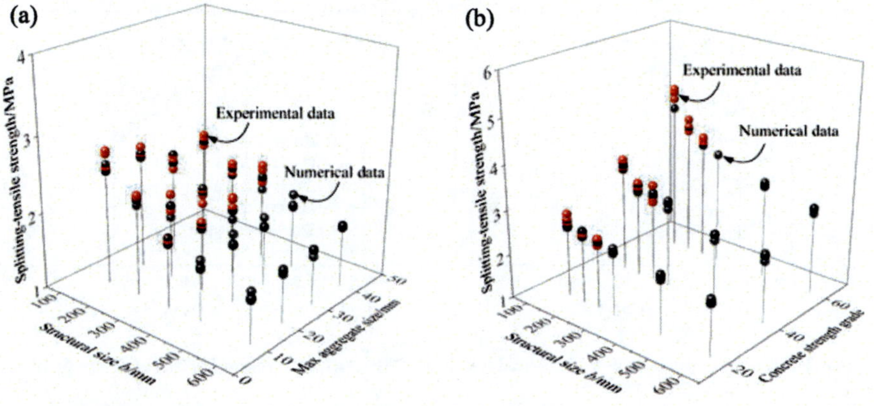

Fig. 4.18 Comparison of splitting-tensile strength between numerical and experimental results: **a** For different MAS; and **b** For different CSG

$$\sigma_{Nu} = \frac{Bf_t'}{\sqrt{1 + D/D_0}} \tag{4.35}$$

where σ_{Nu} represents the size-related nominal strength at failure for specimens similar in 2D; f_t' represents the average splitting-tensile strength of standard specimen (here the structural size is 150 mm); D represents the characteristic structural size of specimen (here D is side length b); B and D_0 represent empirical parameters determined by optimum fitting of nominal strengths σ_N covering a broad enough size range, D_0 is also a physical quantity related to the brittleness of the material. For linear regression analysis, Eq. (4.35) could be rearranged to

$$\left(\frac{f_t'}{\sigma_{Nu}}\right)^2 = \left|\frac{1}{D_0 B^2}\right| D + \frac{1}{B^2} \tag{4.36}$$

Equation (4.36) can be depicted as a linear form:

$$Y = AX + C \tag{4.37}$$

where $Y = (f_t'/\sigma_{Nu})^2, X = D, A = 1/D_0B^2$, and $C = 1/B^2$. A and C are obtained from a regression analysis according to strength data. Based on the regression analysis, the bi-logarithmic plots of $\lg(\sigma_{Nu}/B\ f_t')$ versus $\lg(D/D_0)$ for nominal splitting-tensile strength can be obtained, which are given in Fig. 4.19. The linear elastic fracture mechanics (LEFM, for pure brittle materials) curve and the strength criterion (for plastic materials) curve are also plotted for the comparison with the numerical results.

One could note that both the numerical strength data points basically vary along the curve of Bažant's SEL, illustrating that the Bažant's SEL could well characterize the

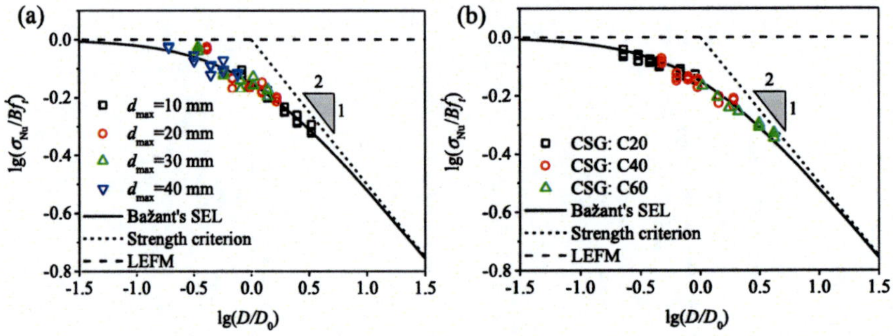

Fig. 4.19 Comparison between numerical results and SEL: **a** For different MAS; and **b** For different CSG

numerical results. Furthermore, with the decrease in the MAS and the improvement in CSG, the data points of splitting-tensile strengths are away from the curve of strength criterion and get closer to the curve of LEFM (slope = −1/2). This indicates that splitting fracture pattern of concrete becomes more brittle and the size effect on splitting-tensile strength is more obvious.

4.3 Flexural-Tensile Strength of Concrete Materials

Flexural-tensile strength is one of the basic mechanical properties of concrete materials and large quantities of researchers have carried out lots of experimental and theoretical studies. These studies have promoted the understanding of the damage mechanism and macro-mechanical properties of concrete materials under flexural load. However, there are limited studies on the influence of aggregate size on the flexural-tensile failure behavior of concrete and the corresponding size effect. Therefore, this section shows physical tests and meso-scale simulation results on flexural-tensile failure behavior of concrete materials, and it also provides the corresponding size effect (Du et al. 2016; Du 2016).

4.3.1 Experimental Analysis

In the physical experiment, a total of 27 plain concrete beams having three kinds of MAS (d_{max} = 10, 20 and 30 mm) and three kinds of structural sizes (height h × width b × length l = 100 mm × 100 mm × 400 mm, 150 mm × 150 mm × 550 mm and 200 mm × 200 mm × 700 mm) were designed for four-point bending tests. The details about the production and introduction of tested concrete beams can refer to this literature (Du et al. 2016). All specimens were tested in the hydraulic universal testing machine with loading velocity of 0.02–0.05 MPa/s according to the Chinese code GB/T 50081–2002 ((Ministry of construction of People's Republic of China 2003)). Four-point bending setup of concrete beam is shown in Fig. 4.20.

1. *Failure pattern*

Figure 4.21 shows the flexural-tensile failure patterns of concrete beams (with $h =$ 150 mm) having different MAS. It can be seen that flexural-tensile cracks appear first in the bottom of concrete beam. When the load is further applied, the main cracks near the mid-span develops and penetrates rapidly towards to the middle part of beam. Small cracks were generated near the main cracks. Finally, the mid-span cracks almost penetrate the entire concrete beam. In addition, the flexural-tensile failure patterns of concrete beams having different MAS are slightly different. The larger the MAS, the more tortuous flexural-tensile cracks. The flexural-tensile failure patterns of concrete beams with different beam heights are basically the same.

2. *Flexural-tensile strength*

According to the Chinese code GB 50010–2010 (Ministry of construction of People's Republic of China 2011), the measured load can be converted to the nominal flexural-tensile strength f_f as

$$f_f = \frac{Fl}{bh^2} \tag{4.38}$$

where f_f represents the nominal flexural-tensile strength; F represents the applied peak load; l represents the length of concrete beam; b represents the width of concrete beam; and h represents the height of concrete beam, $l = 3\ h$.

The nominal flexural-tensile strength f_f, standard deviation S and variation coefficient C_v are shown in Table 4.4. In Table 4.4 the specimen numbers are in the form of BX-Y–Z, in which B represents concrete beam, X represents beam height h, Y represents MAS and Z represents the serial number of specimen. For example, B100-10–1 represents that the concrete beam with beam height $h = 100$ mm, MAS $d_{max} = 10$ mm and serial number No. 1. The test results in Table 4.4 show that the

Fig. 4.20 Four-point bending setup of concrete

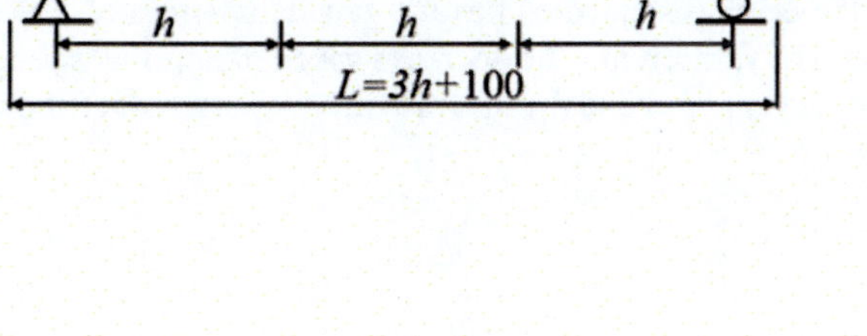

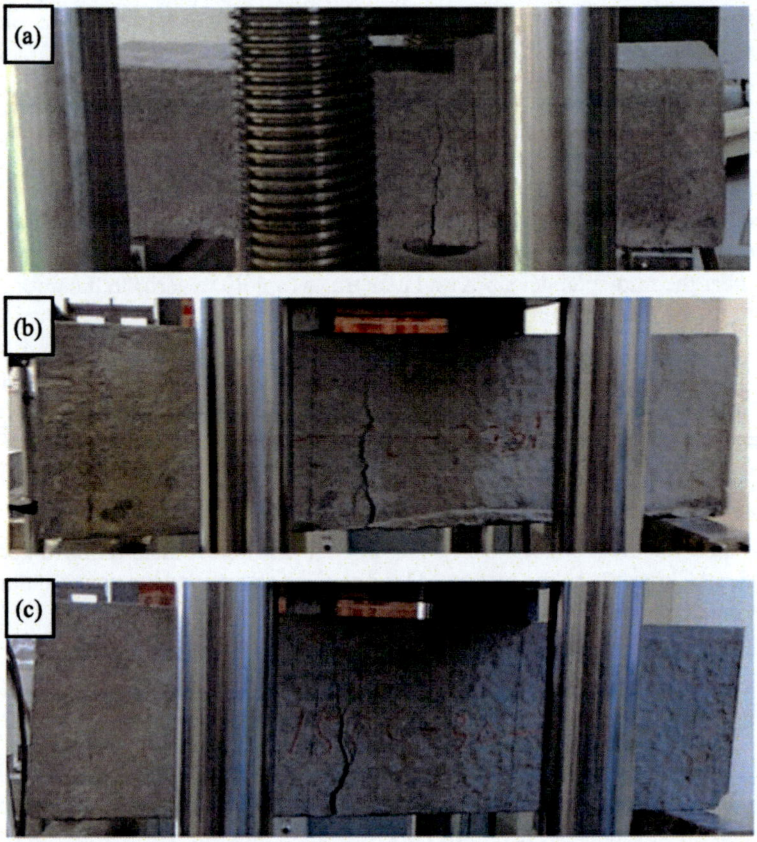

Fig. 4.21 Failure patterns of concrete beams having different MAS (beam height $h = 150$ mm): **a** MAS 10 mm; **b** MAS 20 mm; and **c** MAS 30 mm

variation coefficient C_v of flexural-tensile strength of concrete beams are all less than 10%, indicating that the dispersion of the test results is acceptable.

Figure 4.22 shows the relationship between flexural-tensile strength of concrete specimens and MAS. It can be seen that the flexural-tensile strength of concrete specimens is affected by MAS and the nominal strength basically decreases with the increasing MAS.

Figure 4.23 presents the relationship between flexural-tensile strength of concrete specimens and structural size (beam height). One can note that the nominal flexural-tensile strength of concrete beams decreases with the increasing beam height under the same MAS, showing a significant size effect. Herein, the decrease speed can be described by the linear fitting slope. According to the slope, it can be found that the decrease speed of flexural-tensile strength of concrete having different MAS is slightly different. The decrease speed of nominal strength in specimens with smaller MAS is larger and the increase in MAS can reduce the sensitivity to the size effect on

Table 4.4 Test data of concrete beams

Specimen number	Peak load (kN)	f_f (MPa)	Average f_f (MPa)	Standard deviation S	Variation coefficient C_v (%)
B100-10-1	16.30	4.89	4.61	0.34	7.5
B100-10-2	15.76	4.73			
B100-10-3	14.06	4.20			
B150-10-1	33.82	4.51	4.28	0.22	5.1
B150-10-2	31.50	4.20			
B150-10-3	32.12	4.15			
B200-10-1	52.80	3.96	4.04	0.26	6.6
B200-10-2	57.82	4.34			
B200-10-3	50.82	3.81			
B100-20-1	15.00	4.50	4.57	0.12	4.6
B100-20-2	15.00	4.50			
B100-20-3	15.70	4.71			
B150-20-1	29.77	3.97	4.18	0.18	4.3
B150-20-2	31.93	4.26			
B150-20-3	32.25	4.30			
B200-20-1	51.98	3.90	4.01	0.1	2.5
B200-20-2	54.48	4.09			
B200-20-3	53.98	4.05			
B100-30-1	14.61	4.38	4.48	0.1	2.2
B100-30-2	14.93	4.48			
B100-30-3	15.28	4.58			
B150-30-1	30.23	4.03	4.05	0.15	3.7
B150-30-2	28.90	3.86			
B150-30-3	31.12	4.15			
B200-30-1	54.65	4.10	3.96	0.13	3.2
B200-30-2	52.38	3.93			
B200-30-3	51.32	3.85			

flexural-tensile strength. When concrete beam is applied load, flexural-tensile cracks tend to bypass small aggregate particles. Once internal cracks occur, they can rapidly expand and the failure pattern shows a greater brittleness, causing that size effect is relatively more significant.

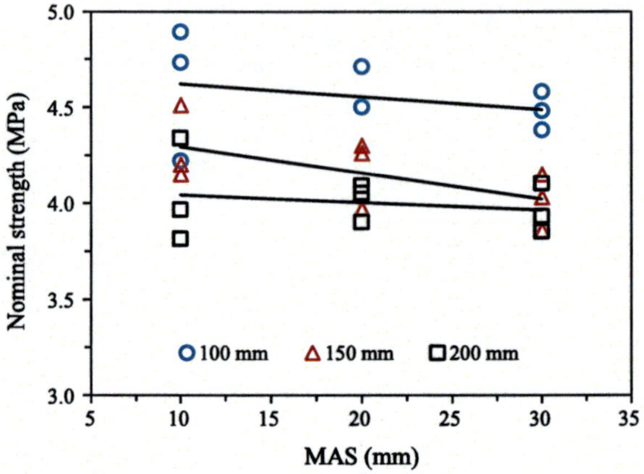

Fig. 4.22 Relationship between flexural-tensile strength of concrete specimens and MAS

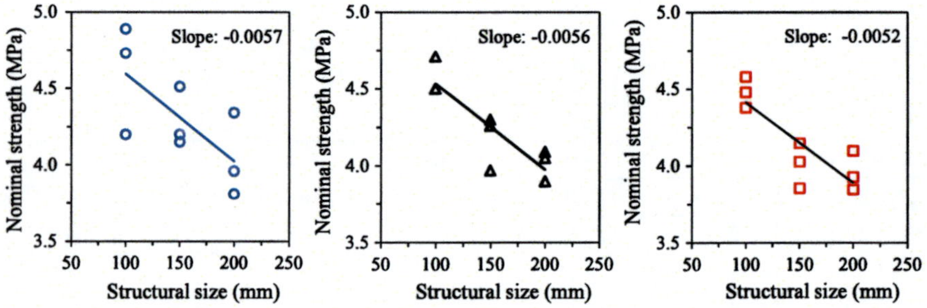

Fig. 4.23 Relationship between flexural-tensile strength of concrete specimens and structural size (beam height): **a** MAS: 10 mm; **b** MAS: 20 mm; and **c** MAS: 30 mm

4.3.2 Numerical Analysis

Similarly, for investigating the influence of MAS on the flexural-tensile failure mechanical behavior of concrete and the corresponding size effect from numerical perspective, the flexural-tensile failure behavior of concrete beam models with three kinds of MAS ($d_{max} = 10$, 20 and 30 mm) and three kinds of structural sizes (height $h \times$ width $b \times$ length $l = 100$ mm $\times$ 100 mm $\times$ 400 mm, 150 mm $\times$ 150 mm $\times$ 550 mm and 200 mm $\times$ 200 mm $\times$ 700 mm) were simulated by the meso-scale numerical method. Figure 4.24 depicts typical 2D plain concrete beam models utilized in this section and coarse aggregate particles were modeled as circles with random distribution enfolded by ITZ. The numerical model was meshed with four-node isoparametric element and the average element size is approximately 1 mm. In addition, the damaged plasticity model described in Sect. 3.3.1of Chap. 3 has been adopted to simulate the mechanical failure behavior of mortar matrix and ITZ. As for

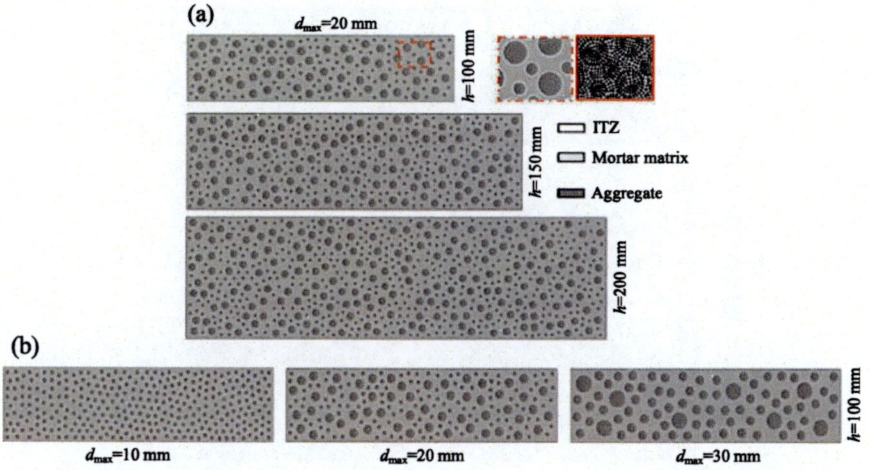

Fig. 4.24 2D meso-scale numerical models of concrete beam for flexural-tensile simulation: **a** With $d_{max} = 20$ mm for different beam heights; and **b** With beam height 100 mm for different MAS

aggregate particles, they were assumed to follow a linear elastic material behavior in this section. The mechanical parameters of three-phase components set in meso-scale simulation are listed in Table 4.5. Detailed modeling introduction can refer to the work (Du 2016).

1. *Failure pattern in simulation*

Figure 4.25 shows the damage cloud diagram of the meso-mechanical model of concrete beam (with $h = 100$ mm, $d_{max} = 20$ mm) under flexural-tensile load. It can be seen that flexural-tensile cracks appear first in the bottom of concrete beam (the damage are first formed from the ITZ). When the load is further applied, the main cracks near the mid-span develops and penetrates rapidly towards to the middle part

Table 4.5 Mechanical parameters of meso components for flexural-tensile simulation

	Aggregate	Mortar matrix	ITZ
Elastic modulus E (GPa)	50[b]	30[a]	25[a]
Poisson's ratio υ	0.16[b]	0.20[a]	0.22[a]
Dilation angle ψ (°)	–	18	15
K	–	0.667	0.667
Compressive yield stress σ_c (MPa)	–	30[a]	28.4[c]
Tensile yield stress σ_t (MPa)	–	3.0[a]	2.8[c]

Note [a]Data were quoted from the literature (Du et al. 2015); [b]Data were obtained from tests; [c]Data were obtained by repeated trial; and other parameters were default value

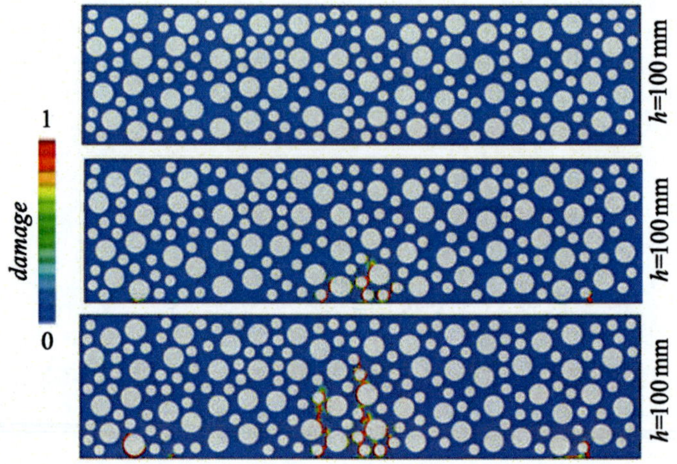

Fig. 4.25 Flexural-tensile damage process of concrete specimens in simulation

of beam. Small cracks were generated near the main cracks. Finally, the mid-span cracks almost penetrate the entire concrete beam.

Figure 4.26 presents flexural-tensile patterns of concrete specimens (beam height $h = 150$ mm) under different MAS obtained from simulation. The flexural-tensile patterns of concrete beams with different MAS are slightly different. The aggregate particles affect the propagation path of damaged cracks and the concrete beam with

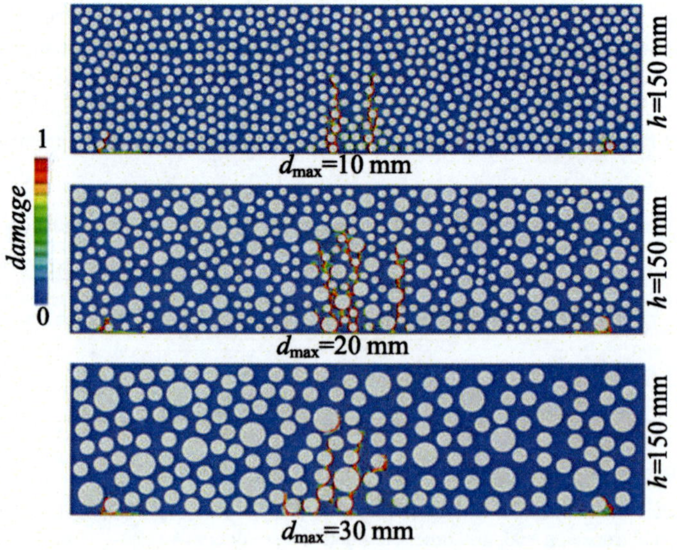

Fig. 4.26 Flexural-tensile patterns of concrete specimens under different MAS in simulation

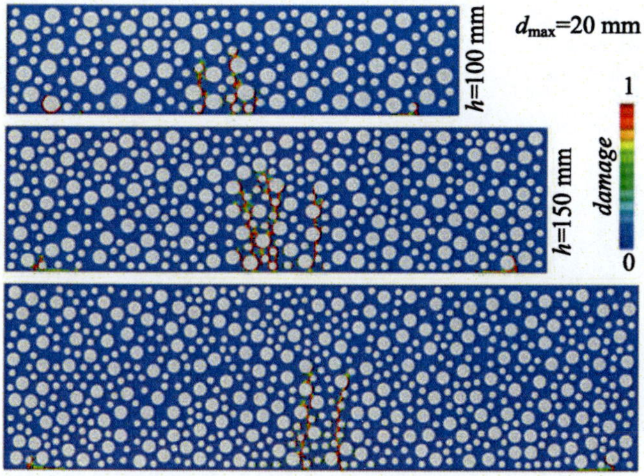

Fig. 4.27 Flexural-tensile patterns of concrete specimens under different structural sizes in simulation

larger-sized aggregate particles has more branched cracks and more tortuous main cracks. Figure 4.27 gives flexural-tensile patterns of concrete specimens (d_{max} = 20 mm) under different structural sizes obtained from simulation. One can note that the flexural-tensile failure patterns of concrete beams with different beam heights are basically the same. These above numerical flexural-tensile patterns are consistent with ones in test results in Sect. 4.3.1.

2. *Flexural-tensile strength in simulation*

Figure 4.28 shows the comparisons of flexural-tensile strength between numerical results and test results. It can be seen that the flexural-tensile strengths obtained from simulation approach the test results, showing that the nominal strength decreases with the increasing structural size. For comparing with Bažant's SEL, the bi-logarithmic plot of $\lg(\sigma_{Nu}/B f_f')$ versus $\lg(D/D_0)$ (here D = beam height h) for numerical flexural-tensile strength are given in Fig. 4.29. One can note that the numerical strength data points basically vary along the curve of Bažant's SEL, illustrating that the Bažant's SEL could well characterize the numerical results. Furthermore, with the decrease in the MAS, the data points of flexural-tensile strengths are away from the curve of strength criterion and get closer to the curve of LEFM (slope = -1/2). This indicates that flexural-tensile fracture pattern of concrete beam becomes more brittle and the size effect on flexural-tensile strength is more obvious.

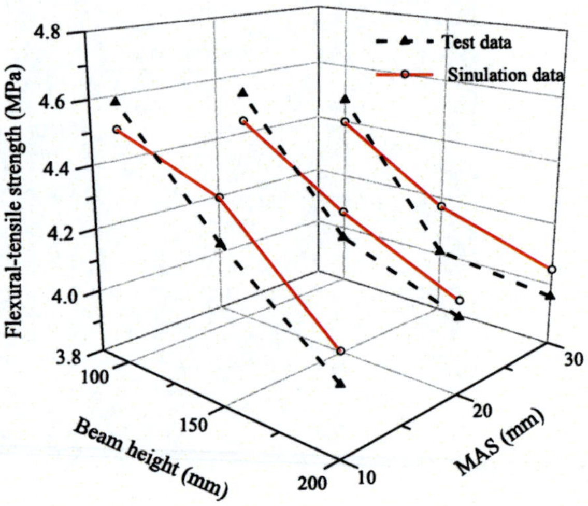

Fig. 4.28 Comparisons of flexural-tensile strength between numerical results and test results

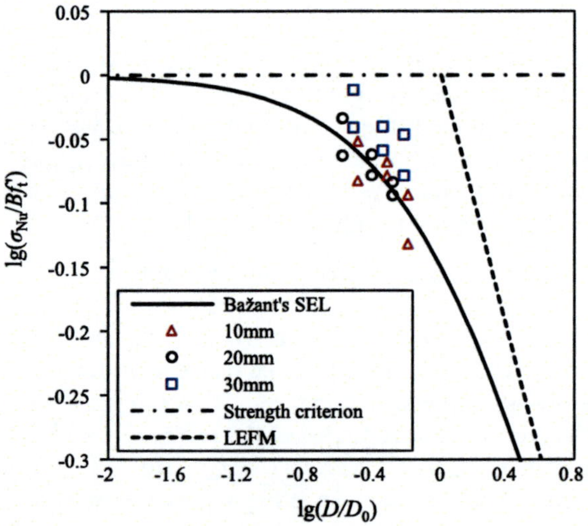

Fig. 4.29 Comparison between numerical results and Bažant's SEL

4.4 Compressive Strength of Concrete Materials

The compressive strength is also one of the main parameters in the design of concrete structures and it is an important indicator reflecting the load bearing capacity of concrete materials. On the whole, the investigations on the static size effect on compressive strength of normal concrete have been conducted relatively completely. However, the investigations on the static size effect on compressive strength of high-performance concrete materials (lightweight aggregate concrete, recycled aggregate concrete, fiber reinforced concrete, etc.) and the size effect on compressive strength of concrete materials under complex loads (bi-axial load, tri-axial load, dynamic load, etc.) are limited, which need more in-depth investigations.

4.4.1 Size Effect of Lightweight Aggregate Concrete

In recent years, a large number of high-rise buildings and large-span structures have emerged, leading to an increase in the weight of the structures. Lightweight aggregate concrete (LWAC) has become an irreplaceable building material for the construction industry because of its light weight. As a new building material used widely, it is of great importance to investigate the mechanical failure behavior of LWAC and the corresponding size effect under static and dynamic loads.

The compressive failure behavior of LWAC and the corresponding size effect under static loads were investigated based on the 3D meso-scale modelling method (Jin et al. 2020a). A total of 18 LWAC cubic specimens having different structural sizes (side length b = 100, 150, 200, 300, 450 and 600 mm) under quasi-static load were simulated by Jin et al. (2020a). For comparison, each case contains three parallel cubic specimens of normal concrete and a total of 18 normal concrete cubic specimens were also simulated.

Figure 4.30 depicts the 3D LWAC cubic specimen utilized in simulation and coarse aggregate particles were modeled as circles with random distribution enfolded by ITZ. The volume fraction of aggregate particles is about 40% and two-graded aggregate concrete was adopted with lightweight aggregates with diameter 12 and 30 mm. The numerical model was meshed with an 8-nodded hexahedral element and the average element size is approximately 2 mm. In order to implement a static load (quasi-static load), a very low loading velocity v (constant velocity) was set on the top surface of model. Herein the strain rate $\dot{\varepsilon} = v/b$ (b = the side length) in numerical models was controlled as 1×10^{-5}/s for obtaining the quasi-static load. The bottom of model was fixed by normal constraint and one corner point of bottom was horizontally

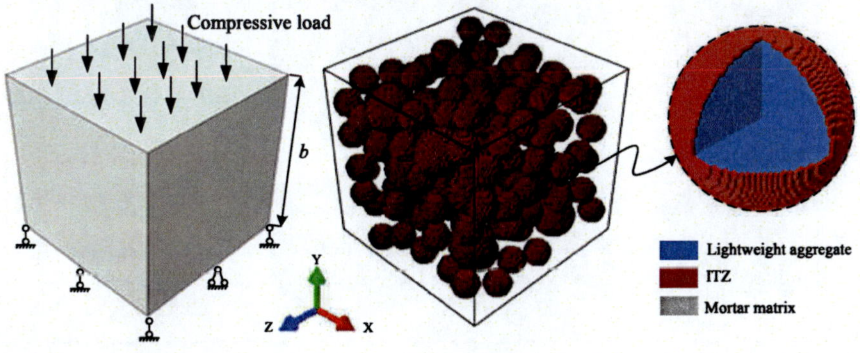

Fig. 4.30 3D meso-scale numerical model of LWAC cubic specimen for compression

and vertically constrained. The other edges were free boundaries. In addition, the damaged plasticity model described in Sect. 3.3.1 has been adopted to simulate the mechanical failure behavior of concrete three-phase components (lightweight aggregate, mortar matrix and ITZ) and the mechanical parameters of three-phase components used in meso-scale simulation are listed in Table 4.6. Detailed modeling introduction can refer to this literature (Jin et al. 2020a).

1. *Failure patterns*

Figure 4.31 shows the failure patterns of aggregate particles, ITZ and mortar matrix at the peak compressive stress of normal concrete and LWAC. It can be seen that the failure patterns of the internal components of normal concrete and LWAC are significantly different, especially for the aggregate particles. When concrete stress reaches peak stress, the damage of LWAC is more serious. Moreover, the damage degree of aggregate particles is similar to that of ITZ and mortar matrix. However,

Table 4.6 Mechanical parameters of concrete meso-components

	Lightweight aggregate	Mortar matrix	ITZ
Compressive yield stress σ_c (MPa)	18.0[b]	25.0[c]	20.0[c]
Tensile yield stress σ_t (MPa)	2.8[b]	2.5[c]	2.0[c]
Dilation angle ψ (°)	18[c]	18	15
Poisson's ratio υ	0.22[b]	0.20[a]	0.20[c]
Elastic modulus E (GPa)	16[b]	38[a]	26[c]
Mass density ρ (kg/m^3)	1400[b]	2750[a]	2750[c]

Note [a]Data were based on the Dilger et al.'s study (Dilger et al. 1984); [b]Data were based on Wang and Chen's study (Wang and Chen 2008); [c]Data were got by repeated trial; and other parameters are default value

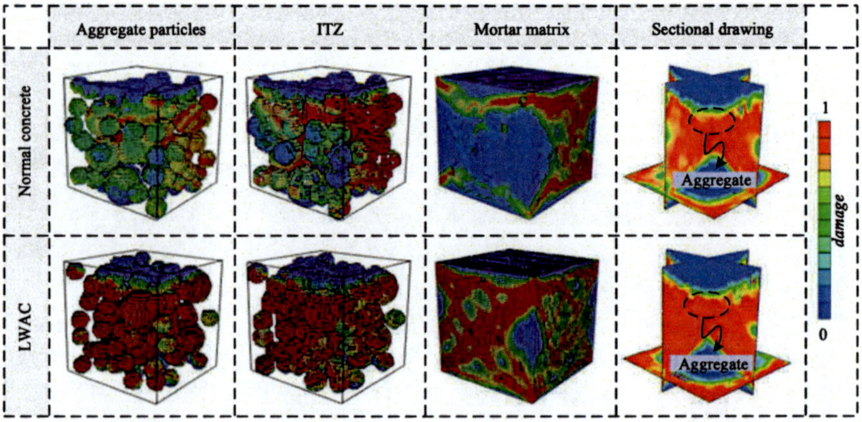

Fig. 4.31 Failure patterns of aggregate particles, ITZ and mortar matrix at the peak compressive stress of normal concrete and LWAC

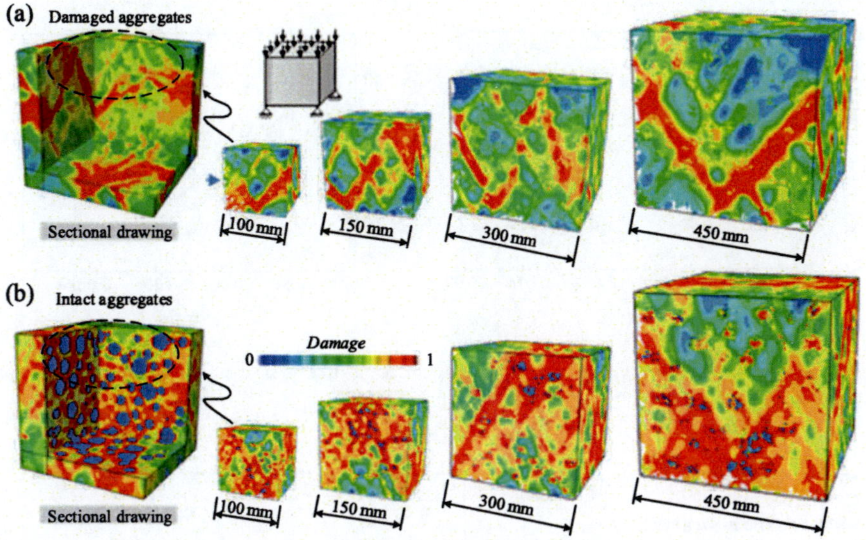

Fig. 4.32 Final failure patterns of concrete with different structural sizes: **a** For LWAC; and **b** For normal concrete

there is little damage in the aggregate particles of normal concrete. It can be found clearly from sectional drawing that the damage degree of aggregate is the main difference between LWAC and normal concrete. When the concrete gets damaged, the lightweight aggregate also reaches damaged while the normal aggregate is almost intact. The results are consistent with those of Cui et al. (2012).

Figure 4.32 exhibits the final failure patterns of normal concrete and LWAC with different structural sizes. Due to the heterogeneity of the concrete materials, the internal damages of concrete specimens are inhomogeneous. The cracks run obliquely through the concrete specimens and the brittle shear failure patterns can be clearly observed. When the concrete gets broken, the aggregate of LWAC is damaged because of its weak mechanical properties, while the aggregate of normal concrete remains intact. On the whole, the failure patterns of concrete with different structural sizes have little difference.

2. *Stress–strain curves*

Figure 4.33 plots the compressive stress–strain relationship of LWAC and normal concrete. Obviously, the changing tendency of stress–strain curves for LWAC with different structural sizes is quite similar. The ascent stage is almost coincident, and the falling section is approximately parallel. With the increasing structural size, the peak stress of LWAC decreases, showing an obvious size effect. In addition, the compressive stress–strain relationship of LWAC under same structural size is similar to that of normal concrete.

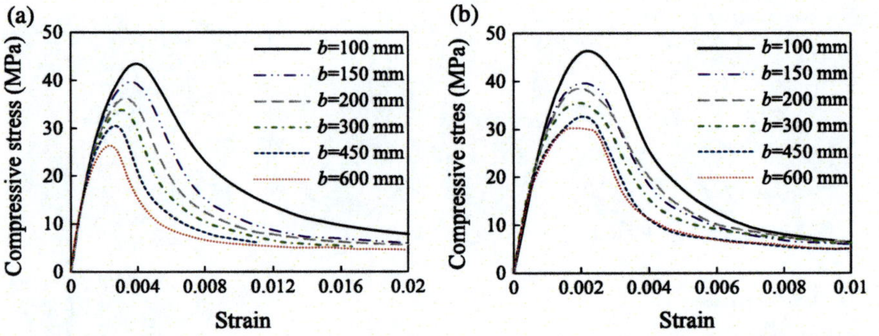

Fig. 4.33 Stress–strain curves of concrete having different structural sizes: **a** LWAC; and **b** Normal concrete

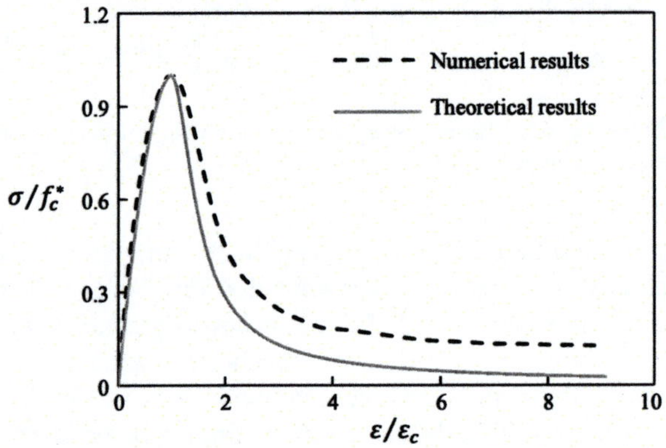

Fig. 4.34 Comparison between the numerical results and theoretical results of Guo et al. (1982)

Figure 4.34 presents the comparison between the numerical stress–strain relationship of LC30 (standard specimen, $b = 150$ mm) and the theoretical curve proposed by Guo et al. (1982). From Fig. 4.34, it can be seen that the numerical results are in good agreement with the theoretical ones, indicating the rationality of the numerical results. The theoretical stress–strain curve equation of LC10-LC45 for LWAC could be expressed as follows:

$$\frac{\sigma}{f_c^*} = \alpha_a \frac{\varepsilon}{\varepsilon_c} + (3 - 2\alpha_a)(\frac{\varepsilon}{\varepsilon_c})^2 + (\alpha_a - 2)(\frac{\varepsilon}{\varepsilon_c})^3, \left(\frac{\varepsilon}{\varepsilon_c} \le 1\right) \tag{4.39}$$

$$\frac{\sigma}{f_c^*} = \frac{\frac{\varepsilon}{\varepsilon_c}}{\alpha_{d1}\left(\frac{\varepsilon}{\varepsilon_c} - 1\right)^2 + \frac{\varepsilon}{\varepsilon_c}}, \left(\frac{\varepsilon}{\varepsilon_c} \ge 1\right) \tag{4.40}$$

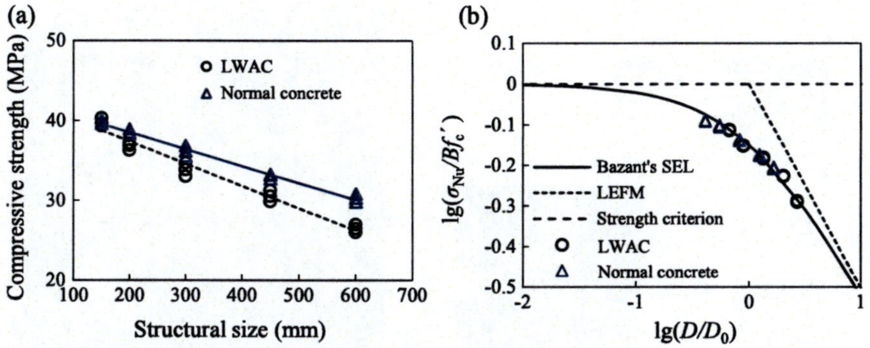

Fig. 4.35 Comparison of size effect between LWAC and normal concrete: **a** Strength versus structural size; and **b** Compared with Bažant's SEL

where ε is the compressive strain of LWAC; ε_c is the peak compressive strain of LWAC; σ is compressive stress; f_c^* is the compressive strength (peak stress) of LWAC; α_a and α_{d1} are constants.

3. *Size effect on compressive strength*

Figure 4.35 presents the comparison of size effect on the compressive strength of LWAC and normal concrete. As displayed in Fig. 4.35a, the compressive strength significantly decreases as the structural size increases for LWAC and normal concrete, showing an obvious size effect. However, the linear fitting slope (i.e. the decrease speed of the compressive strength as the structural size increases) of LWAC specimens is obviously greater than that of normal concrete. This indicates that the LWAC exhibit more obvious size effect than normal concrete. As shown in Fig. 4.35b, the numerical results are compared with Bažant's SEL. It can be seen that the compressive strength data points of LWAC and normal concrete obtained by numerical simulation agree well with Bažant's SEL curve. Moreover, the data points of LWAC are closer to the LEFM curve, indicating that the LWAC is more brittle and the size effect of LWAC is more obvious than that of normal concrete.

4.4.2 Size Effect on Bi-axial Compressive Behavior

In fact, most concrete structures are under complex stress conditions. The strength theory of concrete based on uniaxial stress state is difficult to generalize the stress characteristics under complex stress state. A large number of studies conducted by scholars on the strength characteristics of concrete under biaxial stress indicate that the stress state has an obvious influence on concrete strength (Buyukozturk et al. 1971; Subramaniam and Shah 2003; Shang and Song 2006; Chen and Leung 2014). At present, most studies on the size effect of concrete materials are conducted based on the uniaxial stress state while the understanding of the mechanism of size effect

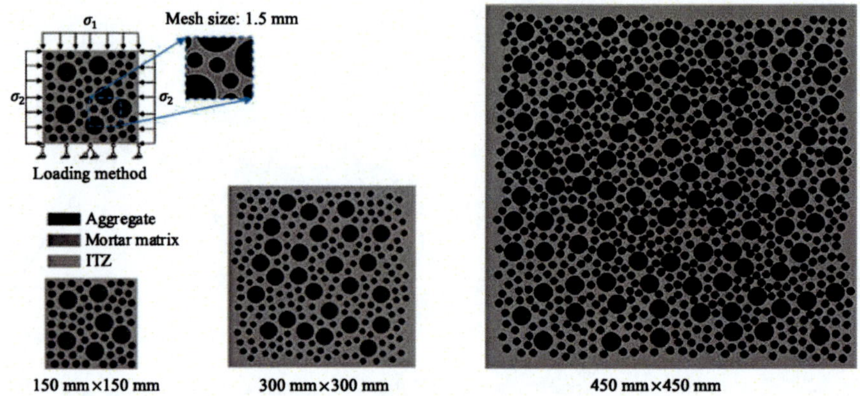

Fig. 4.36 Meso-scale mechanical model of concrete for bi-axial compression

on multi-axis strength is far from enough (Rossi and Ulm 1997). Here, the failure behavior of concrete materials and the corresponding size effect under biaxial load would be introduced.

Two-grade concrete was adopted and 2D square concrete meso-scale models with different structural sizes ($b = 150$, 300 and 600 mm) are presented in Fig. 4.36. The vertical displacement of the bottom of model is limited and other sides of the model are boundaries with displacement loading. Displacement loading on concrete model with different stress ratios (σ_1/σ_2) was adopted in simulation. The numerical model was meshed with four node isoparametric linear element and a fine mesh with a basic element size of 1.5 mm. In addition, the damaged plasticity model described in Sect. 3.3.1 has been adopted to simulate the mechanical failure behavior of mortar and ITZ. As for aggregate particles, they were assumed to follow a linear elastic material behavior in this section. The mechanical parameters of three-phase components set in meso-scale simulation can also refer to Table 4.5. One can see more detailed modeling information in (Li et al. 2017).

1. *Failure pattern*

Figure 4.37 shows the biaxial compressive failure process of concrete specimen (side length $b = 150$ mm) when the stress ratio σ_1/σ_2 is 1.0. It can be seen that the damage first occurs in the ITZ where the mechanical properties are relatively weak. Damage

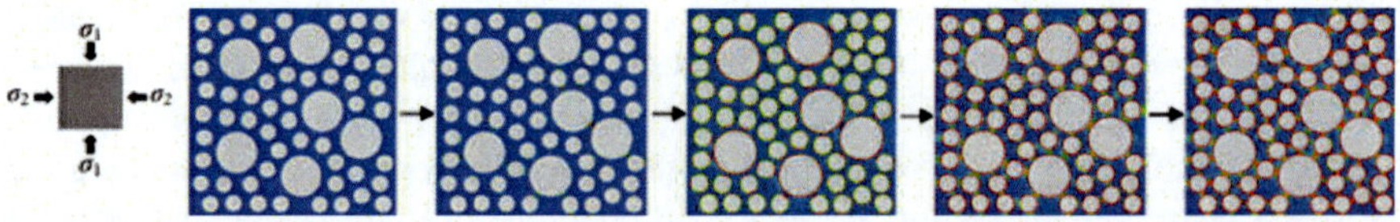

Fig. 4.37 Failure process of square concrete specimen ($b = 150$ mm) under bi-axial compression ($\sigma_1/\sigma_2 = 1.0$)

cracks continue to develop along the ITZ, gradually expanding from the ITZ around large aggregate to that around small aggregate. Then the cracks spread to the mortar matrix and the specimens form one or more macro cracks finally.

Figures 4.38, 4.39 and 4.40 show the failure patterns of square concrete specimens with different structural sizes under biaxial compression when the stress ratios are 0.2, 0.5 and 1.0 respectively. At the meso-scale level, the cracks are mainly distributed in the ITZ between the aggregate particles and the mortar matrix. The fine cracks roughly spread in a direction parallel to the larger principal stress, forming vertically crossing and crisscrossing main cracks. The crack distribution pattern is closely related to the stress ratio. When the stress ratio is small ($\sigma_1/\sigma_2 = 0.2$, as visible in Figs. 4.38a, 4.39a and 4.40a), there are cracks mainly with transverse direction (larger principal stress σ_2 direction), which is basically the same as the failure pattern under uniaxial compressive load (Buyukozturk et al. 1971). However, due to the constrained principal stress applied from the σ_1 direction limits the expansion of micro-cracks in the concrete and the generation of lateral deformation, the biaxial compressive strength gets an enhancement. As the stress ratio gradually increases and approaches 1.0, the restraining effect around the concrete increases gradually. As shown in Figs. 4.38c, 4.39c and 4.40c, crack propagation changes into vertical intersections and crisscrosses, forming a uniform biaxial compression failure patterns. In

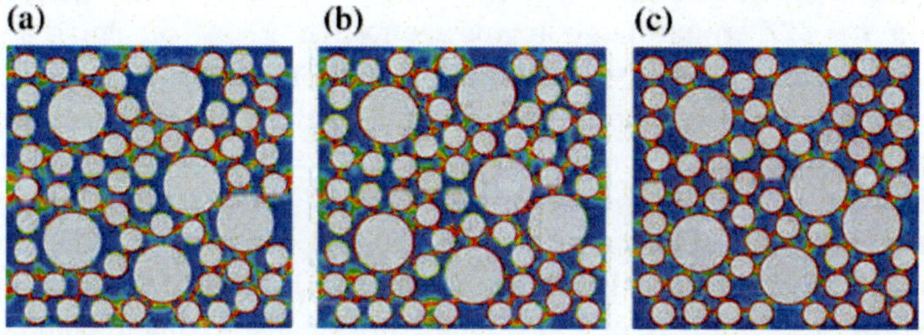

Fig. 4.38 Failure patterns of square concrete specimen ($b = 150$ mm) with different σ_1/σ_2: **a** $\sigma_1/\sigma_2 = 0.2$; **b** $\sigma_1/\sigma_2 = 0.5$; and **c** $\sigma_1/\sigma_2 = 1.0$

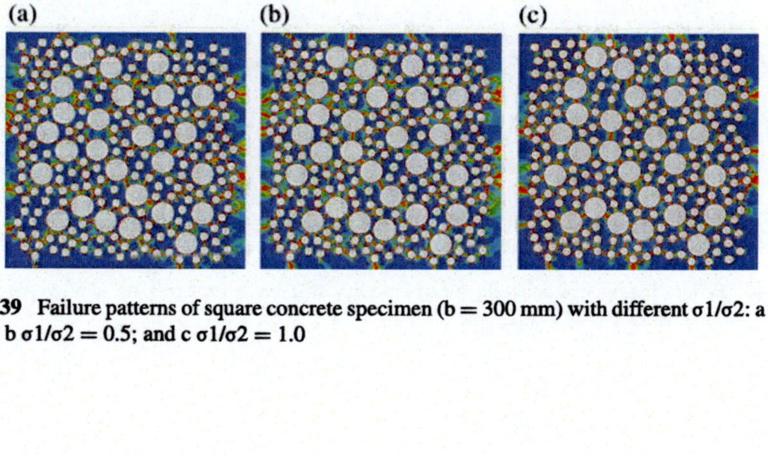

Fig. 4.39 Failure patterns of square concrete specimen (b = 300 mm) with different σ1/σ2: a σ1/σ2 = 0.2; b σ1/σ2 = 0.5; and c σ1/σ2 = 1.0

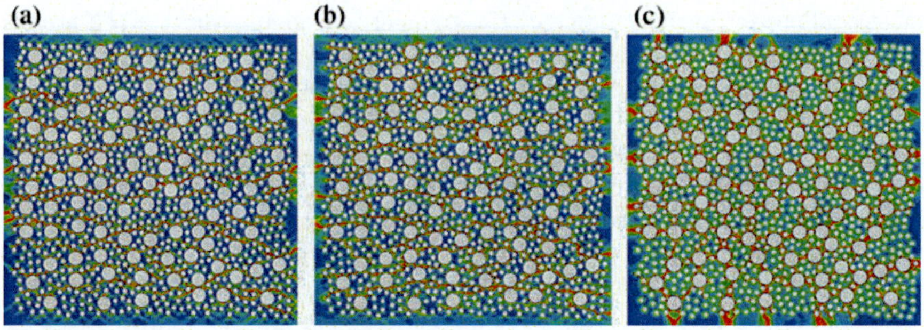

Fig. 4.40 Failure patterns of square concrete specimen ($b = 600$ mm) with different σ_1/σ_2: **a** $\sigma_1/\sigma_2 = 0.2$; **b** $\sigma_1/\sigma_2 = 0.5$; and **c** $\sigma_1/\sigma_2 = 1.0$

addition, it can be also seen from Figs. 4.38, 4.39 and 4.40 that the damage distribution of 2D numerical specimens is basically similar with that of test results (Shang and Song 2006; Chen and Leung 2014) under stress ratios of 0.25, 0.50, and 0.75, respectively (as shown in Fig. 4.41).

It can be found from the failure patterns of concrete specimen (stress ratio is 1.0) that the failure form of concrete specimen changes with the increasing structural size. The damage cracks of small-sized specimens (structural size $b = 150$ mm) gradually expand from the ITZ around large aggregates to that around the small aggregates and all ITZ are damaged finally, which is shown in Fig. 4.38c. The damage cracks of large-sized specimens (structural size $b = 600$ mm) mainly extend along the ITZ around large aggregates, and the damage cracks in the ITZ around small aggregates are not fully developed when specimens get damaged, which is shown in Fig. 4.40c. As shown in Fig. 4.42 (structural size b = 150 and 600 mm, and the stress ratio is 1.0), the failure patterns of local area (Domain-1, 2 and 3) in square concrete specimens with structural size of 150 and 600 mm are analyzed and compared from the perspective of energy release. It can be found that the large-sized specimens are destroyed after the ITZ around large aggregates reach damaged. While small-sized specimens are destroyed when all ITZ get damaged. In other words, when the

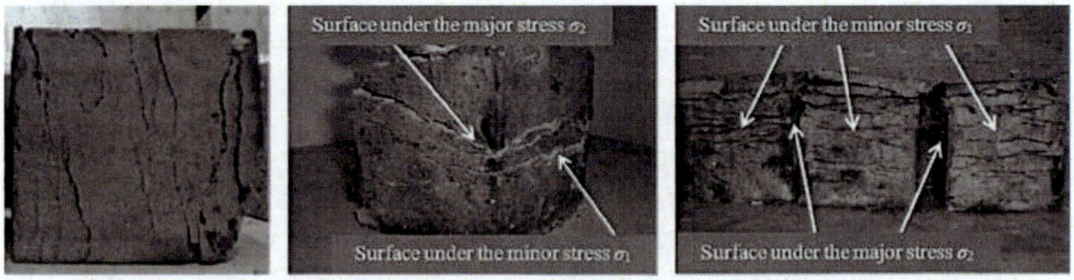

Fig. 4.41 Failure patterns of concrete specimen with different σ_1/σ_2 in tests: **a** $\sigma_1/\sigma_2 = 0.25$; **b** $\sigma_1/\sigma_2 = 0.50$; and **c** $\sigma_1/\sigma_2 = 0.75$

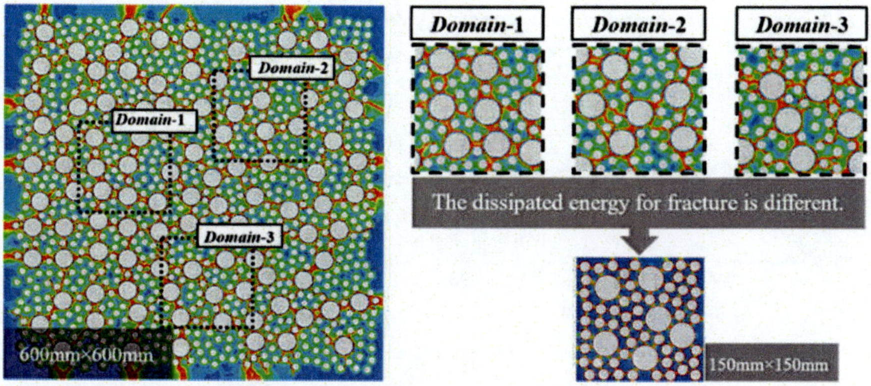

Fig. 4.42 Mechanism analysis on the size effect of concrete under biaxial compression

concrete in unit area gets damaged, the small-sized specimens require more energy and have a higher strength.

In short, due to the heterogeneity of the internal composition and stress concentration in concrete materials, the failure pattern changes with the increasing structural size. The main reason for the size effect on the biaxial compressive strength of concrete is the different fracture energy consumed when the specimen gets damaged.

2. *Biaxial compressive strength*

Figure 4.43 shows the biaxial compressive stress–strain curves of concrete specimens with different structural size under different stress ratios (here c150-0.6 indicates that the structural size of concrete specimen is 150 mm and the stress ratio σ_1/σ_2 is 0.6; the peak strength in the larger principal stress direction (σ_2 direction) is taken as the biaxial compressive strength). It can be seen that the initial stiffness of each specimen is basically the same regardless of whether it is under uniaxial load or biaxial load. When the concrete strength reaches the peak, the concrete specimens having the same structural size exhibit different peak strengths under different stress ratios, which are higher than those under uniaxial load. Furthermore, the softening curves under different stress ratios have some differences.

The small principal stress applied in the σ_1 direction limits the propagation of micro-cracks and the generation of lateral deformation in concrete materials. As a result, the biaxial compressive strength of concrete is higher than the uniaxial compressive strength. These are consistent with the test results of Buyukozturk et al. (1971). In addition, it can be seen from Fig. 4.43 that as the specimen size increases, both the uniaxial compressive strength and biaxial compressive strength of concrete materials decrease, showing a significant size effect.

Figure 4.44a shows the relationship between the biaxial compressive strength and the stress ratio of concrete specimens with different sizes. It can be seen that the biaxial compressive strength of concrete has different increases under different stress ratios. The biaxial compressive strength of concrete reaches its maximum when the

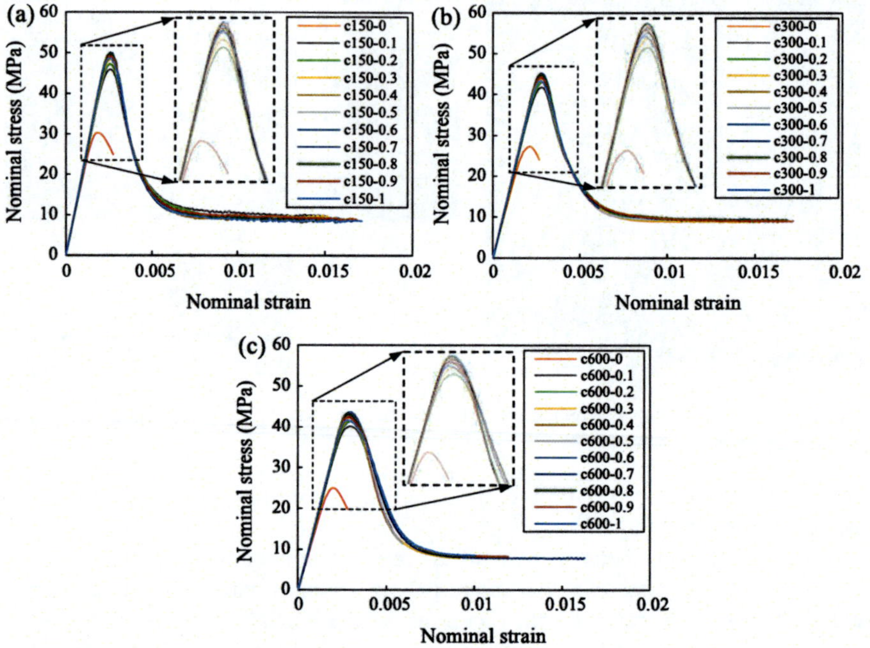

Fig. 4.43 Stress–strain curves of square concrete specimens under different stress ratios: **a** $b =$ 150 mm; **b** $b = 300$ mm; and **c** $b = 150$ mm

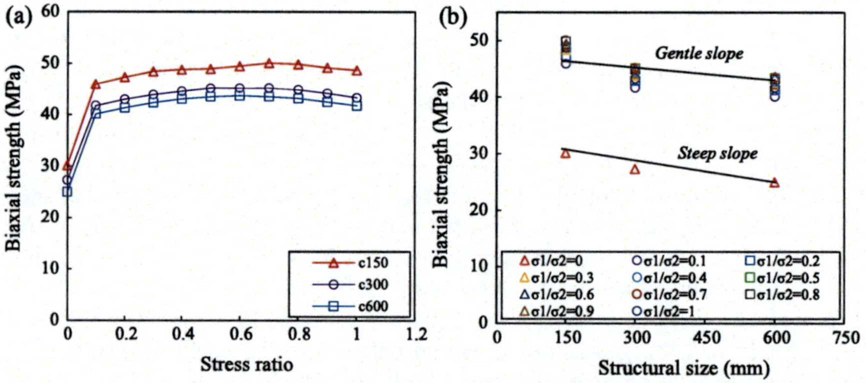

Fig. 4.44 Variation of biaxial compressive strength with stress ratio and structural size: **a** Strength versus stress ratio; and **b** Strength versus structural size

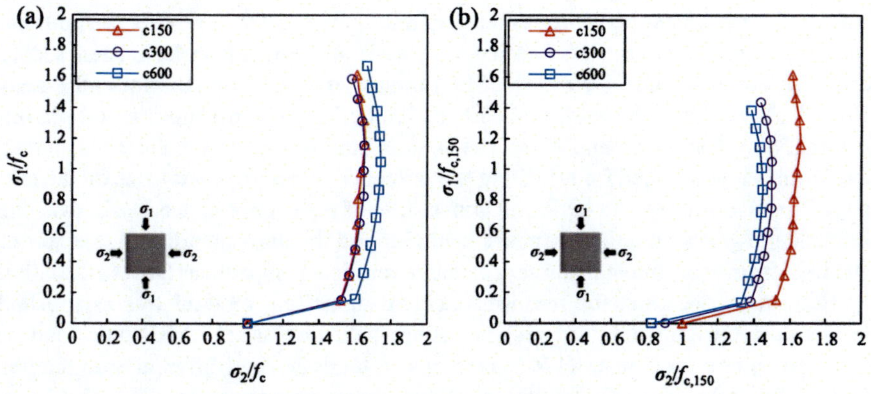

Fig. 4.45 Envelopes of biaxial compressive strength of concrete: **a** σ versus f_c; and **b** σ versus $f_{c,150}$

stress ratio is $\sigma_1 / \sigma_2 = 0.7$. The biaxial compression strength of small-sized specimens under different stresses is larger than that of large-sized specimens. Figure 4.44b shows the change trend of the biaxial compressive strength with the structural size of concrete specimens under different stress ratios. It can be seen that the biaxial compressive strength gradually decreases as the structural size increases and the size effect is obvious. But the trend (the biaxial compressive strength decreases with the increasing structural size) is slower than that of the uniaxial compressive strength. This indicates that under the uniaxial compressive load, concrete materials show stronger brittleness and the size effect is more obvious.

Figure 4.45 shows the envelope of the biaxial compressive strength of concrete specimens having different specimen sizes (here $f_{c,150}$ is the uniaxial compressive strength of specimen sized by 150 mm). It can be seen from Fig. 4.45a that under the biaxial compressive load, the increasing amplitude of concrete compressive strength increases with the addition of structural size, which is consistent with the test results of Rossi and Ulm (1997). It can be seen from Fig. 4.45b that there is a significant size effect on the biaxial compressive strength of concrete and the strength envelope of large-sized specimens shrink inward.

4.5 Novel Size Effect Law Considering MAS

There is no denying that composite materials like concrete behaves significant size effect, which can be attributed to the heterogeneity of internal components. The aggregate size is also an important factor affecting the size effect on nominal strength of concrete materials. The investigations on the influence of aggregate size on size effect in concrete materials are relatively limited and the influence mechanism has not formed a unified cognition, especially for compressive strength. As for theoretical studies, there is a mass of theoretical investigations to reflect the size effect on nominal strength of concrete materials and several classic size effect laws have been widely used, including the size effect law (SEL) proposed by Bažant (Bažant and Planas 1997; Bažant 1984), the statistical SEL of Weibull (1939), the multifractal scaling law (MFSL) (Carpinteri and Ferro 1994), etc., but among these theoretical studies the formulas undertaken for reflecting the influence of heterogeneity size on the size effect are also limited. As reviewed above, it is of great importance to discover the influence of MAS on uniaxial failure behavior and the corresponding size effect of concrete from experimental, numerical and theoretical perspectives (Jin et al. 2020c).

Here in this section, the meso-mechanical modelling method was established to simulate the uniaxial compressive failure behavior under quasi-static load of concrete square specimens (144 specimens in total) having different side lengths (D = 100, 150, 200, 300 and 450 mm), maximum aggregate sizes (MAS $d_{\max}$ = 24, 30, 36 and 42 mm) and concrete strength grades (C30 and C50, according to the Chinese code GB/T 50010–2010 (Ministry of construction of People's Republic of China 2011)). The influence of MAS on nominal compressive strength and the corresponding static size effect has been in discussion from the numerical perspective. Moreover, from theoretical perspective, a modified *Type-2* size effect law considering the MAS for the nominal compressive strength has been proposed (Jin et al. 2020c).

In the meso-scale simulation, the shape of aggregate particle is often simplify assumed as circle. These aggregate particles would be placed randomly inside a volume base on the classic "Take and Place" method (Monte-Carlo method). The ITZ phase and mortar matrix phase are modelled surrounding the already placed aggregates. The volume fraction of aggregate particles is about 50% and two-graded aggregate concrete was adopted with aggregates with diameter 13 mm and maximum diameter 24 mm (30, 36 and 42 mm). The numerical model was meshed with constant strain triangle element and a fine mesh with a basic element size of 2 mm (with refinements around ITZ of 1 mm) was applied. Figure 4.46 depicts 2D meso-scale models of square specimens with different side lengths and MAS in this section.

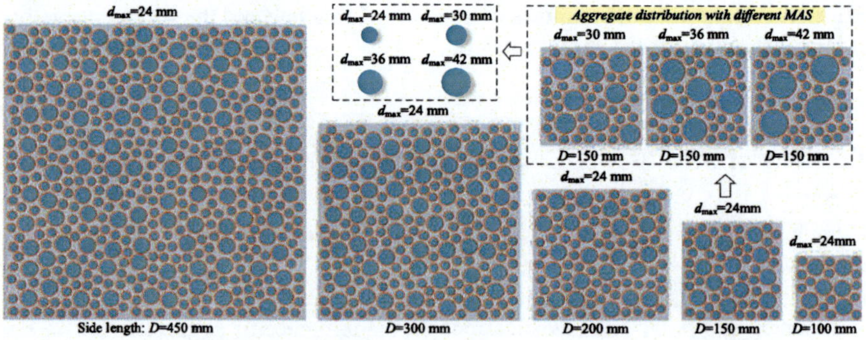

Fig. 4.46 2D meso-scale models of concrete specimens with different side lengths and MAS

Figure 4.47 also depicts the loading scheme and boundary conditions of 2D meso-scale model. The vertical displacement of the bottom of model was limited and one corner of model was horizontally and vertically limited. Both sides of the model were free boundaries. Similar to the Sect. 4.4, a very low loading velocity (constant velocity) was also set on the top surface of model to implement a static load (quasi-static load). Herein the strain rate $\dot{\varepsilon}=v/D$ (D = the side length) in all models was controlled as 1×10^{-5}/s (quasi-static load).

In addition, the damaged plasticity model described in Sect. 3.3.1 has been also adopted to simulate the mechanical failure behavior of concrete three-phase components (aggregate, mortar matrix and ITZ) and the mechanical parameters of three-phase components set in meso-scale simulations are listed in Table 4.7.

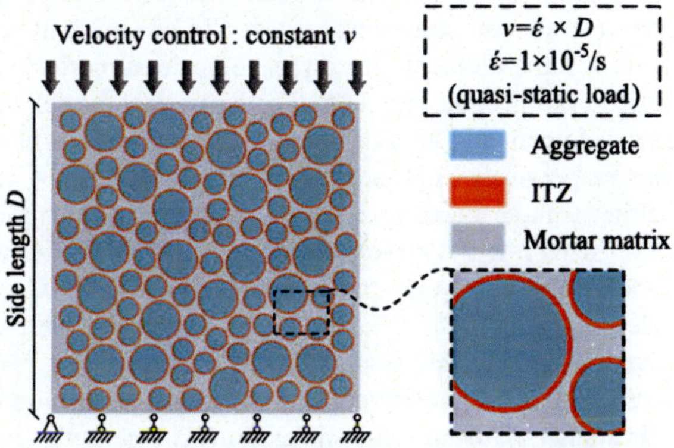

Fig. 4.47 Loading schematic diagram and boundary conditions for uniaxial compression

Table 4.7 Mechanical parameters of the meso-components

	Aggregate	Mortar matrix	ITZ
Young's modulus E (GPa)	70.0[a]	32.5[a]	24.4[b]
Poisson ratio ν	0.16[a]	0.20[a]	0.22
Eccentricity η (%)	0.1	0.1	0.1
Stress ratio $f_{b0} = f_{c0}$	1.16	1.16	1.16
K	0.667	0.667	0.667
Compressive strength σ_c (MPa)	80.0[a]	C30: 31.6[a]	C30: 23.7[b]
		C50: 52.1[c]	C50: 39.1[b]
Tensile strength σ_t (MPa)	8.0[a]	C30: 2.9[a]	C30: 2.2[b]
		C50: 5.2[c]	C50: 3.9[b]

Note [a]Data were referenced to the tests (Jin et al. 2018); [b]Data were set as 75% mechanical parameters of mortar matrix according to Song and Lu (Song and Lu 2012); and [c]Data were determined by the inversion method

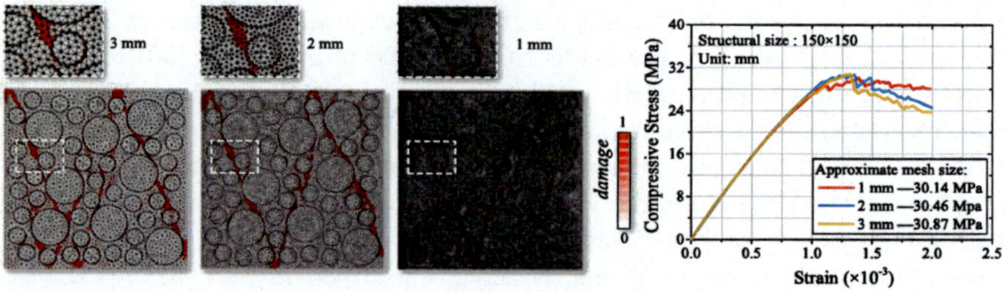

Fig. 4.48 Effect of finite element size on failure pattern and stress–strain curve of concrete square specimen

In order to calibrate and verify the above-established modelling method, the concrete uniaxial compressive test (Jin et al. 2018) was selected. Firstly, the influence of finite element size on meso-scale numerical results has been in discussion, which is showed in Fig. 4.48. The numerical model was meshed with triangle plane stress element and the fine mesh with a basic element size of 1 mm, 2 mm and 3 mm was respectively simulated. As observed in Fig. 4.48, although numerical model having smaller mash size have a more meticulous failure pattern, the influence of mesh size (1, 2 and 3 mm) on failure patterns is within an acceptable range. Similar conclusion can be also found in the comparison of stress–strain curves. The mesh size only affects the descending branch while the stress increase phase is almost identical and it also has an acceptable influence on the peak stress. Therefore, a fine mesh with a basic element size of 2 mm (with refinements around ITZ of 1 mm) was applied in meso-scale simulation. The compressive test (Jin et al. 2018) provide the Young's modulus, Poisson ratio, compressive strength, tensile strength, etc. of aggregate and mortar matrix, which can be applied in simulation. As the mechanical parameters of meso-components listed in Table 4.7 utilized in meso-scale models, the uniaxial compressive failure pattern and stress–strain curves of four concrete specimens (having different aggregate distributions) under static load have been obtained. The comparison with test results (Jin et al. 2018) is given in Fig. 4.49. From Fig. 4.49 it

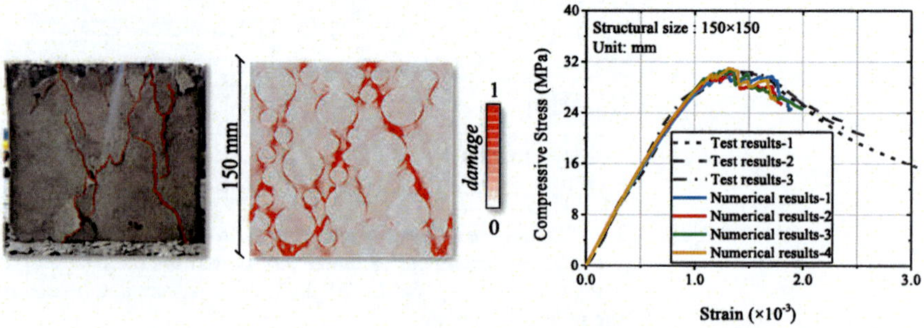

Fig. 4.49 Comparison of failure pattern and stress–strain curves between test and numerical results

can be found that all of four numerical stress–strain curves basically match well with ones of test and the crack propagation form under uniaxial compressive load in simulation shows a good agreement with the one in tests. This comparison results proved that the above-established modelling method and the applied mechanical parameters could yield relatively acceptable and reliable predictions of the mechanical behavior of concrete materials under static compressive load.

1. *Failure pattern*

Figure 4.50 reports the typical final uniaxial compressive failure patterns of square specimens (side length: 150 mm; C30) with four kinds of MAS. As visible in Fig. 4.50, under compressive load damage in the internal components of concrete is not uniform. Many small cracks show up in the areas with weaker mechanical properties (like ITZ and its surrounding areas) and these cracks continuously expand and converge to form several main cracks. These main cracks are distributed obliquely to form a shear failure surface and eventually run through the entire specimen. Most of tortuous cracks take a detour around the aggregate particles, showing an intergranular failure pattern. In addition, the MAS affects the crack propagation form in the internal components of concrete. Compared to the concrete specimen with smaller MAS, the number of main cracks in concrete specimen with larger MAS is more but the width is narrower; and crack distribution is more dispersed. Figure 4.51 depicts the typical final compressive failure patterns of square specimens (C30) having different side lengths (d_{max} = 24 and 36 mm). It can be seen that under the same MAS the failure patterns (having several oblique main cracks) are basically the similar for the specimens having different side lengths. However, there are more oblique cracks run through the entire specimen and more fine cracks around the aggregate particles in large-sized concrete. The crack distribution in the internal components of concrete is also more dispersed.

2. *Stress–strain curves*

Figure 4.52 gives typical compressive stress–strain curves of concrete having different side lengths and MAS. As presented in Fig. 4.52, the structural size and MAS has no effect on the increase phase of compressive stress–strain curves, but has a slight effect on the peak stress and the descending part. Moreover, the strain

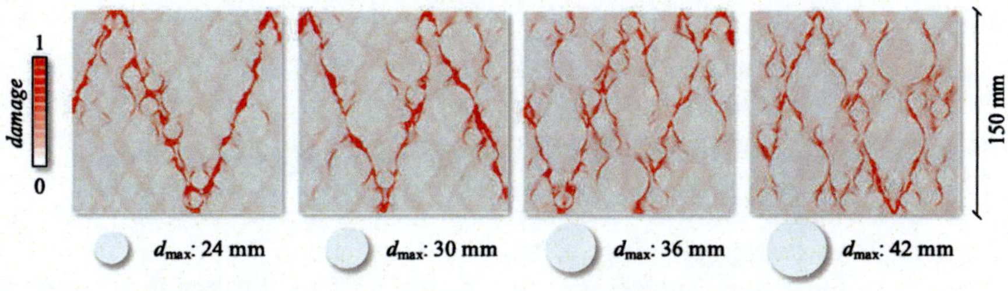

Fig. 4.50 Uniaxial compressive failure patterns of square specimens (C30) with different d_{max}

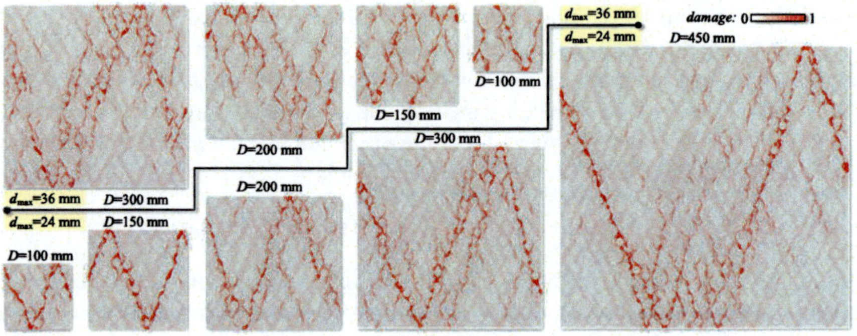

Fig. 4.51 Final compressive failure patterns of square specimens (C30) having different side lengths

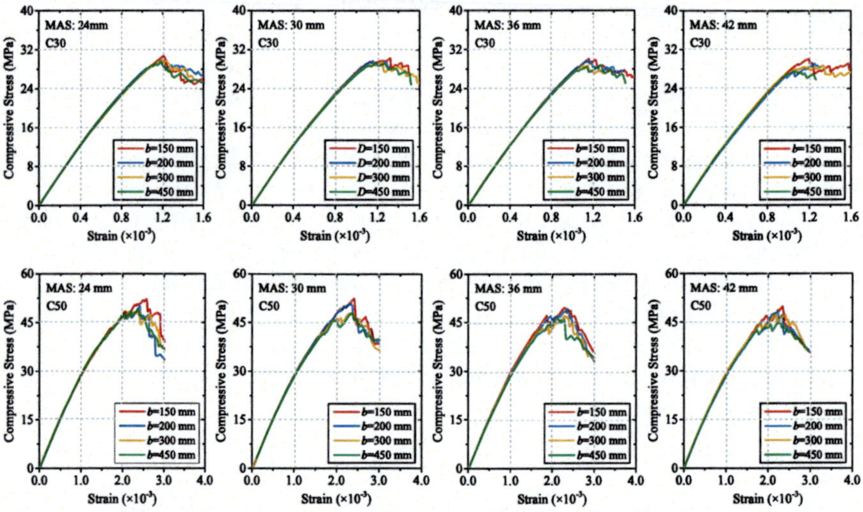

Fig. 4.52 Typical compressive stress–strain curves of concrete having different side lengths and MAS

corresponding the peak stress basically is independent with the structural size and MAS. Namely, the ductility is insensitive with the structural size and MAS.

3. *Influence of MAS on compressive strength*

The nominal compressive strengths of concrete having different MAS are sorted out in Fig. 4.53. As detailed in Fig. 4.53, whether it is C30 or C50 concrete, under the same side length, the nominal compressive strength decreases roughly with the increasing MAS (within the range of MAS in this section). The reason for this is not clear but it may have something to do with the crack distribution in the internal components of concrete, which has been described in Fig. 4.50. Concrete specimen

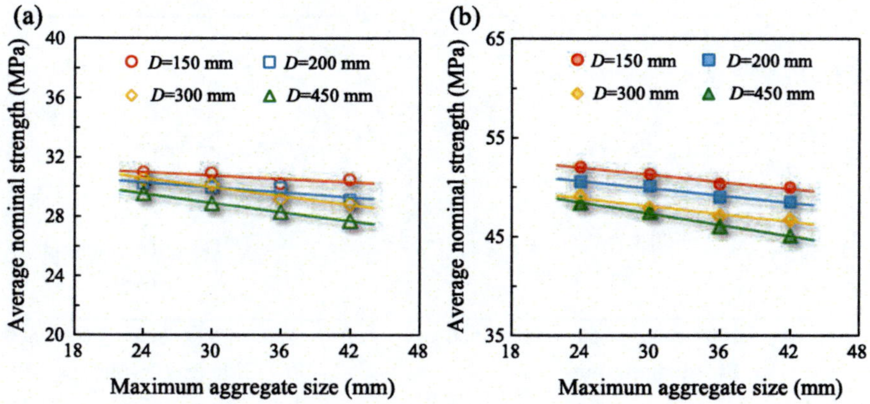

Fig. 4.53 Nominal compressive strengths of concrete having different MAS: **a** C30; and **b** C50

with larger MAS has more fine and narrow cracks with tortuous paths and there is more stress redistribution phenomenon, resulting in that the local stress concentration get alleviated and the internal stored energy get released. The above findings basically coincide with the research results of Refs. (Li and Gao 2013; Uddin et al. 2017).

4. *Influence of MAS on size effect*

Figure 4.54 sorts out the size effect on compressive strength of concrete material having different MAS. In order to intuitively reflect the tendency of nominal strength to decrease with increasing structural size, the slope k is adopted and it could be defined as $\Delta f_c/\Delta D$. As observed in Fig. 4.54, whether it is C30 or C50 concrete, the MAS affects the slope k and the slope k get an improvement with the increasing MAS (see Fig. 4.54e clearly), which indicates that the increasing MAS (within the range of MAS in this section) can enhance the sensitivity of strength-to-size and makes the size effect more obvious. Furthermore, under the same MAS, the slope k of concrete with C50 is generally higher than that of concrete with C30, illustrating that concrete with C50 is more brittle and has more obvious size effect.

5. *Regression analysis based on Bažant's SEL*

The *Type-2* size effect law proposed by Bažant (Bažant and Planas 1997; Bažant 1984) was also selected to analyze the size effect on compressive strength of concrete material having different MAS and it can be written as

$$\sigma_{\mathrm{N}} = \frac{Bf_c'}{\sqrt{1 + D/D_0}} \tag{4.41}$$

where σ_{N} represents the size-related nominal strength at failure for specimens similar in 2D; f_c' represents the average compressive strength of standard specimen (here the

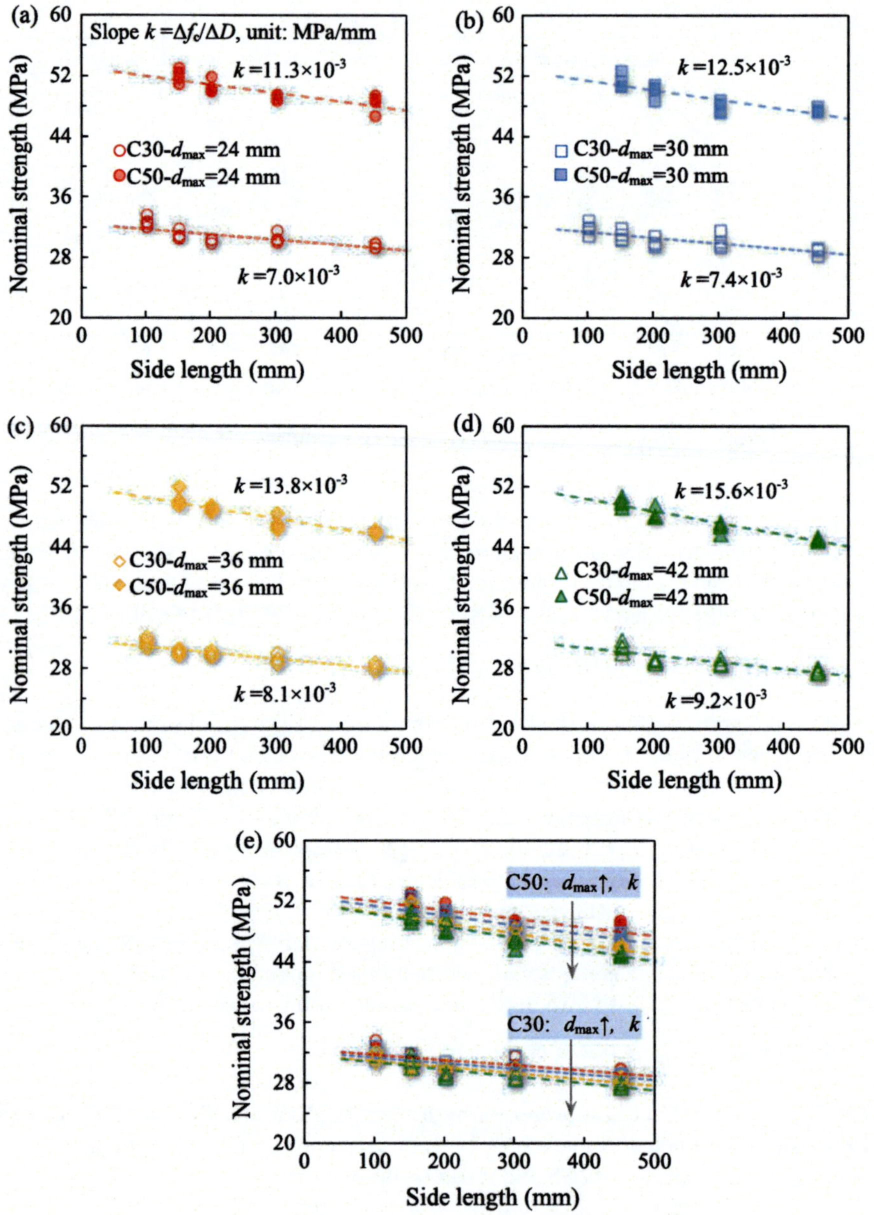

Fig. 4.54 Nominal compressive strengths of concrete having different side lengths: **a** MAS: 24 mm; **b** MAS: 30 mm; **c** MAS: 36 mm; **d** MAS: 42 mm; and **e** all MAS

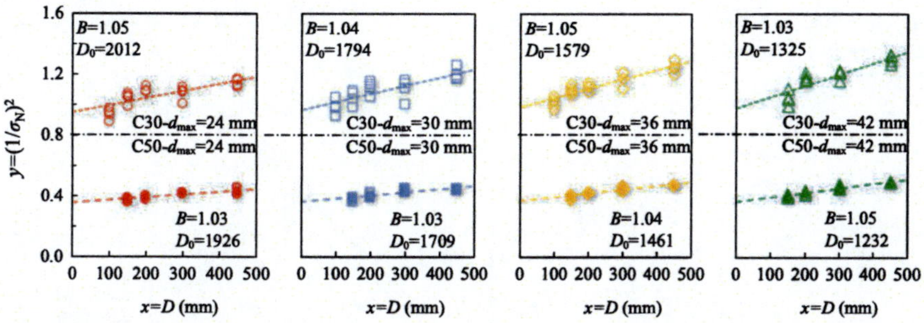

Fig. 4.55 Linear regression analysis of compressive strength data under different MAS according to Bažant's SEL

side length is 150 mm); D represents the characteristic structural size of specimen; B and D_0 represent empirical parameters determined by optimum fitting of nominal strengths σ_N covering a broad enough size range, D_0 is also a physical quantity related to the brittleness of the material. For linear regression analysis, Eq. (4.41) could be rearranged to

$$\frac{1}{(\sigma_N)^2} = \frac{1}{D_0\left(Bf_c'\right)^2}D + \frac{1}{\left(Bf_c'\right)^2} \tag{4.42}$$

In Eq. (4.42), $1/(\sigma_N)^2$ is defined as y; D is defined as x; $1/D_0\left(Bf_c'\right)^2$ is defined as the slope A; and $1/\left(Bf_c'\right)^2$ is defined as the intercept C. Therefore, through linear regression analysis of Eq. (4.42), the empirical parameters B and D_0 of Bažant's SEL under different MAS and concrete strength grades can be obtained based on the above numerical results, which are showed in Fig. 4.55. As displayed in Fig. 4.55, one can clear seen that the empirical parameters B is almost the same and fluctuates in the range of 1.03–1.05 for all series, i.e., it is not affected by the aggregate size. On the contrary, it is important to highlight that the MAS has an obviously influence on the empirical parameters D_0. Whether it is C30 or C50 concrete, as the MAS increases the parameters D_0 basically shows a downward trend, indicating that the concrete with larger MAS is more brittle and has more obvious size effect.

6. *Modified Type-2 size effect law in compression*

Alam et al. (2020) have applied a new size effect model considered the heterogeneity size (aggregate size) to describe the size effect on nominal tensile strength and the quantitative relationship between the parameters D_0 and d_{max} can be assumed as $D_0 = \gamma d_{max}$ (γ is a material constant which is independent of the heterogeneity size or the structural size). Unfortunately, although this new size effect model may have many advantages over the classical Bažant's SEL, it could only reflect the influence of heterogeneity size on size effect on nominal tensile strength and the positive correlation between the parameters D_0 and d_{max}. As for the size effect on

Fig. 4.56 Quantitative relationship of D_0 and d_{max}

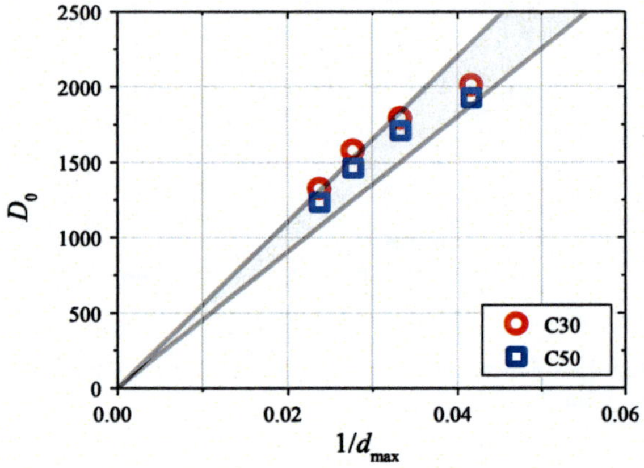

nominal compressive strength, it is not appropriate for characterizing the negative correlation of d_{max} increases as D_0 decreases. Accordingly, referring to the Alam et al. (2020), particular emphasis is placed on finding a quantitative relationship between the parameters D_0 and d_{max} for the size effect on compressive strength considering the MAS. Based on the above meso-scale numerical results and empirical parameter D_0 sorted out in Fig. 4.55, several functions have been attempted to describe this relationship. When the data points (D_0 against $1/d_{max}$) are plotted in Fig. 4.56, it can be found that the relation between the parameters D_0 and $1/d_{max}$ may approach a linear relation (a direct proportion relation). Therefore, according to this finding, it could be reasonably assumed that:

$$D_0 = \xi \cdot \frac{1}{d_{max}} \tag{4.43}$$

where ξ is also an empirical constant which is independent of the heterogeneity size or the structural size and the value of ξ is basically in the range of 4.5–5.5 × 10^4. In this assumption, combining the Eq. (4.41), the classical Bažant's SEL could now be converted to:

$$\sigma_N = \frac{Bf_c'}{\sqrt{1 + D \cdot d_{max}/\xi}} \tag{4.44}$$

Equation (4.44) is a modified *Type-2* size effect law considering the heterogeneity size (MAS) for the nominal compressive strength of concrete material and in this new size effect formula one can quantitatively predict the size effect behavior of compressive strength based on known physical parameters (i.e., the characteristic structural dimension D and the maximum heterogeneity size d_{max}) instead of D/D_0. However, this modified *Type-2* size effect law is an assumption at present base on

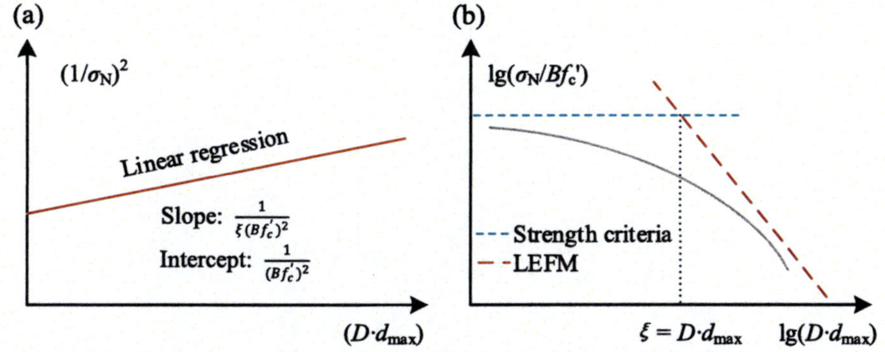

Fig. 4.57 Illustration of analysis applying modified size effect law: **a** Regression curve; and **b** New size effect curve

the direct proportion of D_0 and d_{max}, which needs to be verified and calibrated by tests and simulations.

Referring the regression analysis in Ref. (Alam et al. 2020), the above modified size effect law can also be expressed by the following form of linear equation in terms of $1/(\sigma_N)^2$ and Dd_{max}:

$$\frac{1}{(\sigma_N)^2} = \frac{1}{\xi\left(Bf_c'\right)^2} Dd_{max} + \frac{1}{\left(Bf_c'\right)^2} \tag{4.45}$$

In Eq. (4.45), Dd_{max} is regarded as a global variable and the regression analysis should be conducted by varying the Dd_{max} (scaling D or scaling d_{max}) to obtain the constants Bf_c' and ξ under different cases (see Fig. 4.57a). The modified size effect law can also be plotted as a new size effect curve (see Fig. 4.57b), which is transitional between plasticity (strength based theories, ductile behavior) and linear elastic fracture mechanics (i.e., LEFM, brittle behavior). Table 4.8 lists the regression analysis parameter Bf_c' and ξ based on the modified size effect law. From Table 4.8 one can see that the empirical constant ξ at different maximum aggregate size series has nothing to do with concrete strength grade and it fluctuates with the MAS in the range of 4.5–5.5 × 10^4, which is consistent with the fitting slope in Fig. 4.56. Moreover, the parameter Bf_c' basically does not vary with the change of MAS under each concrete strength grade, illustrating that the parameter B could be regard as a constant.

Figure 4.58 depicts the comparison between the nominal compressive strengths of concrete specimens for each MAS series obtained from meso-scale simulation and the new size effect curve according to the modified size effect law. Similar to the size effect curve of Bažant and the size effect model of Alam et al. (2020), the numerical results present a transitional behavior as varying the Dd_{max} (scaling D or scaling d_{max}) under each concrete strength grade. The numerical data points are basically distributed near the double logarithmic curve and vary along with the

Table 4.8 Regression analysis parameters based on the modified size effect law

Parameters	Size series-C30				Size series-C50			
	d_{max} 24 mm	d_{max} 30 mm	d_{max} 36 mm	d_{max} 42 mm	d_{max} 24 mm	d_{max} 30 mm	d_{max} 36 mm	d_{max} 42 mm
Bf_c'(MPa)	32.5	32.3	31.9	31.9	53.4	52.9	52.3	52.5
$\xi(\times 10^4)$	4.8	5.4	5.2	5.5	4.6	5.1	5.3	5.2

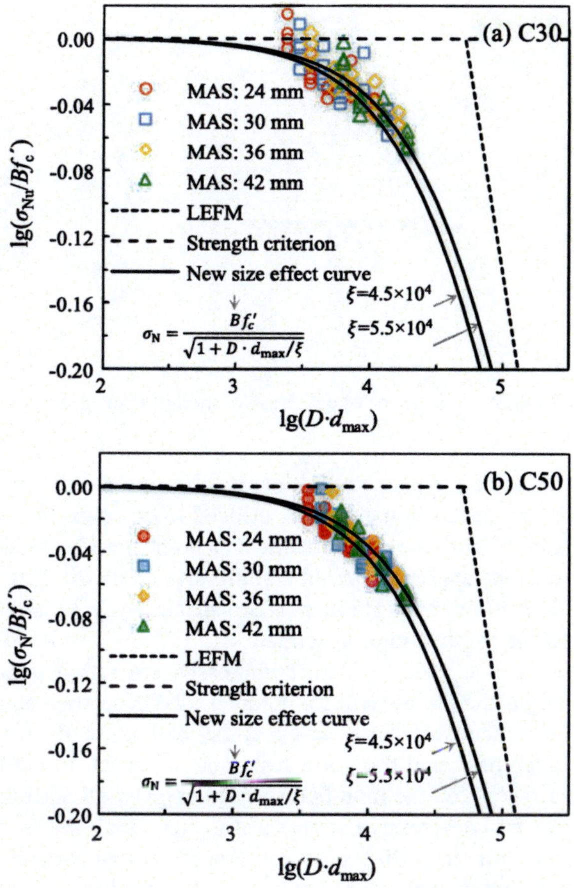

Fig. 4.58 Comparison of numerical results and new size effect curve in present model: **a** C30; and **b** C50

curve, exhibiting that the new curve according to the modified size effect law can well characterize the size effect bahavior on compressive strength consider the MAS. In addition, with the increasing MAS the data points are getting closer to the LEFM, that is to say, the new curve can reflect the compressive brittleness of concrete material having different MAS and the increasing MAS can enhance the size effect.

In addition to the verification from above numerical simulation perspective, the verification from physics experiment perspective is urgently desired. However, due to the limitation of test condition, the appropriate test results used to compare and verify the theoretical formula Eq. (4.44) are really limited, especially the uniaxial compressive size effect tests with large-sized aggregates. Here in this part, the aggregate size effect test of cube concrete (uniaxial compression, $d_{\max} = 10$, 20, 40 and 60 mm) conducted by Li and Gao (2013) and the compressive size effect test of cube concrete ($d_{\max} = 20$ mm, max size length = 200 mm) conducted by Su et al.

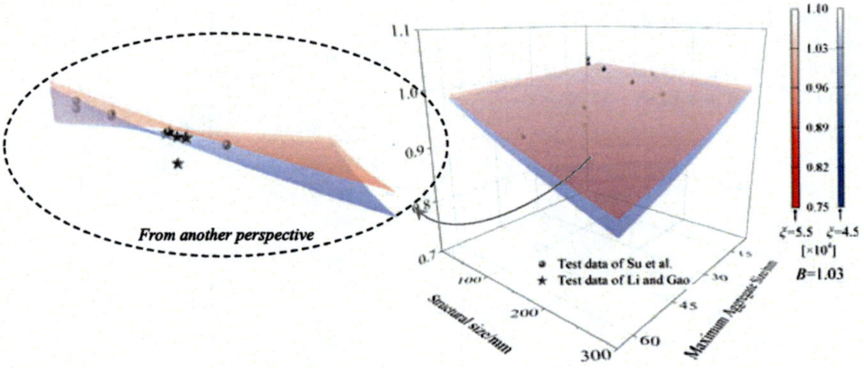

Fig. 4.59 Comparison of compressive strengths between test data and theoretical results

(2015) have been selected for verification. In order to be more persuasive, a dimensionless ratio of σ_N/f_c' was utilized to compare the theoretical surface of nominal strength with the experimental data points (as observed in Fig. 4.59). Here, σ_N represents the applied nominal compressive strength; f_c' represents the average nominal compressive strength of standard specimen (in Li and Gao's test, f_c' is the compressive strength of cube specimen with 150 mm size length and 20 mm MAS at 28-day; in Su et al.'s test, f_c' is the compressive strength of cube specimen with 150 mm size length and 20 mm MAS). Figure 4.59 depicts the comparison between test data and theoretical surfaces ($\xi = 4.5 \times 10^4$ and 5.5×10^4) of the modified size effect law. It can be found that most test data points fall in the range surrounded by the two surfaces, i.e., the modified size effect law well describes the influence of aggregate size on compressive strength in Li and Gao's test and the size effect on compressive strength in Su et al.'s test. The above comparisons between test data, numerical data and theoretical results have preliminarily verified the rationality of the modified size effect law set up based on some assumptions. Applying the modified *Type-2* size effect law considering the MAS, once the structural dimensions and MAS are fixed, it can predict the nominal compressive strength of concrete material from a quantitative perspective.

4.6 Summary

This chapter mainly focus on static size effect on nominal strength of concrete materials. Firstly, the mechanism analysis of size effect on nominal tensile strength was introduced from theoretical perspective. Secondly, the static size effect on tensile strength, flexural strength and compressive strength of concrete materials were provided from experimental perspective and numerical perspective. The material parameters (such as MAS, aggregate type, concrete strength grade, etc.) on size effect of concrete materials were also discussed. In addition, the size effect law considering the MAS for nominal compressive strength was presented from theoretical perspective.

(1) Splitting-tensile strength for concrete cubic specimens decreases with increasing the structural size. The increase in MAS can reduce the sensitivity to the size effect on the splitting-tensile strength. Concrete specimens with high(er) strength grades have more obvious downtrend of strength than the ones with low(er) strength grades.
(2) There is an obvious size effect in the flexural-tensile strength of concrete and the increase in MAS can reduce the sensitivity to the size effect on flexural-tensile strength.
(3) The aggregate type has an important influence on the size effect on nominal strength of concrete, and the size effect on nominal strength of lightweight aggregate concrete is more obvious than that of normal concrete.
(4) The size effect law can well consider the influence of MAS on size effect on nominal compressive strength.

References

Akçaoğlu T, Tokyay M, Çelik (2004) Effect of coarse aggregate size and matrix quality on ITZ and failure behavior of concrete under uniaxial compression. Cement Concr Compos 26(6):633–638

Alam SY, Zhu R, Loukili A (2020) A new way to analyse the size effect in quasi-brittle materials by scaling the heterogeneity size. Eng Fract Mech 225:106864

Bažant Z (1984) Size effect in blunt fracture: concrete, rock, metal. J Eng Mech 110(4):518–535

Bažant Z, Planas JP (1997) Fracture and size effect in concrete and other quasibrittle materials. CRC Press, USA

Beygi MHA, Kazemi MT, Nikbin IM, Amiri JV, Rabbanifar S, Rahmani E (2014) The influence of coarse aggregate size and volume on the fracture behavior and brittleness of self-compacting concrete. Cem Concr Res 66:75–90

Buyukozturk O, Nilson AH, Slate FO (1971) Stress-strain response and fracture of a concrete model in biaxial loading. ACI Struct J 68(8):590–599

Carmona S (2009) Effect of specimen size and loading conditions on indirect tensile test results. Materiales De Construcción 59(294):7–18

Carpinteri A, Ferro G (1994) Size effects on tensile fracture properties: a unified explanation based on disorder and fractality of concrete microstructure. Mater Struct 27(10):563–571

Chen E, Leung CKY (2014) Effect of uniaxial strength and fracture parameters of concrete on its biaxial compressive strength. J Mater Civ Eng 26(6):06014001

Cui HZ, Lo TY, Memon SA, Xu WT (2012) Effect of lightweight aggregates on the mechanical properties and brittleness of lightweight aggregate concrete. Constr Build Mater 35:149–158

Dilger WH, Koch R, Kowalczyk R (1984) Ductility of plain and confined concrete under different strain rates. J Proc 81(1):73–81

Du M, Du X, Jin L, Lu A (2015) Meso-heterogeneity mechanism of the concrete size effect of tension and compression strength. J Civ Environ Eng 37(3):11–18 (in Chinese)

Du M, Jin L, Li D, Du X (2016) Experimental study of the influence of coarse aggregate size on the size effect of concrete flexural strength. J Beijing Univ Technol 42(6):912–918 (in Chinese)

Du M (2016) Research on size effect of concrete and stirrup-confined concrete columns. Beijing University of Technology, Beijing (in Chinese)
Du X, Jin L, Li D (2017a) A state-of-the-art review on the size effect of concretes and concrete structures (I): concrete materials. Chin Civ Eng J 50(9):28–45 (in Chinese)
Du X, Jin L, Li D (2017b) A state-of-the-art review on the size effect of concretes and concrete structures (II): RC members. Chin Civ Eng J 50(11):24–44 (in Chinese)
Fuller WB, Thompson SE (1907) The laws of proportioning concrete. Trans Am Soc Civ Eng 59:67–143
Grassl P, Wong HS, Buenfeld NR (2010) Influence of aggregate size and volume fraction on shrinkage induced micro-cracking of concrete and mortar. Cem Concr Res 40(1):85–93
Guo Z, Zhang X, Zhang D, Wang R (1982) Experimental investigation of the complete stress-strain curve of concrete. J Build Struct 1:1–12 (in Chinese)
Ince R, Gör M, Eren ME, Alyamaç KE (2015) The effect of size on the splitting strength of cubic concrete members. Strain 51(2):135–146
Jin L, Yang W, Yu W, Du X (2020a) Size effect on dynamic compressive strength of lightweight aggregate concrete: Meso-scale investigation. Cement Concr Compos (in preparation)
Jin L, Yu W, Du X (2020b) Size effect on static splitting tensile strength of concrete: experimental and numerical studies. J Mater Civ Eng 32(10):4020308
Jin L, Yu W, Li D, Du X (2020c) Numerical and theoretical investigation on the size effect of concrete compressive strength considering the maximum aggregate size. Int J Mater Sci (in preparation)
Jin L, Zhang S, Li D, Xu H, Du X, Li Z (2018) A combined experimental and numerical analysis on the seismic behavior of short reinforced concrete columns with different structural sizes and axial compression ratios. Int J Damage Mech 27(9):1416–1447
Kadlеček V, Modrý S (2002) Size effect of test specimens on tensile splitting strength of concrete: general relation. Mater Struct 35:28–34
Khaloo AR, Mohamadi Shooreh MR, Askari SM (2009) Size influence of specimens and maximum aggregate on dam concrete: compressive strength. J Mater Civ Eng 21(8):349–355
Li D, Jin L, Du X, Du M (2017) Mechanism of size effect of concrete under biaxial compression. J Beijing Univ Technol 43(08):1190–1198 (in Chinese)
Li D, Jin L, Du X, Liu J (2018) A theoretical method to determine the tortuous crack length and the mechanical parameters of concrete in direct tension-A particle size effect analysis. Eng Fract Mech 197:128–150
Li S, Gao D (2013) Experimental research on the influence of coarse aggregate size on boulder concrete compressive strength. Concrete 2:59–61+64 (in Chinese)
Lilliu G, Van Mier JGM (2003) 3D lattice type fracture model for concrete. Eng Fract Mech 70(7–8):927–941
Malhotra VM (1976) Are 4×8 inch concrete cylinders as good as 6×12 inch cylinders for quality control of concrete. Aci J 73(1):33–36
Meddah MS, Zitouni S, Belâabes S (2010) Effect of content and particle size distribution of coarse aggregate on the compressive strength of concrete. Constr Build Mater 24(4):505–512
Ministry of construction of People's Republic of China (2003) Standard for test method of mechanical properties of ordinary concrete GB/T 50081–2002. China Architecture & Building Press, Beijing (in Chinese)
Ministry of construction of People's Republic of China (2011) Code for design of concrete structures GB 50010–2010 [S]. China Architecture & Building Press, Beijing (in Chinese)
Ministry of construction of People's Republic of China (2015) Code for mix design of hydraulic concrete DL/T 5330–2015. China Architecture & Building Press, Beijing (in Chinese)
Neville AM (1995) Properties of concrete. Longman, London
Qian J, Yang Z, Huang Y, Tang Z (2004) Experimental study of the size effect on the strength of high-strength concrete. J Civ Eng Manage (Urban Science Edition) 21(1):1–4 (in Chinese)
Rocco C, Guinea GV, Planas J, Elices M (1999) Size effect and boundary conditions in the Brazilian test: experimental verification. Mater Struct 32(3):210–217

Rosselló C, Elices M, Guinea GV, 2006. Fracture of model concrete: 2. Fracture energy and characteristic length. Cement Concr Compos 36(7):1345–1353
Rossi P, Ulm FJ (1997) Size effects in the biaxial tensile-compressive behavior of concrete: physical mechanisms and modelling. Mater Struct 30(4):210–216
Shah SP, Chandra S (1968) Critical stress, volume change and microcracking of concrete. J Proc 9:770–781
Shang H, Song Y (2006) Experimental study of strength and deformation of plain concrete under biaxial compression after freezing and thawing cycles. Cement Concr Compos 36(10):1857–1864
Shi J, Xu Y, Ye Z (2009) Meso analysis of concrete behavior due to effect of aggregate gradation. Eng Mech 26(4):134–138 (in Chinese)
Sim JI, Yang KH, Jeon JH (2013) Influence of aggregate size on the compressive size effect according to different concrete types. Constr Build Mater 44:716–725
Song Z, Lu Y (2012) Mesoscopic analysis of concrete under excessively high strain rate compression and implications on interpretation of test data. Int J Impact Eng 46:41–55
Su J, Ye J, Fang Z, Zhao M (2015) Size effect on cubic and prismatic compressive strength of cement paste. J Central South Univ 22(10):4090–4096
Subramaniam KV, Shah SP (2003) Biaxial tension fatigue response of concrete. Cement Concr Compos 25(6):617–623
Uddin MT, Mahmood AH, Kamal MRI, Yashin SM, Zihan ZUA (2017) Effects of maximum size of brick aggregate on properties of concrete. Constr Build Mater 134:713–726
Walraven JC, Reinhardt HW (1981) Theory and experiments on the mechanical behaviour of cracks in plain and reinforced concrete subjected to shear loading. Heron 26(1A):68
Wang L, Chen G (2008) Mesoscopic simulation for mechanical properties of lightweight aggregate concrete by rigid body spring model. J Hydraul Eng 39(5):588–595 (in Chinese)
Wang L, Xing L, Song Y (2014) Mesoscale modelling on size effect of splitting tensile strength and flexural compressive strength of concrete. Eng Mech 31(10):69–76 (in Chinese)
Weibull W (1939) A statistical theory of the strength of materials. Proc Ame Math Soc 151(5):1034
Zhou FP, Balendran RV, Jeary AP (1998) Size effect on flexural, splitting tensile, and torsional strengths of high-strength concrete. Cement Concr Compos 28(12):1725–1736
Zhou H, Che Y, Chen G, Song Y (2010) Size effect on tensile splitting strength of concrete cubes and cylinders. Concrete 8(1):13–20 (in Chinese)

Chapter 5
Dynamic Size Effect in Concrete Materials

As known, there is obvious size effect and strain-rate effect in concrete materials. Therefore, how concrete materials behave under the combined influences of size effect and strain-rate effect should be an interesting issue. Great progresses have been made in static size effect while almost no efforts have been conducted in dynamic size effect of concrete. Due to the limitations of the laboratory test device, limited experimental studies (Krauthammer et al. 2003; Elfahal et al. 2005; Elfahal and Krauthammer 2005; Bindiganavile and Banthia 2006; Hu et al. 2015; Wang et al. 2018; Li et al. 2018) have been conducted on the size effect of concrete materials under dynamic loads. These experimental efforts have demonstrated that the enhancement of material strength under impact loads has been proved to be size-dependent and the size effect of concrete under dynamic loads is more complicated compared to that under static load. However, the research conclusions on dynamic size effect are diversified, and the understanding of its production mechanism is still far from sufficient. In addition, there is lack of theoretical formula for accurately capturing the size effect on dynamic strength. Moreover, due to the limitations of test equipment and conditions, it is really difficult to explore the dynamic size effect of concrete only by physical tests, especially dynamic tests with large structural sizes under high strain rates. The meso-scale modelling method has provided an effective measure for investigating the mechanical failure behavior of concrete materials and the corresponding size effect under dynamic loads.

In this chapter, the spitting-tensile/tensile/compressive failure behavior of concrete materials having different structural sizes under different strain rates are provided by the meso-scale modelling method and the dynamic size effect on nominal strengths (spitting-tensile/tensile/compressive strength) is given. The influence of meso-structure and initial loads on the dynamic size effect is also presented. Moreover, the Static and Dynamic unified Size Effect Law (i.e. **SD-SEL**) for nominal strengths (spitting-tensile/tensile/compressive strength) of concrete materials is presented considering the contribution of strain rate on the dynamic size effect (Jin et al. 2019a, b, c, d, e, 2020a, b, c).

X. Du and L. Jin, *Size Effect in Concrete Materials and Structures*,
https://doi.org/10.1007/978-981-33-4943-8_5

5.1 Dynamic Size Effect on Splitting-Tensile Strength

For investigating the influence of strain rate on the dynamic splitting-tensile failure behavior and the corresponding size effect, the failure behavior of concrete cubic specimens having different structural sizes (side length $b = 100$, 150, 300 and 450 mm) under different strain rates (the applied strain rates are $\dot{\varepsilon}= 10^{-5}$/s (quasi-static), 10^{-3}/s, 10^{-1}/s, 1/s, 10/s, 100/s and 200/s, respectively) were simulated by the meso-scale modelling method. To reduce the discreteness of numerical results as much as possible, three concrete cubic specimens (only the aggregate distribution is different) were conducted in each case and a total of 84 concrete cubic specimens were simulated. The 3D cubic concrete model applied in this section is showed in Fig. 5.1 and the modeling details (including loading method and boundary condition) can refer to those in Sect. 3.6.1 of Chap. 3.

The numerical model was separated with an 8-nodded hexahedral element (type C3D8R: Continuum, 3-D, 8-node, reduce integration) and the average element size is around 2 mm. In addition, the damaged plasticity model described in Sect. 3.3.1 of Chap. 3 has been adopted to simulate the mechanical failure behavior of concrete three-phase components (aggregate, mortar matrix and ITZs) and the mechanical parameters of three-phase components used in meso-scale simulation of splitting-tensile behavior are listed Table 5.1. Only the strength enhancement behavior was considered and the strain rate effect for meso-components on material strength was incorporated into the plastic damage model. The DIF for compressive strength recommended by CEB (1993) and the DIF for tensile strength recommended by Malvar and Ross (1998) (described in Sect. 3.4.1 of Chap. 3) were set in ABAQUS. Detailed modeling introduction can refer to this literature (Jin et al. 2019b).

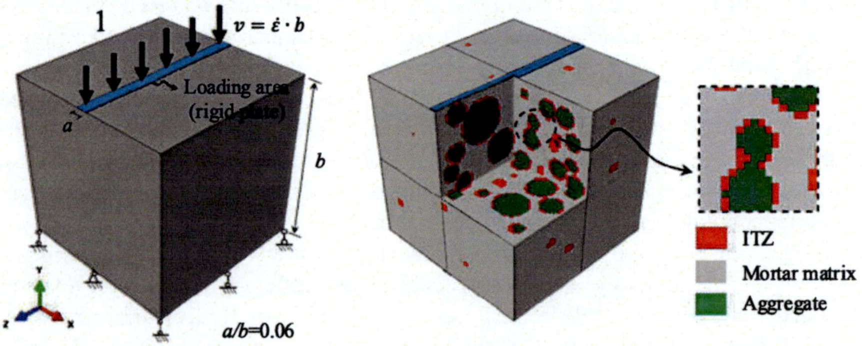

Fig. 5.1 3D meso-scale model of concrete cubic specimens for splitting-tensile simulation

Table 5.1 Mechanical parameters of concrete meso-components for splitting-tensile simulation

	Aggregate particle	Mortar matrix	ITZ
Young's modulus E (GPa)	50[a]	30[a]	26[b]
Poisson ratio ν	0.16	0.20	0.22
Dilatant angle ψ (°)	30	18	15
Eccentricity η (%)	0.1	0.1	0.1
Stress ratio $f_{b0} = f_{c0}$	1.16	1.16	1.16
K	0.667	0.667	0.667
Fracture energy G_c (J/m^2)	60	50	30
Compressive strength σ_c (MPa)	130[a]	30[a]	28.4[c]
Tensile strength σ_t (MPa)	10[a]	3[a]	2.8[c]

Note: [a]Data were obtained from the tests (Du et al. 2017); [b]Data were got from the literatures (Du et al. 2013); and [c]Data were obtained by the inversion method

5.1.1 Dynamic Failure Behavior

1. *Failure process*

Figure 5.2 depicts the splitting-tensile damage process of concrete cubic specimen (structural size b = 100 mm) under different strain rates in simulation. It can be clearly secn that with the increase of strain rate the time required for the specimen reaching damage is getting shorter. The damage generated firstly at the internal region having relatively weak mechanical properties (ITZ), and the damage propagated to mortar matrix or aggregates as time goes on. Under the strain rates of $\dot{\varepsilon}= 10^{-3}$/s and $\dot{\varepsilon}=$ 1/s, the crack started to appear along the central line below load areas and then extended continuously up and down. Finally, many cracks existed throughout the whole specimen and separated the concrete specimen into approximately equal halves. Furthermore, some cracks occurred on the surfaces parallel with the central line later under strain rate $\dot{\varepsilon}=$ 1/s in the end.

2. *Failure pattern*

Typical final failure patterns of concrete cubic specimen (structural size b = 100 mm) under different strain rates in simulation are presented in Fig. 5.3. From Fig. 5.3a it can be seen that the area near upper and lower surfaces (Area A) is in a state of tri-axial compression while Area B is in a state of tension–compression, resulting in that much damage appeared in Area B first and specimens begin to fail in this area. As given in Fig. 5.3b, damage gradually develops from the outside of specimen to the

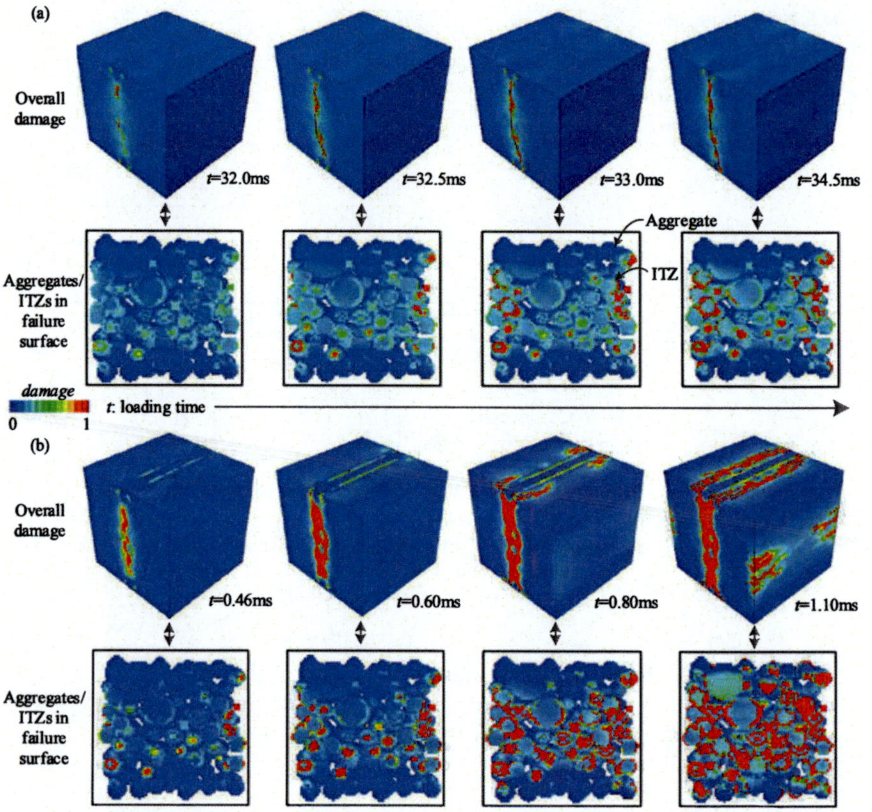

Fig. 5.2 Splitting-tensile damage process of concrete cubic specimen (structural size $b = 100$ mm) under different strain rates: **a** $\dot{\varepsilon} = 10^{-3}$/s; and **b** $\dot{\varepsilon} = 1$/s

inside and the damage outside the specimen is more serious than one inside. Viewed from the lateral view, the number and width of cracks near central line increase with the increasing strain rate. After the strain rate $\dot{\varepsilon}$ reaches 1/s, a fracture belt is formed gradually in the middle of specimen. The cracks begin to appear on both sides of the specimen and continue to expand. One can find from Fig. 5.3d that under low(er) strain rates few aggregate particles gets damaged. But large amount of aggregate particles begins to be pulled off when the applied strain rate reaches 1/s or exceeds 1/s, this may be because that the cracks could not bypass the aggregates in a short time and extend along the shortest path under high strain rates.

Figure 5.4 presents typical failure patterns of concrete cubic specimens having different structural sizes under different strain rates. One can note that the failure patterns are closely similar for the specimens having different sizes under the same strain rate. The fracture belt composed of many cracks near central line widen gradually with the increasing strain rate. The splitting-tensile fracture belt has also been observed in the available test of Wu et al. (2014), which is also presented in Fig. 5.4.

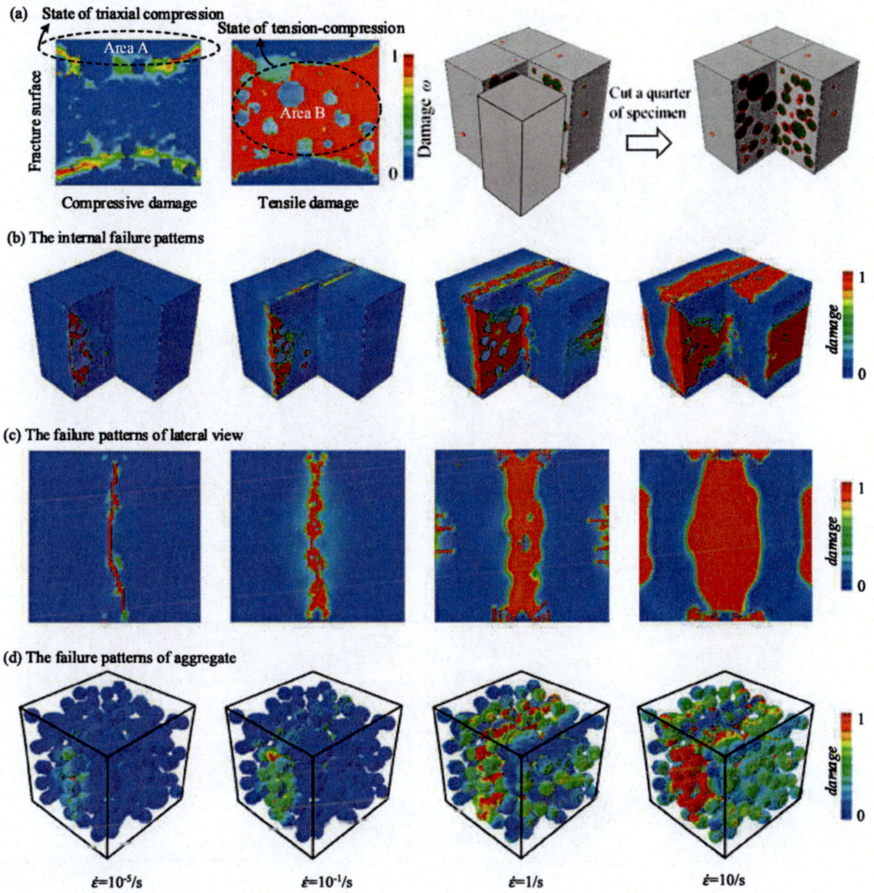

Fig. 5.3 Typical failure patterns of concrete cubic specimen (structural size $b = 100$ mm) under different strain rates: **a** fracture surface; **b** internal failure patterns; **c** failure patterns of lateral view; and **d** failure patterns of aggregates

In addition, due to large vertical deformation in the middle region of the specimens, bending-tensile cracks also occurred on the side surfaces parallel with the central line when the applied strain rate exceeds 1/s. Damage occurred in most areas of the concrete specimen and even local comminuted damage shows up finally under high strain rates.

3. *Load–displacement relations*

Figure 5.5 plots load (P)- displacement (δ) curves of concrete specimens having different structural sizes under different strain rates. The load is the force applied to the loading area and displacement (δ) is the deformation at the loading area. It can

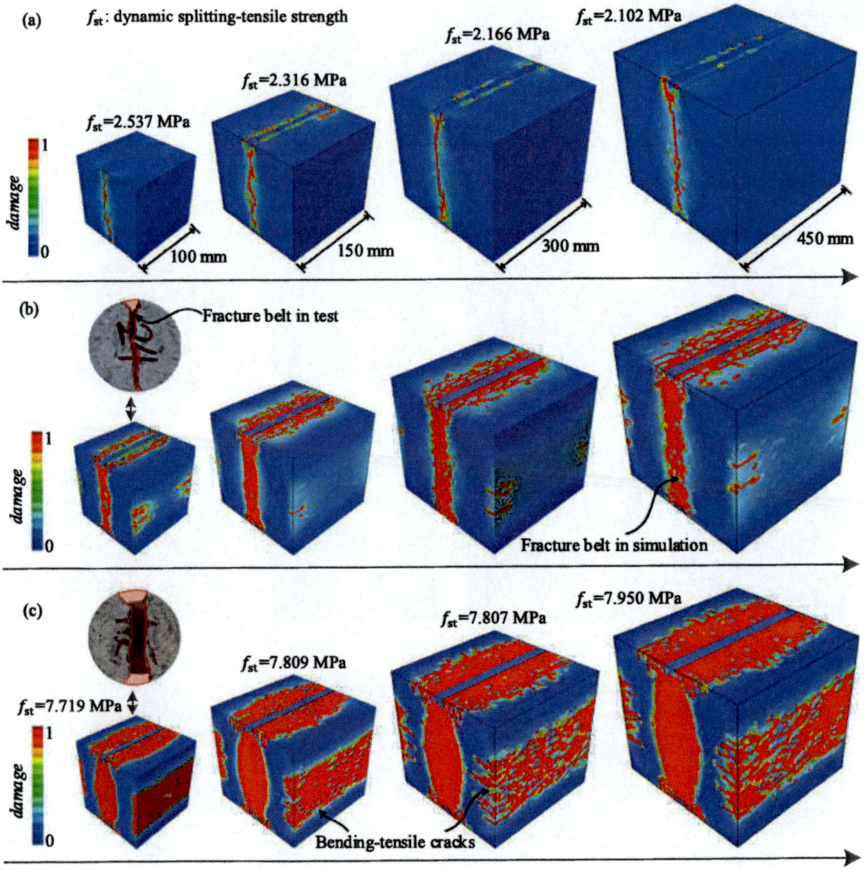

Fig. 5.4 Typical failure patterns of concrete cubic specimens having four kinds of structural sizes under different strain rates: **a** $\dot{\varepsilon}= 10^{-3}$/s; **b** $\dot{\varepsilon}= 1$/s; and **c** $\dot{\varepsilon}= 10$/s

be seen from Fig. 5.5 that with the increase of structural size the peak load increases obviously under different strain rates. As shown in Fig. 5.6, the peak load also enhances and the displacement corresponding to the peak load grows as strain rates increases. Furthermore, the higher strain rate is, the greater stiffness of specimens at early stage of loadings is.

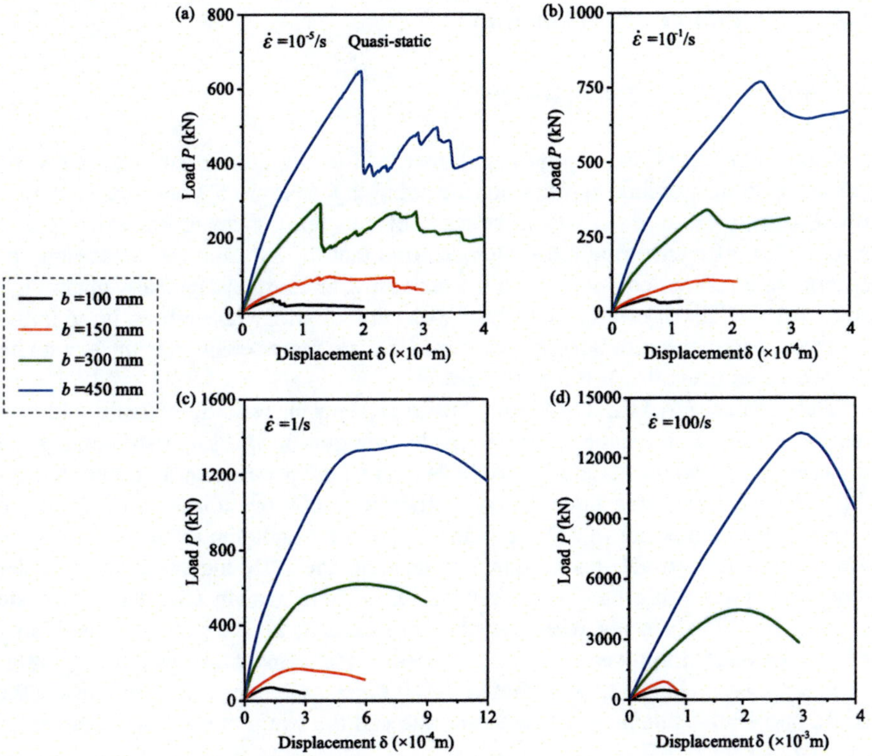

Fig. 5.5 Load (*P*)- displacement (*δ*) curves of loading area of concrete specimens having different structural sizes under different strain rates: **a** $\dot{\varepsilon}= 10^{-5}$/s; **b** $\dot{\varepsilon}= 10^{-3}$/s; **c** $\dot{\varepsilon}=$ 1/s; and **d** $\dot{\varepsilon}=$ 100/s

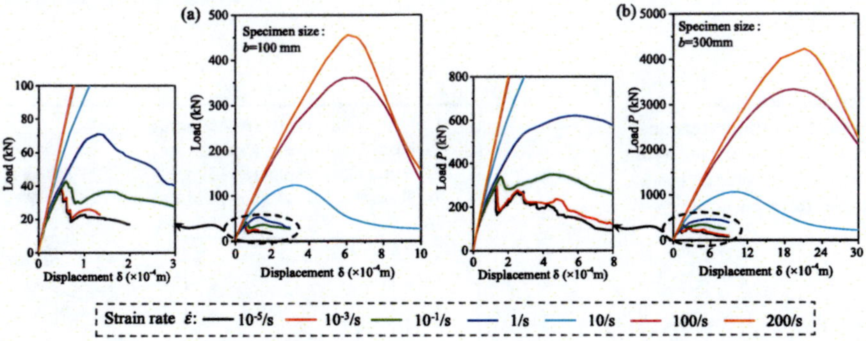

Fig. 5.6 Load (*P*)- displacement (*δ*) curves of loading area of concrete specimen under different strain rates: **a** structural size $b = 100$ mm; and **b** structural size $b = 300$ mm

5.1.2 *Influence of Strain Rate*

1. *Dynamic splitting-tensile strength*

According to Eq. (4.34), the measured load of loading area in simulation can be converted to the nominal splitting-tensile strength f_{st}. Figure 5.7 describes the relationship between splitting-tensile strength of concrete having different structural sizes under different strain rates. One can find that the splitting-tensile strength of specimens enhances obviously with the increasing strain rate, especially under high strain rates. This shows the significant behavior of strain rate effect. In addition, concrete specimens with large(r) structural sizes are more sensitive to the strain rate than those with small(er) structural sizes.

The comparison of the TDIF (the TDIF represents the ratio of dynamic splitting-tensile strengths and static splitting-tensile strengths in the same structural size) between numerical results and the available test results for concrete specimens having different structural sizes under different strain rates is showed in Fig. 5.8. It can be seen that the data points of TDIF obtained from the simulation are basically within the range of the available test results. Furthermore, the TDIF increases slowly under low strain rates while the TIDF shows a remarkable growth after the strain rate exceeds 1/s. That is to say, when the strain rate exceeds 1/s the strain rate hardening effect obviously improved, resulting in great enhancement in dynamic splitting-tensile strength. Herein the strain rate $\dot{\varepsilon}$=1/s is taken as the critical strain rate $\dot{\varepsilon}_{cr}$, the strain rate $\dot{\varepsilon}$<1/s is defined as low strain rate and the strain rate $\dot{\varepsilon}$>1/s is defined as high strain rate.

Figure 5.9 presents the relationship between the strain rate (lg $\dot{\varepsilon}$) and the TDIF of concrete standard specimens (sized by 100 mm). According to the fitting curves, one can get the fitting segmented formula for the increase of dynamic splitting-tensile strength as:

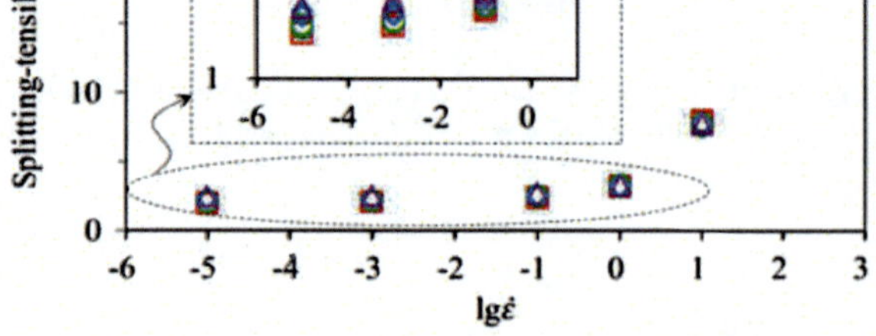

Fig. 5.7 Relationship between splitting-tensile strength of concrete having different structural sizes under different strain rates

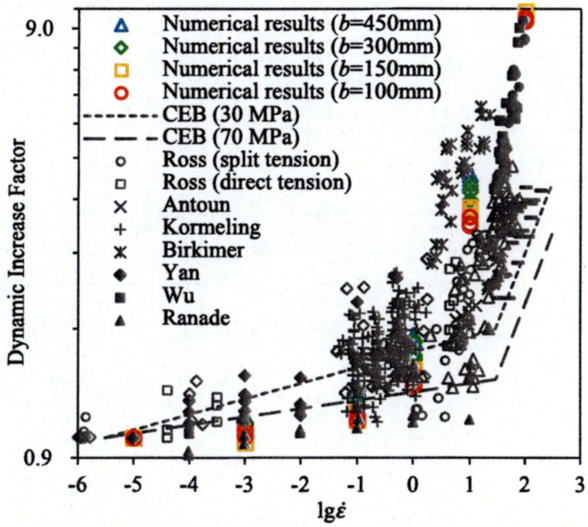

Fig. 5.8 Comparison of TDIF between numerical results and existing experimental results

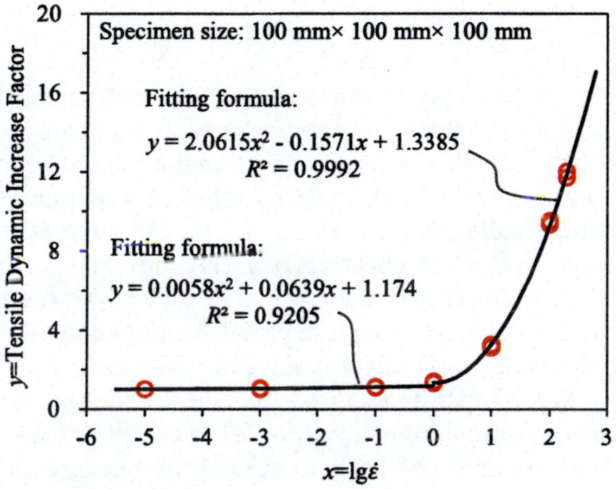

Fig. 5.9 Fitting relationship between the strain rate (lg $\dot{\varepsilon}$) and the TDIF of splitting-tensile strength

$$TDIF = 0.0058(\lg\dot{\varepsilon})^2 + 0.06391\lg\dot{\varepsilon} + 1.1740, (\dot{\varepsilon} \leq \dot{\varepsilon}_{cr} = 1/\text{s}). \quad (5.1)$$

$$TDIF = 2.0615(\lg\dot{\varepsilon})^2 - 0.1571\lg\dot{\varepsilon} + 1.3385, (\dot{\varepsilon} > \dot{\varepsilon}_{cr} = 1/\text{s}) \quad (5.2)$$

1. *Dynamic size effect*

Figure 5.10 gives the change of dynamic splitting-tensile strength of concrete specimens having different structural sizes under different strain rates. One can note that under low strain rates with the addition of structural size the splitting-tensile strength gradually decreases and the concrete shows a size effect behavior. However,

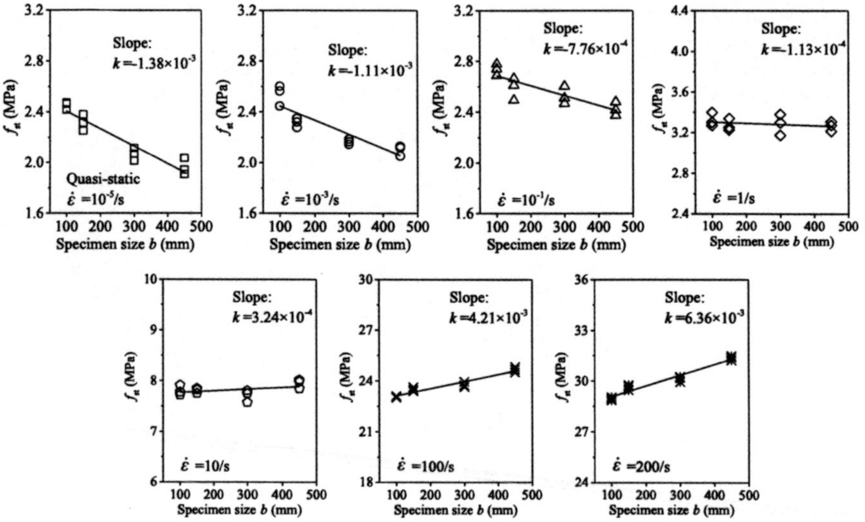

Fig. 5.10 Relationship between dynamic splitting-tensile strength and structural size under different strain rates

the decrease speed of the splitting-tensile strength (described by the linear fitting slope k [MPa/mm]) is getting slower as the strain rate increases, which illustrates that the size effect behavior of splitting-tensile strength is weakened gradually. Even under the critical strain rate $\dot{\varepsilon}_{cr}$ the slope k approaches to zero, indicating that the corresponding size effect behavior is totally suppressed. Moreover, under high strain rates, the size effect behavior of concrete shows up again and it is completely different with that under low strain rates. The splitting-tensile strength enhances as the strain rate improves and the discrepancy in splitting-tensile strength of specimens having different structural sizes tends to increase gradually.

According to above discussion, the dynamic splitting-tensile strength of concrete materials shows significantly different size effect behavior under different strain rates. The mechanism of dynamic size effect is under preliminary analysis as follows. As known, under static load, the size effect on nominal strengths of concrete is essentially a kind of material scientific issues. It is therefore closely influenced by the micro- and meso-structure of concrete, i.e. the aggregate sizes, aggregate spatial distribution, aggregate shape, initial defects, etc. A large number of studies have illustrated that the strength of concrete materials tends to decrease with the increasing structural size. However, under dynamic loads, the structural effect becomes an important influencing factor that affect the dynamic size effect besides the material effect. The material effect contains the concrete heterogeneities and the inherent strain rate effect, and the so-called structural effect involves the lateral confinement and the end friction effect.

Under low strain rates, especially under quasi-static condition, internal stress distribution of concrete is relatively uniform under loads. The weakest region of mechanical properties (like the ITZ) first reaches the ultimate strength and cracks appear. Then cracks have enough time to expand by bypassing aggregates and splitting-tensile damage cracks through the specimen form finally. The larger the specimen size is, and the larger region of poor mechanical properties in concrete is, causing the obvious size effect. As the strain rate increases, the pore-water pressure of concrete enhances and the viscous effect of free water slows down the propagation and expansion of cracks. The lateral restraint effect also enhances gradually and puts concrete specimens be in a complex stress state, leading to suppressing or delaying the expansion of macro cracks. In addition, more and more cracks form and expend. The crack path and number increase as the specimen size improves. The stress distribution within concrete becomes uneven progressively and the effect of internal defects on the dynamic splitting-tensile strength gradually declines. At same time, the effect of inertial force on the enhancement of splitting-tensile strength gradually increases. These factors commonly cause the piecemeal suppression of dynamic size effect behavior of concrete materials under low strain rates and the size effect disappears under the critical strain rate.

Under high strain rates, concrete specimens receive a huge impact and enough energy in a short time, making that the failure pattern has great changes. The high-speed impact brings more energy, which needs to be released in a short time and internal cracks of concrete have not enough time to develop. The restriction effect of concrete significantly enhances, making concrete in a highly constrained, highly complex state of stress. Therefore, the splitting-tensile strength improves obviously as the strain rate exceeds the critical strain rate. Moreover, the internal defects of concrete basically have no effect on the dynamic size effect and the inertial force of concrete greatly enhances as the increasing strain rate. The magnitude of the inertial force is related to the structural size of specimens, and the larger the size, the stronger the inertia. That is to say, the inertial force of concrete plays a leading role in the influence of dynamic size effect, resulting that with the addition of structural size the splitting-tensile strength gradually enhances under high strain rates.

5.2 Dynamic Size Effect on Tensile Strength

In order to investigate the influence of strain rate on the dynamic tensile failure behavior and the corresponding size effect, the failure behavior of double-edge notched concrete specimens having different structural sizes (the widths b of the concrete specimens are 100 mm, 150 mm, 200 mm, 300 mm and 400 mm, respectively) under different strain rates (the applied strain rates are $\dot{\varepsilon}= 10^{-5}$/s (quasi-static), 10^{-4}/s, 10^{-3}/s, 10^{-2}/s, 10^{-1}/s, 1/s, 10/s, 50/s and 100/s, respectively) were simulated by the meso-scale modelling method. To reduce the discreteness of numerical results as much as possible, three double-edge notched concrete specimens (only the aggregate distribution is different) were conducted in each case and a total of 135 double-edge notched concrete specimens were simulated. Figure 5.11 gives the 2D double-edge notched concrete utilized in this section and coarse aggregate particles were modeled

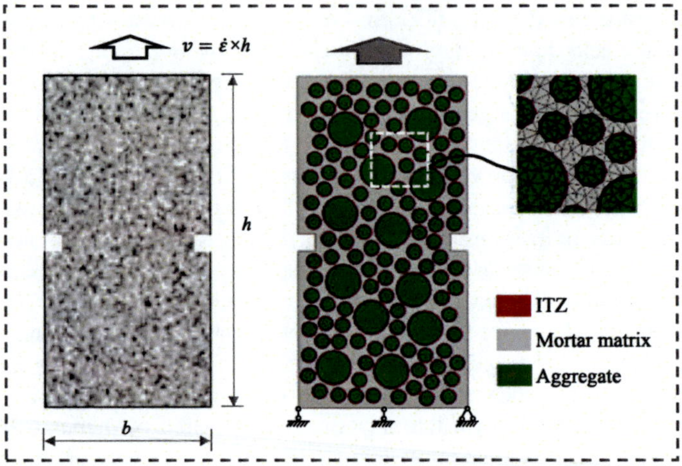

Fig. 5.11 2D meso-scale simulation model of double-edge notched concrete

as circles with random distribution enfolded by ITZ. The volume fraction of aggregate particles is about 47% and two-graded aggregate concrete was adopted with aggregates with diameter 12 mm and 30 mm. The height-width ratio (*h*/*b*) of concrete specimen is 2 (i.e. $h/b = 2$) and the notched area is a square having the side length of *b*/10. The model was meshed with constant strain triangle element and the average element size is approximately1 mm.

The loading method and boundary condition for concrete model were set as follows:

(1) The left and right sides of the model were free boundaries;
(2) The vertical displacement of the bottom of model was limited and one corner of model was horizontally and vertically limited; and
(3) The constant vertical velocity v was applied on the top surface of model, and the nominal strain rate $\dot{\varepsilon}$ is defined as the applied velocity divided by the height of the model, $\dot{\varepsilon}=v/h$.

In addition, the damaged plasticity model described in Sect. 3.3.1 has been also adopted to simulate the mechanical failure behavior of concrete three-phase components (aggregate, mortar matrix and ITZ) and the mechanical parameters of three-phase components set in meso-scale simulation of tensile behavior are listed in Table 5.2. Only the strength enhancement behavior was considered and the strain rate effect for meso-components on material strength was incorporated into the plastic damage model. The DIF for compressive strength recommended by CEB (1993) and the DIF for tensile strength recommended by Malvar and Ross (1998) (described in Sect. 3.4.1) were set in ABAQUS. Detailed modeling information can refer to the work (Jin et al. 2019a).

Table 5.2 Mechanical parameters of the concrete meso components used in the study

	Aggregate	Mortar matrix	ITZ
Mass density ρ (kg/m^3)	2880[a]	2750[a]	2750[a]
Elastic modulus E (GPa)	73[a]	38[a]	26
Poisson's ratio υ	0.16[a]	0.2[a]	0.2[a]
Dilation angle ψ (°)	18	18	15
Compressive yield stress σ_c (MPa)	80.0[b]	25.0[b]	16.0[b]
Tensile yield stress σ_t (MPa)	6.0[b]	2.5[b]	1.6[b]

Note: [a]Data were quoted from Zhou and Hao (2008); [b]Data were obtained by repeated trial; and other parameters were default value

5.2.1 Dynamic Failure Behavior

1. *Failure pattern*

Figure 5.12 gives the final failure patterns of the double-edge notched concrete specimens having different structural sizes under the strain rate $\dot{\varepsilon}$=10^{-2}/s. As shown in Fig. 5.12, the damage of their internal components is not uniform on account of the meso-scale heterogeneity of concrete specimens. The damage concentrates at the notched areas and the cracks propagate along the weaker zones. Then the cracks continued to expand and penetrate the entire specimen near the notched areas finally.

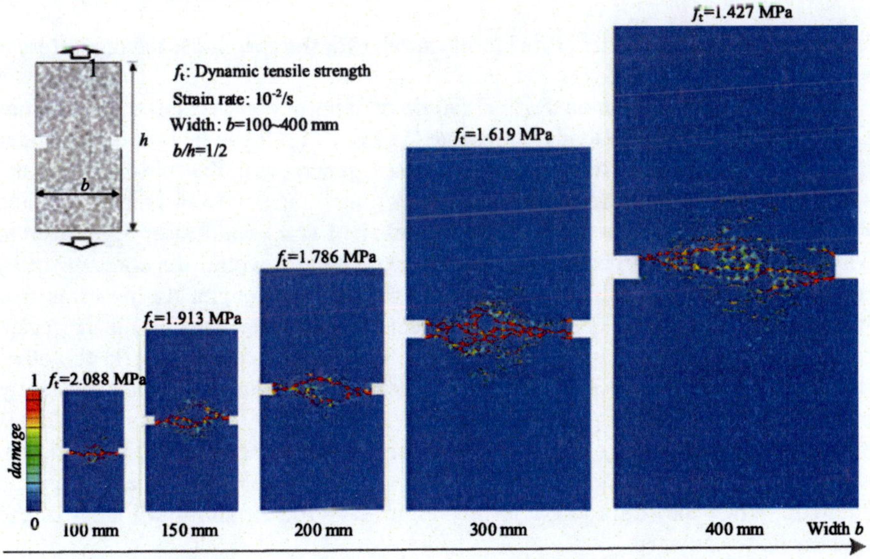

Fig. 5.12 Failure patterns of the notched specimens having different structural sizes under the strain rate $\dot{\varepsilon}$=10^{-2}/s

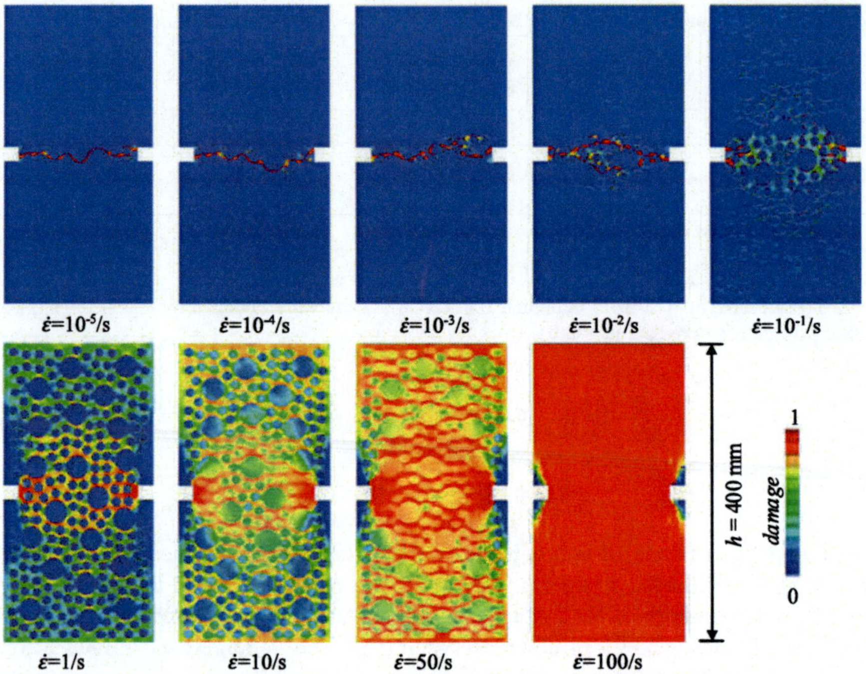

Fig. 5.13 Final failure patterns of concrete specimens having same structural size under different strain rates

In general, the failure patterns are closely similar for the specimens having different sizes.

The effect of strain rate on the final failure patterns of concrete (sized by 200 mm $\times$ 400 mm) under tensile loads is presented in Fig. 5.13. It can be found that, under the quasi-static load ($\dot{\varepsilon}$=10^{-5}/s), only one crack generated at the mid-height of the specimen. As the strain rate increases ($\dot{\varepsilon}$<1/s), the cracks become more tortuous, the damaged region increases, and the number of cracks increases significantly. This definitely causes the increase of dissipated energy within the concrete specimen, resulting in the enhancement of nominal tensile strength. As the strain rate increases ($\dot{\varepsilon} \geq$ 1/s), cracks begin to penetrate the aggregate, causing the aggregate to be damaged. Large number of transverse cracks gradually appear in the other parts of the specimen besides near the notches and the damaged region constantly expands from the middle part of the specimen to the upper and lower sides. That is to say, the internal damage area is further enlarged, spreading to the entire specimen, and further dissipating more energy, resulting in the large enhancement of tensile strength. When the strain rate $\dot{\varepsilon}$ is very high ($\dot{\varepsilon} \geq$ 100/s), almost all areas of the specimen are damaged.

2. *Stress–strain relations*

The obtained tensile stress–strain curves of concrete having different strain rates and different structural sizes are plotted in Fig. 5.14. For the case of same structural sizes, the peak tensile stress (dynamic tensile strength) increases with the increasing strain rate. The trend of increasing strength is relatively slow as the strain rate $\dot{\varepsilon}$<1/s (low strain rates) while the magnitude of strength enhancement significantly increases as the strain rate $\dot{\varepsilon}$>1/s (high strain rates). The peak strain corresponding to the peak tensile stress (dynamic tensile strength) of concrete also has a similar pattern. The peak tensile strain gradually increases with increasing strain rate, especially under high strain rates.

For the same strain rate, the tensile strengths (i.e. the peak tensile stress) of the notched concrete specimens with different widths show a great discrepancy. Especially, the discrepancy turns out to be obvious under lower strain rate (e.g. $\dot{\varepsilon}$=10^{-5}/s). Moreover, the discrepancy decreases as the strain rate increases. When the strain

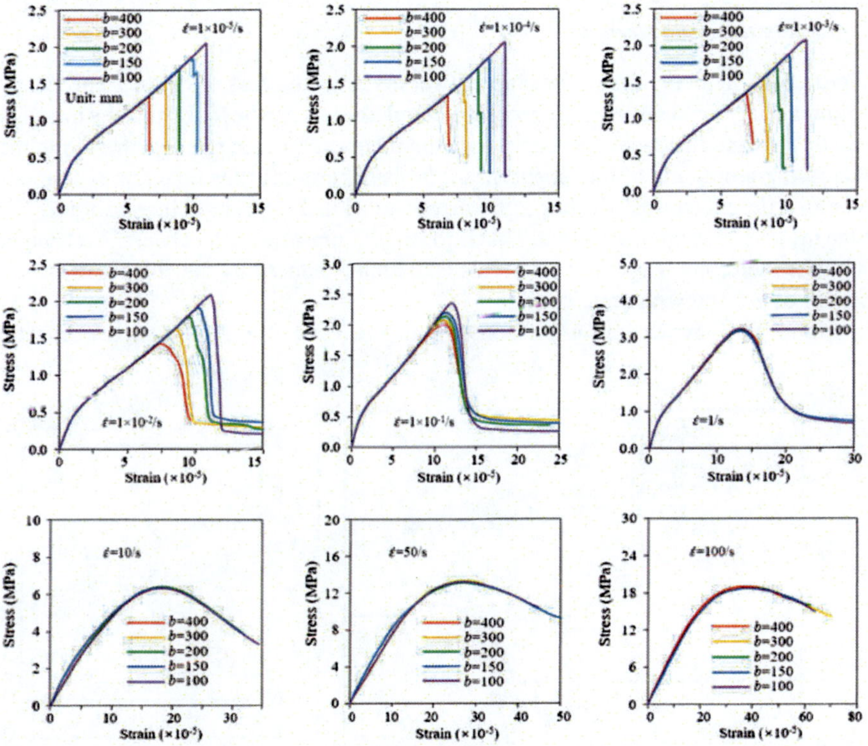

Fig. 5.14 Tensile stress–strain curves of double-edge notched concrete specimens having different structural sizes under different strain rates

rate reaches the critical strain rate $\dot{\varepsilon}_{cr}(=1/\text{s})$, the obtained peak tensile stress, the peak strain and the tensile stress–strain curves for different specimen sizes are basically the same. Under high strain rates, the tensile stress–strain curves are almost overlapping and the tensile strength has no dependence on specimen size. That is to say, under low strain rates, the nominal tensile strength and the peak tensile strain of concrete present obvious size effect, namely the tensile strength and the peak tensile strain decrease as the specimen size increases. Nevertheless, the discrepancy in the tensile strengths and peak tensile strain of concrete having different structural sizes tends to be reduced with the increase of strain rate. Under the critical strain rate $\dot{\varepsilon}_{cr}$ the size effect in tensile strength is completely suppressed. The tensile strength of concrete is not affected by changes in structural size, and the size effect completely disappears under high strain rates.

5.2.2 *Influence of Strain Rate*

1. *Dynamic tensile strength*

Figure 5.15 depicts the relationship between strain rate and the numerical tensile strengths of concrete having different structural sizes. It can be found that the nominal tensile strength increases obviously with the increasing strain rate. Furthermore, the increasing amplitude of the tensile strengths is different for the concrete specimens having different widths. The larger the specimen width is, the larger increasing amplitude the tensile strengths present. That is to say, the strain rate effect becomes stronger with the increasing structural size. This is definitely caused by the combined effects of size effect and strain rate effect.

Figure 5.16 shows the comparison of dynamic TDIF (the TDIF denotes the ratio

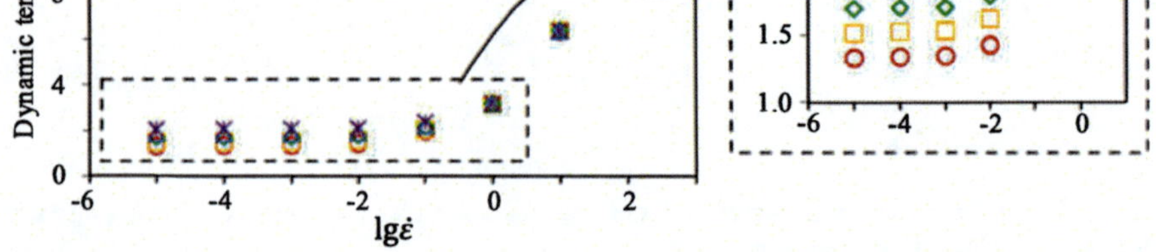

Fig. 5.15 Relationship between strain rate and tensile strengths of concrete having different structural sizes

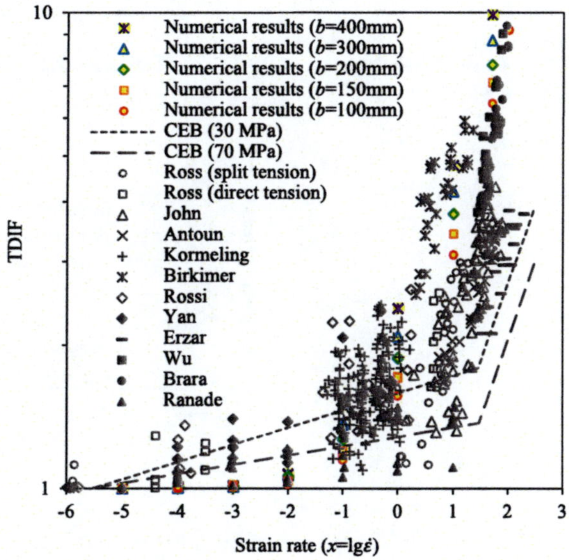

Fig. 5.16 Comparison of TDIF between numerical results and existing experimental results

of dynamic tensile strength and static tensile strength) between numerical results and existing experimental results for concrete specimens having different structural sizes under different strain rates. From Fig. 5.16 one can see that the TDIF obtained from simulation is basically within the range of the dynamic strength increase factor obtained by the tensile experiments of concrete, especially under low strain rates, which further verifies the rationality of the numerical results. Under low strain rates, when the strain rate $\dot{\varepsilon}<10^{-3}$/s, the TDIF basically remains unchanged as the strain rate improves. The TDIF increases slowly with increasing strain rate as the magnitude of the strain rate is between 10^{-3}/s and 1/s, and the increasing trend is much smaller than that under high strain rates. When the applied strain rate is more than the critical strain rate $\dot{\varepsilon}_{cr}$ (here the $\dot{\varepsilon}_{cr}$ is also 1/s), the TDIF shows a marked increasing trend along with the increasing strain rate, and the magnitude of the enhancement continues to increase. In other words, the dynamic tensile strength of concrete has obvious strain rate effect, and the strain rate corresponding to the variable inflection point is near the critical strain rate $\dot{\varepsilon}_{cr}$. Under low strain rates, the strain rate hardening effect is relatively weak, and when the applied strain rate begins to exceed the critical strain rate $\dot{\varepsilon}_{cr}$, the strain rate hardening effect is significantly improved, resulting in an great enhancement in dynamic tensile strength.

Figure 5.17 describes the relationship between strain rate (lg $\dot{\varepsilon}$) and the TDIF of concrete sized by 100 mm × 200 mm. According to the fitting curves, one can get the fitting segmented formula for the increase of dynamic tensile strength as

$$TDIF = 0.0163(\lg\dot{\varepsilon})^3 + 0.1681(\lg\dot{\varepsilon})^2 + 0.547\lg\dot{\varepsilon} + 1.5627, \ (\dot{\varepsilon} \le \dot{\varepsilon}_{cr} = 1/\mathrm{s}) \tag{5.3}$$

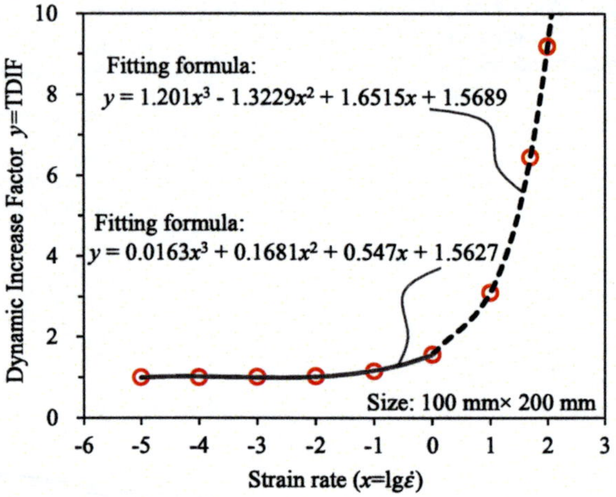

Fig. 5.17 Fitting relationship between the strain rate (lg $\dot{\varepsilon}$) and the TDIF of tensile strength

$$TDIF = 1.201(\lg\dot{\varepsilon})^3 - 1.3229(\lg\dot{\varepsilon})^2 + 1.6515\lg\dot{\varepsilon} + 1.5689, \ (\dot{\varepsilon} > \dot{\varepsilon}_{cr} = 1/\text{s}) \quad (5.4)$$

2. *Dynamic size effect*

The relationship between the structural sizes (i.e. width b) and the numerical dynamic tensile strength of concrete under different strain rates is given in Fig. 5.18. It can be clearly noted that the nominal tensile strength increases obviously with the increasing strain rate. Under low strain rates the obtained dynamic tensile strength of concrete decreases linearly as the specimen size increases. However, the decrease speed of the tensile strength is significantly distinct. Herein, the decrease speed can be described by the linear fitting slope k. In general, the decline trend of the strength slows down as the strain rate increases, indicating that the size effect of tensile strength is weakened or suppressed. Under the critical strain rate $\dot{\varepsilon}_{cr}$, the magnitude of the slope k begins to be close to zero, indicating that the corresponding size effect is completely suppressed. When the strain rate continues to increase, the magnitude of the slope k is stable and fluctuates around zero, Therefore, the dynamic tensile strength of concrete is not influenced by the variation of structural size, and the size effect basically disappears.

According to numerical results and analyses above, the dynamic tensile strength of concrete exhibits significantly distinct size effect under different strain rates. Significant size effect on tensile strength of concrete under static loads can be observed. The main reasons for the phenomena are presented as follows.

Under the low(er) strain rates ($\dot{\varepsilon}$<1/s), as the strain rate increases, the pore-water pressure inside the concrete material increases, and the viscous effect of free water slows down the propagation and expansion of cracks. The lateral restraint effect caused by the Poisson effect get enhanced progressively and the horizontal inertial force makes concrete specimens be in a complex stress state, which suppress or

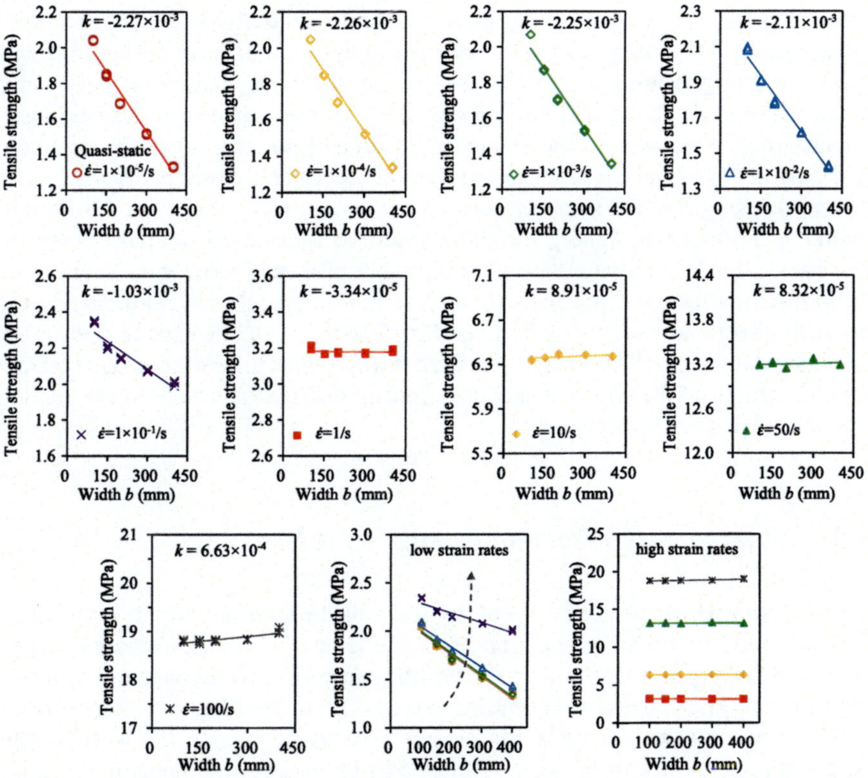

Fig. 5.18 Relationship between dynamic strength and structural size under different strain rates

delay the generation of macro cracks in concrete specimens. Concrete specimens require more energy to be destroyed and the nominal strength of concrete increases gradually. Furthermore, the inhomogeneity of stress distribution in concrete materials progressively increases with the increasing strain rate. When the stress in the weakest region has not reached the ultimate strength, the stress in some regions may have reached ultimate strength. This causes that the influence of factors such as internal defects of materials on the dynamic nominal strength gradually reduces. A larger number of cracks generated and also the crack branching occurred. The larger the specimen size is, the longer the crack path is and the more complicated the crack expansion is. Meanwhile, the influence of vertical inertial force on the enhancement of nominal strength progressively increases and the inertial force is positively related to the specimen size. These lead to the fact that the size effect on dynamic strength of concrete is weakened gradually. When the applied strain rate reaches the critical strain rate $\dot{\varepsilon}_{cr}$, as the strain rate increases, the lateral confinement and the horizontal inertial force enhance to ultimate stability, making concrete specimens in a highly constrained, highly complex state of stress. It is difficult for cracks to further expand around the aggregate particles in a short time after the internal cracks of the specimen are generated. Even the cracks penetrate the aggregate particles in a short time and cause the aggregate to be pulled off. That is to say, the stress increases rapidly under higher strain rates that the cracks do not have sufficient time to propagate along the path of least resistance. The propagating path becomes straighter, the tortuosity of cracks reduces and the cracks propagate along several direct paths with more fractured aggregates. The stress is redistributed inside the specimen and the damage region further expands. Finally, the entire specimen is damaged and more energy is dissipated, resulting in the obvious enhancement of dynamic strength. In addition, the vertical inertial force gradually improves as the strain rate increases, putting the material effect and structural effect reaching a state of stability, mutual restraint, and mutual influence. This causes that there is no obvious difference in the dynamic tensile strength of the concrete specimen having different structural sizes, and the size effect completely disappears.

5.3 Dynamic Size Effect on Compressive Strength

In order to investigate the influence of strain rate on the dynamic compressive failure behavior and the corresponding size effect, the failure behavior of square concrete specimens having different structural sizes (the width b of square concrete specimens are 100, 150, 300, 450 mm and 600 mm, respectively) under different strain rates (the applied strain rates are $\dot{\varepsilon}= 10^{-5}$/s (quasi-static), 10^{-3}/s, 10^{-2}/s, 10^{-1}/s, 1/s, 10/s, 30/s, 50/s and 100/s, respectively) were simulated by the meso-scale modelling method. To reduce the discreteness of numerical results as much as possible, three square concrete specimens (only the aggregate distribution is different) were conducted in each case and a total of 135 concrete specimens were simulated. Figure 5.19 depicts

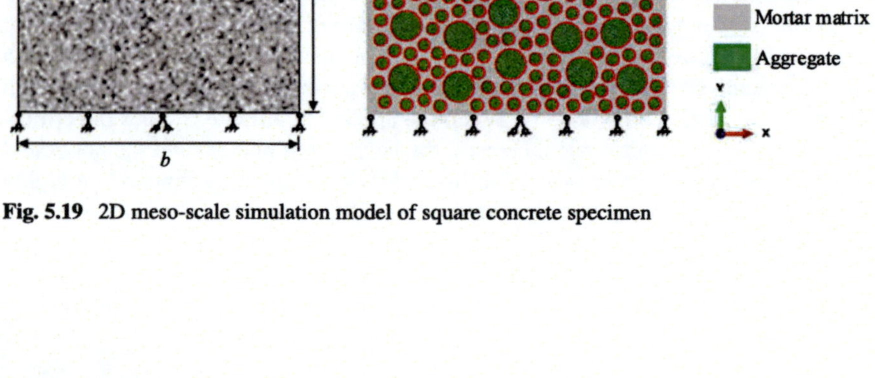

Fig. 5.19 2D meso-scale simulation model of square concrete specimen

Table 5.3 Mechanical parameters of meso components in compressive simulation

	Aggregate	Mortar matrix	ITZ
Mass density ρ(kg/m^3)	2880[a]	2750[a]	2750[b]
Elastic modulus E (GPa)	73[a]	38[a]	26[b]
Poisson's ratio υ	0.16 [a]	0.2[a]	0.2[b]
Dilation angle ψ (°)		18	15
Compressive yield stress σ_c (MPa)		25.0[b]	16.0[b]
Tensile yield stress σ_t (MPa)		2.25[b]	1.6[b]

Note [a]Data were quoted from Dilger et al. (1984); [b]Data were obtained by repeated trial; and other parameters were default value

the 2D square concrete utilized in this section and coarse aggregate particles were modeled as circles with random distribution enfolded by ITZs. The volume fraction of aggregate particles is about 47% and two-graded aggregate concrete was adopted with aggregates with diameter 12 mm and 30 mm. The numerical model was divided with constant strain triangle element and the average element size is approximately1 mm. The loading method and boundary condition of square concrete model were consistent with those of double-edge notched concrete model, as shown in Sect. 5.2.

In addition, the damaged plasticity model described in Sect. 3.3.1 of Chap. 3 has been also adopted to simulate the mechanical failure behavior of mortar and ITZ. As for aggregate particles, they were assumed to follow a linear elastic material behavior in this section. The mechanical parameters of three-phase components set in meso-scale simulation of compressive behavior are listed in Table 5.3. Only the strength enhancement behavior was considered and the strain rate effect for meso-components on material strength was incorporated into the present plastic damage model. The DIF for compressive strength recommended by CEB (1993) and the DIF for tensile strength recommended by Malvar and Ross (1998) (described in Sect. 3.4.1of Chap. 3) were set in ABAQUS. Detailed modeling introduction can refer to this literature (Jin et al. 2019c).

5.3.1 Dynamic Failure Behavior

1. *Failure pattern*

Typically, the final failure patterns of the square concrete specimens having different structural sizes under the strain rate $\dot{\varepsilon}$=1/s are shown in Fig. 5.20. It can be seen that, due to the meso-scale heterogeneity of concrete specimens, the damage of their internal components is not uniform. Damages arise firstly in the ITZs of concrete specimens and then the damage area extends to the mortar matrix gradually. Damages of all the specimens bypass the aggregate and tortuous cracks are formed. The damage area expands and eventually a diagonally coherent plastic deformation zone appears.

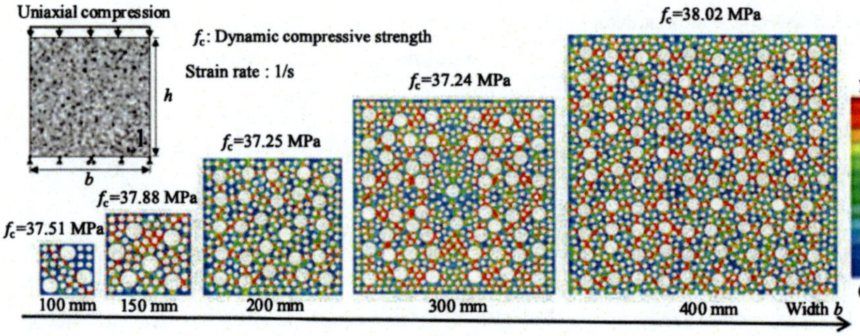

Fig. 5.20 Failure patterns of square concrete specimens having different structural sizes under strain rate $\dot{\varepsilon}$=1/s

The plastic deformation zone intersects and the overall stiffness of concrete specimen decreases until failure. On the whole, the failure patterns are closely similar for the specimens having different sizes.

Figure 5.21 presents the influence of strain rates on the final failure patterns of concrete (sized by 150 mm × 150 mm) under compressive loads. One can find that when the strain rate is small (10^{-5}/s < $\dot{\varepsilon}$ < 10^{-2}/s), only one or two diagonal cracks generate and penetrate the specimen. As the strain rate increases (10^{-1}/s < $\dot{\varepsilon}$ < 1/s), the path of cracks presents a complex and diversified trend, and the number of cracks increases significantly. The cracks branch and cross, and the damaged region increases. This definitely causes the increase of dissipated energy within the concrete specimen, resulting in the enhancement of nominal compressive strength. When the strain rate is relatively high (the strain rate > $\dot{\varepsilon}$50/s), large areas of the specimen are damaged, and concrete specimens are almost crushed under compressive loads. Similar results were also reported in the study of Hao et al. (2013).

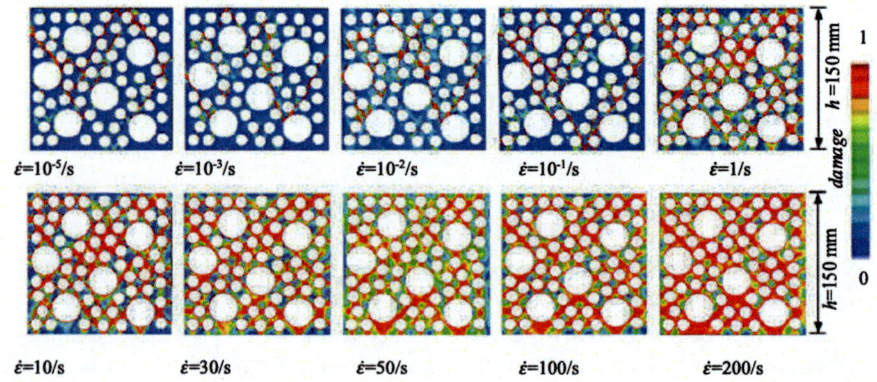

Fig. 5.21 Failure patterns of square concrete specimens having same structural size under different strain rates

2. *Dynamic stress–strain relations*

Figure 5.22 shows the obtained macroscopic compressive stress–strain curves of concrete having different structural sizes and different strain rates. It can be seen that, for the case of same structural sizes, as the strain rate increases, the magnitudes of the dynamic compressive strength and the corresponding peak strain increase. In addition, one can see that the softening curves tend to be less steep as the strain rate increases and the strain rate also affects the peak strain of concrete dynamic compression. At the same structural size, with the increasing strain rate, the damage area of concrete increases, and the dynamic compression peak strain increases.

Under the same strain rate, the difference in peak strain of the specimens having different structural sizes is small while the peak compressive stresses (dynamic compressive strength) of the specimens having different structural sizes are significantly different. That is, there is an obvious size effect phenomenon on the compressive strength of concrete under dynamic loads.

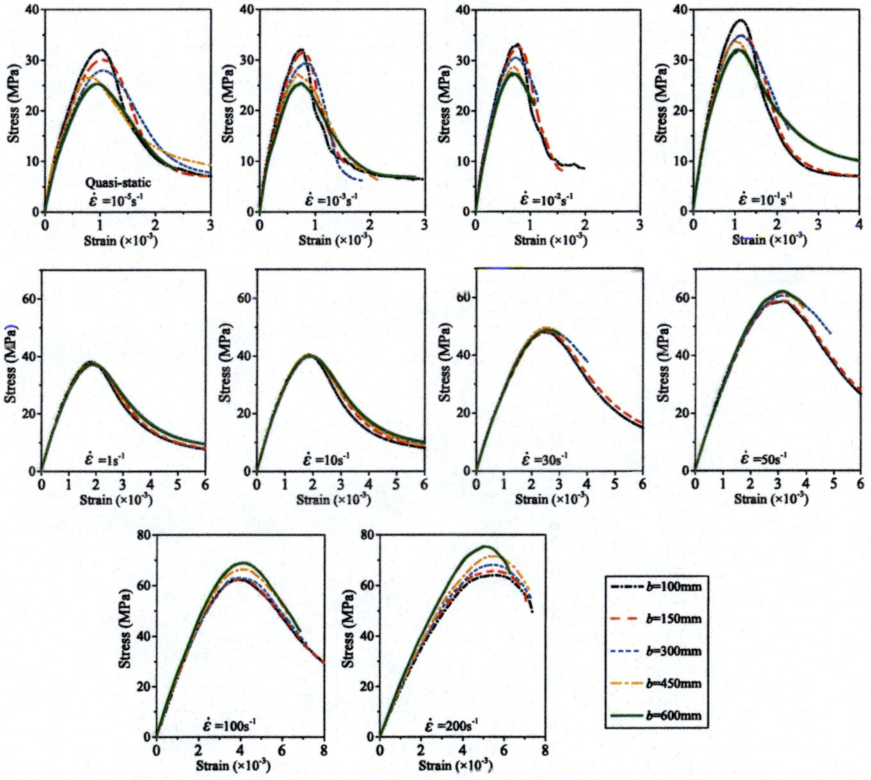

Fig. 5.22 Compressive stress–strain curves of concrete specimens having different structural sizes under different strain rates

5.3.2 *Influence of Strain Rate*

1. *Dynamic size effect*

Figure 5.23 plots the relationship between the structural sizes (i.e. width b) and the simulated dynamic compressive strength of concrete under different strain rates. One can clearly note the following findings:

The nominal compressive strength increases obviously with the increasing strain rate.

When the strain rate $\dot{\varepsilon}$ equals to be 10^{-5}/s (quasi-static), the compressive strengths (i.e. the peak compressive stress) of the square concrete specimens with different widths show a great discrepancy and the size effect is significant. Moreover, when the strain rate $\dot{\varepsilon}$ is between 10^{-5}/s and 1/s, the obtained dynamic compressive strength of concrete decreases linearly as the specimen size increases. However, the decrease

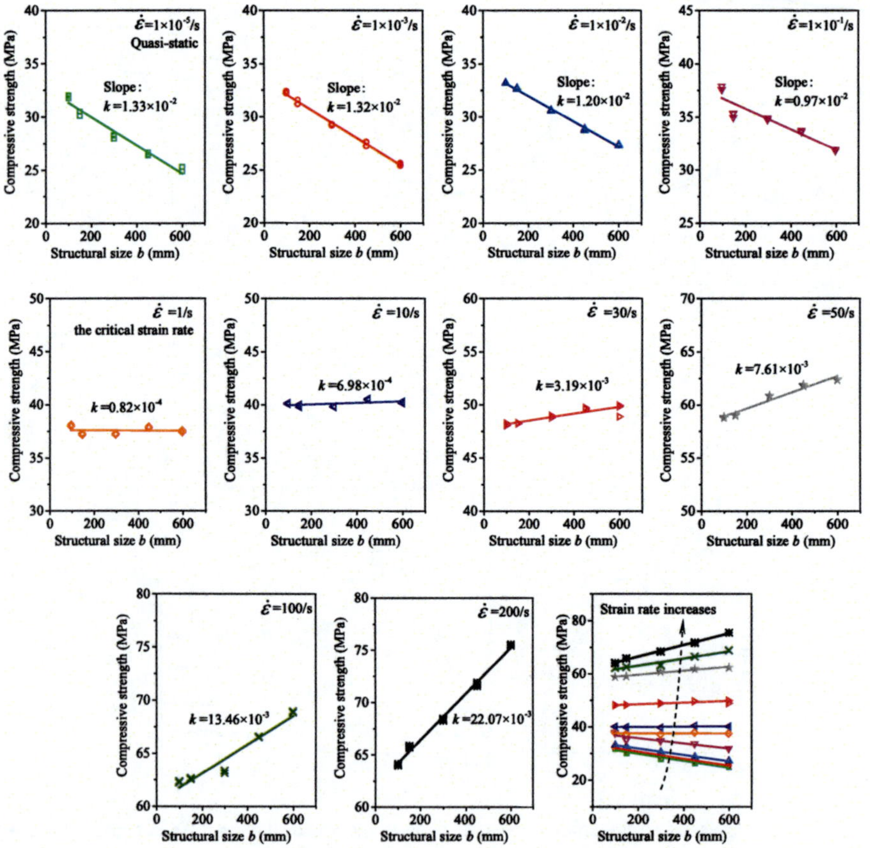

Fig. 5.23 Relationship between dynamic compressive strength and structural size

speed of the strength is significantly different. Herein, the decrease speed can be described by the linear fitting slope k. On the whole, the decline trend of the strength slows down as the strain rate increase, indicating that the size effect of compressive strength is weakened.

When the strain rate $\dot{\varepsilon}$ equals to be 1/s, the magnitude of the slope k is close to zero. The obtained peak compressive stress, the peak strain and the compressive stress–strain curves for different specimen sizes are basically the same. The compressive strength has no dependence on specimen sizes. That is to say, the size effect in compressive strength under the critical strain rate $\dot{\varepsilon}_{cr}$ ($\dot{\varepsilon}_{cr}$=1/s) is completely suppressed.

When the applied strain rate exceeds the critical strain rate $\dot{\varepsilon}_{cr}$ (i.e. 1/s), the obtained dynamic compressive strength of concrete specimens with different widths show a discrepancy again and the nominal compressive strength of concrete increases as the structural sizes increase. With the increasing strain rate, the discrepancy in the compressive strengths of concrete having different structural sizes tends to be enhanced gradually, showing that a distinct size effect behavior on compressive strength under high strain rates.

2. *Dynamic compressive strength*

Figure 5.24 plots the relationship between the strain rate and the numerical compressive strengths of concrete having different structural sizes. One can note that the nominal compressive strength increases obviously with the increasing strain rate. Moreover, for the concrete specimens having different structural sizes, the increasing amplitude of the compressive strengths is distinct. The larger the specimen size is, the larger increasing amplitude the compressive strengths present. In other words, the strain rate effect becomes stronger as the specimen size increases. This is mainly caused by the combined effects of strain rate effect and size effect.

Figure 5.25 presents the relationship between the strain rate (lg $\dot{\varepsilon}$) and CDIF of

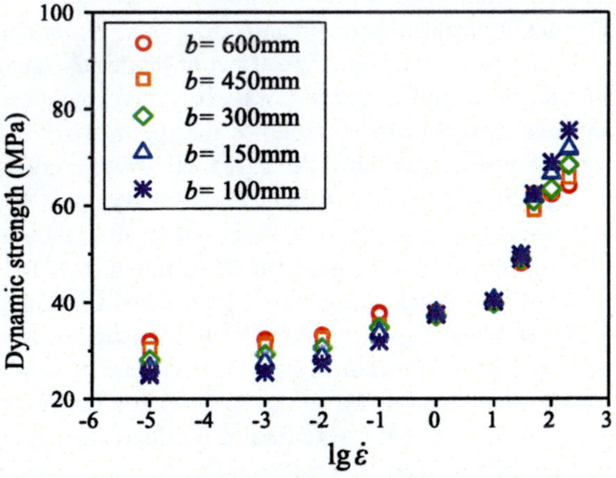

Fig. 5.24 Relationship between strain rate and compressive strengths of concrete with different structural sizes

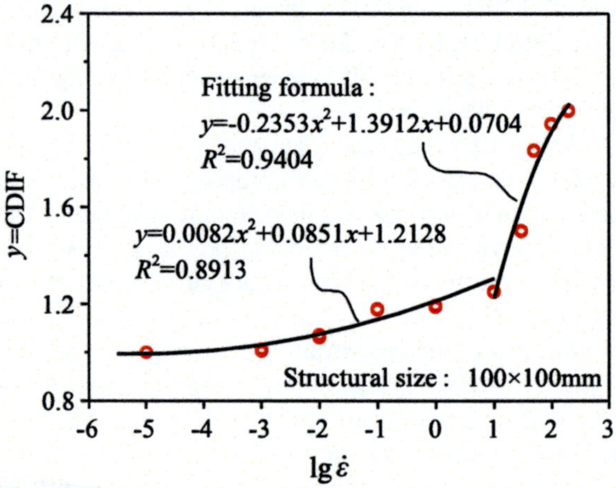

Fig. 5.25 Fitting relationship between strain rate (lg $\dot{\varepsilon}$) and CDIF

concrete sized by 100 × 100 mm. According to the fitting curves, one can get the fitting segmented formula for the increase of dynamic compressive strength as

$$CDIF = 0.0082(\lg\dot{\varepsilon})^2 + 0.0851\lg\dot{\varepsilon} + 1.2128, (\dot{\varepsilon} \leq \dot{\varepsilon}_{cr} = 1/\mathrm{s}) \quad (5.5)$$

$$CDIF = -0.2353(\lg\dot{\varepsilon})^2 + 1.3912\lg\dot{\varepsilon} + 0.0704, (\dot{\varepsilon} > \dot{\varepsilon}_{cr} = 1/\mathrm{s}) \quad (5.6)$$

3. *Preliminary discussion on dynamic size effect*

As shown above, the size effect in compressive strength of concrete under different strain rates is obviously different. Significant size effect in compressive strength of concrete under static load can be observed from the above simulation. Under dynamic loads, as the strain rate increases, the dynamic compressive strength increases, while the size effect on compressive strength is weakened first and then enhanced reversely. The main reasons for the phenomena can be explained as follows.

Under static load, the size effect on nominal strength of concretes is essentially a kind of material scientific issues. It is therefore closely associated with the micro- and meso-structure of concrete, i.e. the aggregate sizes, aggregate spatial distribution, aggregate shape, initial defects, etc. However, under dynamic loads, in addition to the material effect, the structural effect becomes an important influencing factor that affect the dynamic size effect. Here, the material effect contains the concrete heterogeneities and the inherent strain rate effect, and the so-called structural effect involves the lateral confinement and the end friction effect.

When the applied strain rate is less than the critical strain rate $\dot{\varepsilon}_{cr}$ ($\dot{\varepsilon} \leq 1/\mathrm{s}$), as the strain rate and the specimen sizes increase, the influence of viscosity effect on strength enhancement tends to be weaker and weaker. However, the lateral confinement tends to be stronger and the structural effect becomes greater. These make the concrete specimens have a complex state of stress, and the generation and expansion of cracks are suppressed or delayed, causing that the dynamic compressive strength increases and the size effect turns out to be weakened gradually.

When the applied strain rate exceeds the critical strain rate $\dot{\varepsilon}_{cr}$ ($\dot{\varepsilon}$ > 1/s), as the strain rate and the specimen sizes increase, the increasing lateral confinement makes the concrete specimens in a highly constrained and extremely complicated state of stress. Cracks are difficult to spread in a short time and more energy is dissipated, leading to the dynamic compressive strength greatly enhanced. Meanwhile, the inertial effect dominates the strength improvement gradually, the larger the structural sizes is, the greater the inertial effect is, this causes that the dynamic compressive strength shows an increasing trend with increasing structural sizes, indicating a distinctly different size effect compared with that under low strain rates.

On the whole, under dynamic loads, both the material effect and the structural effect contributes to the size effect of concrete materials.

5.4 Influence of Meso-Structure

Concrete is a very important and highly applicable material whose nonlinear mechanical behavior can change with varying percentages of concrete components (e.g. aggregate particles, cement, water, pores additives, etc.). Aggregate particles are granular materials that play a skeleton. The properties of the aggregates (type, shape, particle size, grading, and surface conditions, etc.) play an important role in mechanical failure behavior of concrete materials. Therefore, here in this section, applying the meso-scale modelling method, the influences of aggregate content, maximum aggregate size and aggregate type on the failure behavior (uniaxial compressive failure and splitting-tensile failure) and the corresponding size effect under dynamic loads are introduced.

5.4.1 Influence of Aggregate Content

This part focuses on revealing the influence of aggregate content on the dynamic failure behavior and the corresponding size effect of concrete materials based on meso-scale simulation (Jin et al. 2020b).

1. *Dynamic compressive behavior*

Firstly, applied the compressive model established in Sect. 5.3, the compressive failure behavior of concrete square specimens with four kinds of side lengths (b = 100,150, 300 and 450 mm) and three kinds of aggregate contents (V_{agg} = 45%, 30% and 15%) under seven kinds of strain rates ($\dot{\varepsilon}$ = 10^{-5}/s, 10^{-3}/s, 10^{-1}/s, 1/s, 10/s, 100/s and 200/s) were simulated at the meso-scale level. In each case, three concrete square specimens (having different aggregate distributions) were conducted

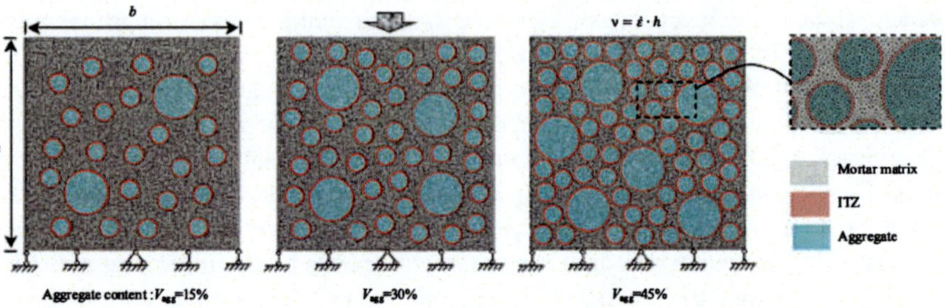

Fig. 5.26 2D Meso-scale models of square specimens having different aggregate contents

and 252 concrete square specimens were simulated in total. Figure 5.26 depicts 2D meso-scale models of square specimens (side length $b = 150$ mm) having different aggregate contents. The mechanical failure behavior of concrete three-phase components (aggregate, mortar matrix and ITZ) were simulated by the plastic damage model.

Figure 5.27 represents the final compressive failure patterns of concrete specimens (size length $b = 300$ mm) with different aggregate contents under different strain rates. Under low strain rates, aggregate content affects the propagation and distribution of macro cracks. The more aggregate content, the denser the arrangement of aggregate particles. A compact space is formed between the aggregate particles, so the cracks need to bypass more aggregates, causing the crack path of concrete specimen with high aggregate content is more tortuous. Under high strain rates, there are more damaged aggregate particles in concrete specimens with larger aggregate content.

Figure 5.28 shows the compressive stress–strain curves of square concrete specimens with different aggregate contents and different sizes under different strain rates. As visible in Fig. 5.27, aggregate content affects the dynamic compression peak stress

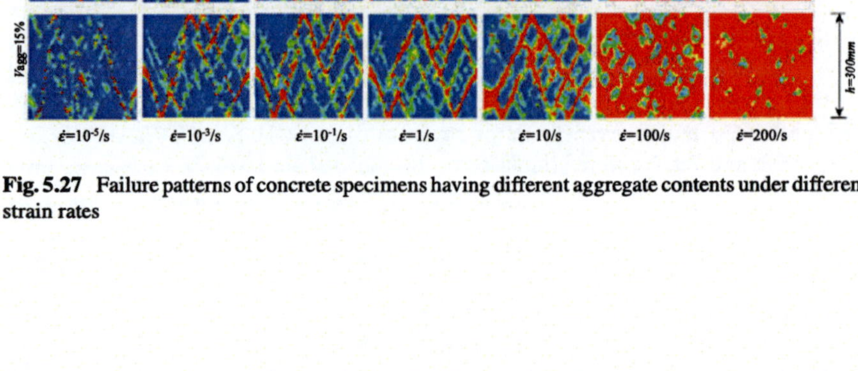

Fig. 5.27 Failure patterns of concrete specimens having different aggregate contents under different strain rates

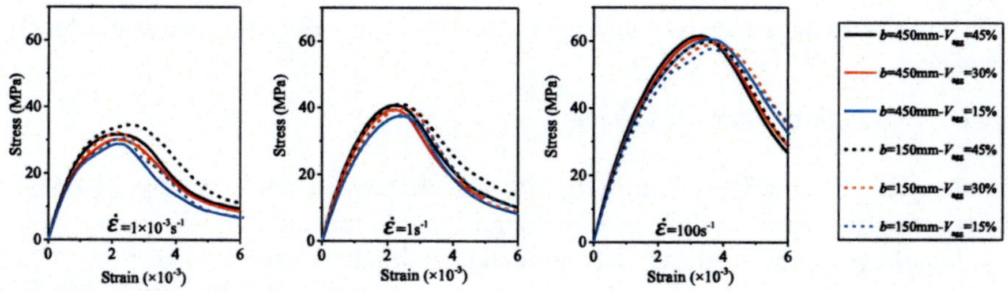

Fig. 5.28 Compressive stress–strain curves of square concrete specimens having different structural sizes and aggregate contents under different strain rates

of concrete. As the aggregate content increases, the dynamic compression peak stress gets an enhancement. Under the same strain rate and structural size, the peak strain and stress–strain curves of concrete specimens with different aggregate contents are basically the same. That is to say the aggregate content has basically no effect on the dynamic compressive stress–strain curve of concrete.

Figure 5.29 shows the relationship between dynamic compressive strength and structural size of concrete specimens with different aggregate contents under different strain rates. It can be seen that aggregate content has some influences on dynamic compressive strength of concrete materials under different strain rates. Dynamic compressive strength increases with the increasing aggregate content under the same strain rate, but concrete specimens with different aggregate contents have little difference in the slope of strength degradation or enhancement with the increasing size.

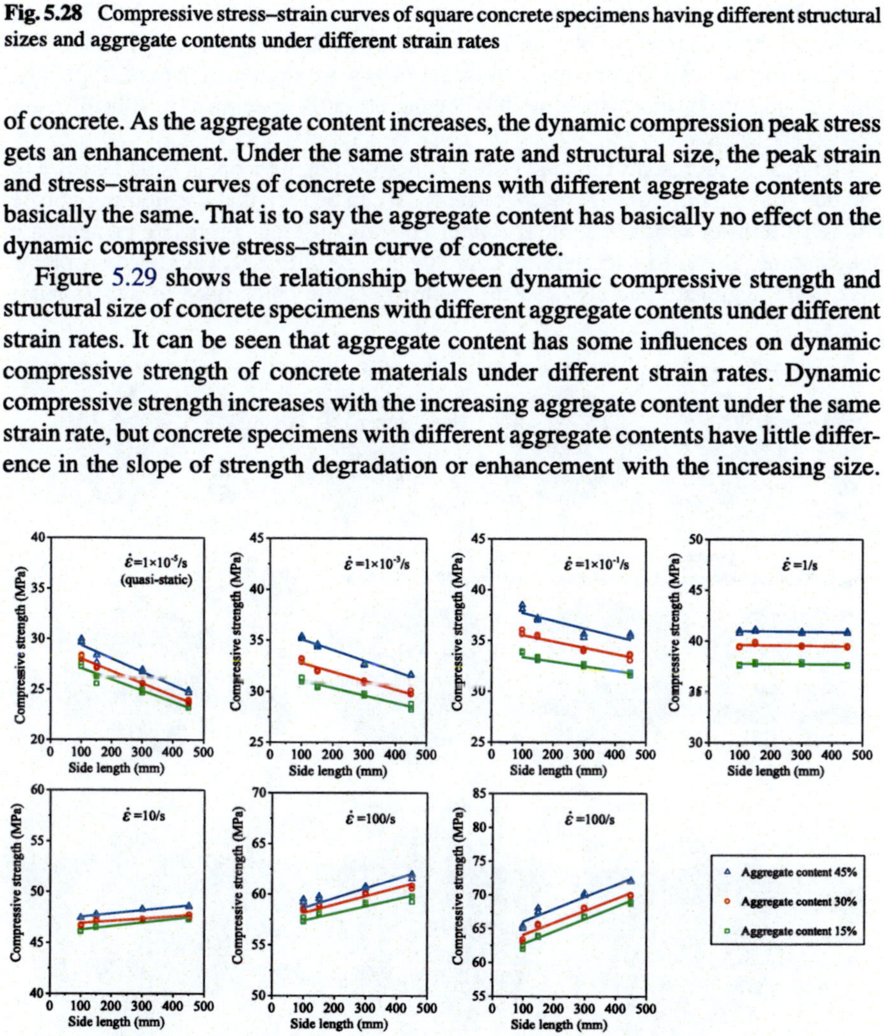

Fig. 5.29 Relationship of dynamic compressive strength between different structural sizes of concrete specimens having different aggregate contents under different strain rates

That is to say the aggregate content has an ignorable influence on the dynamic strength size effect.

2. *Dynamic splitting-tensile behavior*

Similarly, applied the 3D splitting-tensile model established in Sect. 5.1 (see Fig. 5.2), the splitting-tensile failure behavior of concrete cubic specimens with three kinds of side lengths ($b = 150$, 300 and 450 mm) and three kinds of aggregate contents ($V_{agg} = 40\%$, 30% and 20%) under six kinds of strain rates ($\dot{\varepsilon} = 10^{-5}$/s, 10^{-3}/s、10^{-1}/s、1/s、10/s and 100/s) were simulated at the meso-scale level (Jin et al. 2020b). In each case, three concrete cubic specimens (having different aggregate distributions) were conducted and 162 concrete cubic specimens were simulated in total. Typically, Fig. 5.30 depicts 3D meso-scale models of concrete cubic specimens having different aggregate contents.

Figure 5.31 represents the final failure patterns of cubic models (side length $b =$ 150 mm) having different aggregate contents. It can be seen that the splitting-tensile failure patterns of concrete cubic specimens having different aggregate contents are not the same. Under low strain rates, cubic specimens with larger aggregate contents have more aggregate particles and the splitting-tensile crack need to bypass more

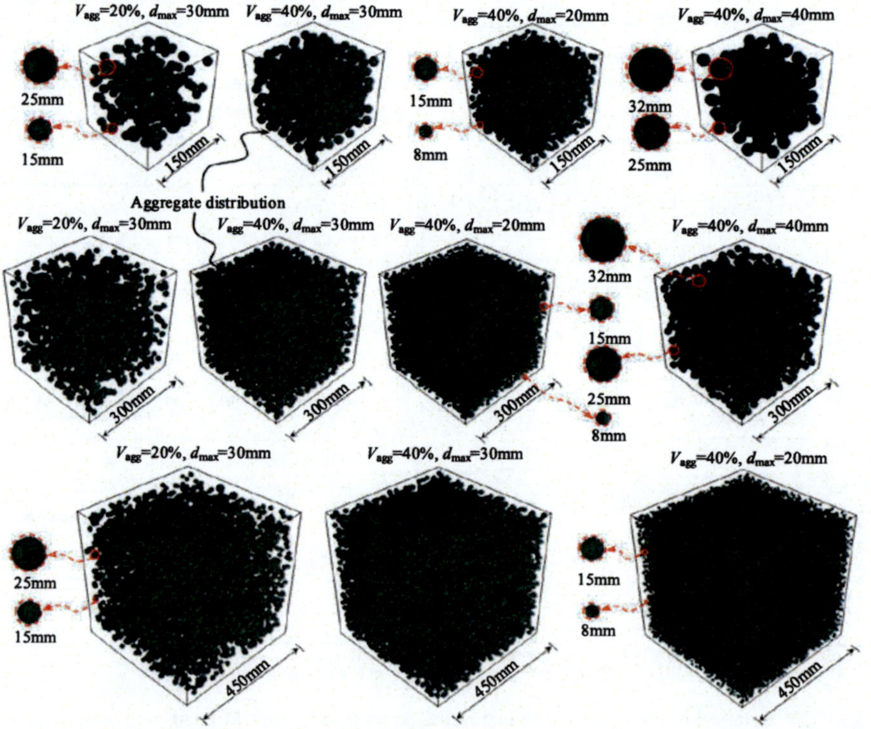

Fig. 5.30 3D meso-scale models of cubic specimens having different aggregate contents and MAS

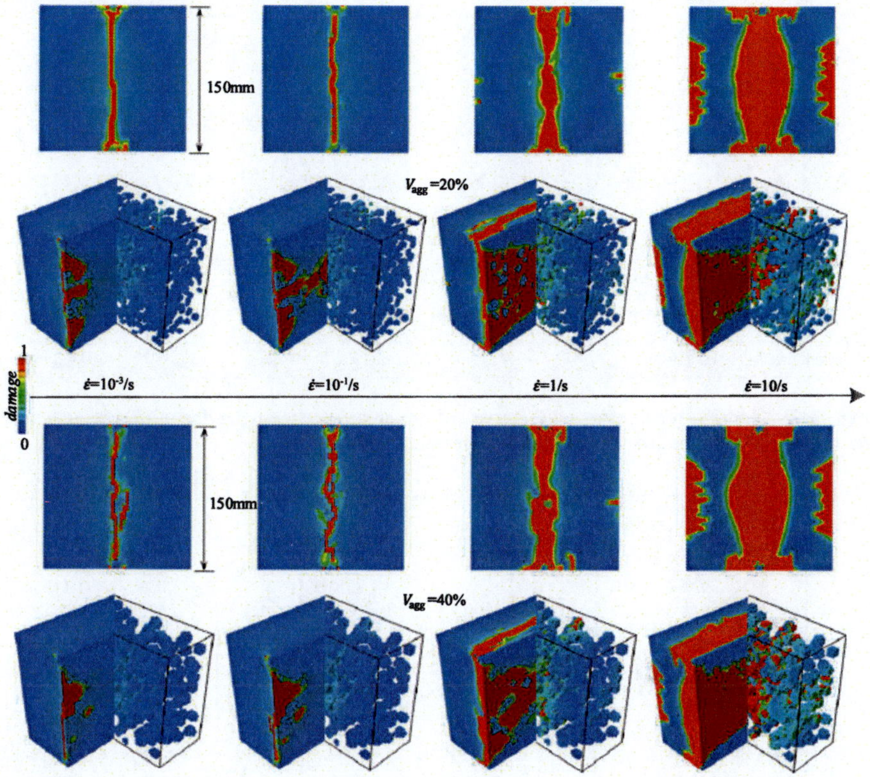

Fig. 5.31 Typical failure patterns of concrete specimens having different aggregate contents under dynamic loads

aggregates, causing the path of crack is more zigzag. Under high strain rates, there are more damaged aggregate particles and more bending tensile cracks in the specimens with larger aggregate contents.

Load (P)—displacement (δ) curves of cubic models with different side lengths having different aggregate contents under low strain rates and high strain rates are depicted in Fig. 5.32. The splitting-tensile peak load gets an enhancement and the displacement of loading area corresponding to the peak load increases with the increasing strain rate. The P-δ curves of cubic specimens (with same side length) having different aggregate contents are basically similar, and there are only some differences in the peak loads under the same strain rate. With the same structural size, the higher the aggregate content is, the larger the peak load is (see Fig. 5.32).

Figure 5.33 depicts the size effect on dynamic splitting-tensile strength of concrete having different aggregate contents under different strain rates. One can note that the aggregate content also has an important influence on the dynamic splitting-tensile strength and the corresponding dynamic size effect: (1) Under low strain rates, in the cubic specimens with high(er) aggregate content, the aggregate distribution is

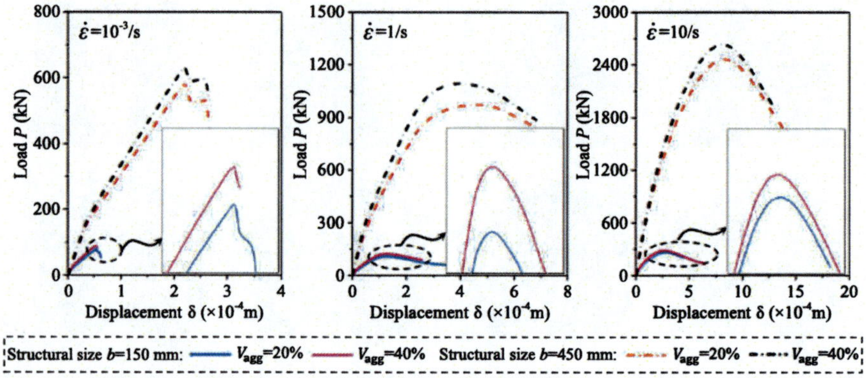

Fig. 5.32 Load(P)—displacement(δ) curves of concrete specimens having different aggregate contents (b = 150 mm and 450 mm) under dynamic loads

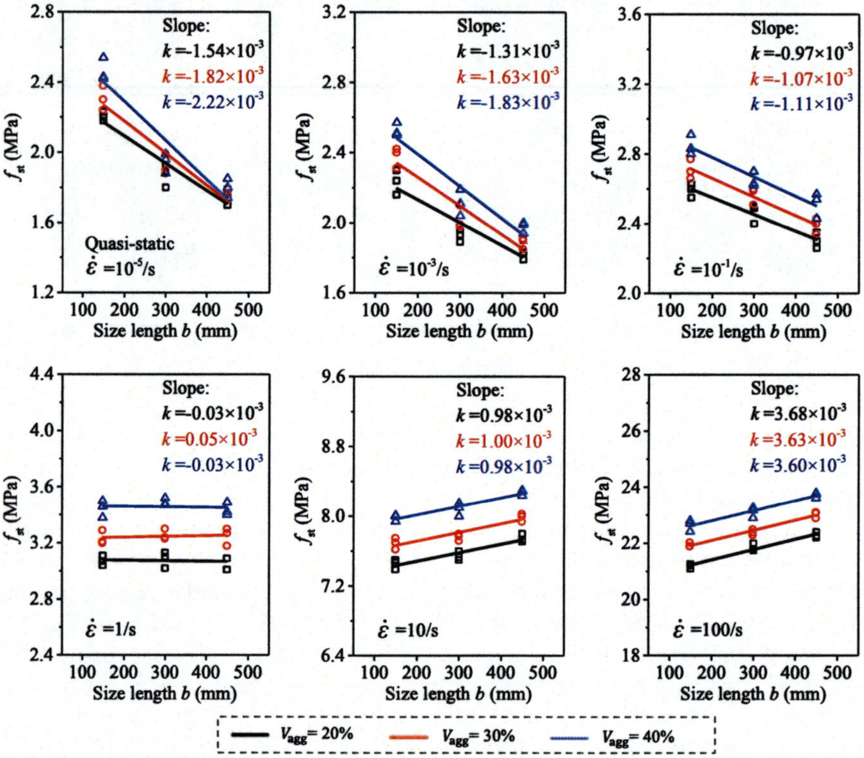

Fig. 5.33 Size effect on dynamic splitting-tensile strength of concrete having different aggregate contents under different strain rates

denser and a more compact space is formed between the aggregates. These cause that cracks need to bypass more aggregates during their expending process and the paths of cracks are more tortuous. Moreover, cubic specimens with higher aggregate content has stronger lateral restraint effect. These factors make that the splitting-tensile strengths enhance with the addition of aggregate content under low strain rates. In addition, there are more defect and region with weak mechanical properties in cubic specimens with higher aggregate contents, and the heterogeneity of meso-scale composition plays an essential role in affecting the size effect on splitting-tensile strengths. Therefore, under low strain rates, the dynamic splitting-tensile strengths get an enhancement with the addition of aggregate content and the size effect on nominal strengths shows more obviously in specimens with higher aggregate content. (2) Under high strain rates, cubic specimens with higher aggregate content have more aggregate particles penetrated and damaged under high-speed impact. The aggregate particles getting damaged need more energy than mortar matrix or the ITZs getting damaged. Therefore, the dynamic splitting-tensile strengths get an enhancement with an addition of aggregate content under high strain rates.

5.4.2 *Influence of Maximum Aggregate Size*

This part focuses on investigating the influence of maximum aggregate size on the dynamic failure behavior and the corresponding size effect of concrete materials based on meso-scale simulation.

Similarly, applied the 3D splitting-tensile model established in Sect. 5.1 (see Fig. 5.2), the splitting-tensile failure behavior of concrete cubic specimens with three kinds of side lengths ($b = 150$, 300 and 450 mm) and three kinds of maximum aggregate sizes (MAS $d_{max} = 20$ mm, 30 mm and 40 mm) under six kinds of strain rates ($\dot{\varepsilon} = 10^{-5}$/s, 10^{-3}/s, 10^{-1}/s, 1/s, 10/s and 100/s) were simulated at the meso-scale level (Jin et al. 2020b). In each case, three concrete cubic specimens (having different aggregate distributions) were conducted and 162 concrete cubic specimens were simulated in total. Typically, 3D Meso-scale models of cubic specimens having different MAS are also depicted in Fig. 5.30.

Figure 5.34 gives the final failure patterns of cubic models (side length $b =$ 150 mm) having different MAS under different strain rates. As displayed in Fig. 5.34, there are also some differences between specimens having different MAS. Cubic specimens having aggregate particles with larger sizes have more tortuous splitting-tensile cracks near the central line of specimen under low strain rates, which is consistent with the test results of Du et al. (2017). Furthermore, under high strain rates (e.g. $\dot{\varepsilon} = 10$/s), there are more aggregates getting damaged in the fracture surface of cubic specimens having larger-sized aggregates.

Load (*P*)—displacement (δ) curves of cubic models with different side lengths having different MAS under low strain rates and high strain rates are also depicted in Fig. 5.35. The *P*-δ curves of cubic specimens (with same side length) having different MAS are basically similar, and there are only some differences in the peak

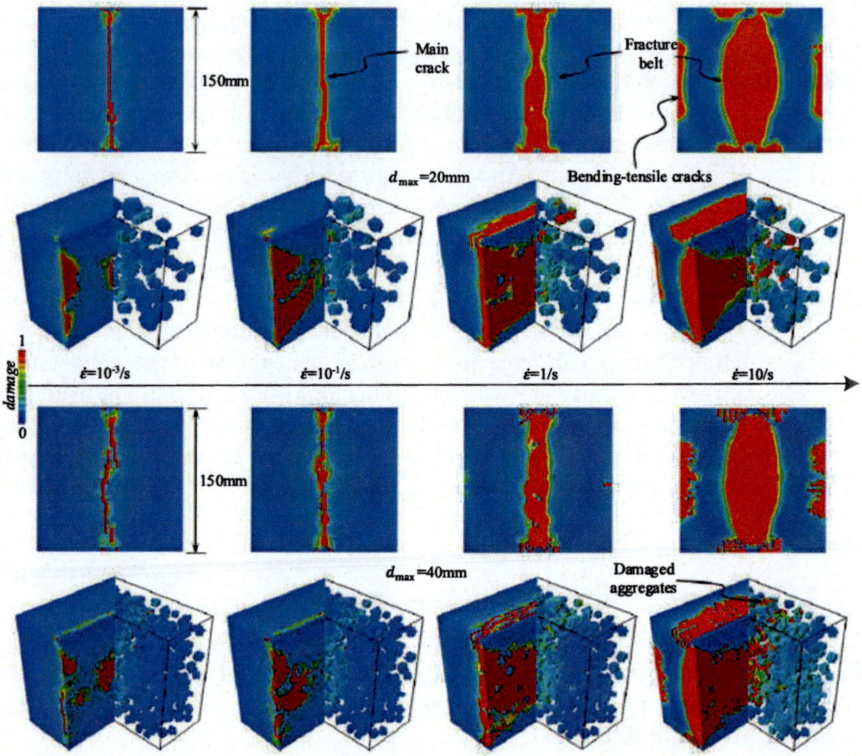

Fig. 5.34 Typical final failure patterns of concrete specimens having different MAS under dynamic loads

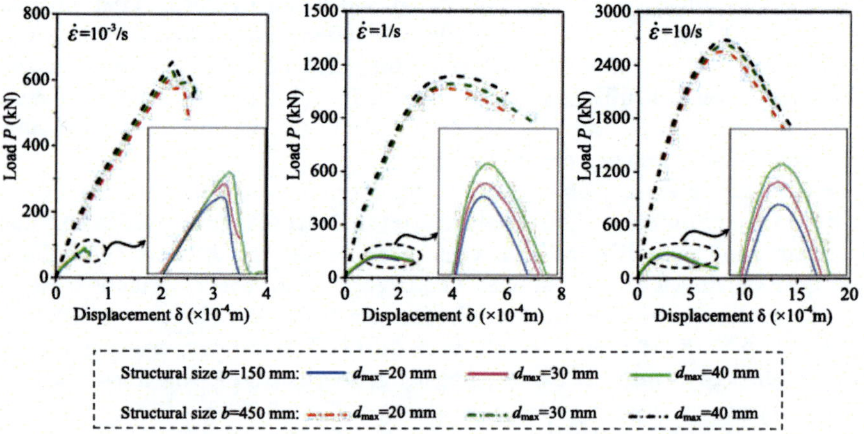

Fig. 5.35 Load (*P*)—displacement (*δ*) curves of concrete specimens having different MAS (b = 150 mm and 450 mm) under dynamic loads

loads under the same strain rate. As visible in Fig. 5.35, the peak load of specimens with same structural size enhances with the increasing MAS under both low and high strain rates.

Figure 5.36 presents the size effect on dynamic splitting-tensile strength of concrete having different MAS under different strain rates. It can be found that the MAS has an important influence on the dynamic splitting-tensile strengths and the corresponding dynamic size effect: (1) Under low strain rates, particularly under the quasi-static loads, most cracks generate in the region with weak mechanical properties and have relatively enough time to bypass aggregate for expending. The aggregate particles having different sizes provide different degrees of obstruction for cracks to develop. The lager the aggregate size is, the stronger the degrees of obstruction is, causing that the splitting-tensile cracks of cubic specimens with larger MAS have more tortuous and longer paths. There are more amount of aggregate particles and more defects in the concrete specimens having aggregates with smaller MAS, and their failure patterns are more brittle. Therefore, under the same strain rate, the

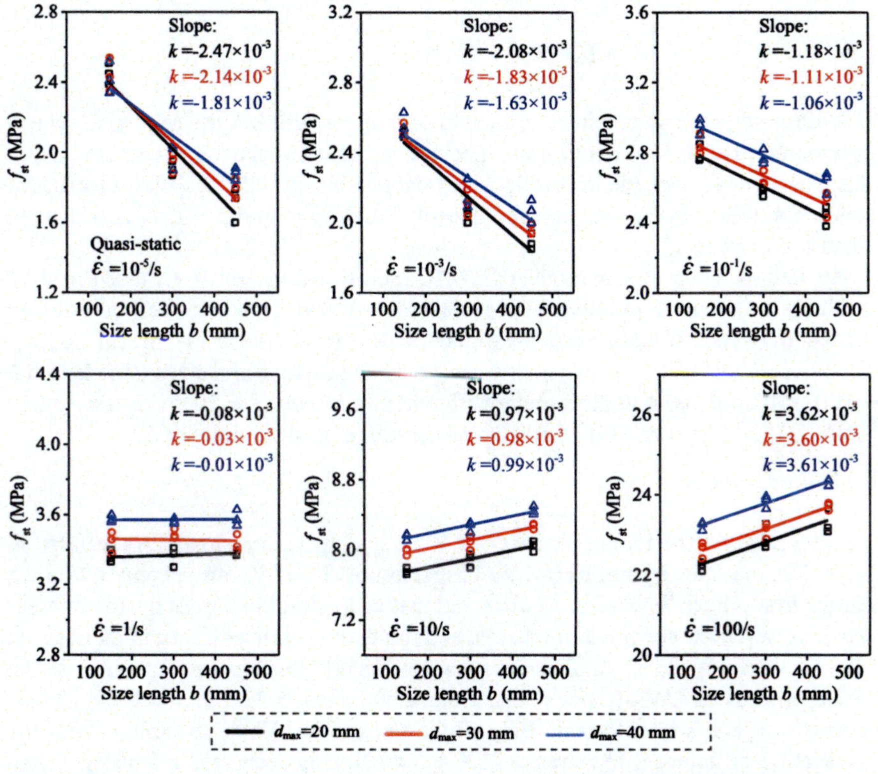

Fig. 5.36 Size effect on dynamic splitting-tensile strength of concrete specimens having different MAS under different strain rates

splitting-tensile strengths of concrete enhance as the MAS increases and the corresponding size effect of concrete specimens having small(er) MAS is more obvious. (2) As the strain rate continues to increase, when the strain rate exceeds $\dot{\varepsilon}_{cr}$, the splitting-tensile failure pattern of cubic specimen absolutely changes. Under high strain rates, the specimens suffer from high-speed impact loads and the cracks have to expand in a short time. More and more aggregate particles are pulled off or crushed near the splitting surface. Compared with specimens having lager aggregate particles, the specimens with small aggregate particles have more damaged aggregates and have more opportunity for the stress redistribution. Furthermore, there are more tortuous cracks and branching cracks in specimens having larger aggregates. That is to say, the paths of cracks in specimens having larger aggregates are more complex and there is more damaged area near the splitting surface, which can be seen from the failure patterns shown in Fig. 5.34. These factors lead to the results that splitting-tensile strengths enhance with the increasing MAS in cubic specimens under high strain rates.

5.4.3 Influence of Aggregate Type

In addition to aggregate content and maximum aggregate size, the aggregate type is an important factor for affecting the mechanical failure behavior of concrete materials. Here in this part, the lightweight aggregate concrete (LWAC) was selected for discussion, which is become an irreplaceable building material widely used for the construction industry.

Applied the compressive model of LWAC established in Sect. 4.4.1 (see Fig. 4.30 and Table 4.6), the compressive failure behavior of LWAC and the corresponding size effect under dynamic loads were investigated at the meso-scale level (Jin et al. 2020a). A total of 72 LWAC cubic specimens having different structural sizes (side length b = 100, 150, 300 and 450 mm) under different strain rates ($\dot{\varepsilon}$= 10^{-5}/s (quasi-static), 10^{-3}/s, 10^{-1}/s, 1/s, 10/s, 100/s and 200/s, respectively) were simulated.

1. *Failure pattern*

Figure 5.37 gives the failure process of LWAC under dynamic loads. As detailed in Fig. 5.37a, the failure patterns of LWAC specimens (b = 150 mm) changes with the loading time. It can be obviously observed that as loading time goes by, the number of cracks in LWAC specimen increases, and the damage intensifies until its ultimate failure occurs. Figure 5.37b plots the stress–strain relationship corresponding to the failure process of LWAC specimens. Comparing Figs. 5.37a, b, when the LWAC reaches the peak stress, there is only slight damage in LWAC specimen. With the aggravation of damage, the stress of LWAC specimens decreases gradually. When the stress drops down to the point c in Fig. 5.37b, the stress of concrete almost drops to the minimum value and the specimens form ultimate failure patterns.

2. *Stress–strain relationship*

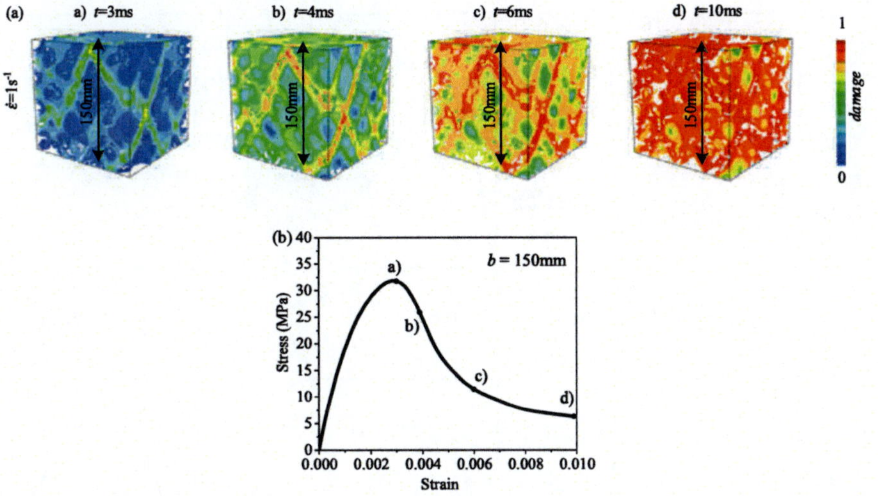

Fig. 5.37 Compressive failure of LWAC under the strain rate of 1/s: **a** failure process of LWAC specimens; **b** the corresponding stress–strain curve

Figure 5.38 plots the numerical stress–strain curve of LWAC specimens having different structural sizes under different strain rates. Under low strain rates (i.e. $\dot{\varepsilon}= 10^{-5}$/s, 10^{-3}/s and 10^{-1}/s), the compressive strengths of LWAC with different structural sizes have obvious differences. For the case with the same strain rate, the compressive strength of LWAC decreases with the increasing structural size, especially under the strain rate of $\dot{\varepsilon}$ $=10^{-5}$/s. When the applied strain rate reaches 1/s, the peak stresses of LWAC specimens with different structural sizes are almost the same. This illustrates that the size effect on the compressive strength of the LWAC under dynamic loads is completely inhibited. Under high strain rates, the stress–strain relationships of LWAC with different structural sizes are almost similar.

3. *Dynamic compressive strength*

Figure 5.39 depicts the relationship between the dynamic compressive strength and the structural size of LWAC under different strain rates. It can be found that: (1) as mentioned above, with the increasing strain rate, the compressive strength of LWAC increases obviously. (2) When the applied strain rate reaches 1/s, the strength degradation slope k is about zero. This phenomenon indicates that under the critical strain rate, the compressive strength of LWAC hardly changes with the structural size, and the corresponding size effect on compressive strength of LWAC is inhibited. Under low strain rates, the compressive strength of LWAC decreases with the structural size increases. The strength degradation slope k decreases with the increasing strain rate, meaning that the size effect on compressive strength of LWAC is gradually inhibited with the increasing strain rate. Under high strain rates, the slope k fluctuates around

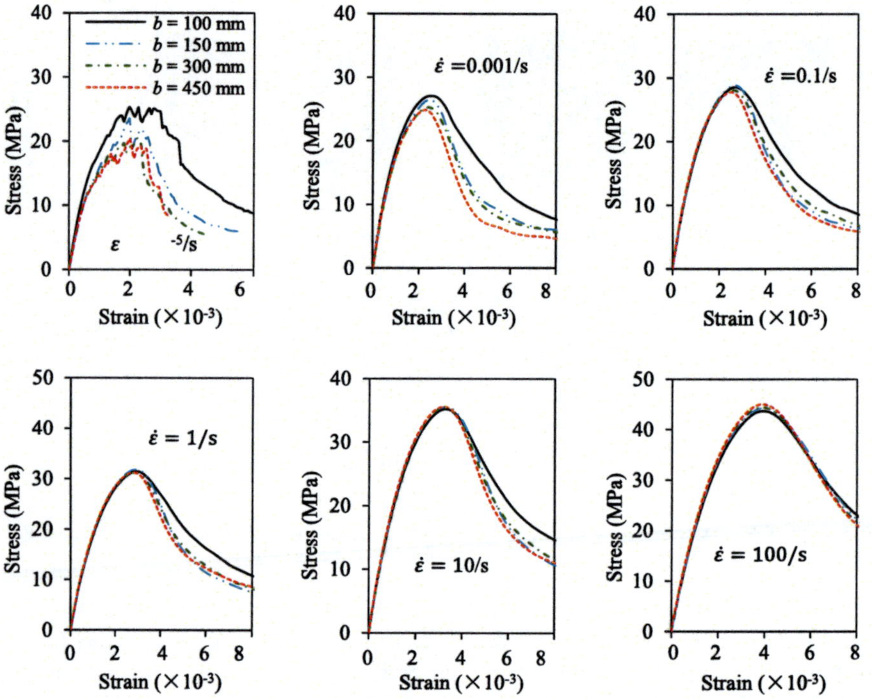

Fig. 5.38 Compressive stress–strain curve of LWAC under different strain rates

zero regardless of the change of strain rate, that is to say the compressive strength of LWAC remains almost unchanged.

The relationship between compressive strength of LWAC and the logarithm of strain rate is visible in Fig. 5.40. Overall, the compressive strength of LWAC increases as the strain rate increases. However, the amplitude of enhancement in compressive strength is different for the LWAC with different structural sizes. The larger the structural size, the larger the amplitude of increase in the dynamic strength with the strain rate increases. This indicates that the large-sized LWAC specimens have stronger sensitivity to strain rate than small-sized ones because of the inertial effect, which is consistent with the conclusion of Wang et al. (2018).

5.5 Influence of Initial Loads

In addition to withstanding static loads, most concrete materials and engineering structures inevitably suffer from dynamic loads, e.g. wind loads, seismic loads and sudden dynamic loads (explosions, shocks, etc.). These dynamic loads become the most critical control factors in concrete structure design due to their unpredictability

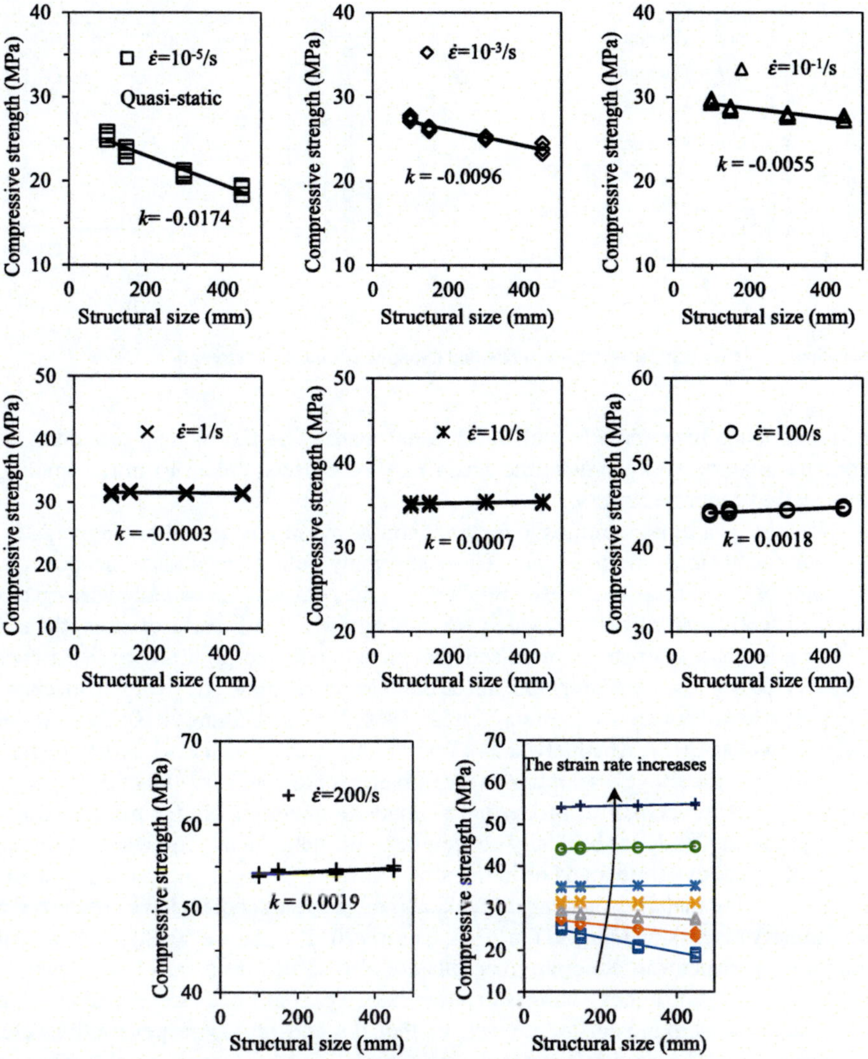

Fig. 5.39 Size effect on dynamic compressive strength of LWAC under different strain rates

and powerful destructiveness. Concrete is a strain rate sensitive material. Lots of scholars have carried out numerous physical experimental studies (Li et al. 2018; Chen et al. 2013; Erzar and Forquin 2011; Yan and Lin 2006) on dynamic mechanical properties of concrete materials since Adrams (1917) first researched the dynamic compression of concrete materials. However, most of these studies were conducted without initial static loads or initial dynamic loads. In fact, concrete engineering structures often suffer from multiple dynamic loads, or prior to being subjected to

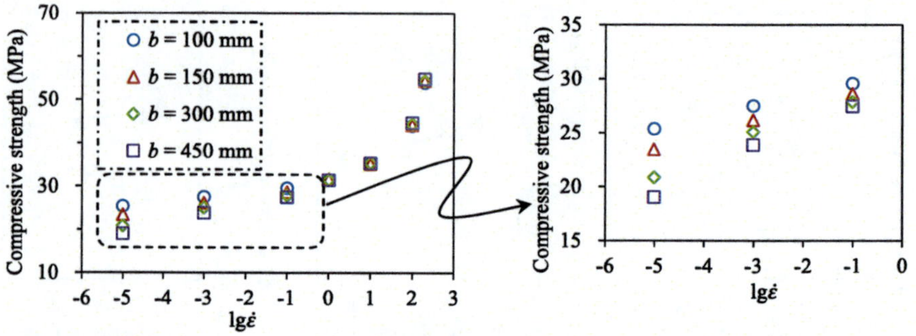

Fig. 5.40 Relationship between strain rate and dynamic compressive strength of LWAC

dynamic loads have already withstood initial loads. The mechanical properties of concrete materials under dynamic loads may be closely linked to initial loading history that concrete suffers.

Concrete is a brittle composite material composed of coarse and fine aggregates, cement and various additives, etc. There are many defects such as micro-pores in the interior of concrete and these initial defects may cause damage and micro cracks under different loading history, which presents significant influences on strengths and deformation characteristics. Some scholars have conducted preliminary researches on the effects of loading history and initial damage on the static and dynamic mechanical properties of concrete materials (Kaplan 1980; Xiao and Zhang 2010; Yan and Lin 2008; Yan et al. 2017; Khandelwal and Ranjith 2013, 2017; Fan et al. 2017). Generally speaking, the above researches on the influence of load history and initial damage on the dynamic mechanical properties of concrete materials are far from consummate and initial loads are basically concentrated on static loads. The research on the effect of initial dynamic loads or initial strain rates on dynamic mechanical properties of concrete materials is almost blank, and only a few scholars have conducted a preliminary discussion (Jin and Du 2014; Cusatis 2011; Bažant et al. 2000). In actual engineering structures, concrete materials are also subject to a variety of dynamic loads or encounter sudden changes in strain rate, e.g. continuous wave load or wind load, continuous explosion or impact, earthquake and projectile penetration, etc. Therefore, it is of great importance to investigate an in-depth study on the effect of different initial loading strain rates (especially initial dynamic loads) on dynamic mechanical behavior of concrete materials and the corresponding size effect.

5.5.1 *Dynamic Compressive Failure*

In order to investigate the effects of initial static load and initial dynamic load on dynamic compressive failure of concrete materials, firstly, the dynamic compressive failure behavior of concrete cubic specimens under different strain rates ($\dot{\varepsilon}= 10^{-5}$/s

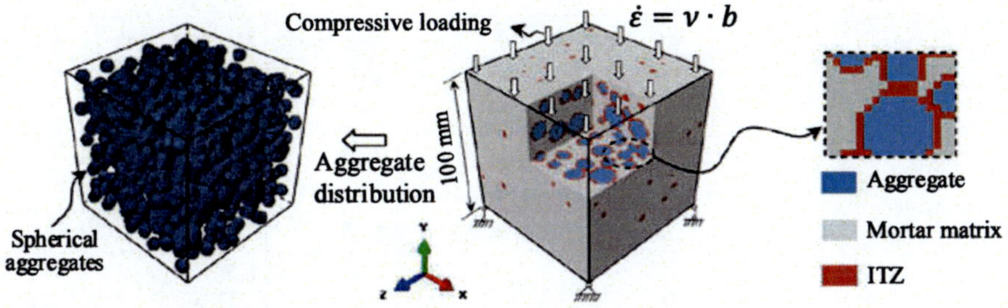

Fig. 5.41 3D meso-scale numerical model of concrete cubic specimen for compression

(quasi-static), 10^{-4}/s, 10^{-3}/s, 10^{-2}/s, 10^{-1}/s, 1/s, 10/s and 100/s) were simulated based on 3D meso-scale mechanical modelling method. Secondly, the dynamic compressive failure behavior of concrete cubic specimens with three kinds of initial static loads (f_0= 30%f_{sc}, 60%f_{sc}, and 90%f_{sc}) and seven kinds of changed strain rate $\dot{\varepsilon}_2$ (from 10^{-4}/s to 100/s) were also simulated. Then the dynamic compressive failure behavior of concrete cubic specimens with six kinds of initial dynamic loads and three kinds of changed strain rate $\dot{\varepsilon}_2$ (10^{-1}/s, 10/s and 100/s) were simulated. (Jin et al. 2020c). Herein the applied load (or strain rate) at the beginning is defined as initial load (or initial strain rate $\dot{\varepsilon}_1$); the applied load (or strain rate) changed later is defined as changed load (or changed strain rate $\dot{\varepsilon}_2$). Particularly, the applied strain rate $\dot{\varepsilon}$= 10^{-5}/s at the beginning is defined as initial static strain rate $\dot{\varepsilon}_{st}$. The peak stress under $\dot{\varepsilon}_{st}$ is defined as f_{sc}. The static stress as load starts to change is defined as f_0. The peak stress (strength) under initial strain rate $\dot{\varepsilon}_1$ is defined as f_{c1}; the dynamic compressive stress as load starts to change is defined as f_1. In all cases $= f_1 60\% f_{c1}$. Furthermore, the cases that dynamic load suddenly increases at f_1 in hardening stage (before reaching peak strength f_{c1}) and in softening stage (after reaching peak strength f_{c1}) are considered. There are three concrete specimens in each loading case and a total of 147 specimens has been simulated.

Figure 5.41 depicts 3D meso-scale numerical model of concrete cubic specimen (side length of b = 100 mm) for compression in this section and coarse aggregate particles were modeled as spheres with random distribution enfolded by ITZ. The volume fraction of aggregate particles is about 45% and two-graded aggregate concrete was adopted with aggregates with diameter 8 mm and 16 mm. The numerical model was meshed with an 8-nodded hexahedral element (type C3D8R: Continuum, 3-D, 8-node, reduce integration) and the average element size is approximately1 mm. Loading and boundary conditions of in meso-scale models could also be observed in Fig. 5.41. The loading method and boundary condition of concrete cubic model are consistent with those in Fig. 5.36.

In addition, the damaged plasticity model described in Sect. 3.3.1 of Chap. 3 has been also adopted to simulate the mechanical failure behavior of concrete three-phase components (aggregate, mortar matrix and ITZ). The mechanical parameters of three-phase components set in meso-scale simulation are listed in Table 5.4. Only

Table 5.4 Mechanical parameters of concrete meso-components

	Aggregate	Mortar matrix	ITZ
Young's modulus E (GPa)	40[d]	24[d]	18[c]
Poisson ratio	0.20	0.20	0.22
Eccentricity η (%)	0.1	0.1	0.1
Stress ratio $f_{b0} = f_{c0}$	1.16	1.16	1.16
K	0.667	0.667	0.667
Fracture energy G_f (N/mm)	0.128[b]	0.049[b]	0.025[b]
Compressive strength σ_c (MPa)	80.0[a]	20.0[a]	15.0[c]
Tensile strength σ_t (MPa)	6.0[a]	2.0[a]	1.5[c]

Note [a]Data were referenced to the tests of Yan and Lin (2006); [b]Data were referenced to CEB-FIP MC90 (1993); [c]Data were set as 75% mechanical parameters of mortar matrix according to Song and Lu (2012); and [d]Data were determined by the inversion method

the strength enhancement behavior was considered and the strain rate effect for meso-components on material strength was incorporated into the present plastic damage model. The CDIF and TDIF recommended by Hao et al. (Hao 2011, 2013), which have removed the inertia confinement effect, were adopted for modelling the strain rate effect of meso-components. Detailed modeling introduction can refer to the study of Jin et al. (2020c).

Yan and Lin's test results (Yan and Lin 2006, 2008) were adopted for validation, which is given in Fig. 5.42. It can be found that stress–strain curves of meso-scale simulation basically match well with ones of test, especially the uniaxial compressive peak stress (strength). The comparison and verification have shown that the modelling method could yield relatively reliable predictions of the mechanical behavior of concrete materials under dynamic loads.

Figure 5.43 reports dynamic compressive stress–strain curves obtained from simulation under different strain rates. As it can be seen, increasing loading strain rates can enhance the peak compressive stress (i.e. compressive strength) and its corresponding peak strain, showing that there is an obvious rate-dependent behavior in compressive stress–strain curves of concrete materials, especially under high strain rates.

The rate-dependent behavior can be also clearly found in dynamic compressive failure patterns, which is displayed in Fig. 5.44. As detailed in Fig. 5.44, more and more damage appears in the concrete cubic specimens with the increasing strain rate. Under quasi-static load ($\dot{\varepsilon}= 10^{-5}$/s), the stress inside concrete specimen grows relatively slowly. The micro-cracks have sufficient time to bypass aggregate particles and expand along areas with weaker mechanical properties. There are several macro main cracks (oblique cracks) appear in concrete specimen surface. As the strain rate

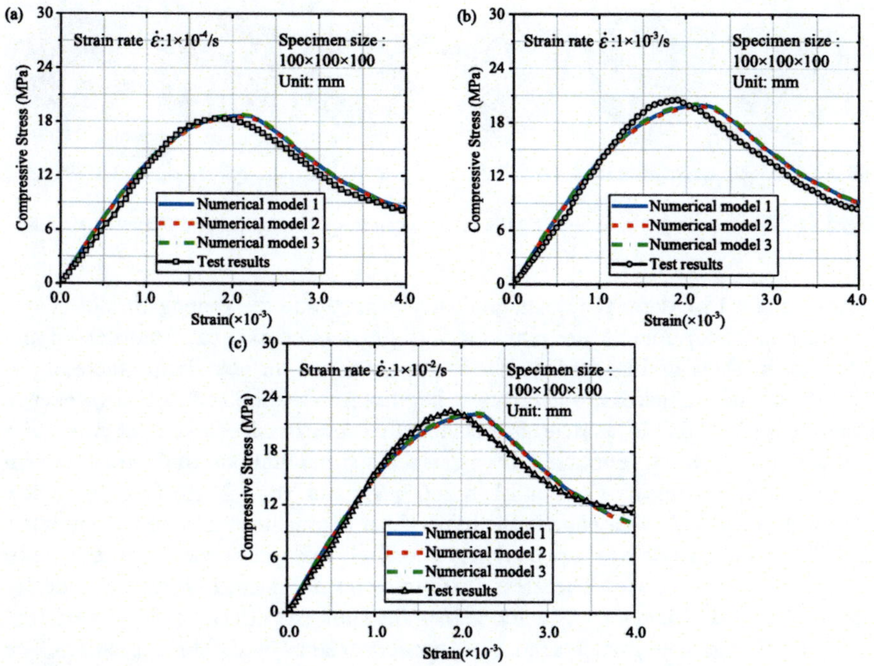

Fig. 5.42 Comparison of stress–strain curves obtained from simulation and test results (Yan and Lin 2008): **a** $\dot{\varepsilon}=10^{-4}$/s; **b** $\dot{\varepsilon}=10^{-3}$/s; and **c** $\dot{\varepsilon}=10^{-2}$/s

Fig. 5.43 Typical compressive stress–strain curves under different strain rates

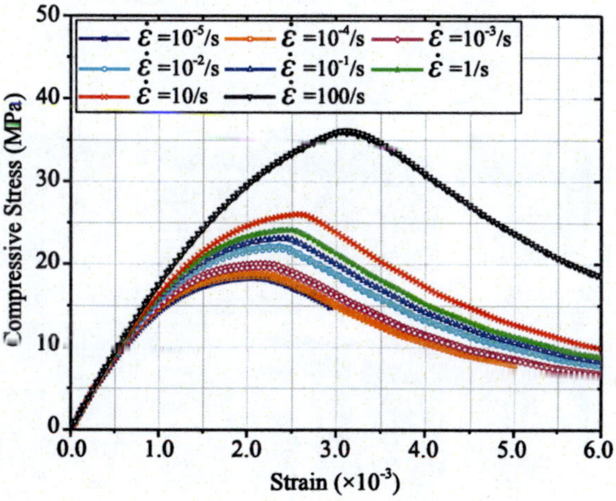

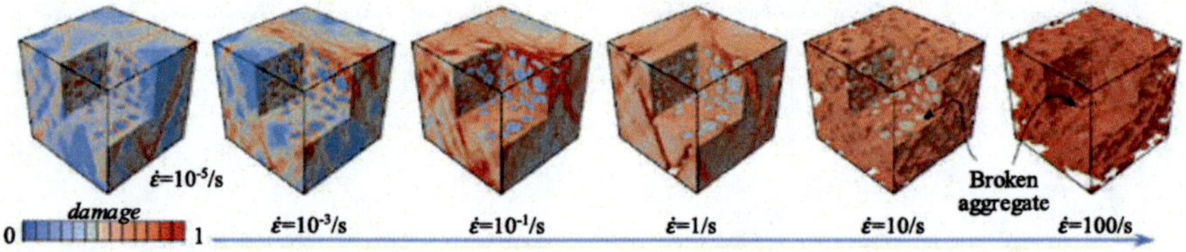

Fig. 5.44 Typical compressive failure patterns of concrete specimens under different strain rates

increases ($\dot{\varepsilon}$< 1/s), damaged areas gradually expand and the propagation of cracks becomes more tortuous. More main cracks (diagonal cracks) can be observed and cross cracks show up like spider webs on the specimen surface. Furthermore, crack width continues to increase and concrete begins to be spalled near the surface, even concrete is crushed in local areas. It is evident that compressive failure patterns under low strain rates ($\dot{\varepsilon}$< 1/s) obtained by meso-scale simulation are in good agreement with existing test results (Yan and Lin 2006; Shi and Song 2018) (see Fig. 5.45). Moreover, there is few damage in the majority of aggregates under low strain rates. In addition, compressive failure patterns of concrete materials under high strain rates ($\dot{\varepsilon}$> 1/s) have changed a lot. The stress distribution inside concrete specimen gradually becomes uneven. Many cracks extend along the shortest path (for releasing energy) and pull off aggregate particles in a short time, instead of bypassing aggregates. The areas where have reached ultimate strength (or damage) at the same time significantly increases. More and more concrete is spalled near the surface, even concrete in local areas is almost smashed into small pieces. These factors may commonly result in that compressive strength of concrete materials get remarkable enhancement under high strain rates.

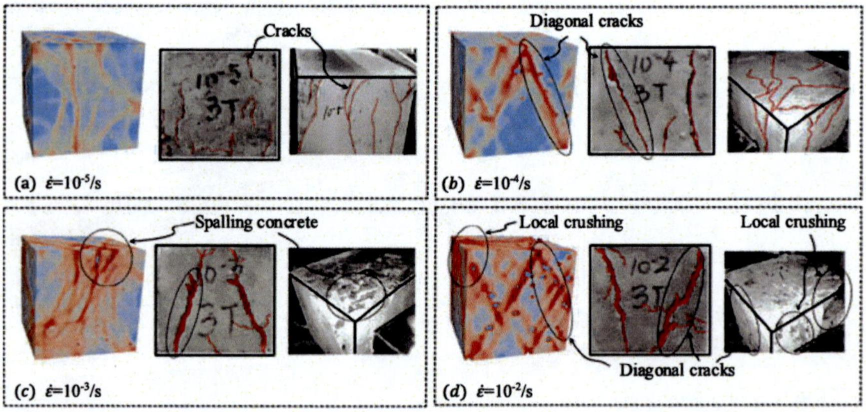

Fig. 5.45 Comparison of compressive failure patterns under low strain rates obtained by meso-scale simulation and existing test results (Shi and Song 2018; Yan and Lin 2006): **a** $\dot{\varepsilon}=10^{-5}$/s; **b** $\dot{\varepsilon}=10^{-4}$/s; **c** $\dot{\varepsilon}=10^{-3}$/s; and **d** $\dot{\varepsilon}=10^{-2}$/s

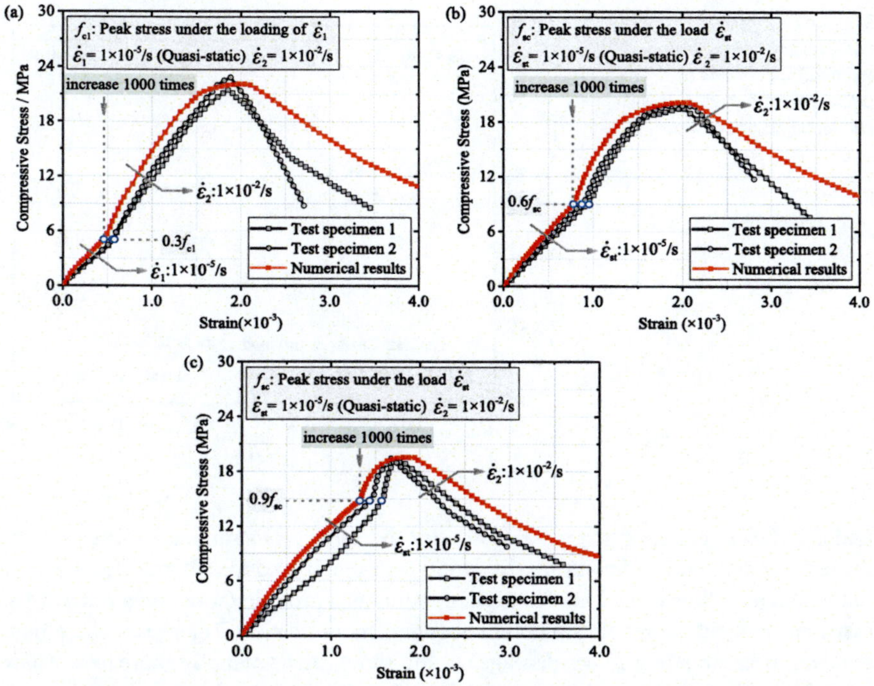

Fig. 5.46 Comparison of dynamic compressive stress–strain curves with different initial static stresses obtained by present simulation and test results of Yan and Lin (2008): **a** initial stress: $30\%f_{sc}$; **d** initial stress: $60\%f_{sc}$; and **c** initial stress: $90\%f_{sc}$

1. *Effect of initial static load*

Figure 5.46 compares dynamic compressive stress–strain curves with different initial static stresses ($f_0 = 30\%f_{sc}$, $60\%f_{sc}$, and $90\%f_{sc}$) obtained by meso-scale simulation and test results of Yan and Lin (2008). One can observe that compressive stress–strain curves obtained from simulation are basically in line with ones obtained from tests. Furthermore, when changed strain rate $\dot{\varepsilon}_2$ ($=10^{-2}$/s) begins to replace static load, the slope in stress–strain curve has changed, i.e., tangent modulus has a sudden increase. This is consistent with research results obtained in the literature (Khandelwal and Ranjith 2013).

Figure 5.47 summaries dynamic compressive strengths under the combined loads of different initial static stresses and changed strain rates. Data in Fig. 5.47 suggest that dynamic compressive strength under different changed strain rates gradually decreases with the increasing initial static load. This may be explained by that different initial static loads have a plastically unrecoverable effect on the development of micro-cracks in concrete, which making the internal structure of concrete damaged. This could lead to deterioration in dynamic mechanical properties of concrete materials. However, the decreasing magnitudes of compressive strength

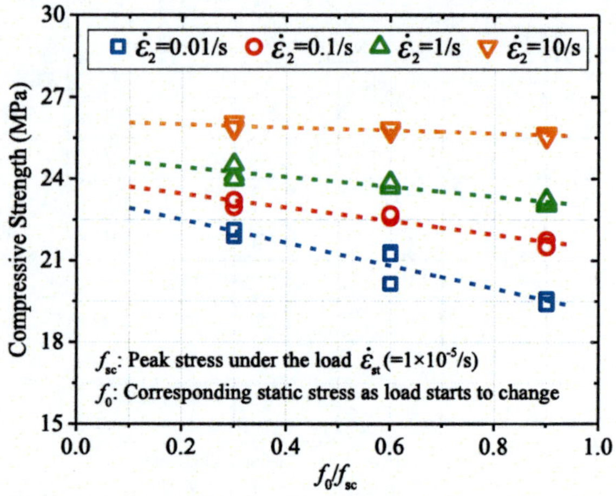

Fig. 5.47 Dynamic compressive strength under the combined loads of different initial static stresses and changed strain rates

under different changed strain rates are not the same. As strain rate increases, the decreasing magnitude of compressive strength gradually reduces. It may be attributed to the fact as follows: the loads of higher strain rates provide more energy and stress distribution inside concrete specimen becomes more uneven. The areas where have reached ultimate strength (or damage) at the same time gradually increases. These factors could weaken the influence of initial static load on dynamic compressive strength of concrete. The above findings are in general agreement with experimental studies (Yan et al. 2017; Chen et al. 2016; Xiao and Zhang 2010).

Figure 5.48 plots dynamic compressive strengths under the combined loads of initial static stress ($f_0 = 60\% f_{sc}$) and different changed strain rates. It is evident from details in Fig. 5.48 that dynamic compressive strength enhances as strain rate

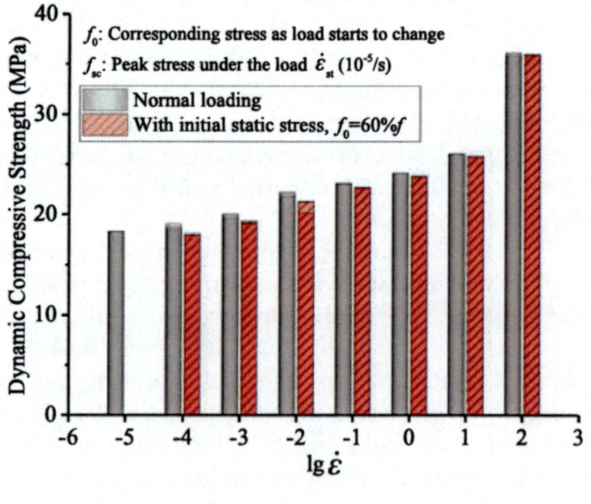

Fig. 5.48 Dynamic compressive strength under the combined loads of initial static stress ($f_0 = 60\% f_{sc}$) and different changed strain rates

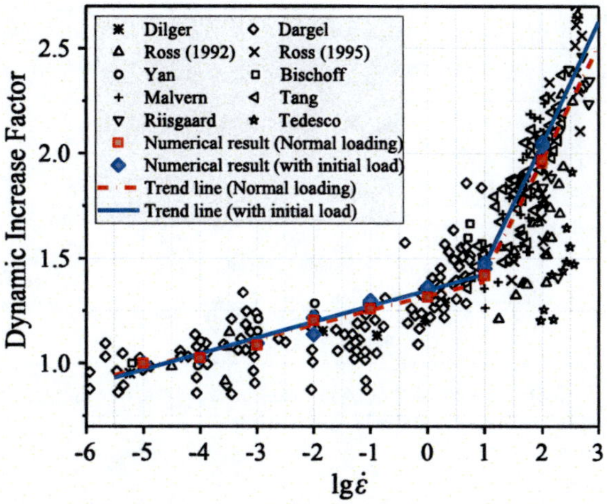

Fig. 5.49 CDIF of compressive strength under the combined loads of initial static loads ($f_0 = 60\% f_{sc}$) and different changed strain rates

increases both under the loads of without initial static stress (normal loads) and with initial static stress. Especially under high strain rates, dynamic compressive strength is significantly sensitive to strain rate and get an obvious enhancement. In addition, dynamic compressive strength obtained without initial static stress is higher than one obtained with initial static stress. However, the gap between them decreases with the addition of strain rate, even under high strain rates ($\dot{\varepsilon}$> 1/s) the initial static load almost has no effect on dynamic compressive strength.

CDIF of compressive strength under the combined loads of initial static loads ($f_0 = 60\% f_{sc}$) and different changed strain rates is given in Fig. 5.49. Comparing numerical results and experimental results one can note that the numerical data points of CDIF are basically within the range of existing test results. Moreover, the trend line of CDIF with initial load is above the one without initial load, illustrating that initial load could enhance the strain rate effect in compressive strength of concrete materials, which is consistent with the conclusion in the study of Yan et al. (2017).

2. *Effect of initial dynamic load*

Typically, effects of sudden increase in strain rate on compressive stress–strain curves are depicted in Fig. 5.50. It is apparent to seen that significant changes in compressive stress–strain curves of concrete have happened as strain rate suddenly increases. When dynamic load changes in hardening stage, the slope in stress–strain curve has changed, i.e. tangent modulus has a sudden increase. The compressive stress reaches a new peak f_{c2} strength under changed strain rate, which is obviously higher than f_{c1}. The corresponding peak strain has been greatly improved. On the other hand, a sudden increase of dynamic load in softening stage reverses the post-peak softening to post-peak hardening. The re-hardening is then followed by a second peak strength (the second peak strength is defined as f_{c3}) and develops a new curve which is approaching the stress–strain curve obtained under the combined loads of $\dot{\varepsilon}_1$ and $\dot{\varepsilon}_2$. The second peak strength f_{c3} may be lower or higher than first peak strength

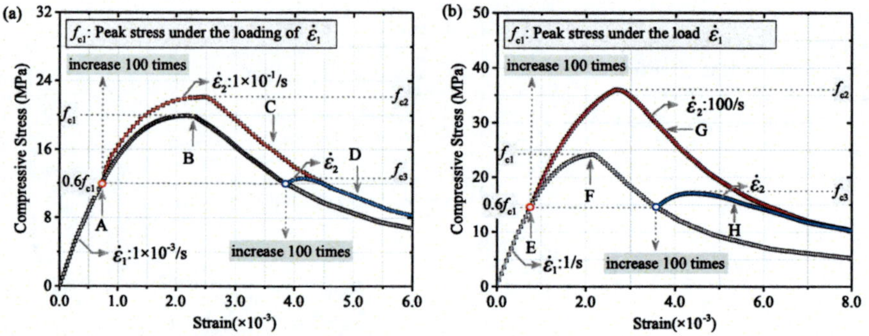

Fig. 5.50 Effects of sudden increase in strain rate on compressive stress–strain curves: **a** under low strain rates; and **b** under high strain rates

f_{c1}, which is depending on the changed strain rate and the residual stress f_1 as load starts to change. These findings are generally in line with the studies in literatures (Tandon et al. 1995; Jin and Du 2014).

Figure 5.51 shows failure patterns of concrete cubic specimens corresponding to different stages according to Fig. 5.50. Linking Stage ABCD or Stage EFGH one can observe the development process of dynamic compressive damage in concrete cubic specimen under different strain rates. Under low strain rates (see Stage ABC) damage first appears in the areas with weaker mechanical properties (like ITZs), and cracks bypass aggregate particles. As the stress increases many micro-cracks expand and confluence to form several macro main cracks (oblique cracks) in concrete specimens. As detailed in the Stage EFG, there is much damage in ITZ phase and some aggregates are pull off at the beginning (see Stage E). As strain rate changes to high strain rate, the ITZ phase and most of the aggregate particles have gotten damaged (see Stage GH), even plastic damage occurs in entire cubic specimen. The majority of concrete

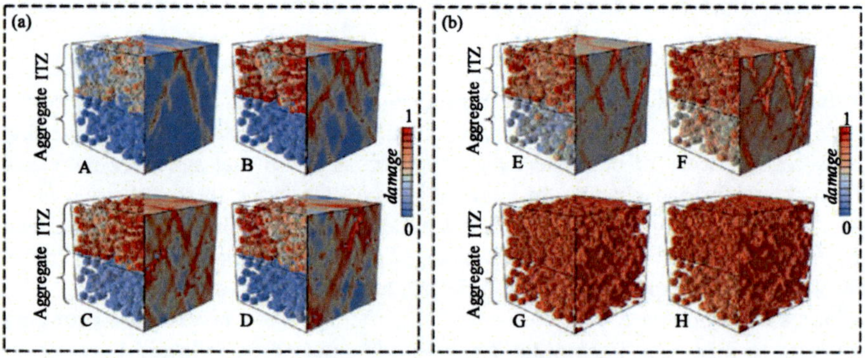

Fig. 5.51 Failure patterns of concrete cubic specimens corresponding to different stages according to Fig. 5.50: **a** under low strain rates; and **b** under high strain rates

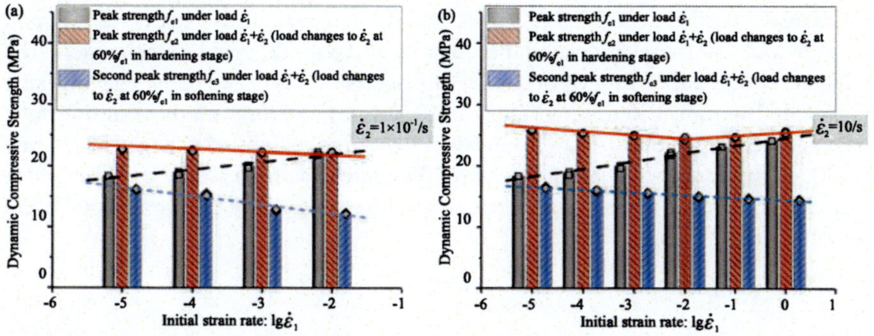

Fig. 5.52 Dynamic compressive strength under different initial dynamic loads and changed strain rates: **a** changed strain rate $\dot{\varepsilon}_2$=0.1/s; **b** changed strain rate $\dot{\varepsilon}_2$=10/s

is spalled or smashed into small pieces near the surface. Comparing Stage B and Stage CD can be found that sudden increase in strain rate provides more energy for crack propagation. The number and width of cracks reach a rapid increase and much more damage shows up in concrete internal areas. Moreover, the failure pattern in Stage C is more serious than one in Stage D. This may be link with following factors: when dynamic load suddenly increases in post-peak softening stage macro-cracks continue to develop along the original extended path and some new cracks are generated in areas close to reach ultimate strength, which performs a re-hardening behavior. However, before strain rate suddenly increases, most of crack propagation is approaching the end and plastic damage generated results deterioration in mechanical properties of concrete materials. Therefore, the peak strength f_{c2} is obviously higher than the second peak strength f_{c3} and there is more damage in failure pattern of Stage C. Similar observation can be visible in Fig. 5.51b.

Typically, dynamic compressive strengths under different initial dynamic loads and changed strain rates are given in Fig. 5.52. Data in both Fig. 5.52a, b demonstrate that the second peak strengths f_{c3} gradually decrease with the increasing initial dynamic load. However, the peak strengths f_{c2} under low strain rates and under high strain rates are a little different. The peak strength f_{c2} shows a slight downward trend with the growth in initial dynamic load under low strain rates, which can be seen in Fig. 5.52a. When changed load is the load of high strain rates, the peak strength f_{c2} first decreases and then slightly increases as initial dynamic load increases, which can be seen in Fig. 5.52b. In addition, the second peak strengths f_{c3} are lower than the first strength f_{c1} in all cases.

5.5.2 *Dynamic Size Effect*

In order to investigate the influence of initial static load on dynamic size effect on compressive failure of concrete materials, based on the 3D meso-scale numerical model established in Sect. 5.5.1, the structural size (side length b) expands to 150 mm, 300 mm and 450 mm. Other modeling details of the model remain unchanged. The dynamic compressive failure behavior of concrete cubic specimens (having side lengths of 100 mm, 150 mm, 300 mm and 450 mm) with four kinds of initial static loads (f_0= 20%f_{sc}, 40%f_{sc}, 60%f_{sc} and 80%f_{sc}) and two kinds of changed strain rate $\dot{\varepsilon}_2$ (10^{-2}/s represents low strain rate and 10/s represents high strain rate) were also simulated. There are three concrete specimens in each loading case and a total of 96 specimens has been simulated.

1. *Effect of Different Initial Static Loads*

Figure 5.53 gives the influence of initial static load on dynamic compressive stress–strain curves of cubic specimens having different side lengths under low and high strain rates ($\dot{\varepsilon}_2$=10^{-2}/s and 10/s). It can be seen that when concrete specimens (having different side lengths) with different initial static loads suddenly suffer dynamic loads, the slopes in stress–strain curves (tangent modulus) have a sudden increase. Under low strain rate, there are some differences in the peak stress and corresponding peak strain in dynamic compressive stress–strain curve of concrete with different initial static loads. As initial static load increases, the peak stress decreases while the peak strain increases. This phenomenon happens in concrete having different structural sizes. Under high strain rate, no matter what structural size of the concrete, the influence of initial static load on the peak stress and corresponding peak strain in dynamic compressive stress–strain curve of concrete can be negligible.

Figure 5.54 summaries the relationship of initial static load f_0 and dynamic compressive strength of cubic specimens having different side lengths under different strain rates. One can note that the influence of initial static load on dynamic compressive strength is different under low strain rate and high strain rate. Under low strain rate, the dynamic compressive strength decreases as initial static load withstood by concrete increases. This downward trend of dynamic compressive strength becomes greater in concrete specimen with larger structural size. However, under high strain rate, the dynamic compressive strength has little to do with the static load withstood by concrete.

Figure 5.55 presents the relationship of structural size (side length) and dynamic compressive strength of concrete specimens withstanding different initial static loads under low strain rates and high strain rates. It can be found that under low strain rate, the dynamic compressive strength of concrete decreases with the increasing structural size, showing an obvious size effect. Under high strain rate, the dynamic compressive strength of concrete increases with the increasing structural size, showing a reverse size effect. These findings are consistent with the conclusions in Sect. 5.3. In addition, the initial static load affects the dynamic size effect on compressive strength under low strain rates. With the addition of initial static load withstood by concrete, the downward trend of dynamic strength decreasing with the increasing structural size becomes greater and the dynamic size effect is more obvious. As mentioned in Sect. 5.3, the size effect is weakened gradually with the increasing strain rate under low strain rates. The addition of initial static load withstood by concrete can postpone

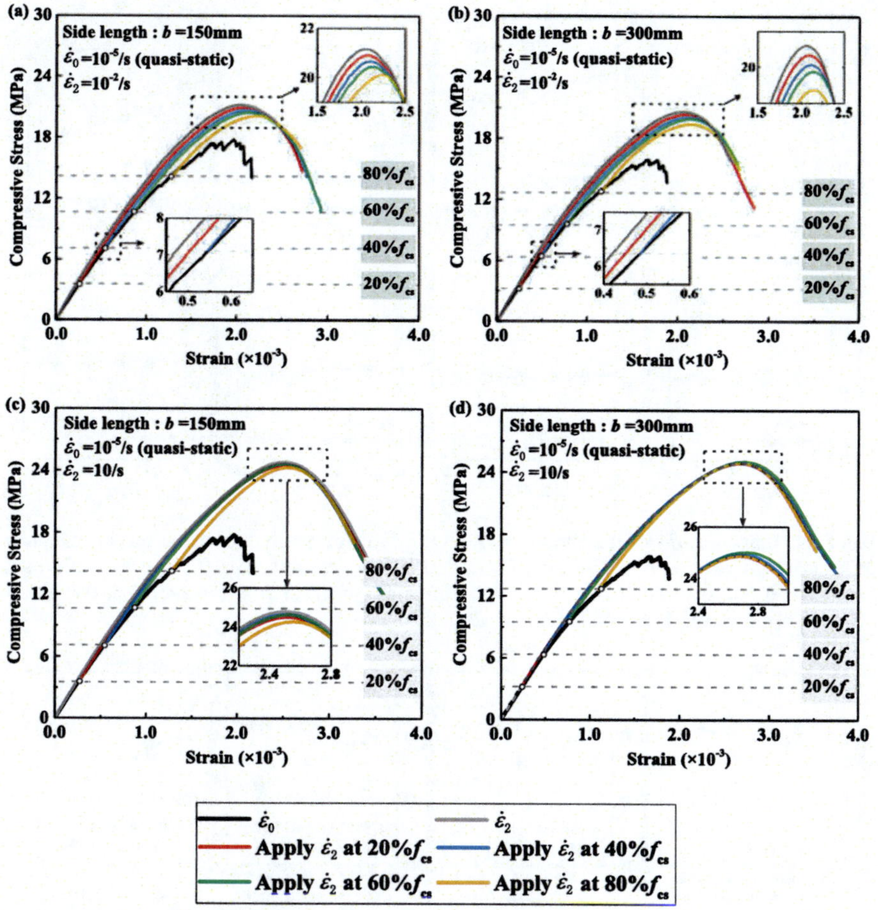

Fig. 5.53 Influence of initial static load f_0 on dynamic compressive stress–strain curves of cubic specimens: **a** b = 150 mm, $\dot{\varepsilon}_2$=10^{-2}/s; **b** b = 300 mm, $\dot{\varepsilon}_2$=10^{-2}/s; **c** b = 150 mm, $\dot{\varepsilon}_2$=10/s; **d** b = 300 mm, $\dot{\varepsilon}_2$=10/s

this weakening process. However, things become different under high strain rates. The influence of initial static load on dynamic size effect on compressive strength can be ignored under high strain rates.

5.6 Static-Dynamic Unified Size Effect Law

There are many theoretical analyses (Bažant and Planas 1997; Bažant 1984; Carpinteri and Ferro 1994; Weibull 1939) on size effect of concrete material based on different study frameworks (e.g. statistical theories of random strength, stress-redistribution and energy release, theory of crack fractality, etc.), which are widely

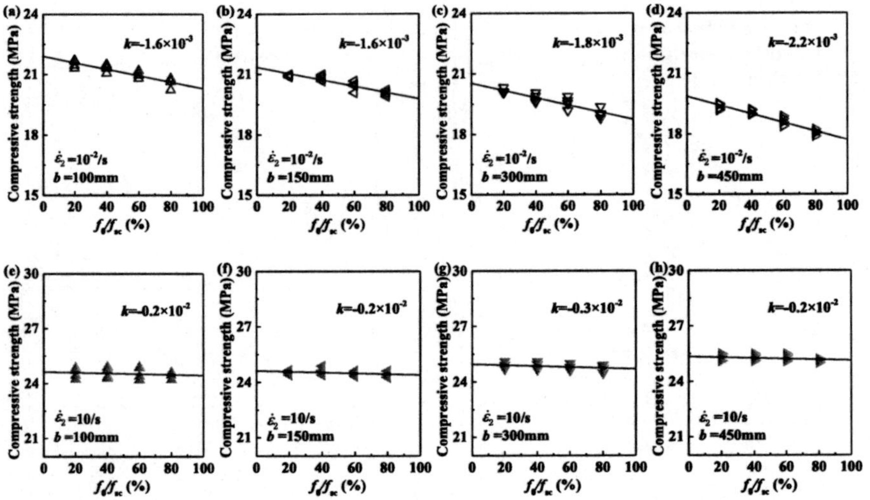

Fig. 5.54 Influence of initial static loads f_0 on the dynamic strength of cubic specimens under different strain rates: **a** b = 100 mm, $\dot{\varepsilon}_2$=10^{-2}/s; **b** b = 150 mm, $\dot{\varepsilon}_2$=10^{-2}/s; **c** b = 300 mm, $\dot{\varepsilon}_2$=10^{-2}/s; **d** b = 450 mm, $\dot{\varepsilon}_2$=10^{-2}/s; **e** b = 100 mm, $\dot{\varepsilon}_2$=10/s; **f** b = 150 mm, $\dot{\varepsilon}_2$=10/s; **g** b = 300 mm, $\dot{\varepsilon}_2$=10/s; and **h** b = 450 mm, $\dot{\varepsilon}_2$=10/s

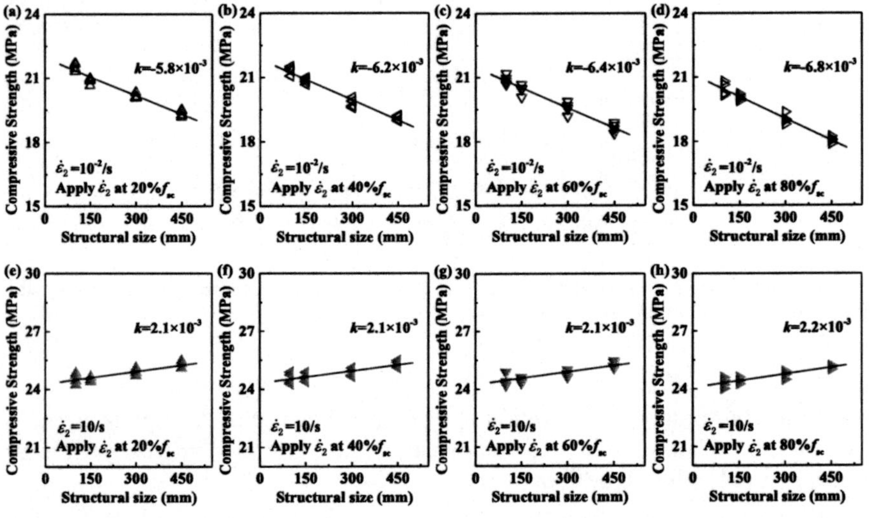

Fig. 5.55 Influence of initial static loads f_0 on the dynamic size effect of cubic specimens under different strain rates: **a** f_0/f_{cs} = 20%, $\dot{\varepsilon}_2$=10^{-2}/s; **b** f_0/f_{cs} = 40%, $\dot{\varepsilon}_2$=10^{-2}/s; **c** f_0/f_{cs} = 60%, $\dot{\varepsilon}_2$=10^{-2}/s; **d** f_0/f_{cs} = 80%, $\dot{\varepsilon}_2$=10^{-2}/s; **e** f_0/f_{cs} = 20%, $\dot{\varepsilon}_2$=10/s; **f** f_0/f_{cs} = 40%, $\dot{\varepsilon}_2$=10/s; **g** f_0/f_{cs} = 60%, $\dot{\varepsilon}_2$=10/s; and **h** f_0/f_{cs} = 80%, $\dot{\varepsilon}_2$=10/s

approbated by lots of scholars. Among these three theories, the size effect law (SEL) based on the linear elastic fracture mechanics proposed by Bažant (Bažant and Planas 1997; Bažant and Pang 2007; Bažant 2004) has been demonstrated to be able to well describe the size effect of concrete. Here in this section, *Type-2* size effect is also appropriate for characterizing the size effect on nominal strength obtained by above simulation and it has the following form:

$$\sigma_{\mathrm{Nu}} = \frac{Bf'}{\sqrt{1 + D/D_0}} \tag{5.7}$$

where σ_{Nu} represents the static nominal strength of concrete; D represents the structural size; and f' represents the nominal strength of concrete which is borrowed only for convenience (f_c' for compressive strength, f_t' for tensile strength and f_{st}' for splitting-tensile strength); B represents the dimensionless geometry-dependent parameter and D_0 represents the size-dependent parameter called transitional size. The B and D_0 are two empirical parameters to be identified by optimum fitting of measured σ_{Nu} values covering a broad enough size range.

Equation (5.7) can be written by mathematical transformation as:

$$\left(\frac{f'}{\sigma_{Nu}}\right)^2 = \left|\frac{D}{D_0 B^2}\right| + \frac{1}{B^2} \tag{5.8}$$

Equation (5.8) can be depicted as a linear form:

$$Y = AX + C \tag{5.9}$$

where $Y = (f'/\sigma_{\mathrm{Nu}})^2, X = D, A = 1/D_0 B^2$, and $C = 1/B^2$. A and C are obtained from a regression analysis based on numerical results.

According to the above numerical results of dynamic size effect on compressive strength in Sect. 5.3, the determined parameters of Bažant's *Type-2* SEL for compressive strength under different strain rates are listed in in Table 5.5. One can note from Table 5.5 that when the applied strain rate is less than the critical strain rate, with the increasing strain rate, the magnitude of the parameter B almost keeps constant, while the value of the parameter D_0 increase significantly. In reality, the parameter D_0 is strongly related to the applied strain rate. When the applied strain rate exceeds the critical strain rate, the parameter B increases as the strain rate increases while the parameter D_0 is no longer a positive value. As the strain rate increases, the absolute value of D_0 gets smaller. In other words, under high strain rates, Bažant's SEL cannot describe size effect behavior of dynamic compressive strength and it needs to be modified.

Based on the fitting parameters, the bi-logarithmic plots of $\log(\sigma_{\mathrm{Nu}}/B f_c')$ versus $\log(D/D_0)$ for static and dynamic loads are described in Fig. 5.56. The linear elastic fracture mechanics (LEFM, for pure brittle materials) curve and the strength criterion

Table 5.5 Parameters of B and D_0 under different strain rates in compressive simulation

Strain rates (/s)	A	C	Bf_c' (MPa)	B	D_0 (mm)
10^{-5}	1.19×10^{-6}	8.87×10^{-4}	33.57952	1.05166	744
10^{-3}	1.12×10^{-6}	8.43×10^{-4}	34.43357	1.07841	754
10^{-2}	8.76×10^{-7}	8.11×10^{-4}	35.11856	1.09986	926
10^{-1}	4.76×10^{-7}	6.88×10^{-4}	38.11645	1.25466	1447
1	2.65×10^{-9}	7.07×10^{-4}	37.59953	1.17756	266,485
10	-2.15×10^{-8}	6.28×10^{-4}	39.89670	1.24951	−29,201
30	-5.46×10^{-8}	4.36×10^{-4}	47.89431	1.49998	−7989
50	-6.85×10^{-8}	2.95×10^{-4}	58.23944	1.82397	−4304
100	-9.66×10^{-8}	2.71×10^{-4}	60.78926	1.90383	−2801
200	-1.32×10^{-8}	2.54×10^{-4}	62.76889	1.96583	−1929

Note f_c' is the compressive strength of concrete specimen with width of 100 mm under the strain rate of 1×10^{-5}/s, here f_c' =31.93 MPa

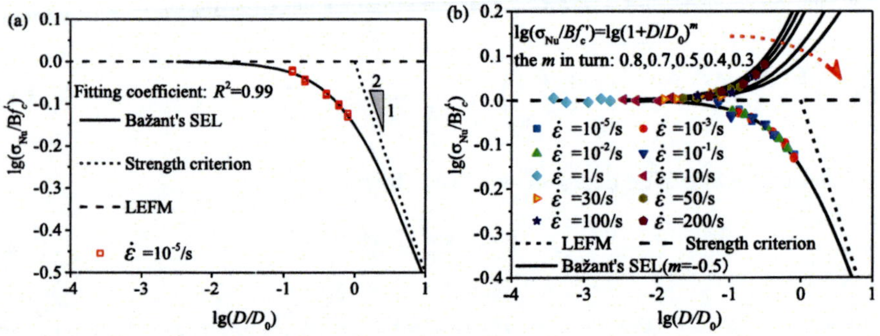

Fig. 5.56 Comparison between numerical results and typical size effect law: **a** under the strain rate of $\dot{\varepsilon}$=10^{-5}/s; and **b** under different strain rates

(for plastic materials) curve are also plotted for the comparison with the numerical results. From Fig. 5.56a one can find that the numerical compressive strength data of concrete under quasi-static load vary along Bažant's SEL curve. The numerical compressive strength data under quasi-static load are the closest to the LEFM (its slope is −1/2), showing an obvious size effect behavior.

In addition, it can be seen from Fig. 5.56b that: (1) when the applied strain rate is less than the critical strain rate, with the increasing strain rate, the strength data points approach closer to the strength criterion curve, namely the size effect is weakened; and (2) when the strain rate reaches the critical strain rate $\dot{\varepsilon}$=1/s, the numerical compressive strength data are almost overlapped with the strength criterion curve, indicating that the size effect is completely suppressed. (3) when the applied strain rate exceeds the critical strain rate, the dynamic compressive strength enhances with the increasing structural size. The numerical compressive strength data under high strain rates are far from the strength criterion curve, demonstrating that as the strain rate increases the size effect behavior opposed to static becomes gradually obvious. Refer to Bažant's SEL, a novel curve was discussed as follows.

5.6.1 Basic Assumptions

As shown in Fig. 5.56, the Bažant's SEL (double logarithmic curve expression: $\lg(\sigma_{\mathrm{Nu}}/Bf_c') = \lg(1 + D/D_0)^{-1/2}$) characterizes the decrease in compressive strength as structural size increases. Under the critical strain rate the compressive strength approaches the strength criterion based on plasticity theory; under quasi-static load the compressive strength approaches the strength limit (its slope is -1/2) based on the LEFM. That is, the compressive strength is at the transition between plasticity and LEFM. Similarly, for the case where the strain rate exceeds the critical strain rate, the compressive strength data points may also vary along a certain double logarithmic curve, this curve can also characterize the size effect behavior (the compressive strength increases with the increasing sizes) and the compressive strength under the ultimate strain rate approaches the strength limit with a slope of *m*. Herein, referring to the formula of Bažant's SEL, one can assume that the double logarithmic curve is expressed as:

$$\lg(\sigma_{\mathrm{Nu}}/Bf_c') = \lg(1 + D/D_0)^m \tag{5.10}$$

where *m* is the approximation coefficient and it can reflect the slope of the strength limit that the curve is asymptotic with. The *m* in the formula of Bažant's SEL is -1/2, and under high strain rates *m* needs to be determined by repeated trials and comparisons with data points. Figure 5.56 gives five types of double logarithmic curves under different *m* ($m = 0.8, 0.7, 0.5, 0.4, 0.3$). From Fig. 5.56 it can be seen that when *m* is in the range of 0.5 to 0.7, the compressive strength data points (the applied strain rate exceeds the critical strain rate) are in good agreement with the novel curve, indicating that the novel curve can well reflect the size effect of compressive strength under high strain rates. It should be noted that because of the discreteness of the simulated data, it is difficult to accurately determine the specific value of *m*, and only the corresponding range can be given. In order to facilitate the following research on the size effect formula, *m* can be selected as 1/2, that is, under the ultimate strain rate the compressive strength approaches the strength limit with a slope of 1/2. Thus, Eq. (5.10) can be rearranged as

$$\sigma_{\mathrm{Nu}} = Bf_c' \cdot \sqrt{1 + D/D_0} \tag{5.11}$$

According to the numerical results and analysis in Sect. 5.3, the contribution of strain rate can be divided into two parts:

(1) Strength enhancement. As the strain rate $\dot{\varepsilon}$ increases, the nominal compressive strength improves. This behavior can be described by strength enhancement coefficient $\varphi_{\dot{\varepsilon}} = CDIF$.
(2) Weakening and reverse enhancement of size effect. As the strain rate increases, the size effect on dynamic nominal strength is weakened or even suppressed before the strain rate reaches the critical strain rate $\dot{\varepsilon}_{cr}$. When the strain rate exceeds the critical strain rate $\dot{\varepsilon}_{cr}$ the size effect of strength-enhancing gradually increases. This behavior can be described by the strain rate influence coefficient $\beta_{\dot{\varepsilon}}$.

5.6.2 Dynamic Size Effect Law for Concrete

1. *Compressive strength*

Based on the above analysis, a **S**tatic and **D**ynamic unified **S**ize **E**ffect **L**aw (i.e., **SD-SEL**) on nominal compressive strength $\sigma_{\mathrm{Nc}}^{\dot{\varepsilon}}$ can be explicitly expressed as

$$\sigma_{\mathrm{Nc}}^{\dot{\varepsilon}} = \frac{Bf_c'}{\sqrt{1 + D/D_0}} \cdot \varphi_{\mathrm{c}}^{\dot{\varepsilon}} \cdot \beta_{\mathrm{c}}^{\dot{\varepsilon}} \tag{5.12}$$

where f_c' denotes the static nominal compressive strength of standard specimen; the $\varphi_{\mathrm{c}}^{\dot{\varepsilon}}$ denotes the strength enhancement coefficient for compressive strength, which can be determined by the fitting formula $CDIF_{\mathrm{fit}}$ referring to Eq. (5.5) (under low strain rates) and Eq. (5.6) (under high strain rates); and $\beta_{\mathrm{c}}^{\dot{\varepsilon}}$ denotes the strain rate influencing coefficient for compressive strength. It is to be noted that, B and D_0 are also two empirical parameters determined by regression analysis. One can clearly note from Eq. (5.12) that, the theoretical formula can not only describe the size effect law under dynamic loads, but also the one under static load. Therefore, it is a kind of static and dynamic unified size effect law.

The determination of the strain rate influencing coefficient $\beta_{\mathrm{c}}^{\dot{\varepsilon}}$ is a complex issue. Figure 5.57 plots the relationship between $\beta_{\mathrm{c}}^{\dot{\varepsilon}}$ and $\lg(\dot{\varepsilon})$.

For the quasi-static condition ($\dot{\varepsilon}_{static}$), i.e., $\dot{\varepsilon} \leq 10^{-5}$/s, the strain rate should have no influence on the size effect, thus $\beta_{\mathrm{c}}^{\dot{\varepsilon}}$=1.

For the critical strain rate $\dot{\varepsilon}_{cr}$ (i.e., $\dot{\varepsilon}$=1/s), the size effect on nominal strength is totally suppressed. Accordingly, under the critical strain rate $\dot{\varepsilon}_{cr}$, the dynamic strength should be

$$\sigma_{\mathrm{Nc}}^{\dot{\varepsilon}} = f_c' \cdot \varphi_{\mathrm{c}}^{\dot{\varepsilon}} \tag{5.13}$$

According to Eqs. (5.12) and (5.13), one can get the coefficient $\beta_{\mathrm{c}}^{\dot{\varepsilon}}$ (> 1) under the critical strain rate $\dot{\varepsilon}_{cr}$ as

Fig. 5.57 Determination of the coefficient $\beta_{\dot{\varepsilon}}$

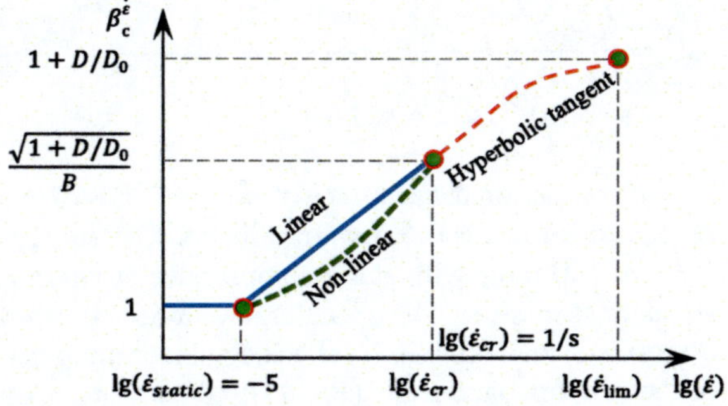

$$\beta_c^{\dot{\varepsilon}}\Big|_{\dot{\varepsilon}=\dot{\varepsilon}_{cr}} = \frac{\sqrt{1 + D/D_0}}{B} \tag{5.14}$$

When the strain rate $\dot{\varepsilon}$ is between $\dot{\varepsilon}_{static}$ and $\dot{\varepsilon}_{cr}$, linear or nonlinear curves can be utilized to describe the effect of strain rate on the size effect law. Herein a linear curve and a nonlinear curve were both assumed and adopted, and these two curves were compared to choose a batter one.

First, the linear relationship was adopted to describe the effect of strain rate. Based on the assumptions, the coefficient $\beta_c^{\dot{\varepsilon}}$ can be fixed as

$$\beta_c^{\dot{\varepsilon}} = 1, (\dot{\varepsilon} \leq \dot{\varepsilon}_{static} = 10^{-5}/\text{s}) \tag{5.15}$$

$$\beta_c^{\dot{\varepsilon}} = \frac{\frac{\sqrt{1+D/D_0}}{B} - 1}{5 + \lg(\dot{\varepsilon}_{cr})} \cdot (lg\dot{\varepsilon} + 5) + 1, (10^{-5}/\text{s} < \dot{\varepsilon} \leq \dot{\varepsilon}_{cr}) \tag{5.16}$$

As the critical strain rate $\dot{\varepsilon}_{cr} = 1/\text{s}$, Eq. (5.16) can be simplify to:

$$\beta_c^{\dot{\varepsilon}} = \frac{1}{5} \cdot (\frac{\sqrt{1 + D/D_0}}{B} - 1) \cdot (lg\dot{\varepsilon} + 5) + 1, (10^{-5}/\text{s} < \dot{\varepsilon} \leq 1/\text{s}) \tag{5.17}$$

When a nonlinear relationship was adopted to describe the effect of strain rate, herein assumed a quadratic curve, similarly, based on the assumptions, the coefficient $\beta_c^{\dot{\varepsilon}}$ can be fixed as:

$$\beta_c^{\dot{\varepsilon}} = \frac{\left(\frac{\sqrt{1+D/D_0}}{B} - 1\right) \cdot (lg\dot{\varepsilon})^2 + 10 \cdot \left(\frac{\sqrt{1+D/D_0}}{B} - 1\right) \cdot lg\dot{\varepsilon} + 25 \cdot (\frac{\sqrt{1+D/D_0}}{B} - 1)}{(\lg(\dot{\varepsilon}_{cr}))^2 + 10\lg(\dot{\varepsilon}_{cr}) + 25} + 1, 10 - 5/\text{s} < \dot{\varepsilon} \leq \dot{\varepsilon}_{cr} \tag{5.18}$$

As the critical strain rate $\dot{\varepsilon}_{cr} = 1/\text{s}$, Eq. (5.18) can be simplify to:

$$\beta_c^{\dot{\varepsilon}} = \frac{\left(\frac{\sqrt{1+D/D_0}}{B}-1\right)\cdot(lg\dot{\varepsilon})^2 + 10\cdot\left(\frac{\sqrt{1+D/D_0}}{B}-1\right)\cdot lg\dot{\varepsilon} + 25\cdot\left(\frac{\sqrt{1+D/D_0}}{B}-1\right)}{25}$$
$$+1,\ 10^{-5}/\mathrm{s} < \dot{\varepsilon} \leq 1/\mathrm{s} \tag{5.19}$$

For comparing the accuracy of the two kinds of expressions for $\beta_c^{\dot{\varepsilon}}$, the numerical results given in Sect. 5.3 were utilized. The square concrete specimens (sized of 100 × 100 mm) with static compressive strength $f_c^{'}$ =31.93 MPa were taken as standard specimens. The $\varphi_c^{\dot{\varepsilon}}$ under low strain rate refers to Eq. (5.5). Moreover, under quasi-static load ($\dot{\varepsilon} = 10^{-5}$/s), the obtained fitting parameters of B and D_0 were used for benchmark parameters ($B = 1.05$, $D_0 = 744$). With these parameters determined, one can predict the magnitude of the dynamic compressive strength of concrete having larger structural sizes. When the applied strain rate is less than the critical strain rate, the predicted results based on the proposed theoretical formula (using two types of expressions of $\beta_c^{\dot{\varepsilon}}$) are respectively plotted against with the simulated results in Fig. 5.58. It can be noted from Fig. 5.58 that using the quadratic curve to describe the strain rate influencing coefficient $\beta_c^{\dot{\varepsilon}}$ is more accurate than using a linear curve, and the theoretical value is closer to the simulated results. Meanwhile, the SD-SEL can well describe the size effect on the compressive strength of concrete under low strain rates.

Similarly, under the critical strain rate, the size effect is completely suppressed, and the dynamic compressive strength is the same as Eq. (5.13). When the strain rate approaches the ultimate strain rate $\dot{\varepsilon}_{\lim}$, according to Eq. (5.11), the data points of dynamic compressive strength approach the strength limit with a slope of 1/2, the strain rate has almost no effect on the size effect, and $\beta_c^{\dot{\varepsilon}}$ infinitely closes to $(1 + D/D_0)$. When the strain rate $\dot{\varepsilon}$ is between $\dot{\varepsilon}_{cr}$ and $\dot{\varepsilon}_{\lim}$, The influence coefficient $\beta_c^{\dot{\varepsilon}}$ is a curve asymptotic to $(1 + D/D_0)$. The hyperbolic tangent function can describe the effect of strain rate temporarily, as shown in Fig. 5.57. Based on the assumptions and the above analysis, combined Eq. (5.14), the coefficient $\beta_c^{\dot{\varepsilon}}$ can be fixed as:

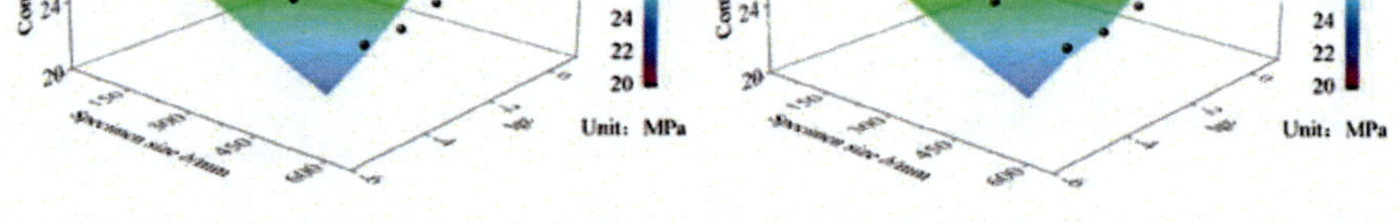

Fig. 5.58 Comparison of the theoretical and numerical results under low strain rates: **a** $\beta_c^{\dot{\varepsilon}}$= linear expression; and **b** $\beta_c^{\dot{\varepsilon}}$= quadratic expression

$$\beta_c^{\dot{\varepsilon}} = \left[(1 + D/D_0) - \frac{\sqrt{1 + D/D_0}}{B}\right] \cdot \tanh(\alpha_c \cdot lg\dot{\varepsilon}) + \frac{\sqrt{1 + D/D_0}}{B}, (\dot{\varepsilon} > 1/\mathrm{s}) \tag{5.20}$$

where α_c is the adjustment coefficient for compressive strength, reflecting the change speed of coefficient $\beta_c^{\dot{\varepsilon}}$ with the strain rate. In reality, Eq. (5.20) is a transcendental equation and it is difficult to obtain its analytical solution. A trial algorithm was employed to approximate the numerical solution in the following part.

In order to determine approximately the range of the adjustment coefficient α_c by trial calculation, and calibrate the proposed SD-SEL under high strain rates, similarly, based on the basic conditions above, combined Eqs. (5.5), (5.12) and (5.20), one can predict the dynamic compressive strength of concrete for more dimensions under high strain rates ($\dot{\varepsilon}$ ˙ > 1/s).

The predicted results based on the developed SD-SEL (the adjustment coefficient α_c is 0.10 and 0.15 respectively) under high strain rates are respectively plotted against with the simulated results in Fig. 5.59. It can be found from Fig. 5.59 that under high strain rates, when the adjustment coefficient α_c is in the range of 0.10 to 0.15, the theoretical value is in good agreement with the simulated results. That is, the SD-SEL can well describe the size effect on the compressive strength of concrete under high strain rates.

On the whole, the developed SD-SEL can reflect the relation between dynamic size effect and static size effect, and it can well describe the size effect on the compressive strength of concrete under different strain rates (10^{-5} ~ 200/s).

2. *Dynamic splitting-tensile strength*

According the numerical results of dynamic splitting-tensile failure behavior in Sect. 5.1, the size effect on nominal splitting-tensile strength of concrete materials under low strain rates and high strain rates is similar with that on nominal compressive strength: the size effect is weakened gradually with the increasing strain rate under low strain rates while the dynamic strength enhances as the structural size increases

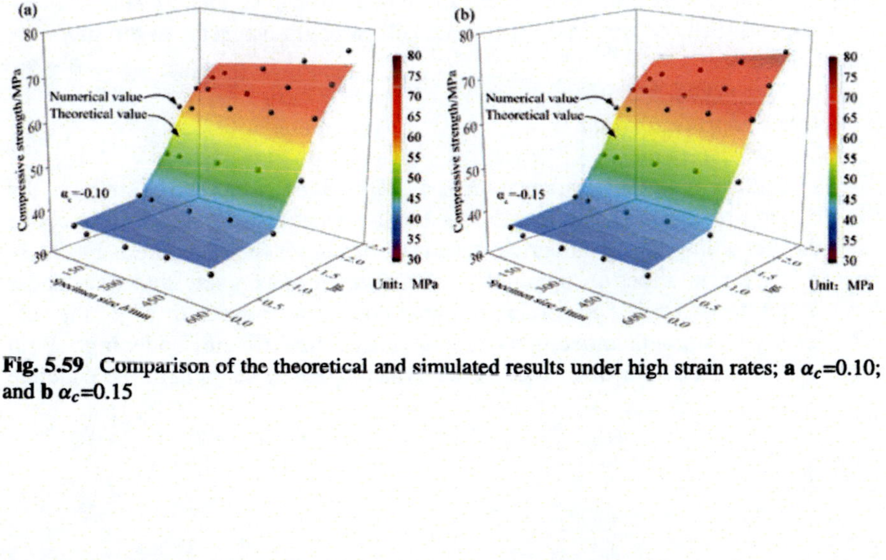

Fig. 5.59 Comparison of the theoretical and simulated results under high strain rates; **a** α_c=0.10; and **b** α_c=0.15

under high strain rates. Therefore, referring to the developed ideas of SD-SEL on compressive strength, the SD-SEL on nominal splitting-tensile strength $\sigma_{\mathrm{Nst}}^{\dot{\varepsilon}}$ can be also established based on the above analysis:

$$\sigma_{\mathrm{Nst}}^{\dot{\varepsilon}} = \frac{Bf'_{\mathrm{st}}}{\sqrt{1 + D/D_0}} \cdot \varphi_{\mathrm{st}}^{\dot{\varepsilon}} \cdot \beta_{\mathrm{st}}^{\dot{\varepsilon}} \tag{5.21}$$

where f'_{st} denotes the static nominal splitting-tensile strength of standard specimen; the $\varphi_{\mathrm{st}}^{\dot{\varepsilon}}$ denotes the strength enhancement coefficient for splitting-tensile strength, which can be determined by the fitting formula $TDIF_{\mathrm{fit}}$ referring to Eq. (5.2) (under low strain rates) and Eq. (5.3) (under high strain rates); and $\beta_{\mathrm{st}}^{\dot{\varepsilon}}$ denotes the strain rate influencing coefficient for splitting-tensile strength.

Similarly, the coefficient $\beta_{\mathrm{st}}^{\dot{\varepsilon}}$ for splitting-tensile strength can be also be fixed as:

$$\beta_{\mathrm{st}}^{\dot{\varepsilon}} = 1, (\dot{\varepsilon} \le \dot{\varepsilon}_{static} = 10^{-5}/\mathrm{s}) \tag{5.22}$$

$$\beta_{\mathrm{st}}^{\dot{\varepsilon}} = \frac{\left(\frac{\sqrt{1+D/D_0}}{B} - 1\right) \cdot (lg\dot{\varepsilon})^2 + 10 \cdot \left(\frac{\sqrt{1+D/D_0}}{B} - 1\right) \cdot lg\dot{\varepsilon} + 25 \cdot \left(\frac{\sqrt{1+D/D_0}}{B} - 1\right)}{(\lg(\dot{\varepsilon}_{cr}))^2 + 10\lg(\dot{\varepsilon}_{cr}) + 25} + 1, 10^{-5}/\mathrm{s} < \dot{\varepsilon} \le \dot{\varepsilon}_{cr} \tag{5.23}$$

$$\beta_{\mathrm{st}}^{\dot{\varepsilon}} = \left[(1 + D/D_0) - \frac{\sqrt{1 + D/D_0}}{B}\right] \cdot \tanh(\alpha_{\mathrm{st}} \cdot lg\dot{\varepsilon}) + \frac{\sqrt{1 + D/D_0}}{B}, (\dot{\varepsilon} > \dot{\varepsilon}_{cr}) \tag{5.24}$$

Herein α_{st} is the adjustment coefficient for splitting-tensile strength, reflecting the change speed of coefficient $\beta_{\mathrm{st}}^{\dot{\varepsilon}}$ with the strain rate.

Combining with Eqs. (5.1), (5.2), (5.21), (5.22), (5.23) and (5.24), one can get the SD-SEL on nominal splitting-tensile strength of concrete materials and the SD-SEL can predict the dynamic splitting-tensile strength of concrete materials having different structural sizes under different strain rates. The detailed derivation process about this SD-SEL on nominal splitting-tensile strength can refer to the literature (Jin et al. 2019b).

3. *Dynamic tensile strength*

Similarly, referring to the developed ideas of SD-SEL on compressive strength, the SD-SEL on nominal tensile strength can be also established based on the above analysis. According to the numerical results of dynamic tensile failure behavior in Sect. 5.2, the size effect on nominal tensile strength of concrete materials under low strain rates is similar to that on nominal compressive strength while the size effect on tensile strength under high strain rates is a little different. Under high strain rates, there is no obvious difference in the dynamic tensile strength of the concrete

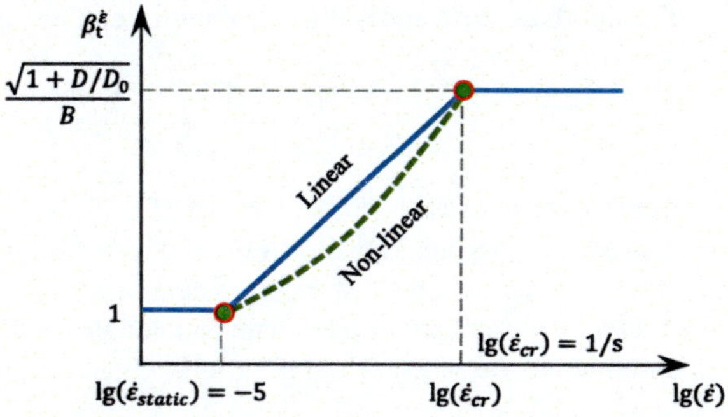

Fig. 5.60 Determination of the coefficient $\beta_t^{\dot{\varepsilon}}$

specimen having different structural sizes and the size effect completely disappears, which affects the determination of coefficient $\beta_t^{\dot{\varepsilon}}$ under high strain rates. Therefore, the SD-SEL on nominal tensile strength $\sigma_{Nt}^{\dot{\varepsilon}}$ can be also established based on the above analysis:

$$\sigma_{Nt}^{\dot{\varepsilon}} = \frac{Bf_t^{'}}{\sqrt{1+D/D_0}} \cdot \varphi_t^{\dot{\varepsilon}} \cdot \beta_t^{\dot{\varepsilon}} \tag{5.25}$$

where $f_t^{'}$ denotes the static nominal tensile strength of standard specimen; the $\varphi_t^{\dot{\varepsilon}}$ denotes the strength enhancement coefficient for tensile strength, which can be determined by the fitting formula $TDIF_{fit}$ referring to Eq. (5.3) (under low strain rates) and Eq. (5.4) (under high strain rates); and $\beta_{st}^{\dot{\varepsilon}}$ denotes the strain rate influence coefficient for tensile strength. Figure 5.60 shows the determination of the coefficient $\beta_t^{\dot{\varepsilon}}$ for tensile strength. It can be observed that the coefficient $\beta_t^{\dot{\varepsilon}}$ under low strain rates is similar to that for compressive strength, which can be also expressed as:

$$\beta_t^{\dot{\varepsilon}} = 1, (\dot{\varepsilon} \le \dot{\varepsilon}_{static} = 10^{-5}/s) \tag{5.26}$$

$$\beta_t^{\dot{\varepsilon}} = \frac{\left(\frac{\sqrt{1+D/D_0}}{B} - 1\right) \cdot (lg\dot{\varepsilon})^2 + 10 \cdot \left(\frac{\sqrt{1+D/D_0}}{B} - 1\right) \cdot lg\dot{\varepsilon} + 25 \cdot (\frac{\sqrt{1+D/D_0}}{B} - 1)}{(\lg(\dot{\varepsilon}_{cr}))^2 + 10\lg(\dot{\varepsilon}_{cr}) + 25} + 1, 10^{-5}/s < \dot{\varepsilon} \le \dot{\varepsilon}_{cr} \tag{5.27}$$

As for $\beta_t^{\dot{\varepsilon}}$ under high strain rates, it needs a further derivation: As the applied strain rate exceeds the critical strain rate $\dot{\varepsilon}_{cr}$, the dynamic tensile strength is insensitive to the structural size and the size effect completely disappears. That is to say, the dynamic tensile strength under high strain rates is only related to the strain rate. Accordingly, under the critical strain rate and high strain rates, the dynamic tensile strength $\sigma_{Nt}^{\dot{\varepsilon}}$ should be expressed as:

$$\sigma_{Nt}^{\dot{\varepsilon}} = f_t^{'} \cdot \varphi_t^{\dot{\varepsilon}} \tag{5.28}$$

The coefficient $\beta_{t}^{\dot{\varepsilon}}$ under high strain rates can be written as:

$$\beta_{t}^{\dot{\varepsilon}} = \frac{\sqrt{1 + D/D_0}}{B}, \dot{\varepsilon} > \dot{\varepsilon}_{cr} \tag{5.29}$$

Combining with Eqs. (5.3), (5.4), (5.25), (5.27), (5.28) and (5.29), one can get the SD-SEL on nominal tensile strength of concrete materials and the SD-SEL can predict the dynamic tensile strength of concrete materials having different structural sizes under different strain rates. The detailed derivation process about this SD-SEL on nominal tensile strength can refer to the literature (Jin et al. 2019a).

4. *Lightweight concrete*

According to the numerical results in Sect. 5.4.3, under low strain rates the size effect on compressive strength of LWAC is weakened gradually with the increasing strain rate while there is no obvious difference in the dynamic compressive strength of LWAC having different structural sizes and the size effect completely disappears under high strain rates, which are similar with the size effect on dynamic tensile strength in Sect. 5.2. Therefore, referring to the developed ideas of SD-SEL on tensile strength above, the SD-SEL on compressive strength of LWAC can be also established as follows:

$$\sigma_{\mathrm{Nlc}}^{\dot{\varepsilon}} = \frac{Bf_{\mathrm{lc}}^{'}}{\sqrt{1 + D/D_0}} \cdot \varphi_{\mathrm{lc}}^{\dot{\varepsilon}} \cdot \beta_{\mathrm{lc}}^{\dot{\varepsilon}} \tag{5.30}$$

where $f_{\mathrm{lc}}^{'}$ denotes the static nominal compressive strength of LWAC standard specimen; the $\varphi_{\mathrm{lc}}^{\dot{\varepsilon}}$ denotes the strength enhancement coefficient for compressive strength of LWAC; and $\beta_{\mathrm{lc}}^{\dot{\varepsilon}}$ denotes the strain rate influence coefficient for compressive strength of LWAC and it can be determined according to Fig. 5.60. The coefficient $\beta_{\mathrm{lc}}^{\dot{\varepsilon}}$ for compressive strength of LWAC can be also expressed as:

$$\beta_{\mathrm{lc}}^{\dot{\varepsilon}} = 1, (\dot{\varepsilon} \le \dot{\varepsilon}_{static} = 10^{-5}/\mathrm{s}) \tag{5.31}$$

$$\beta_{\mathrm{lc}}^{\dot{\varepsilon}} = \frac{\left(\frac{\sqrt{1+D/D_0}}{B} - 1\right) \cdot (lg\dot{\varepsilon})^2 + 10 \cdot \left(\frac{\sqrt{1+D/D_0}}{B} - 1\right) \cdot lg\dot{\varepsilon} + 25 \cdot (\frac{\sqrt{1+D/D_0}}{B} - 1)}{(\lg(\dot{\varepsilon}_{cr}))^2 + 10\lg(\dot{\varepsilon}_{cr}) + 25} + 1, 10^{-5}/\mathrm{s} < \dot{\varepsilon} \le \dot{\varepsilon}_{cr} \tag{5.32}$$

$$\beta_{\mathrm{lc}}^{\dot{\varepsilon}} = \frac{\sqrt{1 + D/D_0}}{B}, \dot{\varepsilon} > \dot{\varepsilon}_{cr} \tag{5.33}$$

Similarly, combining with Eqs. (5.30)-(5.33), one can get the SD-SEL on compressive strength of LWAC and the SD-SEL can predict the dynamic compressive strength of LWAC having different structural sizes under different strain rates. The detailed derivation process about this SD-SEL on compressive strength of LWAC can refer to the literature (Jin et al. 2020a).

5.6.3 Validation of the Theoretical Formula

1. *Dynamic compressive strength*

In addition to the validation of SD-SEL on compressive strength from numerical perspective in Sect. 5.6.2, here in this section the validation of SD-SEL on compressive strength from experimental perspective was conducted.

Figure 5.61 plots the comparison of experimental results in Hu et al. (2015)'s study (under low strain rates) and theoretical results based on the proposed SD-SEL. In order to be more persuasive, f_c/f_c^R ratio was utilized to compare the theoretical results with the experimental ones, where f_c represents the dynamic strength, and f_c^R is the compressive strength of concrete reference specimens under static load (in Hu et al.'s tests, f_c^R is the peak compressive stress of cubic concrete specimen with width of 150 mm under the strain rate of 1×10^{-5}/s. In theoretical results f_c^R is the compressive strength of square concrete specimen with width of 150 mm under the strain rate of 1×10^{-5}/s). From Fig. 5.61 one can note that, although some experimental values are

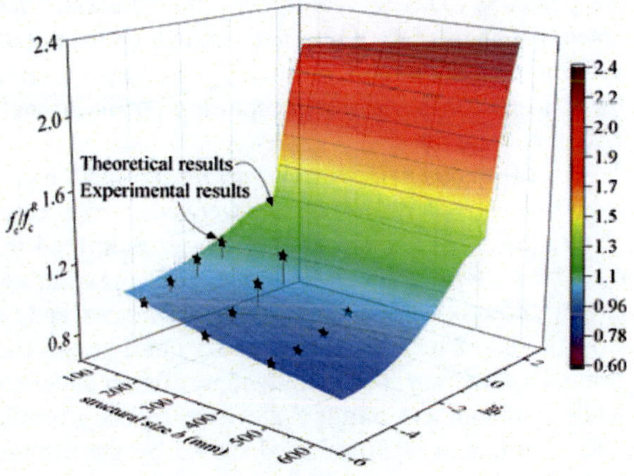

Fig. 5.61 Comparison of the experimental and theoretical results of compressive strength

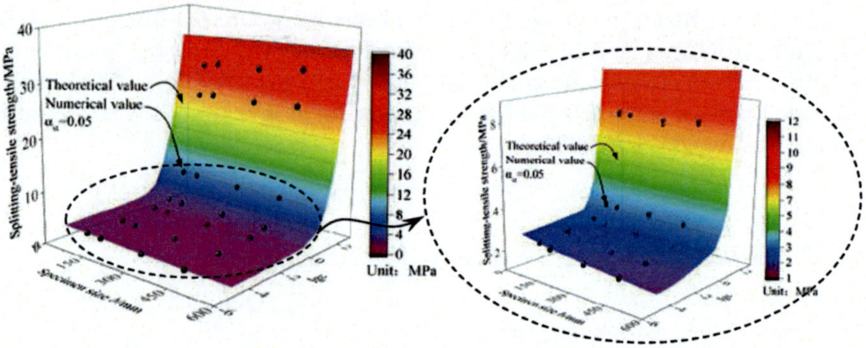

Fig. 5.62 Comparison of the numerical and theoretical results of splitting-tensile strength

slightly larger, most of the test data fall near the theoretical surface. This preliminarily verified the rationality of the proposed SD-SEL on compressive strength.

2. *Dynamic splitting-tensile strength*

The numerical results obtained in Sect. 5.2 were adopted to calibrate the proposed SD-SEL on splitting-tensile strength of concrete. Under the strain rate $\dot{\varepsilon} = 10^{-5}/\text{s}$ (quasi-static), these two empirical parameters B and D_0 ($B = 1.07139$, $D_0 = 541$) are determined and utilized. After a large number of trials when the adjustment coefficient α is defined as 0.05, one can predict the magnitude of the dynamic splitting-tensile strength of concrete having different structural sizes from theoretical perspective, which is shown in Fig. 5.62. In Fig. 5.62 it describes the predicted results based on the proposed SD-SEL against with the numerical data points. It can be obviously noted that, most of the numerical strength points fall on the surface of the theoretical splitting-tensile strength, illustrating that the proposed SD-SEL can well characterize the size effect behavior on the splitting-tensile strength of concrete materials under different strain rates.

In order to compare with the test results and the proposed SD-SEL, experimental data in Chen et al. (2018)'s study were adopted to verify the theory. The comparison of test results and theoretical results is described in Fig. 5.63. For convenience, f_{st}/f_{st}^{R} ratio was applied, where f_{st} denotes the splitting-tensile strength of specimens having different sizes under different strain rates, and f_{st}^{R} denotes the splitting-tensile strength of referent specimens under quasi-static load (f_{st}^{R} = the splitting-tensile strength of referent specimen sized by 30 mm under static load in Chen et al.'s study while in theoretical results f_{st}^{R} = the splitting-tensile strength of cubic specimen with side length of 30 mm calculated by the proposed SD-SEL). It can be seen that, although some test data are slightly below the theoretical surface based on

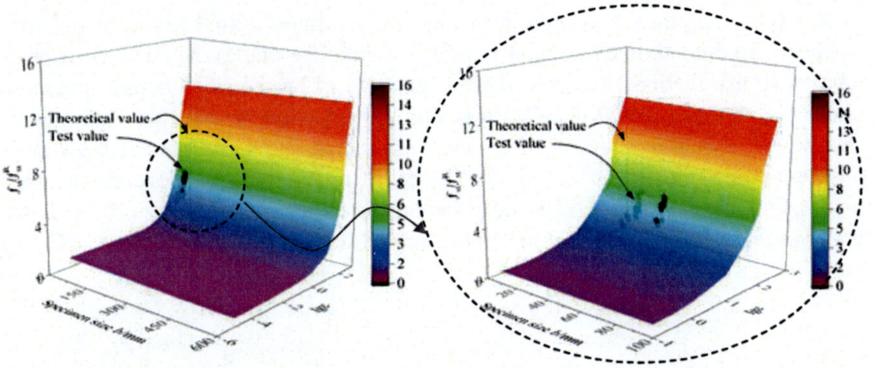

Fig. 5.63 Comparison of the experimental and theoretical results of splitting-tensile strength

the proposed SD-SEL, most tests results approach or fall on the theoretical surface, which preliminarily verifies the proposed SD-SEL.

3. *Dynamic tensile strength*

To calibrate the proposed SD-SEL, the numerical results given in Sect. 5.2 were utilized. The double-edge notched concrete specimens sized of 100 mm × 200 mm with a static tensile strength f_t' of 2.046 MPa were taken as examples. Moreover, under quasi-static load ($\dot{\varepsilon}$=10^{-5}/s), the obtained fitting parameters of B and D_0 (B = 1.34, D_0 = 127) were utilized. With these parameters determined, one can predict the magnitude of the dynamic tensile strength of concrete having larger structural sizes.

The predicted results based on the proposed theoretical formula are plotted against with the numerical results in Fig. 5.64. It can be obviously noted that, most of the numerical strength points fall on the surface of the theoretical tensile strength, indicating that the proposed SD-SEL can well describe the size effect on the tensile strength of concrete under different strain rates.

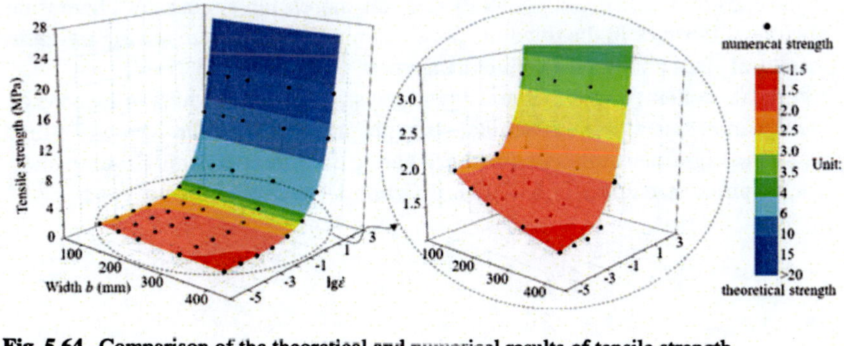

Fig. 5.64 Comparison of the theoretical and numerical results of tensile strength

It is to be noted that, as mentioned previously, quite limited experimental work has been conducted on dynamic size effect of concrete due to the limitations of the laboratory test devices and even there is no tests of large-size concrete specimens under high strain rates. It is difficult to find suitable experimental data of dynamic size effect (dynamic strength of concrete specimens having different sizes under different strain rates) for comparison. Therefore, relevant tests of concrete materials on the dynamic size effect (especially for concrete specimens having different sizes under high strain rates) should be conducted to further verify the proposed SD-SEL in the future studies.

5.7 Summary

This chapter mainly focus on dynamic size effect on nominal strength of concrete materials. Applying the meso-scale modelling method, the spitting-tensile/ tensile/ compressive failure behavior of concrete materials having different structural sizes under different strain rates were simulated and the influence of strain rate on size effect on spitting-tensile/ tensile/ compressive strength was analyzed. Moreover, the influence of meso-structure (aggregate content, maximum aggregate size, aggregate type, etc.) and initial loads on the dynamic size effect were also provided. In addition, the SD-SEL for spitting-tensile/ tensile/ compressive strength of concrete materials is respectively introduced considering the contribution of strain rate on the dynamic size effect. The following conclusions can be drawn:

(1) Under dynamic loads, the size effect on nominal strength (spitting-tensile/ tensile/ compressive strength) of concrete materials is obviously different from that under static loads.
(2) There is a critical strain rate ($\dot{\varepsilon}_{cr}$=1/s) in dynamic nominal strength of concrete materials. Under low strain rates, the size effect is weakened gradually with the increasing strain rate. Under high strain rates, the dynamic nominal spitting-tensile strength and compressive strength increase as the structural size increases and the discrepancy in nominal strengths of concrete having different structural sizes tends to be enhanced gradually with the increasing strain rate. There is no obvious difference in the dynamic tensile strength of concrete having different structural sizes and the size effect disappears.
(3) Aggregate content and maximum aggregate size present obvious influences on the dynamic size effect on nominal strengths under low strain rates while they perform an ignorable influence on the dynamic size effect under high strain rates. The dynamic size effect of LWAC is different from that of normal concrete.

(4) Considering the contribution of strain rate on the dynamic size effect, the developed SD-SEL can well characterize the size effect on nominal strengths of concrete material under different strain rates.

References

Abrams D (1917) Effect of rate of application of load on the compressive strength of concrete [J]. Proc Am Soc Testing Mater 364–377

Bažant ZP (1984) Size effect in blunt fracture: concrete, rock, metal [J]. J Eng Mech 110(4):518–535

Bažant ZP (1997) Fracture and size effect in concrete and other quasibrittle materials [M]. CRC Press, USA

Bažant ZP (2004) Probability distribution of energetic-statistical size effect in quasibrittle fracture [J]. Probab Eng Mech 19(4):307–319

Bažant ZP, Pang S (2007) Activation energy based extreme value statistics and size effect in brittle and quasibrittle fracture [J]. J Mech Phys Solids 55(1):91–131

Bažant Z, Caner F, Adley M, Akers S (2000) Fracturing rate effect and creep in microplane model for dynamics [J]. J Eng Mech 126(9):962–970

Bindiganavile V, Banthia N (2006) Size effects and the dynamic response of plain concrete [J]. J Mater Civ Eng 18(4):485–491

Carpinteri A, Ferro G (1994) Size effects on tensile fracture properties: a unified explanation based on disorder and fractality of concrete microstructure [J]. Mater Struct 27(10):563–571

Chen X, Wu S, Zhou J (2013) Experimental and modeling study of dynamic mechanical properties of cement paste, mortar and concrete [J]. Constr Build Mater 47:419–430

Chen J, Xiang D, Wang Z, Wu G, Wang G (2018) Dynamic tensile strength enhancement of concrete in split Hopkinson pressure bar test [J]. Adv Mech Eng 10(6):2072047318

Comité euro-international du béton, Fédération Internationale de la Précontrainte (1993) CEB-FIP model code 1990: design code [M]. Thomas Telford Publishing, Lausanne, Switzerland

Cusatis G (2011) Strain-rate effects on concrete behavior [J]. Int J Impact Eng 38(4):162–170

Dilger WH, Koch R, Kowalczyk R (1984) Ductility of plain and confined concrete under different strain rates [J]. J Proc 81(1):73–81

Du X, Jin L (2014) Effects of loading rate and its sudden change on concrete compressive failure [J]. J Vibr Shock 19:187–193 (in Chinese)

Du X, Jin L, Ma G (2013) A meso-scale analysis method for the simulation of nonlinear damage and failure behavior of reinforced concrete members [J]. Int J Damage Mech 22(6):878–904

Du M, Jin L, Li D, Du X (2017) Mesoscopic simulation study of the influence of aggregate size on mechanical properties and specimen size effect of concrete subjected to splitting tensile loading [J]. Eng Mech 34(09):54–63 (in Chinese)

Elfahal M, Krauthammer T (2005) Dynamic size effect in normal-and high-strength concrete cylinders [J]. ACI Mater J 102(2):77–85

Elfahal MM, Krauthammer T, Ohno T, Beppu M, Mindess S (2005) Size effect for normal strength concrete cylinders subjected to axial impact [J]. Int J Impact Eng 31(4):461–481

Erzar B, Forquin P (2011) Experiments and mesoscopic modelling of dynamic testing of concrete [J]. Mech Mater 43(9):505–527

Fan X, Hu S, Lu J, Chen Q (2017) Effects of initial static loads on the tensile strength of concrete [J]. J Vibr Shock 36(02):83–88 (in Chinese)

Hao Y, Hao H, Li Z (2013) Influence of end friction confinement on impact tests of concrete material at high strain rate [J]. Int J Impact Eng 60:82–106

Hu W, Zou R, Peng G, Zou S (2015) Research on energy absorption characteristics and size effect of concrete under different strain rates [J]. J Changjiang Acad Sci 05:132–136 (in Chinese)
Jin L, Yu W, Du X, Yang W (2019a) Dynamic size effect of concrete under tension: A numerical study [J]. Int J Impact Eng 132:103318
Jin L, Yu W, Du X, Yang W (2019b) Mesoscopic numerical simulation of dynamic size effect on the splitting-tensile strength of concrete [J]. Eng Fract Mech 209:317–332
Jin L, Yu W, Du X, Zhang S, Yang W, Li D (2019e) Research on size effect of concrete dynamic compressive strength based on meso-simulation [J]. Eng Mech 36(11):50–61 (in Chinese)
Jin L, Yu W, Du X, Zhang S, Li D (2019c) Meso-scale modelling of the size effect on dynamic compressive failure of concrete under different strain rates [J]. Int J Impact Eng 125:1–12
Jin L, Yu W, Du X, Zhang S, Li D (2019d) Meso-scale simulation of size effect of dynamic tensile failure of concrete under low strain rate [J]. Eng Mech 36(08):59–69 (in Chinese)
Jin L, Yu W, Du X, Yang W (2020b) Meso-scale simulations of size effect on concrete dynamic splitting tensile strength: Influence of aggregate content and maximum aggregate size [J]. Eng Fract Mech 230:106979
Jin L, Yu W, Du X (2020c) Effect of initial static load and dynamic load on concrete dynamic compressive failure [J]. J Mater Civ Eng 32(12):4020351
Jin L, Yu W, Du X, (2020a) Size effect on dynamic compressive strength of lightweight aggregate concrete: Meso-scale investigation [J]. Cement Concr Compos (in preparation)
Kaplan S (1980) Factors affecting the relationship between rate of loading and measured compressive strength of concrete [J]. Magazine Concrete Res 32(111):79–88
Khandelwal M, Ranjith PG (2013) Behaviour of brittle material in multiple loading rates under uniaxial compression [J]. Geotech Geol Eng 31(4):1305–1315
Khandelwal M, Ranjith PG (2017) Study of crack propagation in concrete under multiple loading rates by acoustic emission [J]. Geomech Geophys Geo-Energy Geo-Res 3(4):393–404
Krauthammer T, Elfahal MM, Lim J, Ohno T, Beppu M, Markeset G (2003) Size effect for high-strength concrete cylinders subjected to axial impact [J]. Int J Impact Eng 28(9):1001–1016
Li M, Hao H, Shi Y, Hao Y (2018) Specimen shape and size effects on the concrete compressive strength under static and dynamic tests [J]. Constr Build Mater 161:84–93
Malvar LJ, Ross CA (1998) Review of strain rate effects for concrete in tension [J]. Mater J 95(6):735–739
Shi L, Song Y (2018) Effect of initial static stress on the dynamic behavior of larger aggregate concrete under uniaxial compression [J]. Concrete 06:35–38 (in Chinese)
Wang L, Chen G (2008) Mesoscopic simulation for mechanical properties of lightweight aggregate concrete by rigid body spring model [J]. J Hydraulic Eng 39(5):80–87 (in Chinese)
Wang X, Zhang S, Wang C, Song R, Shang C, Fang X (2018) Experimental investigation of the size effect of layered roller compacted concrete (RCC) under high-strain-rate loading [J]. Constr Build Mater 165:45–57
Wang X, Zhang S, Wang C, Song R, Shang C, Tang X (2018) Experimental investigation of the size effect of layered roller compacted concrete (RCC) under high-strain-rate loading [J]. Constr Build Mater 165:45–57
Weibull W (1939) A statistical theory of the strength of materials [J]. Proc Am Math Soc 151(5):1034
Wu M, Qin C, Zhang C (2014) High strain rate splitting tensile tests of concrete and numerical simulation by mesoscale particle elements [J]. J Mater Civ Eng 26(1):71–82
Xiao S, Zhang J (2010) Experiment study on effect of load histories on dynamic compressive damage behaviors of concrete [J]. J Hydraulic Eng 41(08):943–952 (in Chinese)
Yan D, Lin G (2006) Dynamic properties of concrete in direct tension [J]. Cem Concr Res 36(7):1371–1378
Yan D, Lin G (2008) Influence of initial static stress on the dynamic properties of concrete [J]. Cem Concr Res 30(4):327–333

Yan D, Liu K, Fan L, Yang Z (2017) An experimental investigation of pre-loading effects on the dynamic behaviour of concrete [J]. Magazine Concrete Res 69(11):586–594
Zhou X, Hao H (2008) Modelling of compressive behaviour of concrete-like materials at high strain rate [J]. Int J Solids Struct 45(17):4648–4661

Chapter 6
Size Effect in Shear and Flexure Failure of Concrete Beams

The failure modes of concrete beams are mainly divided into shear failure and flexural failure. For RC beams with the small shear-span ratios, the shear failure usually occurs under shear. In general, shear failure is a typical brittle failure, and its size effect behavior is significant. In addition, shear capacity is an important index to measure the bearing capacity of RC beams, which is related to the safety design of the building structures. Due to the high complexity of size effect in shear failure, the shear capacity and shear failure behavior of RC beams have been closely concerned in the engineering field. This chapter mainly introduces the shear failure and size effect behavior of ordinary concrete beams, lightweight-aggregate concrete beams and CFRP-wrapped concrete beams.

For RC beams with the large shear-span ratios, the flexural failure usually occurs. Generally speaking, compared with shear failure, the size effect of flexural failure of RC beams is not obvious. However, under cyclic loading, the flexural failure of beams also has size effect. This chapter mainly introduces the flexural failure and size effect behavior of RC concrete beams.

6.1 Shear Failure in Reinforced Concrete Beams Without Stirrups

Relatively speaking, the mechanics performance of beams without stirrups is relatively simple. Therefore, the study on the shear failure size effect behavior of beams without stirrups becomes the basis of the study on the shear failure size effect behavior of concrete beams. Here, the shear failure behaviors of ordinary RC beams and lightweight-aggregate RC beams without stirrups are presented, according to Jin et al. (2019a, 2020a).

X. Du and L. Jin, *Size Effect in Concrete Materials and Structures*,
https://doi.org/10.1007/978-981-33-4943-8_6

6.1.1 Failure of Ordinary Concrete Beam

Adopting the 3D meso-scale numerical simulation method provided in Chap. 3, the influences of shear-span ratio and longitudinal reinforcement ratio on the shear failure and size effect behavior of ordinary concrete beams without stirrups are given.

Figure 6.1 presents the 3D meso-scale numerical analysis model of RC cantilever beams. The coarse aggregate particles were set as high-strength elastomer, which would not produce large deformation under static loading. The plastic damage constitutive model was used to describe the mechanical behavior of mortar matrix and ITZ. The ideal elastic-plastic constitutive model was used to describe the mechanical behavior of steel bars. The mechanical parameters of concrete meso-components and reinforcement were listed in Table 6.1.

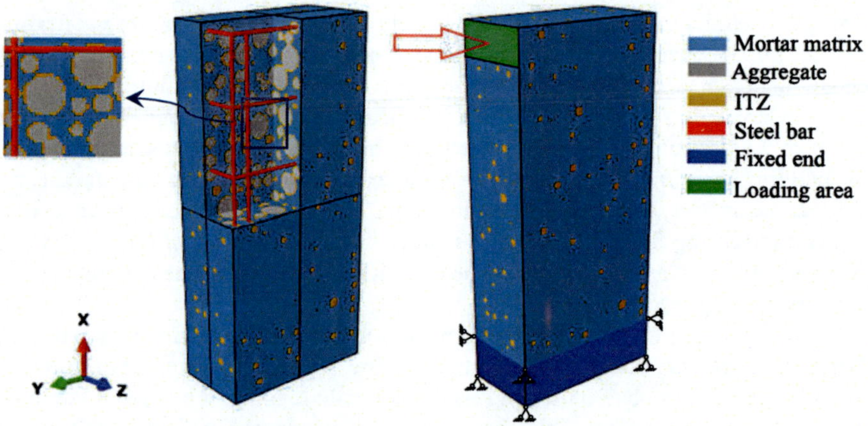

Fig. 6.1 3D meso-mechanical analysis model and loading form

Table 6.1 Mechanical parameters of concrete and reinforcements utilized

Mechanical parameters	Aggregate	Mortar matrix	ITZ	Longitudinal bar	Stirrups
Compressive strength σ_c (MPa)		39.9[a]	31.9[b]		
Tensile strength σ_t (MPa)		3.9[a]	3.2[b]		
Fracture energy G_c (J/m^2)		50	30		
Elastic modulus E (GPa)	70	42.8[a]	38.7[b]	196[a]	210[a]
Poisson ratio ν	0.2	0.2	0.2	0.3[a]	0.3[a]
Yield strength f_y (MPa)				392.1[a]	297.9[a]

Note [a]Data were obtained from the tests (Jin et al. 2016); [b]Data were trial results; and other mechanical parameters (fracture energy and Poisson ratio) are taken from work (Jin et al. 2019b)

1. *Shear-span ratio*

For beams without stirrups or with a small stirrup ratio, the shear-span ratio has a significant effect on shear failure mode and shear bearing capacity (Słowik and Smarzewski 2012). In order to reveal the quantitative influence of shear-span ratio on the shear mechanical properties of beams without stirrups, the 3D numerical analysis models of RC cantilever beams with different shear-span ratios ($\lambda = 1.0$, 1.5, 2.0) were set up by the modelling methods mentioned in Chap. 3. In addition, different section sizes (the beam depths were 200–800 mm) were also considered in these tests to study the size effect on shear strength.

Table 6.2 lists the relevant geometric parameters of some components. The name of the simulated component is "section size-shear-span ratio-longitudinal reinforcement ratio". For example, "S-1.0-1.0" means that the section size of the smaller cantilever beam is 100 mm × 200 mm, the shear-span ratio is 1.0, and the longitudinal reinforcement ratio is 1.0%.

Figure 6.2 presents the ultimate failure forms of RC beams with different shear-span ratios and different section sizes when the longitudinal reinforcement ratio is 1.0%. When the shear-span ratio is smaller (i.e., $\lambda = 1.0$), the RC beam under monotonic concentrated load formed a diagonal crack extending obliquely from the loading point to the fixed end at the bottom, and the RC beam eventually formed several obliquely short columns, showing obvious characteristics of diagonal compression failure. With the increase of shear-span ratio ($\lambda \geq 1.5$), the small RC beam (beam depth was 200 mm) first produced obvious transverse bending cracks under concentrated load. Then the number and extension length of these bending cracks increased. With the further increase of load, these bending cracks began to develop and extended obliquely, and gradually merged with the newly generated obliquely cracks. Finally, shear diagonal cracks extending obliquely from the loading point area to the bottom of

Table 6.2 Geometrical parameters of the RC cantilever beams with different shear-span ratio

Component name	Section size $b \times D$ (mm^2)	Effective depth d_0 (mm)	Shear-span ratio λ	Beam length $l = \lambda \times d_0$ (mm)	Reinforcement ratio ρ (%)
S-1.0-1.0	100 × 200	175	1.0	175	1.0
M-1.0-1.0	200 × 400	370	1.0	370	1.0
L-1.0-1.0	400 × 800	760	1.0	760	1.0
S-1.5-1.0	100 × 200	175	1.5	270	1.0
M-1.5-1.0	200 × 400	370	1.5	560	1.0
L-1.5-1.0	400 × 800	760	1.5	1140	1.0
S-2.0-1.0	100 × 200	175	2.0	350	1.0
M-2.0-1.0	200 × 400	370	2.0	740	1.0
L-2.0-1.0	400 × 800	760	2.0	1520	1.0

Note S—small; M—medium; and L—large

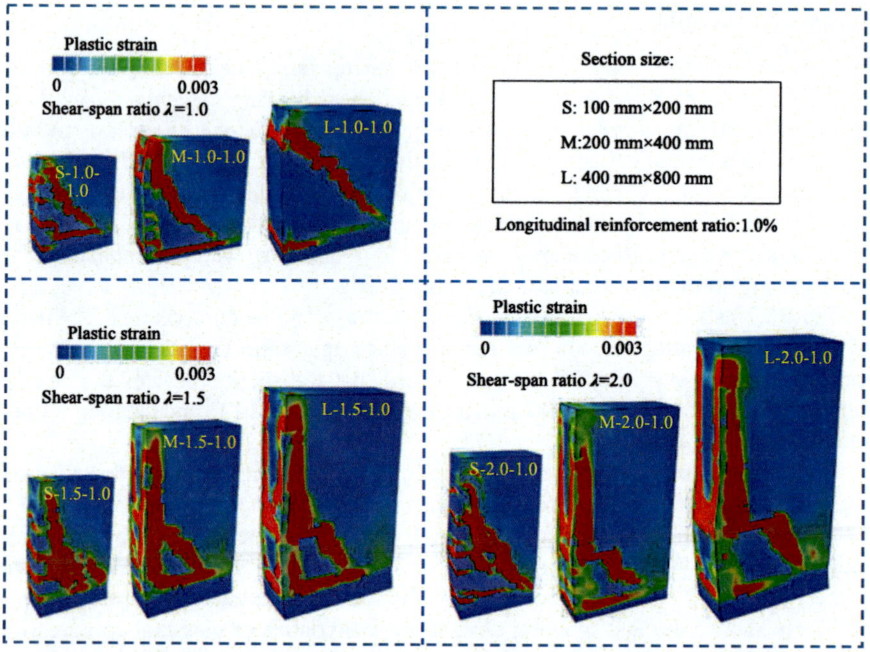

Fig. 6.2 Shear failure modes of the cantilever beams with longitudinal reinforcement ratio of 1.0%

the fixed end were formed, which indicated obvious characteristics of shear compression failure. Compared with the smaller beams, the larger beams had fewer bending cracks when they were damaged, and the starting point of the through oblique cracks were far from the loading point. That was because the larger size beams had greater stiffness and the concrete part near the loading point was difficult to crack when they were sheared.

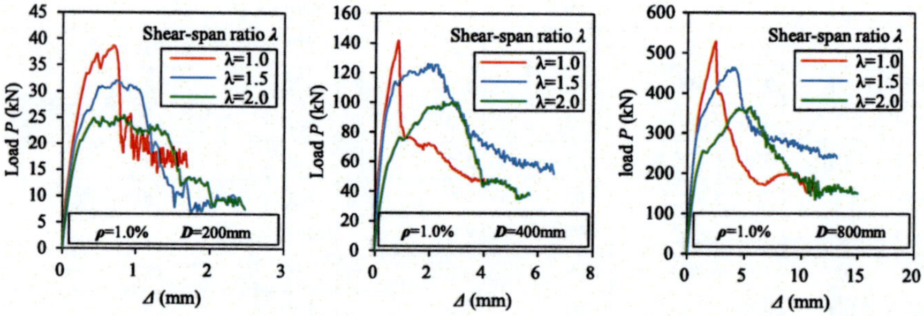

Fig. 6.3 *P-Δ* curves of beams with longitudinal reinforcement ratio of 1.0% under different shear-span ratios

Figure 6.3 plots the load-displacement (P-Δ) curve of each size cantilever beam with the longitudinal reinforcement ratio of 1.0% under different shear-span ratios. It can be seen from Fig. 6.3 that for RC beams with the same section size, the shear-span ratio has a significant impact on the shear capacity which decreases with the increase of shear-span ratio. With the increase of shear-span ratio from 1.0 to 2.0, the shear capacities of the RC beams with the depth of 200 mm were 38.7 kN, 32.1 kN and 25.5 kN, respectively. Compared with the beam with shear-span ratio of 1.0, the shear capacity of beams with shear-span ratio of 1.5 and 2.0 were reduced by 17% and 34%, respectively, which revealed that shear-span ratio has a very significant impact on the shear capacity of RC cantilever beam without stirrups. In addition, the softening sections of the P-Δ curve of the same size cantilever beams were different. When the shear-span ratio was small (shear-span ratio was 1.0), the P-Δ curve of RC cantilever beam dropped suddenly when it reached the peak load. With the increase of the shear-span ratio, the softening stage of the P-Δ curves becomes to be gentle. This indicates that when the shear-span ratio is small, the beam breaks suddenly when it reaches the peak load, and the damage is severe. With the increase of shear-span ratio, the damage of RC cantilever beams gradually eases, which also illustrates that the beams with larger shear-span ratio have better ductility.

The nominal shear strength τ is defined as:

$$\tau = P/(bD_0) \tag{6.1}$$

where, P is the peak value of shear capacity of RC cantilever beam, b is the beam width, and d_0 is the effective depth of beam section.

Figure 6.4 presents the nominal shear strength of RC cantilever beam with different shear-span ratios and section sizes when the longitudinal reinforcement ratio ρ is 1.0%. It can be seen from Fig. 6.4 that the nominal shear strengths of RC beams with the same size decreases with the increase of shear-span ratio. When the beam depth is

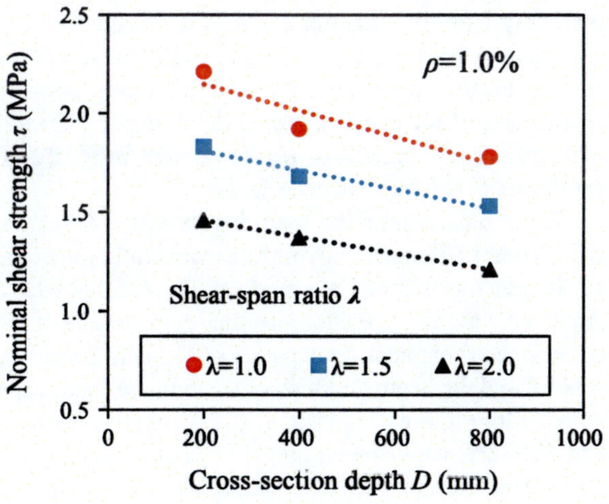

Fig. 6.4 Nominal shear strength of the beams with different shear-span ratios

200 mm, the nominal shear strength of the cantilever beams with different shear-span ratios (1.0, 1.5 and 2.0) are 2.21 MPa, 1.83 MPa and 1.46 MPa respectively. That is to say, compared with the beams with smaller shear-span ratio (1.0), the nominal shear strength of the beams with larger shear-span ratio (1.5 and 2.0) decreased by 17.2% and 33.9% respectively, indicating that the shear-span ratio had a significant impact on the nominal shear strength of RC cantilever beams.

In addition, shown in Fig. 6.4, under the same shear-span ratio, the nominal shear strength of RC cantilever beams decreases with the increase of beam depth. When the shear-span ratio is 1.0, the nominal shear strengths of different beams are 2.21 MPa, 1.92 MPa and 1.78 MPa with the increase of beam depth, respectively. The nominal shear strength of larger size beam is nearly 20% lower than that of smaller size beam, showing a significant size effect. However, the three trend lines of the nominal shear strength of cantilever beams with different sizes are basically the parallel, which indicates that the shear-span ratio had little influence on the size effect of shear strength, that means, the quantitative relationship of the size effect of the nominal shear strength is basically the same under different shear-span ratios.

2. *Longitudinal reinforcement ratio*

In order to study the influence of longitudinal reinforcement ratio on shear failure and size effect of shear strength of RC beams, based on the analysis of the above-mentioned influence of shear-span ratio on shear failure of RC beams, three groups of shear failure simulation tests of RC beams with different longitudinal reinforcement ratio ρ of 0.5, 1.0 and 2.0% were designed under several different shear-span ratios (1.0, 1.5 and 2.0). The geometric parameters and naming method of the test beams are the same as those in Table 6.2.

Figure 6.5 presents the shear failure modes of RC cantilever beams with different longitudinal reinforcement ratio and different cross-section sizes for the shear-span ratio of 1.0. It can be seen from Fig. 6.5 that each RC cantilever beam will produce obvious shear diagonal cracks when it was destroyed. Since the shear-span ratio is relatively small, the concrete beams formed through diagonal cracks extending obliquely from the loading point to the fixed end at the bottom. The failure mode is basically the same as that in Fig. 6.2, because of the large stiffness of the large-size beam, the concrete part near the loading point was difficult to crack, and the crack point of the shear diagonal crack was far from the loading point. In addition, with different longitudinal reinforcement ratio, the ultimate failure mode of each cantilever beam was basically the same.

Figure 6.6 exhibits the load-displacement (P-Δ) curves of RC cantilever beams with different longitudinal reinforcement ratios and beam depths under three different shear-span ratios. It can be seen from Fig. 6.6 that when the shear-span ratio and beam depth were the same, the longitudinal reinforcement ratio had a very significant influence on the shear bearing capacity of RC cantilever beam. When the shear-span ratio was 1.0 and the beam depth was 200 mm, the shear capacities of RC cantilever beams with the longitudinal reinforcement ratios from small to large were 29.0 kN, 38.7 kN and 48.8 kN respectively, and the shear capacities of beams with the longitudinal reinforcement ratio of 1.0 and 2.0% were 33.4 and 68.3% higher than that of beams

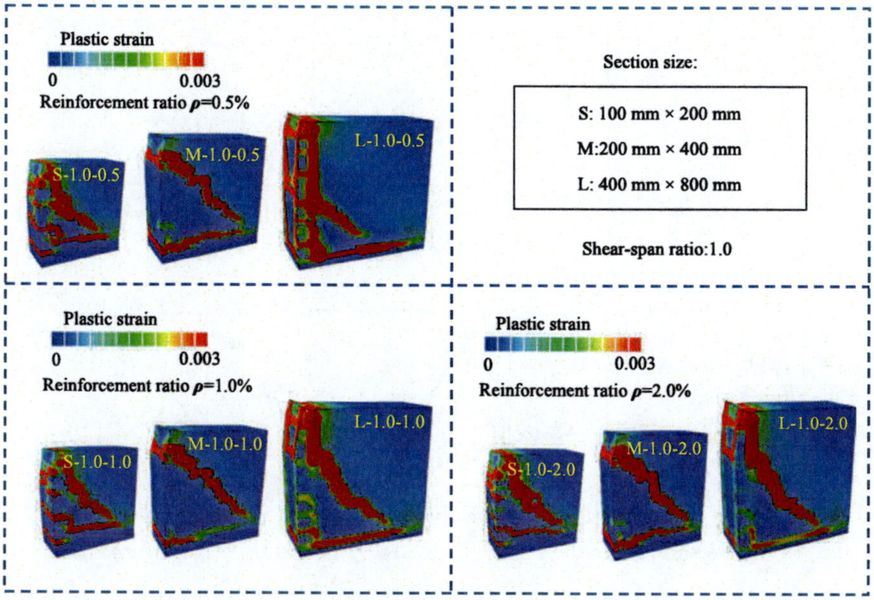

Fig. 6.5 Shear failure modes of beams with different longitudinal reinforcement ratios at shear-span ratio of 1.0

with the longitudinal reinforcement ratio of 0.5%. This indicates that the longitudinal reinforcement ratio has a significant impact on the shear capacity of RC beams. In addition, the longitudinal reinforcement ratio also affects the ultimate displacement of RC beams and the softening behavior of the *P*-*Δ* curves. It can be found that when the longitudinal reinforcement ratio is 2.0%, the ultimate displacements of reinforced concrete beams are small, and the *P*-*Δ* curves suddenly drop and the brittleness increases after reaching the peak value. The reason should be that the shear stiffness of RC beam increases with the increase of longitudinal reinforcement ratio, which leaded to the decrease of its ultimate displacement and deformation ability, and the increase of its brittleness.

Figure 6.7 depicts the nominal shear strength values of the RC cantilever beams with different longitudinal reinforcement ratios and beam depths under three distinct shear-span ratios. It can be seen from Fig. 6.7 that when the shear-span ratio and longitudinal reinforcement ratio are kept the same, the nominal shear strength of RC cantilever beams decreases significantly with the increase of beam depth, exhibiting a significant size effect. Under the same shear-span ratio, the nominal shear strength dramatically increased with the increase of the longitudinal reinforcement ratio. For the beams with the shear-span ratio of 1.0 and the beam depth of 200 mm, the nominal shear strengths of the beams with different longitudinal reinforcement ratios are 1.66 MPa, 2.21 MPa and 2.69 MPa respectively. The nominal shear strength of the RC cantilever beam with longitudinal reinforcement ratio of 2.0% is 62.0% higher than that of the beam with longitudinal reinforcement ratio of 0.5%. In addition, for

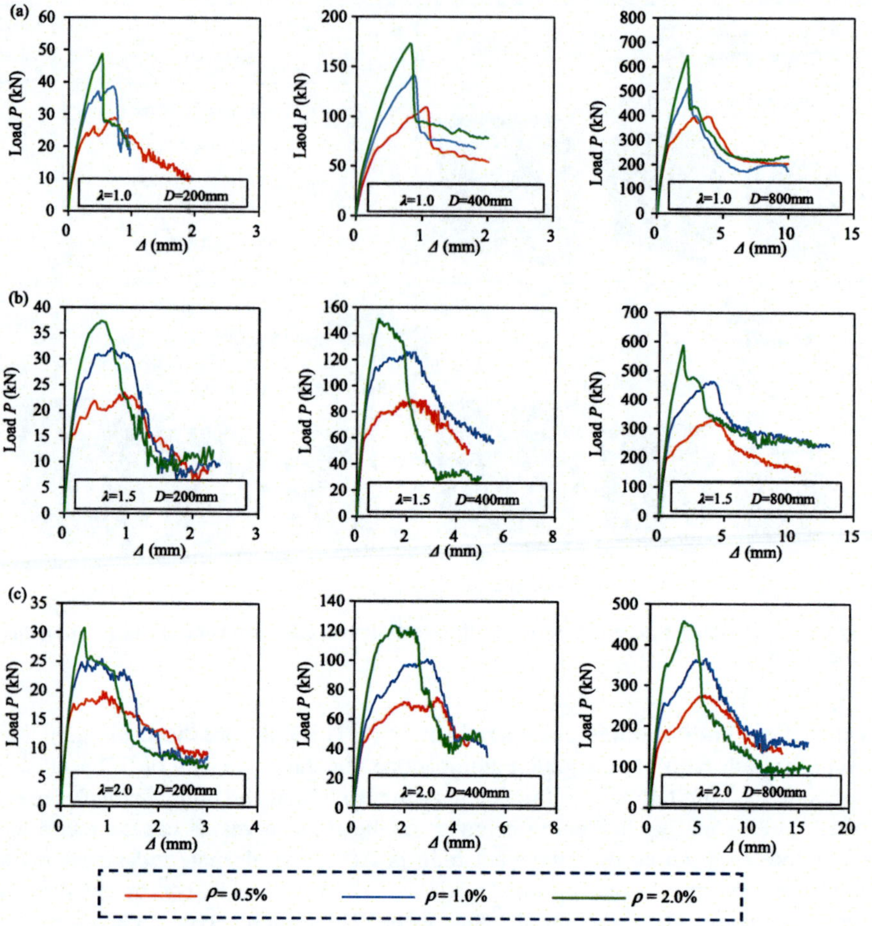

Fig. 6.6 *P-Δ* curves of cantilever beams with different longitudinal reinforcement ratios and sizes under different shear-span ratios: **a** $\lambda = 1.0$; **b** $\lambda = 1.5$; and **c** $\lambda = 2.0$

the same shear-span ratio, the change trends of the nominal shear strength with the section size of RC cantilever beams with different longitudinal reinforcement ratios are basically the same, as observed in Fig. 6.7. This means that the three trend lines are basically the parallel, which indicates that the longitudinal reinforcement ratio only affects the nominal shear strength of RC beams, but has little influence on the size effect of nominal shear strength of RC beams.

Figure 6.8 reveals the effect of longitudinal reinforcement ratio on the shear capacity of RC cantilever beams with different shear-span ratios when the beam depth is 200 mm. It can be seen from Fig. 6.8 that the shear capacity of beams with a smaller shear-span ratio is greater than that of beams with a larger shear-span ratio. In addition, the influence of longitudinal reinforcement ratio on shear capacity of beams with different shear-span ratio is different. When the shear-span ratios are 1.0, 1.5 and 2.0, the shear capacity of the beams with the longitudinal reinforcement ratio of 2.0% is 68.3, 59.9 and 55.8% higher than that of beams with 0.5% longitudinal reinforcement ratio respectively. This indicates that the longitudinal reinforcement ratio has a greater influence on the shear capacity of beams with smaller shear-span ratio. This is because when the shear-span ratio is small, the dowel action of the longitudinal reinforcement is more obvious, which makes the influence of the longitudinal reinforcement ratio on shear capacity greater.

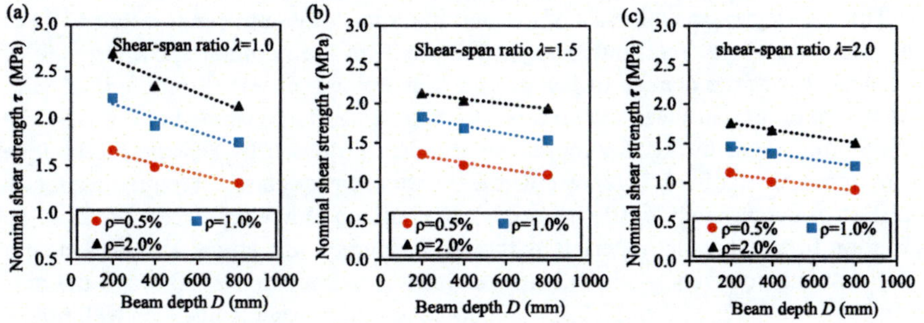

Fig. 6.7 Nominal shear strength of cantilever beams with different longitudinal reinforcement ratios and sizes under different shear-span ratios: **a** $\lambda = 1.0$; **b** $\lambda = 1.5$; and **c** $\lambda = 2.0$

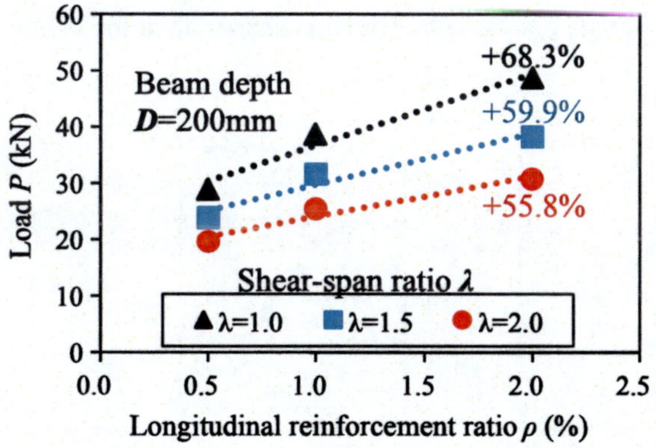

Fig. 6.8 Effect of longitudinal reinforcement ratio on shear capacity of beams with different shear-span ratios

Based on the theory of fracture mechanics of materials, the size effect law (SEL) for predicting the nominal shear strength was proposed by Bažant and Kim (1984) as follows:

$$V_u = \frac{V_0}{\sqrt{1 + D/D_0}} \tag{6.2}$$

where, V_u is the actual shear strength of the concrete member, V_0 and D_0 are empirical coefficients, and D is the characteristic dimension of the concrete member (here is the cross-section depth of the RC beam). The above formula is transformed into a linear equation as follows:

$$Y = AX + C \tag{6.3}$$

where $Y = (1/V_u)^2$, $X = D$, $C = 1/(V_0^2)$, $A = C/D_0$.

Through regression analysis of the nominal shear strength values obtained from the simulation tests, the double logarithm curve of the nominal shear strength of RC beams with the change of beam section size under different longitudinal reinforcement ratios and shear-span ratios was obtained as demonstrated in Fig. 6.9. The logarithmic curve in Fig. 6.9 represents the SEL proposed by Bažant, the oblique line (with a slope of −1/2) represents the theory of linear elastic fracture mechanics (LEFM: for completely brittle materials), the horizontal line represents the strength criterion for the plastic materials without considering size effect. One can clearly note that the correlation coefficients between the simulation results and the size effect law are 0.97, 0.98 and 0.99, which indicates that Bažant's SEL can well reflect the quantitative relationship of size effect of nominal shear strength for RC beams Bažant and Kazemi (1991). In Fig. 6.9, the positions of data points with different longitudinal reinforcement ratios are similar or basically coincident, indicating that the longitudinal reinforcement ratio has little influence on size effect. In addition, the distribution of data points with different shear-span ratios in Fig. 6.9 is basically

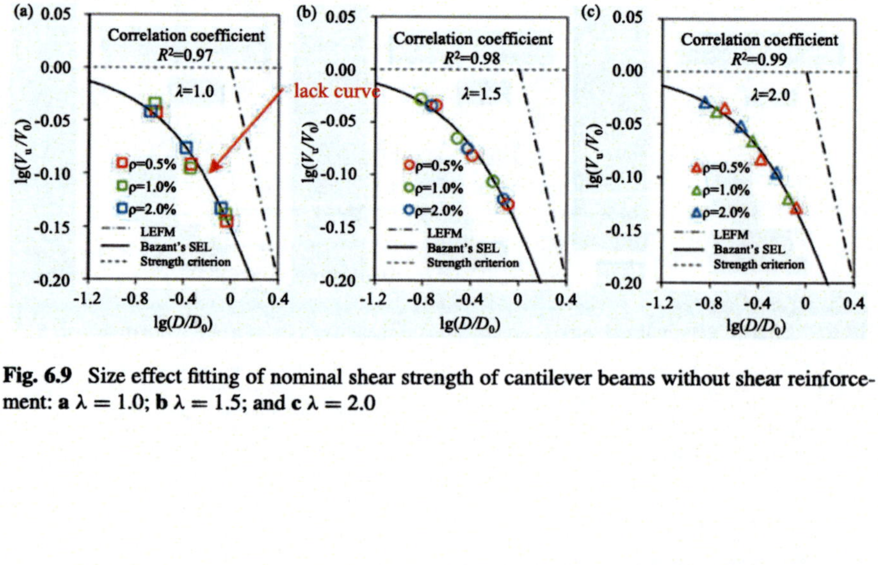

Fig. 6.9 Size effect fitting of nominal shear strength of cantilever beams without shear reinforcement: **a** $\lambda = 1.0$; **b** $\lambda = 1.5$; and **c** $\lambda = 2.0$

the same, indicating that the shear-span ratio has little influence on the size effect of shear failure.

6.1.2 Failure of Lightweight-Aggregate Concrete Beams

Lightweight-aggregate concrete (LWAC) is a superior material which has been widely used in bridge engineering, high-rise buildings and other concrete engineering structures, due to its advantages of light weight, high temperature resistance, and good resistance to fire.

Reinforced concrete beam is a complex component composed of steel and concrete. It is one of the common shear members in civil engineering structures. Shear failure is one of the most dangerous failure modes of reinforced concrete members, which usually shows strong brittleness hence exhibit obvious size effect behavior.

In this section, the shear failure behavior of ordinary and LWAC cantilever beams without stirrups having different section size is presented by the 3D meso-scale numerical simulation method provided in Chap. 3.

In the simulations, the geometrical dimensions (i.e., length L, width b and depth D), cover thickness c, effective beam-depth d_0, shear-span ratio λ and longitudinal reinforcement ratio ρ of the tested beams are given in Table 6.3. It should be noted that NW (normal weight) refers to ordinary aggregate, LW (light weight) refers to lightweight-aggregate, the Numbers "1, 2, 3, 4" refer to the increasing size of the cross section, shear-span ratio is 2.0, longitudinal reinforcement ratio is 1.6%.

Because of LWAC has the characteristics of low aggregate density, low elastic modulus and lower aggregate strength than mortar matrix, the failure characteristics of LWAC and NWC (normal-weight concrete) beams are substantially different. Taking "LW4" specimen as an example, i.e., the LWAC cantilever beam with the

Table 6.3 Geometrical parameters of the concrete cantilever beams

Specimen	Section size $b \times D$ (mm^2)	Cover thickness c (mm)	Effective depth d_0 (mm)	Shear-span ratio λ	Beam length $l = \lambda \times d_0$ (mm)	Reinforcement ratio ρ (%)
NW1	100 × 200	20	180	2.0	360	1.6
LW1	100 × 200	20	180	2.0	360	1.6
NW2	200 × 400	40	360	2.0	720	1.6
LW2	200 × 400	40	360	2.0	720	1.6
NW3	300 × 600	60	540	2.0	1080	1.6
LW3	300 × 600	60	540	2.0	1080	1.6
NW4	400 × 800	80	720	2.0	1440	1.6
LW4	400 × 800	80	720	2.0	1440	1.6

section size of 400 mm × 800 mm. Figure 6.10 shows the load-displacement curve of its shear bearing capacity. Five points are selected on the curve, and the corresponding failure modes are shown in Fig. 6.11a. It can be seen that in the early stage of loading, the specimen presents transverse bending cracks first, and then spreads along the cracks. The stiffness degradation of the beam was not significant, that's point "a".

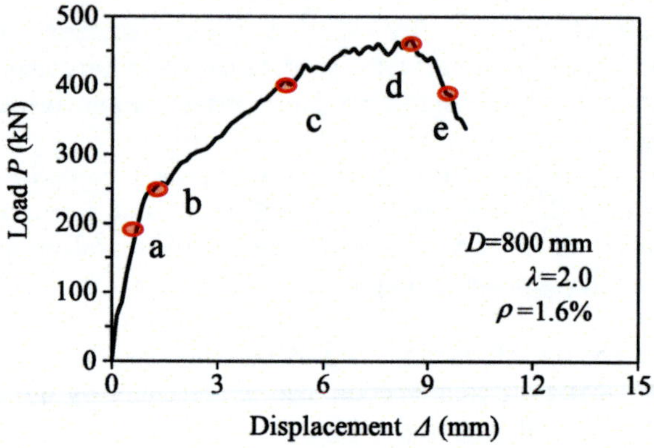

Fig. 6.10 *P-Δ* curves of specimens of LWAC beam at cross-section depth of 800 mm

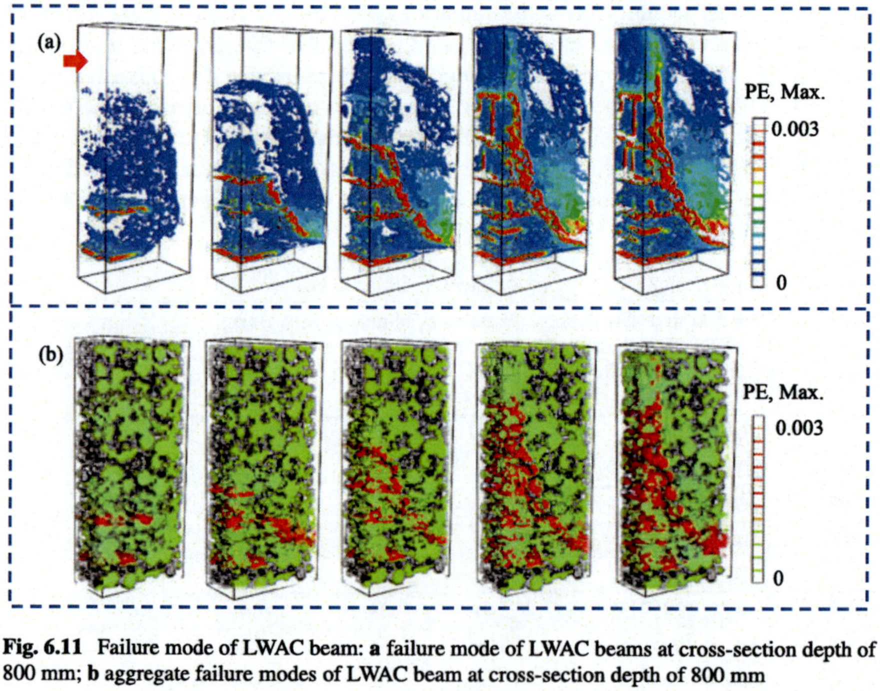

Fig. 6.11 Failure mode of LWAC beam: **a** failure mode of LWAC beams at cross-section depth of 800 mm; **b** aggregate failure modes of LWAC beam at cross-section depth of 800 mm

As the load increases, new cracks appear in the inclined section, the stiffness of the specimen degrades. At point "c", as the crack expands along the aggregate, and penetrates with new cracks appearing in the inclined section, the stiffness of the specimen is unstable and degraded. At point "d", the load reaches its peak, and an inclined crack is formed in the beam that runs through the aggregate from the loading point to the bottom of the fixed end. Until point "e", the inclined crack develops into a critical crack, and the concrete beam is completely destroyed.

As can be seen from Fig. 6.11b, due to the lower strength of lightweight-aggregate particles, internal cracks in the LWAC beams generate cross the interface area and penetrate the lightweight-aggregate particles. Compared with that of NWC beams, the inclined crack surface of LWAC beams under shear failure is smoother, making that the occlusal action of aggregate is less significant relatively, and the brittleness of shear failure will be increased significantly compared with that of NWC beams. This leads to a more obvious size effect in LWAC beams of shear failure inevitably.

The final failure patterns of normal and lightweight-aggregate concrete cantilever beams (shear-span ratio = 2) in different sizes (beam depth: 200–800 mm) are shown in Fig. 6.12. It can be seen that both LWAC and NWC beams have obvious shear brittle failure. The failure modes of the LWAC beams with different sizes are basically the same as that of the NWC beams.

Figure 6.13 shows the load-displacement curves of the numerical models for the mesoscale mechanical analysis of NWC and LWAC cantilever beams with different sizes under displacement loading. It can be seen from Fig. 6.13 that the load-displacement curves of two kinds of aggregate concrete cantilever beams show a linear relationship at the early stage of loading. The curves of two kinds of specimens under the same size are highly coincident, which indicates that the shear force at this stage is basically borne by concrete. When the load reaches 40–50% of the ultimate load, the slope of the curve decreases, and an obvious turning point occurs. The stiffness of the cantilever beam begins to degenerate and enters the inelastic working stage. When the specimens approach the shear capacity (i.e., peak loads), the slope of the curves do not decrease significantly. Finally, the curves drop rapidly after exceeding the shear capacity and the specimen fails.

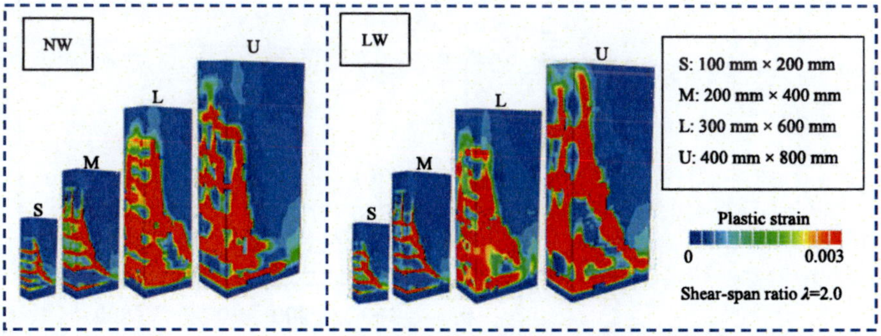

Fig. 6.12 Failure modes of specimens of the concrete cantilever beams at different sizes

Fig. 6.13 *P-Δ* curves of concrete cantilever beams at different sizes

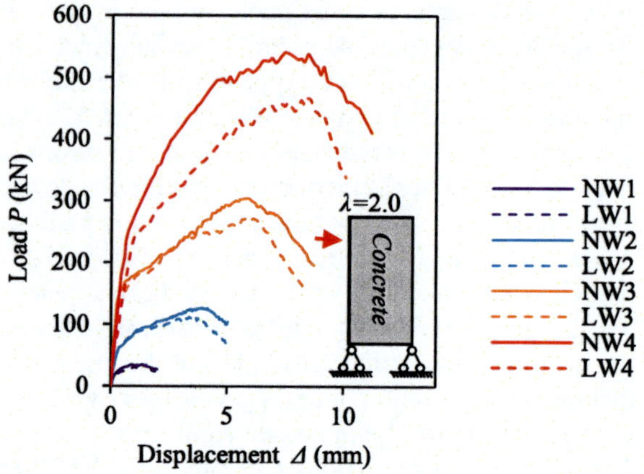

In addition, from Fig. 6.13 one can note that with the increase of the beam size, the LWAC beams gradually fails to reach the shear bearing capacity of the NWC beams. Furthermore, the curvatures of the descending section for LWAC beams are much steeper than that of ordinary ones, indicating that the LWAC beams exhibit more brittle failure patterns than the NWC beams. This, definitely, would lead to a more obvious size effect in shear failure of LWAC beams.

In Fig. 6.14, the nominal shear strengths of the beams having different aggregate types (i.e., lightweight and normal weight aggregates) are compared. It can be seen that the nominal shear strength of LWAC beams is lower than that of NWC beams. For both LWAC and NWC beams, the nominal shear strength decreases with the increase of beam depth, indicating that concrete beams with two types of aggregates exhibit obvious size effect. One can also note that, compared with that of NWC beams, the

Fig. 6.14 The relationship between nominal shear strength and size of cantilever beams

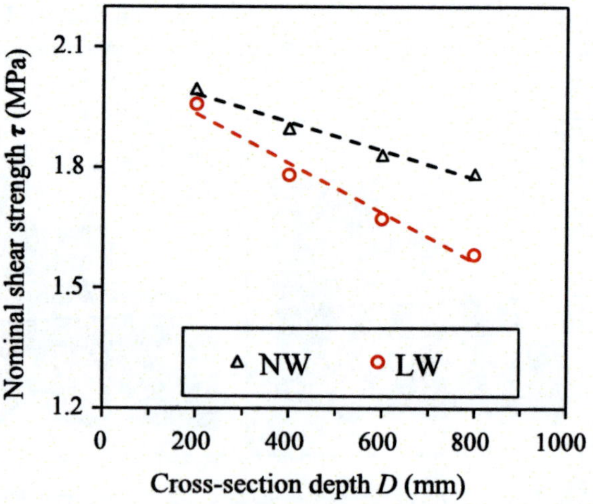

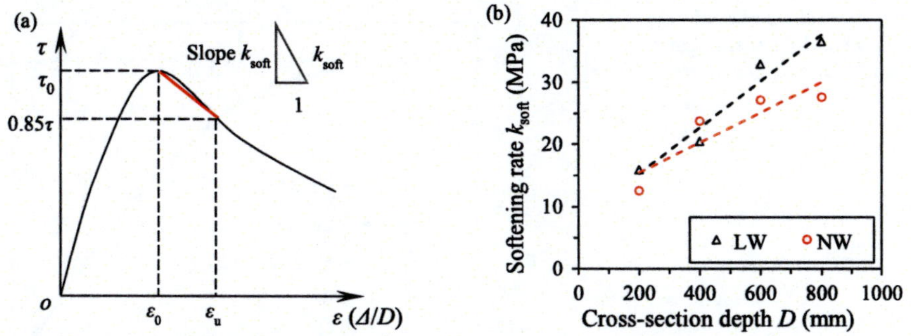

Fig. 6.15 Softening rate k_{soft} of the beams under the CFRP ratios of 0 and 0.0444%: **a** definition of softening rate k_{soft}; and **b** softening rate of the beams with different size

magnitude of the shear strength of LWAC beams decreases faster with the increase of beam depth. This indicates that the LWAC beam presents more significant size effect in shear than the NWA beam. This should be caused by the fact that the LWAC beams exhibit more brittle failure pattern than the NWC ones.

In addition, the size effect is also related to the specimen brittleness, expressed by the softening rate or the loading work during deformation. In this section, the effect of beam depth on the softening behavior at the descending stage was also discussed. Here, the softening behavior is described by a parameter called as softening rate k. In fact, there is more than one way to evaluate ductility of specimens, such as, the ratio of ultimate displacement (the displacement of 85% P_{max} on the descending section of the load displacement curve) to yield displacement. However, there is no unified method for determining the yield point as well. Similar to the work of Imanzadeh et al. (2020), in Fig. 6.15a, the softening rate k_{soft} (unit: MPa) represents the absolute value of slope from the peak nominal shear stress τ_0 to 85% τ_0 in the τ-ε curve (here, the nominal strain ε is calculated as the ratio of displacement Δ to beam length L). The higher the slope (softening rate k_{soft}) value, the steeper the curve of the descending section, and the worse the ductility of the specimens. Actually, these methods are in the characterization of the plastic deformation capacity of specimens, they can all analyze the ductility of specimens qualitatively in essence.

Figure 6.15b presents the relationship between the softening rate k_{soft} and beam depth D for all simulated concrete beams. From Fig. 6.15b, one can note larger-sized beams have a larger softening rate, and thus present a worse ductility. For LWAC beams, the trends that characterize the softening rate curve are above that of NWC beams, and the brittleness of LWAC beams is thus more obvious. However, the strengthening of CFRP would enhance the ductility of concrete beams, as can be concluded from Fig. 6.15b. This is because the FRP pasted on the outer surface of concrete could limit the development of cracks, similar to the effect of stirrups.

The fitting comparison between the simulation results and the Bažant size effect law (SEL) is shown in Fig. 6.16. The fitting correlation coefficient R^2 is 0.99, indicating that the Bažant size effect law can well describe the size effect law of the

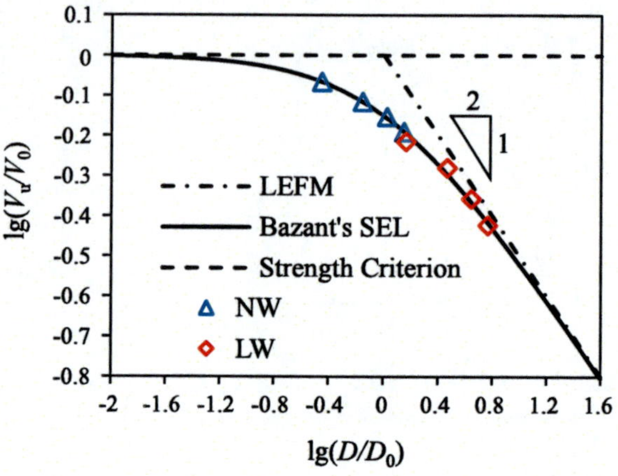

Fig. 6.16 Size effect fitting of nominal shear strength of cantilever beams without stirrups

concrete cantilever beam. In addition, As can be seen from the distribution of data points in Fig. 6.16, the data points of the LWAC specimen are generally below the data points of the ordinary concrete specimen, indicating that the LWAC specimen has great brittleness and low ductility.

6.2 Shear Failure in Reinforced Concrete Beams with Stirrups

Compared with the beam without stirrups, the existence of stirrups in the RC beams limits the development of the cracks, enhances the aggregate interlock behavior, and makes the shear failure behavior of the beam with stirrups more complex. Here, the seismic tests on shear failure of RC beams (Jin et al. 2016) are given. In addition, more simulation works and size effect analysis (Jin et al. 2019b) have been presented.

6.2.1 Seismic Tests on Shear Failure of RC Beams

1. *Test introduction*

Five groups of geometrically similar RC cantilever beams, having different cross-sectional sizes, subjected to low cyclic loading were designed. The RC beams shown in Fig. 6.17 were designed to fail as shear brittle failure. In the present tests, the shear-span ratios λ ($\lambda = l_0/d_0$) of all the RC beams were set as 2, as presented in Fig. 6.17a. The corresponding bending moment and shear distributions along the beam are depicted in Fig. 6.17b.

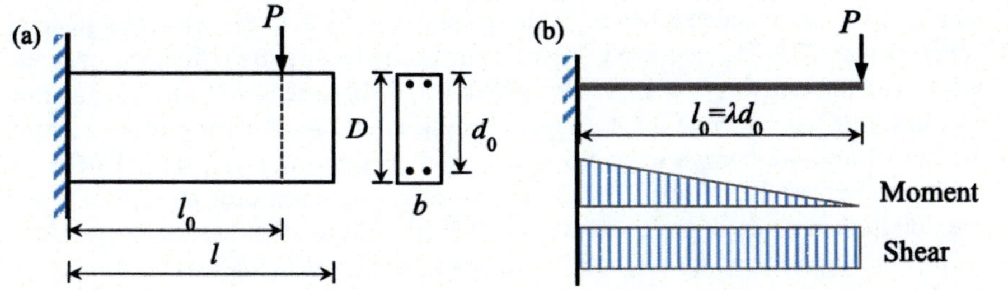

Fig. 6.17 **a** RC cantilever beam design; and **b** bending moment and shear distributions along the beam

The naming of the five groups of specimens was done as XW-1, XW-2, XW-3, XW-4 and XW-5. The cross-sectional sizes of the beams are tested range from 80 mm × 200 mm to 400 mm × 1000 mm, and the total length of the five groups of beams ranges from 490 to 2360 mm. All the specimens were cast from the same batch of continually mixed concrete to minimize statistical scatter of the results, which ensured the material properties of all the specimens to be virtually the same. Plain concrete cube samples with the sizes of 150 mm × 150 mm × 150 mm were also made together with the RC cantilever beams in order to obtain the uniaxial compressive and tensile strengths of the used concrete. The mix consisted of medium sands as the fine aggregate (i.e., the average diameter is between 0 and 5 mm) and crushed pebble stones as the coarse aggregate (i.e., the average diameter is between 5 and 30 mm). It is to be noted that, the concrete mixture components (i.e., aggregate size and distribution) in the present tests were not scaled to the characteristic structure size *D*, and the maximum size of the coarse aggregate particles was about 25 mm for all the specimens. The cover thicknesses of the RC beams of XW-2, XW-3, XW-4 and XW-5 were all 30 mm. While, the thickness of the RC beam of XW-1 with the cross-sectional sizes 80 mm × 200 mm was 15 mm. Since the cross-sectional width for the beam XW-1 is relatively smaller, we did not design a large(r) sized cover thickness. In reality, the cover thickness should be larger than the maximum aggregate size according to design requirements. However, the present study is concerned with the effect of structure size. This is a structural mechanics problem rather than materials science problem. Scaling the aggregate size would complicate evaluation of the test results and prevent isolating the structural size effect from other influences (Li et al. 2016; Barbhuiya and Choudhury 2015)

For the beams of XW-1, XW-2 and XW-3, the structural sizes are relatively small, there were two specimens designed for the tests. While for the beams of XW-4 and XW-5, the structural sizes are much larger, only one specimen was designed for each group due to the economic consideration. In summary, a total of 8 RC cantilever beams were designed and tested, and they are respectively named as XW-1-i, XW-1-ii, XW-2-i, XW-2-ii, XW-3-i, XW-3-ii, XW-4 and XW-5. The detailed structural sizes of the eight specimens, including the cross-sectional sizes, the cross-sectional effective depth d_0, the effective loading length l_0 (i.e., $l_0 = \lambda \times d_0$) and the total length of the beams, are shown in Table 6.4 and Fig. 6.18. In the tests, the grade of concrete was C35. The average value of the splitting tensile strength of the concrete utilized in the tests was $f_t = 1.74$ MPa, which was determined by 150 mm × 150 mm × 150 mm plain concrete cube samples testing after 28-day curing process. And, the actual average compressive strength of the concrete was $f_{cu} = 37.1$ MPa. In addition, the elastic modulus E_c of the concrete used was measured as 27.5 GPa, and the Poisson's ratio ν_c of the concrete was 0.2. In addition, the detailed stirrups and stirrup spacing utilized for all the RC specimens are listed in Table 6.4.

Two material regions of reinforcement have been distinguished. namely, the longitudinal and transverse. The ribbed steel bar of grade HRB335 (i.e., the Hot-rolled Ribbed-steel Bar) was used as the longitudinal reinforcement. To ensure the shear failure pattern of the RC beams, the plain-steel bar of grade HPB235 (i.e., the Hot-rolled Plain-steel Bar) was adopted as stirrups, and the stirrup reinforcement ratios ρ_{sv} of the specimens are all 0.14%. The mechanical parameters of the reinforcing bars used in the tests are listed in Table 6.5, and the loading apparatus are presented in Fig. 6.19.

2. *Shear failure patterns*

Taking the failure behavior of the specimen XW-4 (the cross-sectional sizes are 320 mm × 800 mm) as an example, the development of crack or crack evolution within the RC beam under different levels of loading can be observed from Fig. 6.20. Figure 6.20a, f respectively show the fracture status (i.e., the crack distribution) of the XW-4 specimen, under the loading levels of three to eight. One can see from Fig. 6.20 that the failure process of the specimen is of typical brittle fracture.

For all the test RC cantilever beams, a flexural crack first developed from the bottom surface of the RC cantilever beam in the region near the support, where the

Table 6.4 Physical parameters of the tested RC cantilever beams

Specimen name	XW-1	XW-2	XW-3	XW-4	XW-5
Total number	2	2	2	1	1
Cross-sectional sizes ($b \times D$) (mm^2)	80 × 200	160 × 400	240 × 600	320 × 800	400 × 1000
Longitudinal reinforcement ratio ρ_w (%)	1.66	1.14	1.24	1.07	1.06
Number of stirrup legs	2	2	2	3	3
Stirrup spacing (mm)	220	110	70	80	65
Diameter of stirrup (mm)	4	4	4	4	4
Stirrup reinforcement ratio ρ_{sv} (%)	0.14	0.14	0.14	0.14	0.14
Shear-span ratio λ	2.0	2.0	2.0	2.0	2.0
Effective section depth d_0 (mm)	170	345	540	735	930
Effective loading length $l_0 = \lambda \times d_0$ (mm)	340	690	1080	1470	1860
Total length of beam l (mm)	490	940	1430	1920	2360

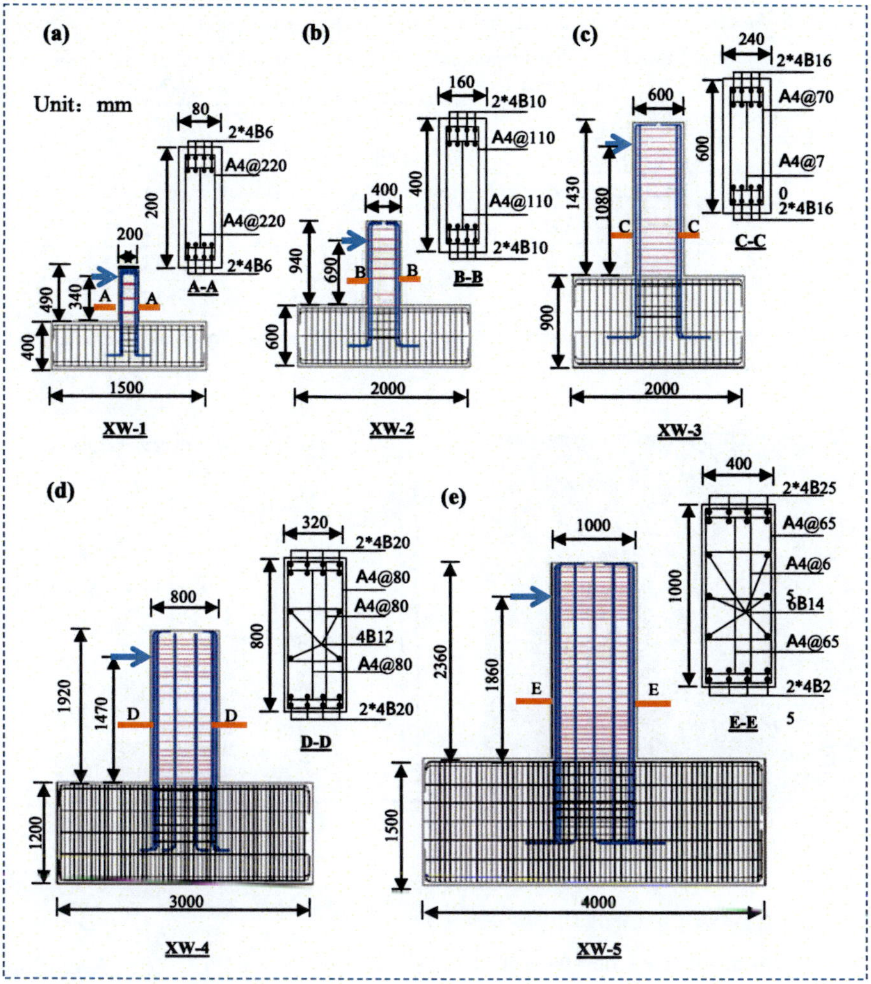

Fig. 6.18 Details of the tested RC cantilever beams

maximum moment occurred. As the applied load continued, flexural cracks spread out toward the loading point. When the load was sufficient to impose diagonal tension that was greater than the concrete's tensile strength, an inclined crack appeared, immediately followed by the failure of the RC beam.

Before the yielding of the XW-4 beam, the number of cracks generated within the specimen was relatively rare, and the corresponding crack width was still small, as presented in Fig. 6.20a, b. When the applied loading approached the yielding load, the width of cracks increased drastically, and the main diagonal cracks formed, as shown in Fig. 6.20c, e. In addition, it can be noted that lots of hairline diagonal cracks generated with the appearance of the main diagonal cracks. As the loading continued,

Table 6.5 Mechanical parameters of the reinforcing bars used in the tests

Diameter (mm)	Grade	Elastic modulus (GPa)	Yield strength (MPa)	Ultimate strength (MPa)
4	HPB235	210	297.89	425.74
6	HRB335	196	405.97	529.64
10	HRB335	198	392.10	564.63
16	HRB335	202	399.13	530.93
20	HRB335	193	402.00	542.51
25	HRB335	195	374.84	536.29

Note "HRB" means "Hot-rolled Ribbed-steel Bar"; and "HPB" means "Hot-rolled Plain-steel Bar"

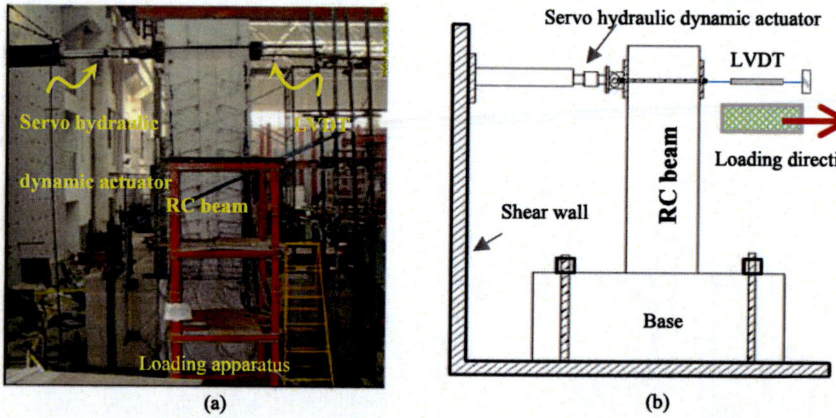

Fig. 6.19 Loading apparatus and schematic diagram

two cross main cracks generated within the RC specimen (see Fig. 6.20f). According to the final failure mode of the specimen, the most severely damaged regions should be the ones where the positive main crack and the negative main crack intersected, as can be described in Fig. 6.20f. In the severely damaged region, pieces of concrete fall off due to concrete crushing. At this time, the cover concrete peeled off, and the coarse aggregate particles as well as the steel reinforced bars can be clearly seen. As the width of cracks increased continuously, the stirrups near the two main inclined cracks may reach their tensile strength and then the necking fracture occurred. In reality, the stirrups with the concrete specimens can bear the shear actions directly and limit the propagation of the diagonal cracks effectively. With the existence of the stirrups, the diagonal cracks cannot open freely, which enhances the friction between the coarse aggregate particles. However, it is to be noted that, when the stirrup ratio exceeds a critical value, the increase of stirrup ratio has a negligible influence on the shear capacity of RC beams. Besides that, one can note from Fig. 6.20f that, severely

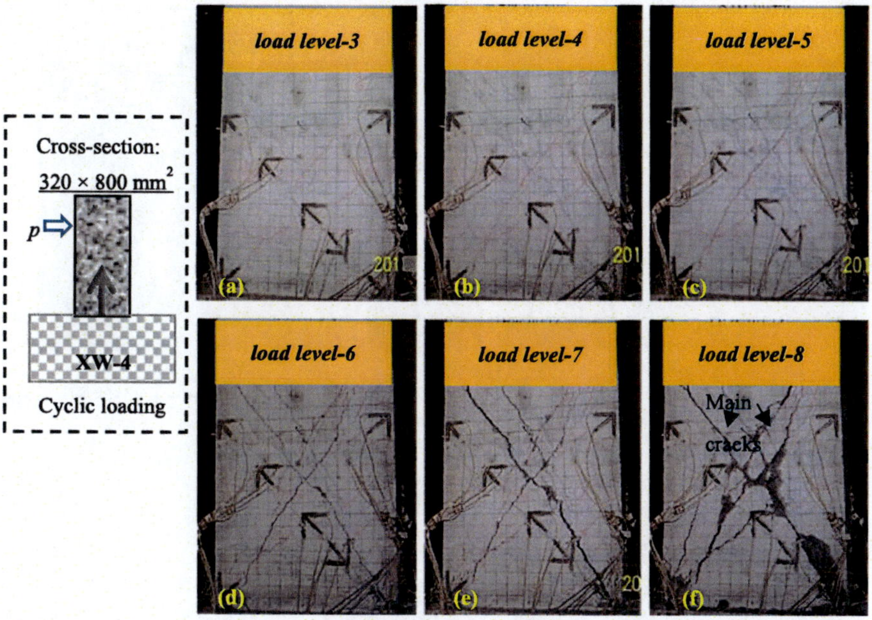

Fig. 6.20 Damage evolution process of the RC beam of XW-4 under different levels of loading

damage also generates at the root areas of the RC cantilever beam under repeated cyclic compressive loadings.

The failure mechanism of the RC cantilever beams of XW-4 can be described as: after appearance of two localized cracks within the specimens, the two diagonal localized cracks started to develop in a shear mode.

Under the low cyclic fatigue loading, the final failure patterns of the eight RC cantilever beams are presented in Fig. 6.21. It can be found clearly from Fig. 6.21 that, the failure patterns of the eight specimens are very similar. Almost all the specimens failed in an “X”-shape, which is of typical brittle shear failure. Nevertheless, the damage distributions of the smaller-sized specimen have some differences with those of the larger-sized specimens. For instance, comparing the XW-1 specimen with the XW-5 specimen, one can see clearly from Fig. 6.21 that the damage of the XW-1 specimen was much more severely than the XW-5 specimen. Compared with that of larger-sized specimens, the distribution of the cracks generated within the smaller-sized specimen was more uniform, and the concretes fall off at most regions of the specimen. However, for larger-sized specimens, e.g. for the XW-4 and XW-5 specimens, the failure localized at the intersection region of the two main diagonal cracks, and the distribution of cracks was relatively focused.

For the smaller-sized specimens, the widths of the main diagonal cracks were small(er), and only the stirrups at the middle-part of the RC beam are broken due to necking fracture, as can be observed in Fig. 6.22. While for the larger-sized specimens, the corresponding crack widths were much larger, and most of the stirrups

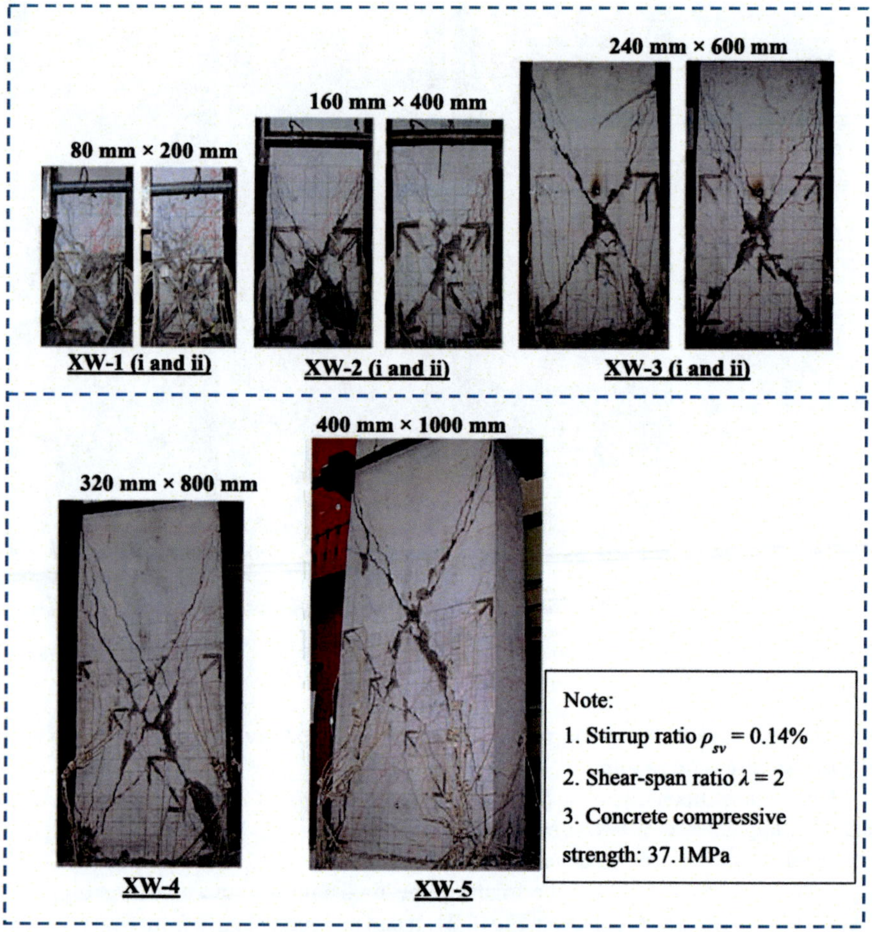

Fig. 6.21 Failure patterns of the eight RC beams with different structure sizes

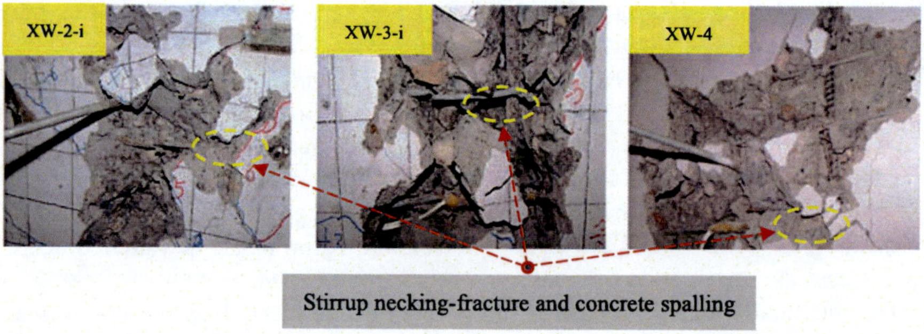

Fig. 6.22 Typical stirrups break due to necking-fracture within the RC cantilever beams

crossing the two main diagonal cracks are snapped. As stated in the work of Bažant and Kim (1984), the most important consequence of wider cracks is diminished residual tensile stress and aggregate interlock. It is to be noted that, the evolution of crack width was not recorded in the tests. Nevertheless, one can see clearly from the broken photographs of the five groups of RC beams that, the corresponding crack widths shows significant differences. In general, the larger the size of the RC specimens is, the larger diagonal crack width the specimens possess when they failed. With a larger crack width, the friction of the shear forces is reduced. Consequently, so also is the ultimate capacity. Furthermore, the diagonal cracks at failure had similar paths and relative lengths at the maximum load independently of the beam size. Consequently, this kind of size effect in failures occurs after large stable crack growth, and it should be caused by the energy release associated with stress redistribution caused by a large crack (herein two main diagonal cracks). This should be the major reason for the size effect on the shear strength of the RC cantilever beams.

3. *Shear-displacement hysteretic behavior and envelopes*

Hysteretic curves of RC beams subjected to low cyclic loading provide important information for the evaluation of seismic behavior, and they can reflect the ductility, energy-dissipating capacity and degradation of stiffness of RC beams. The recorded hysteretic curves of load-displacement relation of the eight RC beams are plotted in Fig. 6.23, in which the "shear" in the Y-axis is equal to the applied load on the beam. From the blue curves plotted in Fig. 6.23, it can be found that, there is an initial elastic response for all specimens and the load-displacement relationship can be considered roughly to be linear. In general, for the eight RC specimens, the shapes of the recorded hysteretic curves are all plump. This illustrates that the RC beams with various cross-sectional sizes have a good energy dissipation capacity.

As illustrated in Fig. 6.23a, h, the pinching behavior within the eight hysteretic curves can be found obviously. The pinching behavior of hysteretic curves is associated with the following factors: the interaction of steel rebar and surrounding concrete (especially the bond-slip behavior), the diagonal crack width within concrete, the plastic deformation accumulation of concrete and the deformation of steel rebar, etc. Under the cyclic loadings, the diagonal cracks within the RC beams open and close continually, causing the decrease of the shear capacity of the RC beams. The larger the shear deformation is, the more serious the shear stiffness degrades. This should be the main reason that the pinching phenomena generates.

In the early stage of the test, under the applied loads, although some diagonal cracks are generated within the RC beams, the global deformation of the RC beams is relatively small. At this time, the beams are in the elastic stage, the bond between steel bars and the surrounding concrete is intact, and the residual deformation is small, leading to a smaller stiffness degradation. In general, at this stage, the corresponding hysteretic hoop obtained exhibits an arcuate shape, not an ideal fusiform shape. Moreover, the areas covered by the hysteretic hoops are relatively small, as can be noticed from curves in Fig. 6.23a, h. In the middle-stage of the test, as the applied shear load increases, the RC beams enter the plastic stage, and plastic deformation

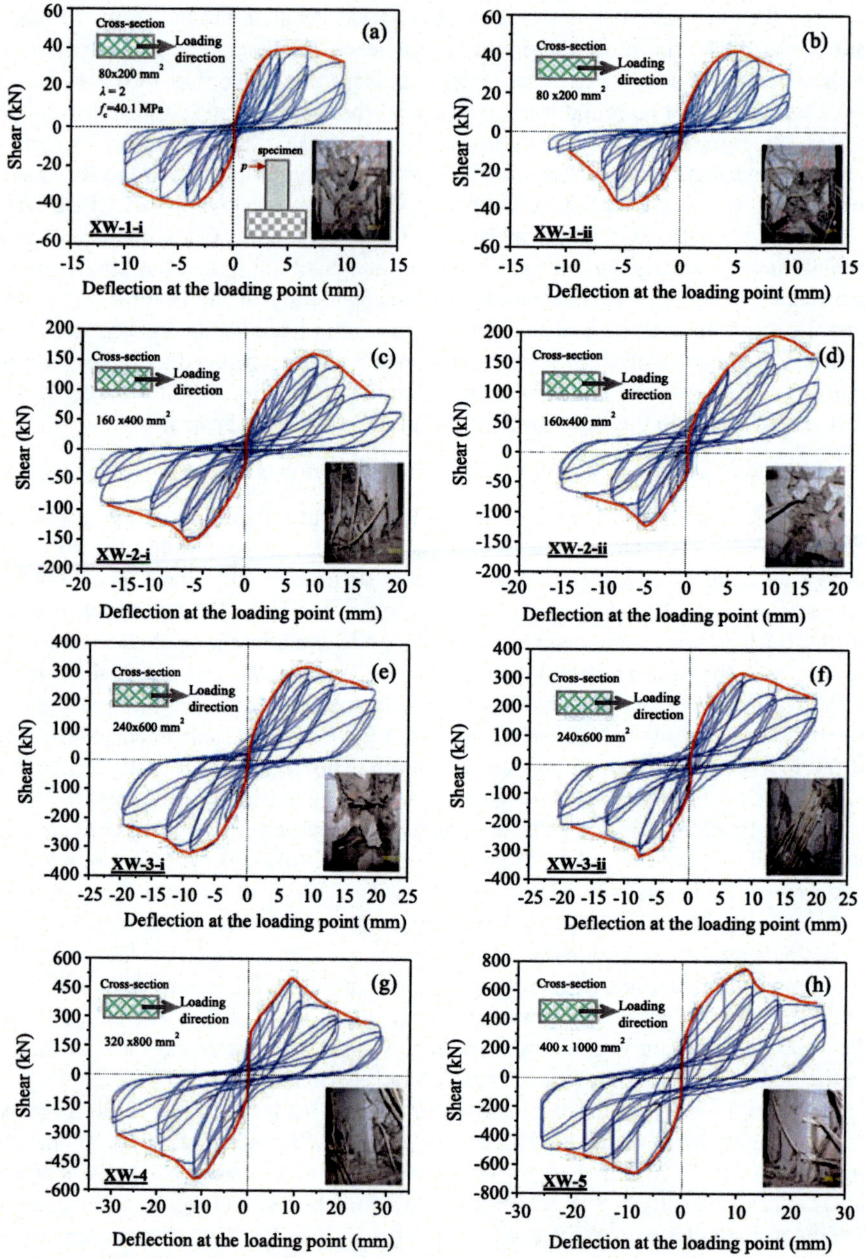

Fig. 6.23 Tested hysteretic curves and the corresponding envelope curves of the eight RC cantilever beams

occurs in the beams. At this stage, the residual deformation of the beams after loading becomes large(r), making that the shear stiffness degrades fast(er).

Exactly for this, the corresponding hysteretic hoops possess a reversed "S-shape", and the areas covered by the hysteretic hoops become large(r). In the post-stage of the test, much larger plastic deformation is generated with the increase of applied displacement, the deformation is thus difficult to recover. The obvious slip between steel bars and surrounding concrete would generate. Moreover, the stirrups within the RC beams may reach their yield strength. And, with the increase of prescribed displacement at the beams, the necking fracture of stirrups could occur, as mentioned previously in Fig. 6.22. For this, the residual deformation of the RC beams is significantly large, which makes the shear stiffness degrade sharply. At this stage, the typical "Z-shaped" hysteretic hoops emerge.

In order to evaluate the seismic performances of the RC beams, the envelope curves (the maximum points of the shear-deflection hysteretic curves) were plotted for the eight RC cantilever beams subjected to low cyclic loading, as the red lines plotted in Fig. 6.23a, h, respectively. One can notice that the shear load-drift curves for all the eight specimens evolved similarly and they all experienced a softening stage after reaching the ultimate loads. Moreover, it is to be noted that the load carrying capacities of each RC cantilever beam are different in the positive and negative loading cycles, which should be caused by the locking and unlocking effects of the spirals.

4. *Loading carrying capacity*

It is well known that the shear strength of a beam is affected by the inclusion of steel reinforcements such as tension and shear reinforcements. According to the Code for design of concrete structures of China (GB 50010 2010), the theoretical load carrying capacity V for the shear performance of RC beams can be calculated by Eq. (6.4),

$$V = \frac{1.75}{\lambda + 1} f_t b h_0 + f_{yv} \frac{A_{sv}}{s} h_0 \tag{6.4}$$

in which, λ is the shear-span ratio, b is the cross-section width of beam, f_t is the tensile strength of concrete, h_0 is the cross-sectional effective depth of beam, A_{sv} is the total area of the stirrup bars in the same section, f_{yv} is the tensile strength of stirrup bars, and s is the stirrup spacing. In addition, the shear reinforcement ratio is defined as $\rho_{sv} = A_{sv}/bs$. The corresponding shear strength v_n can be evaluated as $v_n = V/bh_0$.

Equation (6.4) neglects the size effect obviously, and thus it cannot describe the influence of structural size of the shear strength. Furthermore, one can clearly note from Eq. (6.4) that, the stirrups contribute to the enhancement of the shear capacity and the corresponding shear strength of RC beams.

Herein, the parameter—"safety reserve coefficient η", defined as the ratio of the tested bearing capacity and the theoretical bearing capacity, was utilized to describe the security of the RC components. The bearing capacity measured in the present tests and the computed theoretical bearing capacity, as well as the safety reserve coefficient are shown in Table 6.6. It can be noted from Table 6.6 that, with increasing the structural size, the safety reserve coefficient decreases pronouncedly. When the cross-sectional size increases from 80 mm × 200 mm to 40 mm × 1000 mm, the corresponding safety reserve coefficient η is deduced by 47.7% (i.e., (2.37 − 1.24)/2.37 = 47.7%). That is to say, if the theoretical formula in Eq. (6.4) is utilized to compute the bearing capacity of RC beam with larger cross-section sizes, the obtained bearing capacity will be overestimated. This would make that the results of calculation are not safe.

5. *Ductility capacity*

In the seismic design of composite structures, ductility is one of the most important parameters. Similar with that of Rodrigues et al. (2013), the displacement ductility coefficient, defined as $\mu = \Delta_u/\Delta_y$, was utilized to quantitatively examine the ductility capacity of the RC components. And, it was thus adopted to assess the seismic performance of the RC beams with various cross-sectional sizes. Herein, Δ_u means the displacement when the applied load reaches the ultimate load carrying capacity, and Δ_y characterizes the displacement corresponding to yield strength, namely the yielding displacement.

The method proposed by Rodrigues et al. (2013) was employed to determine the yielding displacement and ductility. The detailed procedure of the method involves four steps: (1) evaluation of the maximum strength of the specimen in both test directions; (2) identification of the cycle for which the strength is lower than three quarters of the previously evaluated maximum strength of the specimen; (3) calculation of the secant stiffness (K_y) for the cycle identified in step 2; and (4) adjustment of the branch corresponding to the post-yield stiffness (K_{pl}). It is to be noted that, determination of the intersection points of K_y and K_{pl}, for each direction, leading to the yield displacement in each direction. The mean of the yield displacements obtained in the two directions provides the reference yield displacement Δ_y. The method adopted to

Table 6.6 The obtained bearing capacity and the displacement ductility coefficient

Specimen name	Size ($b \times D$) (mm^2)	Tested P_{max} (kN)	Theoretical P_{max} (kN)	η	Δ_y (mm)	Δ_u (mm)	μ
XW-1-i	80 × 200	40.03	20.72	1.93	2.92	10.09	3.45
XW-1-ii	80 × 200	40.38	20.72	1.95	3.39	10.38	3.06
XW-2-i	160 × 400	157.30	67.10	2.34	4.30	11.61	2.70
XW-2-ii	160 × 400	159.15	67.10	2.37	6.59	15.49	2.35
XW-3-i	240 × 600	321.63	201.23	1.60	8.95	19.61	2.19
XW-3-ii	240 × 600	320.21	201.23	1.59	7.73	19.79	2.56
XW-4	320 × 800	519.63	362.87	1.43	8.71	19.07	2.19
XW-5	400 × 1000	705.89	570.33	1.24	10.01	17.61	1.76

Note η: safety reserve coefficient; Δ_y: displacement at yield load; Δ_u: displacement at 15% strength degradation from peak; and μ: displacement ductility coefficient

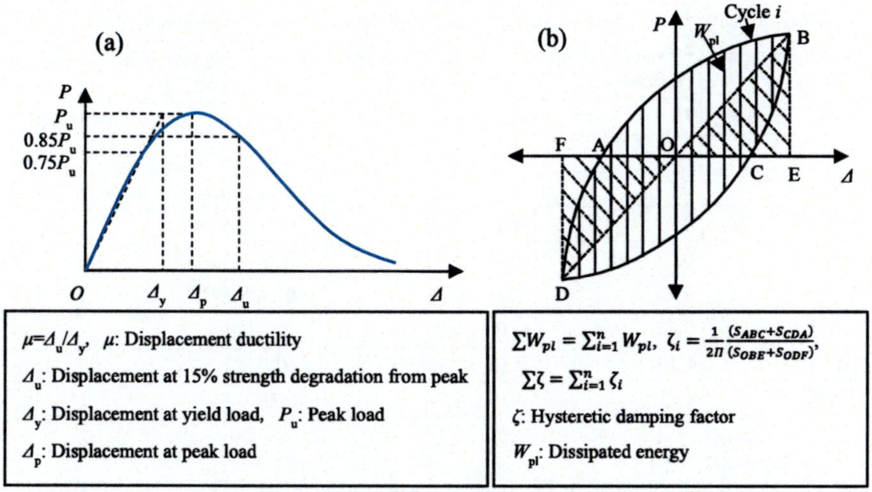

Fig. 6.24 Definition of: **a** the imposed cyclic displacement ductility factor μ; and **b** the dissipated energy W_{pl} and the equivalent hysteretic damping factor ζ

evaluate the yield displacement has a clear advantage in this analysis (in comparison to estimation by simplified formulae) because it is based on experimental test results. The method in details refers to Rodrigues et al. (2013). For the ultimate displacement, its value is assumed to be equivalent to the displacement corresponding to the 85% bearing strength in the decline period (Han and Yang 2005; Lu et al. 2015), as shown in Fig. 6.24a.

The values of Δ_y, Δ_u and μ for the five groups of RC beams with different cross-sectional sizes were calculated and listed in Table 6.6. Herein the ductility coefficients μ are the average values of each set of specimens with same structural sizes. One can note from Table 6.6 that, as the sizes of the cross-section of beam increase, the tested ductility coefficient μ decreases obviously. When the cross-sectional size increases from 80 mm × 200 mm to 400 mm × 1000 mm, the corresponding ductility coefficients μ is deduced by 49% (i.e., (3.45 − 1.76)/3.45 = 49%). Therefore, it can be concluded that the ductility capacity coefficient decreases with increasing the size of RC specimens. The larger the cross-sectional size is, the poor ductility capacity the RC beam possesses. This leads to a weakened deformation capacity and poor seismic performance. In summary, with increasing the structural size, the ductility of the RC beams decreases.

6. *Energy dissipation*

The energy dissipation capacity of the five groups of RC beams with different structural sizes was explored. The load versus deflection relationship can be described as presented in Fig. 6.24b. The area of the hysteretic loop is a very important factor for evaluating the elastoplastic energy dissipation capacity of the specimen. The larger the hysteretic loop is, the better the energy dissipation capacity will be. Equation (6.5)

Table 6.7 Energy dissipation coefficient E_e and equivalent viscous damping ratio ζ_e

Specimen name	Hysteretic hoop (yielding)		Hysteretic hoop (peak loading)		Hysteretic hoop (ultimate loading)	
	E_e	ζ_e	E_e	ζ_e	E_e	ζ_e
XW-1-i	0.53	0.08	0.55	0.09	0.54	0.09
XW-1-ii	0.77	0.12	0.86	0.14	0.74	0.12
XW-2-i	0.31	0.05	0.84	0.13	1.10	0.17
XW-2-ii	0.57	0.09	1.01	0.16	1.02	0.16
XW-3-i	0.55	0.09	0.89	0.14	1.05	0.17
XW-3-ii	0.62	0.10	0.93	0.15	1.72	0.27
XW-4	1.13	0.18	1.79	0.29	1.16	0.18
XW-5	1.21	0.19	1.74	0.28	1.31	0.21

can be utilized to calculate the equivalent damping coefficient (ζ_e). Herein, S_{ABC} and S_{CDA} are the areas under the curves of ABC and CDA, as shown in Fig. 6.24b, respectively. And, S_{OBE} and S_{ODF} are the areas within triangles of OBE and ODF, as shown in Fig. 6.24b, respectively. ζ_e has the form as (Wang et al. 2014):

$$\zeta_e = \frac{1}{2\Pi}\frac{(S_{ABC} + S_{CDA})}{(S_{OBE} + S_{ODF})} \tag{6.5}$$

The dissipated energy capability (E_e) of each hysteretic loop can be computed as

$$E_e = 2\Pi\zeta_e \tag{6.6}$$

The energy dissipated coefficients and the equivalent damping coefficients of the hysteretic hoops, corresponding to those when reaching the yielding loads, the maximum loads and the ultimate loads, are calculated and listed in Table 6.7. One can see from Table 6.7 that, both the energy dissipation coefficient (E_e) and the equivalent damping coefficient (ζ_e) of the RC specimens have a trend of increasing as the structural size becomes larger at each state of the loading history.

Figure 6.25 presents the plots of the dissipated energy W_{pl} and the cumulative dissipated energy $\sum W_{pl}$ versus the number of cycles *n*, respectively. From Fig. 6.25a one can see that in each hysteretic loop, the quantity of dissipated energy gets larger as the number of cycles *n* increases, and from Fig. 6.25b it indicates that the larger the specimen is, the more energy will be dissipated.

Figure 6.26 illustrates the relationships between the hysteretic damping factor ζ for each of the loop or the cumulative hysteretic factor $\sum\zeta$ and the number of cycles *n*. Figure 6.26a shows that the equivalent hysteretic damping factor for each loop is getting lower as the number of cycles arises. This is due to the fact that the progressive damage evolves in the specimen as the load gets to be larger. One can see from Fig. 6.26b that the cumulative damping factor of each specimen becomes larger as the number of cycles arises, and the larger the structural size is, the larger the

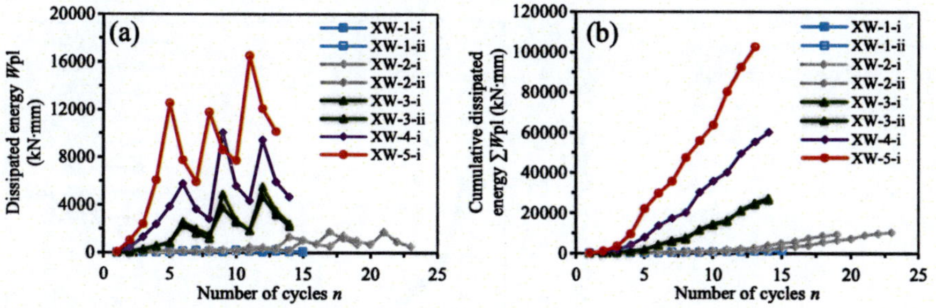

Fig. 6.25 **a** The dissipated energy W_{pl}; and **b** the cumulative dissipated energy $\sum W_{\text{pl}}$ curves for the eight RC cantilever beams

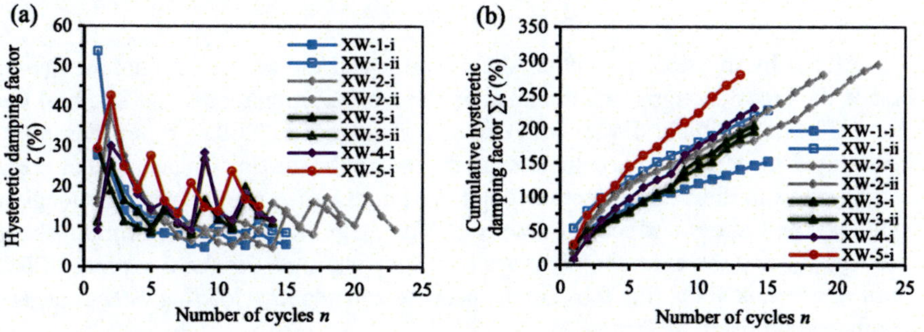

Fig. 6.26 **a** The hysteretic damping factor ζ; and **b** the cumulative hysteretic damping factor $\Sigma\zeta$ curves for the eight RC cantilever beams

cumulative damping factor will be. This definitely indicates that a larger specimen pronounces a greater energy dissipation capacity.

7. *Stiffness degradation*

The stiffness degradation of the specimen plays an important role in the energy dissipation capacity. The relationship between the stiffness and the number of cycles influences the area of the hysteretic loop greatly. Figure 6.27 presents such relation for normalized stiffness E/E_0 with the number of cycles n. E is the secant stiffness at different cycles and E_0 is the initial tangent stiffness. From Fig. 6.27, one can see that the normalized stiffness evolves similarly for all the eight RC specimens. That is to say, the stiffness degradations of the RC beams having different structure sizes are very close. Nevertheless, one can note that, on the whole, compared to that of smaller-sized specimens, the stiffness of the larger-sized RC specimens degrades faster due to their higher brittleness.

8. *Nominal shear stress-strain relations*

As the relationship of the applied shear and the drift is obtained, the nominal shear stress and the nominal shear strain can be evaluated. Herein, the nominal shear stress

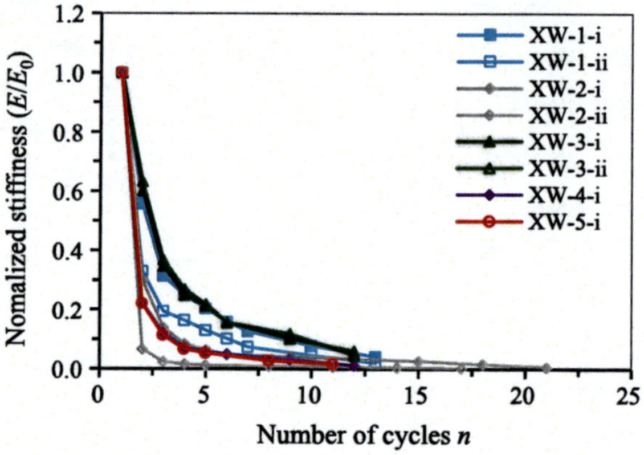

Fig. 6.27 Stiffness degradation curves for the eight RC cantilever beams

τ is defined by division of applied shear load p and the cross-sectional area ($b \times d_0$, b is the cross-sectional width and d_0 is the cross-sectional effective depth of the beam). The definition is similar with that of Bažant and Yu (2005a, b). Furthermore, the nominal shear strain γ is calculated as the ratio of beam-tip displacement Δ at the loading point to the beam effective length l_0 (i.e., the distance between the loading point and the bottom-surface) of the beam. The magnitudes of the maximum shear loading p_{max} and the corresponding nominal shear strength for the five types of RC cantilever beams, including those in the positive and negative loading directions, are calculated and listed in Table 6.8.

The obtained quantitative relationships between the nominal shear stress and shear strain are plotted in Fig. 6.28a, for the eight RC beams tested. From Fig. 6.28a, one can note that, with the increase of structural size, both the nominal shear strength and the displacement ductility coefficient decrease obviously. Moreover, the failure

Table 6.8 Load carrying capacity and the corresponding nominal shear strengths for the eight RC beams

Specimen name	Cross-sectional width b (mm)	Effective beam depth d_0 (mm)	Load carrying capacity p_{max} (kN) (positive/negative)	Nominal shear strength $\tau_{max} = p_{max}/(b \times d_0)$ (MPa) (positive/negative)
XW-1-i	80	170	40.33/39.73	2.96/2.92
XW-1-ii	80	170	42.53/38.22	3.13/2.81
XW-2-i	160	345	161.57/153.03	2.92/2.77
XW-2-ii	160	345	195.63/122.66	3.54/2.22
XW-3-i	240	540	321.63/321.62	2.48/2.48
XW-3-ii	240	540	320.41/320.01	2.47/2.47
XW-4	320	735	499.07/540.19	2.12/2.30
XW-5	400	930	755.42/656.35	2.03/1.76

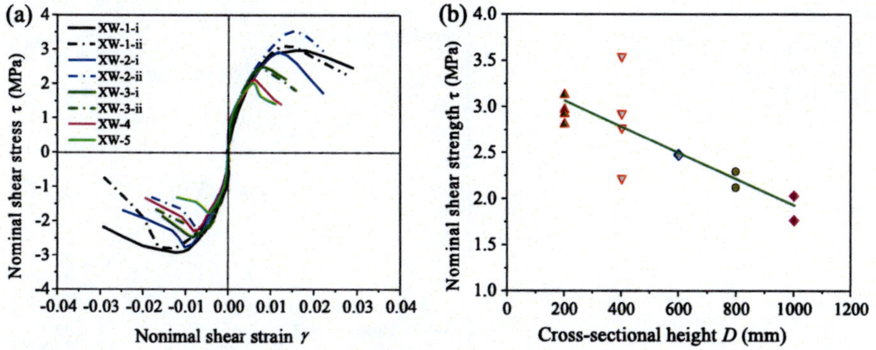

Fig. 6.28 **a** Nominal shear stress-strain curves of the eight RC beam subjected to low cyclic loading; and **b** nominal shear strength versus cross-sectional depth

becomes more brittle as the structural size increases, as can be observed from the post-peak declining curves. Figure 6.28b shows the decrease in the nominal shear strength (corresponding to the peak stress plotted in Fig. 6.28a) as the structural size increases. It is to be noted that, for each RC beam, two nominal shear strengths were obtained, involving the peak shear stresses in the positive loading and the one in the negative loading directions. For the tested beams named as XW-1, XW-2 and XW-3, there are two same specimens with same sizes for each beam. Therefore, there are four discrete data on the nominal shear strength. While, for the XW-4 and XW-5 RC specimens, there are only two values for each beam on the nominal shear strength. It can be found from Fig. 6.28a, b that the shear strength of RC cantilever beam exhibits a strong size effect. One can observe clearly that as the structural size of the RC specimen increases, the tested nominal shear strength decreases.

As observed previously in Fig. 6.21 the diagonal crack at failure in the present tests had essentially similar paths and relative lengths at the maximum load independently of the beam size.

Subsequently, the tested nominal shear strengths will be utilized to validate the available size effect law (SEL) proposed by Bažant (1984).

9. *Comparison with Bažant's SEL*

For the size effect in RC members, it is a structural mechanics problem including not only the most basic material size effect, but also the size effect contributed by the influence of mutual effect between reinforcing steel and surrounding concrete. Simply to say, the size effect in the nominal strength of RC members would be caused and affected by (1) the inherent features of concrete and steel, and (2) the mutual effect between concrete and steel rebar. Herein the following analysis, the detailed size effect law for the nominal shear strength of the RC cantilever beams was explored.

The bi-logarithmic plot was drawn in Fig. 6.29. Moreover, the size effect law (SEL) proposed by Bažant (1984), the strength criterion (for plastic materials) without considering the influence of size effect and the classic linear elastic fracture

Fig. 6.29 The normalized nominal shear strength of the present test data fitted by Bažant's SEL

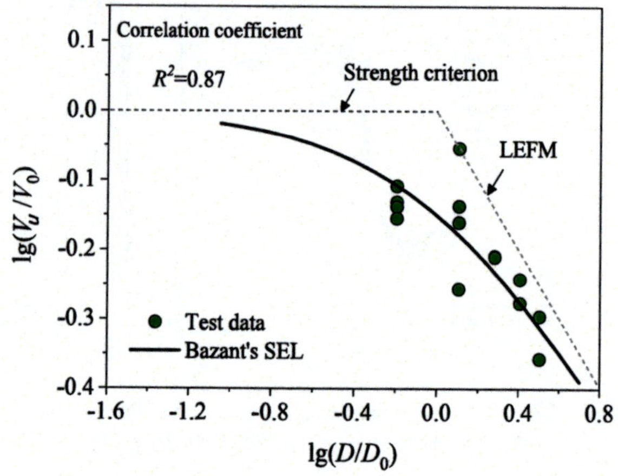

mechanics (LEFM) were also plotted in Fig. 6.29 for the comparison. According to the calculations, one can get the correlation coefficient to be 0.87. This means that the SEL curve fits the present tested data satisfactorily, as illustrated in Fig. 6.29. According to the above analyses, it can be concluded that the plot shows the presence of size effect on the shear strength of RC beams in accordance with Bažant's size effect law.

10. *Preliminary discussion on the source of size effect*

In general, for RC beams, because of the presence of transversal reinforcement, the size effect in shear strength should be mitigated, weakened or even suppressed, as observed in Angelakos et al. (2001) and Lubell et al. (2004). Stirrups or other shear reinforcement are known to significantly enhance the shear strength of deep beams, thanks to strengthening of the so-called compression strut. However, a series of test observations, e.g. in the efforts of Bažant (2005) and Bažant and Xiang (1997), as well as Walraven and Lehwalter (1994) have reported that a marked size effect is documented for geometrically similar deep beams with stirrups heavier than the minimum requirement. In addition, the simulation results of Yu and Bažant (2011) on slender beams presented that the stirrups cannot suppress the size effect completely regardless of stirrup ratio.

Similarly, just as the present test observations show, a strong size effect on shear strength is still observed in the RC cantilever beams with stirrups subjected to cyclic fatigue loading. The source for such a strong size effect may be caused by the following reasons.

(1) Under low-cycle fatigue loadings, even with the transversal reinforcement, the failure of the RC cantilever beams is still brittle. Simply to say, it attributes to the cyclic nature of fatigue loading.
(2) The detailed mechanism of shear failure of the RC cantilever beams is very complicated, however it consists of some complex form of cohesive softening

fracture, as can be observed previously from the broken photographs. In addition, the concrete softening due to distributed fracturing also generated, and it localized into narrow bands obviously.

(3) From the comparison of the crack widths of the eight RC beams, one can note that for larger-sized RC beams, the corresponding crack widths as the beams failed are obviously larger than those for smaller-sized beams. With a larger crack width, the bite force between aggregates and the friction of shear forces should be reduced, therefore, so also is the ultimate capacity.

(4) From the viewpoint of mechanism, the size effect should be attributed by the fact that, the compression-shear failure generated at the compressive zone of concrete leads to the final shear failure of RC beams with stirrups before the stirrups yielded. Simply to say, the failure of the RC beams with stirrups is controlled by the compression-shear failure of concrete, but not the yield of stirrups. For the present tests on the RC beams, most of the shear force should be carried by a single strut due to the arching action.

6.2.2 *Simulations on Shear Failure of RC Beams*

Here, the influences of shear-span ratio, stirrup ratio and stirrup spacing on shear failure and size effect of RC beams with stirrups are introduced.

1. *Shear-span ratio*

In order to reveal the influence of shear-span ratio on shear strength and size effect of RC beams, three groups of shear failure simulation tests (the shear-span ratios are 1.0, 1.5, 2.0, respectively) of cantilever beams under monotonic loading were designed. Each group contains eight beams with different section sizes (S: 100 mm × 200 mm, M: 200 mm × 400 mm, L: 400 mm × 800 mm, U: 800 mm × 1600 mm) and different stirrup ratios (0 and 0.2%). The geometric parameters of the models are listed in Table 6.9.

Take "CB-1.0-0-S" as an example to illustrate the naming method, the letter "CB" stands for Concrete Beam, the number "1.0" stands for the shear-span ratio of 1.0, the number "0" stands for the stirrup ratio of zero, and the letter "S" stands for the cross-sectional size of the cantilever beam is small, which is 100 mm × 200 mm. Because of the large number of models, Table 6.9 only lists the geometric parameters of some specimens. The names and parameters of other specimens can be obtained by analogy according to the naming method.

The ultimate failure modes of the RC cantilever beams without stirrups under different shear-span ratios ($\lambda = 1.0, 1.5, 2.0$) are shown in Fig. 6.30. According to the characteristics of the ultimate failure modes, all specimens are regarded as shear failure. At the case of small shear-span ratio and cross-sectional size, such as "CB-1.0-0-S", the principal diagonal crack of the cantilever beam extended obliquely from the loading point to the bottom of the fixed end, and the concrete formed several obliquely compressed short columns and was crushed. The failure mode showed

Table 6.9 Geometrical parameters of the RC cantilever beam

Specimen name	Section size $b \times D$ (mm × mm)	Effective sectional depth d_0 (mm)	Shear-span ratio λ	Stirrup ratio ρ_{sv} (%)
CB-1.0-0-S	100 × 200	175	1.0	0
CB-1.0-0-M	200 × 400	370	1.0	0
CB-1.0-0-L	400 × 800	760	1.0	0
CB-1.0-0-U	800 × 1600	1550	1.0	0
CB-1.5-0-S	100 × 200	175	1.5	0
CB-1.5-0-M	200 × 400	370	1.5	0
CB-1.5-0-L	400 × 800	760	1.5	0
CB-1.5-0-U	800 × 1600	1550	1.5	0
CB-2.0-0.2-S	100 × 200	175	2.0	0.2
CB-2.0-0.2-M	200 × 400	370	2.0	0.2
CB-2.0-0.2-L	400 × 800	760	2.0	0.2
CB-2.0-0.2-U	800 × 1600	1550	2.0	0.2

Note S—small; M—medium; L—larger; and U—ultra-large

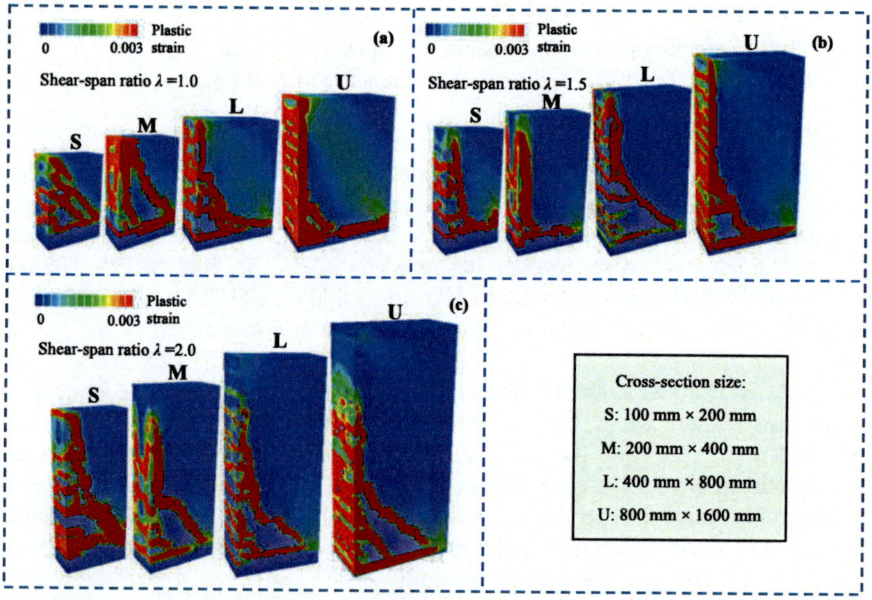

Fig. 6.30 Failure modes of RC beams with different shear-span ratios: **a** λ = 1.0; **b** λ = 1.5; and **c** λ = 2.0

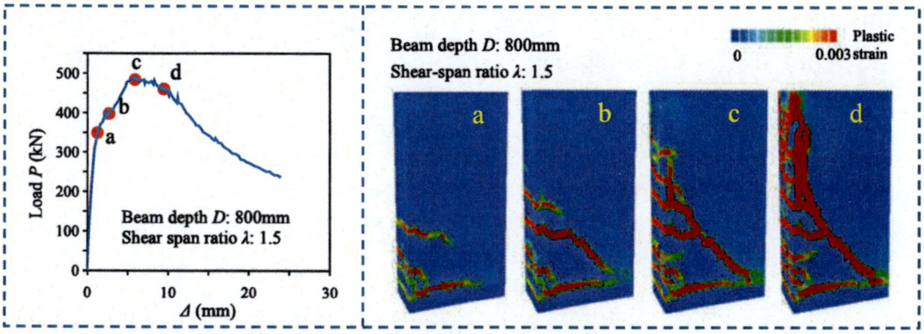

Fig. 6.31 Crack propagation process for the RC beam (CB-1.5-0-L)

obvious characteristics of diagonal compression failure. With the increase of shear-span ratio and section size, more obvious flexural cracks extensively developed. This indicates that in the members of shear-span ratio less than 2.5, the transfer from a beam action to an arch action is observed (Słowik and Smarzewski 2012). The failure modes of beams were similar.

Taking "CB-1.5-0-L" as an example, Fig. 6.31 presents the expanding process of the cracks. In the initial loading phase, the transverse cracks first appeared on the beam. In this phase, corresponding to point "a", the load-displacement curve was close to a straight line, and the stiffness degradation was not significant. The beam was in an elastic stage. With the load increasing, the transverse cracks began to extend diagonally. Corresponding to point "b", the load-displacement (P-Δ) curve began to fluctuate, which meant that stiffness began to decrease. With the load increasing further, the newly generated cracks developed diagonally and merged gradually with the existing diagonal cracks. At this time, corresponding to point "c", the stiffness of the beam degraded obviously and the applied load gradually reached the peak value. At last, corresponding to point "d", some diagonal cracks developed into a critical diagonal crack, and the concrete beam was destroyed. Figures 6.30 and 6.31 present the extended simulation results in purpose to study the influence of shear-span ratio on shear strength and size effect. It is noted that there is no available experimental data for comparison purposes. More experiments would be conducted to verify the correctness of the simulation results.

Figure 6.32 depicts the load-displacement (P-Δ) curves of the cantilever beams with different sizes and shear-span ratios. In the initial loading stage, the curve is close to a straight line, and the stiffness is basically constant. There are some fine cracks at this time. Then the curve fluctuates to a certain extent, the stiffness begins to degenerate, and the cracks begin to develop diagonally. Continuing loading, the curve is still in the ascending stage, but with the diagonal extension of the cracks, the stiffness degenerates obviously. After the final curve reaches its peak value, the stiffness of the beam decreases sharply, obviously shear failure occurs, and the bearing capacity is completely lost. As presented in Fig. 6.32, the shear capacity of beams decreases significantly with the increase of shear-span ratio.

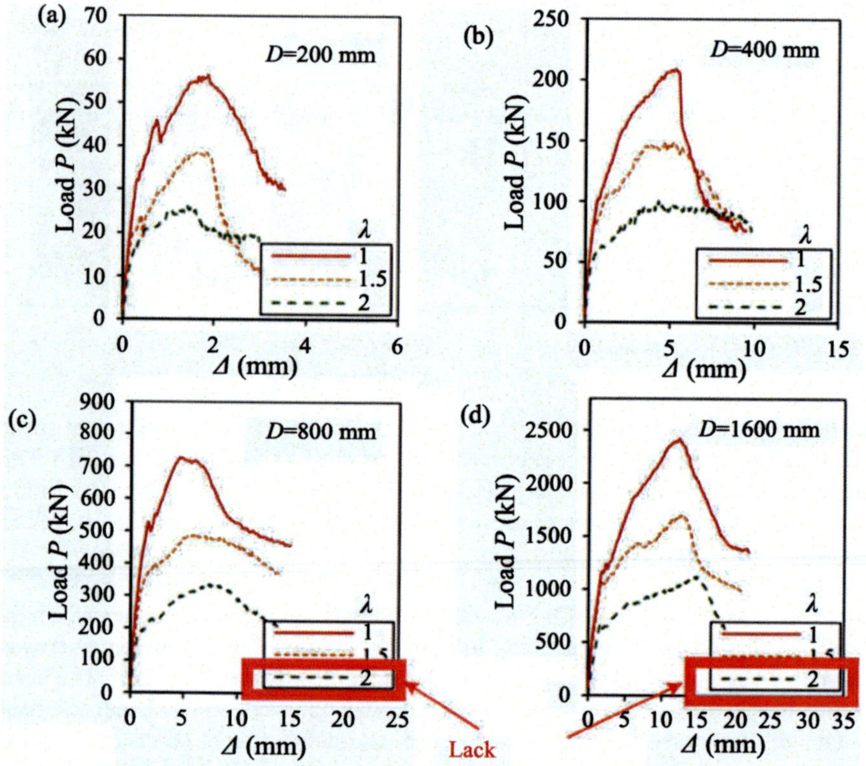

Fig. 6.32 *P-Δ* curves of the RC beams with different sizes and shear-span ratios

Fig. 6.33 Nominal shear strength of cantilever beams without stirrups

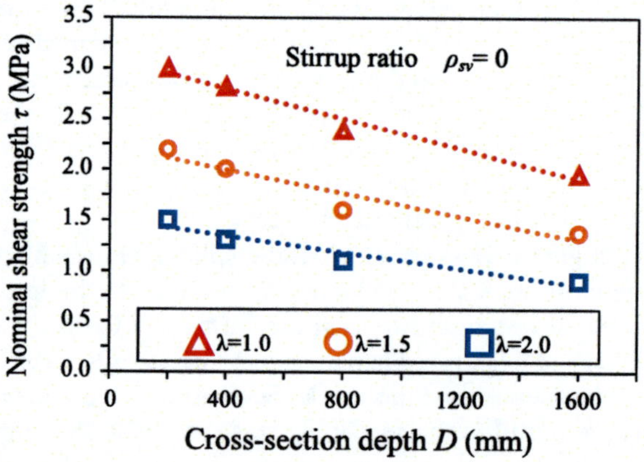

Fig. 6.34 Softening rate of cantilever beams without stirrups

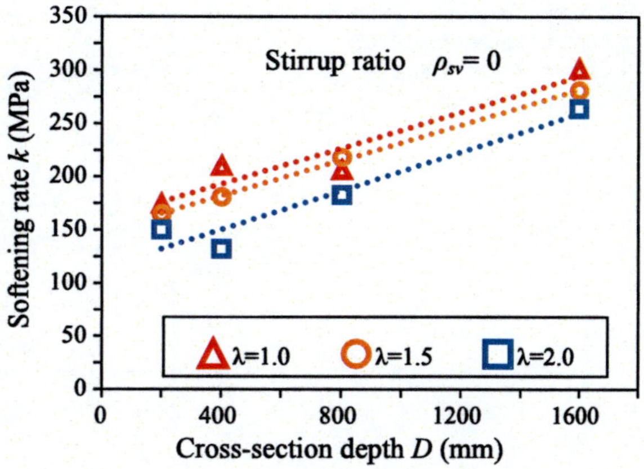

Figure 6.33 presents the relationship between the nominal shear strength and cross-sectional depth of beams. The trend lines of the data were derived by regression analysis. For beams with the same size, the nominal shear strength decreases with the increase of shear-span ratio. The nominal shear strength of beams with the smallest shear-span ratio ($\lambda = 1.0$) is nearly 50% higher than that of beams with the largest shear-span ratio ($\lambda = 2.0$). This shows that the shear-span ratio has a very important influence on the shear capacity. For beams with the same shear-span ratio, the nominal shear strength decreases with the beam depth increasing. The nominal shear strength of beams with 1600 mm in depth decrease by nearly 40% compared with that of beams with 200 mm in depth, indicating the presence of obvious size effect. Nevertheless, the three trend lines are roughly parallel, it can be concluded that the shear-span ratio has little influence on size effect.

The softening rate k can reflect the brittleness of RC beams to a certain extent. Herein, the definition of the softening rate k of the RC beams are presented in Fig. 6.15a. It represents the absolute value of slope from the peak nominal shear strength τ_0 to $85\%\tau_0$ in the τ-ε curve. As it can be seen in Fig. 6.34, with the increase of the cross-section size of RC beams, the softening rate increases. That means larger specimens are more brittle if the same failure mode occurs. Moreover, according to the trend of softening rate with the cross- section size under different shear-span ratios, it can be concluded that the shear-span ratio has little influence on size effect.

Figure 6.35 is the size effect fitting curve comparing the simulation results with the size effect law. The fitting correlation coefficient R^2 for Bažant's SEL and the present simulated strengths is 0.99, indicating that Bažant's SEL can well describe the size effect on shear strength of RC cantilever beams. Moreover, as it can be seen in Fig. 6.35, the data points of the same size and different shear-span ratios roughly coincide with each other, illustrating that the shear-span ratio has little influence on the size effect of shear strength. However, the influence of shear-span ratio on strength value is still needed to be considered in strength design.

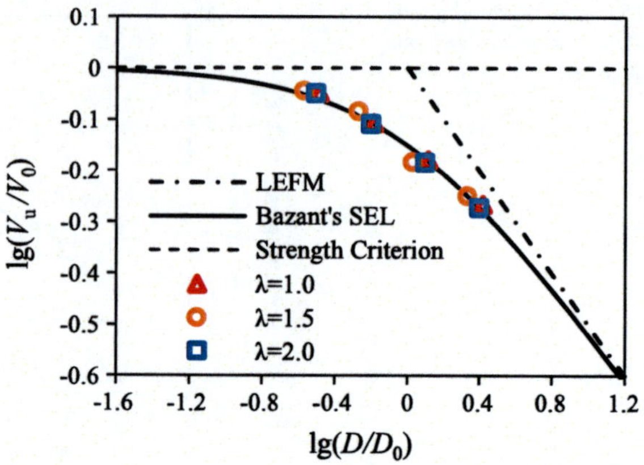

Fig. 6.35 Size effect fitting of nominal shear strength of cantilever beams without shear reinforcement

In the case of low stirrup ratio, the shear-span ratio has a significant influence on the shear failure mode and shear strength of RC beams. Based on the above simulation tests of beams without shear reinforcement, adopting stirrup ratio of 0.2%, the simulation tests of cantilever beams with different shear-span ratios (1.0, 1.5 and 2.0) were designed. The geometric parameters of these cantilever beams are presented in Table 6.9.

The relationship between the nominal shear strength and beam depth is plotted in Fig. 6.36. One can clearly note that for these cantilever beams with the same size, the smaller the shear-span ratio is, the larger the nominal shear strength will be. With the same shear-span ratio, the nominal shear strength will decrease with the increase of the beam depth. In Fig. 6.36, the three trend lines are roughly parallel, indicating that the shear-span ratio has little influence on the size effect of beams with low stirrup ratios. The conclusion is consistent with that of the beams without stirrups.

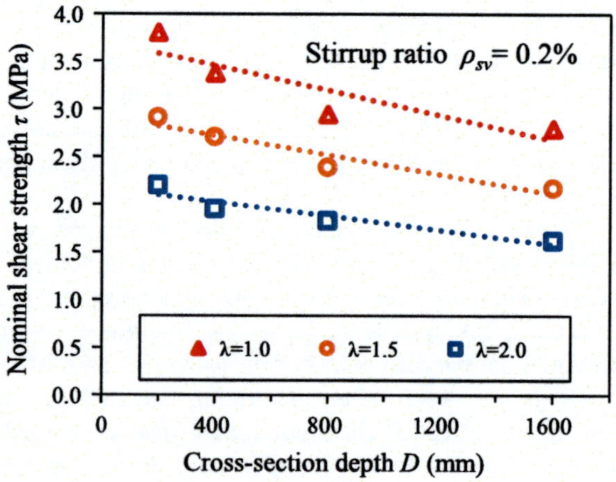

Fig. 6.36 Nominal shear strength of cantilever beams with stirrup

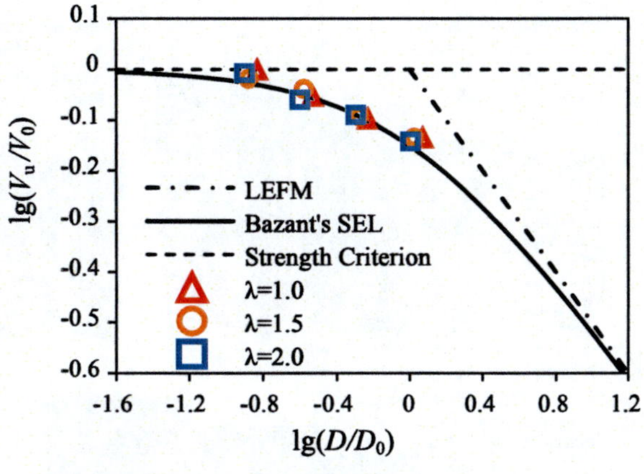

Fig. 6.37 Size effect fitting of nominal shear strength of cantilever beams with stirrup

Compared with Fig. 6.33, the slopes of trend lines decrease, which indicates that the stirrups weaken the size effect of the beams.

Figure 6.37 plots a logarithmic fitting curve of the nominal shear strength of cantilever beams with shear reinforcement and various structural sizes. The fitting correlation coefficient R^2 for Bažant's SEL and the present simulated strengths is 0.97, indicating that Bažant's SEL can well describe the size effect on nominal shear strength. Consistent with Fig. 6.35, the data points of the three shear-span ratios roughly coincide, indicating that shear-span ratio has little influence on the size effect of shear capacity. However, the influence of shear-span ratio on shear strength value cannot be neglected in strength design. Based on the above shear strength change rules of beams with and without shear reinforcement, it can be concluded that the shear-span ratio has little influence on size effect and has a great influence on the shear strength value.

2. *Stirrup ratio*

Based on the shear failure simulation tests of beams without shear reinforcement, more simulation tests of beams with different stirrup ratios (0.2, 0.4, 0.6 and 0.8%) were designed (the shear-span ratio of all simulated test beams is 2). The quantitative influence of the stirrup ratio on shear capacity and size effect is obtained.

Figure 6.38 shows the ultimate failure patterns of cantilever beams with different sizes and stirrup ratios. Compared with the beams without stirrups, with the increase of stirrup ratio, transverse cracks in the beams were more fully developed. Moreover, the damage area and the cracks in the tension zones of the concrete increased. The flexural failure phenomenon was more obvious, which shows that the existence of stirrups improves the ductility of RC beams. On the whole, the ultimate failure modes of beams with different stirrup ratios were roughly the same, only the damage degree of the concrete was slightly different. That means the range of stirrup ratio in this study cannot completely suppress the shear failure mode of the beams, and similar shear failure mode of beams still occurred ultimately. However, compared with the

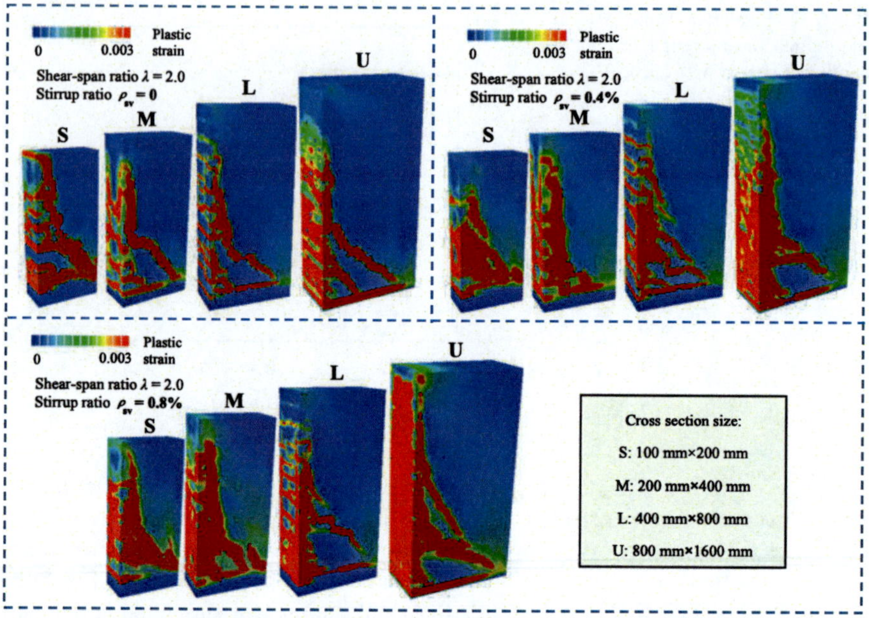

Fig. 6.38 Failure modes of cantilever beams with different structural sizes and stirrup ratios

beams without shear reinforcement, the cracks in the beams with stirrups were more fully developed.

Taking "CB-2.0-0.4-L" as an example, Fig. 6.39 presents the expanding process of the cracks. As CB-1.5-0.0-L in Fig. 6.31, the crack propagation process of CB-2.0-0.4-L can also be divided into the transverse extension stage (point "a"), diagonal extension stage (point "b"), diagonal penetration stage (point "c") and failure stage (point "d"). In addition, the development trend of the load-displacement (P-Δ) curve is also similar to that of CB-1.5-0.0-L. In summary, the failure process of the beams with the two different stirrup ratios (0 and 0.4%) are similar.

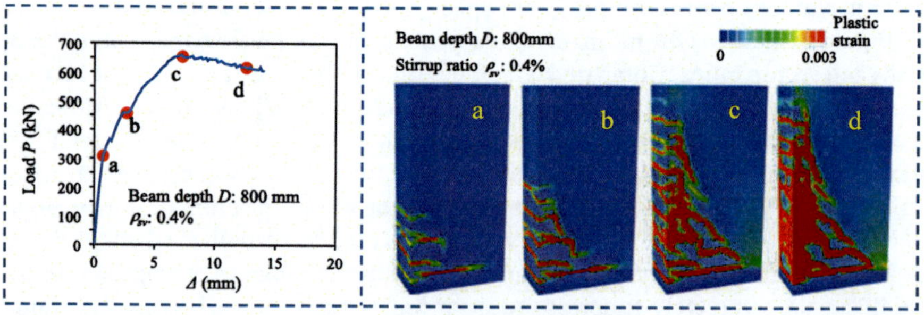

Fig. 6.39 Crack propagation process for the RC beam (CB-2.0-0.4-L)

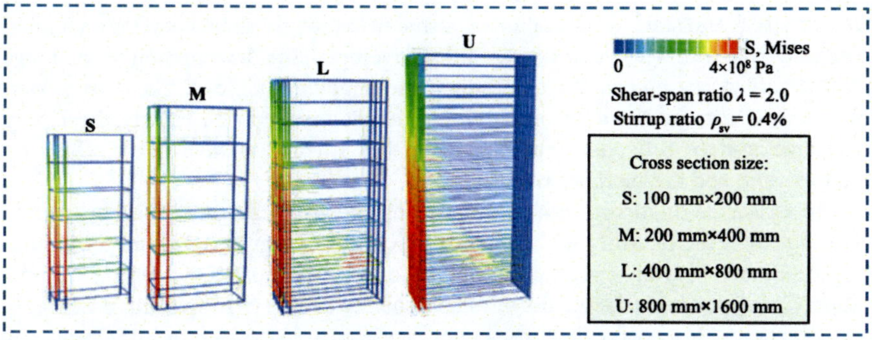

Fig. 6.40 The deformation of the reinforcing bars of different sizes beams at 0.4% stirrup ratio

Under the stirrup ratio of 0.4%, the deformation of steel bars is presented in Fig. 6.40. With the development of cracks in the beams, the deformation of steel bars increased gradually. Near the fixed end of the cantilever beams, the longitudinal bars yielded gradually due to the large strain, and the deformation of longitudinal bars decreased gradually along the longitudinal direction. With the extension of diagonal cracks, the deformation of stirrups increased gradually, and some of them yielded eventually. Comparing with the location and development of the diagonal cracks in Fig. 6.38, the stirrups yielded at the wider diagonal cracks. It can be judged that stirrups played a good restraining effect on concrete and improved the ductility of concrete beams.

Figure 6.41 shows the yielding process of the cantilever beam with the depth of 800 mm and the stirrup ratio of 0.4%. When no load is applied, the steel bars in the beam were not loaded, nor were they deformed. At the initial stage of loading,

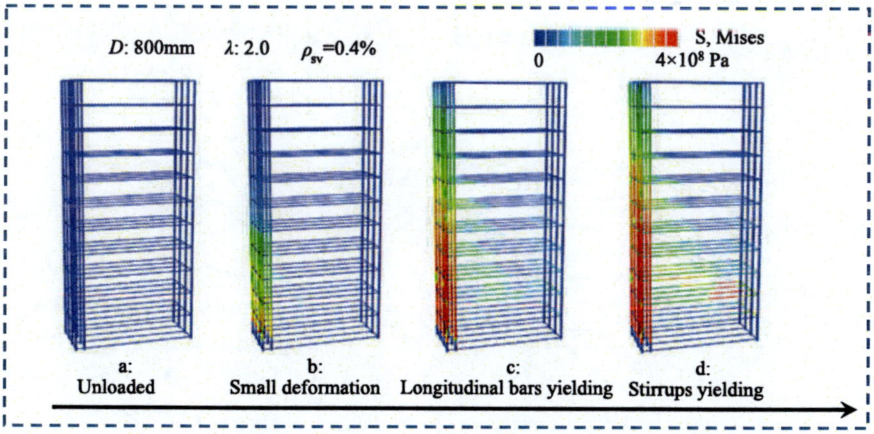

Fig. 6.41 Deformation process of the reinforcing bars of 800 mm cantilever beam at 0.4% stirrup ratio

tension stress and small deformation occurred in the longitudinal bars. When loading continues, the transverse cracks were fully developed, the deformation of the longitudinal bars increased and the longitudinal bars gradually yielded. The stirrups were obviously deformed due to the extension of the diagonal cracks. Finally, these diagonal cracks were fully developed, some of the stirrups yielded due to the large deformation, and the member was destroyed.

The existence of stirrups not only increases the shear capacity of RC beams, but also improves the ductility and deformation capacity. Figure 6.42 shows the load-displacement (P-Δ) curves of cantilever beams with different beam depths and stirrup ratios. Compared with the results of plain concrete beams (stirrup ratio of zero), the shear capacity of beams with stirrups improves in various degrees, and the peak-post

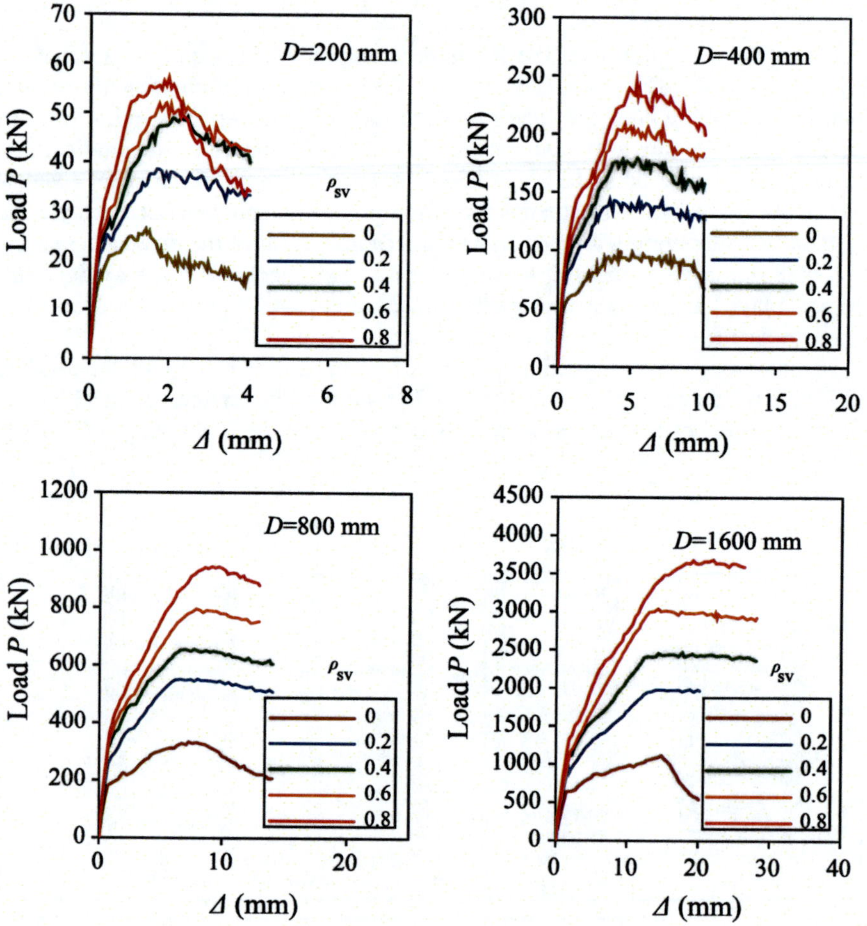

Fig. 6.42 P-Δ curves of the RC beams with different sizes and stirrups ratios

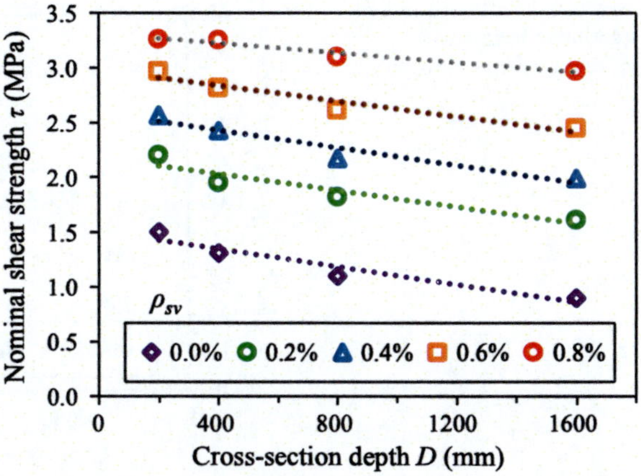

Fig. 6.43 Nominal shear strength of cantilever beams with different stirrup ratios

curves become more and more gentle with the increase of stirrup ratio. This indicates that the ductility of beams has been improved significantly. Moreover, there is no significant change in the displacements at the beam ends corresponding to the peak loads under different stirrup ratios.

Figure 6.43 presents the relationship between the nominal shear strength and the beam depth. The trend lines of the data were derived by regression analysis. One can clearly note that the nominal shear strength of beams increases with the stirrup ratio increasing, indicating that the existence of stirrups improves the shear capacity of beams. In addition, the nominal shear strength decreases with the increase of beam depth, indicating the presence of size effect. However, with the increase of stirrup ratio, the downward trend decreases and the trend lines become gentle, which indicates that the existence of stirrups weakens the size effect on shear strength. As presented in Fig. 6.43, even at a large stirrup ratio (e.g. 0.8%), the nominal shear strength of large-sized beams still tends to decrease, which indicates that the increase of stirrup ratio will weaken the size effect. Nevertheless, the increase of stirrup ratio cannot completely eliminate the size effect, which is consistent with the conclusion of Yu and Bažant (2011).

In addition, the softening rate of the RC beams are presented in Fig. 6.44. One can clearly note that with the increase of stirrup ratio, the softening ratio decreases gradually. Moreover, the softening ratio increases with the increase of the cross-section depth of the beams, that means the brittleness of RC beams enhances with the increase of the cross-section depth. The largest specimens were the most brittle when the same failure mode occurred. However, with the increase of stirrup ratio, the slopes of each trend line in Fig. 6.44 decreases gradually, which illustrates that the existence of stirrups weakens brittleness of RC beams.

In order to analyze the contribution of stirrups to the shear capacity of RC cantilever beams, the incremental shear strength of beams with adjacent stirrup ratios was quantitatively compared and analyzed. Figure 6.45 shows the incremental shear strength of beams with adjacent stirrup ratios. The abscissa 0.2 in the figure is the

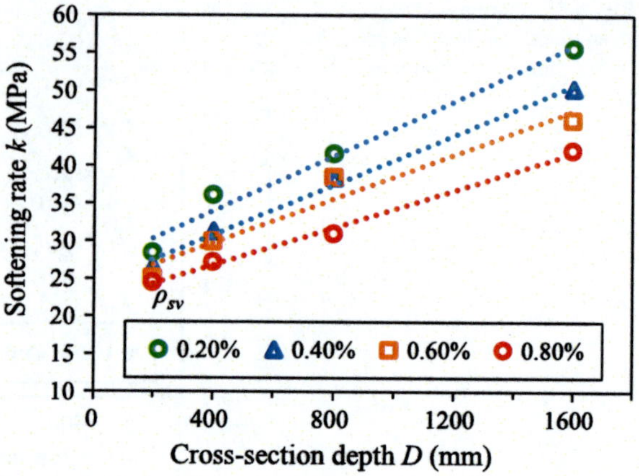

Fig. 6.44 Softening rate of cantilever beams with different stirrup ratios

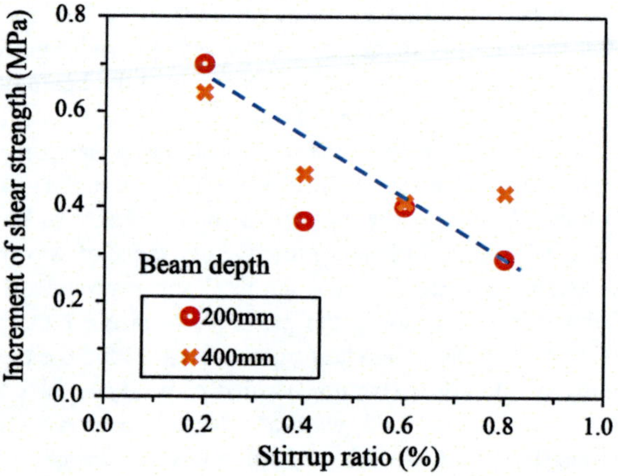

Fig. 6.45 Increment of shear strength between adjacent stirrup ratios

increment of shear strength of cantilever beams with 0.2% stirrup ratio compared with the beams without shear reinforcements, the abscissa 0.4 is the increment of shear strength of the beams with 0.4% stirrup ratio compared with that of 0.2% stirrup ratio, and etc. According to the trend line in Fig. 6.45, in the case of small stirrup ratio, the effect of stirrups on shear strength is greater. However, with the increase of stirrup ratio, the value of shear strength increment begins to decrease. In the case of high stirrup ratio, the effect of further increasing the stirrup ratio on the shear strength is no longer obvious, which indicates that the shear strength cannot increase indefinitely with the increase of stirrup ratio. This should be caused by the fact that the stirrups in beams with high stirrup ratio do not yield when beams are destroyed by shear, in other words, the stirrups do not play a full role. In this case, the shear failure of the beam mainly depends on when the concrete is crushed, and the failure mode is mostly diagonal compression failure. Therefore, considering both

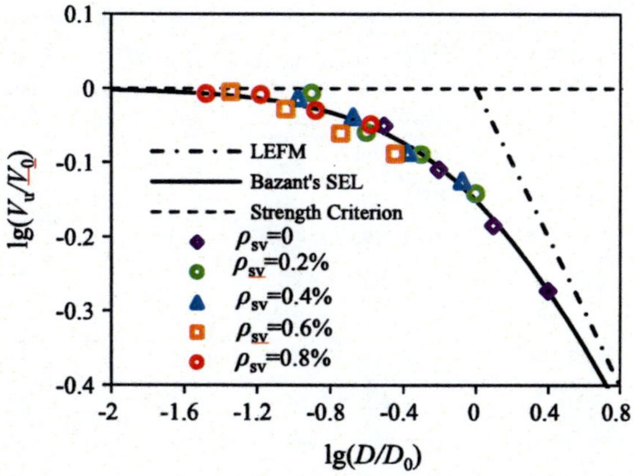

Fig. 6.46 Size effect fitting of nominal shear strength of RC beams with different stirrup ratios

safety and the economic benefits, the stirrup ratio should ensure that the beam can bear the load required by the design, and should not be too large. In the case of small stirrup ratio, diagonal tension failure often occurs in beams subjected to shear, which means, once the diagonal cracks appear in the beam, the stirrups intersecting with the diagonal cracks will yield rapidly because they cannot restrict the development of the diagonal cracks. In order to prevent the diagonal tension failure, the minimum stirrup ratio ($\rho_{sv,min} = 0.24f_t/f_{yv}$) of RC beams is stipulated in the Chinese code (GB 50010 2010).

Also, under the same stirrup ratio, the simulated failure patterns of the RC members having different structural sizes are geometrically similar. Therefore, the SEL proposed by Bažant could be adopted. Figure 6.46 plots a logarithmic fitting curve of nominal shear strength of the beams with different stirrup ratios and different section sizes. The classic SEL proposed by Bažant was selected for the comparison. The fitting correlation coefficient R^2 for Bažant's SEL and the present simulated strengths is 0.99, indicating that Bažant's SEL can well represent the size effect on shear capacity of RC beams. It can be seen from Fig. 6.46 that with the increase of stirrup ratio, the strength data points move up along the Bažant's SEL curve and approach the horizontal line of plastic strength, indicating that the size effect on the nominal shear strength is weakened to a certain extent with the increase of the stirrup ratio.

From the above analysis, it can be concluded that for RC beams with stirrups, on the one hand, stirrups can directly bear shear force, thereby improving the shear capacity of RC beams; on the other hand, the existence of stirrups can limit cracks, thereby improving the ductility of beams and weakening the size effect on shear strength.

3. *Stirrup spacing*

In order to study the influence of stirrup spacing on shear failure and size effect of RC cantilever beam, several groups of RC cantilever beams with different cross-section sizes (S: 100 mm × 200 mm, M: 200 mm × 400 mm, L: 400 mm × 800 mm,

Table 6.10 Geometrical parameters of the RC cantilever beams with different stirrup spacing

Component name	Section size $b \times D$ (mm × mm)	Stirrup spacing s (mm)	Stirrup section area A_{sv1} (mm^2)	Effective depth d_0 (mm)	Shear-span ratio λ	Longitudinal reinforcement ratio ρ (%)
S-I-0.4	100 × 200	30	6	170	2	1.0
S-II-0.4	100 × 200	60	12	170	2	1.0
S-III-0.4	100 × 200	90	18	170	2	1.0
S-IV-0.4	100 × 200	120	24	170	2	1.0
M-I-0.4	200 × 400	30	12	370	2	1.0
M-II-0.4	200 × 400	60	24	370	2	1.0
M-III-0.4	200 × 400	90	36	370	2	1.0
M-IV-0.4	200 × 400	120	48	370	2	1.0
L-I-0.4	400 × 800	30	24	760	2	1.0
L-II-0.4	400 × 800	60	48	760	2	1.0
L-III-0.4	400 × 800	90	72	760	2	1.0
L-IV-0.4	400 × 800	120	96	760	2	1.0
U-IV-0.4	800 × 1600	30	48	1550	2	1.0
U-IV-0.4	800 × 1600	60	96	1550	2	1.0
U-IV-0.4	800 × 1600	90	144	1550	2	1.0
U-IV-0.4	800 × 1600	120	192	1550	2	1.0

U: 800 mm × 1600 mm), stirrup spacing (I: 30 mm, II: 60 mm, III: 90 mm, IV: 120 mm) and stirrup ratios (0.2, 0.4, 0.6%) are designed. The relevant geometric parameters of some models (stirrup ratio is 0.4%) are listed in Table 6.10, and the relevant parameters of other models can be obtained by geometric multiple analogy.

The ultimate failure modes of RC cantilever beams with different stirrup spacing and different section sizes under monotonic shear loading are shown in Fig. 6.47. According to the development of cracks, all specimens emerged obvious through oblique cracks at the ultimate failure, which represents significant characteristics of shear failure. The ultimate failure modes of cantilever beams with different sizes are similar, which appeared that a large number of transverse bending cracks develop obliquely into through diagonal cracks, and finally extended to the fixed end of the beam bottom. In addition, for the beams with the same section size, it is found that when the stirrup spacing and stirrup diameter changed correspondingly, the trend of diagonal crack development was approximately the same, and the final failure modes are closely similar.

Figure 6.48 presents the crack development process of the cantilever beam CB-L-I-0.4 (beam depth is 800 mm, and stirrup spacing is 30 mm). At the initial stage of displacement loading (point "a"), the bottom side of the beam first cracked laterally, and the bending cracks gradually developed and extended. At this stage, the rigidity degradation of the beam was not obvious, and the load-displacement curve was close to a straight line. As the load continued to increase (point "b"), the transverse bending

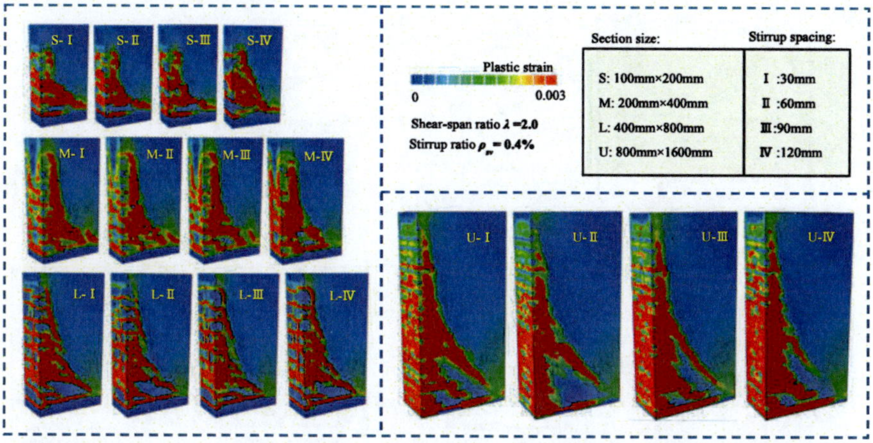

Fig. 6.47 Ultimate failure modes of the cantilever beams with different sizes and different stirrup spacing

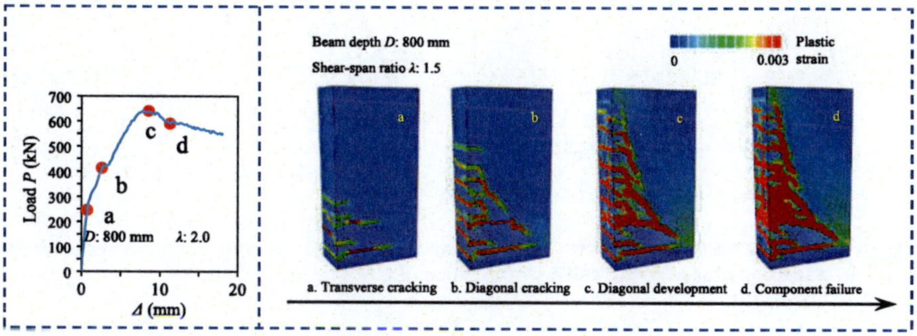

Fig. 6.48 Crack development process of the beam CB-L-I-0.4

cracks began to develop obliquely towards the fixed end of the beam bottom, and obvious obliquely cracks appeared. At this stage, the slope of load-displacement curve began to decrease and the rigidity of RC beam began to degenerate. With the further increase of the loading displacement (point "c"), the inclined crack continued to extend, and the new bending crack developed obliquely at the same time, and gradually converged with the existing crack as the main diagonal crack. At this stage, the rigidity of the beam degenerated obviously, and the load gradually reached the peak value. At the last stage (point "d"), the diagonal cracks further extended and penetrated, and finally formed a through diagonal crack extending from the loading point to the fixed end of the beam bottom, and the beam was destroyed.

Figure 6.49 presents the load-displacement (P-Δ) curves of each RC cantilever beam, in which Fig. 6.49a, c represents the results for the stirrup ratios of 0.2%, 0.4% and 0.6% respectively. The trend of P-Δ curves of the cantilever beams with

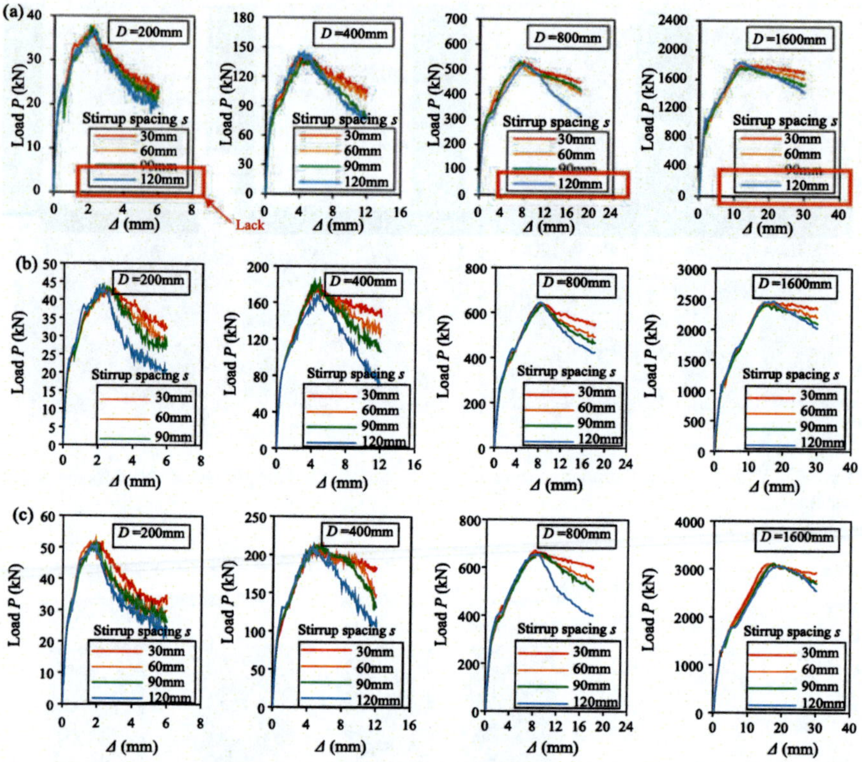

Fig. 6.49 *P*-*Δ* curves of the beams with different sizes and different stirrup spacing: **a** stirrup ratios of 0.2%; **b** stirrup ratios of 0.4%; and **c** stirrup ratios of 0.6%

the same size and different stirrup spacing are basically the same, and the peak load and corresponding displacement are almost the same. This indicates that in the case of the same stirrup ratio, the corresponding change of stirrup spacing and stirrup diameter have an ignorable effect on the shear capacity of RC beams.

Compared with the *P*-*Δ* curves of beams with the same size and different stirrup spacing in Fig. 6.49, the elastic stages roughly coincide, that means, in the elastic stage, the shear of the beam is mainly borne by the concrete part. Continued loading until the transverse cracks in the beam develops gradually, the diagonal cracks begin to appear, and the *P*-*Δ* curves of beams with different stirrup spacing no longer coincide, however, the trends are still roughly the same, indicating that there is no obvious difference on the mechanical properties of each concrete beam at this stage. However, one can see that the softening stages for different *P*-*Δ* curves are significantly distinct. Compared with those of the beam with larger stirrup spacing, the *P*-*Δ* curves of the beams with smaller stirrup spacing are gentler, and the rigidity degradation are slower, that is to say, the reduction of stirrup spacing can improve the shear ductility of the beam. In addition, with the increase of stirrup ratio, the

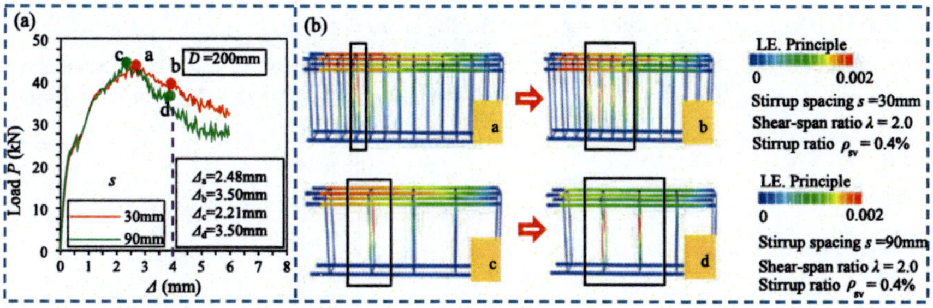

Fig. 6.50 Comparison of yield process of stirrups in beams with different stirrup spacing: **a** load-deflection curves; and **b** the strain of web reinforcement

shear capacity of the beams also increases correspondingly. With the stirrup ratio increasing from 0.2 to 0.6%, the shear capacity of beams with four cross-section sizes increases by more than 30%, indicating that compared with the stirrup spacing, the stirrup ratio has more significant influence on the shear capacity of RC beams.

Figure 6.50 reveals the yield process of reinforcements of two RC beams with different stirrup spacing (i.e., 30 and 90 mm). In Fig. 6.50, the two points "a" and "c" correspond to the peak load of the two RC beams, "b" and "d" correspond to the displacement of 3.5 mm, and the yielded stirrups are remarked with the black frames. It can be seen from Fig. 6.50 that when the load reached the peak value ("a" and "c"), some stirrups in the beams yielded, which means that during the shear failure process of RC beams, the stirrups played an important role that effectively limited the crack development and increased the shear bearing capacity of the RC beams. Continued to load until the displacement was 3.5 mm ("b" and "d"), it is found that compared with small stirrup spacing, the stirrups of larger stirrup spacing yielded in a larger range (in the black frame), and the *P*-*Δ* curve dropped faster. This is due to the increase of stirrup spacing, the scope of concrete directly restrained by each stirrup becomes to be larger. Therefore, after the stirrup yielded, the larger range of concrete cannot be effectively restrained in the process of cracking, the cracks would expand rapidly, resulting in more sudden damage, and the *P*-*Δ* curve would drop more sharply. This indicates that compared with the larger stirrup spacing, the smaller stirrup spacing makes the stirrup have more uniform restraining effect on the concrete and more obvious restraining effect on the crack development.

Displacement ductility factors of cantilever beams with different stirrup spacing are listed in Table 6.11. In Table 6.11, the yield displacements for the beams with the same size and different stirrup spacing are approximately the same, but the ultimate displacements decrease gradually with the increase of stirrup spacing. This is because the beam with large stirrup spacing dropped more rapidly after the *P*-*Δ* curve reached the peak value. In addition, the change trend of displacement ductility factors of each RC beam is depicted in Fig. 6.51 also indicates that the shear ductility of beams with the same section size decreases with the increase of stirrup spacing. However, with the further increase of stirrup spacing, the decrease of displacement ductility factor was no longer obvious, and the trend line turns to be gradually gentle, which illustrates

Table 6.11 Displacement ductility factors of the RC cantilever beams

Component name	Stirrup spacing *s* (mm)	Yield displacement Δ_y (mm)	Ultimate displacement Δ_u (mm)	Displacement ductility factor Δ_u/Δ_y
S-I-0.4	30	1.35	4.24	3.14
S-II-0.4	60	1.31	3.66	2.79
S-III-0.4	90	1.28	3.47	2.71
S-IV-0.4	120	1.22	2.73	2.24
M-I-0.4	30	2.90	8.88	3.06
M-II-0.4	60	2.86	7.02	2.45
M-III-0.4	90	2.84	6.72	2.37
M-IV-0.4	120	2.81	6.47	2.30
L-I-0.4	30	4.81	17.53	3.64
L-II-0.4	60	5.15	14.57	2.83
L-III-0.4	90	4.96	12.77	2.57
L-IV-0.4	120	4.98	11.03	2.21
U-IV-0.4	30	11.1	32.1	2.89
U-IV-0.4	60	12.4	31.5	2.54
U-IV-0.4	90	12.8	30.1	2.35
U-IV-0.4	120	12.9	27.8	2.16

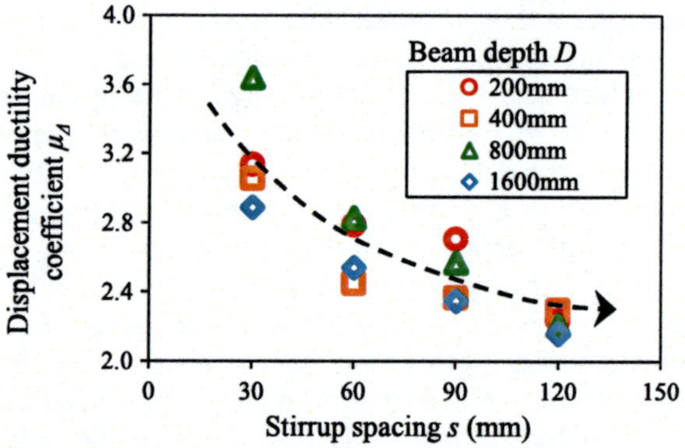

Fig. 6.51 Variation trend of displacement ductility factors

that the shear ductility of RC beams would not infinitely reduce with the increase of stirrup spacing, but would gradually tend to the shear ductility of RC beams without stirrup.

Figure 6.52 shows the trend of softening rate *k* changing with beam depth. It can be seen from Fig. 6.52 that the softening rate *k* of RC beam increases with the increase of section size, which indicates that the brittle behavior of large-size beam is more significant. When the stirrup spacing is large, the softening rate of the beam

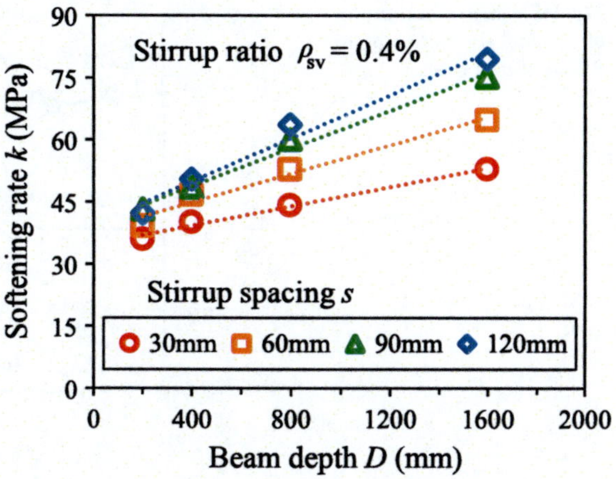

Fig. 6.52 Trend of softening rate k with beam depth

is more sensitive to the section size. With the reduction of stirrup spacing, the size sensitivity of the softening rate is reduced, indicating that the reduction of stirrup spacing weakens the sensitivity of the softening rate to the section size, and the brittleness behavior of the large-size beam is restrained to a certain extent.

Figure 6.53 presents the nominal shear strength of RC cantilever beams with different stirrup spacing under the same stirrup ratio. The nominal shear strength values of RC beams with the same section size under different stirrup spacing are approximately the same, which indicates that stirrup spacing has an ignorable influence on the nominal shear strength values of RC beams. However, with the increase of the section size, the nominal shear strength of the beam obviously decreases, presenting a significant size effect on shear strength.

Through regression analysis of the simulation data, the double logarithm curve of the nominal shear strength of RC cantilever beam with the change of section size can be obtained, as shown in Fig. 6.54. the simulation results are compared with Bažant's SEL, and the fitting correlation coefficient is 0.98, which illustrates that

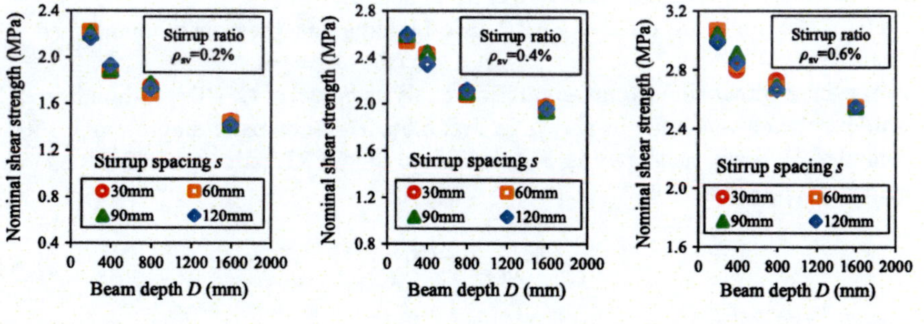

Fig. 6.53 Nominal shear strength of cantilever beams with different stirrup spacing for the stirrup ratio of: **a** $\rho_{sv} = 0.2\%$; **b** $\rho_{sv} = 0.4\%$; and **c** $\rho_{sv} = 0.6\%$

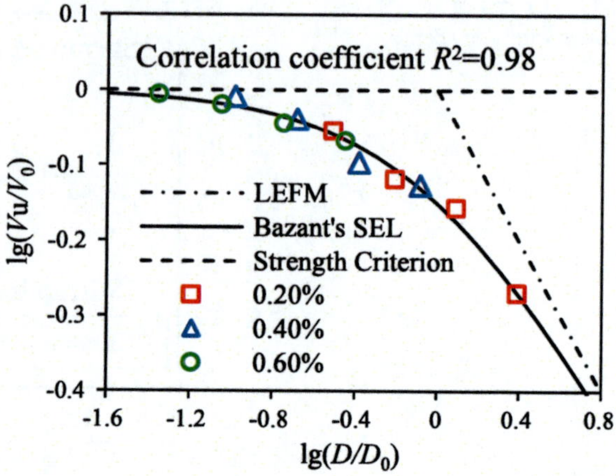

Fig. 6.54 Size effect fitting of nominal shear strength of RC cantilever beams

Bažant's SEL can well reflect the quantitative relationship of size effect of nominal shear strength for RC beams. In addition, one can clearly note from Fig. 6.54 that, as the stirrup ratio increases, the strength data approaches more closely to the horizontal curve (i.e., the strength criterion curve). This indicates that the corresponding size effect is weakened due to the increase of stirrup ratio.

6.3 Shear Failure in CFRP-Wrapped Concrete Beams

The role of CFRP in shear strengthening of RC beams is to share or replace the shear force supported by stirrups and improve the stress state of RC beams (Godat et al. 2010; Matta et al. 2013). In addition, the presence of CFRP restricts the expansion speed of oblique cracks in RC beams, and therefore would improve the displacement ductility. The objective of this section is mainly to present the influence of CFRP ratio and shear-span ratio on the shear capacity and size effect of CFRP-strengthened RC beams having different sizes (Jin et al. 2020a, b).

Here, the focus is on the CFRP side-bonding, as previously mentioned in Fig. 6.55a.

The three-dimensional meso-numerical analysis model of CFRP side-bonding RC cantilever beam was built, as shown in Fig. 6.56. The schematic diagram of CFRP-strengthened beam, as shown in Fig. 6.57. and the CFRP ratio ρ_f of CFRP can be defined as (Bae et al. 2012)

$$\rho_f = \frac{2w_f t_f}{b s_f} \tag{6.7}$$

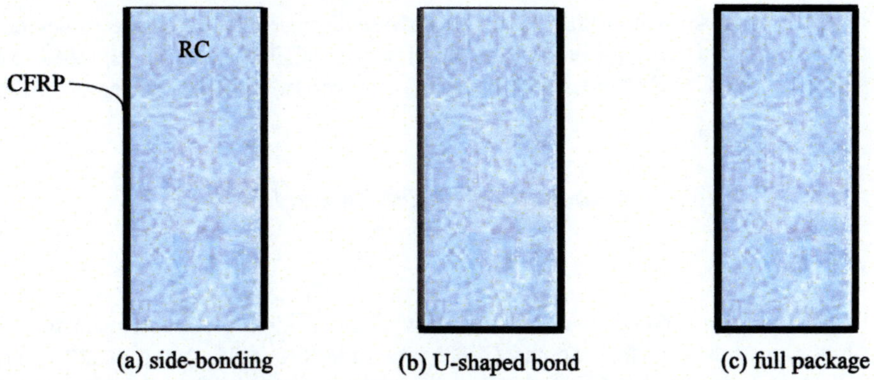

Fig. 6.55 The bonding method of CFRP-strengthened concrete beams

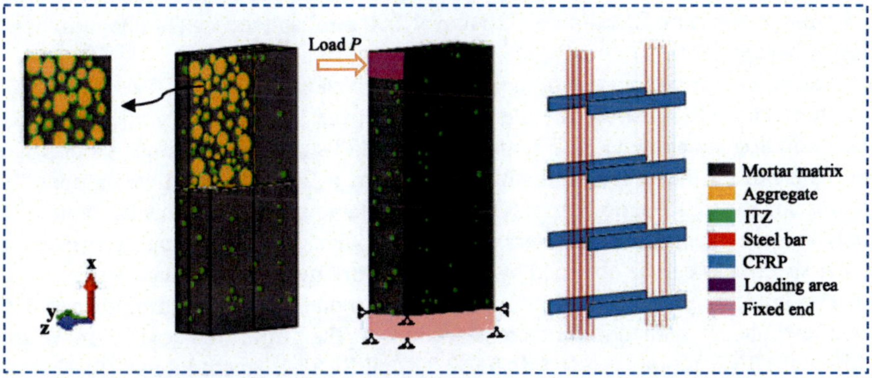

Fig. 6.56 3D meso-scale numerical model of CFRP-strengthened RC cantilever beam

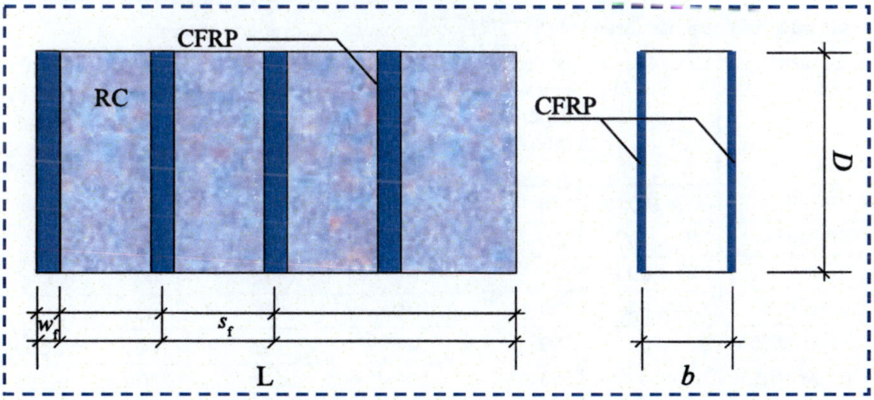

Fig. 6.57 Schematic diagram of CFRP-strengthened beam

in which w_f presents the width of the CFRP sheets; b presents the cross-section width of the reinforced beam; t_f presents the thickness of the CFRP sheets; and s_f means the spacing of the CFRP sheets, i.e., the spacing between two adjacent sheets of CFRP.

6.3.1 CFRP-Strengthened Ordinary Concrete Beams

1. *CFRP ratio*

Based on the verified meso-scale simulation methods, eight groups of shear failure simulation tests (i.e., the adopted CFRP ratios are 0, 0.0555, 0.111, 0.222, 1.11, 2.22, 4.44 and 8.88%) of CFRP side-bonding strengthened cantilever beams without web reinforcement were designed and modeled. Each group contains four different cross-sectional sizes (Small: 80 mm × 200 mm, Medium: 160 mm × 400 mm, Large: 240 mm × 600 mm, Ultra-large: 320 mm × 800 mm) and the shear-span ratio of all tested beams is 2.0.

The geometric parameters of the models are given in Table 6.12. Take "CBF-S-0" as an example to illustrate the naming method, the letter "CBF" indicates CFRP side-bonding strengthened RC beam, the letter "S" stands for the cross-section size of the cantilever beam is small, which is "80 mm × 200 mm", and the number "0" represents the CFRP ratio is zero. Owing to the large number of models, Table 6.12 only lists the geometric parameters of several models. The names and parameters of other specimens can be obtained by analogy according to the naming method.

Figure 6.58 presents the ultimate failure mode of CFRP-strengthened RC cantilever beam with different cross-section depths (from 200 to 800 mm) and different CFRP ratios (i.e., 0, 0.0555, 2.22 and 8.88%). According to the characteristics of the ultimate failure modes of the simulation results, it can be judged that

Table 6.12 Physical parameters of the CFRP-strengthened RC cantilever beam

Specimen name	Cross-section size $b \times D$ (mm)	Effective cross-section depth d_0 (mm)	Beam length $l = \lambda \times d_0$ (mm)	CFRP layers n	The width of the CFRP w_f (mm)	The spacing of the CFRP s_f (mm)	CFRP ratio ρ_f (%)
CBF-S-0	80 × 200	180	360	0	20	100	0.0
CBF-M-0	160 × 400	360	720	0	40	200	0.0
CBF-L-0	240 × 600	540	1080	0	60	300	0.0
CBF-U-0	320 × 800	720	1440	0	80	400	0.0
CBF-S-0.0555	80 × 200	180	360	1	20	100	0.0555
CBF-M-0.0555	160 × 400	360	720	2	40	200	0.0555
CBF-L-0.0555	240 × 600	540	1080	3	60	300	0.0555
CBF-U-0.0555	320 × 800	720	1440	4	80	400	0.0555

(continued)

Table 6.12 (continued)

Specimen name	Cross-section size $b \times D$ (mm)	Effective cross-section depth d_0 (mm)	Beam length $l = \lambda \times d_0$ (mm)	CFRP layers n	The width of the CFRP w_f (mm)	The spacing of the CFRP s_f (mm)	CFRP ratio ρ_f (%)
CBF-S-0.111	80 × 200	180	360	2	20	100	0.111
CBF-M-0.111	160 × 400	360	720	4	40	200	0.111
CBF-L-0.111	240 × 600	540	1080	6	60	300	0.111
CBF-U-0.111	320 × 800	720	1440	8	80	400	0.111
CBF-S-0.222	80 × 200	180	360	4	20	100	0.222
CBF-M-0.222	160 × 400	360	720	8	40	200	0.222
CBF-L-0.222	240 × 600	540	1080	12	60	300	0.222
CBF-U-0.222	320 × 800	720	1440	16	80	400	0.222

(continued)

Table 6.12 (continued)

Specimen name	Cross-section size $b \times D$ (mm)	Effective cross-section depth d_0 (mm)	Beam length $l = \lambda \times d_0$ (mm)	CFRP layers n	The width of the CFRP w_f (mm)	The spacing of the CFRP s_f (mm)	CFRP ratio ρ_f (%)
CBF-S-1.11	80 × 200	180	360	20	20	100	1.11
CBF-M-1.11	160 × 400	360	720	40	40	200	1.11
CBF-L-1.11	240 × 600	540	1080	60	60	300	1.11
CBF-U-1.11	320 × 800	720	1440	80	80	400	1.11

Note The shear-span ratio of all beams is 2 (i.e., $\lambda = 2$)

all beams exhibit obvious shear failure. At the case of small CFRP ratio and cross-section size, e.g. the "CBF-L-0", it can be found that the principal diagonal crack of the cantilever beam extended obliquely from the loading point to the bottom of the fixed end. Its destruction process and characteristics are analogous to that presented in Fig. 6.58. This is consistent with the results of the work (Godat et al. 2012).

When the CFRP ratio of CFRP was different, the ultimate shear failure modes of RC beams were as well as different. Compared with the beams without CFRP sheets, with the increase of CFRP ratio, transverse cracks in the beams were more fully developed. Taking "CBF-M-0.0555" as an example, the expanding process of the cracks is illustrated in Fig. 6.59.

At the initial stage of loading, the transverse cracks appeared in the beam first. As the load increased, the crack further extended in the lateral direction is presented in Fig. 6.59a. Figure 6.60 illustrates the load-displacement (P-Δ) curve was close to a straight line, and the stiffness degradation behavior was not obvious. At this moment,

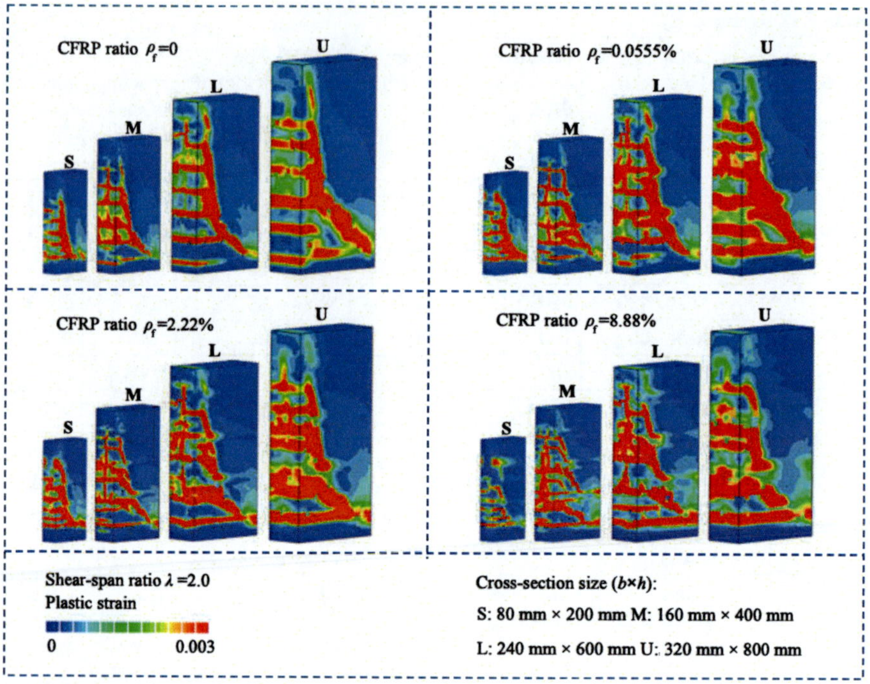

Fig. 6.58 Failure modes of CFRP-strengthened RC beams with different cross-section sizes and CFRP ratios

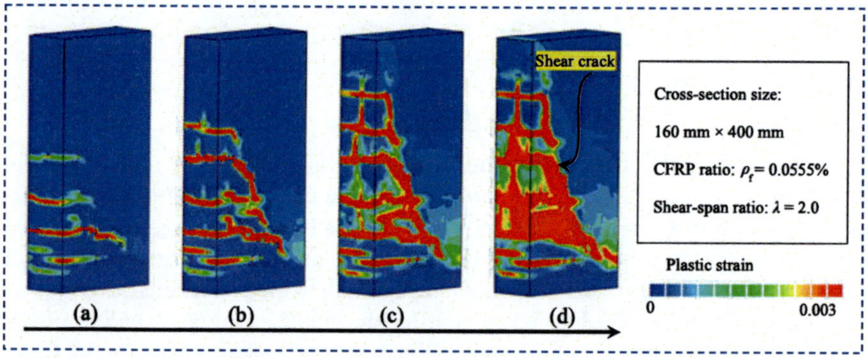

Fig. 6.59 Crack propagation process for CFRP-strengthened RC beam

the beam was in an elastic stage. As the load continued to increase, the transverse crack began to extend obliquely and gradually formed a diagonal crack, as presented in Fig. 6.59b. Due to the restraint of CFRP sheets, more oblique cracks appeared in the shearing zone. The oblique crack continued to propagate, it gradually penetrated and the width of the cracks increased continuously is presented in Fig. 6.59c. At this

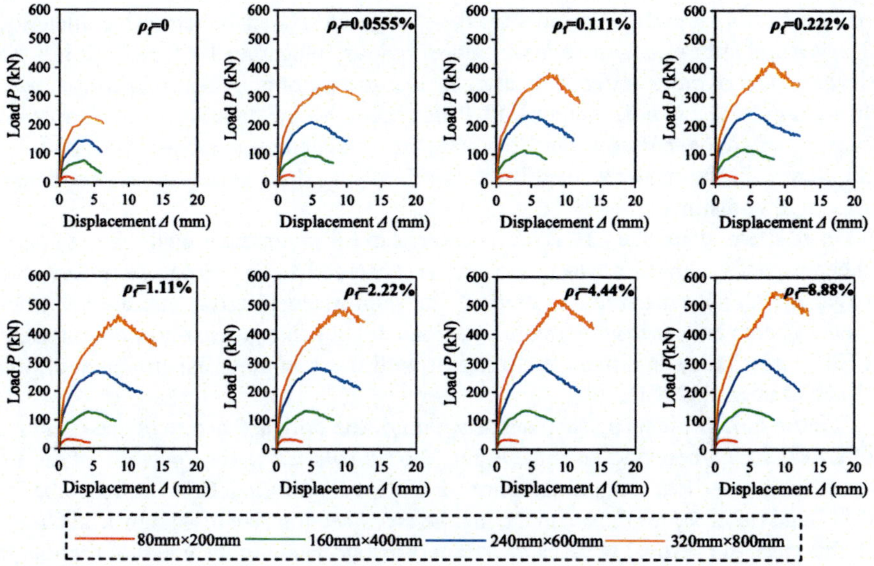

Fig. 6.60 *P-Δ* curves of CFRP-strengthened RC beams with the same CFRP ratios and different cross-section sizes

point, the stiffness of the beam was obviously degraded, and the load reached the peak value gradually. Subsequently, some diagonal cracks developed into a critical diagonal crack and the concrete beam was destroyed eventually, as presented in Fig. 6.59d. The failure process is consistent with the results of Jin et al. (2018).

Figure 6.61 depicts the load-displacement (*P-Δ*) curve of the cantilever beam with the same cross-section size and different CFRP ratios. Compared with the unreinforced concrete beam (with the CFRP ratio of 0), it can be seen that with the increase of CFRP ratio, the shear capacity of cantilever beam was improved

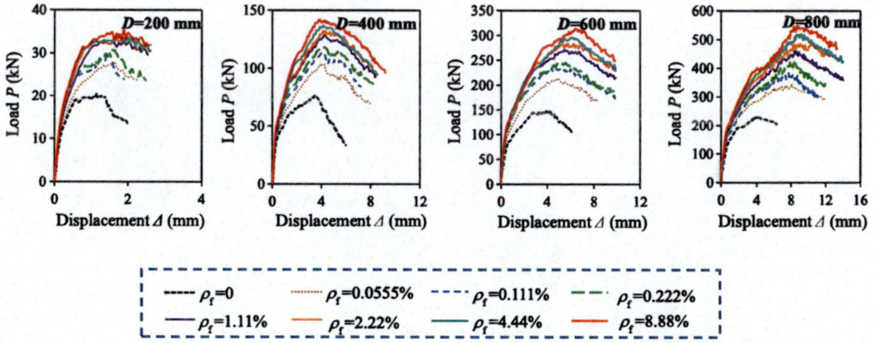

Fig. 6.61 *P-Δ* curves of CFRP-strengthened RC beams with different cross-section sizes and CFRP ratio

obviously and the end displacement of the strengthened beam increased significantly. The flexural failure phenomenon was more obvious, indicating that CFRP inhibits the development of cracks effectively after the occurrence of oblique cracks in concrete. In addition, one can note that the CFRP also improves the ductility and deformation capacity of concrete beams. On the whole, the ultimate failure modes of beams with different CFRP ratios were roughly the same, only the damage degree of the concrete was slightly distinct.

In addition, when the CFRP ratio becomes to be larger, the growth trend of peak shear capacity slowed down obviously, as observed in Fig. 6.61. Therefore, the larger the CFRP ratio, the less obvious the reinforcement effect and the displacement of beam end corresponding to peak load did not change greatly under different CFRP ratio. These findings are consistent with those in other experimental efforts (Kachlakev et al. 2001).

Figure 6.62 presents a schematic diagram of the critical fracture of shear failure of CFRP-strengthened RC beams. Figure 6.63 shows the vertical position of the F2 sheet and the vertical strain distribution of CFRP with different CFRP ratios (0.0555, 0.111 and 0.222%) for the beam having the cross-section size of 80 mm × 200 mm. In this case, the vertical strain is the strain distribution along the depth of section. It

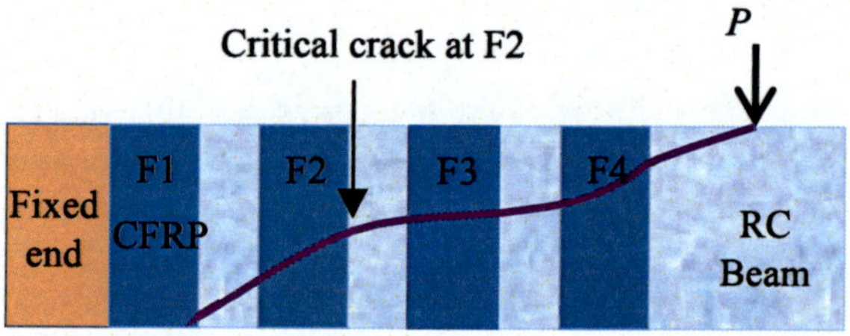

Fig. 6.62 Schematic diagram of critical crack in shear failure

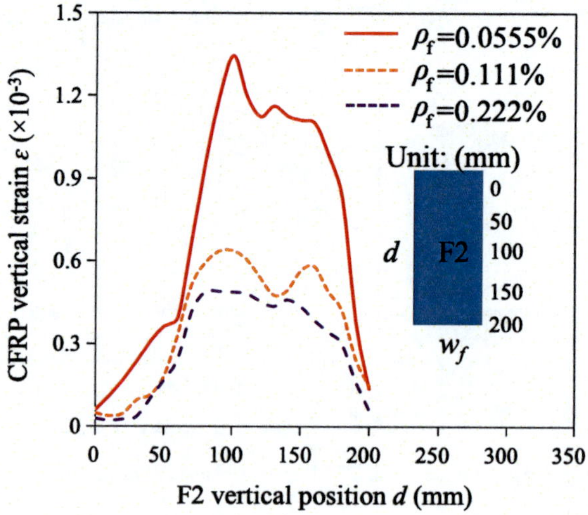

Fig. 6.63 Distribution of vertical strain of CFRP

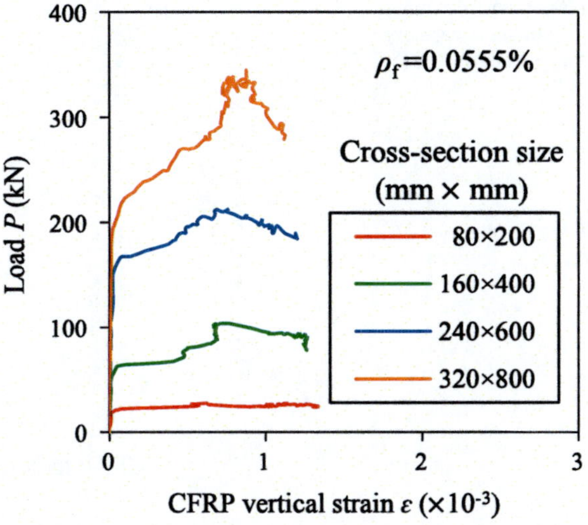

Fig. 6.64 *P-ε* curve of CFRP at F2 critical crack

is noted that the largest vertical strain of CFRP generates in the middle depth near the position of the oblique crack, while the smaller vertical strain of CFRP generates at the ends of the oblique crack near the bottom and top of the beam.

For the CFRP ratio of 0.0555%, the development of CFRP load-strain (*P-ε*) at the critical crack of the F2 sheet in the CFRP shear strengthened RC beam under different cross-section sizes is presented in Fig. 6.64. At the initial stage, the shear force was mainly borne by the concrete. As the loading continued, the CFRP strain changes significantly and increases rapidly after the oblique crack occurs in the beam, indicating that the CFRP plays an important role in limiting the development of cracks. Meanwhile, it can be found that the strain development trend of CFRP is basically the same for the beams having different cross-section sizes. This has also been observed in Godat et al. (2012).

The CFRP maximum strain ε_m and the increment of shear capacity V_{fk} of the F2 sheet of the beams having different cross-section sizes with the CFRP ratio of 0.0555% are plotted in Fig. 6.65. The increment of shear capacity refers to the difference of the maximum shear capacity between the reinforced beam and the unreinforced beam. The horizontal coordinate represented the maximum vertical strain in the CFRP sheet, which was an important parameter for studying the size effect (Godat et al. 2012).

It has been realized that the shear capacity is higher with the increase of the size of beam when the CFRP ratio keeps the same. However, the maximum vertical strain in the CFRP sheet will decrease, which indicates that the shear contribution of CFRP decreases as the beam size increases. It demonstrates that CFRP reinforcement has the largest contribution in small-sized beams and the shear contribution of CFRP decreases with the increase of beam size. Similar observations have been reported by Bae et al. (2012).

Fig. 6.65 Maximum strain-shear capacity increment of CFRP at F2

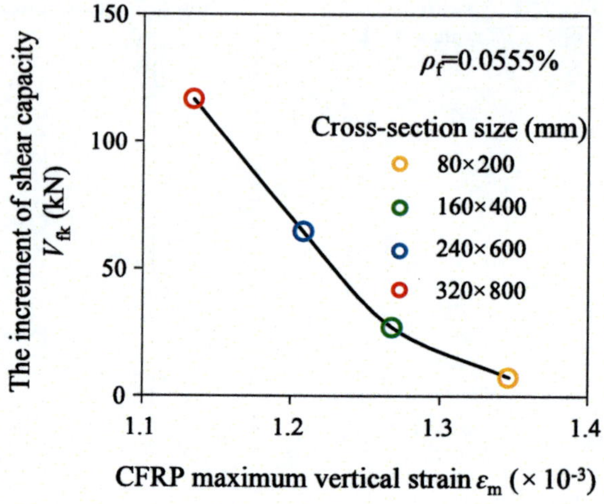

The relationship between the obtained nominal shear strength of CFRP shear-strengthened cantilever beams without web reinforcement and the cross-sectional depth D are presented in Fig. 6.66. One can clearly note that, under the same cross-sectional depth, the nominal shear strength increases with the increase of CFRP ratio, namely the strengthening of CFRP improves the shear capacity of the beam. For the beams with the same CFRP ratio, the nominal shear strength decreases with the increasing size, exhibits a prominent size effect, as can be observed from Fig. 6.66. Moreover, with the increase of CFRP ratio, this downward trend continues to decrease, and the trend line gradually becomes gentle, indicating that the size effect on shear strength of beams is weakened. The nominal shear strength of CFRP-strengthened beams at large CFRP ratios still has a downward trend, indicating

Fig. 6.66 Nominal shear strength of cantilever beams without stirrups

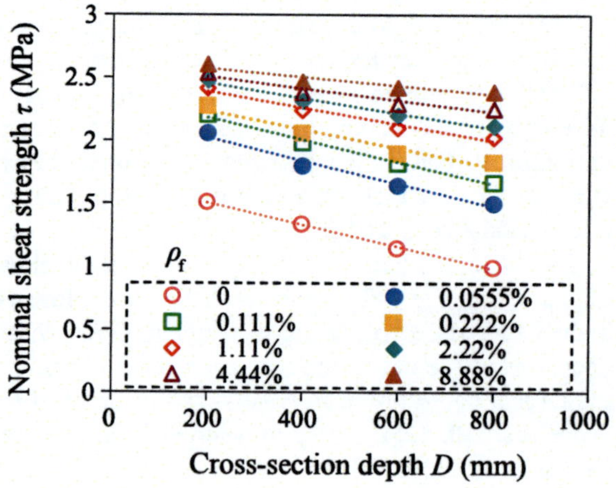

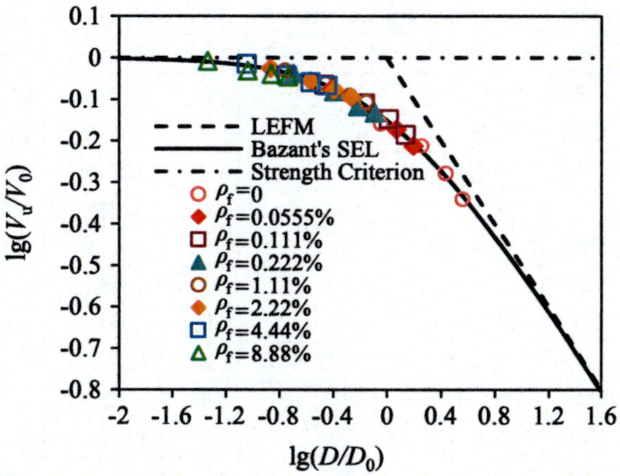

Fig. 6.67 Size effect fittings on the nominal shear strength of cantilever beams without stirrup

that the size effect cannot be completely suppressed. This is consistent with the conclusion of Yu and Bažant (2011). Furthermore, the enhancement effect generated by CFRP on nominal shear strength weakens gradually as the CFRP ratio increases. Therefore, when the CFRP ratio reaches a certain level, the effect of the CFRP shear-strengthened beam is not obvious. The same conclusion is reached in the works of Kachlakev et al. (2001).

According to the regression analysis of the simulation data, the logarithmic fitting curve on the nominal shear strength of the RC cantilever beam without web reinforcement and with different CFRP ratios can be obtained, as plotted in Fig. 6.67. The fitting correlation coefficient R^2 for Bažant's SEL (Bažant 1984) and the present simulated strengths is 0.99, indicating that Bažant's SEL can well describe the size effect on the shear strength of CFRP side-bonding strengthened RC cantilever beams.

2. *Shear-span ratio*

The shear strength and failure mode of concrete beams are obviously affected by the shear-span ratio of CFRP-strengthened beams without stirrups. With the change of the shear-span ratio of beams, the brittleness of strengthened beams may also change, which may have an impact on the size effect of their shear strength. In order to study the influence of shear failure and size effect on CFRP-strengthened beams without stirrups by shear-span ratio, this section adopts the above verified 3-D meso-scale model and method and carries out finite element numerical simulation test based on the previous research (Jin et al. 2020b). Herein, cantilever beams with different cross-section sizes (S: 80 mm × 200 mm, M: 160 mm × 400 mm, L: 320 mm × 800 mm), different shear-span ratios ($\lambda = 1.0, 1.5, 2.0$), and different CFRP ratios (0.0555, 0.111, 0.222%) are designed. Taking the specimen "CBR-S-1.0-0.0555" as an example for the name format. The letter "CBR" means CFRP-strengthened concrete beams. The letter "S" stands for the cantilever beam cross-section size of 80 mm × 200 mm, and the number "1.0" stands for the shear-span ratio of 1.0; The number "0.0555" represents a CFRP ratio of 0.0555%. Due to the large number of

Table 6.13 Physical parameters of the CFRP-strengthened concrete beam

Specimen name	d_0 (mm)	λ	ρ_f (%)	s_f (mm)	w_f (mm)	t_f (mm)
CBR-S-1.0-0.0555	180	1.0	0.0555	100	20	0.111
CBR-S-1.5-0.0555	180	1.5	0.0555	100	20	0.111
CBR-S-2.0-0.0555	180	2.0	0.0555	100	20	0.111
CBR-M-1.0-0.0555	360	1.0	0.0555	200	40	0.222
CBR-M-1.5-0.0555	360	1.5	0.0555	200	40	0.222
CBR-M-2.0-0.0555	360	2.0	0.0555	200	40	0.222
CBR-L-1.0-0.0555	720	1.0	0.0555	400	80	0.444
CBR-L-1.5-0.0555	720	1.5	0.0555	400	80	0.444
CBR-L-2.0-0.0555	720	2.0	0.0555	400	80	0.444
CBR-S-1.0-0.111	180	1.0	0.111	100	20	0.222
CBR-S-1.5-0.111	180	1.5	0.111	100	20	0.222
CBR-S-2.0-0.111	180	2.0	0.111	100	20	0.222
CBR-M-1.0-0.111	360	1.0	0.111	200	40	0.444
CBR-M-1.5-0.111	360	1.5	0.111	200	40	0.444
CBR-M-2.0-0.111	360	2.0	0.111	200	40	0.444
CBR-L-1.0-0.111	720	1.0	0.111	400	80	0.888
CBR-L-1.5-0.111 CBR-L-1.5-0.111	720	1.5	0.111	400	80	0.888
CBR-L-2.0-0.111	720	2.0	0.111	400	80	0.888

Note S: 80 mm × 200 mm, M: 160 mm × 400 mm, L: 320 mm × 800 mm. d_0 means the effective cross-section depth; λ means the shear-span ratio; ρ_f means the CFRP ratio; w_f represents the width of CFRP sheet; t_f represents the thickness of CFRP sheet; and s_f represents the spacing of CFRP sheet

models, the relevant geometric parameters of some specimens are listed in Table 6.13. The names and parameters of other specimens can be obtained by analogy according to the naming method.

Figure 6.68 shows the shear failure of CFRP-strengthened concrete cantilever beams with the same CFRP ratio (0.0555%), different shear-span ratios (1.0, 1.5, 2.0), and different sizes (S: 80 mm × 200 mm, M: 160 mm × 400 mm, L: 320 mm × 800 mm). According to the crack development, all the specimens showed obvious penetrating main oblique cracks at the final failure, which had significant shear failure characteristics. The shear-span ratio has a great influence on the failure pattern and inclination angle of the inclined crack of the strengthened beam. In the same size (cross-section size is 320 mm × 800 mm), with the increase of shear-span ratio, and horizontal to the angle of inclined cracks θ gradually decreases, and change the failure pattern.

Figure 6.69 shows the crack development process of CBR-S-2.0-0.0555% (beam depth is 200 mm) cantilever beam. The deformation of the CFRP of different shear-span ratio beams at 0.0555% CFRP ratio is shown as Fig. 6.70. At the initial stage

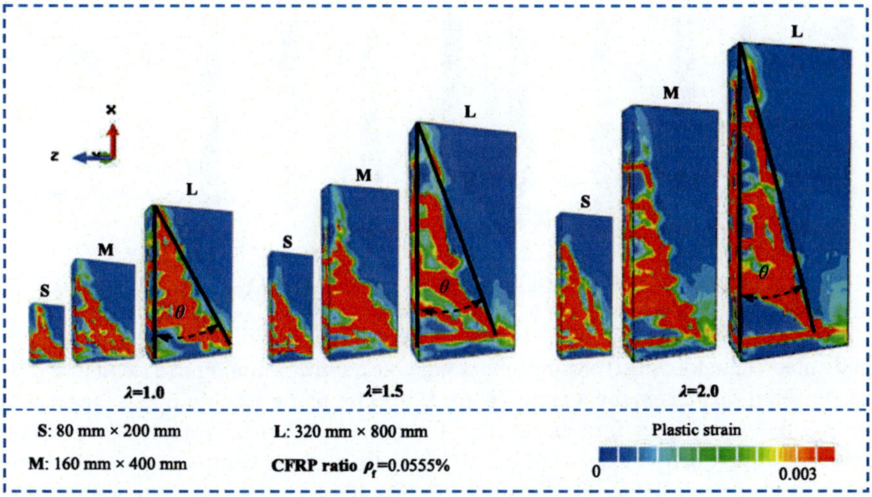

Fig. 6.68 Ultimate failure modes of the cantilever beams with different sizes and shear-span ratios

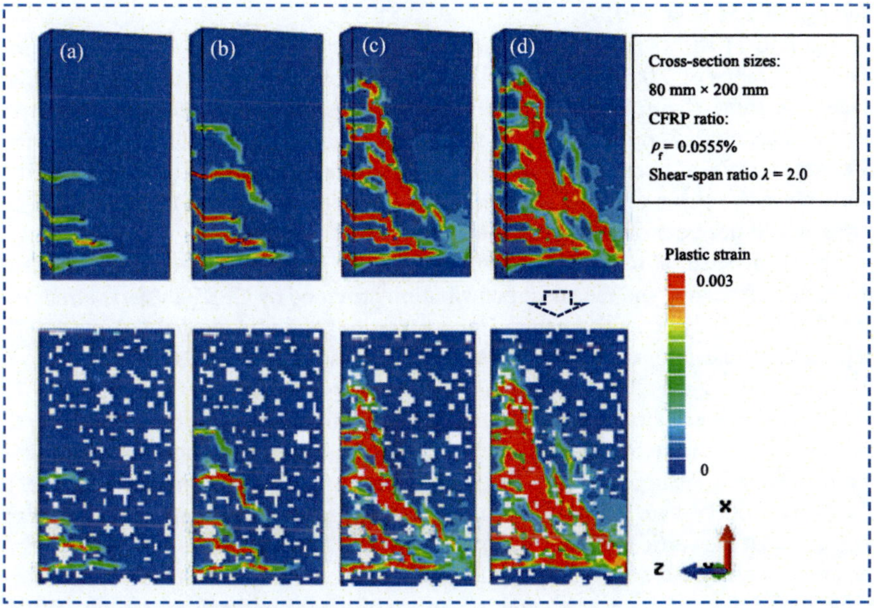

Fig. 6.69 Crack development process of the beam of CBR-S-2.0-0.0555

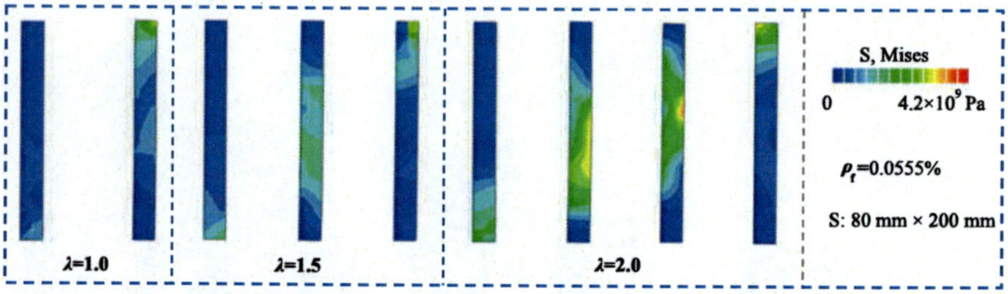

Fig. 6.70 The deformation of the CFRP of different shear-span ratio beams at 0.0555%

of displacement loading, forming a bending crack and continuing to extend laterally, and the load-displacement curve (Fig. 6.71) is close to the straight line, as presented in Fig. 6.69a, b. After further loading, the transverse crack began to extend and developed diagonally to the fixed end of the bottom of the beam and form the main inclined crack is presented in Fig. 6.69c. Eventually forming a penetrating diagonal crack that extends from near the loading point to the fixed end at the bottom of the beam, causing component failure is shown in Fig. 6.69d. This is consistent with the findings of Bae et al. (2012).

Figure 6.71 shows the load-displacement curves (*P*-*Δ*) of the tested beams, for the CFRP ratios of 0.0555%, 0.111% and 0.222%, respectively. It can be found that under the same size CFRP ratio, with the increase of shear-span ratio, the shear bearing capacity of strengthened beam is decreased obviously, illustrating that the shear-span ratio of beam is an important factor affects the shear bearing capacity. Here, since the strain of CFRP cannot reach the ultimate strain of CFRP, the CFRP strips would not be broken at peak load.

The shear-span ratio has a great influence on the shear bearing capacity of shear strengthened beams, and the shear contribution provided by CFRP in shear strengthened beams largely depends on the shear-span ratio (Li and Leung 2016). Figure 6.72 depicts the shear failure loading diagram of CFRP-strengthened RC beams without stirrups, marking the CFRP sheets F1, F2, F3 and F4 from left to right. Figure 6.73 presents the vertical strain distribution of CFRP along the beam cross-section depth at the vertical position of F2 sheet when the beam reaches the bearing capacity under different shear-span ratios ($\lambda = 1.0, 1.5, 2.0$) of the same CFRP ratio ($\rho_f = 0.0555\%$). The vertical coordinate in Fig. 6.73 shows the cross-section depth of the beam. It can be found that the vertical strain distribution of CFRP sheet presents "convex". When the shear-span ratio is small (i.e., $\lambda = 1.0$), the maximum CFRP vertical strain occurs at the top of the beam, i.e., the crack at the top of the beam develops rapidly. When the shear-span ratio is large (i.e., $\lambda = 1.5$ or $\lambda = 2.0$), the maximum CFRP vertical strain basically occurs in the middle of the intersection with the diagonal crack, while the strain at the bottom and top of the beam is the minimum. Meanwhile, it can be found that the smaller the shear-span ratio ($1.0 \le \lambda \le 2.0$), the smaller the CFRP vertical strain, indicating that the smaller the shear contribution of CFRP in the small shear-span ratio, which is consistent with (Li and Leung 2016). In addition, it can

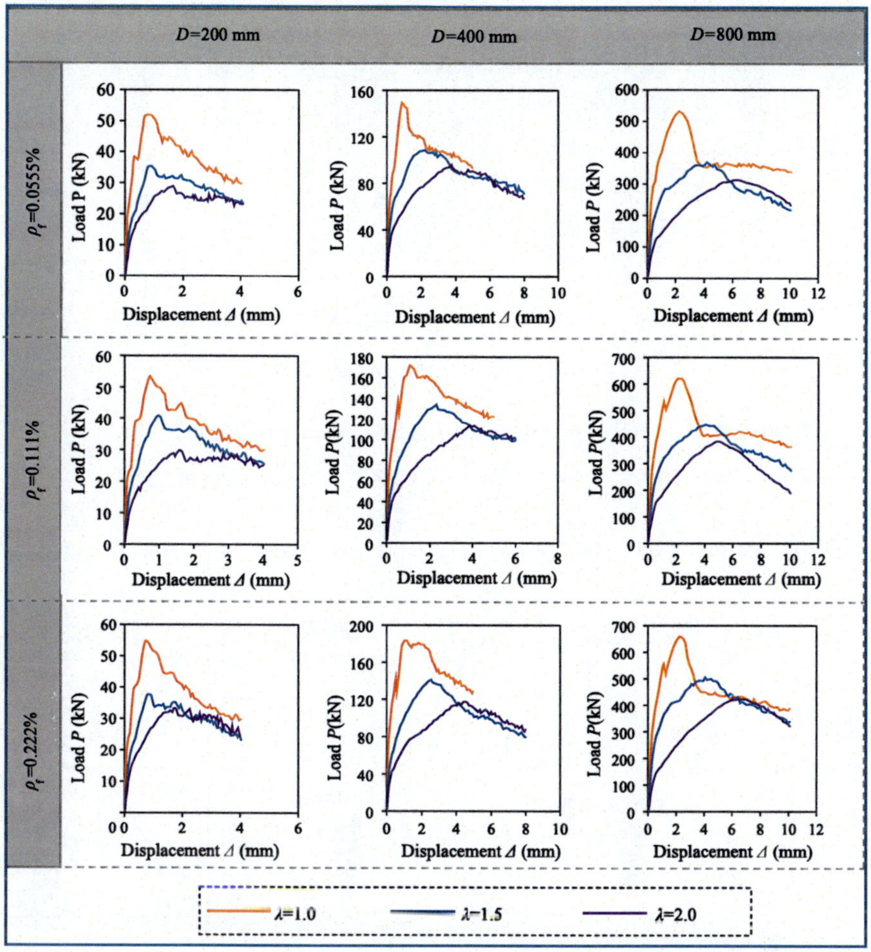

Fig. 6.71 *P-Δ* curves of specimens with different cross-section sizes and different shear-span ratios

be seen that under the same CFRP ratio, the CFRP strains of larger cross-section depth beams are smaller by comparing the CFRP strains of different beam depths. For example, the CFRP strain of 200 mm beam is greater than that of 800 mm beam significantly. In a word, for the smaller size beam, the CFRP sheets has a larger shear contribution, which is consistent with the previous conclusion (Jin et al. 2020b).

In order to further explain the relationship among shear contribution of CFRP sheets, cross-section size and shear-span ratio, the load-strain (*P-ε*) curve at the critical crack of F2 under different shear-span ratios of CFRP-strengthened RC beams without stirrups under the same cross-section size is shown in Fig. 6.74. It can be found that at the initial stage of loading, the concrete bears the shear capacity at the initial stage, and the CFRP strain is close to zero. After further loading, it is

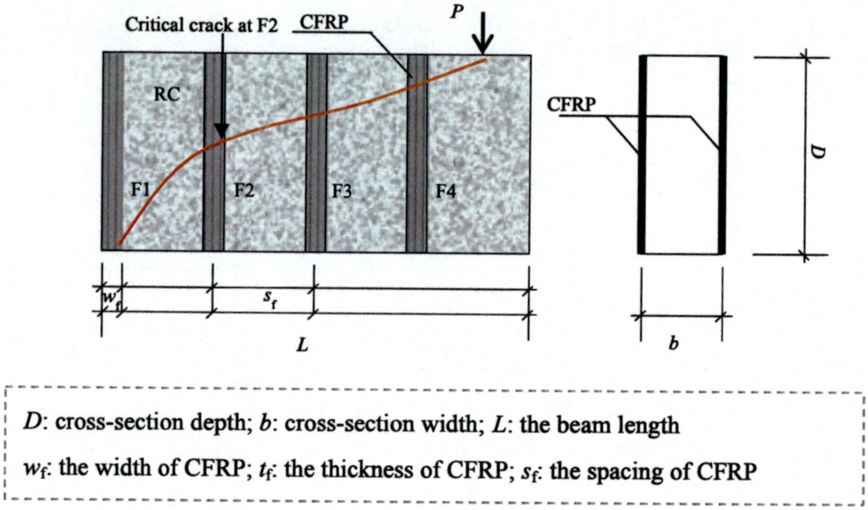

Fig. 6.72 Loading diagram of CFRP-strengthened RC beams without stirrups

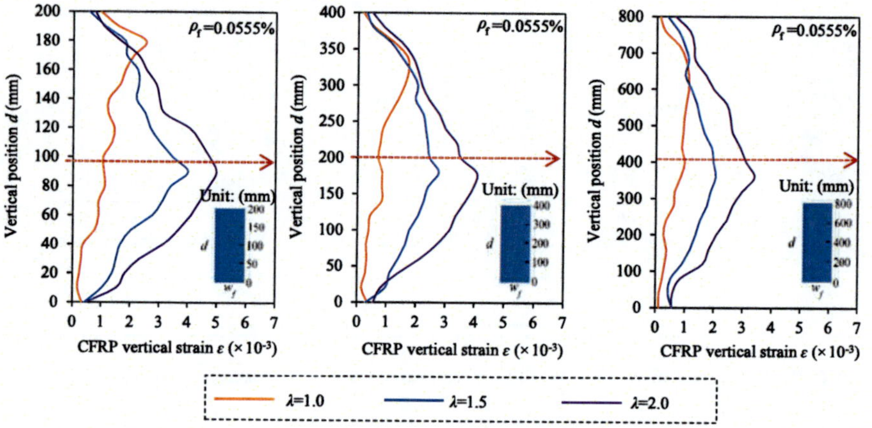

Fig. 6.73 The vertical strain distribution of CFRP in different areas of beams with the same CFRP ratio of 0.0555% (from F2 CFRP sheet in Fig. 6.72)

found that the strain of CFRP sheet increases sharply when diagonal cracks appear in the beam, which indicates that CFRP sheet begins to play the role in limiting the expansion of diagonal cracks. Here, since the strain of CFRP cannot reach the ultimate strain of CFRP, the CFRP strips would not be broken at peak load. With the increase of shear-span ratio, the slower the strain growth trend of CFRP is, the larger the corresponding maximum strain value is. In conclusion, the strengthening effect of CFRP is better under the condition of large shear-span ratio ($1.0 \leq \lambda \leq 2.0$), while the strengthening effect of CFRP under the condition of large size is not as good as

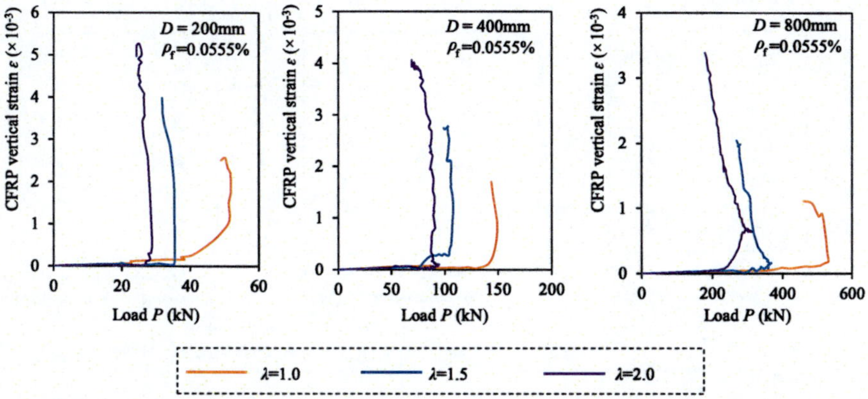

Fig. 6.74 The trend of P-ε curve of CFRP at critical crack (from F2 CFRP sheet in Fig. 6.69)

that under the condition of small size, that is to say, the strengthening effect of CFRP in small size is better.

Figure 6.75 shows the variation trend of nominal shear strength of cantilever beams strengthened by CFRP with different sizes, shear-span ratios and CFRP ratios. For beams with the same size and CFRP ratio, the nominal shear strength decreases with the increase of shear-span ratio. For beams with the same shear-span ratio, the nominal shear strength decreases with the increase of the size, exhibits a prominent size effect.

According to the regression analysis of the simulated data, the logarithmic curve of nominal shear strength of the CFRP-strengthened RC cantilever beams without stirrups with different shear-span ratios is obtained, as shown in Fig. 6.76. At the same time, the simulation results were also fitted with Bažant's SEL and the fitting correlation coefficient R^2 was 0.90, indicating that Bažant's SEL could describe the size effect on shear strength of CFRP-strengthened RC cantilever beams. As can be seen from the distribution of data points in Fig. 6.76, the positions of data points of the same size and different shear-span ratios basically coincide, indicating that shear-span ratio has little effect on the size effect of shear strength, so the influence on the size effect can be ignored, However, it should be noted that for different shear-span ratios and CFRP ratios, the shear failure of CFRP-strengthened concrete beams will be different, and the shear contribution of CFRP will also change. In addition, the influence of shear-span ratio on shear strength must be considered in strength design.

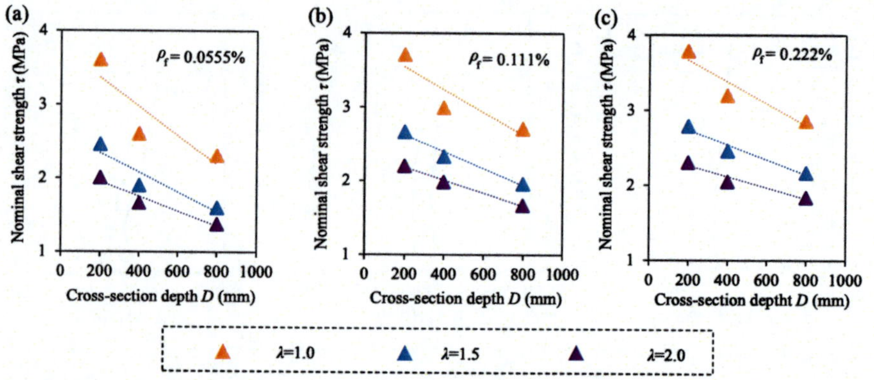

Fig. 6.75 Nominal shear strength of CFRP-strengthened cantilever beams without stirrups, for CFRP ratio of: **a** $\rho_f = 0.0555\%$; **b** $\rho_f = 0.111\%$; and **c** $\rho_f = 0.222\%$

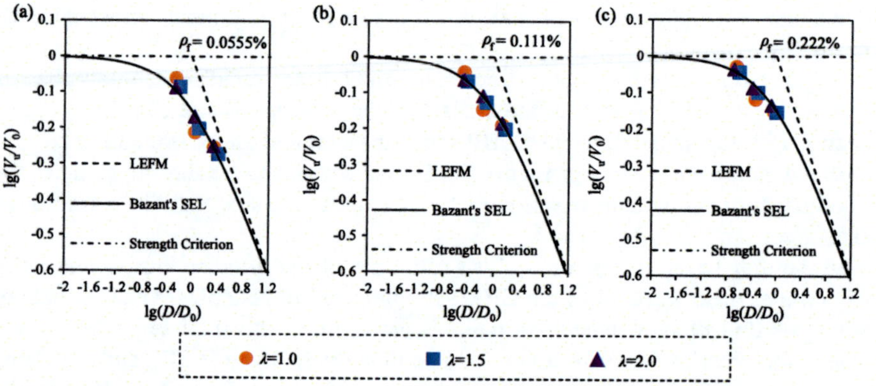

Fig. 6.76 Size effect fitting of nominal shear strength of CFRP-strengthened cantilever beams without stirrups, for CFRP ratio of: **a** $\rho_f = 0.0555\%$; **b** $\rho_f = 0.111\%$; and **c** $\rho_f = 0.222\%$

6.3.2 CFRP-Strengthened Lightweight-Aggregate Concrete Beams

Compared with NWC beams, LWAC beams have lower elastic modulus, more obvious brittleness, and more prone to shear failure. This may lead to the fact that the shear failure of CFRP-strengthened LWAC beams would show more obvious size effect.

This section presents the shear failure behavior of two kinds of concrete beams strengthened with CFRP, explores the size effect law of CFRP-strengthened LWAC beams, and analyzes the shear contribution of CFRP strips in LWAC beams emphatically.

The detailed information on the utilized CFRP, involving CFRP thickness t_f, widths of CFRP sheet w_f, spacing of CFRP sheets s_f and CFRP ratios ρ_f are provided in Table 6.14. According to the name of specimens given in Table 6.14, for example "1-LCBF1", the first character indicates the section size of the specimen in which "1, 2, 3, 4" indicates the beam depth increasing from 200 to 800 mm, and "LCBF" means CFRP side-bonding strengthened LWAC beam. The Numbers "1 to 6" after LCBF in Table 6.14 represent six different CFRP ratios ρ_f. Since the LCBF series specimens are different from the CBF series specimens mentioned in the following text only in aggregate parameter, which are consistent with the specimen parameters as well as the parameters of CFRP. Therefore, the CBF series is not listed separately in Table 6.14.

The final failure modes of CFRP-strengthened ordinary and LWAC cantilever beams (shear-span ratio $\lambda = 2$) at different CFRP ratios (0–0.888%) of different sizes (beam depth: 200–800 mm) are shown in Fig. 6.77, respectively. It can be seen that both LWAC beams strengthened with CFRP and NWC beams strengthened with CFRP have obvious shear brittle failure. The failure mode of CFRP-strengthened LWAC beams with different sizes and CFRP ratio is basically the same as that of CFRP-strengthened NWC beams, and finally, both of them develop a diagonal crack from the loading point to the bottom of the fixed end, leading to the failure of the whole structure.

The load-displacement (P-Δ) curve of the CFRP-strengthened LWAC beams under different sectional dimensions and CFRP ratios as shown in Fig. 6.78.

Table 6.14 CFRP shear-strengthening dimensions and configurations

Number	Specimen	CFRP dimensions			CFRP ratio ρ_f (%)
		t_f (mm)	w_f (mm)	s_f (mm)	
1	1-LCBF1	0	20	100	0
2	1-LCBF2	0.111			0.0444
3	1-LCBF3	0.2775			0.111
4	1-LCBF4	0.555			0.222
5	1-LCBF5	1.11			0.444
6	1-LCBF6	2.22			0.888
7	2-LCBF1	0	40	200	0
8	2-LCBF21.51	0.222			0.0444
9	2-LCBF3	0.555			0.111
10	2-LCBF4	1.11			0.222
11	2-LCBF5	2.22			0.444
12	2-LCBF6	4.44			0.888
13	3-LCBF1	0	60	300	0
14	3-LCBF2	0.333			0.0444

(continued)

Table 6.14 (continued)

Number	Specimen	CFRP dimensions			CFRP ratio ρ_f (%)
		t_f (mm)	w_f (mm)	s_f (mm)	
15	3-LCBF3	0.8325			0.111
16	3-LCBF4	1.1665			0.222
17	3-LCBF5	3.33			0.444
18	3-LCBF6	6.66			0.888
19	4-LCBF1	0	80	400	0
20	4-LCBF2	0.444			0.0444
21	4-LCBF3	1.11			0.111
22	4-LCBF4	2.22			0.222
23	4-LCBF5	4.44			0.444
24	4-LCBF6	8.88			0.888

Compared with the beam with CFRP ratio of zero, the shear capacity of the strengthened beam is increased significantly, and the displacement increases with the increase of CFRP ratio as well, and the curve tends to be flat. This indicates that CFRP plays a good role in restraining the development of oblique fractures. Here, since the strain of CFRP cannot reach the ultimate strain of CFRP, the CFRP strips would not be broken at peak load. Taking the specimens with section depth of 800 mm strengthened by different CFRP ratios as an example, the maximum shear capacity of CFRP-strengthened beams was increased by 11.73%, 18.16%, 22.48%, 26.83%, respectively compared with the control beam ($\rho_f = 0$). However, the efficiency of increasing the ultimate shear capacity of CFRP is 2.74, 1.98, 1.29, and 1.21% between the two adjacent CFRP ratios. The efficiency is decreasing continuously. It indicates that CFRP can improve the shear capacity of the specimens, but the increase is not linear, and the large amount of CFRP does not necessarily mean a better strengthening effect.

Figure 6.79 depicts the critical crack diagram of shear failure of CFRP-strengthened LWAC beams, labelling the strip on one side F1, F2, F3, F4 from left to right, starting near the fixed end. Figure 6.80 presents the CFRP vertical strain distribution along the beam depth at the vertical position of F3 sheet when the LWAC beams reaches the bearing capacity under five different CFRP ratios (0.0444, 0.111, 0.222, 0.444 and 0.888%). The vertical coordinate in Fig. 6.80 denotes the beam depth. It can be found that the vertical strain distribution of CFRP shows a "convex" shape, and the maximum vertical strain is basically generated in the middle of the intersection with the oblique crack, while the strain at the bottom and top of the beam is the minimal. At the same time, it can be found that the smaller the CFRP ratio is, the larger the magnitude of the CFRP vertical strain is, indicating that the shear contribution of CFRP in the small CFRP ratio is larger.

In addition, it is worth noting that the strain distribution of CFRP in different locations is also significantly distinct. The strain distribution of CFRP in F4 (which

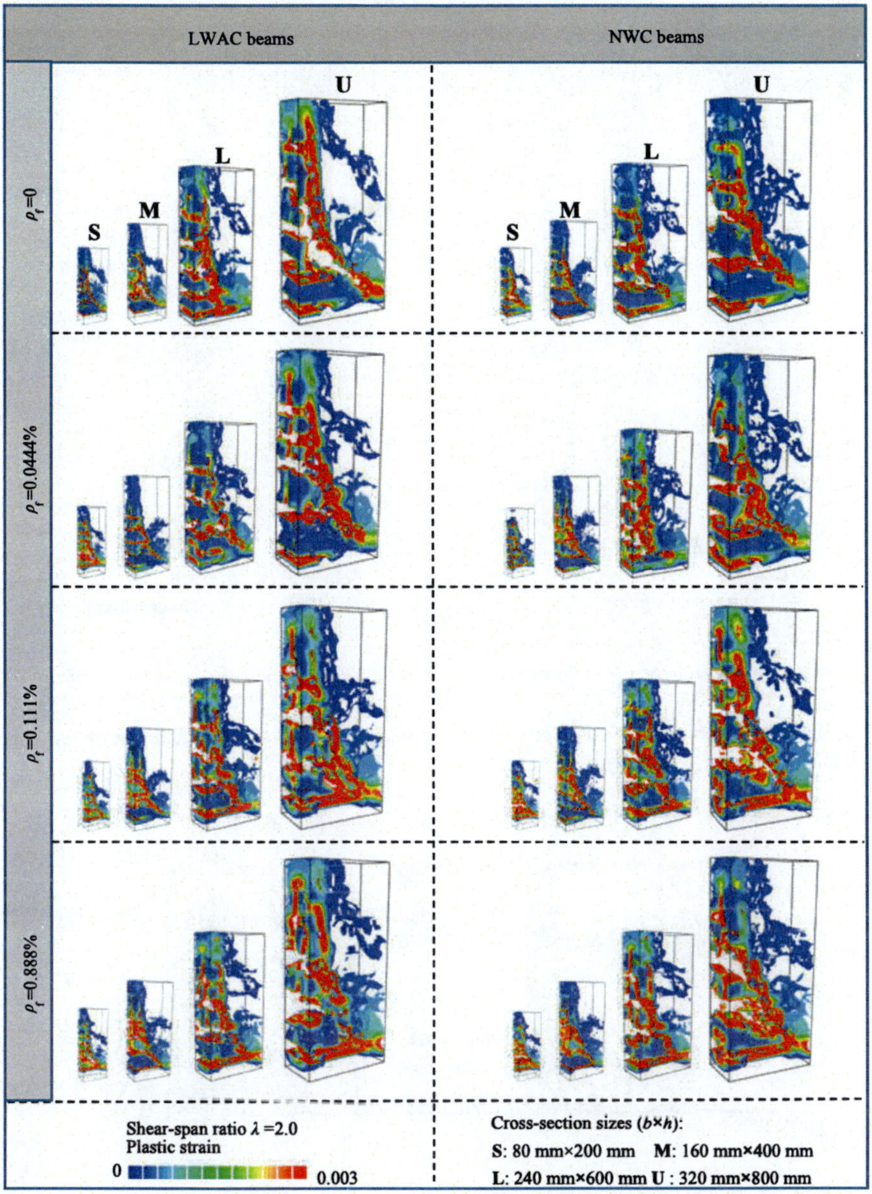

Fig. 6.77 Failure modes of CFRP-strengthened LWAC and NWC beams with different sizes and CFRP ratios

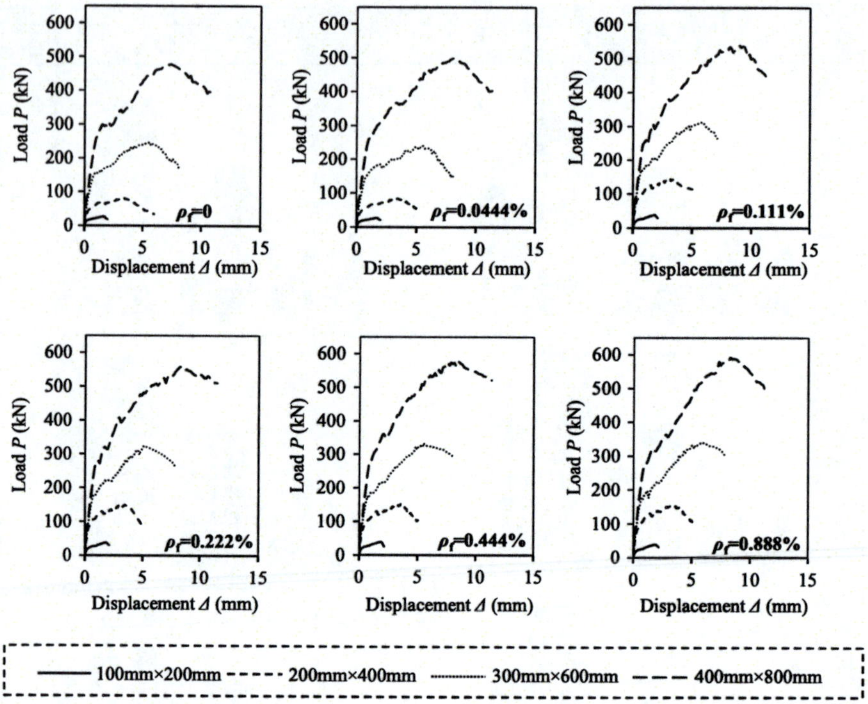

Fig. 6.78 *P-Δ* curves of CFRP-strengthened LWAC beams with different sizes under the same CFRP ratios

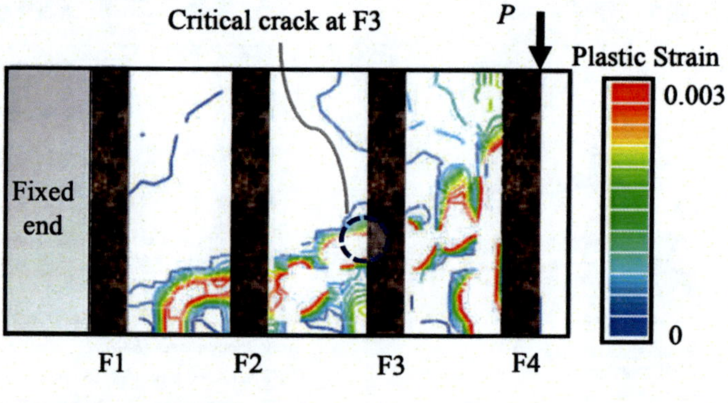

Fig. 6.79 Schematic diagram of critical crack in shear failure

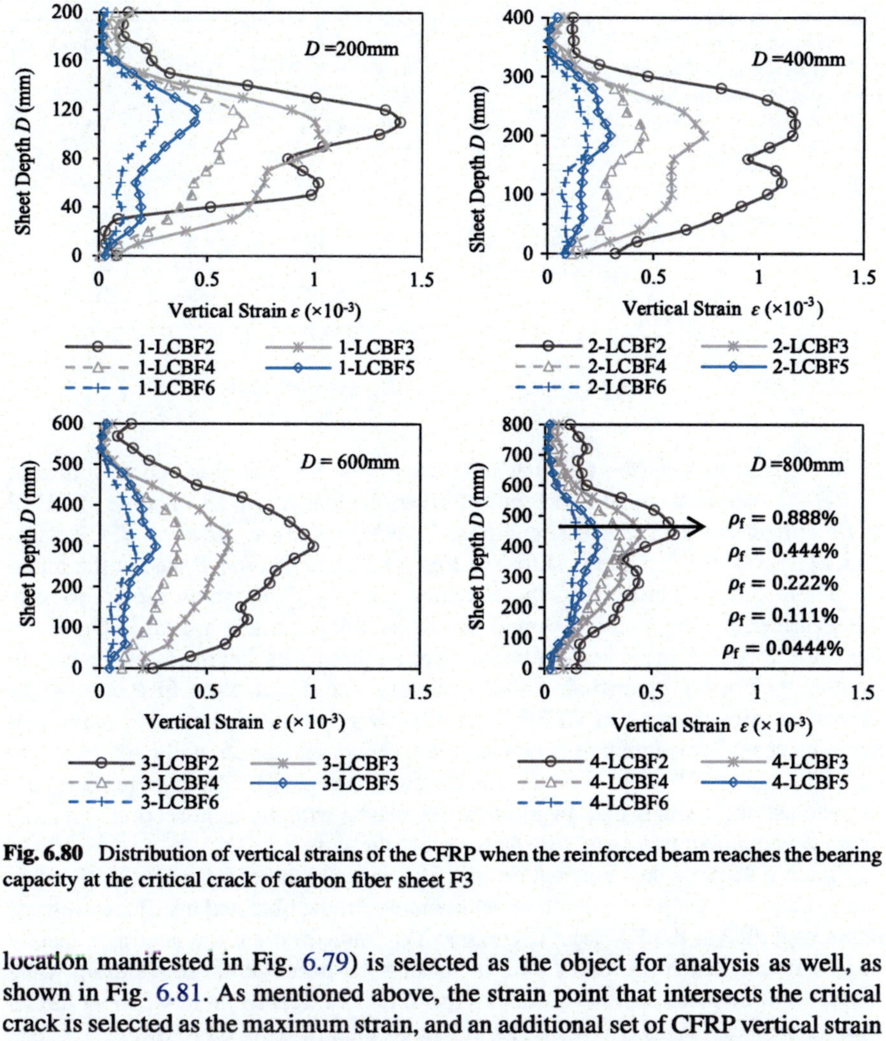

Fig. 6.80 Distribution of vertical strains of the CFRP when the reinforced beam reaches the bearing capacity at the critical crack of carbon fiber sheet F3

location manifested in Fig. 6.79) is selected as the object for analysis as well, as shown in Fig. 6.81. As mentioned above, the strain point that intersects the critical crack is selected as the maximum strain, and an additional set of CFRP vertical strain with a cross-sectional size of 200 mm and the CFRP ratio of 0.0444% is plotted. From the distribution diagram, it can be seen that the maximum strain of CFRP is different under the same CFRP ratio and the same size for different strips which is related to the distribution of cracks. Furthermore, it can be seen from Fig. 6.81 that the maximum strain ε_m of CFRP strip is 5.58×10^{-3} which is about 30% of the ultimate tensile strain. Moreover, one can note from the comparison of the CFRP strains for the beams having different beam depths, under the same CFRP ratio, the larger-sized beam has a smaller CFRP strain. For instance, the magnitude of the CFRP strains for the beam with the depth of 200 mm are obvious greater than that with the depth of 800 mm. In summary, CFRP sheets have larger shear contribution for smaller-sized beams.

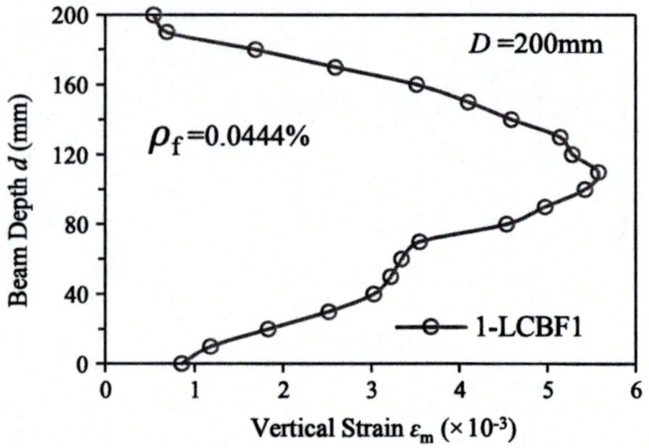

Fig. 6.81 Distribution of vertical strains of the CFRP at the critical crack of carbon fiber sheet F4 with the beam depth of 200 mm (ρ_f = 0.0444%)

In order to express the relationship between the three of shear contribution of CFRP, the section size and CFRP ratio further, the load-strain (P-ε) curve of CFRP at F3 critical crack in CFRP-strengthened LWAC beams at different CFRP ratios under the same section size is shown in Fig. 6.82. It can be found that, at the initial stage of loading, concrete bears shear bearing capacity, CFRP strain is close to zero. After further loading, it can be found that when oblique cracks appear in the beam, the strain of CFRP increases steeply, indicating that CFRP begins to play a role in limiting the crack propagation. At the same time, the smaller the CFRP ratio is, the slower the growth trend of CFRP is, and the larger the corresponding maximum strain value is. Some intuitive conclusions are shown that the strengthening effect of CFRP in small CFRP ratio is better, and the contribution of CFRP in large size is not as good as that in small size. In addition, for beams with different section sizes, the strain development trend of CFRP is basically the same.

Figure 6.83 plots the contribution of CFRP visually under different CFRP ratios and structural sizes. As the depth of beam increases, the obtained maximum vertical strain of CFRP at the F3 sheet decreases. This indicates that the bearing capacity of CFRP undertaken decreases with increasing the beam depth. In addition, when the CFRP ratio increases, the measured maximum vertical strain of CFRP at the F3 sheet decreases, means that the CFRP effect is better in small CFRP ratio.

As can be seen in Fig. 6.84, for LWAC beams, the obtained nominal shear strength decreases with the increase of beam depth when the CFRP ratio is the same. At the same time, the decline angle of CFRP-strengthened LWAC beams is more evident. In addition, with the increase of CFRP ratio, this size effect phenomenon will be weakened, but will not be completely suppressed, that is, the impact of section size on the shear capacity of the beam will be reduced. This is consistent with Mohammad et al. (2018).

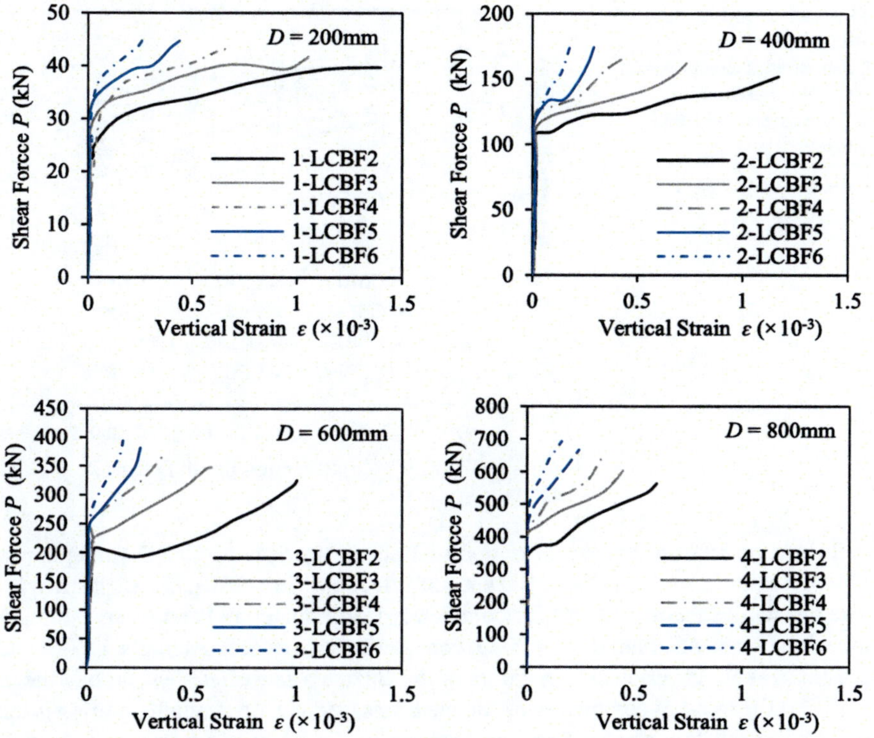

Fig. 6.82 The trend of CFRP vertical strain in cross section at the carbon fiber sheet F3 critical crack

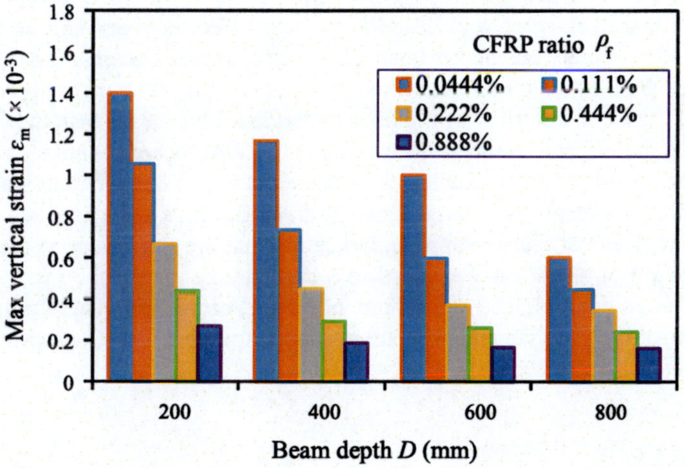

Fig. 6.83 The variation of the maximum vertical strain ε_m of CFRP with different CFRP ratios and section depths

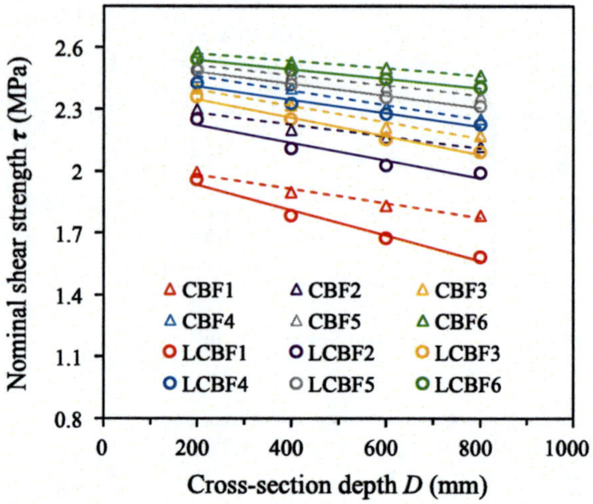

Fig. 6.84 Nominal shear strength of CFRP-strengthened LWAC and NWC beams with different sizes and CFRP ratios

In the regression analysis, the nominal shear strength of CFRP strengthened LWAC beams is compared with Bažant size effect law (i.e., SEL), linear elastic fracture mechanics theory (LEFM for completely brittle materials) and plastic strength criterion (for plastic materials, without considering size effect). Actually, LEFM and plastic strength criterion are the limits of the Bažant's size effect law, among them, LEFM is the trend of completely brittle materials, and plastic strength criterion is the trend of completely plastic. Because concrete is a quasi-brittle material, it changes gradually between the two boundaries. The logarithmic curve of the nominal shear strength of CFRP-strengthened concrete cantilever beams changing with the component size is obtained, as shown in Fig. 6.85. One can see from Fig. 6.85 that, Bažant's size effect law (SEL) can well describe the size effect in shear failure of CFRP-strengthened concrete beams for both LWAC and ordinary beams. It can be noted from Fig. 6.85a, as the CFRP ratio increases, the simulated strength approaches to the line of "strength criterion" (for plastic materials). This means that, the CFRP can weaken the size effect in nominal shear strength of LWAC beams. Similar conclusions can also be found for NWC beams, as presented in Fig. 6.85b. With the increase of CFRP ratio, the strength data points move up along Bažant's SEL curve and approach the horizontal line of plastic strength, indicating that the size effect on the nominal shear strength of LWAC is weakened to a certain extent with the increase of the CFRP ratio. In addition, the data points of LWAC beams are generally below the data points of ordinary concrete specimens, indicating that LWAC specimens are of high brittleness and low ductility.

6.4 Flexural Failure in Reinforced Concrete Beams

Here in this section, the seismic tests on flexural failure of RC beams are introduced, and some simulation works of larger sized beams are provided (Jin et al. 2018).

6.4.1 Seismic Tests on Flexural Failure of RC Beams

1. *Test introduction*

Five groups of geometrically similar RC cantilever beams with different structural sizes, subjected to monotonic and cyclic loading were tested. The dimensions, shape, loading point locations, specimen label, and reinforcement details of specimens used in the experiment are shown in Table 6.15 and Fig. 6.86. The cross-sectional depth of the beams was tested ranged from 200 to 1000 mm. The tested RC beams presented in Fig. 6.86 were designed to fail as flexural failure, and a shear-span ratio (length to depth) of 4 was utilized. There were three specimens designed for every group of tested specimens having the same structural sizes, in which one of them was subjected to monotonic loading, and the other two were designed to bear cyclic loading. The specimens under monotonic loading were named as MB, and the other two specimens under cyclic loading were named as CB. Herein, the letter "M" means the "Monotonic loading", "C" means the "Cyclic loading", and "B" denotes "Beam". A total of 15 RC beams were thus tested.

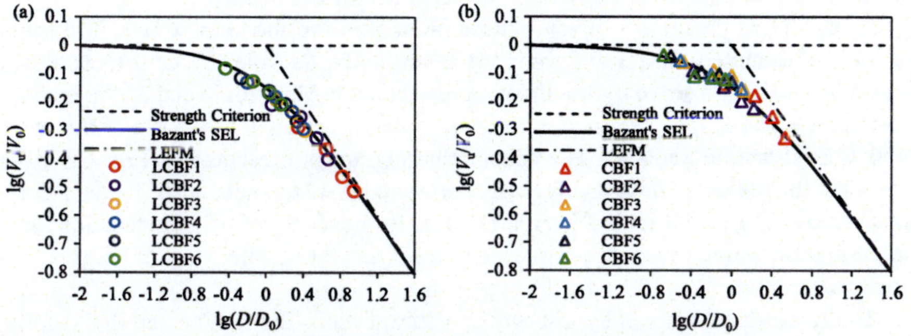

Fig. 6.85 Size effect fitting on nominal shear strength of LWAC beams with different CFRP ratios: **a** LWAC beams; and **b** NWC beams

Table 6.15 Physical parameters of the tested RC beams

Specimen name	MB-1/CB-1	MB-2/CB-2	MB-3/CB-3	MB-4/CB-4	MB-5/CB-5
Total number	3	3	3	3	3
Cross-section size ($b \times D$) (mm^2)	80 × 200	160 × 400	240 × 600	320 × 800	400 × 1000
Reinforcement ratio ρ (%)	1.26	1.23	1.31	1.22	1.23
Stirrup ratio ρ_{sv} (%)	0.14	0.14	0.14	0.14	0.14
Shear-span ratio λ	4.0	4.0	4.0	4.0	4.0
Beam effective depth d_0 (mm)	170	345	540	735	930
Beam effective length $a = \lambda * d_0$ (mm)	6800	1380	2160	2940	3720
Total length of beam l (mm)	830	1630	2510	3390	4220
Concrete tensile/compressive strengths (MPa)	2.3/42.8	2.3/42.8	2.3/42.8	2.3/42.8	2.3/42.8

The compressive strength of concrete is 42.8 MPa, and the splitting tensile strength is 2.3 MPa. Moreover, the detailed information about the parameters of the reinforcing steel bars are listed in Table 6.16, including the diameters, the yield strength and the ultimate strength, etc. In the tests, the strength of concrete, reinforcement ratio, stirrup ratio, and shear-span ratio were kept the same for all RC beams.

Figure 6.87a presents a typical failure photograph of the beam CB-3. The specimen was mounted on a solid component vertically, and the top of the cantilever beam was held by a servo hydraulic dynamic actuator. Monotonic and cyclic loading were employed to the specimen in the horizontal direction with the aid of the servo hydraulic dynamic actuator. The schematic diagram is presented in Fig. 6.87b. In the tests, the collected data involved the horizontal displacement at the loading point measured by Linear Variable Differential Transformer (i.e., LVDT), the reaction force at the loading point, concrete compressive strain, and the tensile strain of longitudinal reinforcement.

The force-displacement hybrid control method was utilized. The detailed loading scheme can be illustrated as follows:

(1) Monotonic loading: firstly, the force control was used until the RC beams yielded. The yielding displacement defined as the displacement when the RC beams yielded was then adopted as the level difference for the loading. It was utilized until the bearing capacity of the beams dropped to 85% peak bearing capacity or significant damage occurred in the beams.
(2) Cyclic fatigue loading: the typical force and displacement history applied to specimen is shown in Fig. 6.88. The applied force at the horizontal direction

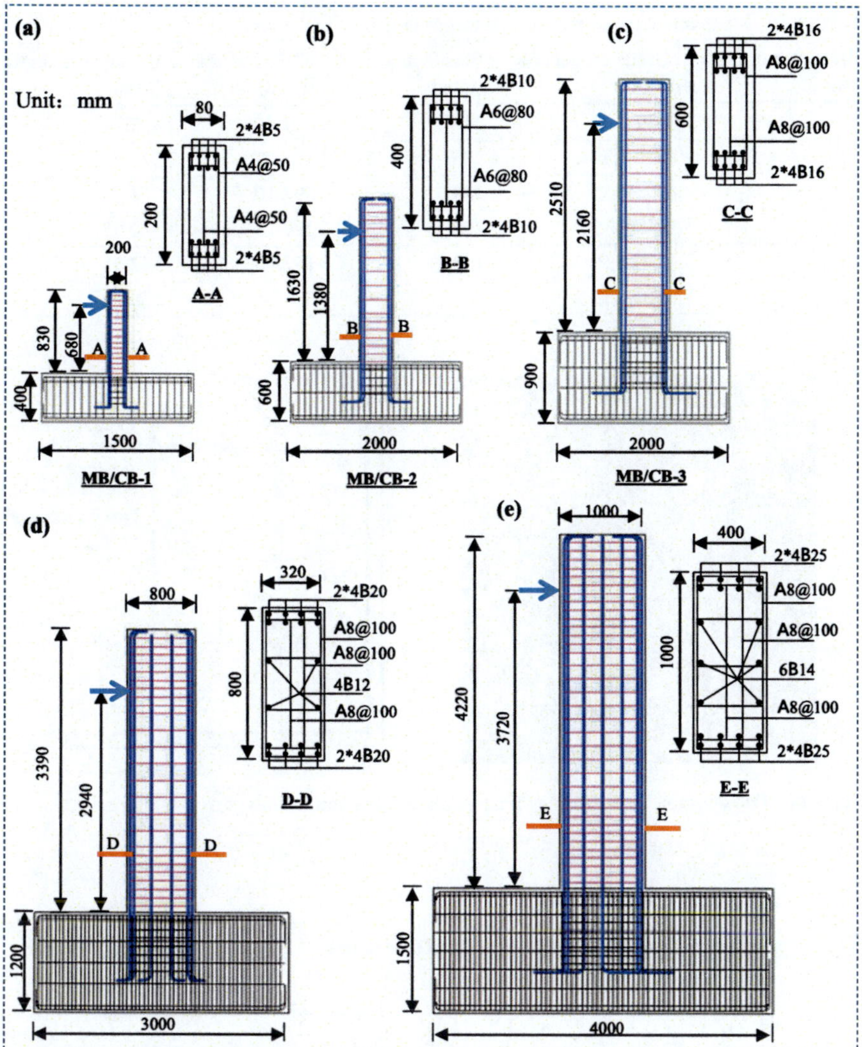

Fig. 6.86 Details of the tested RC cantilever beams

was loaded and only one cycle was applied at each level of loading. After the RC beams yielded, the multiple of yielding displacement was take as the loading control, and three cycles were applied at each level. Similarly, the experiment was stopped at the stage when the load came down in the range of 85% of the ultimate load carrying capacity or significant damage occurred. It is to be noted that, it cost about 10 min for each cyclic to ensure the pseudo-static loading.

Table 6.16 Mechanical parameters of the reinforcing bars in the test

Diameter of rebar (mm)	Grade of steel bar	Elastic modulus (GPa)	Yield strength (MPa)	Ultimate strength (MPa)
4	HPB300	210	297.89	425.74
6	HRB335	196	405.97	529.55
10	HRB335	198	392.10	564.63
16	HRB335	202	399.13	530.93
20	HRB335	193	402.00	542.51
25	HRB335	195	374.84	536.29

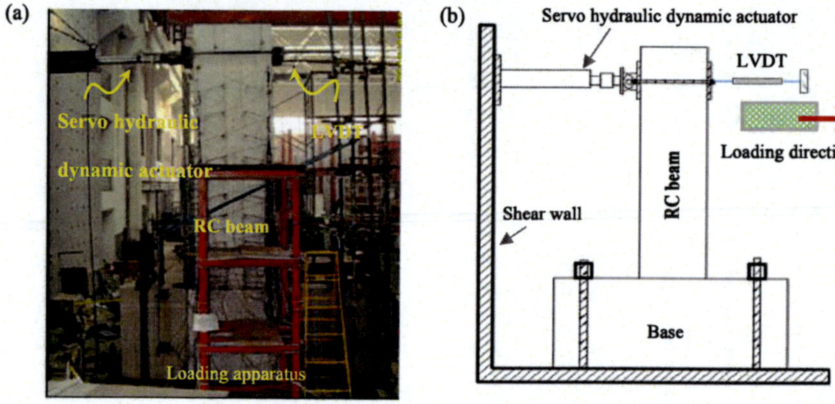

Fig. 6.87 Test equipment: **a** loading apparatus; and **b** schematic diagram

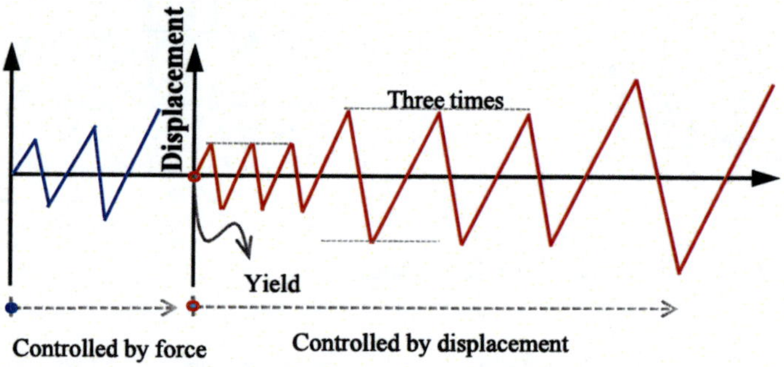

Fig. 6.88 Scheme of cyclic loading

It is to be noted that, fatigue loading is usually divided into two categories, including low-cyclic (e.g. structures subjected to earthquakes with 1–102 loading cycles) and high cyclic loading (Lee and Barr 2004).

2. *Flexural failure patterns*

Under the monotonic loading, the final failure patterns of the five RC cantilever beams having different structural sizes are presented in Fig. 6.89a. It can be found that, the failure patterns of the five groups of beams are closely similar. For the smaller-sized specimens (e.g. MB-1), the widths of the main tensile cracks were small, while for the larger-sized specimens (e.g. MB-5), the corresponding crack widths were much larger, as also observed in the work of Bažant and Kim (1984), the most important consequence of wider cracks is diminished residual tensile stress.

In general, the larger the size of the RC beam is, the larger crack width and more cracks the specimens possess when the beam failed (Syroka-Korol et al. 2013; Syroka-Korol and Tejchman 2014; Che et al. 2012, 2013).

Figure 6.89b shows the final failure pattern of the five groups of RC cantilever beams under cyclic loading. On the whole, the failure patterns of the ten beams are closely similar. The failure of all beams occurred at the bottom part where the concrete crushing generated. At this time, the bond lost efficacy at the interface between the longitudinal steel bars and the surrounding concretes. Moreover, it can be seen from Fig. 6.89b that much more cracks generated as the structural size (cross-sectional depth) of the beams increases.

Compared Fig. 6.89b with Fig. 6.89a, it can be noted that the failure patterns of the beams under monotonic and cyclic loading are significantly distinct. Under low cyclic fatigue loading, both the tensile/compressive strengths of the concrete and the yield strength of reinforcing bars degrade. Moreover, the bearing capacity (or the shear capacity) at the interface of steel bars and surrounding concretes would also decrease due to the low cyclic fatigue behavior. That is to say, the failure definitely takes place in concrete that is a quasi-brittle engineering material. While for monotonically loaded RC beams, the reinforcement could make the failure of the structure more ductile. This finally leads to the fact that, the failure of the beams subjected to cyclic loading is less ductile than the one subjected to monotonic loading.

In addition, the detailed information regarding crack patterns (including crack inclination and crack spacing) are described in Fig. 6.90a, b respectively for the beams under monotonic and cyclic loading. The concrete spalling areas (i.e., the grey region) can also be observed in Fig. 6.90.

3. *Moment-displacement hysteretic behavior and envelopes*

The recorded cyclic hysteretic load-displacement curves (i.e., the solid black lines) and the corresponding skeleton curves (i.e., the blue dotted lines) of the ten RC beams under low cyclic horizontal loading are plotted in Fig. 6.91. Moreover, the moment-displacement curves (i.e., the solid red lines) for the five beams under monotonic loading are also presented in Fig. 6.91 for comparison.

In general, for the ten RC beams subjected to cyclic loading, the shapes of the recorded hysteretic curves are all plump, as can be observed in Fig. 6.91. As the applied loading continued, the RC beams under cyclic loading gradually entered the elastic-plastic stage, and unrecoverable plastic deformation occurred. At this moment, the residual deformation of the beams after loading became larger and

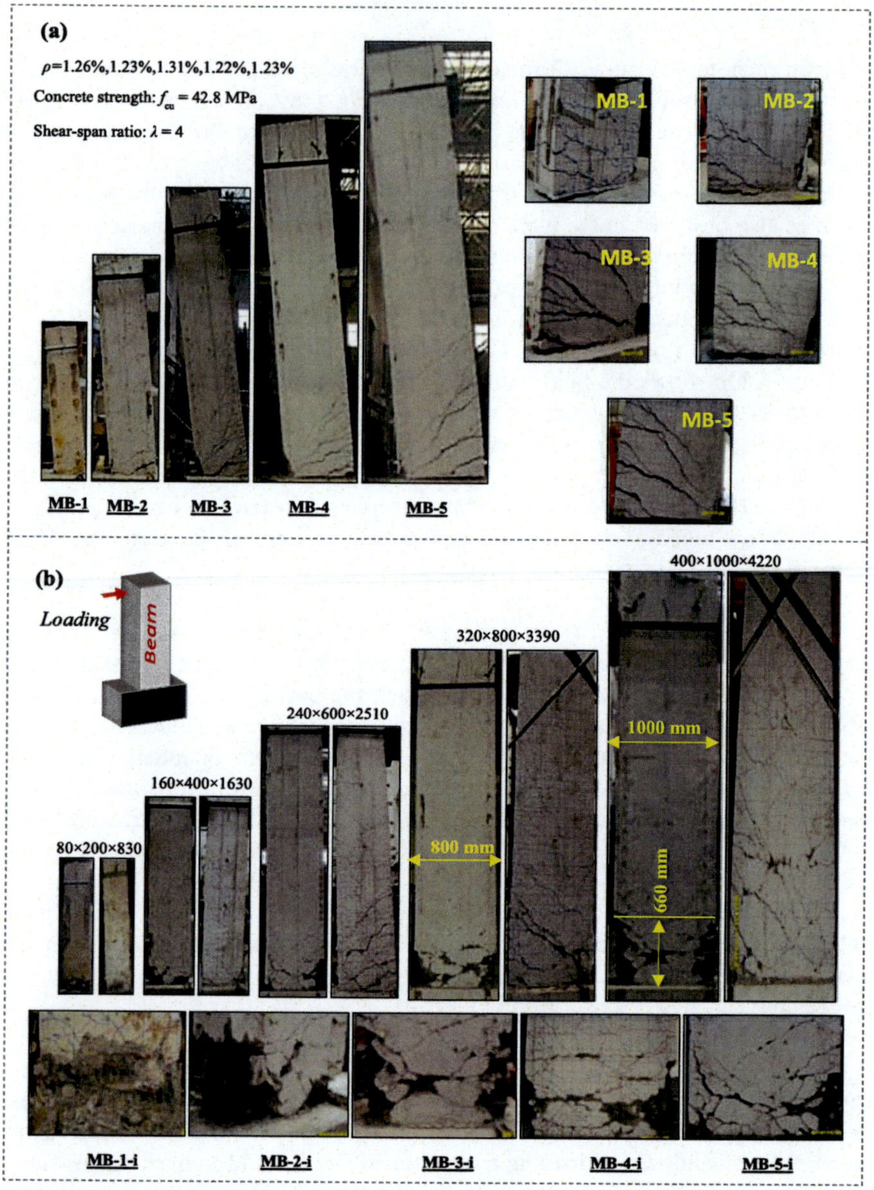

Fig. 6.89 Typical failure patterns for the normal-strength RC beams, subjected to: **a** monotonic loading; and **b** cyclic loading

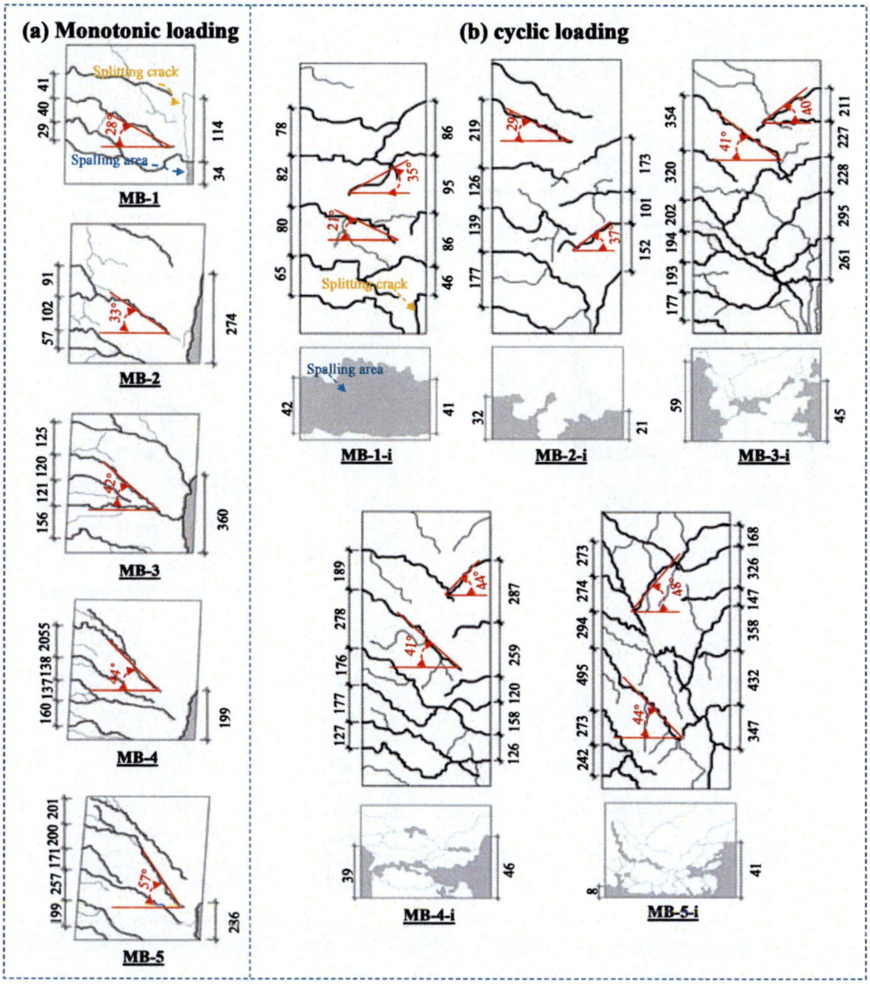

Fig. 6.90 Sketching of the crack distribution and the spalling area of different sizes obtained from the test for the normal-strength RC beams: **a** monotonic loading; and **b** cyclic loading

larger, the cracks developed quickly, making that the stiffness degraded quickly. In the last stage, larger plastic deformation generated, and concrete crushing and buckling of reinforcing bars happened. The corresponding skeleton curves obtained exhibit a reversed 'S-shape', as shown in Fig. 6.91. Under the low cyclic fatigue loadings, the cracks within the RC beams opened and closed continually, resulting into the decrease in the flexural capacity of the RC beams. The larger the flexural deformation is, the more seriously the flexural stiffness degrades. In addition, one can note from Fig. 6.91 that, the softening curve becomes steeper with increasing the beam size, indicating that the brittleness of the specimen increases with the increase of

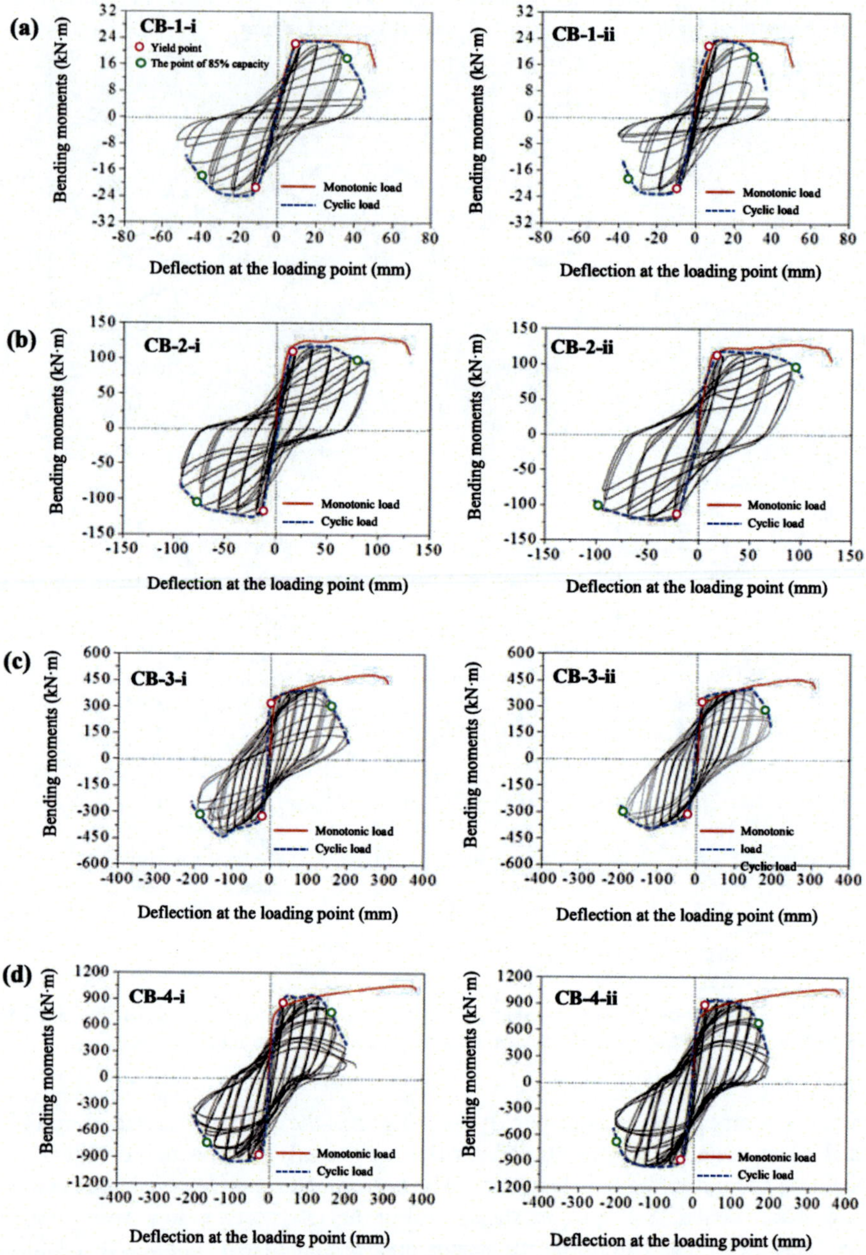

Fig. 6.91 Tested hysteretic curves and the corresponding skeleton curves of the 15 RC cantilever beams subjected to monotonic and low cyclic loading

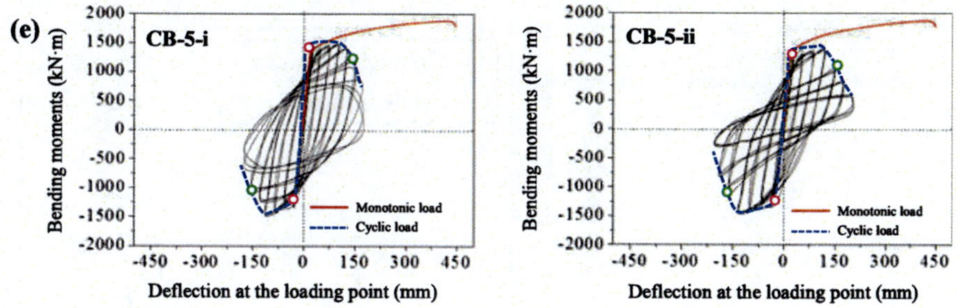

Fig. 6.91 (continued)

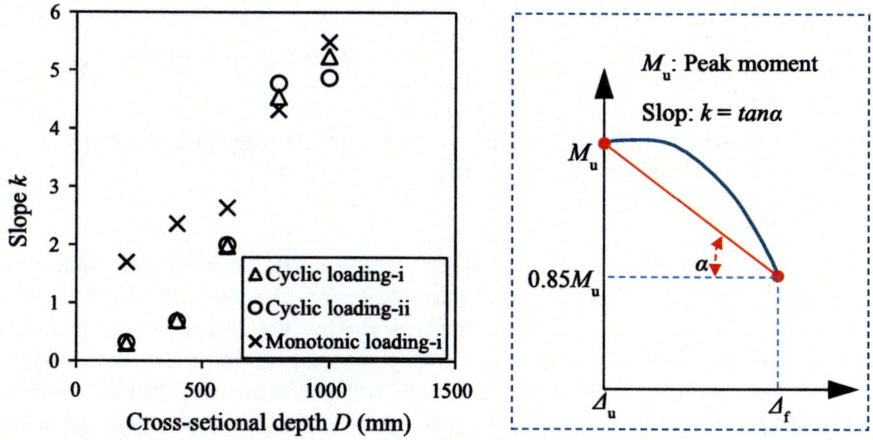

Fig. 6.92 Effect of size on softening rate

the structural dimension. This can be clearly observed from the relationship between the softening rate and structural size presented in Fig. 6.92. Herein, the slope k is utilized to describe the softening rate or the brittleness, as shown in Fig. 6.92. One can note from Fig. 6.92 that, the slope k increases with increasing the cross-sectional depth D.

By comparing with the skeleton curves of the beams under monotonic and cyclic loading, one can clearly find that the bearing capacity of the large-sized beams under monotonic loading is obviously larger than that under cyclic loading. While for the small-sized beams (CB-1-i and CB-1-ii), the bearing capacities are almost the same under monotonic and cyclic loading. Moreover, from the skeleton curves, one can note that under monotonic loading the RC beams exhibit a long yield platform. That is to say, the deformation capacity of the RC beams under cyclic loading is significantly weaker than that under monotonic loading. In reality, compared with that under monotonic loading, not only the tensile/compressive strengths of concrete meso-components and yield strength of the steel bars degrade, but also the bonding capacity

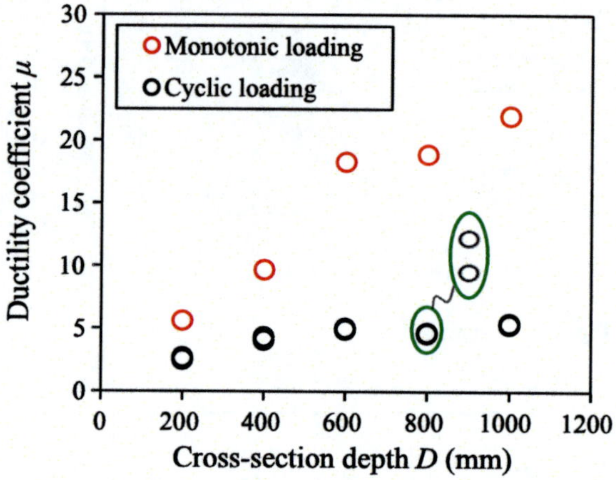

Fig. 6.93 Displacement ductility coefficients of the RC beams having different structural sizes

at the interface of concrete and steel bars weakens due to the damage accumulation within the RC beams under cyclic loading.

4. *Ductility capacity*

The obtained ductility coefficients of the RC cantilever beams under both monotonic and cyclic loading are plotted against the cross-sectional depth h in Fig. 6.93. One can see that the ductility coefficient increases obviously with increasing the cross-sectional depth of the specimens under monotonic loading, while increases slightly for the beams under cyclic loading. In addition, it can be clearly seen that the ductility coefficients for the beams under cyclic loading are much smaller than the one under monotonic loading. This definitely is caused by the fact that under cyclic fatigue loading, the failure of the beams becomes less ductile. In reality, under low cyclic fatigue loading, the bonding between the concrete and the reinforcement would be deteriorated, and the crushing of concretes at the end of the specimen becomes the main reason for the failure of the specimens. Under monotonic loading, for the present balanced-reinforced structure, after the beam yielded, many cracks generated at the flexural-tensile side of the beam, and the longitudinal reinforcements would consume most energy. It is to be noted that, in the present tests, there were almost no concrete crushing at failure for the five beams under monotonic loading, as can be observed from Fig. 6.93. Therefore, the failure of the beams would be more ductile (or less brittle) as compared to that of cyclically loaded RC beams.

5. *Stiffness degradation*

The stiffness degradation of the specimen plays an important role in the energy dissipation capacity (Jin et al. 2016). The secant stiffness K declined with the increase of displacement of the RC beam under the cyclic loading can be calculated by

$$K = \frac{|+P_i| + |-P_i|}{|+f_i| + |-f_i|} \tag{6.8}$$

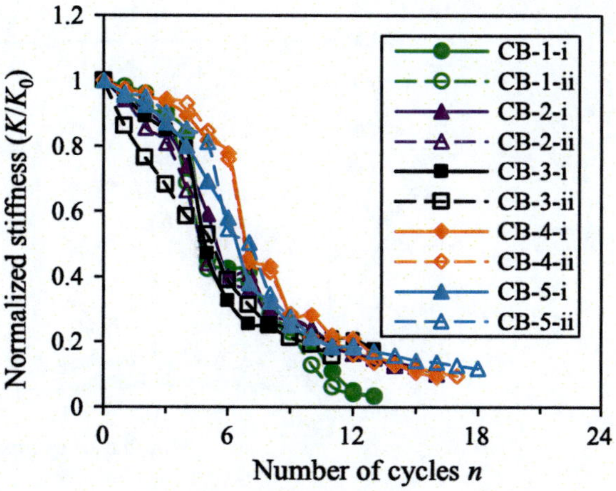

Fig. 6.94 Stiffness degradation curves of the RC cantilever beams subjected to cyclic loading

in which, $+P_i$ is the load in the positive direction, $-P_i$ is the load in the negative direction, $+f_i$ is the displacement in the positive direction, and $-f_i$ is the displacement in the negative direction.

Figure 6.94 presents the relation for normalized stiffness K/K_0 with the number of cycles. Herein, K is the secant stiffness at different cycles, and K_0 is the initial tangent stiffness. It can be noted from Fig. 6.94 that at the early stage of the test, some micro-cracks should generate within the concrete beam. However, the damage was not heavy since there was no macro-crack generated within the concrete beam, resulting into a slight degradation of global stiffness at the early stage. As the cracks developed quickly within the RC beams, some macro-cracks occurred due to the flexural-tensile action, and therefore the stiffness of the beams degraded quickly. One can see from Fig. 6.94 that the normalized stiffness degradations for the ten beams having different structural sizes are similar, indicating that the specimen sizes have a negligible influence on the stiffness degradation.

6. *Energy dissipation*

The dissipated energy W_{pl} and the cumulative dissipated energy ΣW_{pl} versus the number of cycles n for the tested RC beams under cyclic loading are plotted in Fig. 6.95a, b. From Fig. 6.95a, it can be seen that the capacity of dissipated energy becomes larger with the increase of the number of cycles n. This indicates that energy dissipation of the RC beams occurred mostly at the elastic-plastic stage of specimens. At a common level of displacement loading, with increasing the number of cycles, the capacity of dissipated energy becomes to be weaker. In addition, it can be found from Fig. 6.95b that with the increase of the beams size, the cumulated dissipated energy increases.

As mentioned previously, the larger the size of the RC beam is, the larger crack width and more cracks the specimens possess when the beam failed (Syroka-Korol et al. 2013; Syroka-Korol and Tejchman 2014; Che et al. 2013). Simply to say, as the structural size of beam increases, the crack width turns to be larger and the crack

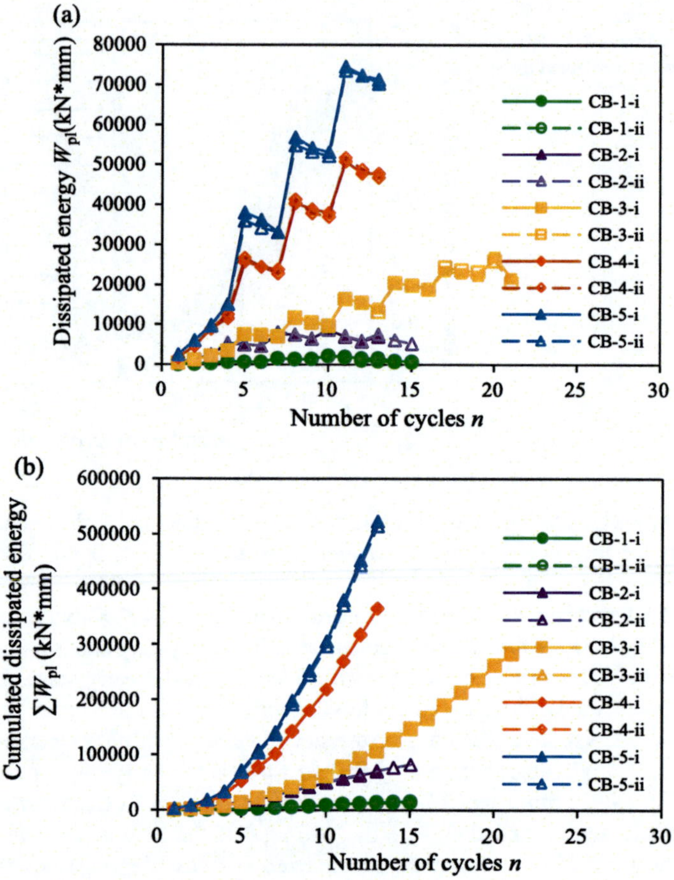

Fig. 6.95 The energy consumption of the RC beams subjected to cyclic loading: **a** dissipated energy of every cyclic; and **b** cumulated dissipated energy

length is also longer when the beam fails, and therefore damage regions becomes to be larger. This finally leads to more energy dissipation of the beam. The work of Jin et al. (2016) also reported the similar test observations on the shear failure of RC beams.

The relationship between the hysteretic damping factor ζ for each of the loop and the number of cycles n for the ten RC beams under cyclic loading is presented in Fig. 6.96. One can see from Fig. 6.95 that the equivalent hysteretic damping factor for each loop gets to be larger first, and then it becomes to be smaller as the number of cycles n increases. This is because the progressive damage generated and developed gradually in the beams as the loading turns out to be larger. Moreover, it can be noted from Fig. 6.96 that the variations of the hysteretic damping factor for all tested beams are similar.

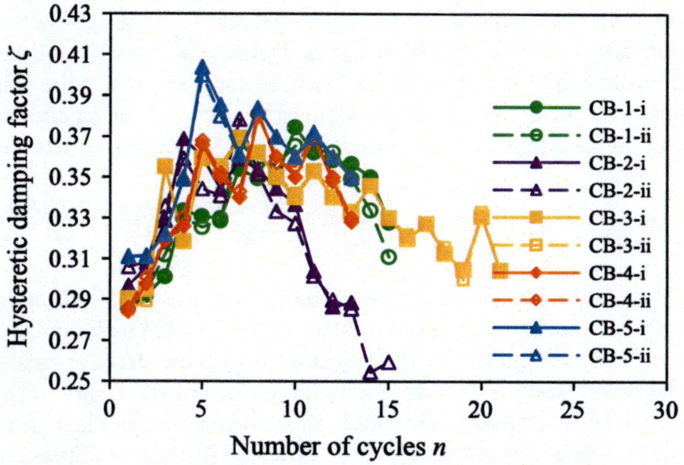

Fig. 6.96 The hysteretic damping factor ζ of the RC beams subjected to cyclic loading

7. *Size effect on the tested nominal flexural strength*

The cyclic loading can make the RC beams exhibit brittle failure behavior because of the low cycle fatigue behavior. Here, the existence of size effect in the flexural strength was examined.

We collected some calculation formulae for the design of flexural capacity from several codes, e.g. EN 1992-1-1, ACI 318-2011 and GB 50010-2010 of China. The comparison of the present tested data and design values calculated by these codes are made and plotted in Fig. 6.97. It can be clearly seen from Fig. 6.97 that all design values are lower than the test values, and the flexural capacity M increases with increasing the structural size (i.e., cross-sectional depth D). In addition, one can note that the calculation results obtained from the three different codes are closely similar.

Fig. 6.97 Relationship between the peak moment M and cross-sectional depth D

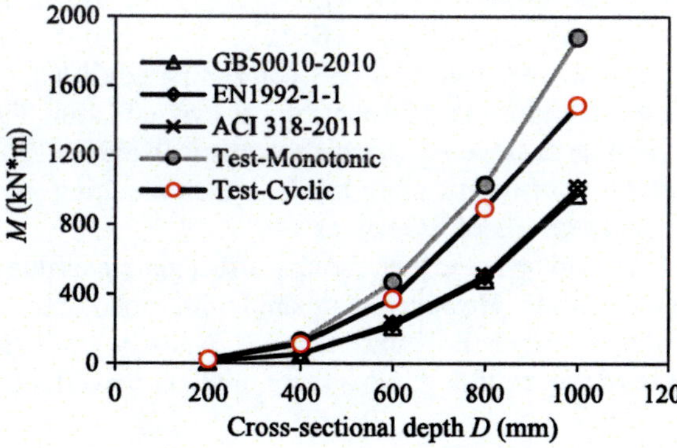

For pure flexural specimens, the normal stress can be calculated by $\sigma = M/W$. W can be evaluated by $W = I_z/y$, where y is the distance between the calculated point and neutral axis, and I_z is the inertial moment of cross-section to the neutral axis. For rectangular cross-section, W has the form as $W = bh^2/6$. The maximum nominal flexural stress (i.e., nominal flexural strength) σ_{Nu} can be easily determined by

$$\sigma_{Nu} = \frac{6M_{max}}{bh_0^2} = \frac{6P_{max}l}{bh_0^2} \tag{6.9}$$

in which M_{max} is the maximum moment generated at the fixed end, $M_{max} = P_{max}$ l. The length l is the distance from the loading point to the fixed end, P_{max} is the maximum applied loading, b is the beam width and h_0 is the effective cross-sectional depth of the beam. The parameters of the five groups of RC beams with different structural sizes, under both monotonic and cyclic loadings are provided in Table 6.17.

The tested nominal flexural strengths of the tested RC beams subjected to monotonic and cyclic loading are plotted in Fig. 6.98a, and the calculated nominal strengths corresponding to the calculated flexural moments plotted in Fig. 6.97 are presented in Fig. 6.98b. One can clearly note from Fig. 6.98a that the obtained flexural strength of the RC beams under cyclic loading is smaller than that under monotonic loading. Moreover, the flexural strength of the RC beams under cyclic loading decreases much faster as the structural size increases. That is to say, the RC beams under cyclic loading presents a stronger size effect than the beams under monotonic loading. However, the utilized formulae in EN 1992-1-1, GB 50010-2010 and ACI 318-2011 could not reveal any size effect, as shown in Fig. 6.98b. It is to be noted in Fig. 6.98b that the nominal strengths for the cross-sectional depths of 200 and 600 mm are slight larger than those of others due to the larger longitudinal reinforcement ratio.

Figure 6.99 shows the bi-logarithmic plot of nominal bending strength. The correlation coefficient R^2 for the present tested data and Bazant's SEL (Bažant and Kim 1984) was also calculated. The correlation coefficient R^2 under monotonic and cyclic loading are respectively 0.92 and 0.93. This means that Bažant's SEL curve fits the present tested data satisfactorily. According to the above analyses, it can be concluded that the plot shows the presence of size effect on the flexural strength of RC beams in accordance with Bazant's SEL.

It can be seen from Fig. 6.99 that compared with the one under monotonic loading, the test data for the RC beams under cyclic loading lies much closer to the inclined asymptote of slope −1/2, which corresponds to the LEFM. The cyclic loading makes the failure of beams more brittle, which resulting into a stronger size effect in the flexural strength of RC beams.

It should be noted that, the size effect generates due to structural geometry but not cyclic loading. However, compared with monotonic loading, the low cyclic fatigue loading changes the failure features. Namely, cyclic loading makes RC beams fail more brittle, resulting into a more obvious size effect.

Table 6.17 Ultimate load carrying capacity and corresponding nominal flexural strength

Specimen name	Cross-sectional width b (mm)	Effective cross-sectional depth d_0 (mm)	Tested p_{max} (kN) (positive/negative)	Tested M_{max} (kN * m) (positive/negative)	Nominal flexural strength σ_{Nu} (MPa) (positive/negative)
MB-1	80	170	32.82	23.32	60.45
MB-2	160	365	94.75	130.75	41.18
MB-3	240	540	217.71	470.25	40.31
MB-4	320	735	350.49	1030.44	35.76
MB-5	400	930	505.45	1880.26	32.61
CB-1-i	80	170	33.26/34.62	22.62/23.54	58.70/61.09
CB-1-ii	80	170	35.29/31.56	22.62/23.54	62.28/55.69
CB-2-i	160	365	83.63/80.30	115.41/110.81	36.36/34.91
CB-2-ii	160	365	84.47/83.64	116.57/115.43	36.73/36.36
CB-3-i	240	540	170.94/179.84	369.23/388.46	31.66/33.30
CB-3-ii	240	540	179.84/160.16	388.46/365.38	33.31/31.33
CB-4-i	320	735	305.82/303.87	894.13/893.39	31.21/31.01
CB-4-ii	320	735	305.63/305.17	898.56/897.20	31.19/31.14
CB-5-i	400	930	402.98/399.11	1487.58/1484.68	26.00/25.75
CB-5-ii	400	930	410.73/400.68	1492.63/1490.53	26.50/25.85

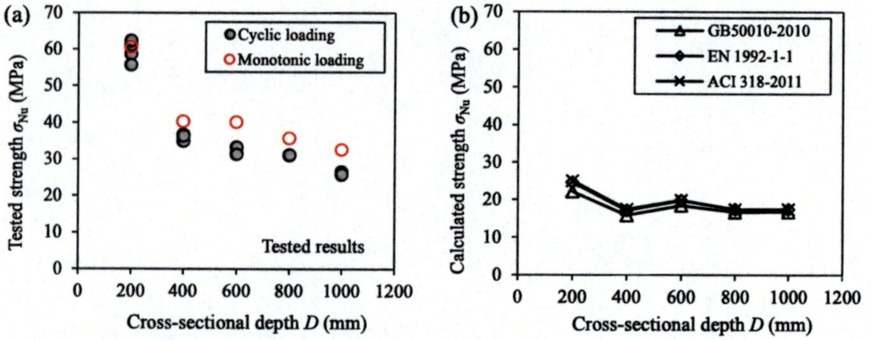

Fig. 6.98 Nominal flexural strengths σ_{Nu} versus cross-sectional depth D for the RC beams: **a** tested strength; and **b** calculated strength by different codes

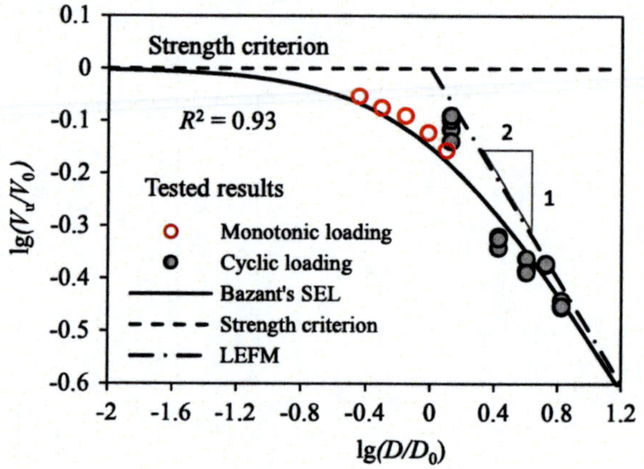

Fig. 6.99 The comparison of nominal flexural strength fitted by Bazant's SEL

6.4.2 Simulations on Flexural Failure of RC Beams

Based on the above-mentioned flexural failure tests of RC cantilever beams, some 3D meso-scale numerical simulation works were given, including the simulation works of test beams and larger sized beams.

The main mechanical parameters for the three meso-components and the reinforcing bars, including elastic modulus E, tensile/compressive strengths σ_t/σ_c, and fracture energy G_c, etc., are listed in Table 6.18. The mechanical parameters of aggregate and mortar matrix are real test ones (labeled by "*"), and the magnitudes of the fracture energy for the three meso-components are quoted from the work of Yılmaz and Molinari (2017). The mechanical parameters of ITZs (labeled by "^") are determined by repeated trials, because it is uneasy to measure by test. Therefore, the parameters of ITZs were obtained by trial calculation, according to Chap. 3 for details.

Table 6.18 Mechanical parameters of the three meso-components of concrete and reinforcing bars utilized

Mechanical parameters	Aggregate	Mortar matrix	ITZ	Longitudinal bar	Stirrups
Compressive strength σ_c (MPa)		*45.3	^37.2		
Tensile strength σ_t (MPa)		*3.8	^3.3		
Elastic modulus E (GPa)	*70.1	*30.8	^24.7	*196	*210
Poisson's ratio ν	*0.2	*0.2	^0.2	*0.3	*0.3
Fracture energy G_c (J/m^2)	60	50	30		
Shear-dilative angle ψ (°)		18	15		
Yield strength f_y (MPa)				*405	*298

1. *Simulation works of the test beams*

The final simulated failure patterns of the RC test beams having different structural sizes under both monotonic and cyclic loadings are presented in Fig. 6.100. It can be found clearly from Fig. 6.100 that the simulated failure patterns have good agreement with the actual tested ones. Nevertheless, one can clearly note from Fig. 6.100 that there are still some differences from the numerical outcomes and the experimental observations, e.g. the crack inclinations and the crack spacing, especially for the RC beams subjected to monotonic loading. This should be caused by the following reasons: (1) the adopted bond-slip relations and the related parameters should be

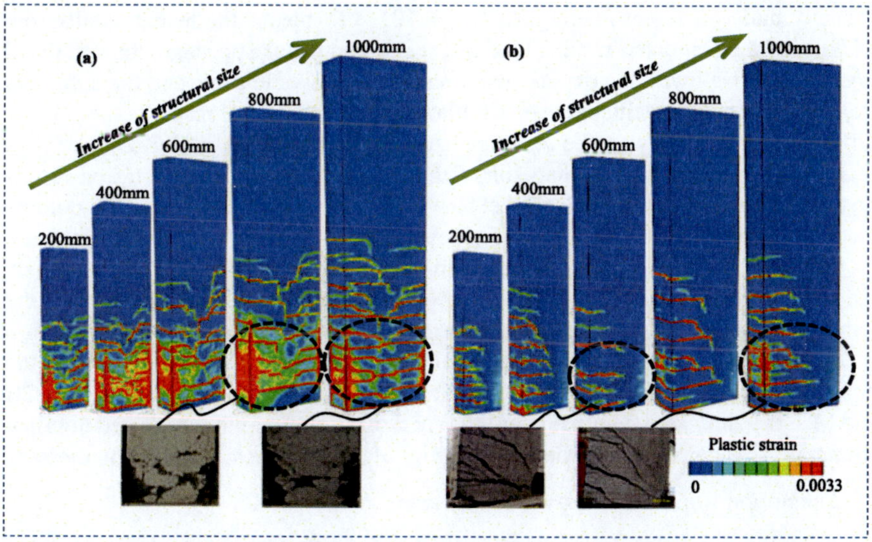

Fig. 6.100 Comparison of simulated and tested failure patterns of the RC beams: **a** under cyclic loading; and **b** under monotonic loading

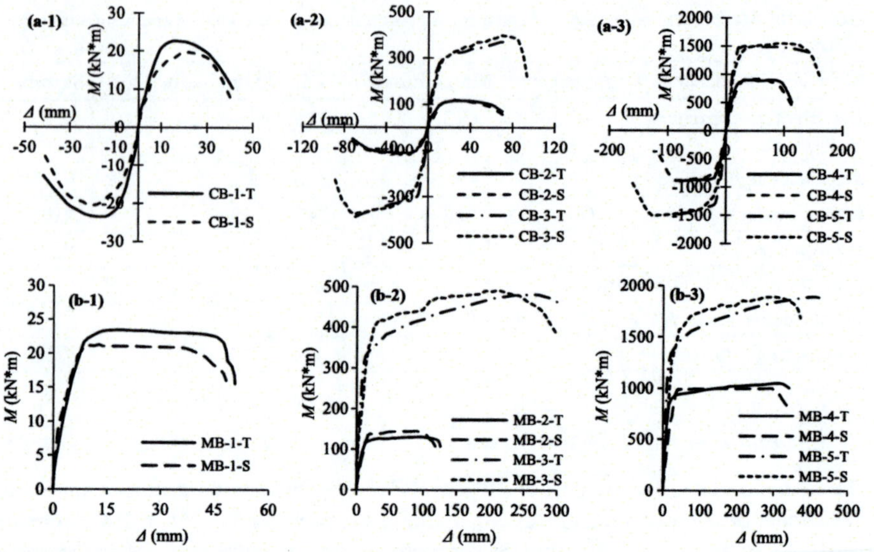

Fig. 6.101 Comparison of the skeleton curves obtained by numerical and experimental results: a under cyclic loading; and b under monotonic loading

different from the real tested ones; (2) in the simulations, the constitutive models for the concrete meso-components cannot avoid the mesh sensitivity totally.

Figure 6.101 shows the comparison between the simulated and tested bearing capacity under different loading. In Fig. 6.101, "T" means the tested results, and "S" means the simulated results. One can see that in the elastic stage, the simulated results agree well with the experimental results. In the yield stage and the softening stage, the bearing capacity between simulated result and the test result has somewhat different, but the errors can be accepted excepted for the beam of CB-1.

Figure 6.102 plots the comparison of the simulated results and the tested results on the strain ε of the longitudinal steel bars at the end of the beams under monotonic load. One can note that the simulation results are consistent with the tested ones. At the initial stage, the beams were all in elastic stage. As the loading continued, the steel strains increased linearly, and the strains increased significantly after the longitudinal steel bars reached their yield strength.

Table 6.19 presents the comparison of the tested and simulated results on the nominal strength of the RC beams. One can clearly note that, for both series CB and MB, the simulated nominal strengths are closely similar with the tested nominal strengths. This demonstrates the reasonability of the meso-scale simulation method.

2. *Simulation works of the larger sized beams*

Based on the above simulation results, the flexural failure simulation tests of larger beams are carried out. The cross-sectional depths of the simulated beams were

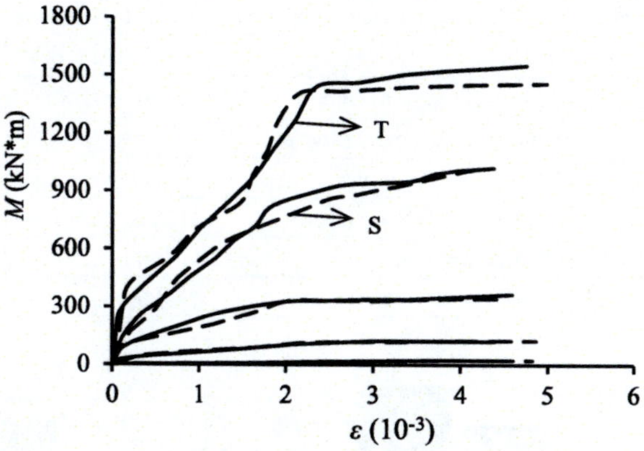

Fig. 6.102 Simulated and tested tensile strains of the longitudinal bars at the fixed bottom

Table 6.19 Comparison of nominal strength obtained from tests and simulations

Specimen label	Tested strength (MPa)	Simulated strength (MPa)	Error (%)
CB-1-i	58.70	51.45	12.35
CB-1-ii	62.28		17.39
CB-2-i	36.36	37.35	2.72
CB-2-ii	36.73		1.69
CB-3-i	31.66	33.27	5.10
CB-3-ii	33.31		0.12
CB-4-i	31.21	32.17	3.01
CB-4-ii	31.19		3.14
CB-5-i	26.00	26.24	0.92
CB-5-ii	26.50		0.98
MB-1	60.45	54.58	9.71
MB-2	41.18	45.86	11.37
MB-3	40.31	43.07	6.85
MB-4	35.76	34.76	2.80
MB-5	32.61	32.72	0.34

1200 mm (CB-6/MB-6), 1600 mm (CB-7/MB-7) and 2000 mm (CB-8/MB-8). The detailed parameters of the beams are listed in Table 6.20.

The failure modes of the three larger-sized RC beams (the depths are 1200, 1600 and 2000 mm) and the Mises stresses distribution within the steel bars are presented in Fig. 6.103. It can be clearly found that the failure modes of the three larger-sized RC beams are closely similar with those of RC beams having smaller sizes, under both monotonic and cyclic loadings. It can also be noticed from Fig. 6.103 that, for larger-sized beams, the horizontal cracks propagate towards the bottom of the specimens, and oblique cracks generate, resulting to the flexural-shear failure pattern.

Table 6.20 Physical parameters of the larger sized RC beams simulated

Specimen name	CB-6/MB-6	CB-7/MB-7	CB-8/MB-8
Cross-section size ($b \times h$) (mm^2)	480 × 1200	640 × 1600	800 × 2000
Longitudinal reinforcement ratio ρ	1.26%	1.24%	1.24%
Hoop reinforcement ratio ρ_{sv}	0.14%	0.14%	0.14%
Effective depth of beam d_0 (mm)	1120	1510	1900
Beam effective length $a = \lambda \times h_0$ (mm)	4480	6040	7200
Total length of beam l (mm)	4900	6500	7700

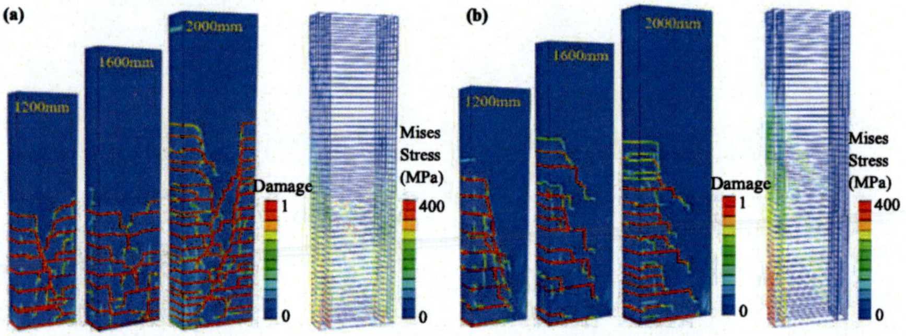

Fig. 6.103 Failure patterns of larger-sized RC beams subjected to: **a** cyclic loading; and **b** monotonic loading

Moreover, one can note from Fig. 6.103 that the Mises stresses contour of the beams with cross-sectional depth of 2000 mm that, the longitudinal steel bars yielded (reached the average yield strength of 400 MPa) at the end of the specimen under both monotonic and cyclic loadings.

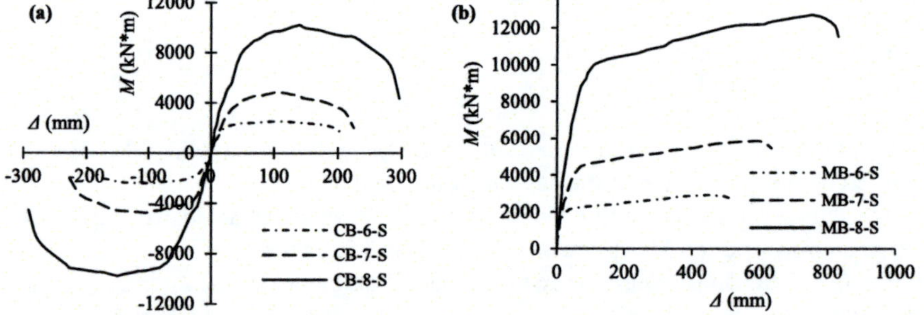

Fig. 6.104 The skeleton curve of larger-sized RC beams subjected to: **a** cyclic loading; and **b** monotonic loading

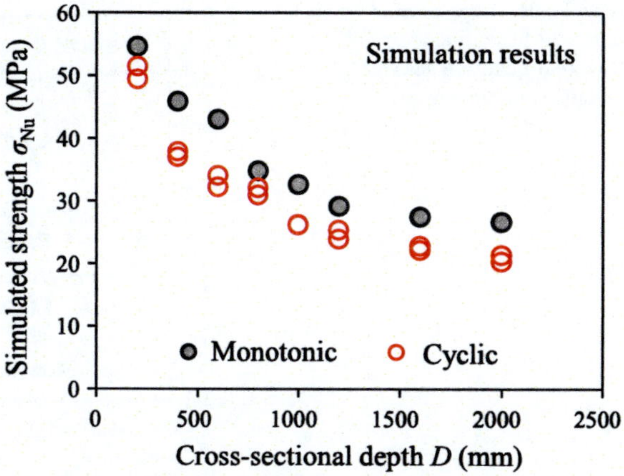

Fig. 6.105 Relationship between the simulated nominal flexural strength and structure size

Figure 6.104a, b plot the skeleton curves (moment M versus deflection Δ) of the three larger-sized RC beams subjected to cyclic and monotonic loadings, respectively. Again, one can note from the comparison of Fig. 6.104a, b that, (1) the peak moments M of the beams under cyclic loading are obviously smaller than those under monotonic loading; (2) the deformation capacity or the ductility capacity of the beams under cyclic loading is weaker than those under monotonic loading. In summary, the mechanical behavior of the larger-sized RC beams is basically similar with that of the smaller-sized RC beams tested.

The relationships between the simulated flexural strengths and the structural size D (i.e., cross-sectional depth D) of the RC beams under both monotonic and cyclic loadings are all plotted in Fig. 6.105. It can be seen that the simulated flexural strengths of the geometrically similar RC beams decrease as the cross-sectional depth increases, indicating a size effect. Moreover, it can be noted from Fig. 6.105 that as the structural size increases, the decrease trend of the obtained flexural strengths slows down. It is believed that as the structural size becomes larger and larger, the flexural strength should be close to a constant.

The correlation coefficient R^2 for the simulated data and Bažant's SEL in Fig. 6.106 was evaluated to be 0.975. This means that, for the simulated larger-sized RC beams, the nominal flexural strength is also in accordance with Bažant's SEL. Moreover, compared with that for monotonic loading, the simulated data for cyclic loading lies closer to the LEFM. This indicates a more pronounced size effect for the beams under cyclic loading.

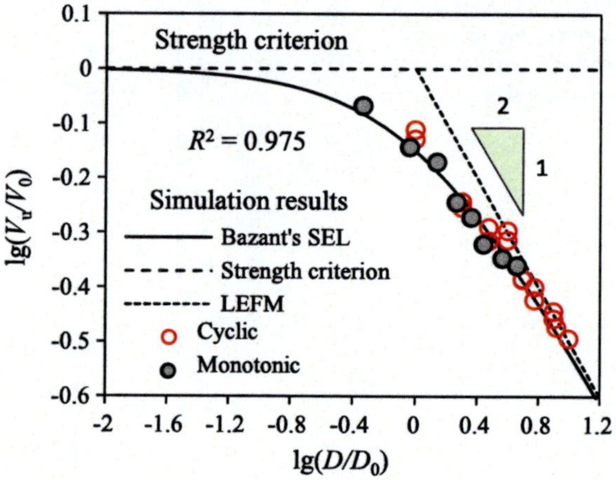

Fig. 6.106 Logarithmic curve of the simulated nominal flexural strength versus structure size

6.5 Size Effect Law for Shear Failure in Concrete Beams

Based on the above test and simulation results, combined with the influence of various factors on the size effect of shear strength of RC beams, the size effect law for shear failure in RC beams is provided.

6.5.1 Basic Assumptions

There are many factors that affect the shear behavior of beams. This chapter mainly considers the influence of some main factors on the shear behavior of beams. For beams without stirrup, factors such as shear-span ratio and longitudinal reinforcement ratio which have great influence on shear strength are considered here. And shear-span ratio, stirrup ratio, longitudinal reinforcement ratio and stirrup spacing are considered for beams with stirrups. Also, the influence of shear-span ratio and CFRP ratio are introduced for CFRP-strengthened ordinary concrete beams, and CFRP ratio is considered for CFRP-strengthened lightweight-aggregate concrete beams.

In addition, the shear capacity of concrete beams is divided into two parts: concrete and stirrups (or fibers). For reinforcement and fiber, it is considered that they are independent to size effect. Therefore, only the size effect of concrete is considered.

6.5.2 *Size Effect Law for Shear Strength*

1. *Ordinary RC beams*

The shear capacity V of RC beam mainly consist of two parts described in Fig. 6.107: (1) the shear force V_{sv} provided by the stirrups which directly bear the shear force; (2) the shear force V_c provided by the concrete beams without shear reinforcement. Therefore, the total shear capacity V of RC beams can be expressed as

$$V = V_{sv} + V_c \tag{6.10}$$

The two shear forces V_{sv} and V_c can be obtained respectively by

$$V_{sv} = f_{yv}\frac{A_{sv}}{s}h_0 \tag{6.11}$$

$$V_c = \tau_c b h_0 \tag{6.12}$$

in which, f_{yv} is the yield strength of stirrups; A_{sv} is the total cross-sectional area of the stirrups configured in the section; s is the stirrup spacing along the length direction; h_0 is the effective cross-sectional depth; τ_c is the nominal shear strength of concrete beams without shear reinforcement; and b is the cross-section width of the beam. Here, the shear strength borne by stirrups is defined as:

$$\tau_{sv} = f_{yv}\frac{A_{sv}}{bs} \tag{6.13}$$

Therefore, the nominal shear strength τ of the RC beam with stirrups is as follows:

$$\tau = \tau_c + \tau_{sv} \tag{6.14}$$

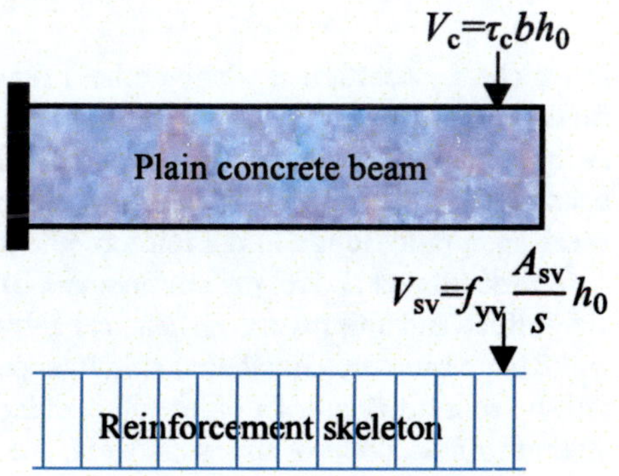

Fig. 6.107 Shear bearing capacity of RC beams

According to Eq. (6.11), one can clearly note that V_{sv} denotes the shear force directly generated by the stirrups, which should be independent of size effect. While, the shear capacity V_c (or nominal shear strength τ_c) should be related to the factors involving cross-section depth, shear-span ratio and stirrup ratio. This means that the nominal shear strength τ_c contributed by the concrete beams without shear reinforcement should exhibit obvious size effect. Herein, it is to be noted that the existence of stirrups ratio, shear-span ratio and longitudinal reinforcement ratio should have significant influences on the size effect of the nominal shear strength of beams without shear reinforcement.

As discussed previously, the shear-span ratio has a significant influence on the shear strength but an ignorable influence on the size effect. The effect of shear-span ratio is first considered here. With the increase of shear-span ratio, the shear strength of beams tends to decrease, which can be characterized by an influence coefficient γ.

Moreover, according to the previous analysis, the stirrups not only bear shear directly, but also hinders the developing and propagation of cracks. Therefore, the stirrups can improve the shear capacity of RC beams and weaken the size effect of shear strength. When the restriction effect of stirrups on concrete is very strong (i.e., the stirrup ratio is large enough), the size effect would be significantly weakened. This behavior can be described by a weakening coefficient β.

In addition, the longitudinal reinforcement ratio ρ has a significant influence on the nominal shear strength of RC beams, and the nominal shear strength tends to increase with the increase of longitudinal reinforcement ratio, which can be characterized by the influence coefficient φ.

Based on the above two viewpoints and Bažant's SEL (Bažant and Kim 1984), considering the influences of shear-span ratio and stirrup ratio on shear capacity and size effect, a semi-empirical and semi-theoretical formula for quantitatively describing the size effect on shear strength of RC beams is given, and the nominal shear strength τ_c is expressed as:

$$\tau_c = \frac{V_0}{\sqrt{1 + D/D_0}} \cdot \gamma \cdot \beta \cdot \varphi \tag{6.15}$$

in which, the coefficient γ should be a function of shear-span ratio, β should be a function of stirrup ratio, and φ should be a function of longitudinal reinforcement ratio. Generally speaking, τ_c (the nominal shear strength borne by the concrete beam without stirrup) is related to size effect, which will be affected by stirrup ratio, shear-span ratio, longitudinal reinforcement ratio, section size and other factors.

As mentioned above, the confinement of stirrups on concrete would weaken the size effect and improve the shear capacity of RC beams. If the shear-span ratio and stirrup ratio are considered simultaneously, the determination of shear strength will be more difficult. In order to consider the effect of shear-span ratio on shear strength of concrete beams separately, the shear strength of beams without shear reinforcements is fitted and analyzed first, and the function form of γ is obtained.

Finally, the accuracy of the function form is verified by the simulation results of beams with stirrups.

As mentioned previously, the formula for calculating shear capacity of RC beams in the Code for Design of Concrete Structures (GB 50010 2010) in China is given as follows:

$$V_{cs} = \alpha_{cv} f_t b h_0 + f_{yv} \frac{A_{sv}}{s} h_0 \tag{6.16}$$

in which, α_{cv} is the shear capacity coefficient of concrete with inclined section, adopting 0.7 for general flexural members, and $1.75/(\lambda + 1)$ for independence beams under concentrated load (including various loads, in which the shear value of bearing section or joint edge caused by concentrated load accounts for more than 75% of the total shear value, λ is the shear-span ratio). Herein, α_{cv} is actually a coefficient considering the influence of shear-span ratio on the shear capacity. From the form of the coefficient α_{cv}, it can be seen that the shear capacity decreases with the increase of shear-span ratio, which is consistent with the previous results.

Herein, the form of the code formula was adopted to determine the coefficient γ, and the relationship between γ and a function form of λ is inversely proportional. The simulation results were fitted and analyzed, and the fitting result is present in Fig. 6.108. The formula of λ can be obtained from Fig. 6.108 as

$$\gamma = \frac{1}{\lambda} \tag{6.17}$$

It should be pointed out that the formula of coefficient γ here is only for the concretes used in this study, and further discussion and verification in other cases are needed.

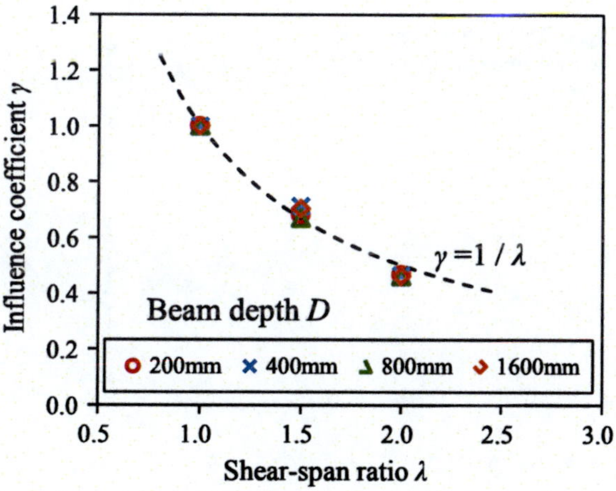

Fig. 6.108 Determination of the influential coefficient γ for shear-span ratio

As mentioned above, stirrups can limit the crack development, improve the ductility of beams and weaken the size effect on shear strength. The specific influence of stirrup ratio on size effect is discussed as follows.

(1) For $\rho_{sv} \leq \rho_{sv,min}$ ($\rho_{sv,min}$ is the lower limit value of stirrup ratio), the stirrups have little influence on size effect. At this case, the coefficient $\beta = 1$.
(2) For $\rho_{sv} \to 1$ (i.e., stirrup ratio is large enough), the confinement of stirrups on concrete is strong. At this case, the size effect in shear strength can be greatly weakened, therefore, the corresponding shear strength of "plain" concrete beams tends to be the shear strength τ_c of the concrete material. The relationship between the shear strength and tensile strength of concrete material proposed by Zhang and Guo (1992) is expressed as follows:

$$\tau_c = 1.20 f_t^{0.89} \tag{6.18}$$

According to Eqs. (6.16)–(6.18), one can get the weakening coefficient β as follows:

$$\beta|_{\rho_{sv}\to 1} = \frac{1.2 f_t^{0.89}\sqrt{1 + D/D_0} \cdot \lambda}{V_0} \tag{6.19}$$

For $\rho_{sv} > \rho_{sv,min}$, the function β becomes a non-linear curve that tends to $\frac{1.2 f_t^{0.89}\sqrt{1+D/D_0}\cdot\lambda}{V_0}$ infinitely from 1. Here, as shown in Fig. 6.109, hyperbolic tangent curve is used to describe the change trend of β, and thus the expression of the coefficient β can be obtained as follows:

$$\beta = \begin{cases} 1, & \rho_{sv} \leq \rho_{sv,min} \\ (\beta_0 - 1) \cdot \tanh\left[\alpha\left(\rho_{sv} - \rho_{sv,min}\right)\right] + 1, & \rho_{sv} > \rho_{sv,min} \end{cases} \tag{6.20}$$

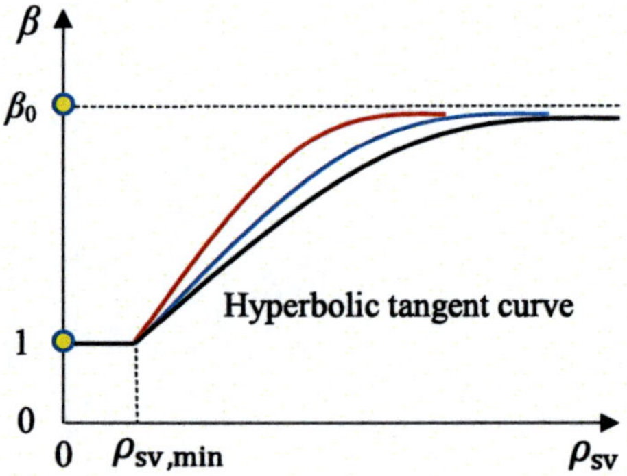

Fig. 6.109 Determination of the weakening coefficient β for stirrup ratio

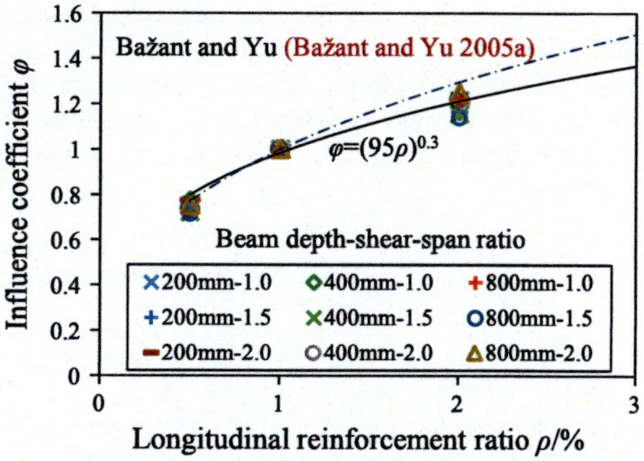

Fig. 6.110 Fitting curve of influence coefficient φ

in which, $\beta_0 = \frac{1.2 f_t^{0.89} \sqrt{1+D/D_0} \cdot \lambda}{V_0}$, α is an adjustment coefficient reflecting the rate of β varies with the stirrup ratio.

It has been concluded from previous conclusion that the nominal shear strength tends to increase with the increase of longitudinal reinforcement ratio. Based on the nominal shear strength of RC beams with the longitudinal reinforcement ratio of 1.0%, a comparative analysis between the formula proposed by Bažant and Yu (2005a) and the test data is shown in Fig. 6.110. It is found that when the longitudinal reinforcement ratio is large, the formula proposed by Bažant and Yu (2005a) will overestimate the influence of the longitudinal reinforcement ratio on the nominal shear strength. Here, the power function with index less than unity is adopted to describe the influence of longitudinal reinforcement ratio on the nominal shear strength, the expression is given as follows:

$$\varphi = (\theta \rho)^m \tag{6.21}$$

where ρ is the longitudinal reinforcement ratio, k and m are undetermined coefficients (m is less than 1.0). Through the fitting analysis of the simulation results, as shown in Fig. 6.110, it is found that when $\theta = 95$, $m = 0.3$, Eq. (6.21) can better reflect the influence of the longitudinal reinforcement ratio on the nominal shear strength.

It should be noted that the standard specimen is the beam with a longitudinal reinforcement ratio of 1.0%, so the coefficient φ of the beam with the longitudinal reinforcement ratio of 0.5% will be less than 1.0 (φ of the standard specimen). In fact, the shear strength of beams is always improved due to the existence of longitudinal reinforcement ratio.

2. *CFRP-wrapped beams*

In CFRP-strengthened concrete beams without stirrups, the enhanced shear capacity of CFRP in beams is similar to that of stirrups. Herein, a novel size effect law that can describe the quantitative influence of CFRP ratio and shear-span ratio on size effect of shear strength of CFRP-strengthened NWC and LWAC beams is given.

In the code of CECS146: 2003, the calculation on the shear capacity of CFRP-strengthened concrete beam is given when CFRP is attached to the axis of vertical beam. Based on the frame of calculating the shear bearing capacity of the stirrups on NWC beams, and combined with the method of calculating the shear bearing capacity of inclined section. The shear capacity V_{cf} provided by CFRP sheets is written as:

$$V_{cf} = kf_{cf}\frac{A_{cf}}{s}h_{cf} \tag{6.22}$$

$$A_{cf} = 2n_{cf}w_{cf}t_{cf} \tag{6.23}$$

in which, k is the correction factor; f_{cf} is the design value of CFRP sheet tensile strength; A_{cf} is the cross-section area of CFRP sheet; s is expressed as the spacing between two adjacent CFRP sheet; h_{cf} is the pasting height of CFRP sheet; n_{cf} presents the number of CFRP pasted layer; w_{cf} presents the width of the CFRP sheet; t_{cf} presents the thickness of the CFRP sheet.

For the CFRP-strengthened LWAC beam, the bearing capacity of the whole member V is jointly borne by the force provided with the concrete beam itself V_c and the bearing capacity provided by CFRP V_{cf}. The expression is given as:

$$V = V_c + V_{cf} \tag{6.24}$$

$$V_c = \tau_c b h_0 \tag{6.25}$$

However, in a practical or experimental situation, the actual shear capacity of the component is smaller than the sum of these two terms. This is due to the interaction between CFRP and concrete. Therefore, the shear capacity of components cannot be characterized by simple superposition.

For CFRP-strengthened NWC beams without stirrups, the shear strength provided by CFRP is independent of the structure size. However, the shear V_c provided by the concrete beam without stirrups should be related to the cross-section depth, CFRP ratio, shear-span ratio and other factors, that is, the nominal shear strength of the concrete beam without stirrups has an obvious size effect. Previous studies have shown that the CFRP ratio has a significant effect on the size effect on nominal shear strength of non-stirrups beams, which is mainly reflected in two aspects: one is the strength enhancement factor φ, represents an increase in the nominal shear strength of the beam; another one is the weakening factor β of size effect, which reflects the effect of CFRP ratio on the size effect of shear strength of RC beams strengthened with CFRP.

According to the conclusion above, the shear strength of the beam decreases with the increase of the shear-span ratio, which can be characterized by the influence coefficient of the shear-span ratio on the shear strength. Based on the above two points, based on the theoretical formula of Bažant's size effect law (Bažant and Kim 1984), the influence of shear-span ratio and CFRP ratio on the shear strength and size effect of CFRP-strengthened beams is considered, and the following expression is given:

$$\tau = \frac{V_0}{\sqrt{1 + D/D_0}} \cdot \varphi \cdot \beta \cdot \gamma \tag{6.26}$$

where τ is shear strength of the whole concrete members, and V_0 and D_0 are empirical coefficients related to concrete types. D is the characteristic size of the concrete member (here is the beam cross-section depth). The parameters γ are functions of the shear-span ratio, and φ and β are functions of the CFRP ratio. Therefore, the τ here actually takes into account the nominal shear strength of RC beams strengthened by CFRP without stirrups under the influence of shear-span ratio, CFRP ratio and member cross-section size.

The CFRP ratios of CFRP-strengthened beams selected in here are 0.0555, 0.111, and 0.222%. Herein, for the sake of simplicity, the relationship between the shear strength enhancement coefficient and the CFRP ratios of CFRP-strengthened concrete beams without stirrups is as follows:

$$\varphi(\rho_f) = A \ln(\rho_f) + B \quad (0 < \rho_f < 1.11\%) \tag{6.27}$$

where the parameters of A and B could be determined according to the fittings. According to the previous studies (Jin et al. 2020b), when the CFRP ratios are 0–1.11%, A and B are respectively recommended to be 0.0749 and 1.6028, as presented in Fig. 6.111, for the present case with the beam shear-span ratio of $1.0 \leq \lambda \leq 2.0$.

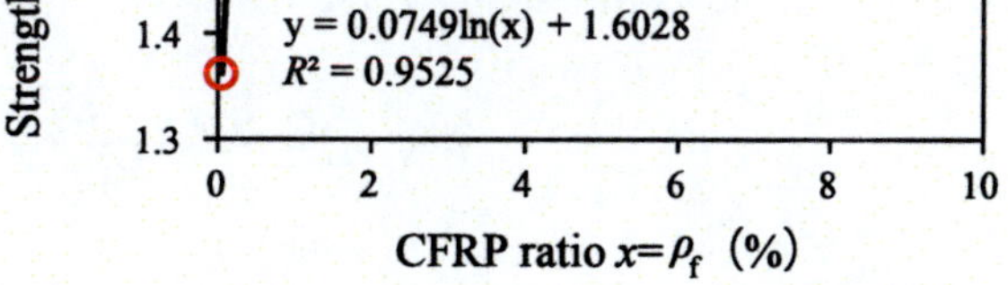

Fig. 6.111 The fitting curve of strength increase coefficient under the influence of CFRP ratios

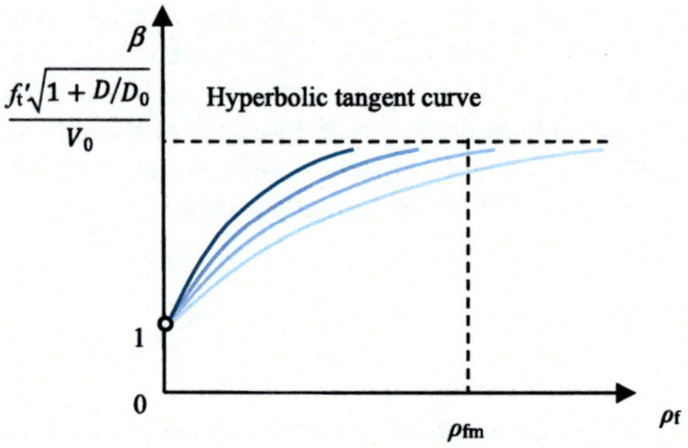

Fig. 6.112 Determination of the weakening coefficient β for the CFRP ratio

In Eq. (6.27), the strength enhancement coefficient φ is determined according to the influence of CFRP ratios on the shear strength of CFRP-strengthened beams.

According to the previous studies (Jin et al. 2020b), with the increase of CFRP ratio, the size effect of shear strength of CFRP-strengthened RC beams gradually weakens, as presented in Fig. 6.112. Herein, the influence coefficient of shear-span ratio is equal to 1 (i.e., $\gamma = 1$).

For $\rho_f = 0$ (i.e., unreinforced concrete beam), since CFRP does not act as a reinforcement, β has no influence on the size effect of nominal shear strength of RC beams, thus one can get:

$$\beta|_{\rho_{f=0}} = 1 \tag{6.28}$$

For $\rho_f \geq \rho_{fm}$ (ρ_{fm} is the ultimate CFRP ratio under the working conditions in this work), due to the excessive CFRP ratio of CFRP under this condition, the existence of CFRP has a strong constraint on concrete. At the moment, the size effect on shear strength of CFRP strengthened RC beams will be greatly weakened. Therefore, the nominal shear strength of the beam having a larger enough CFRP ratio can be considered to be independent on structural size. In this case, the magnitude of the nominal shear strength τ of the beam should be:

$$\tau = f_t' \cdot \varphi \tag{6.29}$$

According to Eqs. (6.26) and (6.29), the size effect weakening coefficient β is obtained for the CFRP ratio $\rho_f \geq \rho_{fm}$:

$$\beta|_{\rho_{f} \to \rho_{fm}} = \frac{f_t'\sqrt{1+D/D_0}}{V_0} \tag{6.30}$$

For $0 < \rho_f < \rho_{fm}$, the magnitude of the size effect weakening coefficient β should be between 1 and $\frac{f_t'\sqrt{1+D/D_0}}{V_0}$. In reality, with the increase of the CFRP ratio, the size effect on shear strength of CFRP strengthened concrete beams is weakened gradually, and the size effect weakening factor β tends to be $\frac{f_t'\sqrt{1+D/D_0}}{V_0}$.

According to the above analysis, a non-linear asymptote was used to describe the effect of the CFRP ratio on the size effect, as presented in Fig. 6.112. Herein, the hyperbolic tangent curve is used to describe the change trend of β. To summarize, the relation between size effect attenuation coefficient β and CFRP ratio ρ_f is as follows:

$$\beta = \begin{cases} 1, & \rho_f = 0 \\ \left(\frac{f_t'\sqrt{1+D/D_0}}{V_0} - 1\right)\tanh(\alpha\rho_f) + 1, & \rho_f > 0 \end{cases} \tag{6.31}$$

where f_t' is the tensile strength of concrete (i.e., 1.79 MPa, which is the simulated value on tensile strength). α is the adjustment coefficient, representing the rate of β varies with the CFRP ratio, which can be determined by fitting simulation and experimental results.

In China's current design code for concrete structures (GB 50010 2010), the calculation formula of shear strength V_{cs} of oblique section of RC beams mentioned previously in Eq. (6.16). α_{cv} presents the shear bearing capacity coefficient of concrete with inclined section and it is 0.7 for general flexural members. For individual beams under concentrated load, the dimension α_{cv} is $1.75/(\lambda + 1)$, and λ presents the shear-span ratio. The coefficient α_{cv} actually represents the influence coefficient of the shear-span ratio on the shear strength of the beam. Moreover, it can be seen that with the increase of shear-span ratio, the shear capacity decreases, which is consistent with the previous conclusion.

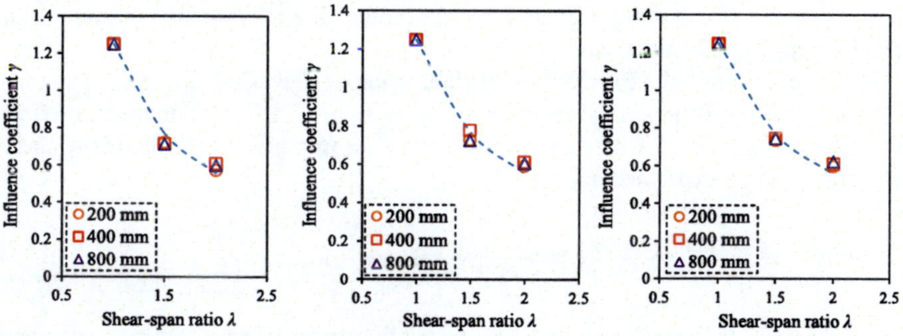

Fig. 6.113 Determination of the influential coefficient γ for shear-span ratio ($\gamma = 1/(\lambda - 2)$)

Here the coefficients γ refer to the form of the design code formula and consider the inverse relationship between the coefficients γ and some functional form of the shear-span ratio λ. The simulation results were fitted for analysis, and the fitting situation was shown in Fig. 6.113.

$$\gamma = \frac{1}{\lambda - 2} \tag{6.32}$$

It should be noted that the coefficient γ here is only for the concrete used in this study, other cases need further discussion and verification.

To sum up, the shear-span ratio and CFRP ratio will affect the shear strength of reinforced beams. The two influencing factors are combined here and expressed as the strength coefficient φ_0, that is:

$$\varphi_0 = \varphi \cdot \gamma \tag{6.33}$$

Combined with Eqs. (6.26) and (6.33), the size effect law can be obtained to quantitatively describe the effect of dimension effect on the shear strength of CFRP-strengthened RC beams with CFRP without stirrups, which can be described as follows:

$$\tau = \frac{V_0}{\sqrt{1 + D/D_0}} \cdot \varphi_0 \cdot \beta \tag{6.34}$$

For CFRP-strengthened LWAC beams, the effect of CFRP ratio mainly includes two aspects. On the one hand, with the raise of CFRP ratio, the nominal shear strength of the beam increasing, and the shear capacity being risen as well. On the other hand, under different structure sizes, with the increase of CFRP ratio, the size effect of nominal shear strength of strengthened beams will be weakened, that is to say, CFRP ratio also has a greater impact on the size effect of shear strength of CFRP-strengthened LWAC beams.

Here, combining with the classical Bažant material size effect law (SEL), a semi-empirical and semi-theoretical formula of the effect of CFRP ratio on the size effect of shear strength of CFRP-strengthened LWAC beams given. The nominal shear strength τ can be expressed as:

$$\tau = \frac{V_0}{\sqrt{1 + D/D_0}} \cdot \varphi \cdot \beta \tag{6.35}$$

where the two coefficients of φ and β represent the strength enhancement coefficient and the size effect weakening coefficient, respectively. Here, the concrete beam and CFRP sheet are treated as a whole, the role of CFRP sheet in the system will be reflected by these two coefficients.

It is noted that, the influence of shear-span ratio on the strength enhancement coefficient is not discussed here. In addition, the factors influencing the shear strength of CFRP-strengthened LWAC beams may include the type of FRP material and the bonding angle, etc., which will be further explored.

With the same section size, the nominal shear strength of the beam will increase with the increase of CFRP ratio. Taking the beam with the cross-section depth of 200 mm as an example, six beams with 100 mm × 200 mm cross section size and different CFRP ratios (0, 0.0444%, 0.111%, 0.222%, 0.444% and 0.888%, respectively) were selected. The nominal shear strength of the beam corresponding to the CFRP ratio of zero is set as the initial reference value, and the strength enhancement coefficient of the beam under different CFRP ratio is obtained. The relationship between shear strength enhancement coefficient φ and CFRP ratio ρ_f is shown in Fig. 6.114. Herein, for the sake of simplicity, the relationship between the strength enhancement coefficient φ and CFRP ratio ρ_f is fitted by a logarithmic function as:

$$\varphi(\rho_f) = A\ln(\rho_f) + B \quad (0 < \rho_f < 0.888\%) \tag{6.36}$$

where the parameters of A and B could be determined according to the fittings. According to the fitting results, A and B are respectively recommended to be 0.027 and 1.175, for the present case with the beam shear-span ratio of 2.0. In Eq. (6.36), the strength enhancement coefficient φ is determined according to the influence of CFRP ratio on the shear strength of CFRP-strengthened LWAC beams.

Similar to the effect of stirrups, with the increase of CFRP ratio, the size effect of nominal shear strength of strengthened beams will be gradually weakened, but the size effect cannot be completely eliminated.

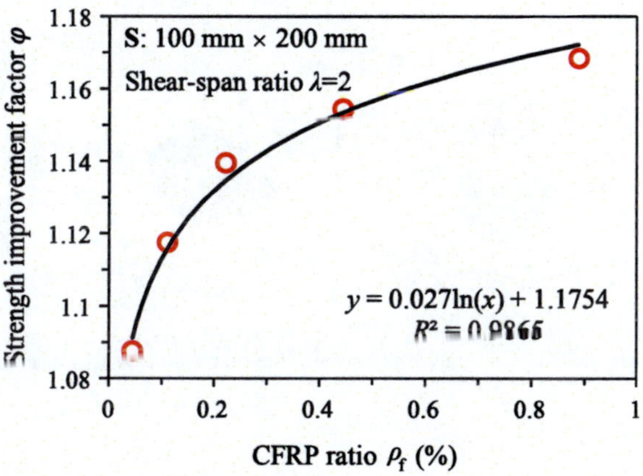

Fig. 6.114 The fitting curve of strength increase coefficient for CFRP-strengthened LWAC beams

The determination method of β is shown in Fig. 6.112. The expression for the coefficient β is as follows:

$$\beta = \begin{cases} 1, & \rho_f = 0 \\ \left(\frac{f_t'\sqrt{1+D/D_0}}{V_0} - 1\right)\tanh^{\frac{1}{2}}(\alpha\rho_f) + 1, & \rho_f > 0 \end{cases} \tag{6.37}$$

6.5.3 Validation of the Theoretical Formula

1. *Ordinary RC beam*

There are three parameters required to be determined in Eqs. (6.16), (6.17), (6.20) and (6.21), i.e., V_0, D_0 and the adjustment coefficient α. The cantilever beam without stirrup and having a shear-span ratio of 1.0, longitudinal reinforcement ratio of 1.0% was used as the reference specimen. In the process of regression analysis, $V_0 = 2.388$ and $D_0 = 876$ are obtained. For the adjustment coefficient α, it can be determined by comparing with the above simulation results.

According to the Chinese code, the lower limit value of stirrup ratio $\rho_{sv,min}$ for RC beams equals to be $0.24f_t/f_{yv}$. Herein, the minimum stirrup ratio $\rho_{sv,min}$ can be calculated to be 0.23% according to the two known parameters (f_t and f_{yv}). Combining with the given parameters above, it is found that when the magnitude of the adjustment coefficient is between 8 and 12, i.e., $\alpha = 8$–12, the obtained theoretical value would have a good agreement with the simulated value. Herein, α is recommended to be 10 for the sake of simplicity.

Fig. 6.115 Comparisons between the theoretical values and numerical results

On this basis, more simulation results are compared with the size effect law. As presented in Fig. 6.115, when the shear-span ratio are 1.0, 1.5 and 2.0, the average values of the ratio of the simulation results to the theoretical value are 0.98, 1.21 and 1.31 respectively, and the coefficients of variation are 0.039, 0.053 and 0.041 respectively, which indicates that the SEL can well predict the nominal shear strength of RC beams under different shear-span ratios, stirrup ratios, longitudinal reinforcement ratios and section sizes.

In addition, some available test results are adopted to furtherly illustrate the reasonability and accuracy of the SEL. Syroka-Korol and Tejchman (2014) carried out the shear failure test of RC beams with the longitudinal reinforcement ratio of 1.0% and the section depth of 200–400 mm, El-Sayed and Shuraim (2016) designed the shear failure test of RC beams with the longitudinal reinforcement ratio of 1.4% and beam depth of 350–1000 mm, Yu et al. (2013) carried out the shear failure test of RC beams with the longitudinal reinforcement ratio of 1.1% and the beam depth of 500–1200 mm, and Jin et al. (2016) carried out the shear failure test of RC beams with the stirrup ratio of 0.14%, longitudinal reinforcement ratio of 1.2% and the beam depth of 200–1000 mm. All the above tests verified the existence of size effect in shear failure of RC beams. These test results are adopted to compared with the predicted values of the SEL, as shown in Fig. 6.116. One can clearly note that, the predicted values of the SEL are in good agreement with the existing test results, which verifies the rationality and accuracy of the SEL. Nevertheless, it should be noted that more systematic and comprehensive test results of different shear-span ratios, stirrup ratios and longitudinal reinforcement ratios are needed to verify the applicability and accuracy of the SEL.

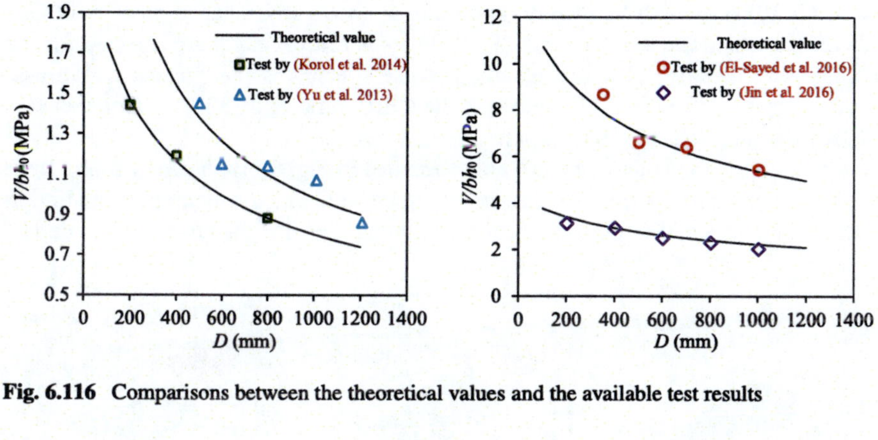

Fig. 6.116 Comparisons between the theoretical values and the available test results

2. ***CFRP-wrapped beam***

In Eqs. (6.31) and (6.34), there are three coefficients that need to be determined: V_0, D_0 and α. In fact, the values of V_0 and D_0 are definitude after determining the strength grade of concrete.

As mentioned above, for CFRP-strengthened NWC beams, the beam with the CFRP ratio of zero is used as the reference specimen and the shear-span ratio is one. The value of V_0 and D_0 are 2.25 and 221.3. As for the adjustment coefficient α, it can be determined by comparing with the above simulation results. When the value of the adjustment coefficient α is between 8.0 and 12.0, the theoretical value and the simulation value will have a good agreement. Here, adopting the value of 10.0 for simplicity, and the 3-D diagram of the comparison between the theoretical values and numerical results is shown in Fig. 6.117. The predicted results (theoretical value) obtained by the theoretical formula are in good agreement with the simulation results.

For CFRP-strengthened LWAC beams, in Eqs. (6.35) and (6.37), the value of V_0 and D_0 are 3.66 and 137.65 respectively by simulation. Also, when the value of the adjustment coefficient α is between 8 and 12, the theoretical value and the simulation value will have a good agreement. Here, adopting the value of 10 for simplicity, and the 3-D diagram of the comparison between the theoretical values and numerical results is shown in Fig. 6.118. The predicted results (theoretical value) obtained by the given theoretical formula are in good agreement with the simulation results.

Furthermore, in order to furtherly illustrate the accuracy and rationality of the SEL presented for CFRP-strengthened NWC and LWAC beams without stirrups, we verified the theoretical formulas of two different aggregate types respectively.

For CFRP-strengthened NWC beams, the available test results reported in Godat et al. (2010) (in which the way to bond CFRP sheets is U-shaped paste) and Luo (2015) (in which the way to bond CFRP sheets is full package) are selected for the comparative analyses. The test results of some specimens (i.e., beams with cross-section depths of 200 mm–400 mm, and the CFRP ratio of 0.133%) in Godat et al.'s (2010) test are selected as comparative analysis. The comparisons of theoretical results and the tested ones are made and presented in Fig. 6.119. It can be found from Fig. 6.119 that the experimental data points agree well with the theoretical prediction curves. In addition, the comparison of experimental results (the cross-section depths

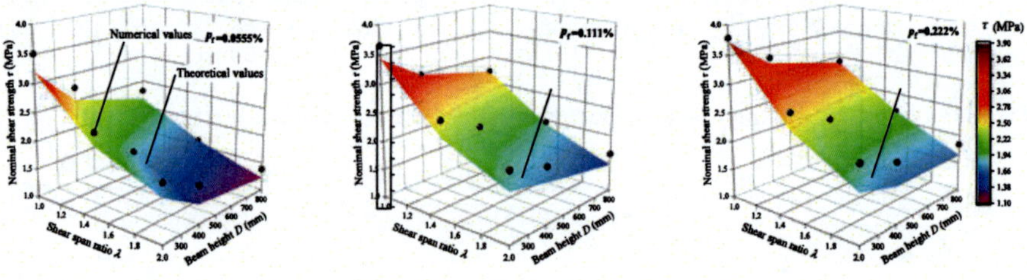

Fig. 6.117 Comparisons between the theoretical values and the present numerical values

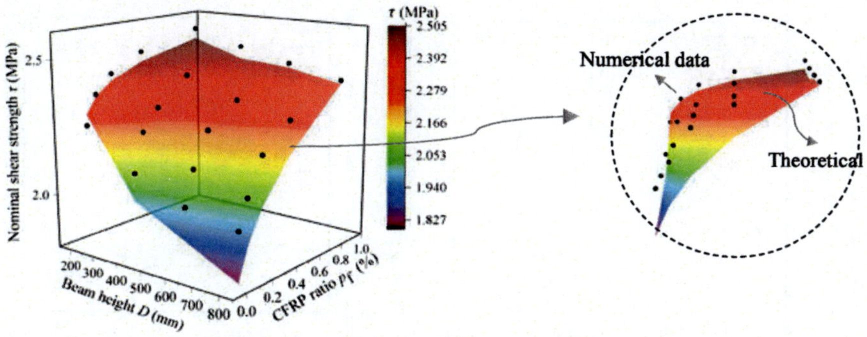

Fig. 6.118 Comparison between the theoretical values and numerical results

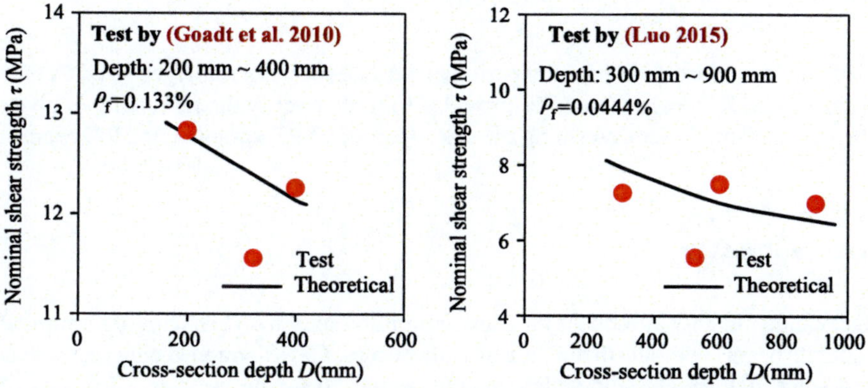

Fig. 6.119 Comparisons between the theoretical results and the available experimental results

of beams are 300 mm, 600 mm and 900 mm respectively and the CFRP ratio of 0.0444% for each specimen) in Luo's (2015) test and theoretical curves presented here. Thus, one can note that good agreement can be observed between the theoretical values of nominal strength and the available tested ones.

For CFRP-strengthened LWAC beams, some scholars have studied the shear properties of CFRP-strengthened LWAC beams as well in recent years, such as Mohammad et al. (2018) and Al-Allaf et al. (2019). Now compare the theoretical formula with the experimental data as shown in Fig. 6.120. One can see from that, the experimental data are in good agreement with the theoretical value, which verifies the correctness and rationality of the given theoretical formula again.

In summary, the novel SEL given in Sect. 6.5.2 presents good consistence with the available experimental results and the meso-scale simulation results, for both CFRP-strengthened NWC and LWAC beams. The theoretical formula considering the influence of structural size, CFRP ratio and shear-span ratio can well predict the nominal shear strength of externally bonded CFRP-strengthened RC beams.

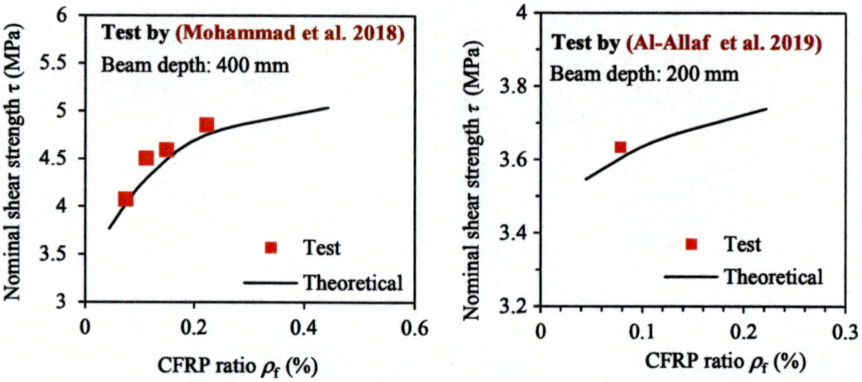

Fig. 6.120 Comparisons between the theoretical results and the available experimental results

Nevertheless, a large number of experimental data, especially for larger-sized CFRP-strengthened RC beams should be provided to further verify the rationality of given theoretical formula for describing the size effect of CFRP-strengthened RC beams.

6.6 Summary

This chapter mainly introduces the shear failure and size effect behavior of reinforced concrete beams without stirrup, beams with stirrups, CFRP-wrapped concrete beams and some lightweight-aggregate concrete beams. Based on the size effect law of Bažant, the size effect laws reflecting various influencing factors are presented. The main conclusions of this chapter are presented as follows:

(1) The effect of stirrups or CFRP on the size effect behavior of shear strength is mainly reflected in two aspects: enhancement on shear strength and weakening of the size effect.
(2) The shear-span ratio and longitudinal reinforcement ratio mainly affect the shear strength of the beam, and their influence on size effect can be ignored.
(3) Different from ordinary concrete beams, LWAC beams are more brittle because of their lower aggregate strength. The shear failure of LWAC beams shows stronger size effect than that of ordinary concrete beams with the same strength grade.
(4) The presented size effect laws can well describe the size effect behavior of shear strength. The simulation results and the available test data demonstrate their accuracy and rationality.

References

Al-Allaf MH, Weekes L, Augusthus NL (2019) Shear behaviour of lightweight concrete beams strengthened with CFRP composite. Mag Concr Res 71(18):949–964

Angelakos D, Bentz EC, Collions MP (2001) Effect of concrete strength and minimum stirrups on shear strength of large members. ACI Struct J 98(3):290–300

Bae SW, Tann BD, Belarbi A (2012) Size effect of reinforced concrete beams strengthened in shear with externally bonded CFRP sheets. In: The 6th international conference on FRP composites in civil engineering, pp 1–8

Barbhuiya S, Choudhury AM (2015) A study on the size effect of RC beam–column connections under cyclic loading. Eng Struct 95:1–7

Bažant ZP (1984) Size Effect in Blunt Fracture: Concrete, Rock. Metal. J Eng Mech 110(4):518–535

Bažant ZP (2005) Scaling of structural strength, 2nd edn. Elsevier

Bažant ZP, Kazemi MT (1991) Size effect on diagonal shear failure of beams without stirrups. ACI Struct J 88(3):268–276

Bažant ZP, Kim JK (1984) Size effect in shear failure of longitudinally reinforced beams. ACI Struct J 81(5):456–468

Bažant ZP, Xiang Y (1997) Size effect in compression fracture: splitting crack band propagation. J Struct Eng 123(2):167–172

Bažant ZP, Yu Q (2005a) Designing against size effect on shear strength of reinforced concrete beams without stirrups: I. formulation. J Struct Eng 131(12):1877–1885

Bažant ZP, Yu Q (2005b) Designing against size effect on shear strength of reinforced concrete beams without stirrups: II. verification and calibration. J Struct Eng 131(12):1886–1897

CECS (2003) Technical specification for strengthening concrete structures with carbon fiber reinforce polymer laminate. China Association for Engineering Construction Standardization, Beijing in Chinese

Che Y, Zhang XF, Wang JJ (2012) Size effect on flexural behavior of reinforced high strength concrete beams subjected to monotonic loading. J Build Struct 33(6):96–102 in Chinese

Che Y, Zhang XF, Wang JJ, Song YP (2013) Experimental study on size effect of flexural behavior of reinforced high-strength concrete beams subjected to cyclic loading. J Build Struct 34(8):100–106 in Chinese

El-Sayed AK, Shuraim AB (2016) Size effect on shear resistance of high strength concrete deep beams. Mater Struct 49:1871–1882

Godat A, Qu Z, Lu XZ, Labossiere P, Ye LP, Neale KW (2010) Size effects for reinforced concrete beams strengthened in shear with CFRP strips. J Compos Construct 14(3):260–271

Godat A, Labossiere P, Neale KW (2012) Numerical investigation of the parameters influencing the behaviour of FRP shear-strengthened beams. Constr Build Mater 32:90–98

Han LH, Yang YF (2005) Cyclic performance of concrete-filled steel CHS columns under flexural loading. J Constr Steel Res 61:423–452

Imanzadeh S, Jarno A, Hibouche A, Bouarar A, Taibi S (2020) Ductility analysis of vegetal-fiber reinforced raw earth concrete by mixture design. Constr Build Mater 239:117829

Jin L, Du XL, Li D, Su X (2016) Seismic behavior of RC cantilever beams under low cyclic loading and size effect on shear strength: An experimental characterization. Eng Struct 122:93–107

Jin L, Yu WX, Su X, Zhang S, Du XL, Han JY, Li D (2018) Effect of cross-section size on the flexural failure behavior of RC cantilever beams under low cyclic and monotonic lateral loadings. Eng Struct 156:567–586

Jin L, Wang T, Du XL (2019a) Size effect theory on shear strength of RC cantilever beams without stirrups. Chin J Comput Mech 1–10

Jin L, Wang T, Jiang XA, Du XL (2019b) Size effect in shear failure of RC beams with stirrups: simulation and formulation. Eng Struct 199:109573

Jin L, Jiang X, Xia H, Chen FJ, Du XL (2020a) Size effect in shear failure of lightweight concrete beams wrapped with CFRP without stirrups: influence of fiber ratio. Compos Part B Eng 199:108257

Jin L, Xia H, Jiang XA, Du XL (2020b) Size effect on shear failure of CFRP-strengthened concrete beams without web reinforcement: meso-scale simulation and formulation. Compos Struct 236:111895

Kachlakev DI, Miller TH, Potisuk T (2001) Finite element modeling of reinforced concrete structures strengthened with FRP laminates. Oregon Department of Transportation. Research Group

Lee MK, Barr BIG (2004) An overview of the fatigue behavior of plain and fibre reinforced concrete. Cem Concr Compos 26:299–305

Li W, Leung CKY (2016) Shear span–depth ratio effect on behavior of RC beam shear strengthened with full-wrapping FRP strip. J Compos Constr 20(3):106–117

Li D, Jin L, Du XL, Fu J, Lu A (2016) Size effect tests of normal-strength and high-strength RC columns subjected to axial compressive loading. Eng Struct 109:43–60

Lu Y, Li N, Li S, Liang H (2015) Behavior of steel fiber reinforced concrete-filled steel tube columns under axial compression. Constr Build Mater 95:74–85

Lubell A, Sherwood T, Bentz EC, Collins MP (2004) Safe shear design of large, wide beams. Concr Int 26(1):67–78

Luo RF (2015) Experimental study on size effect of RC beam shear-strengthened with CFRP wraps. Guangdong University of Technology, Guangdong in Chinese

Matta F, El-Sayed AK, Nanni A, Benmokrane B (2013) Size effect on concrete shear strength in beams reinforced with fiber-reinforced polymer bars. ACI Struct J 110(4):617–628

Ministry of Construction of People's Republic of China (2010) Code for design of concrete structures GB 50010-2010. China Architecture & Building Press, Beijing in Chinese

Mohammad A, Sarsam K, Al-Bayati N (2018) Shear strength enhancement of lightweight aggregate reinforced concrete deep beams by using CFRP strips. MATEC Web Conf 162:04011

Rodrigues H, Arêde A, Varum H, Costa AG (2013) Experimental evaluation of rectangular reinforced concrete column behaviour under biaxial cyclic loading. Earthq Eng Struct Dyn 42(2):239–259

Słowik M, Smarzewski P (2012) Study of the scale effect on diagonal crack propagation in concrete beams. Comput Mater Sci 64:216–220

Syroka-Korol E, Tejchman J (2014) Experimental investigations of size effect in reinforced concrete beams failing by shear. Eng Struct 58:63–78

Syroka-Korol E, Tejchman J, Mróz Z (2013) FE calculations of a deterministic and statistical size effect in concrete under bending within stochastic elasto-plasticity and non-local softening. Eng Struct 48:205–219

Walraven J, Lehwalter N (1994) Size effect in short beams loaded in shear. ACI Struct J 91(5):585–593

Wang B, Wang J, Gong X, Liu B (2014) Experimental studies on circular CFST frames with ALC walls under cyclic loadings. Int J Steel Struct 14(4):755–768

Yılmaz O, Molinari JF (2017) A mesoscale fracture model for concrete. Cem Concr Res 97:84–94

Yu Q, Bažant ZP (2011) Can stirrups suppress size effect on shear strength of RC beams. J Struct Eng 137(5):607–617

Yu L, Che Y, Song YP (2013) Shear behavior of large reinforced concrete beams without web reinforcement. Adv Struct Eng 16(4):653–665

Zhang Q, Guo ZH (1992) Investigation on shear strength and shear strain of concrete. J Build Struct 13(5):17–24 in Chinese

Chapter 7
Size Effect in Compressive Failure Behavior of Concrete Columns

As an important axial loaded component supporting beam, slab and other structures, the mechanical properties of concrete columns are of great important. With the development of large-scale structures, the size of concrete column increases, which results in obvious size effect phenomenon (Bažant and Kwon 1994; Němeček and Bittnar 2004). For size effect of concrete columns, it is a structural mechanics problem including not only the most basic material size effect, but also the size effect contributed by the influence of mutual effect between different materials. Actually, the size effect of concrete columns has been extensively investigated (Bažant and Kwon 1994; Němeček and Bittnar 2004; Wang and Wu 2010; Chen et al. 2018). However, the size of these tested RC columns was still restricted to a relatively small range. In addition, with the increase of the structure size, the safety factor of the existing code equation in predicting the bearing capacity of concrete columns decreases gradually, which has certain risk. Therefore, more tests for practical full-scale components should be conducted for the investigation of size effect of RC columns.

As for the current situation of size effect tests of RC columns mostly located into the small-scale specimens, a series of systematically designed experimental and numerical studies of large-scale RC columns are presented. The analyses of various parameters including the concrete strength, load eccentricity, loading pattern, slenderness ratio, the longitudinal reinforcement ratio, the cross-sectional shape, and the confinement material (stirrup, FRP, and steel tube) are given. Moreover, the size effect law on nominal compressive strength of RC columns is presented, in which the size effect of concrete material and the influence of confinement effect on the size effect are considered.

X. Du and L. Jin, *Size Effect in Concrete Materials and Structures*,
https://doi.org/10.1007/978-981-33-4943-8_7

7.1 Axial Compressive Failure of Normal-Strength RC Column

Herein the present section, the experimental research (Li et al. 2016) on normal-strength RC columns subjected to axial compressive loading with different sizes is described. The influences of reinforcement ratio and slenderness ratio on mechanical performances and size effect of the RC columns are given.

7.1.1 Experimental Program

1. *Specimen description*

All the specimens were cast from the same batch of C30 mixed concrete. The standard plain concrete cube samples with the size of 150 mm × 150 mm × 150 mm were also made together with the RC columns in order to obtain the uniaxial compressive and splitting tensile strengths of the used concrete. Here, the corresponding average values of 28 days compressive and splitting tensile strengths of concrete cube samples were 33.2 MPa and 3.2 MPa. Longitudinal reinforcement was made from the steel rebar of grade HRB335 (i.e. the Hot-rolled Ribbed-steel Bar) and the tested average yield strength was 383 MPa. Hoop reinforcement was made from smooth bars of grade HPB235 and the tested average yield strength was 249 MPa. It has to be noted that, a few number of the RC columns employed the smooth rebar (HPB235) for the longitudinal reinforcement. This was for the purpose of study on the influence of different types of steel rebar on the global mechanical properties of RC columns.

According to different slenderness ratio, longitudinal reinforcement ratio, and stirrup ratio, 19 RC columns are divided into five series, i.e. NA, NB, NC, ND, and NE. The corresponding geometric details of the 19 RC columns utilized in the test are respectively summarized in Table 7.1. All geometric dimensions of the normal-strength RC columns including reinforcement were scaled to obtain geometrically proportional specimens with the ratio 1: 2: 3. Details of the range of the RC columns utilized in the test are presented in Fig. 7.1. To ensure proper load transmission into the RC columns at both ends and to limit the possibilities of bond slip, the RC columns were provided with steel mesh cast into the concrete, as shown in Fig. 7.1.

2. *Test setup and loading arrangement*

All test specimens were loaded by the multifunctional electrohydraulic servo loading test system (MESLTS) at Beijing University of Technology of China, as shown in Fig. 7.2. The RC columns were simply supported at both ends, and the load P was applied axially. The top and bottom ends of the RC columns were located against spherical seating supports.

It was controlled by constant increment of averaged longitudinal forces before the peak load and switched to constant increment control of averaged longitudinal

Table 7.1 Geometric details of the normal-strength RC columns

Series	Specimen label	Cross-sectional size (mm)	Length (mm)	Longitudinal reinforcement ratio (%)	Transverse reinforcement ratio (%)	Slenderness ratio	Cover thickness (mm)
NA	NALR	600 × 600	2700	2.5	0.65	4.5	30
	NAMR	400 × 400	1800	2.5	0.65	4.5	20
	NASR	200 × 200	900	2.5	0.65	4.5	10
	NASP	200 × 200	900	2.5	0.65	4.5	10
NB	NBLR	600 × 600	1800	2.5	0.65	3	30
	NBMR	400 × 400	1200	2.5	0.65	3	20
	NBSR	200 × 200	600	2.5	0.65	3	10
	NBSP	200 × 200	600	2.5	0.65	3	10
NC	NCLR	600 × 600	2700	1.5	0.65	4.5	30
	NCMR	400 × 400	1800	1.5	0.65	4.5	20
	NCSR	200 × 200	900	1.5	0.65	4.5	10
	NCSP	200 × 200	900	1.5	0.65	4.5	10
ND	NDLR	600 × 600	2700	2.5	0.94	4.5	30
	NDMR	400 × 400	1800	2.5	0.94	4.5	20
	NDSR	200 × 200	900	2.5	0.94	4.5	10
	NDSP	200 × 200	900	2.5	0.94	4.5	10
NE	NELR	600 × 600	2700	2.5	1.31	4.5	30
	NEMR	400 × 400	1800	2.5	1.31	4.5	20

(continued)

Table 7.1 (continued)

Series	Specimen label	Cross-sectional size (mm)	Length (mm)	Longitudinal reinforcement ratio (%)	Transverse reinforcement ratio (%)	Slenderness ratio	Cover thickness (mm)
	NESR	200 × 200	900	2.5	1.31	4.5	10

Note "N" stands for the normal-strength concrete; "A, B, C, D and E" correspond to the series; "S", "M" and "L" mean "small", "middle" and "large" which are equal to the cross-sectional size of 200 mm × 200 mm, 400 mm × 400 mm, 600 mm × 600 mm, respectively; "R" represents the "Hot-rolled Ribbed-steel Bar"; and "P" represents the "Hot-rolled Plain-steel Bar"

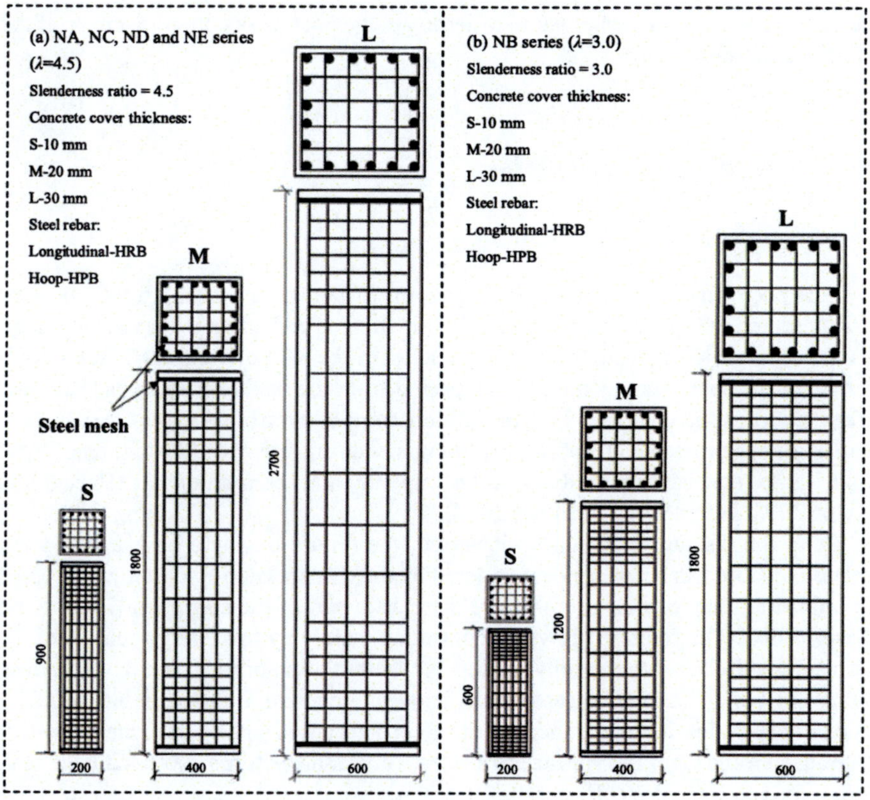

Fig. 7.1 Test RC columns with different slenderness ratios of: **a** 4.5; and **b** 3.0

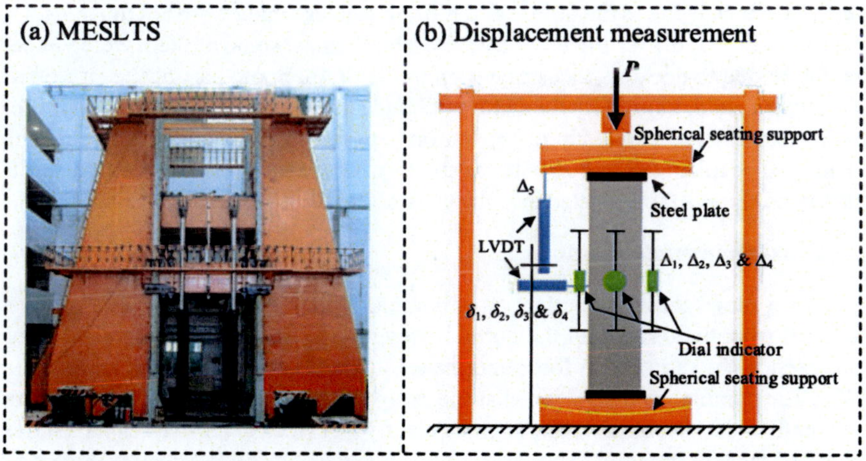

Fig. 7.2 Loading device and schematic diagram: **a** MESLTS; and **b** Displacement measurement

deformations after the peak load in order to get the post-peak softening curve of the stress-strain relationship.

7.1.2 Results and Discussions

1. *Failure patterns*

Typical photographs about the failure patterns of the normal-strength RC columns are presented in Fig. 7.3. One can clearly find that the end effect still exists although having steel mesh cast into the specimen. Most of the RC columns break at the mid-length of the specimen, as might be expected from the buckling analysis (Kim and Yang 1995), as shown in Fig. 7.4a. In addition, some of them break at the top or bottom end of the specimen which definitely pronounces an end effect of RC columns. These findings are consistent with that in other experimental research efforts (Bažant and Kwon 1994; Němeček and Bittnar 2004).

From the fracture mechanics viewpoint, the failure ought to be caused by the release of stored strain energy engendered by the compression fracture, which leads to unloading due to axial extension of the elastic part of the specimen (Bažant and Kwon 1994). The fracture growth is controlled mainly by this energy release of the structure while the strength only matters for the initiation of the fracture. Moreover, failure of bond (i.e. slip or separation) between concrete and steel rebar could be another reason for the appearance of end effect and caused some part of the size effect (Saroushian and Choi 1989; Ruiz et al. 1999; Yerlici and Ozturan 2000; Ichinose et al. 2004; Yu and Bažant 2011).

In the present tests, most of the longitudinal steel rebar were yielded and buckled as can be seen in Fig. 7.4a, some were only partly yielded, some didn't and some were not detected due to the failure of the strain gauges which were not reliable up to the completion of the testing procedure. For the stirrups, some of them were yielded and broke due to necking, as can be seen in Fig. 7.4b. Some part of the size effect of RC members is caused by the failure of bond (Bažant and Kwon 1994; Němeček and Bittnar 2004), i.e. the slip or separation between concrete and steel rebar. This could partly explain the reason why the strains of some steel rebar were not detected and the failure position located into the ends of the RC columns.

2. *Stress-strain relationships*

Figure 7.5 shows the entire collection of nominal stress versus nominal strain curves obtained from the test for all the normal-strength RC columns. The nominal stress is defined as the ratio of the force and the area of the square cross-section, i.e. $\sigma = P/D^2$. The nominal strain is defined as the ratio of the longitudinal displacement and the length of the specimen. It can be noted clearly from the curves that for each series, the initial tangent modulus varies very little while the post-peak softening curves are distinct as the structure size changes. This should be caused by the internal heterogeneities of concrete. As the structural size increases, the aggregate distribution, the

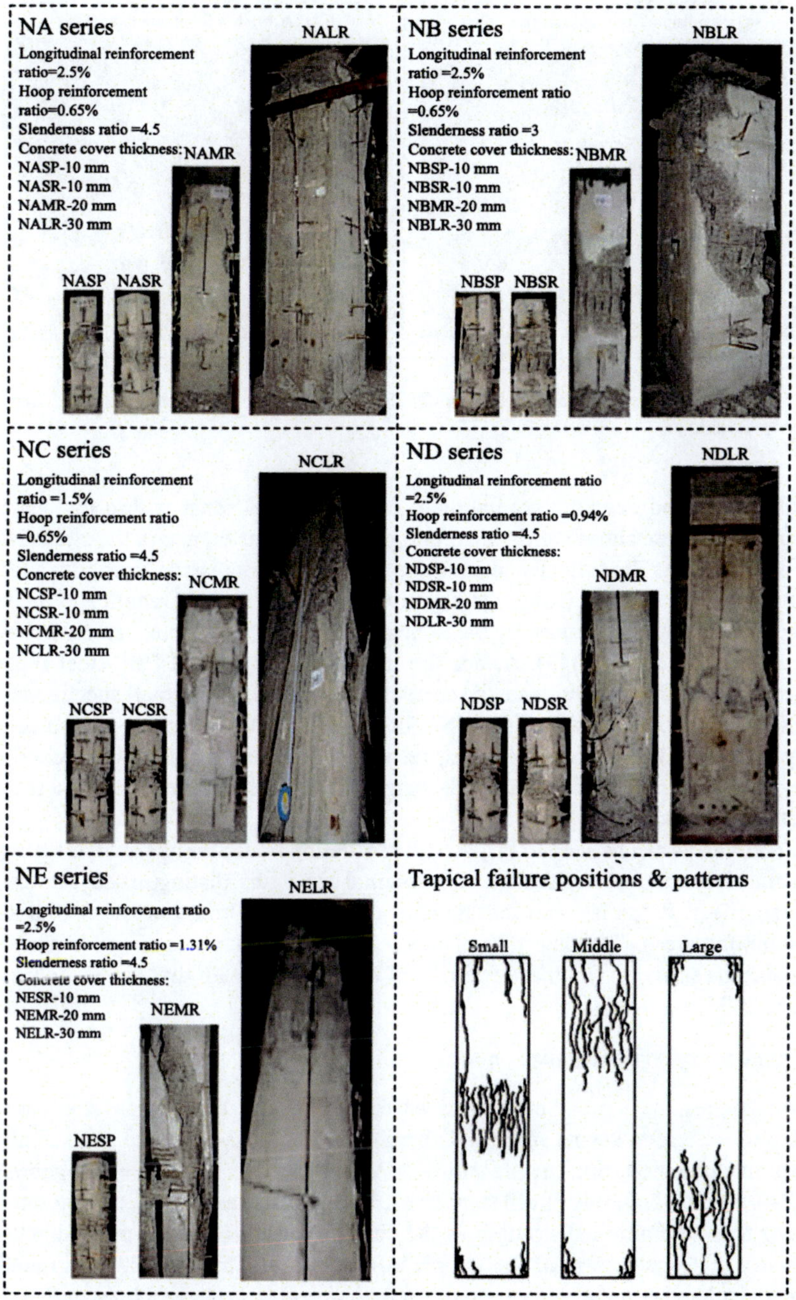

Fig. 7.3 Typical failure photographs for the normal-strength RC columns

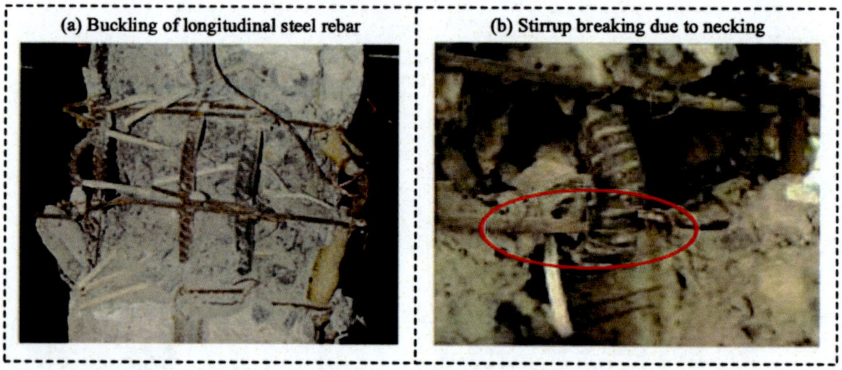

Fig. 7.4 Failure patterns of the steel rebar (NCSR): **a** Buckling of longitudinal steel bar; and **b** Stirrup breaking

initial defects and even the confinement effects of steel rebar within the concrete specimens become significantly different, causing the difference in the stress status and deformation. This, finally, makes the nominal compressive strengths and the softening curves distinct. Simply to say, the differences in the strength and the softening curves should be caused by the nonlinearity of the global mechanical behavior of concrete (Du et al. 2013). As for the use of "HRB" and "HPB" steel rebar in longitudinal reinforcement, it can be seen that the curves of the small specimens (i.e. NASP/NASR, NBSP/NBSR, NCSP/NCSR and NDSP/NDSR) show a good consistency in NA, NB, NC and ND series, respectively. Thus, it can be concluded that the type of steel rebar does not disturb the global mechanical properties and the size effect in RC structures.

Moreover, as can be seen in Fig. 7.5, the softening rates after the peak are roughly the same in each series although the nominal strengths distinguished because of the size effect. It can be seen that most of the softening branches of smaller-sized RC columns are higher than that of lager-sized RC columns and the larger-sized RC columns seems to fail more soon. This is consistent with that in (Němeček and Bittnar 2004).

3. *Influence of structural size on the nominal strength*

Figure 7.6 presents nominal compressive strength σ_{Nu} (i.e. the nominal stress at peak load, $\sigma_{Nu} = P_u/D^2$) versus structural size D (i.e. the cross-sectional size) of all the data points obtained from the present test. As one can see, in each series, geometrically similar RC columns of different sizes, made of the same batch of concrete, are sharing the same nominal strength as the currently used RC structure design codes (Ministry of Construction of the People's Republic of China 2003, 2011) predict. These calculated nominal compressive strengths are clearly straight lines with a slope of zero which are parallel to the axis of structural size D. This is definitely not indicating any size effect. On the other hand, Fig. 7.6 shows the reduction in the measured nominal compressive strengths as the structural size D increases in each series. This

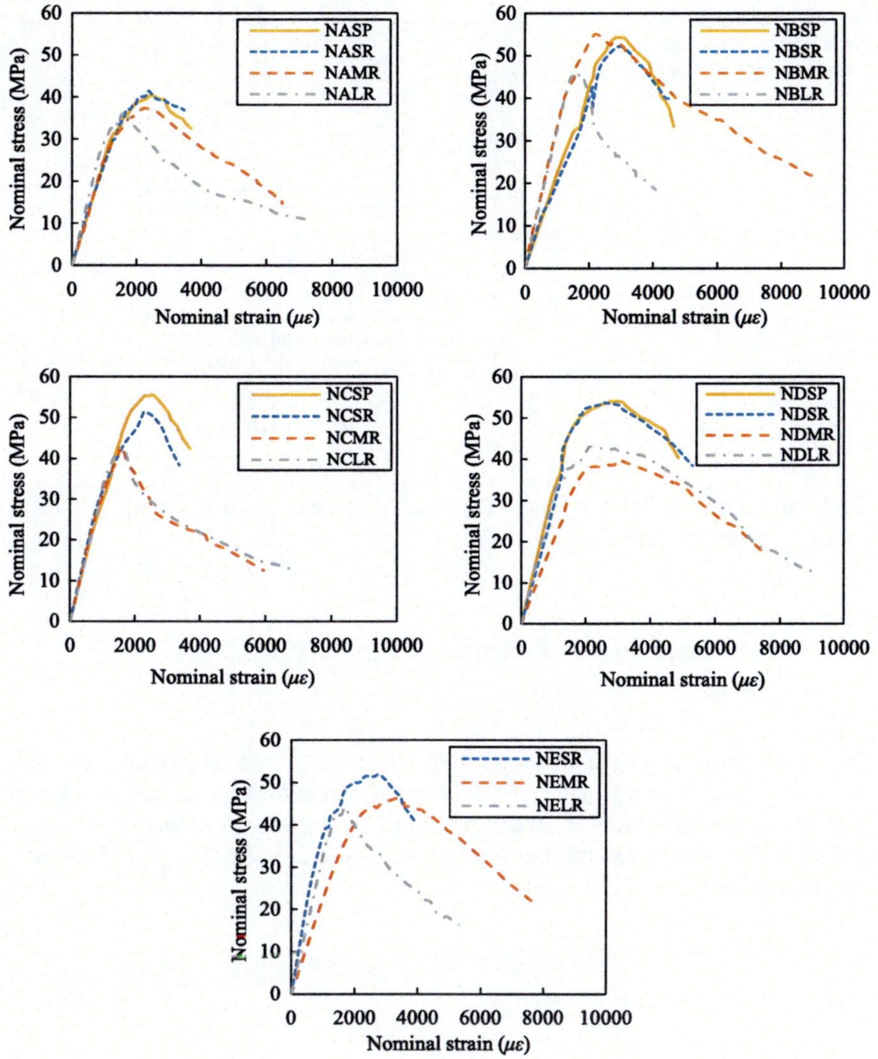

Fig. 7.5 Nominal stress-strain curves for the normal-strength RC columns

diagrams significantly prove the existence of a size effect as theoretically expected. Note that this trend is not as obvious as plain concrete components which especially fail in tension, one possible reason could be the presence of the reinforcement, which contributes to the suppression and disturbance of size effect as can be concluded from previous studies (Ouyang and Shah 1994). Furthermore, some failures were caused by axial splitting cracks in a sudden. In this case, the axial cracks do not disturb the macroscopic field. Therefore, they do not cause any energy release and yield no size

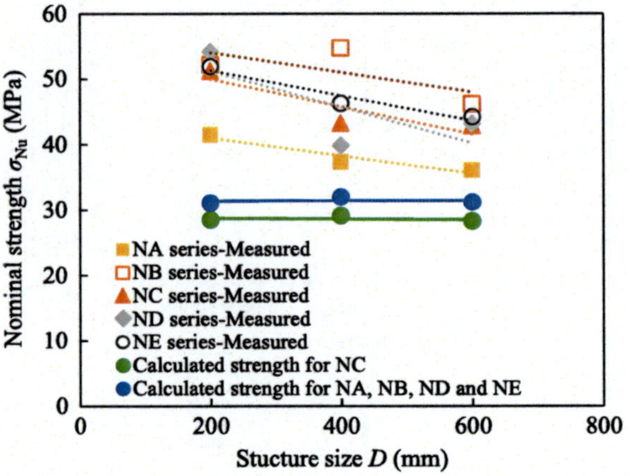

Fig. 7.6 Nominal strength versus structural size for the normal-strength RC columns

effect (Němeček and Bittnar 2004). This should be another possible explanation of the suppression of size effect.

7.2 Axial Compressive Failure of High-Strength RC Column

This section presents the experimental study (Jin et al. 2016) on a series of geometrically similar moderate high-strength RC columns with different cross-sectional sizes under monotonic and cyclic axial compression. The influence of loading pattern on mechanical performances and size effect of the moderate high-strength RC columns is also presented.

7.2.1 Experimental Program

1. *Specimen description*

A total of 16 moderate high-strength RC columns with different sizes (in the ratio 1:2:3:4) were tested. The corresponding geometric details of the high-strength RC columns utilized are respectively summarized in Table 7.2 by two series (i.e. HM and HC). Here, the average values of compressive and splitting tensile strengths of standard concrete cube samples sized of 150 mm × 150 mm × 150 mm were 65.8 MPa and 6.4 MPa, respectively.

The cross-sectional dimensions and the reinforcement details are presented in Fig. 7.7. Additionally, to ensure proper load transmission into the high-strength RC

Table 7.2 Geometric details of the high-strength RC columns

Series	Specimen label	Cross-section (mm)	Length (mm)	Longitudinal reinforcement ratio (%)	Hoop reinforcement ratio (%)	Slender-ness ratio	Cover thickness (mm)
HM	HMG-1/2	800 × 800	2400	1.5	0.65	3	40
	HML-1/2	600 × 600	1800	1.5	0.65	3	30
	HMM-12	400 × 400	1200	1.5	0.65	3	20
	HMS-1/2	200 × 200	600	1.5	0.65	3	10
HC	HCG-1/2	800 × 800	2400	1.5	0.65	3	40
	HCL-1/2	600 × 600	1800	1.5	0.65	3	30
	HCM-1/2	400 × 400	1200	1.5	0.65	3	20
	HCS-1/2	200 × 200	600	1.5	0.65	3	10

Note "H" represents the high-strength RC columns; "M" and "C" stand for the loading patterns of "monotonic loading" and "cyclic loading", respectively; "S", "M" and "L" mean "small", "middle" and "large" size, respectively; "G" means "giant" which is equal to the cross-sectional size 800 × 800 mm; and "1" and "2" correspond to the replicate RC columns

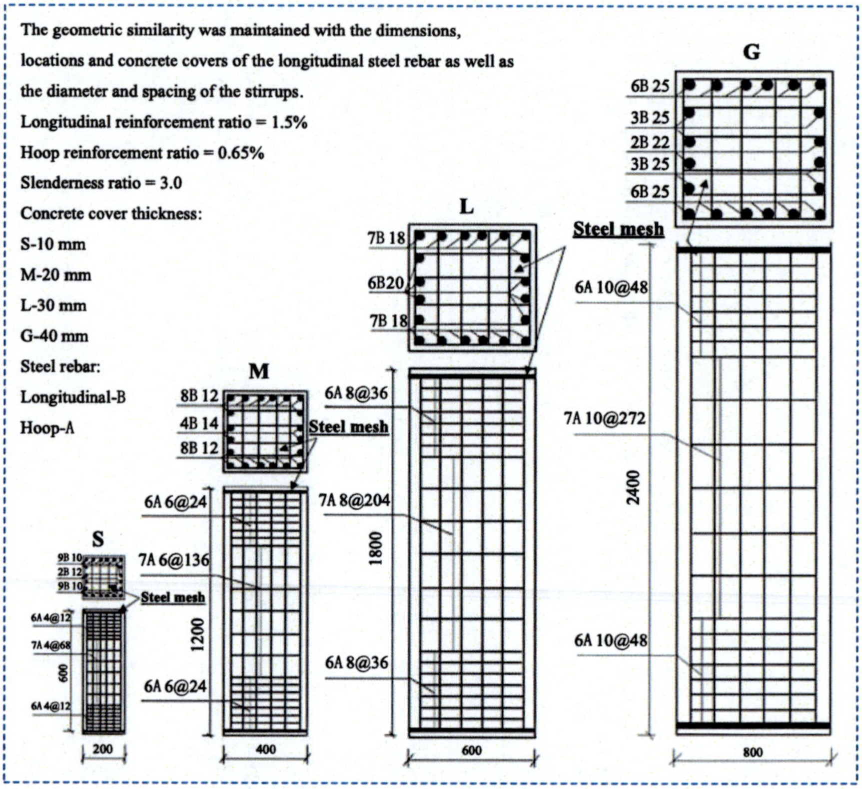

Fig. 7.7 Cross-sectional dimensions and the reinforcement details

columns at both ends, the specimens were provided with steel mesh casted into the concrete, as shown in Fig. 7.7.

2. *Test setup and loading arrangement*

The moderate high-strength RC columns were tested under monotonic and cyclic axial compression using the 40,000 kN capacity multifunctional electrohydraulic servo loading test system (MESLTS). The loading arrangement and displacement measurements used in the test is the same as that in Sect. 7.1.1, as shown in Fig. 7.2b.

For the HM series, the application method of axial loading was the same as that of normal-strength RC columns in Sect. 7.1.1. For the HC series, the RC columns were loaded under repeated cyclic compression involving several unloading and reloading cycles at each prescribed load or displacement level. The repeated cyclic compression was applied at approximately 10% calculated peak load, P_u, intervals. P_u is the calculated peak load according to the currently design codes of RC structures of China (Ministry of Construction of the People's Republic of China 2011). The high-strength RC columns were set to subject to unloading/reloading for a single

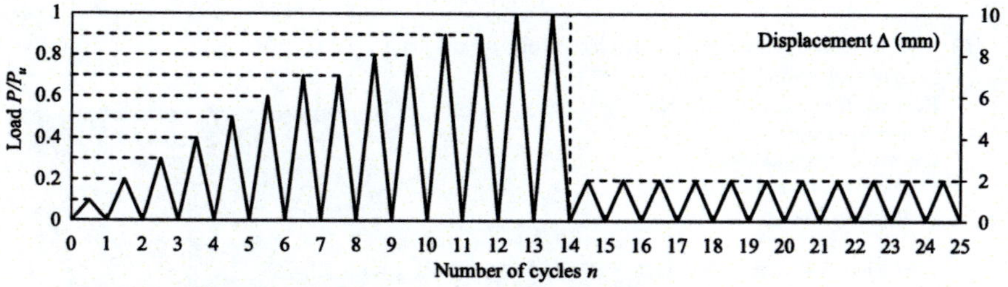

Fig. 7.8 Loading protocol for cyclic loading Number of cycles

cycle before reaching 0.6 P_u, for two cycles after reaching 0.6 P_u and then it was switched to a constant displacement control after the peak load. The loading protocol is presented in Fig. 7.8.

7.2.2 Results and Discussions

1. *Failure patterns*

Typical failures of the test moderate high-strength RC columns are shown in Fig. 7.9. As can be seen in Fig. 7.9, two different types of failure patterns were observed in the specimens. It can be noted that, for the HM series under monotonic axial compression as shown in Fig. 7.9a, most of the RC columns fail by the rupture of concrete and the damage is highly localized around an inclined major shear crack appeared at the surface of the specimen. Most of the RC columns are damaged at mid-length of the specimens, and the damage is apparently a compression-shear failure mode. While, for the HC series under repeated cyclic axial compression as presented in Fig. 7.9b, most of the failures arise at mid-length of the specimen as well, but they should be a compression-induced crushing mode. This should be caused by the repeated loading function. In reality, the deformation should be concentrated at the mid-length of the RC columns for the weak confinement by the stirrups. Under axial compressive loading, the stresses and deformations within the concrete increase and then localize at the mid-length of the RC columns. With the continuous increase of the external applied compressive loading, the concrete cover at the mid-length falls, the longitudinal reinforcements may buckle and the stirrups may fracture due to necking failure, as can be seen from Fig. 7.9b.

In each series, the failure of the specimens evolved in a similar way. Take one of the largest specimens named of HCG-1 as an example, the entire broken surfaces and the local failures involving buckling of longitudinal steel rebar and spalling of concrete cover are shown in Fig. 7.10. The buckling behavior of the longitudinal steel rebar embedded in the concrete and the spalling of concrete cover can be clearly noticed from the global view of the specimen in Fig. 7.10a. Figure 7.10b illustrates the four broken surfaces and the remarkable local failures on the surfaces. There is a small quantity of cracks on surfaces 1 and 2, only a few cases of spalling of concrete cover

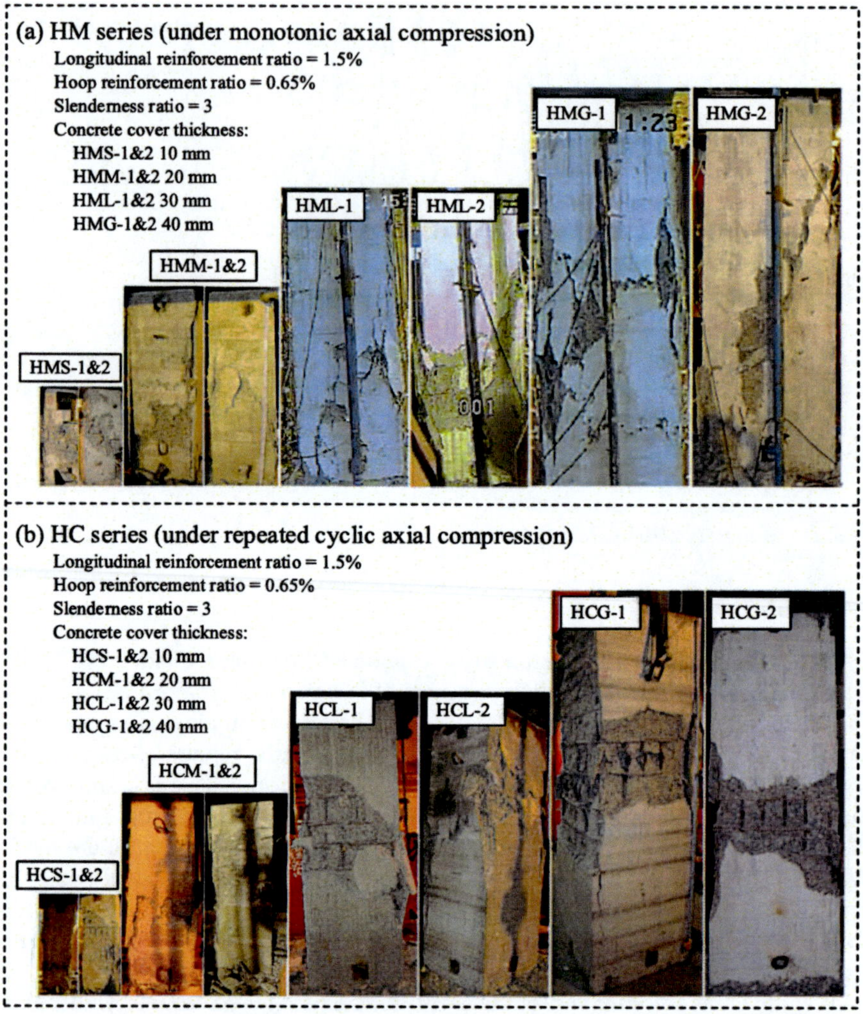

Fig. 7.9 Typical failure patterns of the high-strength RC columns: **a** HM series; and **b** HC series

occur at the edge of the specimen. A large area of spalling of concrete cover happens on the surfaces of 3 and 4, and the buckling of longitudinal steel rebar generated as a consequence. In reality, all the specimens experienced an eccentric compression due to the inevitable eccentricity and the non-uniformity of the columns. The buckling of longitudinal reinforcement in all the specimens was observed in the tests.

2. *Axial load-displacement behavior*

Figure 7.11 presents the entire collection of load versus displacement curves obtained from the test for the RC columns of HM series subjected to monotonic axial compressive loading. It can be found clearly from the curves in Fig. 7.11 that for the replicates

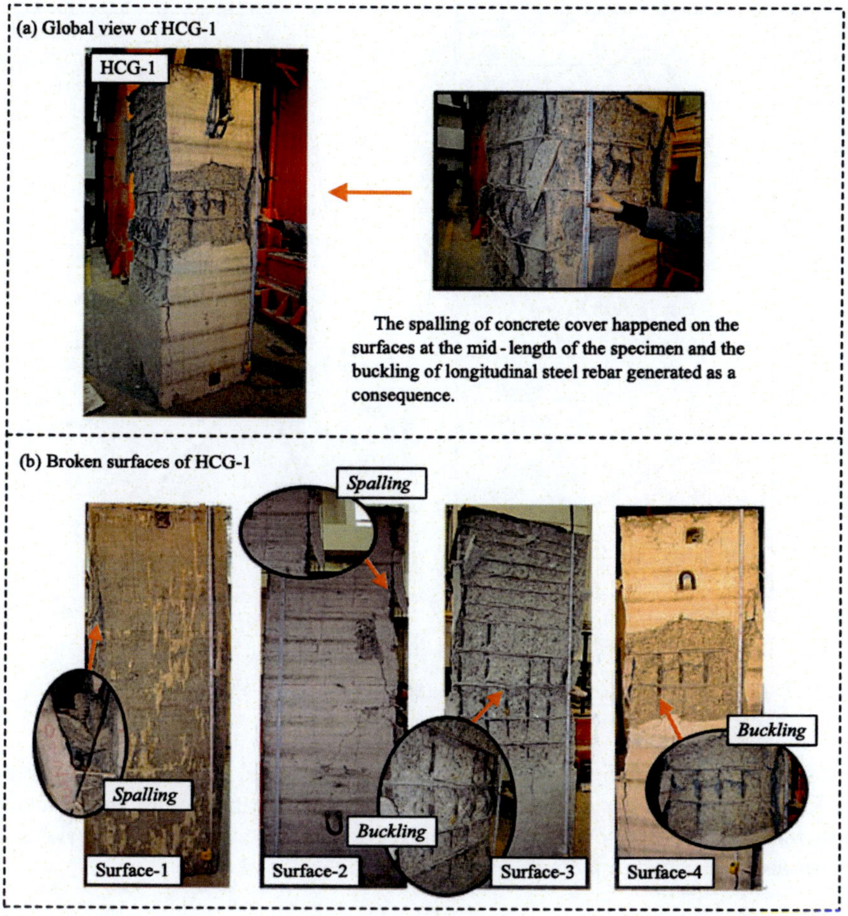

Fig. 7.10 Entire broken surfaces and local failures of HCG-1: **a** Global view; and **b** Local broken surfaces

of each structural size, the initial tangent modulus varies very little while the peak load and the post-peak softening curves are distinct. As described in Sect. 7.1.2, this should be caused by the internal heterogeneities and nonlinearity of mechanical behavior of concrete.

The hysteretic curves of RC members subjected to cyclic loading can reflect the ductility of displacement, the degradations of strength and stiffness, and the energy dissipating capacity. Figure 7.12 plots the entire collection of load versus displacement hysteretic curves obtained from the test for the RC columns of HC series subjected to repeated cyclic axial compressive loading. The skeleton curves (i.e. the link of the maximum points of the load-displacement hysteretic curve) are also plotted Fig. 7.12. It can be found that, similar with that of the HM series subjected

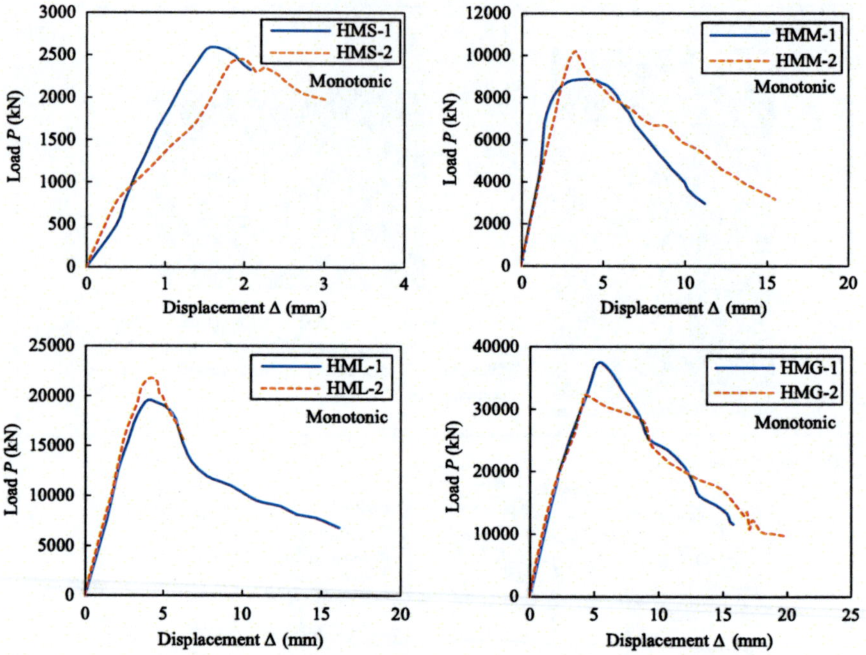

Fig. 7.11 Load versus displacement curves of HM series under monotonic loading

to monotonic axial compressive loading, there is an initial elastic response for all the RC columns, and the load-displacement relationship can be considered roughly to be linear as the skeleton curve shows. Moreover, the load-displacement skeleton curves also experience a softening stage after the peak load.

3. ***Stress-strain response***

Nominal axial compressive stress-strain curves of the test moderate high-strength RC columns are plotted in Fig. 7.13. Here, the nominal compressive stress is the applied compressive load divided by the cross-sectional area of the RC column. In each of the plots shown in Fig. 7.13, the stress-strain curves of two nominally identical monotonically loaded and the two nominally identical cyclically loaded specimens are all included for comparison.

It can be observed in Fig. 7.13 that most of the moderate high-strength RC columns experience a sudden drop in strength starting right at the transition point on the stress-strain curve. This phenomenon should be closely associated to the brittleness nature of the high-strength concrete materials. From fracture mechanics viewpoints, the size effect is closely related with the brittleness nature of concrete. High-strength concrete presents a more obvious size effect than normal-strength concrete due to its inherent high brittleness. Moreover, the size effect in RC columns not only depends upon the characteristic of the concrete materials, but also the interaction of reinforcing

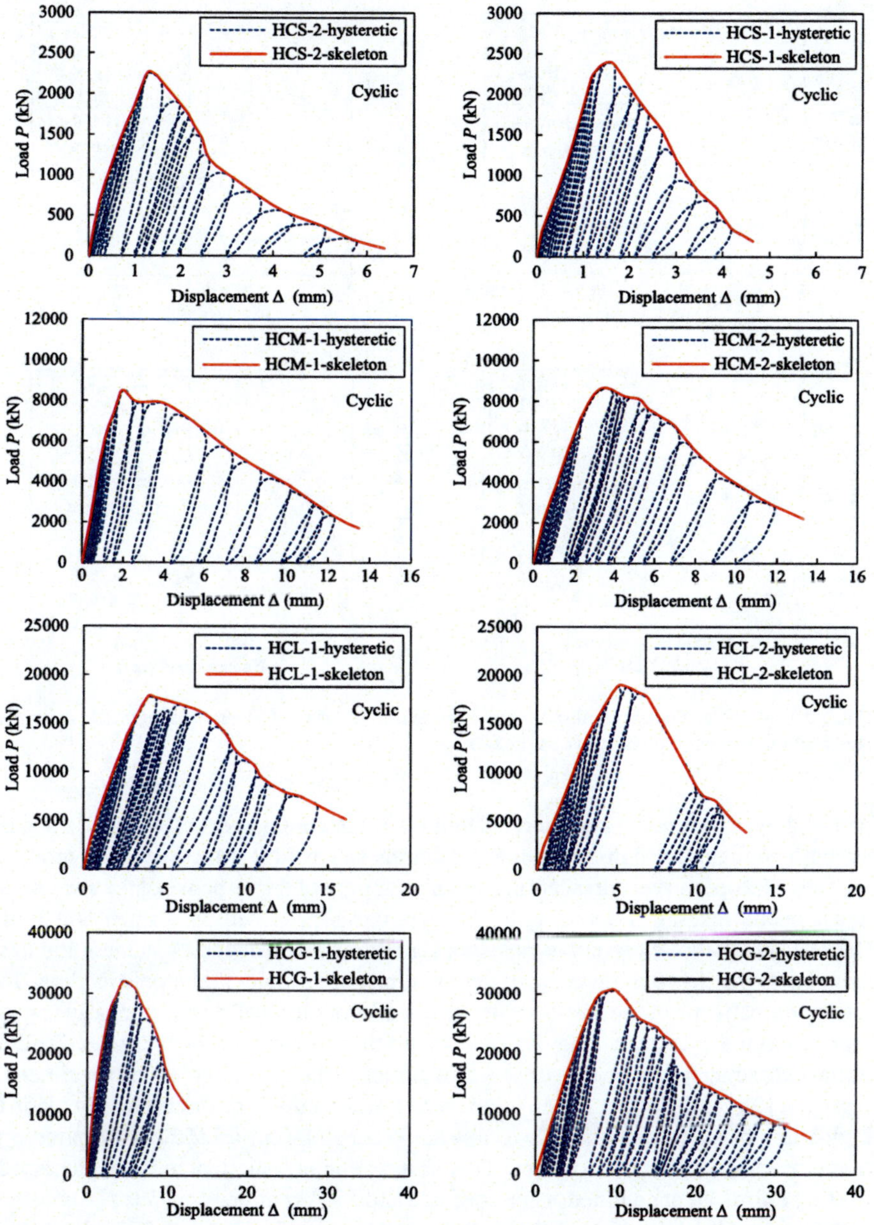

Fig. 7.12 Load versus displacement hysteretic and skeleton curves of HC series under repeated cyclic loading

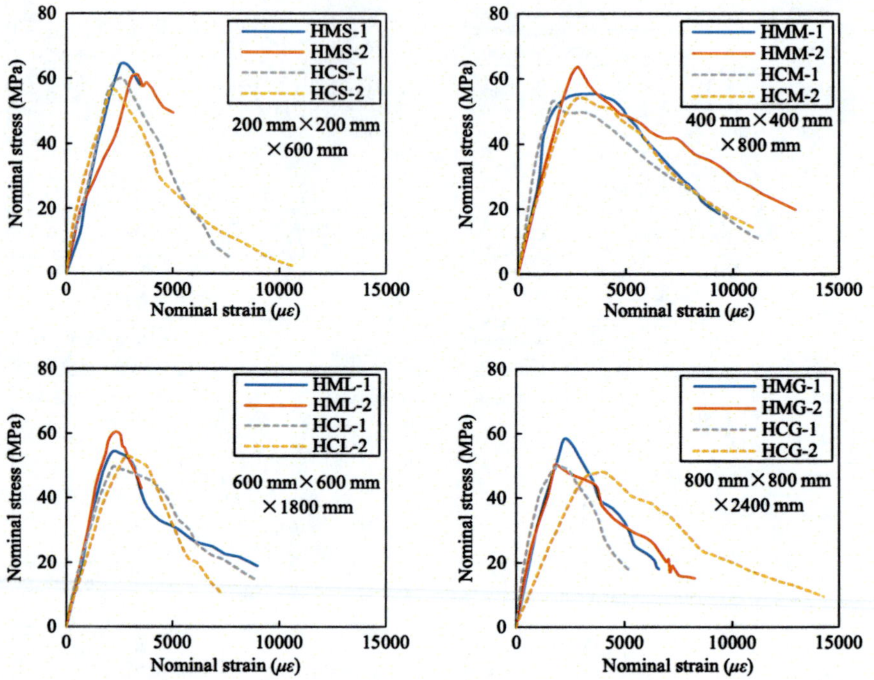

Fig. 7.13 Nominal axial compressive stress-strain curves of the test high-strength RC columns under monotonic and cyclic axial compression

bars and the surrounding concretes. Certainly, if the confinement effect of stirrups is strength enough, the high-strength RC columns can exhibit a ductile failure mode.

The envelope curve represents the upper boundary of the response of the specimen under cyclic loading. According to the hypothesis proposed by Karsan and Jirsa (1969), a unique envelope curve exists for a given cyclically loaded specimen and it is identical to the stress-strain curve of the same specimen under monotonic loading. To allow comparisons of the stress-strain curves of monotonically and cyclically tested high-strength RC columns, Fig. 7.13 also draws the envelope curves by connecting the initial unloading points on the stress-strain curves of the cyclically loaded specimens. Figure 7.13 manifests that the envelope stress-strain curves of the cyclically loaded high-strength RC columns closely follow the stress-strain curves of the corresponding monotonically loaded RC columns. This observation is consistent with that reported in other efforts on unconfined, steel-confined and FRP-confined concretes (Abbasnia and Ziaadiny 2010; Wang et al. 2012; Bahn and Hsu 1998; Lam et al. 2006; Lam and Teng 2009), and it also supports the basic hypothesis of envelope curves proposed by Karsan and Jirsa (1969).

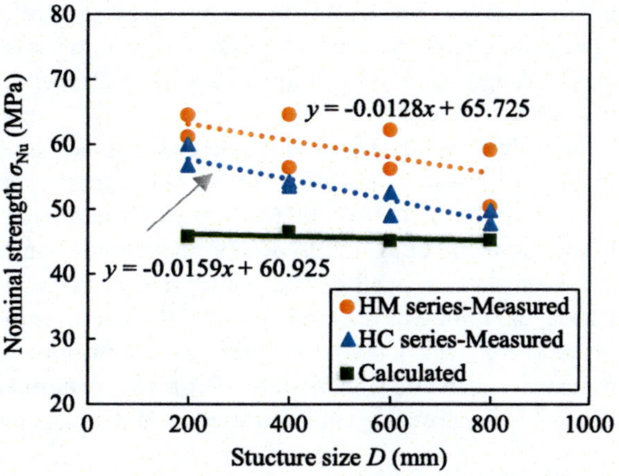

Fig. 7.14 Nominal compressive strength versus structural size of the RC columns

4. *Influence of structural size on the nominal strength*

Figure 7.14 presents the nominal compressive strength σ_{Nu} (i.e. the nominal stress at peak load) versus structural size D (i.e. cross-section dimension) for all the moderate high-strength RC columns. One can clearly see that the calculated nominal compressive strengths of geometrically similar moderate high-strength RC columns of different sizes, made of the same batch of concrete and the same ratio of longitudinal and hoop reinforcement, are sharing the same nominal strength as the codes (Ministry of Construction of the People's Republic of China 2011) predict.

One can see from Fig. 7.14 that, for all the measured nominal compressive strengths of the moderate high-strength RC columns whether under monotonic or under cyclic axial compression, the nominal compressive strength σ_{Nu} decreases with the increase of structural size D. These diagrams significantly prove the existence of size effect in larger-sized RC columns. Additionally, one also can notice from Figs. 7.13 and 7.14 that, the nominal compressive strengths of the RC columns obtained under repeated cyclic compressive loading are slightly lower than the ones under monotonic compressive loading. This phenomenon can be attributed to the low-cycle fatigue behavior (Verderame et al. 2008; Kim and Kim 1996). Moreover, it is to be noted that, compared with the HM series RC columns, the HC series RC columns have a smaller discreteness degree on the nominal compressive strength σ_{Nu}. This may be caused by the fact that, under repeated cyclic loadings, the damage within concrete should be much more concentrated and the energy dissipates much more uniform. In addition, it can be noted from Fig. 7.14 that, the decline degree (i.e. the slope of the fitting curve) of the nominal compressive strength of the RC columns under axial monotonic loading is closely similar with that of columns under cyclic loading, indicating that the detailed laws of size effect should be similar. However, the declining slope of the nominal strengths of the cyclic-loaded RC columns having different structural sizes is slightly larger than that of monotonically loaded RC columns. That is because the low cyclic fatigue loading could make the RC members much more brittle by separating the bond between steel bars and the surrounding concrete under repeated cyclic compression. Certainly, under repeated cyclic compression, the failure of concrete materials and reinforcing bars should also be brittle.

In reality, the size effect of RC members is caused and affected by (1) the inherent features of concrete and steel, and (2) the complex mutual effect between concrete and steel rebar (Li et al. 2016). Actually, the fatigue (or cyclic) loading, compare to monotonic loading, could deteriorate the mechanical properties of both concrete and steel bars, as well as the bond between these two materials. At this circumstance, the failure of the cyclic-loaded RC columns is mainly caused by the fracture of concrete, which definitely presents a more obvious size effect. From fracture mechanics viewpoint, the failure of concrete material is more brittle than that of RC structures considering of the attribution in the plasticity of steel bars.

5. *Hysteretic behavior*

Herein the section, to evaluate the ductility of the moderate high-strength RC columns in terms of displacement, the definition method (Lu et al. 1999) shown in Fig. 7.15 was selected. The displacement at yield δ_y is defined as the point of intersection of a straight line from the origin through the point representing 75% of the peak load with the horizontal line at peak load. The displacement at failure is determined as the corresponding displacement value $\delta_{0.85}$ of the point of intersection of a horizontal line at 85% of the peak load with the descending branch of the skeleton curve. The displacement ductility factor is then defined as $\mu_\delta = \delta_{0.85}/\delta_y$.

The obtained displacement ductility coefficient μ_δ of the 16 RC columns are listed in Table 7.3. Figure 7.16 plots the relationship between the displacement ductility and the structural size of the moderate high-strength RC columns. However, it cannot

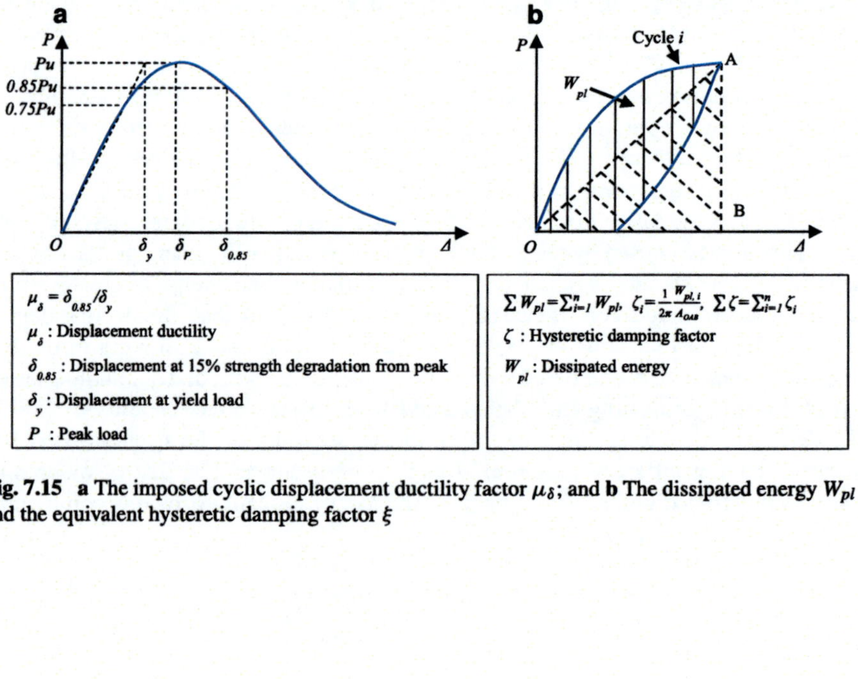

Fig. 7.15 **a** The imposed cyclic displacement ductility factor μ_δ; and **b** The dissipated energy W_{pl} and the equivalent hysteretic damping factor ξ

Table 7.3 Measured displacement response and ductility factors

Dimensions (mm)	Specimen label	δ_y^a (mm)	$\delta_{0.85}^b$ (mm)	μ_δ^c
G (800 × 800 × 2400)	HMG-1	4.89	7.55	1.54
	HMG-2	4.26	9.13	2.14
	HCG-1	2.99	7.18	2.40
	HCG-2	7.13	12.53	1.76
L (600 × 600 × 1800)	HML-1	3.40	6.03	1.77
	HML-2	3.53	5.39	1.53
	HCL-1	3.31	8.16	2.47
	HCL-2	4.27	7.55	1.77
M (400 × 400 × 1200)	HMM-1	2.87	6.26	2.18
	HMM-2	2.90	4.73	1.63
	HCM-1	1.50	5.16	3.44
	HCM-2	2.67	6.21	2.33
S (200 × 200 × 600)	HMS-1	1.46	2.21	1.51
	HMS-2	1.97	2.65	1.35
	HCS-1	1.29	2.07	1.60
	HCS-2	1.14	1.93	1.69

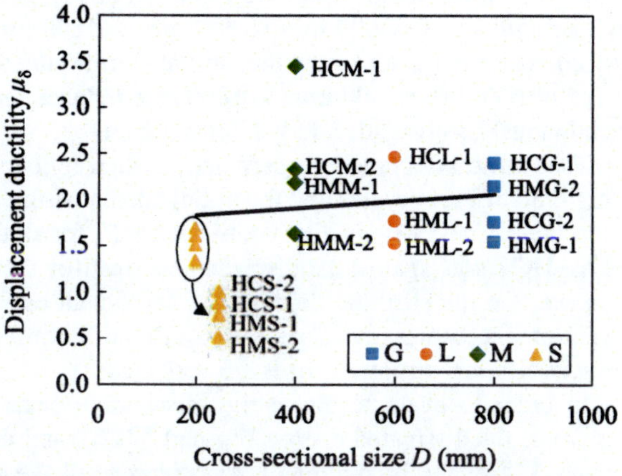

Fig. 7.16 Displacement ductility of the high-strength RC columns with different structural sizes

be found that there are any trends between structural size and ductility of the high-strength RC columns.

The strength degradation regulation of the test specimens can be evaluated by the strength degradation coefficient (defined by p/p_u) at the same loads. Herein, p and p_u are respectively the maximum load at the corresponding displacement ductility factor and the maximum load in the whole test process. Figure 7.17a shows the strength degradation coefficients p/p_u calculated from the repeated cyclic axial compressive

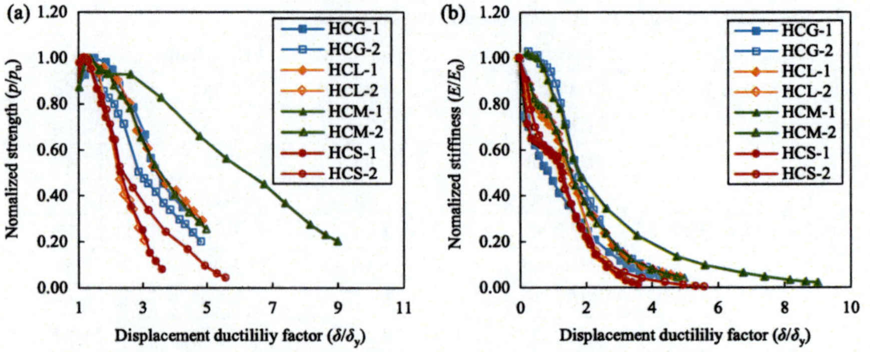

Fig. 7.17 Strength and stiffness degradation curves: **a** p/p_u-δ/δ_y; and **b** E/E_0-δ/δ_y

tests of HC series versus the displacement ductility factor δ/δ_y. It can be concluded that the strength degradation is closely associated with damage state, namely, the initiation of crack, the spalling of concrete cover and the buckling of steel rebar. In all of the tests, at the first two damage states, the strength degradation is limited to less than 20% until the relative displacement $\delta/\delta_y = 2$ is reached. After this, the strength degradation worsens, which should be associated with the damage of concrete and the bond failure between longitudinal steel rebar and the surrounding concrete. Furthermore, it can be found from Fig. 7.17a that, the strengths of the RC column sized of 600 mm × 600 mm × 1800 mm degrades the slowest. This phenomenon indicates that at the ultimate limit state, the strength of the RC column sized of 600 mm × 600 mm × 1800 mm reduces rarely and the RC column still continuously withstands a larger deformation.

The stiffness degradation coefficient is often defined as the ratio of the secant stiffness E at different cycles and the initial tangent stiffness E_0, i.e. E/E_0. Figure 7.17b plots the stiffness degradation coefficient E/E_0 for all the high-strength RC columns of the HC series against the displacement ductility factor δ/δ_y. From Fig. 7.17b one can see that the stiffness degradation coefficient evolves similarly under repeated cyclic axial compression. That is to say, the stiffness degradations of the RC columns having different structural sizes are very close.

In order to evaluate the energy dissipation capacity of the high-strength RC columns, the dissipated energy W_{pl} and $\sum W_{\text{pl}}$, and the hysteretic damping factors ζ and $\sum\zeta$ of the high-strength RC columns of the HC series at various loading cycles "n" are presented in Figs. 7.18 and 7.19, respectively. Herein, the parameters of W_{pl}, $\sum W_{\text{pl}}$, ζ and $\sum\zeta$ are defined in Fig. 7.15b, which is the same. W_{pl} is the corresponding area of each cycle and $\sum W_{\text{pl}}$ is the cumulative dissipated energy of all the cycles. Hysteretic damping factor ζ is defined as:

$$\zeta_i = \frac{1}{2\pi}\frac{W_{pl,i}}{A_{OAB}} \tag{7.1}$$

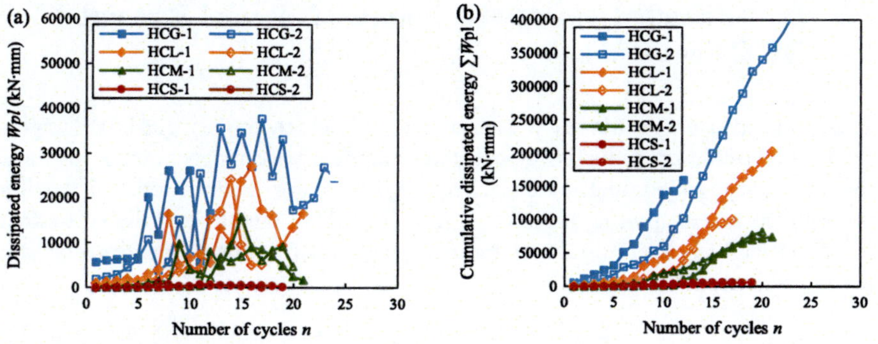

Fig. 7.18 **a** The dissipated energy $W_{\rm pl}$; and **b** The cumulative dissipated energy $\sum W_{\rm pl}$

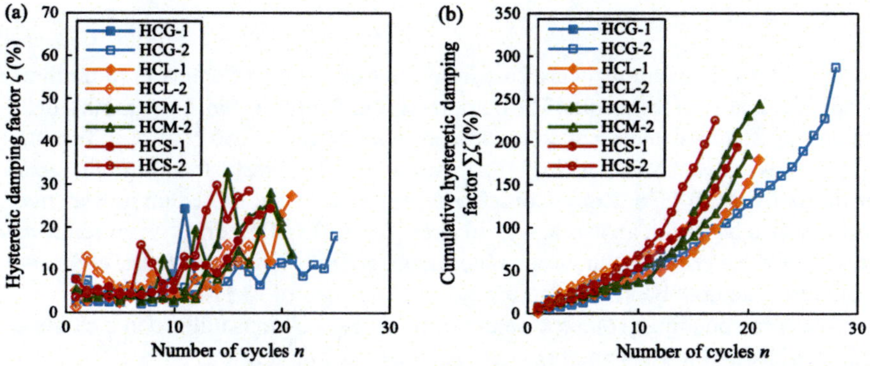

Fig. 7.19 **a** The hysteretic damping factor ζ; and **b** The cumulative hysteretic damping factor $\sum\zeta$

where $A_{\rm OAB}$ is the area of the rectangular triangle sharing the same peak of the corresponding cycle i, and $\sum\zeta$ is the cumulative hysteretic damping factor.

Figures 7.18a and 7.19a respectively illustrate the dissipated energy and the corresponding damping factor of the cyclic loaded RC columns during the generic ith cycle. One can note from Figs. 7.18a and 7.19a that, for the replicates of each size, their energy dissipation capacity is in good consistency. In addition, it can be found from Fig. 7.18a that, the RC columns having larger structural size dissipates more energy at each cycle. However, the hysteretic damping characteristic of the RC columns at different cycles presents different trends, as can be seen in Fig. 7.19a. Figures 7.18b and 7.19b present the cumulative dissipated energy $\sum W_{\rm pl}$ and the cumulative hysteretic damping factor $\sum\zeta$. It can be seen from Fig. 7.18b that, the larger the structural size of the RC column is, the more the energy dissipates. However, there is not any trends between the cumulative hysteretic damping factor and the structural size, as presented in Fig. 7.19b.

7.3 Eccentrically Compressive Failure of Normal-Strength RC Column

The eccentric loading could result from accidental load eccentricity, and initial imperfection of the column. Actually, almost all the columns work under eccentric loadings involving the so called small-eccentric and large-eccentric compression. This section presents the research work of Jin et al. (2017b), which investigated the size effect in eccentrically loaded normal-strength RC columns from both experimental and numerical perspectives.

7.3.1 Experimental Investigations

1. *Experimental program*

A total of 30 RC columns were cast from the same batch of concrete to minimize statistical scatter of the results. The compressive and splitting tensile strengths, determined by standard plain concrete cube samples sized of 150 mm × 150 mm × 150 mm, were 30.7 MPa and 2.4 MPa, respectively. The elastic modulus of concrete measured was 30 GPa. Longitudinal reinforcement in the RC columns was made from ribbed bars with an average yield strength of 335 MPa. Transversal reinforcement, only designed for suppressing local damage from happening, in the corbel was made from smooth bars with an average yield strength of 235 MPa.

The corresponding geometric details of the 30 RC columns utilized in the tests are summarized in Table 7.4 and Fig. 7.20. All geometric dimensions of the RC columns including the reinforcement details are scaled to obtain geometrically proportional specimens with the ratio of 1:2:4.

A schematic view of the loading arrangement utilized in the test is illustrated in Fig. 7.21. The RC columns were simply supported at both ends with different eccentricities of e_0, and the load P was applied axially. The top and bottom ends of the RC columns were located against steel rods. Transversal deflections (Δ) at mid-length of the RC columns were measured by a dial gauge with a 50 mm measuring range.

The loading was controlled based on a mixed constant increment of averaged longitudinal forces and deformations. It was controlled by constant increment of averaged longitudinal forces before the peak load which is for the purpose of efficiency. In order to get the post-peak softening curve of the load-deflection response during the test, it was switched to constant increment control of averaged longitudinal deformations after the peak load.

2. *Failure patterns*

Figure 7.22 present the typical failure patterns of the small-sized, medium-sized and large-sized RC columns, respectively. It can be found that all the RC columns break

Table 7.4 Geometric details of the tested RC columns

Series	Specimen /Amount	Cross-section (mm × mm)	Length (mm)	Longitudinal reinforcement	Transverse reinforcement in corbel	Concrete cover (mm)	e_0 (mm)
EA ($e_0 = 0.1h_0$)	EAL/2	800 × 800	3600	8B32	A10@108/130	40	74
	EAM/2	400 × 400	1800	8B16	A8@70/80	20	37
	EAS/2	200 × 200	900	8B8	A6@50	10	19
EB ($e_0 = 0.25h_0$)	EBL/2	800 × 800	3600	8B32	A10@108/130	40	185
	EBM/2	400 × 400	1800	8B16	A8@70/80	20	93
	EBS/2	200 × 200	900	8B8	A6@50	10	46
EC ($e_0 = 0.4h_0$)	ECL/2	800 × 800	3600	8B32	A10@108/130	40	296
	ECM/2	400 × 400	1800	8B16	A8@70/80	20	148
	ECS/2	200 × 200	900	8B8	A6@50	10	74
ED ($e_0 = 0.6h_0$)	EDL/2	800 × 800	3600	8B32	A10@108/130	40	444
	EDM/2	400 × 400	1800	8B16	A8@70/80	20	222
	EDS/2	200 × 200	900	8B8	A6@50	10	111
EE ($e_0 = 0.9h_0$)	EEL/2	800 × 800	3600	8B32	A10@108/130	40	466
	EEM/2	400 × 400	1800	8B16	A8@70/80	20	233
	EES/2	200 × 200	900	8B8	A6@50	10	167

Note EA, EB, EC, ED and EE series, correspond to eccentricity of $e_0 = 0.1h_0, 0.25h_0, 0.4h_0, 0.6h_0$ and $0.9h_0$, respectively; S, M and L means the cross-sectional sizes of 200 mm × 200 mm, 400 mm × 400 mm, and 800 mm × 800 mm, respectively; e_0 is eccentricity; and A, B present the diamater of transverse reinforcement and longitudinal reinforcement, respectively

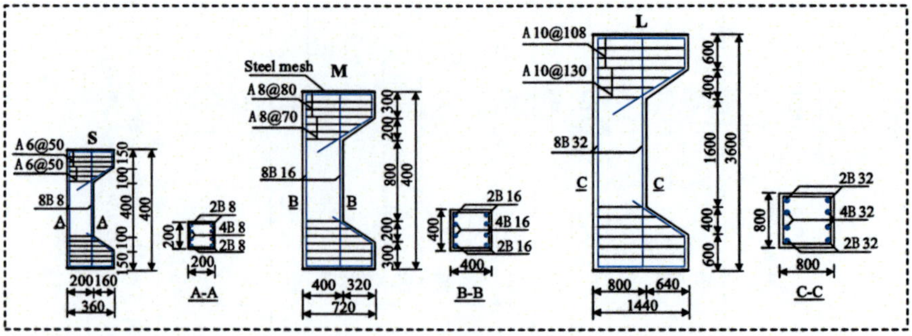

Fig. 7.20 Test RC columns with different dimensions

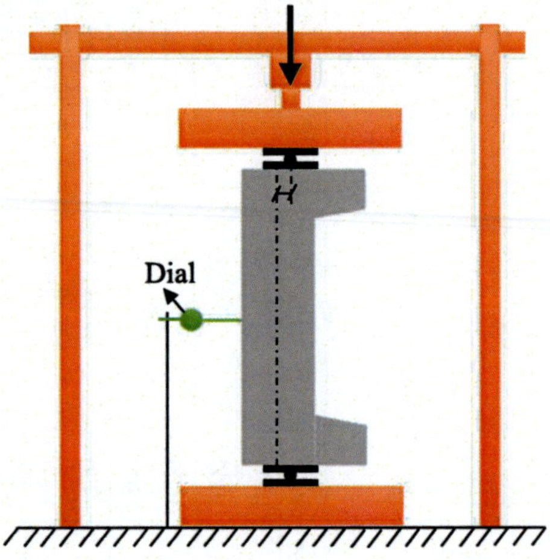

Fig. 7.21 Loading arrangement

right at the mid-length of the specimens, as might be expected from the buckling analysis (Kim and Yang 1995; Guo and Bažant 2006). From the fracture mechanics viewpoint, the failure ought to be caused by the release of stored strain energy engendered by the compression fracture, which leads to unloading due to axial extension of the elastic part of the specimen (Bažant and Kwon 1994; Bažant and Xiang 1997). One can clearly note from Fig. 7.22 that the failure patterns for the eccentrically loaded RC columns with different eccentricities are distinct. Under small(er) eccentric compression, e.g. for the ones with the eccentricities of $0.1h_0$ and $0.25h_0$, the crushing of concrete occurred at the compressive sides of the columns for the three different structural sizes. In addition, almost all the longitudinal steel rebar buckled at the compressive sides. While, under large(r) eccentric compression, e.g. for the ones with the eccentricities of $0.6h_0$ and $0.9h_0$, the flexural tensile fracture happened. For the RC columns under the eccentricity of $0.4h_0$, both the crushing of concrete at the compressive side and the flexural tensile cracking at the tensile side contribute to

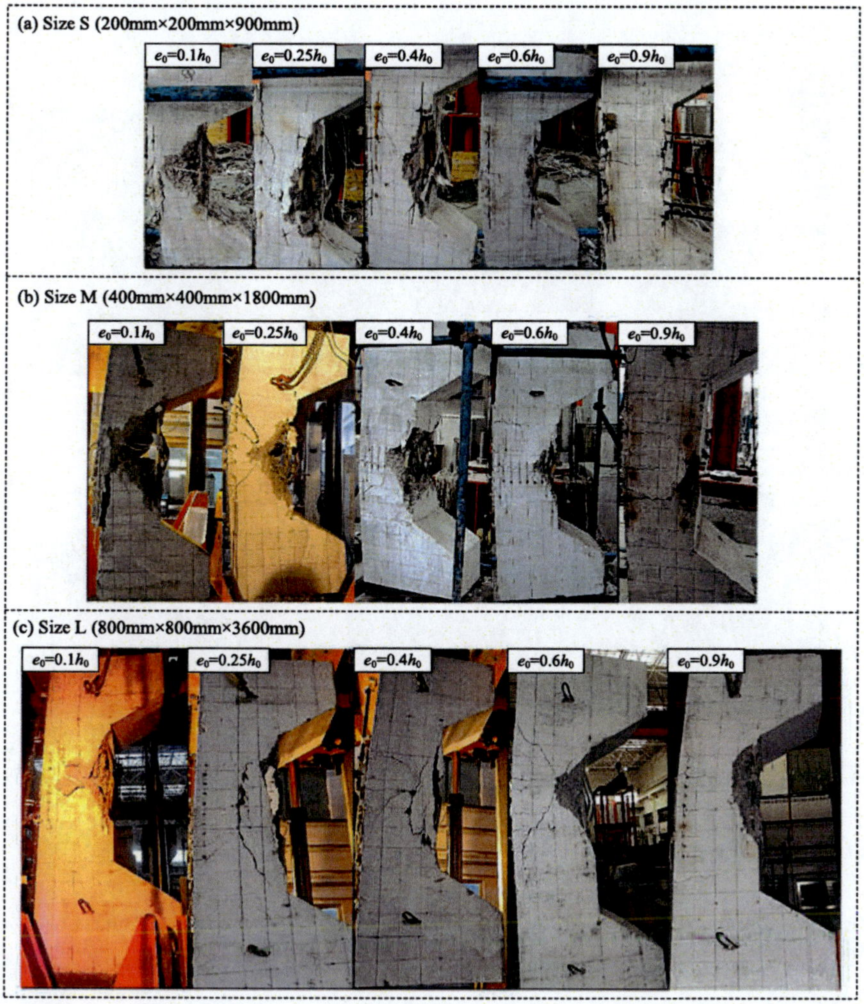

Fig. 7.22 Global failure patterns of the eccentrically loaded RC columns: **a** Size S; **b** Size M; and **c** Size L

the final failure. This should be a brittle-ductile transition behavior, which is similar with that in Porras et al. (2016).

Figure 7.23 presents the local failure patterns of the eccentrically loaded RC columns under the two typical eccentricities of $0.25h_0$ and $0.6h_0$ due to different fracture mechanisms. Brittle failure pattern generates as the RC column is slightly eccentrically loaded. Taking the RC column named of "EBM" as an example, the fracture of specimen is caused by a diagonal shear crack, which is mainly due to the brittle behavior of concrete material as shown in Fig. 7.23a. When the RC column is loaded under a larger eccentricity, taking the RC column named of "EDM" as

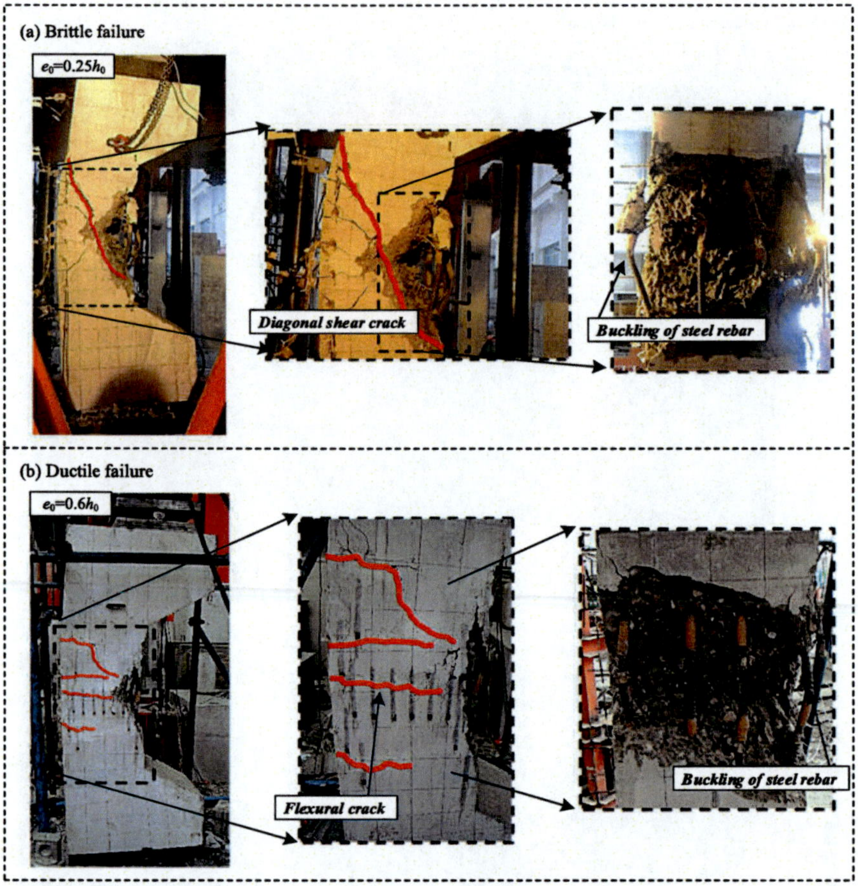

Fig. 7.23 Local failure patterns of the eccentrically loaded RC columns due to different fracture mechanisms

an example, it exhibited a less brittle failure mode, as can be observed from the numerous flexural cracks at the tensile side in Fig. 7.23b. Due to the existence of longitudinal reinforcement within concrete, the failure is closer to ductile fracture. Nevertheless, the buckling of longitudinal steel rebar can also be observed after the spalling the concrete cover at the compressive side.

Figure 7.24 shows a summarized schematic diagram, based on the test observations and the foregoing photographs in Figs. 7.22 and 7.23, describing the influence of eccentricity on the failure pattern of the eccentrically loaded RC columns. Figure 7.24a shows the variation of failure patterns with different eccentricities. It can be clearly seen that the failure pattern presents a brittle-ductile transition behavior as the eccentricity arises (Porras et al. 2016). For the RC columns with smaller eccentricities ($e_0 = 0.1h_0$ and $0.25h_0$), all the specimens exhibited a brittle failure pattern

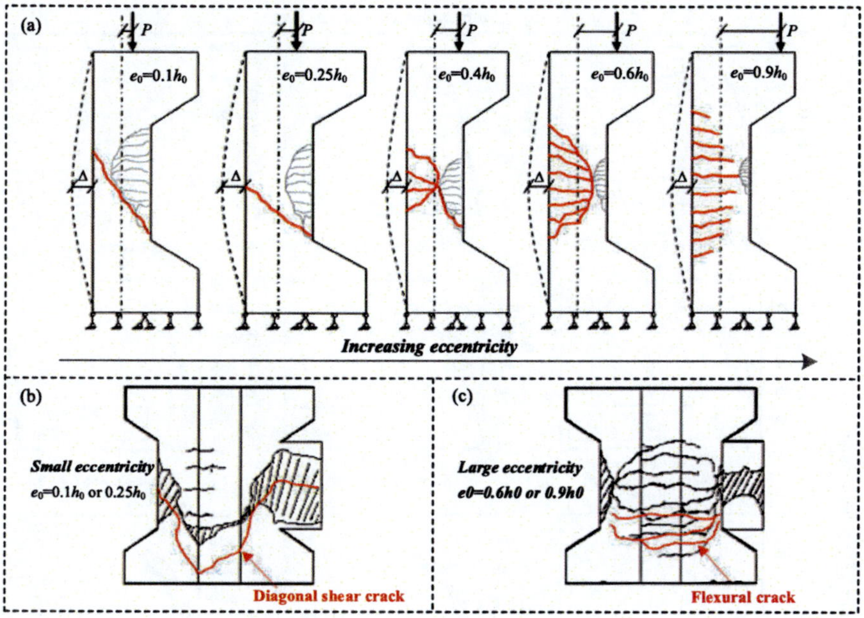

Fig. 7.24 Schematic diagram of failure patterns: **a** RC columns with different eccentricities; **b** An RC column under small eccentricity; and **c** An RC column under large eccentricity

with a diagonal shear crack, as can be seen from Fig. 7.24b. Nevertheless, for the RC columns with larger eccentricities ($e_0 = 0.6h_0$ and $0.9h_0$), the specimens exhibited a ductile failure pattern with flexural-tensile cracks, as shown in Fig. 7.24c. It can be regarded as a transitional eccentricity of $e_0 = 0.4h_0$, since the fracture of the RC column pronounces both the brittle and ductile failure patterns.

3. *Load-deflection response*

Figure 7.25 shows the entire collection of applied load (P) versus lateral deflection (Δ) curves obtained from all tested RC columns. For each series as the structural size increases, the load-deflection curves evolve in a similar way for sharing similar initial tangent modulus and similar post-peak softening branches. Besides, as the eccentricity arises, the load capacity is getting weaker. That is to say, the ultimate load decreases obviously with increasing eccentricity. However, the ductility capacity improves for the lateral deflection is getting larger and larger, as can be observed from the post-peak softening curves. This is definitely due to the brittle-ductile transition behavior of the failure pattern or fracture mechanism (Porras et al. 2016).

For the eccentrically loaded RC columns with smaller eccentricities (i.e. $e_0 = 0.1h_0$ or $0.25h_0$), they present larger load capacities for the using of compressive strength of concrete, but lack ductility with short post-peak softening branches, which definitely pronounces a brittle failure pattern and a remarkable size effect. For the eccentrically loaded RC columns with larger eccentricities (i.e. $e_0 = 0.6h_0$ or $0.9h_0$),

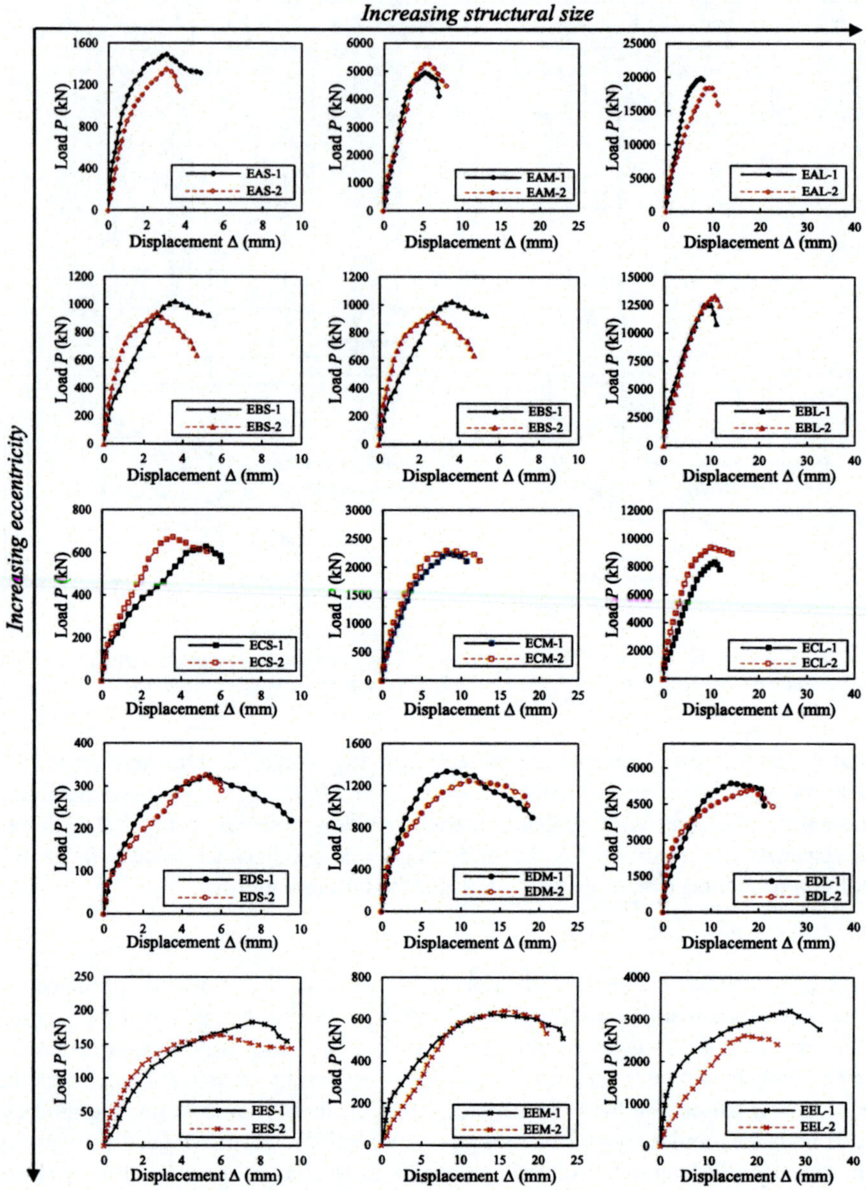

Fig. 7.25 Load-deflection curve for all RC columns with different structural sizes eccentricities

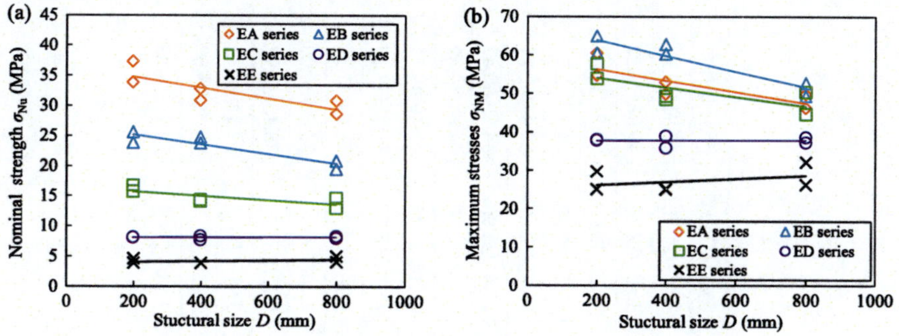

Fig. 7.26 Variation of: **a** Nominal strengths σ_{Nu}; and **b** Maximum stresses σ_{NM}

they clearly exhibit a ductile failure pattern. In Fig. 7.25, the load-deflection curves of the RC columns with larger eccentricities show lower load capacities and longer yielding plateaus. The load capacity of these RC columns is mostly provided by the tensile strength of concrete and steel rebar. In this case, the fracture of the specimens exhibits a ductile failure pattern and this definitely pronounces no size effect or a very weak size effect. For the RC columns with the eccentricity of $e_0 = 0.4h_0$, the failure patterns are sort of an intermediate state between the aforementioned two states, corresponding to the typical photographs of failure pattern in Figs. 7.22, 7.23 and 7.24.

4. *Size effect analysis based on fracture mechanics theory*

Figure 7.26 describes the variation of nominal compressive strengths σ_{Nu} and maximum stresses σ_{NM} with structural size D of all the data points obtained from the present test. The nominal compressive strength is defined as $\sigma_{Nu} = P_u/D^2$, where P_u is the peak load and D is the structural size represented by the cross-sectional width.

When considering the column of sinusoidal initial curvature with initial load eccentricity at mid-span, from the theory of bending, the maximum stress in the column should have the form as

$$\sigma_{NM} = \sigma_{Nu} + \frac{6Pz}{D^3} \tag{7.2}$$

in which, $z = \frac{e_0 D}{1 - P_u/P_{cr}}$, $P_{cr} = \frac{\pi^2 E}{12\lambda^2}$, E is the elastic modulus of concrete and P_{cr} is the Euler load, this yields

$$\sigma_{NM} = \sigma_{Nu}\left(1 + \frac{6e_0}{1 - \frac{12\lambda^2\sigma_{Nu}}{\pi^2 E}}\right) \tag{7.3}$$

Table 7.5 shows all the calculated values of the nominal compressive strengths and the maximum stresses based on the test results. Figure 7.26a presents the plot

Table 7.5 Calculated values of nominal compressive strengths and maximum stresses

Specimen/Amount	Cross-section size (mm)	Length (mm)	Eccentricity ratio e_0/h_0	P (kN)	σ_{Nu} (MPa)	σ_{NM} (MPa)
EAL/2	800 × 800	3600	0.1	19,770/18,366	30.89/28.70	49.91/46.33
EAM/2	400 × 400	1800	0.1	4950/5257	30.94/32.86	49.98/53.12
EAS/2	200 × 200	900	0.1	1495/1356	37.38/33.90	60.51/54.82
EBL/2	800 × 800	3600	0.25	12,500/13,360	19.53/20.88	49.31/52.73
EBM/2	400 × 400	1800	0.25	3816/3970	23.85/24.81	60.34/62.81
EBS/2	200 × 200	900	0.25	1026/956	25.65/23.90	64.95/60.47
ECL/2	800 × 800	3600	0.4	8340/9410	13.03/14.70	44.64/50.42
ECM/2	400 × 400	1800	0.4	2263/2301	14.14/14.38	48.49/49.31
ECS/2	200 × 200	900	0.4	629/673	15.73/16.83	53.96/57.77
EDL/2	800 × 800	3600	0.6	5350/5170	8.36/8.08	38.66/37.35
EDM/2	400 × 400	1800	0.6	1350/1246	8.44/7.79	39.02/36.00
EDS/2	200 × 200	900	0.6	330/328	8.25/8.20	38.15/37.92
EEL/2	800 × 800	3600	0.9	3207/2610	5.01/4.08	32.18/26.17
EEM/2	400 × 400	1800	0.9	621/625	3.88/3.91	24.91/25.07
EES/2	200 × 200	900	0.9	185/156	4.63/3.90	29.70/25.03

of nominal compressive strength σ_{Nu} versus structural size D. From Fig. 7.26a, one can clearly see that the measured nominal compressive strengths decrease with the structural size in EA and EB series, which significantly proves the existence of size effect in RC columns with a small eccentricity. This must be caused by the brittle failure behavior as the micro-mechanism of compression fracture inside the concrete material create a laterally propagating band of axial splitting cracks. The band propagates in a direction either inclined or orthogonal to the direction of the compressive stress of the largest magnitude, which leads to a global compression fracture (Bažant and Xiang 1997). However, the trend lines of ED and EE series are nearly parallel to the structural size axis, which clearly indicates no size effect. The possible reason is that, as the eccentricity arises, the RC columns present a flexural failure pattern. The nominal flexural strength is mainly provided by the longitudinal steel rebar, and steel apparently pronounces no size effect by its inherent plastic features. One also can conclude from Fig. 7.26a that EC series is some sort of an intermediate transition state between brittle failure and ductile failure, because the trend line only exhibits a size effect slightly.

Figure 7.26b shows a plot of maximum stresses σ_{NM} versus structural size D. Note that Fig. 7.26b also presents a reduction in the maximum stresses as the structural size D increases in EA, EB and EC (slightly) series. Nevertheless, the maximum stresses in EB series for the three different structural sizes are generally larger than that in EA series, and this is possibly caused by a pre-stress-state due to the critical eccentricity of $e_0 = 0.25h_0$. When the eccentricity is too small, as $e_0 = 0.1h_0$, the whole failure process of the RC column is due to axially compression, then the maximum stress is only related to the compressive strength of concrete. However, when the eccentricity increases to a relatively larger amount, but not beyond a critical eccentricity (e.g. $e_0 = 0.25h_0$), then the eccentrically loaded RC column would be under full compression at first. The longitudinal steel rebar would be compressed and they could be regard as a pre-stressed steel rebar when the RC column changes to a bending state. Thus, the RC column with a relatively larger eccentricity presents a larger maximum stress.

The bi-logarithmic plots of $\lg(\sigma_{Nu}/Bf_t^{'})$ versus $\lg\beta$ are shown in Fig. 7.27. From Fig. 7.27, one can see that, ED and EE series present no size effect as concluded by the prior analysis. Note that for the RC columns with a smaller eccentricity, the size effect is stronger, with the response lying closer to the LEFM asymptote slope of $-1/2$. This is not surprising, since the RC columns with a smaller eccentricity present much more brittle failure patterns. The correlation coefficients ρ_{XY} was 0.86. These clearly indicate the existence of size effect in eccentrically loaded RC columns, and the theoretical work of Bažant (1984) can well describe the size effect on nominal strength of RC columns under eccentric compression.

7.3.2 Numerical Investigations

1. *Numerical method*

Figure 7.28 presents the 2D mesoscale numerical model for the RC column sized as 200 mm × 900 mm. The horizontal and vertical constraints are adopted at the

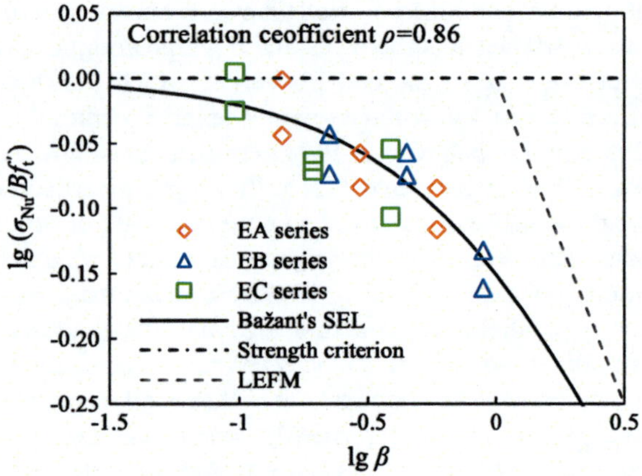

Fig. 7.27 Size effect plots of all eccentrically loaded RC columns

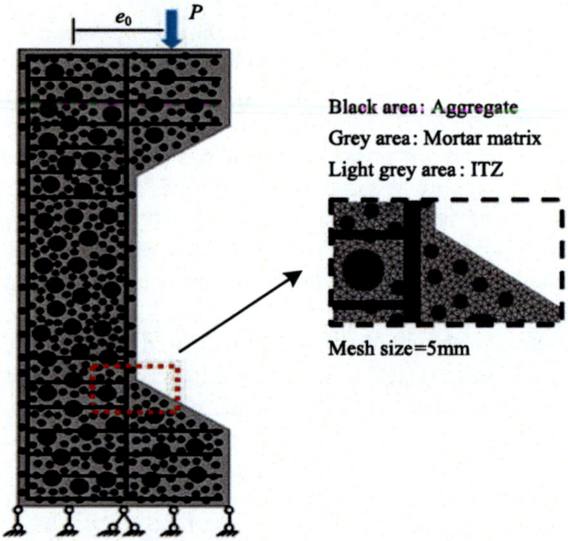

Fig. 7.28 2D mesoscale model for RC column sized of 200 mm × 900 mm

middle-bottom of the numerical model, and vertical constraints are conducted in other places of the numerical model. The left and right sides of the column are free. The prescribed displacement is applied at the top surface of the numerical model. The aggregate parameters used in the numerical models are summarized in Table 7.6.

The mechanical parameters of the three mesoscale components of concrete are listed in Table 7.7. For the concrete part, the corresponding meshing was conducted with triangular elements. Ideal bond between the steel reinforcement and concrete was assumed. The mesh sizes of all the RC column models are set to be 5 mm.

Table 7.6 Aggregate parameters used in the numerical models

Size (mm × mm)	150 × 150	200 × 900	400 × 1800	800 × 3600
Number of small aggregates	49	394	1576	6306
Number of large aggregates	6	52	206	825

Table 7.7 Mechanical parameters of the three meso-components and steel rebar

	Aggregate particle	Mortar matrix	ITZ	Steel rebar
Young's modulus E (GPa)	300[a]	30[a]	26[b]	200
Compressive strength σ_c (MPa)	80[a]	30[a]	22[b]	–
Tensile strength σ_t (MPa)	8.3[a]	3.1[a]	2.2[b]	–
Poisson ratio ν	0.2[a]	0.2[a]	0.2[b]	0.3
Dilatant angle ψ (°)	–	18	15	–
Eccentricity η (%)	–	0.1	0.1	–
Stress ratio f_{b0}/f_{c0}	–	1.16	1.16	–
Yield strength f_y (MPa)	–	–	–	335/235
Reinforcement ratio ρ (%)	–	–	–	1.1
Cross-sectional area A (mm^2)	–	–	–	1.0048

Note [a]Data are obtained from the tests; and [b]Data are given with reasonable assumption

2. *Discussion on 2D and 3D size effects*

It has to be noted that, differences between 2D and 3D size effects should be of particular concern for discussion. Accordingly, the nominal strength is usually defined as

$$\sigma_{Nu} = c_N \frac{P_u}{bD} \tag{7.4}$$

in which, σ_{Nu} is independent of the structural size D which exhibit no size effect. P_u is the peak load, b is the width of the structure or specimen and c_N is a dimensionless constant chosen for convenience. A constant width b makes the specimens at two-dimensional scaling, while a proportional increased width b makes the specimens at three-dimensional scaling, either in experimental models or in numerical models. In 3D models of concrete components, there is a layer, near the side walls, of constant thickness in which the crack must be nonplanar and the crack front curved (which is called the shear-lip phenomenon) (Hoover and Bažant 2014). It would bring about additional problems, because different portions of the specimen thickness b would be occupied by the surface layer as b varies. This could cause the average fracture energy varies with width b and create additional scatter in the size effect trends, although it only slightly affects the results. Therefore, 2D models of concrete components are suitable and reasonable whether in experimental tests or in numerical simulations.

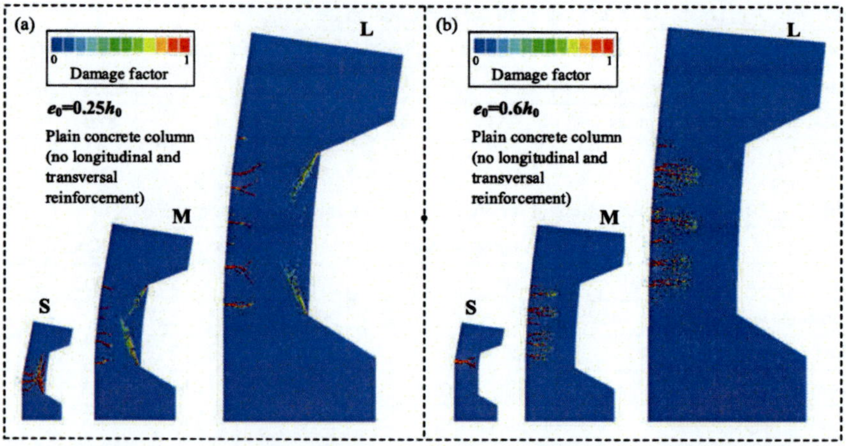

Fig. 7.29 Numerical results of the eccentrically loaded plain concrete columns: **a** Small eccentricity of $e_0 = 0.25h_0$; and **b** Large eccentricity of $e_0 = 0.6h_0$

3. *Simulation results and discussion*

Figure 7.29 shows the numerical results of the eccentrically loaded plain concrete columns with small and large eccentricities of $e_0 = 0.25h_0$ and $e_0 = 0.6h_0$, respectively. One can clearly see that the numerical outcomes are showing a strong brittle failure mode in different dimensions (i.e. S, M and L), which can be expected for size effect analysis, in the plain concrete columns with both the eccentricities, although the failure mechanisms are different.

Figure 7.30 shows the numerical results of the eccentrically loaded RC columns with different sizes. Figure 7.30a presents the final failure pattern of the eccentrically loaded RC column with a small eccentricity of $e_0 = 0.25h_0$. According to the test observations in Sect. 7.3.1, this kind of RC columns exhibit a brittle failure with a diagonal shear crack and pronounce a strong size effect. The numerical outcomes have shown a great accordance with the experimental results as the micro-mechanism of compression fracture inside the concrete material create a laterally propagating band of axial splitting cracks. The band propagates in a direction either inclined or orthogonal to the direction of the compressive stress of the largest magnitude, which leads to a global compression fracture (Bažant and Xiang 1997). Figure 7.30b presents the final failure patterns of the eccentrically loaded RC columns with the eccentricity of $e_0 = 0.6h_0$. It is also in good agreement with the test results.

Furthermore, some additional transversal reinforcement was added inside the mid-length of the models. Stirrup confinement is a vital factor that could strongly influence the size effect behavior in RC members. Figure 7.31 shows the numerical results of the eccentrically loaded columns with both longitudinal and transversal reinforcement. It can be clearly seen from Fig. 7.31 that, the shape and size of the damage zone in concrete were changed because of the existence of the transversal reinforcement. Compare with that in Fig. 7.30a, the damage zone in Fig. 7.31a

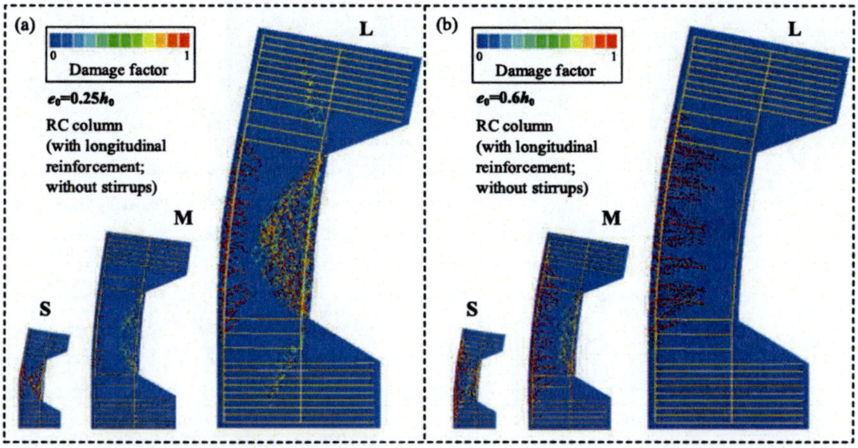

Fig. 7.30 Numerical results of the eccentrically loaded RC columns: **a** Small eccentricity of $e_0 = 0.25h_0$; and **b** Large eccentricity of $e_0 = 0.6h_0$

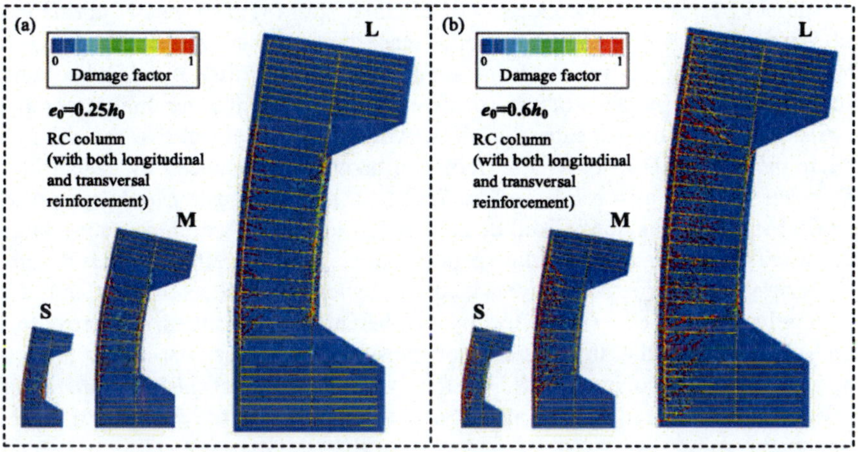

Fig. 7.31 Numerical results of the eccentrically loaded RC columns (with a stirrup ratio of 0.66% at the mid-length): **a** Small eccentricity of $e_0 = 0.25h_0$; and **b** Large eccentricity of $e_0 = 0.6h_0$

apparently reduced from a relatively larger triangular area across the longitudinal steel rebar on the compressive side to a barely seen rectangular band outside of the longitudinal compressive reinforcement. From the comparison of Figs. 7.30a and 7.31a, one can note the damage zone extended to the entire tensile side of the specimen. This is possibly caused by the existence of the stirrups, which make the distribution of stiffness more uniform and therefore no obvious weak domain can be observed in the specimen.

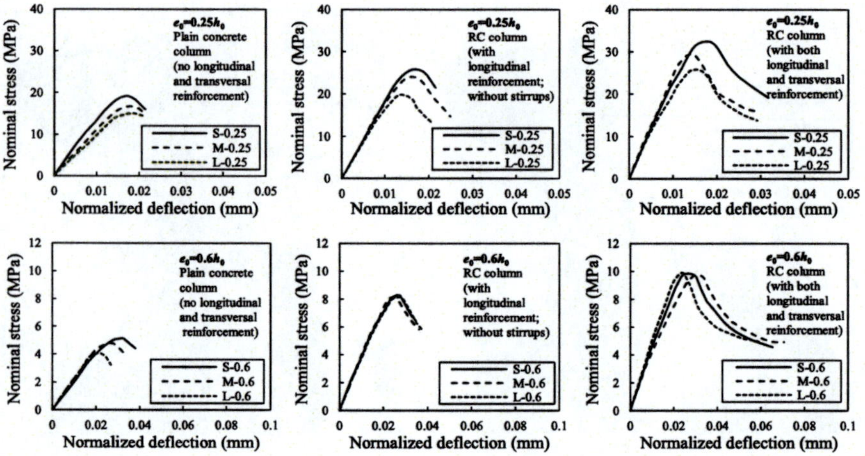

Fig. 7.32 Collection of nominal stress versus normalized deflection of all the mesoscopic numerical tests

Figure 7.32 presents a collection of nominal stress (P/D^2) versus normalized deflection (Δ/D) of all the aforementioned mesoscopic numerical tests. One can note from Fig. 7.32 that, both the nominal strength and the ductility of the RC columns increase with the increase of the stirrup ratio. However, the nominal strength is getting weaker as the eccentricity arises, which is definitely due to the different failure mechanism (Brittle-ductile transition behavior) introduced by eccentricity. Moreover, one can also see from Fig. 7.32 that, it is obvious by looking into the nominal stress versus normalized deflection curves, the RC columns with a larger eccentricity present a ductile failure pattern and yield no size effect. That is because the flexural strength is mainly provided by the longitudinal steel rebar and steel apparently pronounces no size effect by its inherent plastic features. Furthermore, it can be clearly seen that the existence of transversal reinforcement makes the load capacity of the RC columns with both the small and large eccentricities increased. However, the size effect is mitigated as the stirrups improve the ductility of the RC columns.

7.4 Eccentrically Compressive Failure of High-Strength RC Column

The research work of Xu et al. (2016) and Jin et al. (2018) are summarized in this section, in which the mechanical properties and size effect in eccentrically loaded high-strength RC columns are examined.

7.4.1 Experimental Investigations

1. *Experimental program*

A total of 24 geometrically similar RC columns were designed and tested. All geometric dimensions including reinforcement were scaled to obtain geometrically proportional specimens with the ratio of 1:2:4. The slenderness ratios λ of the RC specimens were all 4.5. The tested columns were divided into four series according to whether or not with middle stirrups and the degrees of eccentricity, i.e. HA, HB, HC and HD series, respectively. The corresponding geometric details of the high-strength RC columns tested are summarized in Table 7.8 and Fig. 7.33.

All of the specimens were cast from the same batch of continually mixed concrete. The quantities of ingredients used in the concrete mix are shown in Table 7.9. The reinforcement details are summarized in Table 7.10. The stirrup ratios at the ends of the RC columns were designed much larger than that at the mid-length to prevent local damage at this region.

The experiment of the medium- and large-sized RC columns were carried out in the multifunctional electrohydraulic servo loading test system (MESLTS). The test of the small-sized specimens with a structural size of 200 mm were carried out in another loading equipment, which was a hydraulic jack with a capacity of 3000kN. The top and bottom ends of the high-strength RC columns were located at round-edge hinged supports (See Fig. 7.34). In order to validate the hypothesis that the sections remain in plane strain, three strain gauges were set up at the middle length of the columns and spaced uniformly in Fig. 7.34. Three other strain gauges are employed to measure the strain at both the compressive and the tensile zones along the length of the columns. Five LVDTs were placed along the along the section height of column horizontally to obtain the lateral deformation.

The loading scheme was similar with the experiment of normal-strength RC column in Sect. 7.3.1, and more details were recorded in Xu et al. (2016).

2. *Failure patterns*

The failure patterns of the high-strength RC columns without stirrups at mid-length of the specimen are shown in Fig. 7.35, while that with stirrups at mid-length of the columns are presented in Fig. 7.36. In addition, the corresponding typical failure patterns and eccentricity degrees are also described in Table 7.11 (Fig. 7.37).

The RC columns (i.e. the HA series) under the large eccentricity of $0.6h_0$, with different structural sizes and without middle stirrups broke right at the mid-length of the specimen, as shown in Fig. 7.35a. The breaking of all the 6 RC columns presents a typical flexural-tensile fracture mode, which should be caused by the bending moment. The large bending moment at the mid-length promotes the start of the fracture at mid-length. As the loading continued to increase, the flexural-tensile cracks generated propagate from the outer surface to the inner section. At the end, the flexural-tensile cracks propagate throughout the entire section of the columns with several main cracks at the tensile zones and crushed concrete at the compressed side. The longitudinal steel bars in the tensile zones of the columns yielded, while that in

Table 7.8 Geometric details of the high-strength RC columns

Series	Specimen/Amount	Eccentricity	Cross section (mm × mm)	Length (mm)	Longitudinal reinforcement (%)	Middle stirrup (%)	Slender ratio	Cover thickness (mm)
HA	HAS/2	$e_0 = 0.6h_0$	200 × 200	900	1.11	0	4.5	15
	HAM/2		400 × 400	1800	1.11	0	4.5	30
	HAL/2		800 × 800	3600	1.11	0	4.5	60
HB	HBS /2	$e_0 = 0.25h_0$	200 × 200	900	1.11	0	4.5	15
	HBM/2		400 × 400	1800	1.11	0	4.5	30
	HBL /2		800 × 800	3600	1.11	0	4.5	60
HC	HCS /2	$e_0 = 0.6h_0$	200 × 200	900	1.11	0.66	4.5	15
	HCM/2		400 × 400	1800	1.11	0.66	4.5	30
	HCL/2		800 × 800	3600	1.11	0.66	4.5	60
HD	HDS/2	$e_0 = 0.25h_0$	200 × 200	900	1.11	0.66	4.5	15
	HDM/2		400 × 400	1800	1.11	0.66	4.5	30
	HDL/2		800 × 800	3600	1.11	0.66	4.5	60

Note Series HA: $e_0 = 0.60h_0$, without stirrups in the middle part of the RC columns; series HB: $e_0 = 0.25h_0$, without middle stirrups; series HC: $e_0 = 0.60h_0$, having middle stirrups; series HD: $e_0 = 0.25h_0$, having middle stirrups; and "S", "M", and "L" present the cross-sectional width of 200 mm, 400 mm, and 800 mm, respectively

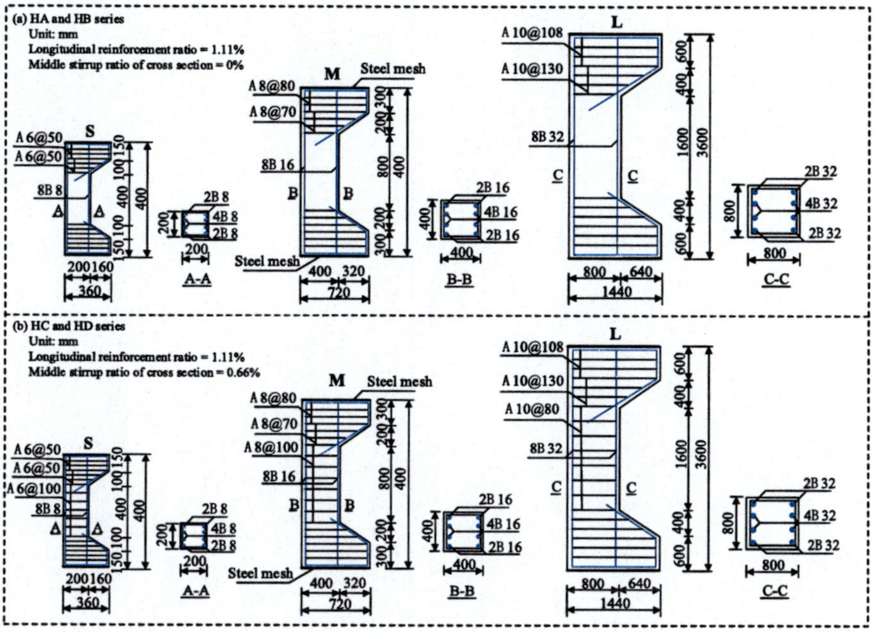

Fig. 7.33 Details of specimens for: a HA and HB series; and b HC and HD series

the compressed side of the cross section, almost all of them buckled. The concrete in the central part of the compression zones cracked and crushed out slowly. These can also be observed from Fig. 7.38a.

The RC columns (i.e. the HB series), under the small eccentricity of $0.25h_0$, with different structural sizes and without middle stirrups broke at the mid-length of the specimen as well, as shown in Fig. 7.35b. The failure patterns of all the 6 columns with different structural sizes are very similar, presenting a compression shear crushing failure mode. At beginning of the test, under the small eccentric loading with the eccentricity of $0.25h_0$, most regions in the concrete cross-section are in compressive status. As the increase of loading displacement, the concrete at the compressed side crushed and flaked, and the longitudinal steel bars without the confinement effects of concrete cover and stirrups may buckle, as can be seen in Figs. 7.35b and 7.38b. Furthermore, as the longitudinal steel rebar at the compressed side buckled, the main shear cracks propagated rapidly and other shear cracks developed indistinctively. Diagonal macro-cracks along the compressed side can be observed in the tests. The crushed areas are obviously larger than those of the columns under large eccentric compression, as can be found from the comparison of Fig. 7.35a and b. For the HB series, the failure of the columns happened in a sudden with crisp noise, and some compressive concrete blocks collapsed off when the RC columns failed, making the failure of the columns be a typical brittle failure pattern. Furthermore, it can be seen from the comparisons of the failure patterns shown in Fig. 7.35a and b that the failure

Table 7.9 High-strength concrete mixture proportions

Aggregate (kg/m^3)				Cement (kg/m^3)	Water (kg/m^3)	Sum (kg/m^3)	*w/c* ratio	Compressive strength (MPa)	Tensile strength (MPa)
0–5 mm	5–10 mm	10–20 mm	20–30 mm						
760	185	526	344	565	170	2550	0.3	62.1	6.06

Table 7.10 Details of the reinforcements used in the RC columns

Grade of steel	Diameter (mm)	Yield strength (MPa)	Ultimate strength (MPa)	Elongation (%)	Elasticity modulus (GPa)
HRB335	8	560	623	24.17	199
	16	383	565	26.00	192
	32	379	517	28.60	197
HPB235	6	269	367	12.8	208
	8	247	340	13.7	202
	10	256	398	11.5	210

Note "HRB" means "Hot-rolled Ribbed-steel Bar"; and "HPB" means "Hot-rolled Plain-steel Bar"

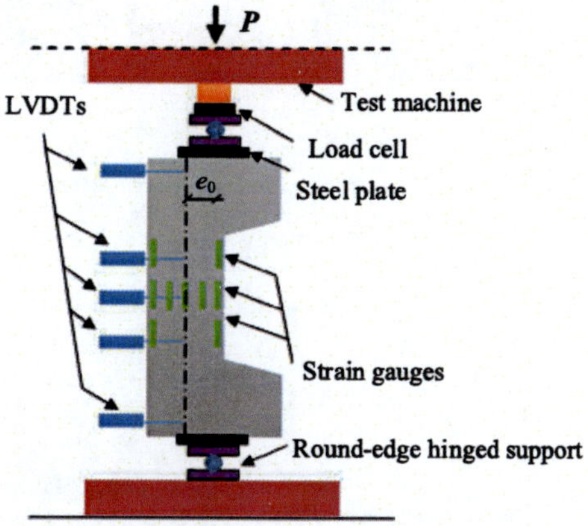

Fig. 7.34 Schematic diagram of loading arrangement

of the small eccentric compressive high-strength RC columns is much more brittle than that of columns under large eccentric compressive loading.

Similar with that in HA series, under the large eccentric loading, one can note from Fig. 7.36a that the failure of the 6 RC columns in HC series is also caused by the large bending moment. In addition, one can notice from Fig. 7.36a that the final failure patterns of the columns with different structural sizes are almost exactly the same. Similar with those without middle stirrups (i.e. the HA series), the flexural-tensile cracks propagated through the whole section of the columns having middle stirrups. In general, the break of the RC columns is still a typical flexural-tensile fracture. Nevertheless, because of the confinement of the transversal reinforcement, the fracture of the RC columns of HC series with the structural sizes of 800 mm × 800 mm is still brittle for the high brittleness number of the high-strength concretes, and the corresponding fracture happened suddenly. Comparing Figs. 7.35a and 7.36a, one can note that, for the RC columns with middle stirrups, the existence of the middle stirrups does not change the failure patterns significantly, and the ductility of the columns having middle stirrups improves slightly.

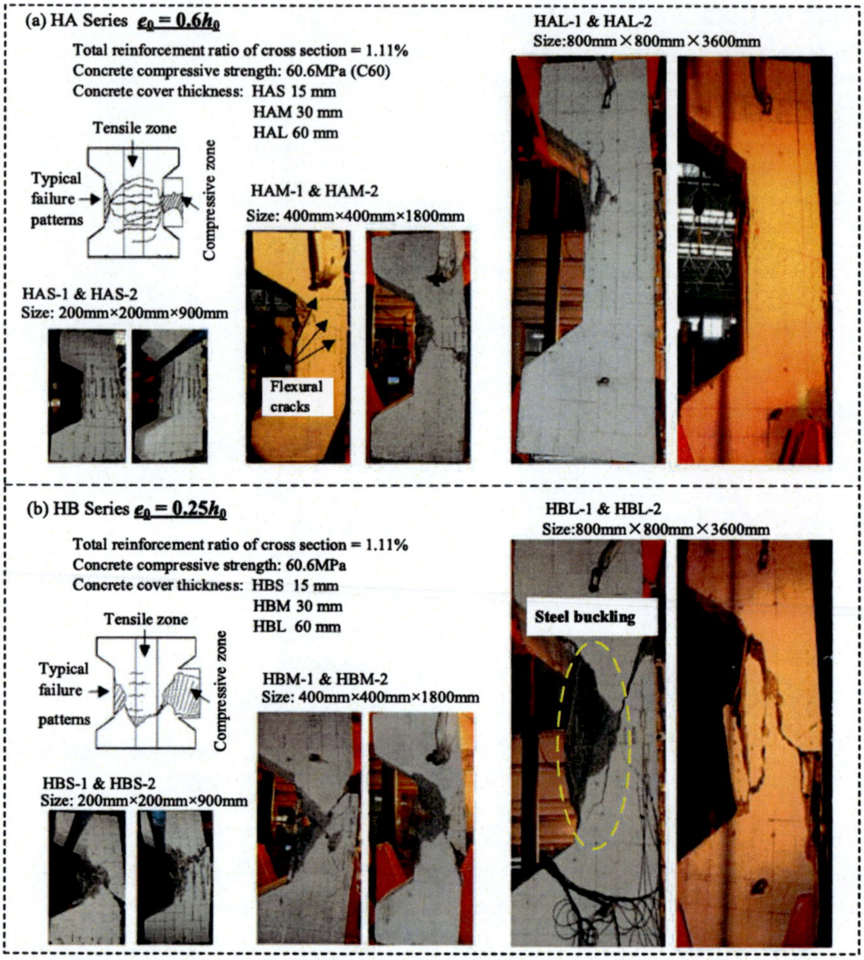

Fig. 7.35 Typical failure patterns for the high-strength RC columns without middle stirrups: **a** Flexural-tensile fracture for HA series; and **b** Compressive-shear fracture for HB series

For the HD series with stirrups at the mid-length and eccentricity of $0.25h_0$, the failure patterns are significantly different with those of the columns of HB series without middle stirrups, as can be found from the comparison of the failure photographs in Figs. 7.35b and 7.36b. The failure of the RC columns with middle stirrups, especially for the larger-sized specimens, is dominated by the compression crushing of concrete at the compressed side and the shear fracture at the tensile zones. It can be seen from Fig. 7.36b that the failure localized at the middle part of the column where a wedge-shaped pattern developed. Under the small eccentric compression loading, shear cracks generated in the columns of the HD series.

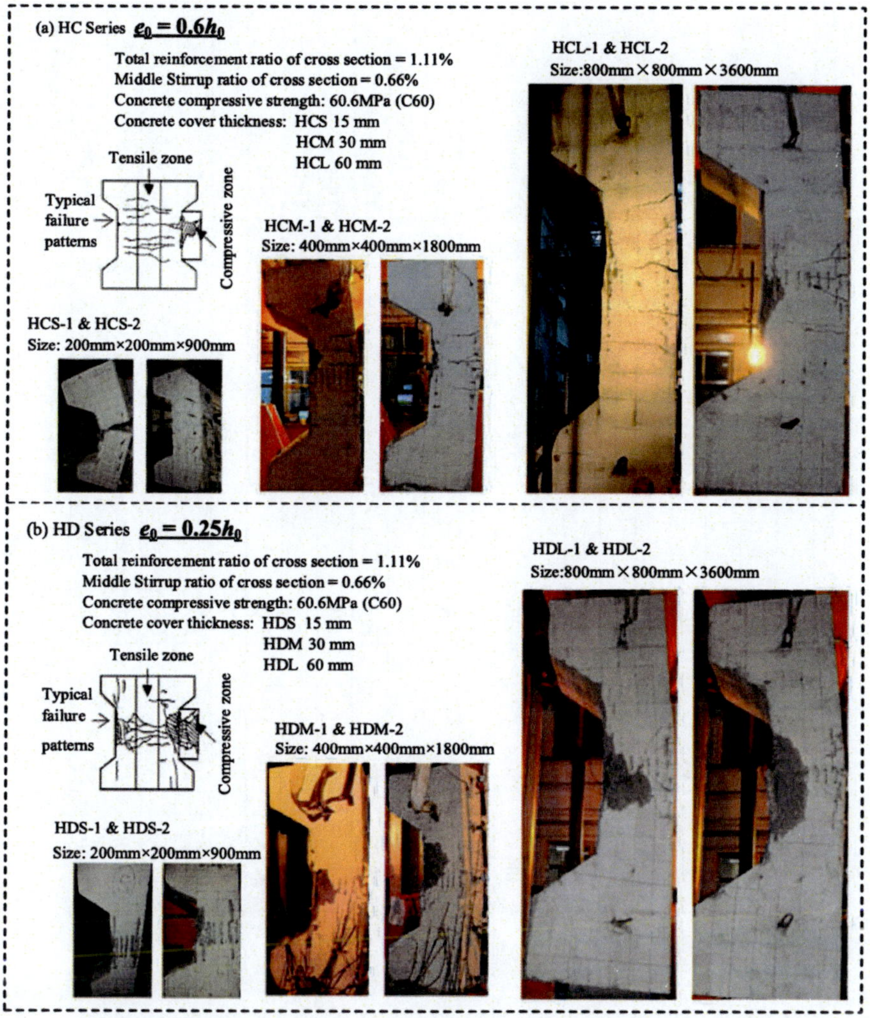

Fig. 7.36 Typical failure patterns for the high-strength RC columns having middle stirrups: **a** Flexural-tensile fracture for HC series; and **b** Compressive-shear fracture for HD series

However, they did not form a main crack. This is due to the fact that, the confinement contributed by the transversal reinforcement limits the evolution of the cracks, which makes the inner concrete be under a tri-axial-compressive status. This, definitely, will cause the improvement of shear bearing capacity. Therefore, the main failure pattern of the columns of HD series is concrete crushing rather than shear fracture. Comparing Figs. 7.35b with 7.36b, one can conclude that, the presence of the middle stirrups could generate the tri-axial confining pressure on the RC columns, causing RC columns to become less brittle and less prone to axial compression/shear

Table 7.11 Peak compressive load and nominal strength

Series	Specimen label	$P_{u,cal}$ (kN)	$\sigma_{Nu,cal}$ (MPa)	$P_{u,exp}$ (kN)	$\sigma_{Nu,exp}$ (MPa)	Failure patterns	Compressive reinforcement	Tensile reinforcement
HA	HAS-1	367	9.2	485	12.1	F/T	Buckling	Yield
	HAS-2	367	9.2	545	13.6	F/T	Buckling	Yield
	HAM-1	1650	10.0	1896	11.9	F/T	Buckling	Yield
	HAM-2	1650	10.0	1650	10.3	F/T	Buckling	Yield
	HAL-1	5320	8.3	6118	9.6	F/T	Buckling	Yield
	HAL -2	5320	8.3	6350	9.9	F/T	Buckling	Yield
HB	HBS-1	648	16.2	1373	34.3	C/S	Buckling	Not yield
	HBS-2	648	16.2	1254	31.4	C/S	Buckling	Not yield
	HBM-1	2801	17.5	4864	30.4	C/S	Buckling	Not yield
	HBM-2	2801	17.5	4735	29.6	C/S	Buckling	Not yield
	HBL-1	11,375	17.8	16,507	25.8	C/S	Buckling	Not yield
	HBL -2	11,375	17.8	16,089	25.1	C/S	Buckling	Not yield
HC	HCS-1	367	9.2	522	13.1	F/T	Yield	Yield
	HCS-2	367	9.2	573	14.3	F/T	Yield	Yield
	HCM-1	1650	10.0	1874	11.7	F/T	Yield	Yield
	HCM-2	1650	10.0	1738	10.9	F/T	Yield	Yield
	HCL-1	5320	8.3	7052	11.0	F/T	Yield	Yield
	HCL -2	5320	8.3	6696	10.5	F/T	Yield	Yield
HD	HDS-1	648	16.2	1426	35.7	Compression	Buckling	Not yield
	HDS-2	648	16.2	1400	35.0	Compression	Buckling	Not yield
	HDM-1	2801	17.5	5656	35.4	Compression	Buckling	Not yield

(continued)

Table 7.11 (continued)

Series	Specimen label	$P_{u,cal}$ (kN)	$\sigma_{Nu,cal}$ (MPa)	$P_{u,exp}$ (kN)	$\sigma_{Nu,exp}$ (MPa)	Failure patterns	Compressive reinforcement	Tensile reinforcement
	HDM-2	2801	17.5	5143	32.1	Compression	Buckling	Not yield
	HDL-1	11,375	17.8	18,496	28.9	Compression	Buckling	Not yield
	HDL-2	11,375	17.8	18,309	28.6	Compression	Buckling	Not yield

Note $P_{u,cal}$ presents calculated peak load; $\sigma_{Nu,cal}$ is calculated nominal strength; $P_{u,exp}$ presents experimental results; $\sigma_{Nu,exp}$ is experimental nominal strength; "F/T" denotes flexural tensile failure; and "C/S" means crushing-shear failure

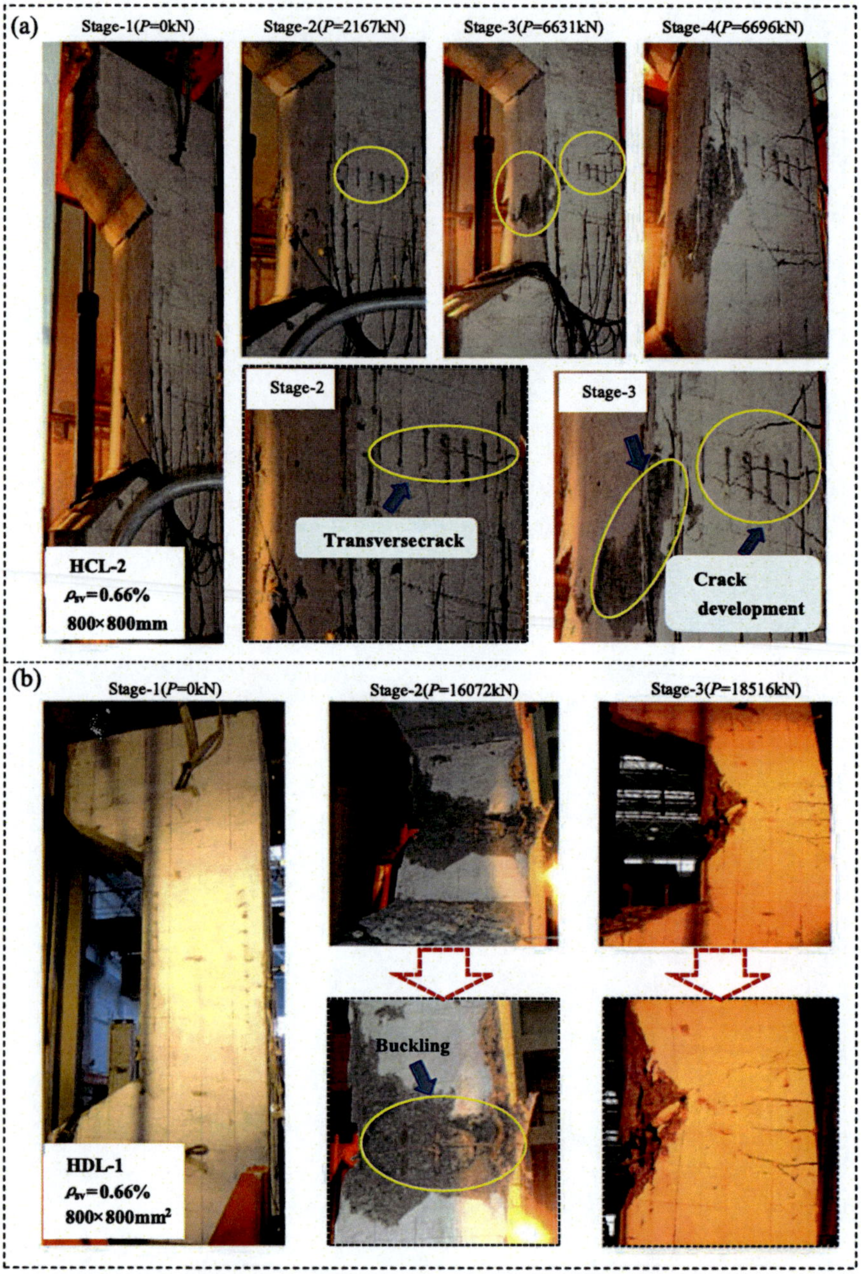

Fig. 7.37 Evolution of fracture in the high-strength RC columns for: **a** HCL-2; and **b** HDL-1

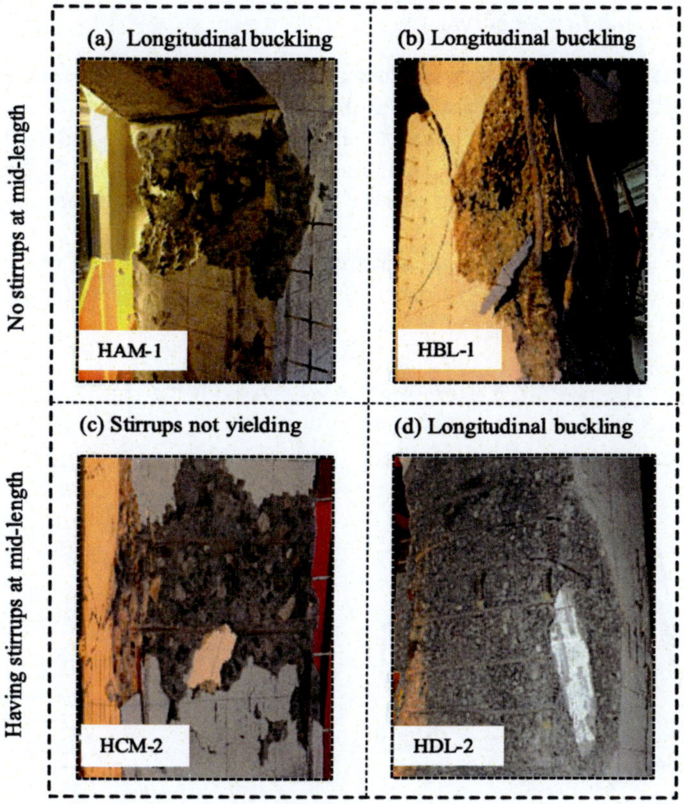

Fig. 7.38 Failure patterns of the steel bars within columns

fracture. This, in reality, will doubtless exhibit a weaker size effect than that of RC columns without the restraint of middle stirrups.

3. *Evolution of cracks within the RC columns*

In order to capture the evolution of cracks generated in the RC columns subjected to large and small eccentric loads, the fracture process of the two large-sized columns (i.e. with cross-sectional size of 800 mm) of the HCL and HDL series was selected and described in Fig. 7.37.

In Fig. 7.37a, the stage-1 denotes the initial state of the columns at the beginning of the test. From stage-2 to stage-3, the evolution of the flexural-tensile cracks of the column of the HCL-1 can be clearly seen from Fig. 7.37a. At the final stage-4, the spalling of concrete cover occurred at the edge of the specimen, and the column was broken due to the flexural-tensile cracks. For the RC column of HDL-1, under small eccentric compressive loads, both the flexural tensile cracks and the concrete crushing contribute to the final failure. At the compressed side of the column, the concrete cover flaked off, and the longitudinal reinforcement buckled.

4. *Nominal stress-strain curves*

The nominal compressive stress-strain curves obtained from the present tests for all the 24 high-strength RC columns with different structural sizes, without and with stirrups at the mid-length, are respectively plotted in Figs. 7.39 and 7.40. The test results of the tested RC columns, including the measured peak load P_u, the nominal strength σ_{Nu}, the failure patterns and the final deformation status of reinforcement are summarized in Table 7.12.

The calculated peak load $P_{u,cal}$ and strength $\sigma_{Nu,cal}$ in Table 7.12 were computed according to the China code (Ministry of Construction of the People's Republic of China 2011), i.e.,

$$P_u e = \alpha_1 f_c b x (h_0 - x/2) + f_t' A_s' \left(h_0 - a_s'\right) \tag{7.5}$$

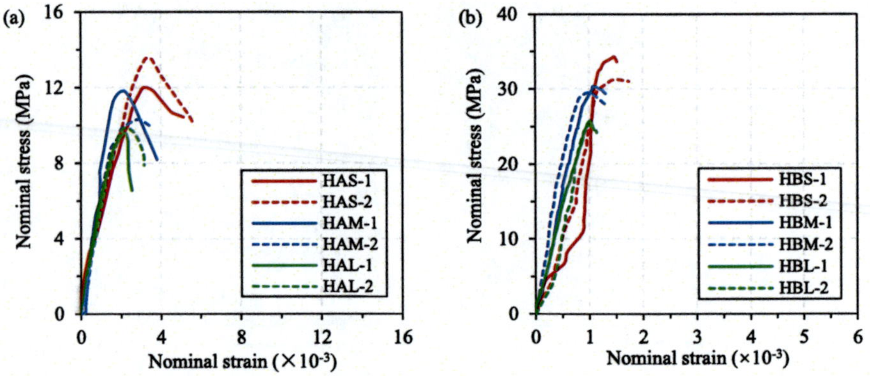

Fig. 7.39 Nominal compressive stress-strain curves for the high-strength RC columns without middle stirrups for: **a** HA series; and **b** HB series

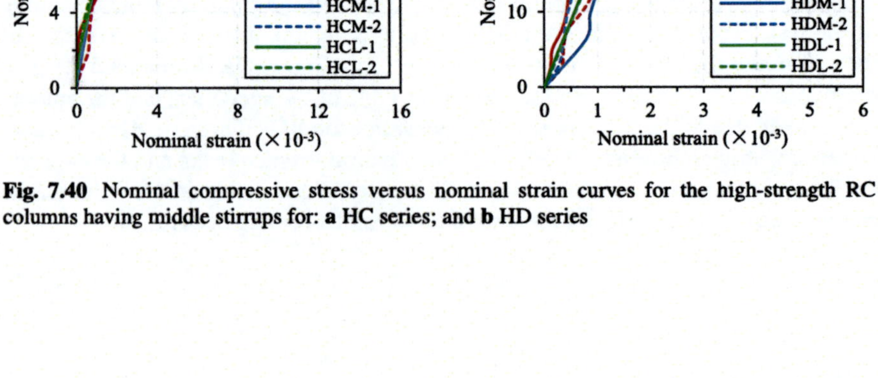

Fig. 7.40 Nominal compressive stress versus nominal strain curves for the high-strength RC columns having middle stirrups for: **a** HC series; and **b** HD series

Table 7.12 Mechanical parameters of concrete

	Mortar matrix	ITZ	Aggregate particle	Steel rebar
Young's modulus E (GPa)	30[a]	25[b]	70[a]	210[a]
Poisson ratio ν	0.20[a]	0.22[b]	0.16[a]	0.3[a]
Dilatant angle ψ (°)	18	15	–	–
K_c	0.667	0.667	–	–
Eccentricity η (%)	0.1	0.1	–	–
Stress ratio σ_{b0}/σ_{c0}	1.16	1.16	–	–
Compressive strength f_c (MPa)	67.6[a]	56.2[b]	80[a]	–
Tensile strength f_t (MPa)	6.5[a]	5.6[b]	8[a]	–
Yield strength f_y (MPa)	–	–	–	335/235[a]

Note [a]Data were obtained from the tests; [b]Data were given with reasonable assumption; and the other parameters are default values or determined by referring to Marzec et al. (2007) and Syroka-Korol et al. (2014)

where, e is the eccentricity; α_1 is a coefficient, $\alpha_1 = 0.98$; f_c and f_t' are respectively the compressive and tensile strengths of concrete; b is the width of the column section; x is the height of concrete in compressed side of cross section; h_0 is the effective sectional height; A_s' is total cross-sectional area of the compression steel bars; and a_s' is the distance between resultant force of all longitudinal reinforcements in compressed side and the tensile side of cross section.

From the post-peak softening curves in Figs. 7.39 and 7.40, one can clearly note that the failure of the RC columns is brittle (as can be demonstrated from the broken photographs, especially for the columns of HC series in Fig. 7.36a). Actually, because of the brittle performance of high-strength concrete, most of the RC columns failed catastrophically and therefore the softening branches could not have been fully recorded in the stress-strain relationships, thus the rate softening change was not considered in this paper.

For the HA and HC series shown in Figs. 7.39a and 7.40a, as discussed previously, the corresponding failure should be caused by bending moment. However, one can note that the failure is obviously brittle. The brittle failure of the HA and HC series under large eccentric load should be attributed to the high brittleness of the concrete material used. Figures 7.39a and 7.40b respectively present the nominal stress-strain curves of the RC columns without stirrups at the mid-length, under the large eccentric compression (HA series, $e_0 = 0.6h_0$) and the small eccentric compression (HB series, $e_0 = 0.25h_0$). One can clearly note that the tested nominal strengths (i.e. the peak stress) of the columns with the same structural sizes show obvious randomness, e.g. for the HAS-1 and HAS-2. Nevertheless, on the whole, the nominal strength decreases with increasing the structural size, illustrating the presence of size effect in RC columns subjected to eccentric loading.

One can note from Figs. 7.39 and 7.40 that some of the RC columns failed catastrophically and therefore the softening branches could not have been fully recorded in the stress-strain relationships, especially for the HB series shown in Fig. 7.39b. It can be noted clearly from the curves in Fig. 7.39a that for each series, the initial tangent modulus varies little while the post-peak softening curves are distinct as the structural size changes. From the post-peak slope in Fig. 7.39a, one can note that the failure of the high-strength RC columns presents more and more brittle as the structural size increases, as can be concluded from the post-peak slope of the stress-strain curves. For the RC columns with the cross-sectional sizes of 800 mm × 800 mm, one can see from the two green curves in Fig. 7.39a that, the nominal stresses decrease rapidly when they exceed the peak stress. Compared with those of the RC columns sized of 200 mm × 200 mm and 400 mm × 400 mm, the global stiffness of the larger-sized RC columns (i.e., HAL-1 and HAL-2 whose cross-sectional sizes are all 800 mm × 800 mm) degrades much faster. This can also be demonstrated and illustrated from the test observations aforementioned in Fig. 7.35a. One can note from Fig. 7.39b that, the obtained softening curves of the columns of HB-series are all very short, or even the post-peak softening curves are almost not presented. This is because that, under the small-eccentric compressive loading, the fracture process of the RC columns without middle stirrups happened at a sudden, and the fracture is brittle. Compared with the columns of HB series in Fig. 7.39b, the columns of HA series in Fig. 7.39a show a less brittleness, as observed by the slope of stress-strain diagram. Figure 7.40a plots the nominal stress-strain curves of the RC columns subjected to the large eccentric compressive loading ($e_0 = 0.6h_0$), and Fig. 7.40b presents the ones of the RC columns subjected to the small eccentric compressive loading ($e_0 = 0.25h_0$). Again, with increasing the structural size the decrease of nominal strength of the RC columns demonstrates the existence of size effect.

Compared Fig. 7.39 with 7.40, it can be found that the obtained nominal strengths of the columns improve slightly due to the constraint effects of transversal reinforcements. This can be also illustrated from the previous analysis on the failure behavior of the RC columns. In addition, comparing Figs. 7.39 with 7.40, one can find that, the failure of the RC columns of HC and HD series with middle stirrups is less brittle than the ones (HA and HB series) without stirrups. The declining slope at the post-peak for the HC and HD series are less steep than those for HA and HB series, which is caused by the confinement effect of transversal reinforcement. Nevertheless, it can be noticed from the comparison of curves in Figs. 7.39b and 7.40b that, for the larger-sized RC columns (i.e. HBL and HDL series whose structural size is 800 mm), the confinement effects of stirrups seem to not improve the ductility of the RC columns. The failure of the columns of HDL-series is still extremely brittle. This should be caused by the high brittleness of high-strength concrete components with large sizes.

5. *Effect of transversal reinforcement on the nominal-strength*

To further illustrate the quantitative effect of transversal reinforcement on the nominal strength σ_{Nu}, let us plot the nominal strength σ_{Nu} versus the structural size D (i.e. the cross-sectional size), which presents all the data points gathered from these tests. Such plots for large and small eccentric compressive loadings are respectively given in Fig. 7.41a and b, in which the value of nominal strength $\sigma_{Nu} = P_u/D^2$ was calculated from the peak loads measured in individual tests. According to the

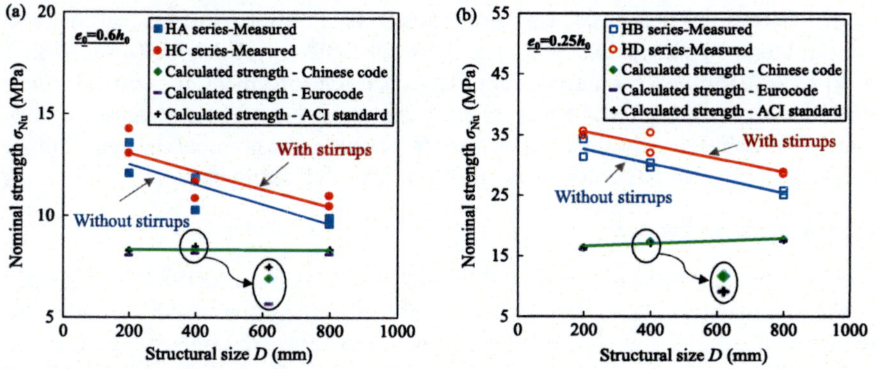

Fig. 7.41 Nominal strength versus structure size for all the high-strength RC columns: **a** HA series and HC series; **b** HB series and HD series

current RC structure design codes of China (Ministry of Construction of the People's Republic of China 2011), these calculated nominal compressive strengths presented in Fig. 7.41 are clearly straight lines with a slope of zero which are parallel to the axis of structural size D. This definitely does not indicate any size effect. Additionally, the comparison of experimental results with Eurocode 2 (CEN 2004a) and ACI 318-14 (ACI 2014) are plotted in Fig. 7.41. The Eurocode 2 (CEN 2004a) happens to be the most conservative standard for the design of RC structures. However, it can be clearly seen from Fig. 7.41 that, the codes, including the Chinese code, the Eurocode 2 (CEN 2004a) and the ACI 318-14 (2014), present approximate design values for the RC specimens, and the calculated nominal strengths are all parallel to the horizontal axis, which definitely presents no size effect. Nevertheless, one can note the reduction in the measured nominal strengths as the structural size D increases in each series. These diagrams clearly indicate the existence of a size effect, as theoretically expected.

It can be clearly seen from Fig. 7.41 that, under both the large and small eccentric compressive loadings, the nominal strength decreases with increasing structural size, indicating the existence of size effect in eccentrically loaded RC columns. Furthermore, one can note that the obtained nominal strengths of the RC columns with transversal reinforcement are significantly larger than those without transversal reinforcement. This phenomenon illustrates the improvement of nominal strength contributed by the confinement effect of middle stirrups. From Figs. 7.41a and b, it should be noted that the declining slopes of the nominal strength of the RC columns with and without transversal reinforcement are significantly distinct. Compared with those without transversal reinforcement (i.e. the HA and HB series), the nominal strengths of the RC columns having middle stirrups (i.e. the HC and HD series) decrease much slower as the structural size increases. This indicates that the existence of the transversal reinforcement could weaken the size effect of eccentrically loaded RC columns. This can be explained by the fact that, increase of tri-axial confining pressure engendered by the transversal reinforcement causes concrete to become less brittle and less prone to axial crushing fracture. As known, for an extremely case, for a concrete column encased in a very strong steel tube, the global mechanical behavior turns out to be plastic and ductile, making the size effect probably be eliminated.

In a word, for the RC columns with transversal reinforcement, the corresponding size effect descending trend becomes less steep than that without transversal reinforcement. Subsequently, the detailed size effect law for the nominal strength of these eccentrically loaded RC columns is provided in the following section.

6. *Size effect on the eccentrically loaded RC columns*

The draw bi-logarithmic plots according to the SEL of Bažant (1984) is presented in Fig. 7.42. Figure 7.42a describe the combined test results of the high-strength RC columns of HA and HC series, and Fig. 7.42b shows the ones of HB and HD series. All these plots and the corresponding correlation coefficients clearly indicate a strong size effect. In addition, the SEL fits the present test data satisfactorily. With increasing the structural size, the brittleness number increases. For small structural sizes these fitting curves approach a horizontal asymptote, and for large structural sizes they approach an inclined asymptote of slope −1/2, which corresponds to linear elastic fracture mechanics.

Comparing the tested data of the large-eccentrically loaded columns with transversal reinforcement (i.e. the HC series) and that of columns without middle stirrups (i.e. HA series), as shown in Fig. 7.42a, one can note that stirrups could obviously mitigate the size effect. The increase of the tri-axial confining pressure engendered by the stirrups makes concrete less brittle. Therefore, the test results of the HA series approach much closer to the LEFM asymptote than those of HC series.

From Fig. 7.42b, one can note that the tendency of size effect in RC columns under small eccentric load is similar with that under large eccentric load. Due to the confinement effect of transversal reinforcement, the failure of the columns having middle stirrups turns out to be less brittle compared to the one without middle stirrups. The middle stirrups, in turn, weaken the size effect in RC columns. For the same structural sizes, the brittleness number of the columns with transversal reinforcement

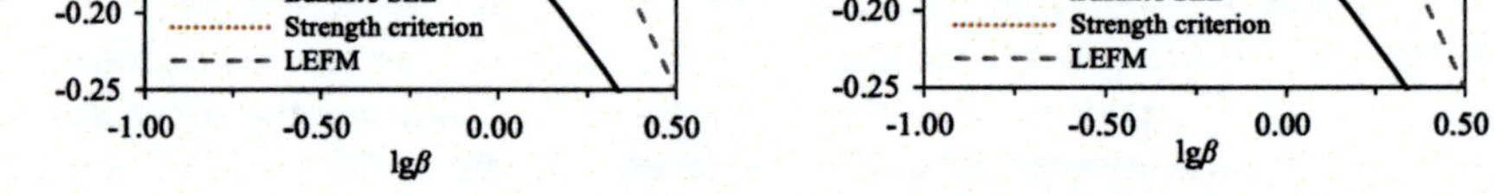

Fig. 7.42 Test results on size effect for the high-strength RC columns: **a** HA and HC series; and **b** HB and HD series

is obviously smaller than that having no middle stirrups, and the test data of the HB series approaches closer to the LEFM asymptote than that of HD series.

7.4.2 Numerical Investigations

Jin et al. (2018) conducted the numerical tests of the RC columns with larger stirrup ratios and eccentricity of $e_0 = 0.25h_0$, to discover the confinement effect on the behavior of high-strength RC columns. Herein, the details of numerical results are presented.

1. *Numerical method*

The generation of 2D mesoscale model (Fig. 7.43) and boundary conditions are similar with that in Sect. 7.3.2. The mechanical properties of concrete and steel bars are listed in Table 7.12. It is to be noted that, the larger stirrup ratios were achieved by reducing the spacing of stirrups. Four different stirrup ratios were set, including $r_{sv} = 0$, 0.66%, 1.2% and 2.4%, corresponding to HB/HD/HE/HF series.

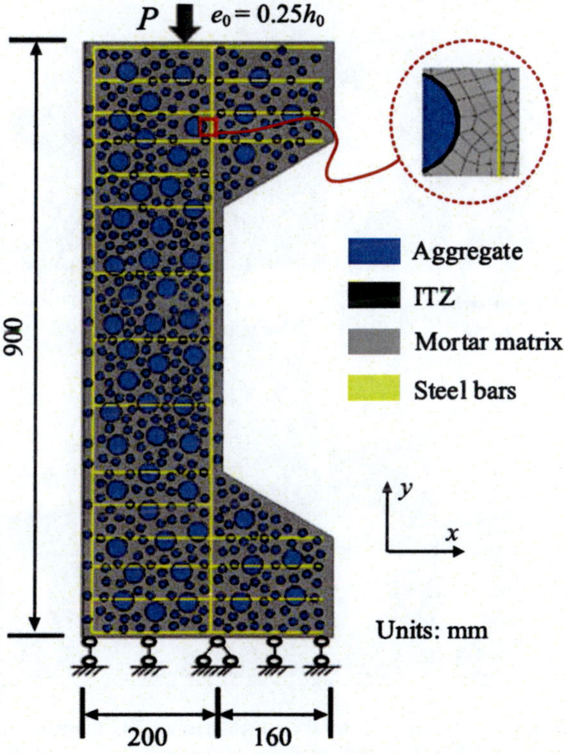

Fig. 7.43 Meso-scale mechanical model for RC column

2. *Numerical results and discussion*

The final failure patterns of the 12 high-strength RC columns are showed in Fig. 7.44. With the improvement of the stirrup ratios, it can be noticed from Fig. 7.44a–d that, the shape and size of the damage zones in concrete change as well, and longitudinal reinforcement buckles. In addition, as the stirrup ratio increases, the load-carrying capacity increases.

3. *Size effect on the nominal strength*

The results of the four series (i.e. HB, HD, HE and HF) of moderate high-strength RC columns, involving the peak load obtained from numerical simulation $P_{u,num}$ at the loading point and the calculated nominal compressive strength $\sigma_{Nu,cal}$ (Eq. 7.1) are summarized in Table 7.13.

To further illustrate the quantitative effect of transversal reinforcement on the nominal compressive strength σ_{Nu}, Fig. 7.45 plots the relationship between the

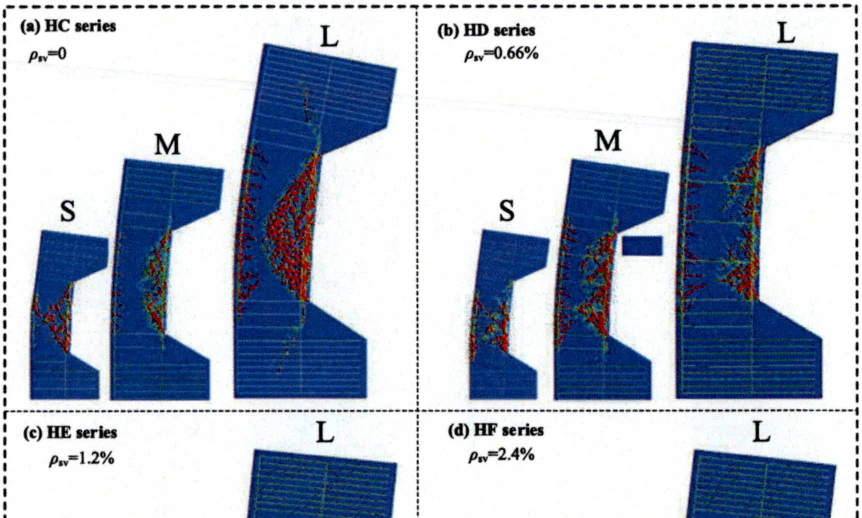

Fig. 7.44 Failure patterns of the high-strength RC columns under the eccentricity of $e_0 = 0.25h_0$: **a** $\rho_{sv} = 0$; **b** $\rho_{sv} = 0.66\%$; **c** $\rho_{sv} = 1.20\%$; and **d** $\rho_{sv} = 2.40\%$

Table 7.13 Peak compressive load and nominal strength

Series	Specimen	$P_{u,cal}$ (kN)	$\sigma_{Nu,cal}$ (MPa)	$P_{u,num}$ (kN)	$\sigma_{Nu,num}$ (MPa)
HB	HCS	648	16.2	1410	35.26
	HCM	2801	17.5	4765	29.78
	HCL	11,375	17.8	17,464	27.29
HD	HDS	648	16.2	1426	35.65
	HDM	2801	17.5	5142	32.14
	HDL	11,375	17.8	18,309	28.61
HE	HES	648	16.2	1426	35.65
	HEM	2801	17.5	5242	32.76
	HEL	11,375	17.8	18,849	29.45
HF	HFS	648	16.2	1426	35.65
	HFM	2801	17.5	5314	33.21
	HFL	11,375	17.8	19,401	30.31

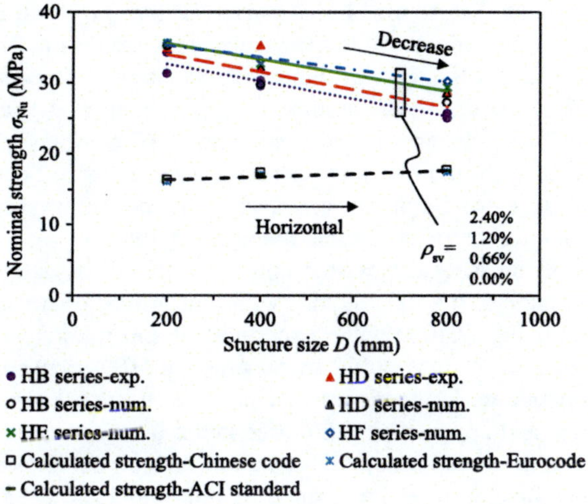

Fig. 7.45 Nominal compressive strength versus structural size

nominal strength σ_{Nu} versus the structural size D for the RC columns with small eccentricity, as well as the calculated value according to Chinese code (Ministry of Construction of the People's Republic of China 2011), Eurocode 2 (CEN 2004a) and ACI 318-14 (ACI 2014). Similar with that in Sect. 7.4.1, all the codes present approximate design values for the RC columns, and the calculated nominal strengths are all parallel to the horizontal axis.

However, the eccentrically loaded RC columns of four stirrup ratios exist distinct size effect. In addition, it can be found that with the improvement of the stirrup ratio, the corresponding nominal compressive strength increase obviously. This indicates

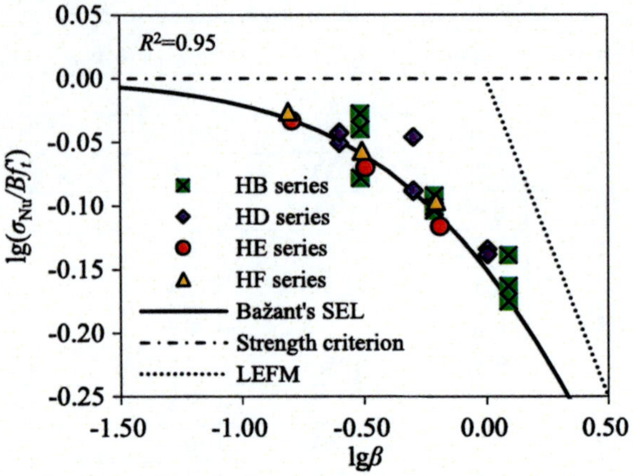

Fig. 7.46 Bi-logarithmic curves of nominal axial compressive strength and structural size

that the existence of stirrups could weaken the size effect of high-strength RC columns under small-eccentric compressive loading, as shown in Sect. 7.4.1.

According to the SEL of Bažant (1984), Fig. 7.46 illustrates the size effect plots of the numerical results and experimental results of HB series. The fitting correlation coefficient (i.e. R^2) for the numerical data and Bažant's SEL curve was 0.95. This illustrates that the theoretical work of Bažant (1984) can well describe the size effect on the nominal compressive strength of RC columns subjected to the small-eccentric loading. Similar with the experimental results in Sect. 7.4.1, the nominal strength of large-sized columns approaches the inclined asymptote, and that of small-sized columns approach the horizontal asymptote.

Comparing the experimental and numerical data on the nominal compressive strength of the RC columns with four groups of stirrup ratios, one can note clearly from Fig. 7.46 that the experimental and numerical data obtained from the RC columns without middle stirrups (i.e. the HB series) approaches closest to the LEFM asymptote. As the stirrup ratio improves, the data keeps away from the LEFM asymptote and approach much closer to the plastic strength criterion. In a word, with a smaller stirrup ratio, the fracture of the column is more brittle, and the decrease of nominal compressive strength is more obvious, exhibiting a more significant size effect.

7.5 Compressive Failure of Stirrups-Confined Concrete Column

In this section, a series of experimental tests (Du et al. 2017; Jin et al. 2017a) and numerical test (Jin et al. 2020a) on the axial-loaded stirrup-confined circular and squared RC columns are given. Moreover, the size-dependent stress-strain models to predict the stress-strain behavior of large-scaled stirrup-confined concrete columns (Jin et al. under review) is also presented.

7.5.1 *Experimental Investigations of Circular Concrete Column*

1. *Experimental program*

The mixture proportions of the concrete utilized in the tests are summarized in Table 7.14. It is to be noted that, the concrete mixture components (i.e. aggregate size and aggregate distribution) were not scaled to the characteristic dimension, and the maximum diameter of the large coarse aggregates was about 25 mm for all the specimens. The corresponding average values of 28 days compressive and splitting tensile strengths of standard plain concrete cube samples, sized of 150 mm × 150 mm × 150 mm, were 42.8 MPa and 3.0 MPa, respectively. The modulus of elasticity of the concrete utilized was 30.8 GPa. The longitudinal and transverse reinforcement of columns are the same type of steel bars and are summarized in Table 7.15.

A total of 18 confined circular RC columns were cast from the same batch of continually mixed concrete, which ensured the material properties of all the specimens to be virtually the same. The corresponding geometric configuration and reinforcement layout of the tested columns are presented in Fig. 7.47 and Table 7.16. The slenderness ratio of all these columns was three. In addition, the longitudinal reinforcement within the columns was mainly used for fixing the transversal reinforcement. Three longitudinal reinforcing bars were adopted in the columns. To avoid the early damage at the ends of the columns during the loading process, a more hooped reinforcement was used at each end of 1/6 full height of specimens.

The experiment was carried out in the multifunctional electrohydraulic servo loading test system (MESLTS), as shown in Fig. 7.48a. All the RC columns were

Table 7.14 Concrete mixture proportions

Material	Aggregate	Sand	Cement	Water	Water-reducing admixtures	Silica fume
(kg/m^3)	1030	760	280	195	3.7	75

Table 7.15 Details of the steel rebar utilized in the concrete columns

Grade of steel	Diameter (mm)	Ultimate strength (MPa)	Yield strength (MPa)	Yield strain ($\mu\varepsilon$)	Elasticity modulus (GPa)
HRB400	8	644	458	2238	205
	12	665	478	2335	205
	18	675	494	2395	206

Note "HRB" means "Hot-rolled Ribbed-steel Bar"

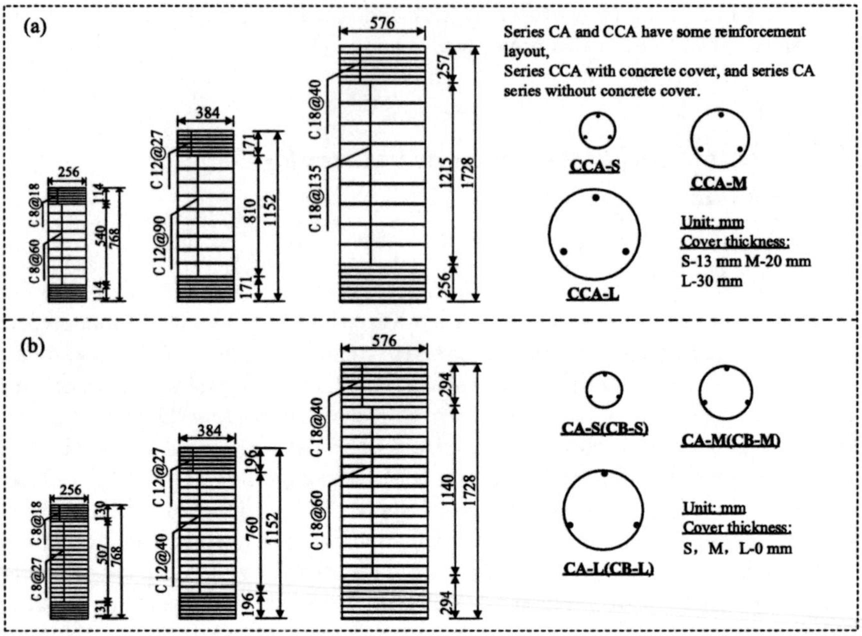

Fig. 7.47 Geometrical configuration and reinforcement layout: **a** Series CA and CCA; and **b** Series CB

Table 7.16 Geometric details of the confined columns

Series	Specimen	Diameter or edge length (mm)	Length (mm)	Stirrup form	Stirrup ratio ρ_{sv} (%)	Cover thickness (mm)
CA	CA-S	256	768	8@ 60	1.26	0
	CA-M	384	1152	12@90	1.26	0
	CA-L	576	1728	18@135	1.26	0
CCA	CCA-S	282	768	8@60	1.26	13
	CCA-M	424	1152	12@90	1.26	20
	CCA-L	636	1728	18@135	1.26	30
CB	CB-S	256	768	8@27	2.89	0
	CB-M	384	1152	12@40	2.89	0
	CB-L	576	1728	18@60	2.89	0

Note Serials CCA denote the ones with concrete cover; Serials CA and CB denote those without concrete cover; and "A" and "B" stand for the stirrup ratios of 1.26% and 2.89%, respectively

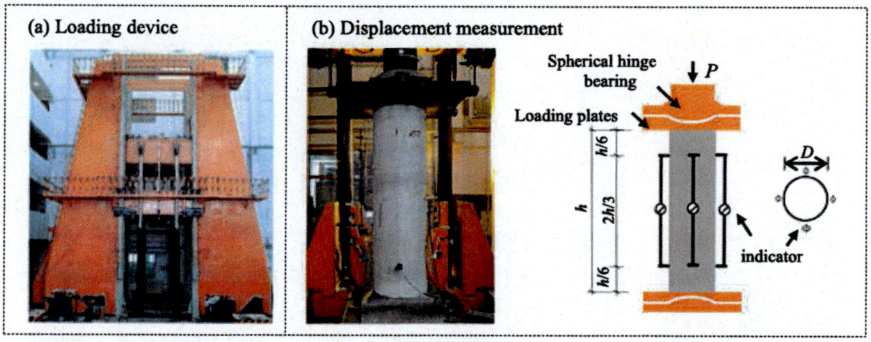

Fig. 7.48 Loading device and schematic diagram: **a** Loading device; and **b** Displacement measurement

simply supported at both ends, and the loading was applied axially. The top and bottom ends of the confined RC columns were located against spherical seating supports, which were designed to minimize the eccentricity of loading, as presented in Fig. 7.48b.

The control was based on a mixed constant increment of averaged longitudinal forces and deformations. It was controlled by constant increment of averaged longitudinal forces before reaching the peak load. To get the post-peak softening curves of the stress-strain relationship (load-deformation curve during the test in fact), it was switched to constant increment control of averaged longitudinal deformations after the peak load.

2. *Failure patterns*

From Figs. 7.49, one can clearly note that almost all the columns failed by compression-shear modes in the gauge length, because of the strengthening behavior at the ends. For the case of same stirrup ratios and cross-section patterns, the failure modes of the RC columns with different structural sizes are closely similar. Moreover, an obvious inclined crack appeared in the failure regions in all the stocky specimens subjected to axial compression, which had also been reported by Barbhuiya and Choudhury (2015) and Syroka-Korol and Tejchman (2014). Furthermore, the failure regions of all specimens are in the middle length of the specimens, where the slip or separation behavior between concrete and steel rebar occurred and the lateral expansion is more obvious than other areas.

Comparing Fig. 7.49a with b, one can see that after the ultimate load, concrete cover cracked and fell off, and cannot work. Therefore, the failure modes of the columns of "CCA" series with concrete cover are almost totally the same with those of "CA" series without concrete cover.

The stirrup ratio has a great influence on the failure behavior of RC columns. With the increase of stirrup ratio, the confinement action improves, and the failure of columns turns out to be more ductile, as can be observed from Fig. 7.49. Moreover, it can be seen that most of the longitudinal steel rebar were buckled, and the lateral

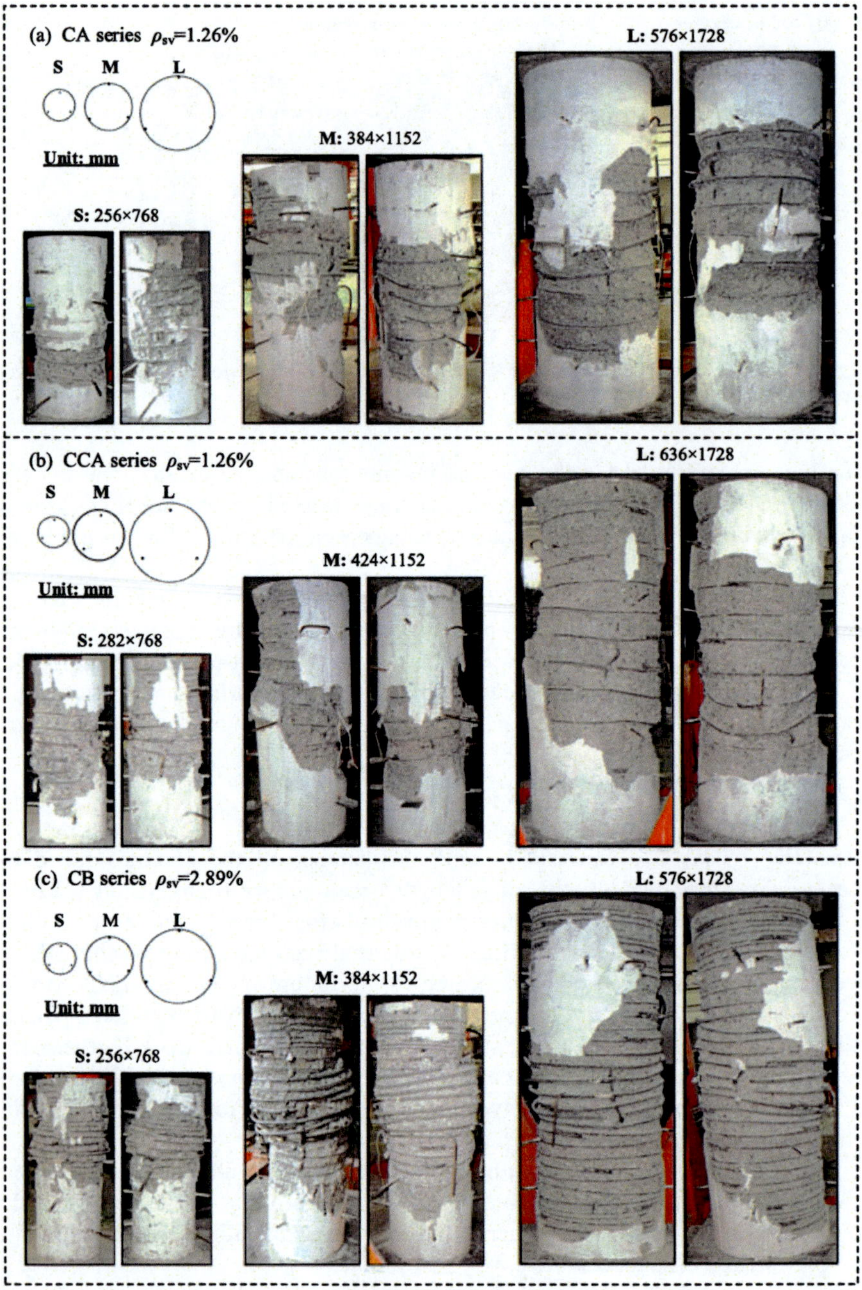

Fig. 7.49 Failure patterns of the circular RC columns: **a** Series CA; **b** Series CCA; and **c** Series CB

deformations of transverse reinforcements are prominent. Moreover, some of stirrups crossed the main cracks are broken due to necking fracture, for both the columns having smaller and larger stirrup ratios. The concrete flaked off at the middle length of the columns.

3. *Nominal stress-strain relation*

The nominal axial compressive stress-strain curves of the tested circular columns are plotted in Fig. 7.50. The nominal stress σ is defined as the ratio of reaction force P at the loading point and the cross-sectional areas, i.e. $\sigma = 4P/\pi D^2$ for the columns with circular section. Herein, D denotes the cross-sectional diameter of the circular RC columns. The nominal compressive strain ε is defined as the ratio of the average vertical displacement Δ at 1/2 height of the column and $L/2$ (L is the height of the RC columns), i.e. $\varepsilon = 2\Delta/L$. Herein, the average vertical displacement Δ is calculated by averaging the displacements measured at the 1/3 and 2/3 height of the column.

The tested load capacity P_u, the nominal compressive strength σ_{Nu}, the nominal compressive strain ε_{cc} at the peak stress, and the displacement ductility coefficient u (i.e. $u = \varepsilon_{85}/\varepsilon_{cc}$) for the RC columns are all listed in Table 7.17. Herein, ε_{cc} means the strain at peak stress, and ε_{85} denotes the strain when the applied loading drops to 85% P_u.

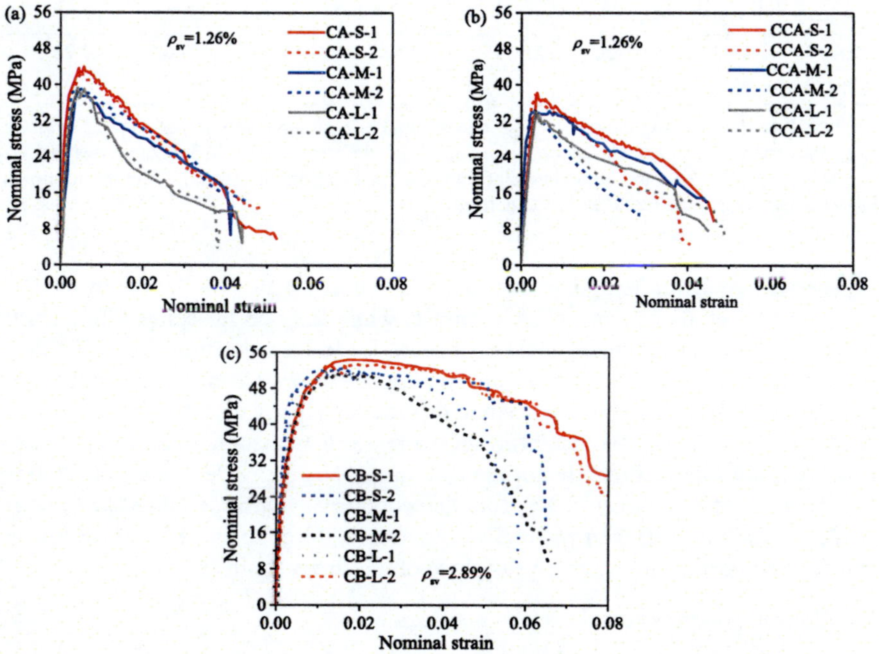

Fig. 7.50 Nominal stress-strain curves of the circular columns: **a** Series CA; **b** Series CCA; and **c** Series CB

Table 7.17 Test results of the confined concrete columns

Specimen label	P_u (kN)	Mean value of P_u (kN)	σ_{Nu} (MPa)	Mean value of σ_{Nu} (MPa)	Mean value of ε_{Nu} (%)	Mean value of ε_{85}(%)	Strain rati u = $\varepsilon_{85}/\varepsilon_{cc}$
CA-S-1	2222	2244	43.20	43.63	0.58	1.43	2.47
CA-S-2	2266		44.05				
CA-M-1	4571	4529.5	39.49	39.13	0.49	1.17	2.39
CA-M-2	4488		38.77				
CA-L-1	9662	9997	37.10	38.39	0.45	0.93	2.07
CA-L-2	10,332		39.67				
CCA-S-1	2380	2392.5	38.12	38.33	0.36	1.23	3.42
CCA-S-2	2405		38.53				
CCA-M-1	4801	4818	34.02	34.14	0.34	1.02	3
CCA-M-2	4835		34.26				
CCA-L-1	10,923	10,864.5	34.40	34.22	0.37	0.95	2.57
CCA-L-2	10,806		34.03				
CB-S-1	2800	2783.5	54.43	54.11	1.69	4.82	2.85
CB-S-2	2767		53.78				
CB-M-1	6081	6055.5	52.53	52.31	1.56	4.76	3.05
CB-M-2	6030		52.09				
CB-L-1	13,382	13,456	51.38	51.67	1.30	3.62	2.78
CB-L-2	13,530		51.95				

Note "P_u" is load-bearing capacity; "f_{cc}" the nominal compressive strength of the RC column; "ε_{cc}" the strain of the peak stress corresponding; "ε_{85}" the strain when the loading drops to 85% P_u; "*u*" the displacement ductility coefficient; and the final number of "1" and "2" in the specimen label corresponds to the replicate of RC columns

The post-peak softening curves in Fig. 7.50 indicate that the failure of the RC columns for serials CA and CCA is brittle, while the one for series CB and PB is quasi-brittle since there is stable post peak. It can be seen from Fig. 7.50c that, the nominal stress-strain curves of the columns of serials CB decrease gently after reaching the nominal compressive strength. One can clearly note from all the stress-strain curves in Fig. 7.50 that, with the increase of structural sizes, the obtained nominal compressive strengths for the columns of any series decreases, illustrating the presence of size effect in RC columns confined by stirrups subjected to axial loading. In addition, as the structural size increases, the columns fail more brittle, as can be concluded from the post-peak slope of the stress-strain curves.

4. *Effect of concrete cover*

From Table 7.17, one can clearly notice that, at the same structural size and cross section pattern, the load-bearing capacity of the RC columns of "CCA" serial is obviously larger than that of "CA" serials. The nominal stress-strain curves for the

"CA" serials decrease a little faster than those of the "CCA" serials. That is to say, the brittleness number of columns without concrete cover is larger than that of columns having concrete cover. Nevertheless, in the present tests, the measured nominal peak stress (i.e. the ratio of load-bearing capacity and total cross-section) of "CCA" serial columns is lower than that of "CA" serial columns. That means that the load-bearing capacity of the RC columns is not synchronous with that of the core area of specimens. When the overall bearing capacity of the specimen gets to the maximum, the bearing capacity of the columns at the core regions is still increasing.

5. *Effect of stirrup ratio*

Comparing Fig. 7.50a with c, one can see that, the confinement effect on the RC columns improves significantly with the increase of stirrups ratio. This, not only makes the enhancement of nominal strengths of the columns, but also the ductility capacity of the columns.

Moreover, the RC columns with the higher stirrup ratio (i.e. the series "CB") failed less brittle than the ones with the lower stirrup ratio (i.e. series "CA"). The declining slopes at the post-peak for "CB" series are obviously less steep than those for "CA" series, which is definitely caused by the confinement effect of transversal reinforcement. From Table 7.17, one can notice, compared with that of RC columns with lower stirrup ratio, the displacement ductility coefficient "u" described by the strain ratios "$\varepsilon_{85}/\varepsilon_{cc}$" of the RC columns with higher stirrup ratio are obviously much larger.

Be different from the others, the failure of the "CB" serials columns (with the circular cross section and the stirrup ratio of 2.89%) shown in Fig. 7.50c seems to be ductile. Especially, the columns having a larger structural size in "CB" serials (i.e. "CB-L-1"and "CB-L-2" specimens) exhibit more brittle failure than the columns having smaller structural sizes.

6. *Size effect on the nominal strength*

All the nominal compressive strengths of the tested circular columns were plotted against with the structural size D (i.e. the diameter or the cross-sectional width of the columns), as presented in Fig. 7.51. One can clearly note from Fig. 7.51 that, with the increase of structural size, all the obtained nominal compressive strength decreases.

For series "CA" and "CCA", when the concrete cover is set, the corresponding nominal compressive strength of columns decreases. In addition, the nominal strength of the "CA" series decreases faster than that of the "CCA" series. For series "CA" and series "CB", by comparing the two fitting trend curves, one can see that, the increase of stirrup ratio will make the nominal strength of RC columns be improved. Moreover, for the series "CB" with a higher stirrup ratio, the declining slope of the strength is obviously less steep than the one for the series "CA" with a lower stirrup ratio, indicating that the size effect is weakened with the increase of stirrup ratio.

Figure 7.52 gathers the data points of the nominal strength of the tested circular columns, as well as the strength criterion and the linear elastic fracture mechanics

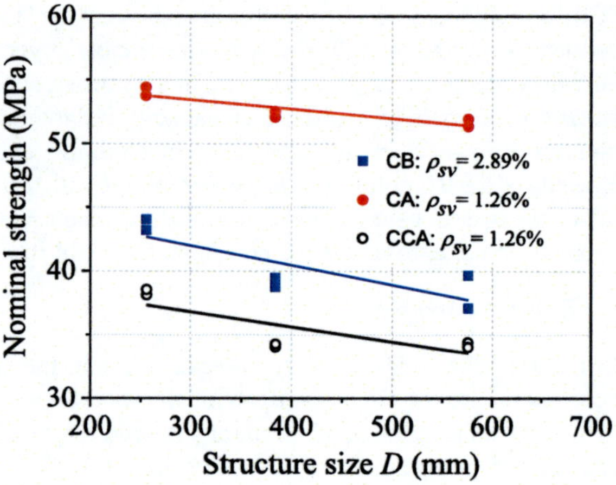

Fig. 7.51 Nominal compressive strength versus structural size of the RC columns tested

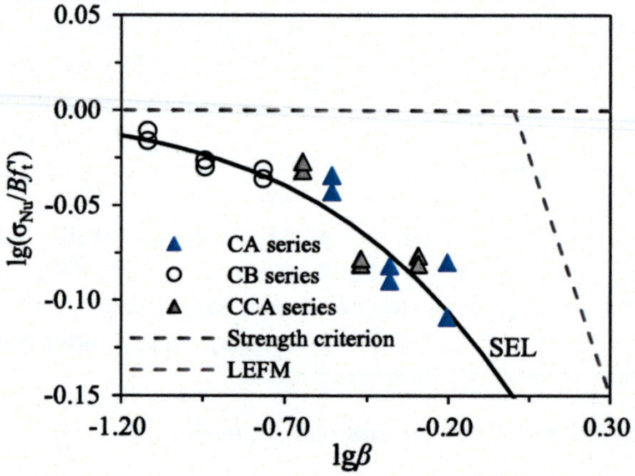

Fig. 7.52 The size effect for the tested RC columns

(LFEM, for pure-brittle materials) are also plotted for the comparisons. All the tested data keep far from the two curves (strength criterion and LEFM), indicating that the failure of the tested RC columns should be between pure ductile and pure brittle.

One can note from Fig. 7.52 that the size effect is stronger for the columns of series "CA" with a lower stirrup ratio, and the tested data lie much closer to the LEFM asymptote slope of $-1/2$. The higher the stirrup ratio is, the larger the triaxial confining pressure is, the less brittle the failure of the RC columns exhibits. This, definitely, is the major reason that the RC columns with a higher stirrup ratio exhibit a weaker size effect.

In addition, the size effect is obviously stronger for the RC columns without concrete cover. Simply to say, when concrete cover is adopted, the transverse deformation of stirrup is restricted. As a result, the cement mortar and coarse aggregate in the core of concrete are restricted well by stirrups, and they are combined more fully.

The RC confined column without concrete cover, having a lower stirrup ratio presents much stronger size effect. One can notice from Fig. 7.52 that, the size effect for the series CA is relatively the strongest in the three series, and the corresponding test data approaches closest to the LEFM. Compared with the series CA, the series CCA exhibit less brittle failure patterns, and therefore the corresponding brittleness number of the series CCA is relatively smaller. The failure the series CB should be the most ductile one in the three series, and thus the corresponding size effect on the nominal strength is the weakest one.

As mentioned above, the tested data keep far from the two curves (strength criterion and LEFM). The analysis of the effect of structural size on the nominal strength of quasi-brittle structures failing in compression was carried out by Bažant and Xiang (1997). In their work, a simple, intuitively clear explanations of the phenomenon of size effect was given, and a simplified model for the global mechanism of propagation of compression fracture was formulated. The modified size effect has the slope of $-2/5$ rather than $-1/2$. The comparison of the theory with the present test data is plotted is Fig. 7.53, for all the tested circular columns. As seen from Fig. 7.53, the simplified law in Bažant and Xiang (1997) fits the present test results equally well. In summary, for large sizes D, the size effect on the nominal strength of the column approaches the power $D^{-2/5}$.

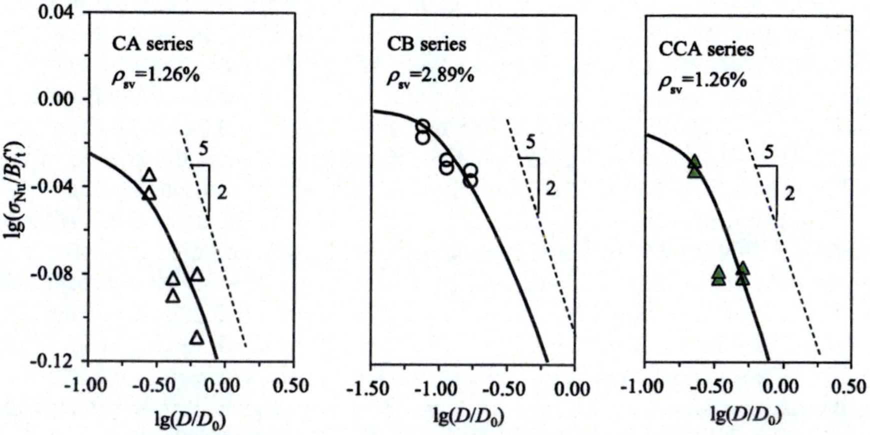

Fig. 7.53 Measured size effect compared to theoretical size effect proposed by Bažant and Xiang (1997)

7.5.2 *Experimental Investigations of Square Concrete Column*

1. *Experimental program*

The mixture proportions of the concrete utilized in the present tests are given in Table 7.18. The corresponding mean values of 28 days splitting tensile and compressive strengths of standard plain concrete cube samples, sized of 150 mm × 150 mm × 150 mm, were 3.0 MPa and 42.8 MPa, respectively. The measured modulus of elasticity of the concrete utilized was 34.8 GPa. To explore the effect of yield strength of stirrups on the failure of RC columns, two types of steel rebar were used, as summarized in Table 7.19.

A total of 26 geometrically-similar RC columns were designed and tested. The corresponding geometric configuration and reinforcement layout are presented in Fig. 7.54. The slenderness λ of all columns is 3. The four longitudinal reinforcements arranged in all specimens were mainly used for fixing the transversal reinforcement. Two groups of stirrups ratios ρ_{sv} were adopted, i.e. $\rho_{sv} = 1.26\%$ and 2.89%. In order to avoid the early damage at the ends of the columns during the loading process, a more hooped reinforcement was used at each end of 1/6 full length of the columns, as can be observed in Fig. 7.54.

As presented in Fig. 7.54, the first letter used in the column labels means stirrups ratio (namely, A and B respectively stand for the stirrups ratio of 1.26% and 2.89%). The second letter "X" represents the number of stirrup legs (e.g. if X = 2, the corresponding number of the stirrup legs is 2). The letter "H" in the series "AXH" means the high-strength steel. The letter "S", "M" and "L" mean "small", "middle" and "large" corresponding to the cross sectional sizes of 267 mm × 267 mm, 400 mm × 400 mm, and 600 mm × 600 mm. The numbers "1" and "2" at the end of all column labels denote the replicate of the columns. Moreover, the cover thicknesses of the columns with "S", "M" and "L" are respectively 10 mm, 20 mm and 30 mm.

To sum up, there are 16 RC columns denoted as series A, including series A2, AX and AXH. While, there are 10 columns denoted as series B (i.e. "B2" and "BH"). All geometric sizes of the RC columns tested, including the reinforcement were scaled to get geometrically proportional samples with the ratio of 1:1.5:2.25. The geometric similarity was kept with the sizes, locations and covers of the reinforcing bars, as well as the diameter and the spacing of the stirrups. All of them were scaled in proportion to the characteristic sizes. The test setup and loading arrangement were similar with that in Sect. 7.5.1. For more details, please see that in Jin et al. (2017b).

2. *Failure patterns*

Figure 7.55 presents the final failure modes of the columns of series A ($\rho_{sv} = 1.26\%$). Figure 7.56 shows the failure modes of series B ($\rho_{sv} = 2.89\%$). Furthermore, the typical failure of stirrups and the surrounding concretes for series B are presented in Fig. 7.57.

From Figs. 7.55 and 7.56, one can clearly note that almost all RC columns failed by compression-shear modes. No failure occurred nearby the two ends of the specimens

Table 7.18 Concrete mixture proportions utilized in the concrete columns

Aggregate (kg/m^3)			Cement (kg/m^3)	Water (kg/m^3)	Water-reducing admixtures (kg/m^3)	Silica fume (kg/m^3)	Sum (kg/m^3)	*w/c* ratio
0–5 mm	5–10 mm	10–20 mm						
1090	185	436	460	184	4.0	75	2434	0.4

Table 7.19 Details of the steel rebar utilized in the concrete columns

Grade of steel	Diameter (mm)	Ultimate strength (MPa)	Yield strength (MPa)	Yield strain ($\mu\varepsilon$)	Elasticity modulus (GPa)
HRB400	8	644	458	2238	205
	10	671	485	2371	204
	12	665	478	2335	205
	18	675	494	2395	206
High-strength steel, ϕ^{PM}	9	1092	1044	4890	200

because of the strengthening behavior at the ends. The failure modes of the tested RC columns with different structural sizes are closely similar. Moreover, an obvious inclined crack appeared in the failure regions in all specimens. The failure regions are located in the middle length of the specimens, where the slip or separation behavior between steel bars and the surrounding concretes occurred and the corresponding lateral expansion is obvious.

From the comparison of the failure modes shown in Fig. 7.55a and b, one can see that tie arrangement type of stirrups has a great influence on the failure behavior of columns confined by reinforcement. This can also be illustrated from the comparison of failure patterns presented in Fig. 7.56a and b. Compared with series AX shown in Fig. 7.55b, the failure of series AXH with higher-strength stirrups is more ductile due to the stronger confinement. The number of broken stirrups within series AX is obviously larger than that within series AXH. In addition, according to the comparison of Figs. 7.55a and 7.56a, as well as that of Figs. 7.55b and 7.56b, it can be discovered that the stirrup ratio has a great influence on the failure behavior of RC columns. As the stirrup ratio increases, the confinement capacity increases, and the failure of columns becomes less brittle.

In Fig. 7.57, as the axial loading continued to increase, the lateral deformation at the mid-length of the columns increased, namely, the concrete at the mid-length would expand outward continuously. This would cause the continuous increase of tensile strain of stirrups, and the necking fracture finally happened for some of the stirrups crossing the main crack. Moreover, after the concrete spalling at the mid-length the columns, most of the longitudinal steel bars buckled due to the lack of constraint of cover concrete.

3. *Nominal stress-strain relation*

The nominal compressive stress-strain curves are plotted in Fig. 7.58. The nominal compressive stress σ is defined as the ratio of reaction force P at the loading point and the cross-sectional areas, i.e. $\sigma = P/D^2$. The nominal compressive strain ε is defined as the ratio of the vertical displacement Δ at the loading point and the column length H, i.e. $\varepsilon = \Delta/H$.

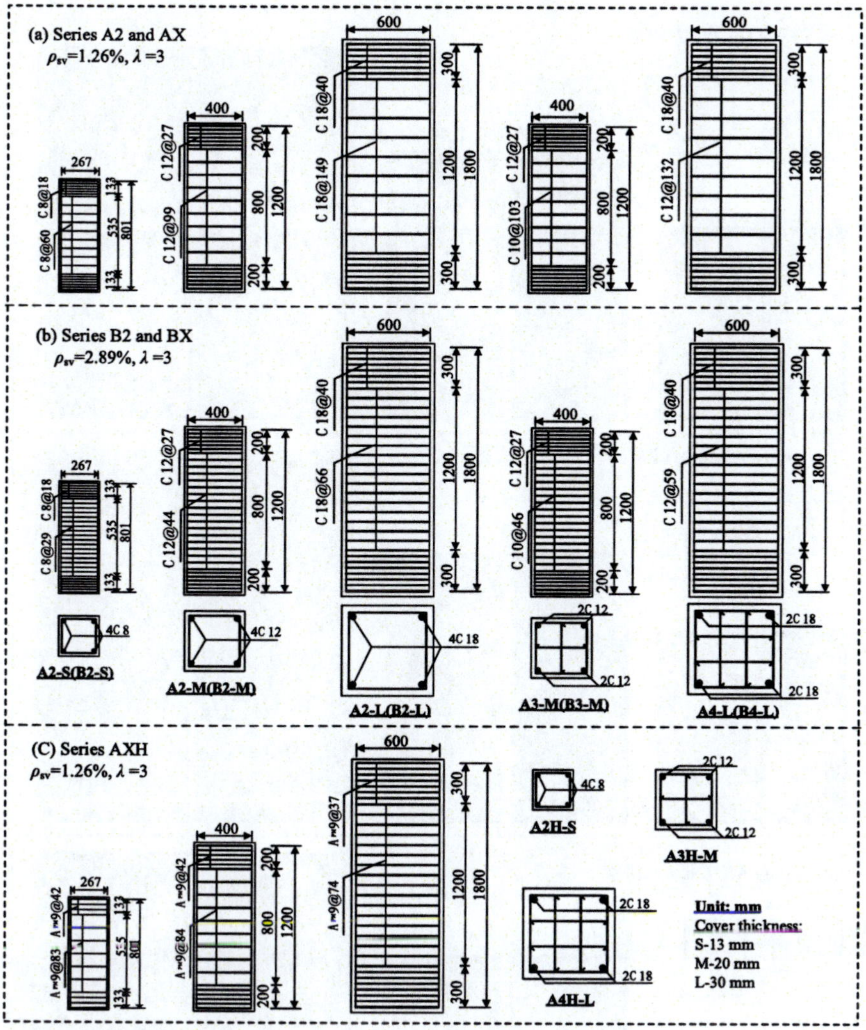

Fig. 7.54 Geometrical configuration and reinforcement layout: **a** Series A2 and AX; **b** Series B2 and BX; and **c** Series AXH

The measured load capacities P_u, the nominal compressive strengths σ_{Nu}, the nominal compressive strains ε_{Nu} at the peak stress, and the displacement ductility coefficients u ($u = \varepsilon_{85}/\varepsilon_{Nu}$) of 26 RC columns are listed in Table 7.20. Herein, ε_{85} denotes the strain when the applied loading drops to 85%P_u after the peak.

One can clearly note from Fig. 7.58 that there is no yield plateau for all stress-strain curves. Therefore, the failure of all columns is not plastic but brittle. In reality, the absence of a plateau implies the existence of softening in the material (or component) due to fracture, cracking, or other damage. This further implies that the failure process

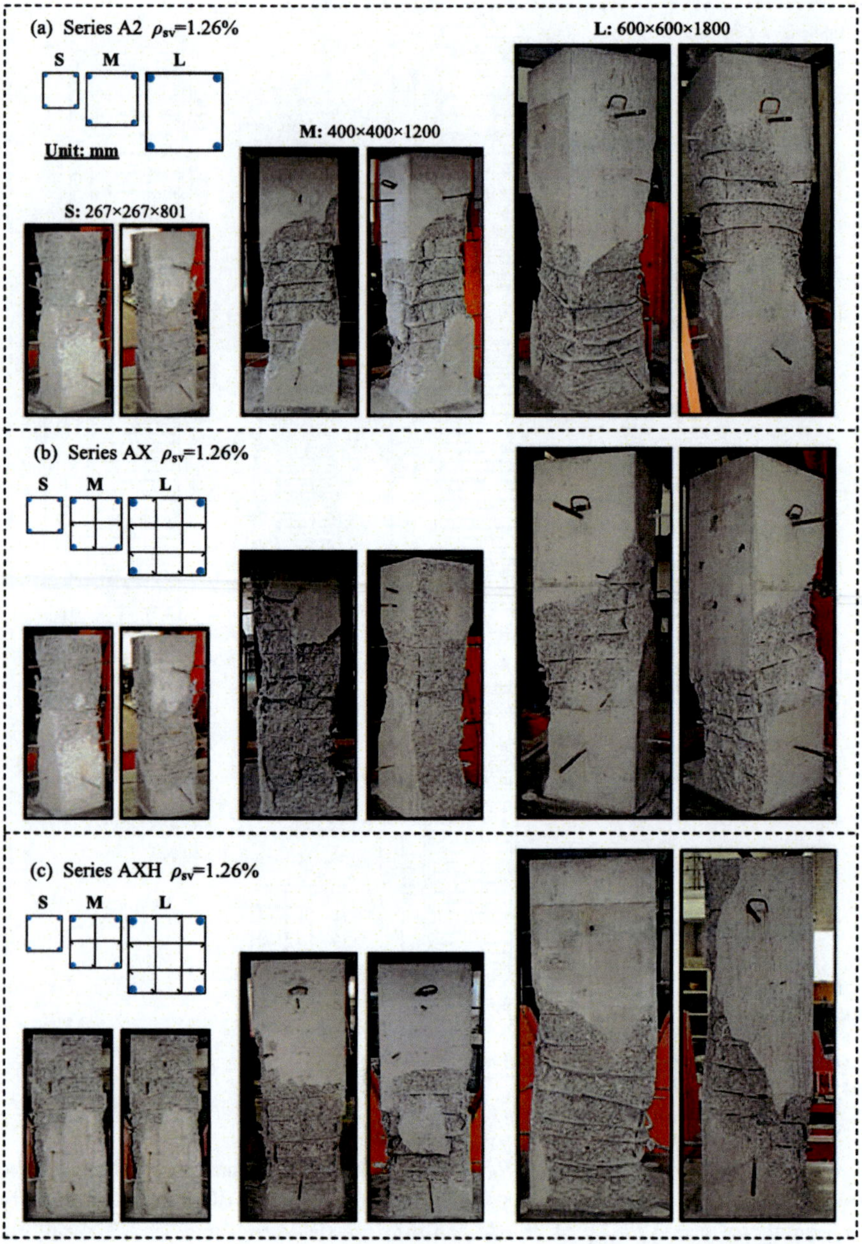

Fig. 7.55 Failure patterns of the RC columns: **a** Series A2; **b** Series AX; and **c** Series AXH

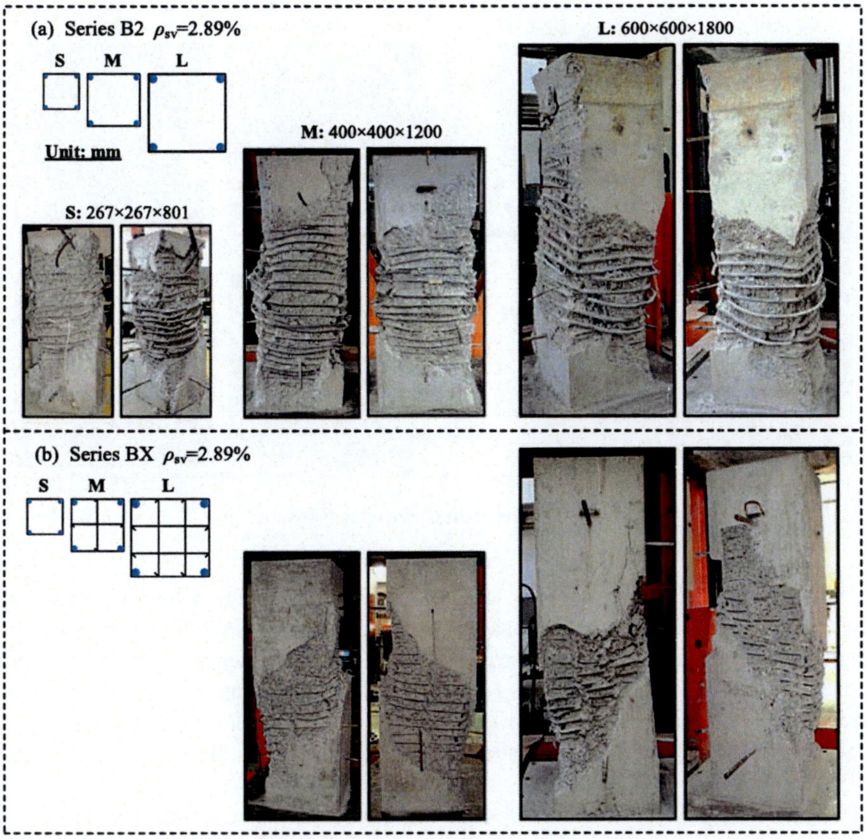

Fig. 7.56 Failure patterns of the RC columns: **a** Series B2; and **b** Series BX

of the tested columns cannot develop a single-degree-of-freedom mechanism but consists of propagation of the failure zones throughout the structure. This kind of brittle failure would cause the existence of size effect.

4. *Effect of stirrup ratio*

Due to the existence of transversal reinforcement, under axial compressive loadings, the RC columns turn out to be in a tri-axial compressive status. Comparing Fig. 7.58b with a, one can see that for the same tie arrangement types, with the increase of stirrup ratio, the confinement capacity on the RC columns improves significantly. This makes the nominal strength of the columns increase obviously (as can also be observed from the test data listed in Table 7.20. This can also be concluded from the comparison of the curves shown in Fig. 7.58c and d.

Moreover, due to the confinement effect of the transversal reinforcement, the failure of columns of series B is less brittle than that of series A. The declining slopes at the post-peak for series B2 and BX (shown in Fig. 7.58b and d, respectively) are

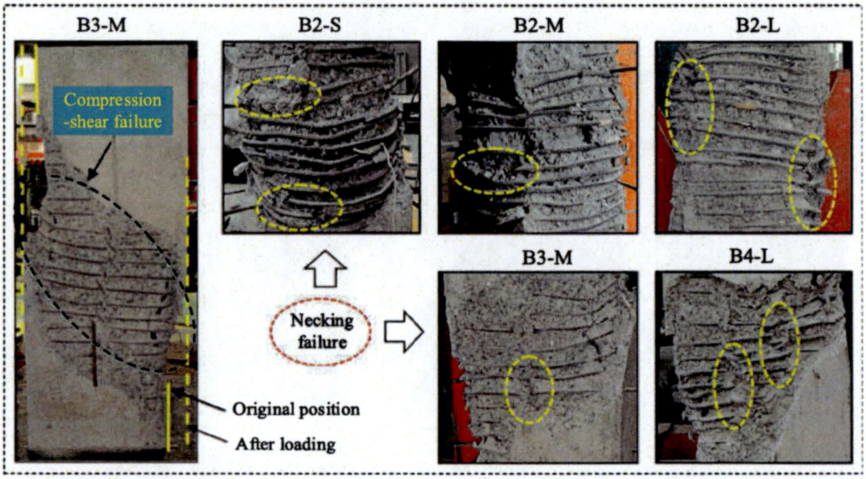

Fig. 7.57 Failure of the stirrups and surrounding concretes (for series B with ρ_{sv} of 2.89%)

obviously less steep than those for series A2 and AX, which is definitely caused by the confinement effect of transversal reinforcement. In Table 7.20, compared with that of RC columns with a lower stirrup ratio, the displacement ductility coefficient u described by the strain ratios $\varepsilon_{85}/\varepsilon_{Nu}$ of the RC columns with a higher stirrup ratio are obviously much larger. To sum up, as stirrup ratio increases, both the nominal compressive strength and the displacement ductility capacity of RC columns improve significantly.

In reality, as stirrup ratio increases, instead of unstable distributed crack, the crack formation is much more stable with finer micro-cracks. This makes an increase of the axial load-carrying capacity and the corresponding axial strain. In addition, this can also lead to a high lateral strain at the peak stress.

Nevertheless, one can note from Fig. 7.58b and d that the columns of series B still exhibit brittle failure, and the corresponding size effect should also exist. However, the stronger confinement effect the column possess, the less brittle the failure of the columns is, and the weaker size effect the failure exhibits.

5. *Effect of the tie arrangement type*

Tie arrangement type should have a great influence on the confinement effect. That is because that the mutual effects between steel bars and the surrounding concretes should be distinct for the columns having different tie arrangement types. Three types of tie arrangement were employed in the tests, and the leg number of ties for the columns having the "S", "M" and "L" structural sizes is 2, 3 and 4, respectively.

From Table 7.20 as well as the comparison of Fig. 7.58a and c, for the columns having the same structural sizes, stirrup ratios and steel strengths, one can find that the obtained nominal compressive strengths of the columns of serials AX are larger than those of the columns of series A2. In addition, one can note from the comparison

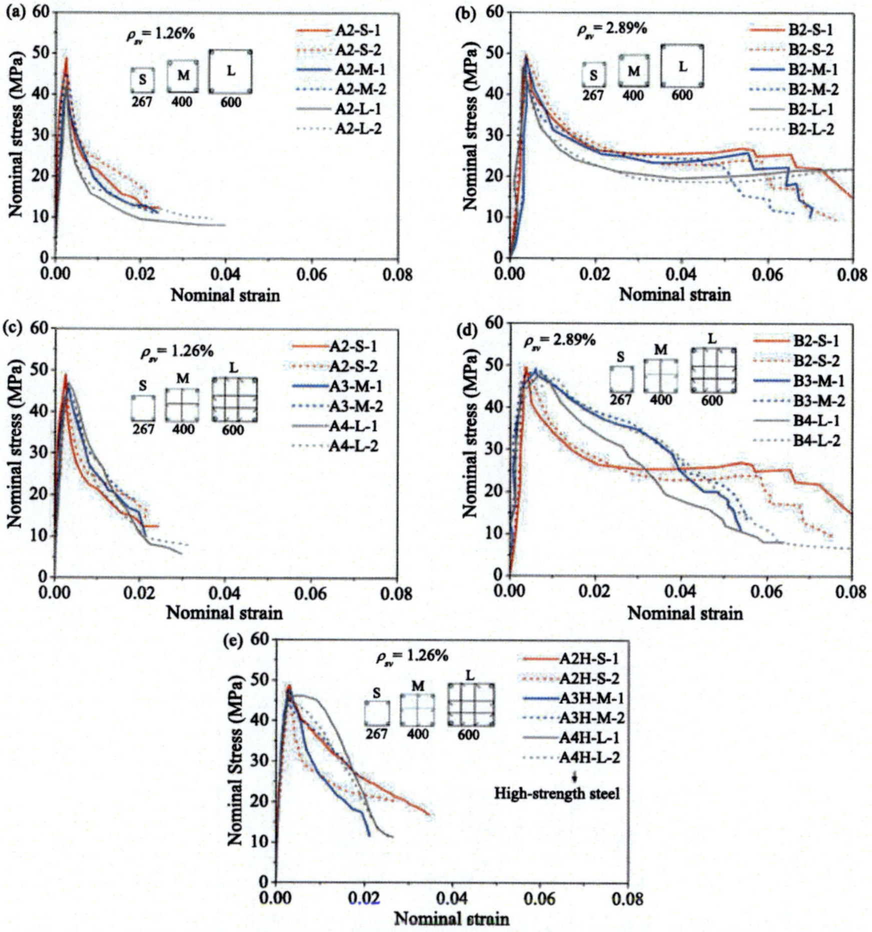

Fig. 7.58 Nominal compressive stress-strain curves of the tested RC columns: **a** Series A2; **b** Series B2; **c** Series AX; **d** Series BX; and **e** Series AXH

of Fig. 7.58b and d that the tie arrangement type has a significant influence on the nominal strength of the columns. Simply to say, the tested data indicates that the larger the leg number of ties is, the larger the constraining force on the columns becomes, resulting into the higher compressive strength of columns. This is because, when the number of stirrup legs increases, the constituent of concrete, such as cement mortar and coarse aggregates, are combined more closely, which makes the concretes not easy to break.

The deformation characteristics of the tested columns with different stirrup configurations are distinct obviously. From Table 7.20, it can be noticed that the mean values of ε_{Nu} (i.e. the peak strain) and ε_{85} for series AX are obviously larger than those for

Table 7.20 Test results of the confined concrete columns

Specimen	Load capacity P_u (kN)	Nominal strength σ_{Nu} (MPa)	Mean load capacity (kN)	Mean strength (MPa)	Mean peak strain ε_{Nu} (%)	Mean ε_{85} (%)	Ductility coefficient $u = \varepsilon_{85}/\varepsilon_{Nu}$
A2-S-1	3491	48.97	3395	47.63	0.25	0.31	1.24
A2-S-2	3300	46.29					
A2-M-1	6674	41.71	6954	43.47	0.25	0.33	1.32
A2-M-2	7235	45.22					
A2-L-1	15,131	42.03	15,334	42.60	0.25	0.28	1.12
A2-L-2	15,538	43.16					
A3-M-1	7218	45.11	7352	45.95	0.3	0.53	1.77
A3-M-2	7486	46.79					
A4-L-1	16,917	46.99	16,236	45.10	0.3	0.58	1.93
A4-L-2	15,554	43.21					
B2-S-1	3527	49.47	3550	49.80	0.37	0.55	1.49
B2-S-2	3573	50.12					
B2-M-1	7875	49.22	7678	47.99	0.36	0.60	1.67
B2-M-2	7482	46.76					
B2-L-1	16,690	46.36	17,003	47.23	0.29	0.37	1.28
B2-L-2	17,316	48.10					
B3-M-1	7816	48.85	7858	49.11	0.61	1.55	2.54
B3-M-2	7899	49.37					
B4-L-1	16,991	47.20	17,161	47.67	0.89	1.56	1.75
B4-L-2	17,330	48.14					

(continued)

Table 7.20 (continued)

Specimen	Load capacity P_u (kN)	Nominal strength σ_{Nu} (MPa)	Mean load capacity (kN)	Mean strength (MPa)	Mean peak strain ε_{Nu} (%)	Mean ε_{85} (%)	Ductility coefficient $u = \varepsilon_{85}/\varepsilon_{Nu}$
A2H-S-1	3498	48.78	3451	48.33	0.28	0.45	1.61
A2H-S-2	3405	47.88					
A3H-M-1	7114	46.03	7119	46.77	0.28	0.58	2.07
A3H-M-2	7123	47.50					
A4H-L-1	15,838	46.81	15,763	45.93	0.30	0.53	1.77
A4H-L-2	15688	45.05					

series A2 having the same structural sizes. For series B2 and BX, the same conclusions can also be drawn. Furthermore, from the comparison of stress-strain curves plotted in Fig. 7.58a and c, one can get that the post-peak softening slopes of the nominal stress-strain curves for series AX are less steep than those for series A2. The above analysis illustrates that with increasing the number of stirrup legs, the confinement effect generated by transversal reinforcement on the columns become much stronger. That can make the increase of the load bearing capacity, and cause the improvement of the ductility capacity of RC columns.

6. *Effect of yield strength of stirrups*

Compared with that of the columns of series AX, the yield strength of the stirrups utilized in series AXH is much larger. It can be obtained from Table 7.20 that compared with those of columns with the lower-strength stirrups, the mean nominal strengths of the columns with the higher-strength stirrups improves by 1.5, 1.8 and 1.8% for the three different structural sizes, i.e. the “S”, “M” and “L” ones. In addition, one can clearly note from the post-peak softening slopes plotted in Fig. 7.58c and e that the deformation capacity of the columns with the higher-strength stirrups is obviously stronger than that with the lower-strength stirrups. The confinement effect can effectively hinder the crack propagation within concrete, leading to the enhancement of load bearing capacity, deformation capacity and the peak strain $\varepsilon_{\mathrm{Nu}}$ at the peak stress. In addition, one can note from Table 7.20 that the ductility coefficient $u = \varepsilon_{85}/\varepsilon_{\mathrm{Nu}}$ increases as the yield strength of stirrups increases. This should be caused by the fact that as the strength of stirrups increases, the confinement effect due to the transverse rebar on the columns becomes much stronger. In other words, both the nominal compressive strength and the displacement ductility of the tested RC columns improve with the increase of yield strength of stirrups.

7. *Size effect on nominal compressive strength*

The nominal compressive strengths of all tested columns are plotted against with structural size D (i.e. the cross-sectional width), as presented in Fig. 7.59. One can clearly note from Fig. 7.59 that with the increase of structural size, all obtained nominal compressive strength decreases, but with different declining slopes. This, obviously indicates the existence of size effect in the RC columns tested in this work. Subsequently, the influences of stirrup ratio, tie arrangement type and yield strength of stirrups on the size effect in the tested RC columns will be explored in details.

Figure 7.59a plots the quantitative relationship of the tested nominal strength σ_{Nu} and structural size for series A2 and B2, to study the influence of stirrup ratio on size effect. One can note that the increase of stirrup ratio makes the nominal strength improve. However, for series B2 with higher stirrup ratio, the declining slope of the strength-size curve is obviously steeper than the one for series A2 with a lower stirrup ratio, indicating that the size effect is weakened. This should be caused by the fact that as the stirrup ratio increases, the confinement effect improves, making that the column becomes less brittle and thus less prone to the axial splitting failure. This is consistent with that in Bažant and Kwon (1994).

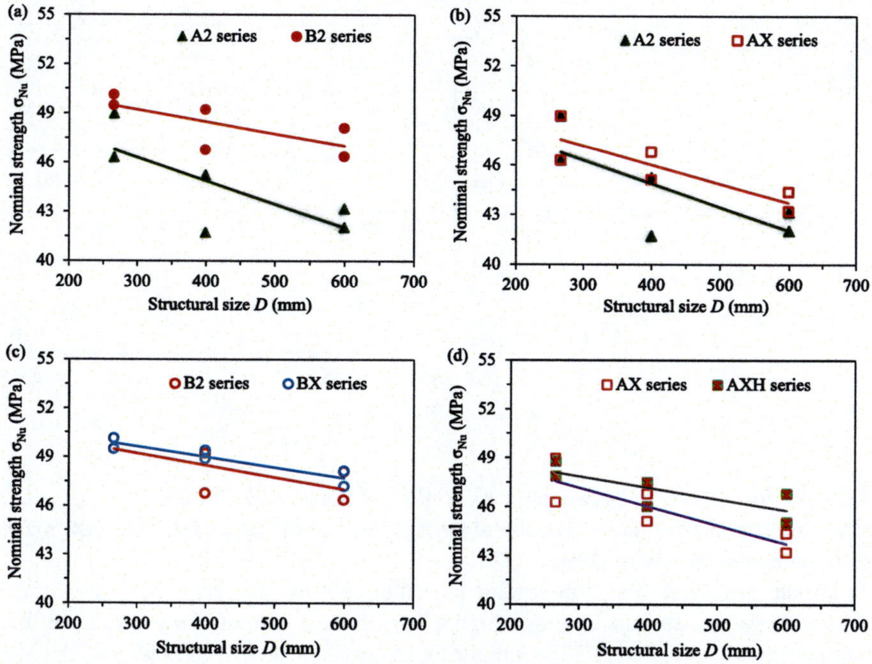

Fig. 7.59 Nominal strength versus structural size of the tested RC columns: **a** Series A2 and B2; **b** Series A2 and AX; **c** Series B2 and BX; and **d** Series AX and AXH

Figure 7.59b and c respectively show nominal strengths versus structural size for series A and B. It can be found from Fig. 7.59b and c that as the leg number of stirrups increases, the tested nominal strength increases. Moreover, the nominal strengths for series A2 decrease slightly faster than those of series AX. The similar tendency can also be observed for series B2 and BX in Fig. 7.59c. That is to say, the columns with complex tie arrangement type will exhibit a weaker size effect than those with simple tie arrangement type.

Figure 7.59d presents the relationship of the nominal strength against with structural size for series AX and AXH, to discuss the influence of yield strength of stirrups. By comparing the two fitting trend curves, one can see that the increase of strength of stirrups causes the improvement of nominal strength but the reduction of size effect. Under axial compression, the triaxial confining pressure engendered by stirrups with higher yield strength should be larger than that with lower yield strength. This, definitely, will make the failure of RC columns with high-strength stirrups less brittle. Thus, the columns present a weaker size effect.

Figure 7.60 gathers all data points of the nominal strength of 26 columns, along with the SEL of Bažant (1984), the strength criterion and the linear elastic fracture mechanics (LFEM). One can see from Fig. 7.60 that the SEL proposed by Bažant (1984) has a good agreement with the test data. All tested data keep far from the two curves (strength criterion and LEFM), indicating that the failure of the tested

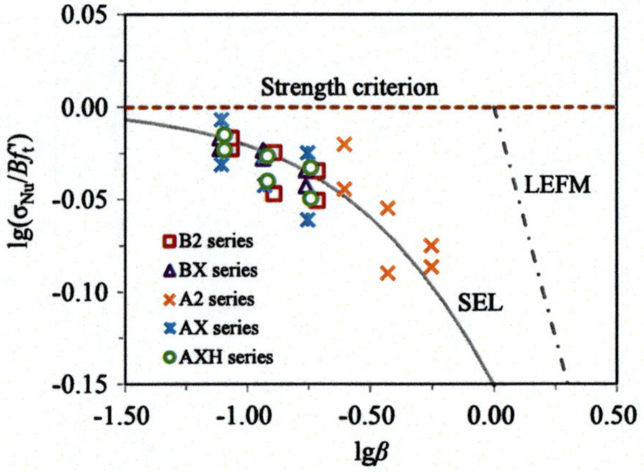

Fig. 7.60 Test results on size effect for all tested RC columns

RC columns should be between pure ductile and pure brittle. In addition, for small sizes, the curves approach a horizontal asymptote, and for larger sizes they approach an inclined asymptote of slope $-1/2$.

Similar with that of stirrup-confined circular columns in Sect. 7.5.1, the stirrup-confined square columns having a higher stirrup ratio exhibit a weaker size effect (comparing series A2 with B2). Comparing series A2 with AX, one can note that the size effect is obviously stronger for the RC columns with a smaller number of stirrup legs, and the tested data for series A2 approaches closer to the LEFM. By comparing series AX with AXH, it can be found that the size effect is a slightly stronger for the RC columns with a lower yield strength of stirrup. In summary, RC columns with a looser tie arrangement type, a lower reinforcement ratio and a lower strength grade of stirrups present a stronger size effect.

7.5.3 Numerical Investigations

This section presents the numerical results and discussions of circular and squared stirrup-confined concrete columns (Jin et al. 2020d).

1. *Stress-strain responses*

Figure 7.61 plots the nominal axial compressive stress-strain responses of circular (Jin et al. 2020c) and square stirrup-confined RC columns (Jin et al. 2020d). From Fig. 7.61a, the ascending part of stress-strain curves with the identical volumetric stirrup ratio almost coincided, and the initial tangent modulus were basically consistent. The results indicated that the size effect of the elastic modulus was not evident. This phenomenon can be explained by the following: the size effect is caused by mechanisms closely associated with the fracture response of concrete, while the

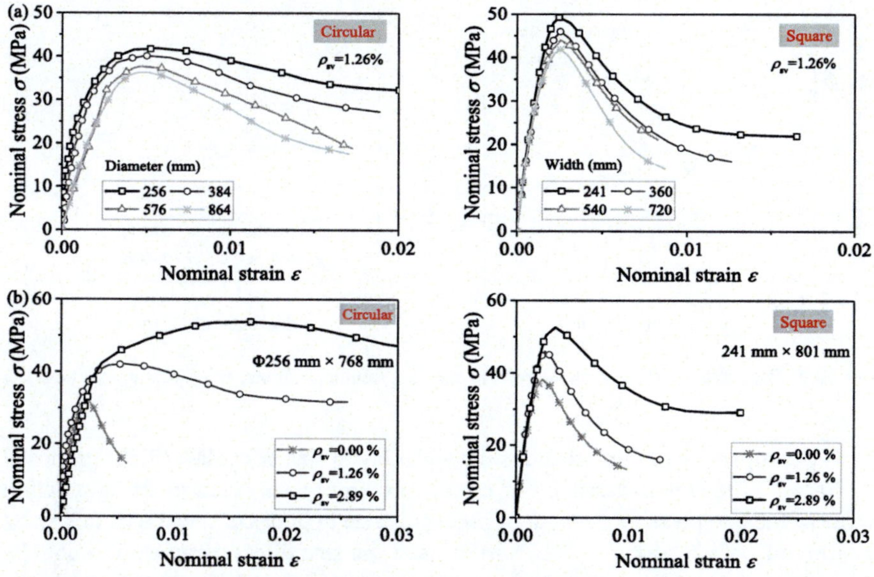

Fig. 7.61 Stress-strain curves of the square and circular stirrup-confined RC columns: **a** Tnfluence of specimen size; and **b** The influence of volumetric stirrup ratio

elastic modulus is the mechanical parameter in the elastic stage which is not influenced by the concrete fracture. Moreover, for columns having same stirrup ratio, with the increase of the section size, the descending part of the curve becomes steeper. That is because the confinement effect of stirrup weakens as the section size increases.

Figure 7.61b depicts the stress-strain curves of the columns with different volumetric stirrup ratios and identical cross-section size. For both square and circular RC columns, as the volumetric stirrup ratio increases, the stress at the peak point increases, and the declining slope at the post-peak are evidently less steep. That is definitely caused by the confinement effect of transversal reinforcement.

2. *Peak axial stress*

The peak axial stress σ_{Nu} of the concrete specimen under axial compression means the nominal compressive strength defined in Sect. 7.5.1. Figure 7.62 presents the peak axial stress σ_{Nu} under different stirrup ratios and section size of square RC columns. In Fig. 7.62, as the section size increased from smallest to largest value, the peak axial stress decreased, which indicated that peak axial stress exhibited evident size effect behavior. In addition, with the increase of stirrup ratio, the confining effect of transverse reinforcements enhanced, and the peak axial stress exhibited an increasing trend. This phenomenon illustrated that the volumetric stirrup ratio can increase the peak axial stress. In Fig. 7.62, it is noteworthy that the nominal strength of the columns having larger stirrup ratios decreased less steep with the increasing section size. This demonstrated that the confining effect of transverse reinforcements can weaken the size effect of the peak axial stress.

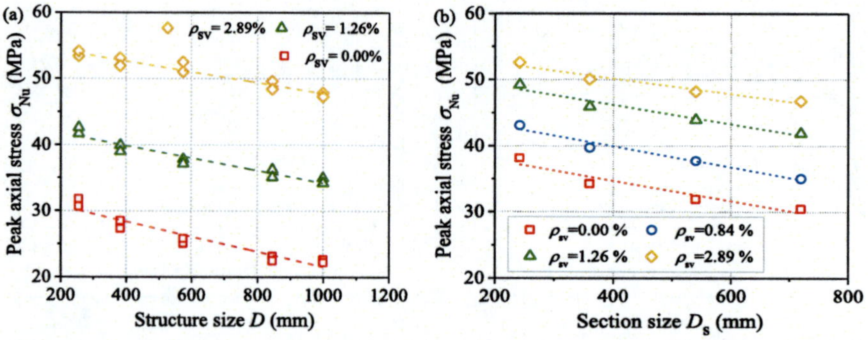

Fig. 7.62 Peak axial stress versus the section size: **a** Circular columns; and **b** Square columns

The comparison of numerical results with classic size effect law (SEL) proposed by Bažant (1984) is presented. For circular column, f_c is taken as the simulation value of the compressive strength of plain concrete cylindrical specimens and sized of 150 mm × 300 mm (i.e., 34.76 MPa), and the size effect parameters B and D_0 were respectively 1.017 and 800. For square columns, f_c is taken as the simulation value of the compressive strength of concrete prism sized of 150 mm × 150 mm × 300 mm (i.e., 39 MPa), and the size effect parameters B and D_0 were respectively 1.176 and 714. Consequently, the numerical data points along with the size effect curve, plastic strength criteria and the linear elastic fracture mechanics (LEFM) were plotted in Fig. 7.63. For both circular and square columns, the numerical results of plain concrete columns generally align with the SEL, illustrating that SEL can well describe the size effect on the nominal compressive strength of axially loaded plain concrete columns.

3. *Peak axial strain*

The peak axial strain ε_{Nu} means the value of axial compressive strain at the peak axial stress σ_{Nu}. Figure 7.64a and b presents the peak axial strain having different stirrup

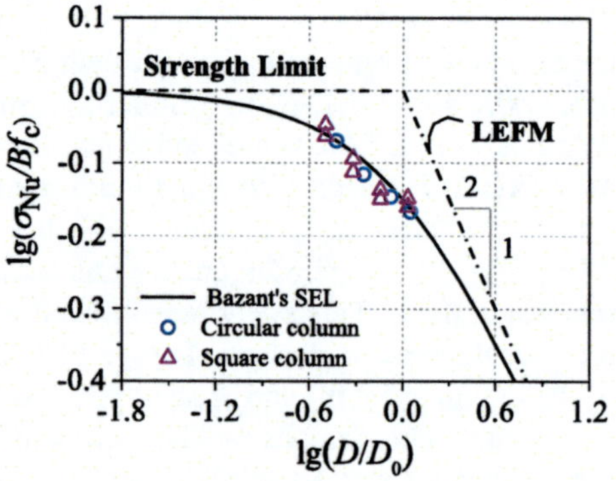

Fig. 7.63 Bi-logarithmic curves for plain concrete columns

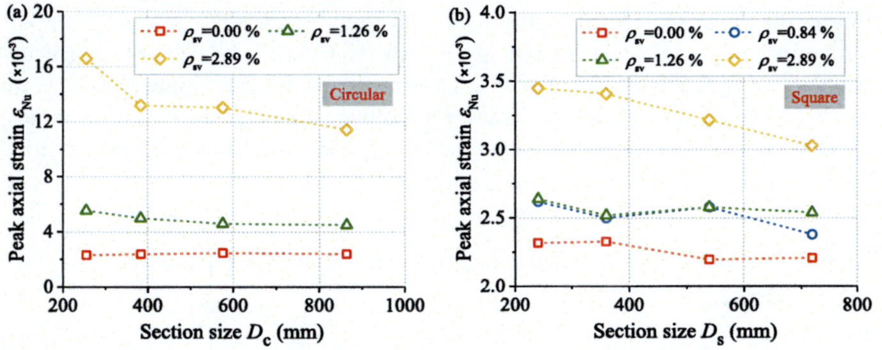

Fig. 7.64 Peak axial strain versus the section size for: **a** Circular columns; and **b** Squared columns

ratios and section size, for the circular and square columns, respectively. In Fig. 7.64b, for square columns with the stirrup ratio of 2.89%, the peak axial strain exhibited a decreasing trend with the increase of section size. In addition, the plain concrete columns showed the same trend. However, it was observed that the columns with the stirrup ratios of 0.84% and 1.26% did not exist an obvious trend. In Fig. 7.64a, for circular concrete columns with the stirrup ratios of 1.26% and 2.89%, the peak axial strain decreased with the increase of section size. Generally, the peak axial strain decreased with the increase of section size for circular and square stirrup-confined concrete columns.

4. *Softening rate of the descending branch*

Figure 7.65 presents the relationship between softening rate K_{soft} and specimen size D of the circular and square columns. Here, the softening rate K_{soft} is defined as the

Fig. 7.65 Softening rate of stirrup-confined RC columns

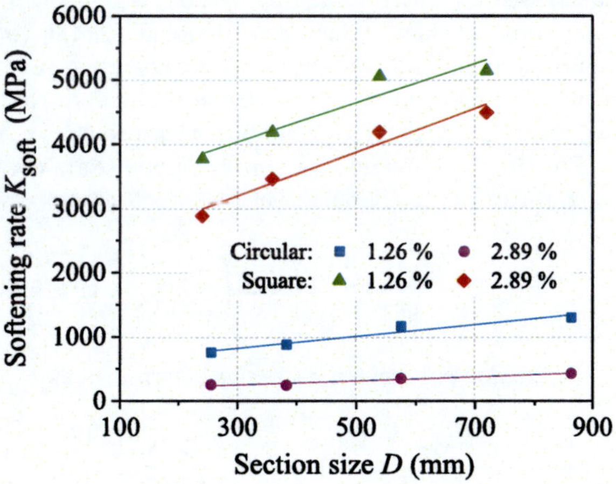

slope of the linear segment from the peak axial stress σ_{Nu} to 85% σ_{Nu} in the stress-strain curve. In Fig. 7.65, the softening rate increases with the decrease of stirrup ratio, indicating that the specimens with a larger stirrup ratio have higher ductility. In addition, the softening rate increases as the specimen size increases, which illustrates that in the identical stirrup ratio, the failure of larger-sized specimens is more brittle.

7.5.4 *Theoretical Analysis on Size-Dependent Stress-Strain Model*

In this section, a size-dependent stress-strain model for circular and squared stirrup-confined RC columns is presented, in which the size effect of peak axial stress, peak axial strain and the softening behavior of descending branch was considered (Jin et al. 2020d).

1. *Size effect equation on peak axial stress*

The size effect equation which was based on classical SEL with considering the confinement effect of stirrups, is described by Eqs. (7.17)–(7.19a) and (7.31) in Sect. 7.8.2.

2. *Size effect equation on peak axial strain*

In reality, only a few research works focus on the size effect on peak axial strain. The study on the size effect of the peak axial strain is inadequate. In addition, the research on the size effect of peak axial strain of concrete confined by stirrups lacks relevant experimental data and theoretical analysis model.

As one of the important parameters in the stress-strain model of confined-concrete under axial compression, the peak axial strain is affected by several aspects, e.g. the concrete strength, specimen dimension, confinement effect, and the loading rate. Therefore, considering the cross-sectional size (D_{c} for circular column, and D_{s} for square column) and the ratio of lateral confining stress to the compressive strength of unconfined concrete (f_l/f_c), the equation for peak axial strain was obtained by regression analysis through nonlinear least squares method (Jin et al. 2020d).

For circular columns, the corresponding strain corresponding to peak stress has the form as

$$\frac{\varepsilon_{Nu,c}}{\varepsilon_c} = \left[1 + 2.3\left(\frac{D_c}{150}\right)^{-0.5}\right]\left[1 + 0.36\left(\frac{10f_l}{f_c}\right)^4\right] \tag{7.6}$$

For square columns, the strain is determined as

$$\frac{\varepsilon_{Nu,s}}{\varepsilon_c} = \left[1 + 0.45\left(\frac{D_s}{150}\right)^{-0.5}\right]\left[1 + 0.21\left(\frac{10f_l}{f_c}\right)^4\right] \tag{7.7}$$

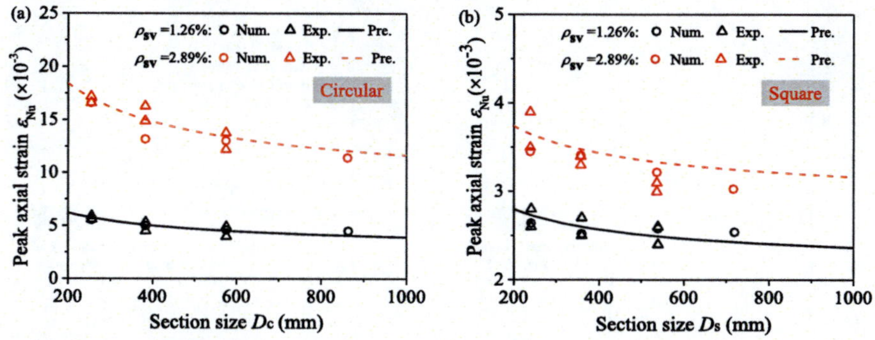

Fig. 7.66 Comparisons between predictions and simulation data of peak axial strain for: **a** Circular columns; and **b** Square columns

in which, $\varepsilon_{Nu,c}$ and $\varepsilon_{Nu,s}$ are the peak axial strain of the confined concrete for circular and square column, respectively; ε_c is the axial strain corresponding to the peak axial stress of unconfined concrete, and is assumed as 0.002 in all data interpretation in this section; f_l is the effective lateral confining stress by stirrup. In addition, the f_l of circular and square column can be determined by Eqs. (7.8) and (7.9), respectively.

In Eqs. (7.6) and (7.7), the peak axial strain of stirrup-confined concrete is in inverse proportion with cross-sectional size and direct proportion with the value of f_l/f_c. The difference between Eqs. (7.6) and (7.7) are considerably caused by the stirrup type, i.e., the confinement effect of circular hoops is higher than square hoops.

Figure 7.66 presents predicted peak axial strain compared with numerical and experimental results for circular and square columns. In Fig. 7.66, the equation [Eqs. (7.6) and (7.7)] can well estimate the size effect of the peak axial strain. Specially, the equation [Eq. (7.7)] overestimate the peak strain of square columns with larger stirrup ratio, indicating the model is less stable in predicting the peak axial strain of square columns. That may be resulting from the discreteness in the measurement of strain. Nevertheless, the comparisons of other specimens can indicate the accuracy of the proposed equation of peak axial strain.

Apart from the numerical results, the comparison of proposed equation with the collected experimental data is plotted in Fig. 7.67. $\varepsilon_{\text{Nu,exp}}$ and $\varepsilon_{\text{Nu,pre}}$ present the experimental and predicted values, respectively. The ratios of the predicted peak axial strain to the experimental data $\varepsilon_{\text{Nu,pre}} : \varepsilon_{\text{Nu,exp}}$ were obtained. The average value of $\varepsilon_{\text{Nu,pre}} : \varepsilon_{\text{Nu,exp}}$ for circular and square columns were 0.94 and 0.93, respectively. The coefficients of variation were 0.15 and 0.20 for circular and square columns, respectively. In Fig. 7.67, for stirrup-confined circular and square columns, the model can provide reasonable close predictions of peak axial strain. Due to the sensitivity of measurement of peak axial strain, the comparison results of peak axial strain are slightly discrete. Nevertheless, the model of peak axial strain needs to be further verified.

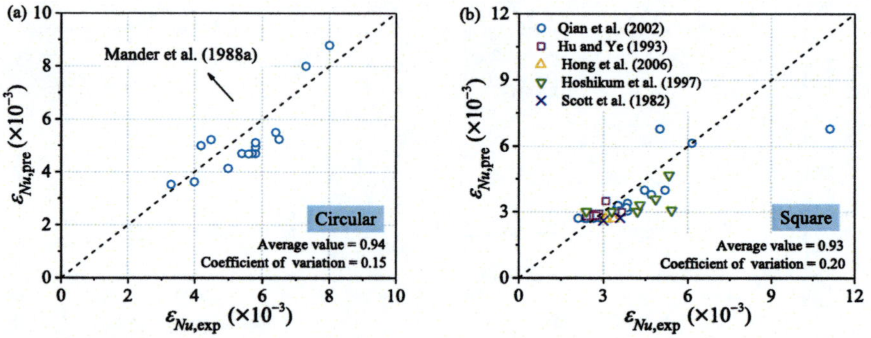

Fig. 7.67 Comparisons between predictions and test data of peak axial strain for: **a** Circular columns; and **b** Square columns

Overall, the predictions are in good agreement with the available experimental results. Nevertheless, more experiments are required to further verify the accuracy of the size effect formula for peak axial strain.

3. *Softening rate considering the effect of specimen size*

According to the preceding analyses, softening rate of descending branch was associated with the stirrup ratio and the specimen size. Therefore, the size effect equation of the softening rate was presented, which was established by regression analysis on the numerical results of the circular and square columns (as plotted in Fig. 7.68). The relations obtained from regression analyses are as follows:

$$K_{soft,c} = 0.3\left(\frac{\rho_{su}f_y}{f_c^2}\right)^{-1.38}\left(\frac{D_c}{150}\right)^{0.5} \tag{7.8}$$

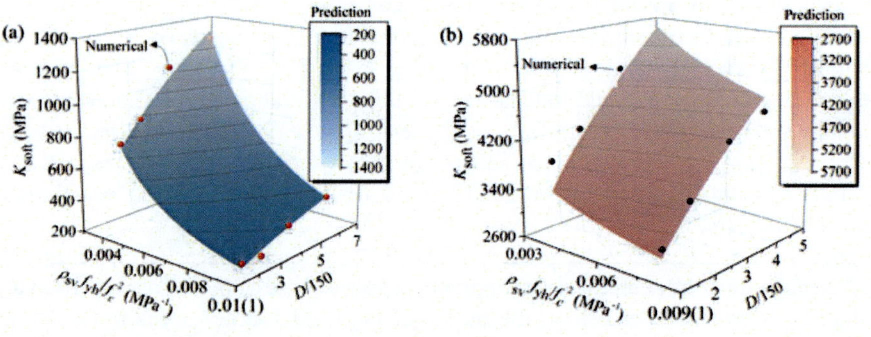

Fig. 7.68 The influence of confinement factor and specimen size on the softening rate for: **a** Circular columns; and **b** Square columns

$$K_{soft,s} = 751\left(\frac{\rho_{su}f_y}{f_c^2}\right)^{-0.22}\left(\frac{D_s}{150}\right)^{0.5} \tag{7.9}$$

where $K_{\text{soft,c}}$ and $K_{\text{soft,s}}$ are the softening rate of the confined concrete for circular and square column, respectively. The influence of stirrup ratio and specimen size was considered in the size effect equation of softening rate.

4. *Stress-strain equations*

Once the size effect formula of peak axial stress and peak strain are given, the following equations can be used to predict the stress-strain response of stirrup-confined circular and square column. The ascending branch adopt the equation initially proposed by Popovics (1973), which was also employed by Mander et al. (1988), Razvi and Saatcioglu (1999), Jiang and Teng (2007), etc. The ascending branch is expressed as

$$\sigma = \frac{\sigma_{Nu} r(\varepsilon/\varepsilon_{Nu})}{r - 1 + (\varepsilon/\varepsilon_{Nu})^r} \varepsilon \leq \varepsilon_{Nu} \tag{7.10a}$$

with

$$r = E_c/(E_c - \sigma_{Nu}/\varepsilon_{Nu}) \tag{7.10b}$$

$$E_c = 4700\sqrt{f_c} \tag{7.10c}$$

where σ and ε are the axial stress and corresponding axial strain of the concrete respectively; r is the coefficient of the ascending branch; E_c is the concrete's initial elastic modulus. In addition, E_c is typically defined at $0.4f_c$ where no micro crack is likely to occur, thus the size effect of concrete is slight and can be ignored.

The descending branch is described by a linear segment from the peak point, which can be written as:

$$\sigma = \sigma_{Nu} - K_{soft}(\varepsilon - \varepsilon_{Nu}) \geq 0.2\sigma_{Nu} \tag{7.11}$$

in which, K_{soft} controls the slope of the descending branch and can be determined by Eqs. (7.8) and (7.9).

The comparisons of predictions of size-dependent stress-strain model with the measured stress-strain curves of circular (Sect. 7.5.1) and square (Sect. 7.5.2) stirrup-confined RC columns, were plotted in Figs. 7.69 and 7.70. The available models proposed by Mander et al. (1988), Hoshikuma et al. (1997), and Legeron and Paultre (2003) were also plotted. It is to be noted that, the vertical axis (Figs. 7.69 and 7.70) represents the axial stress of the core concrete. Obviously, the predicted stress-strain curves generally satisfy for all series of specimens, irrespective of the specimen size, stirrup ratio and the cross-sectional shape. This illustrates the good performance of the model considering size effect. In addition, the models (Mander et al. 1988;

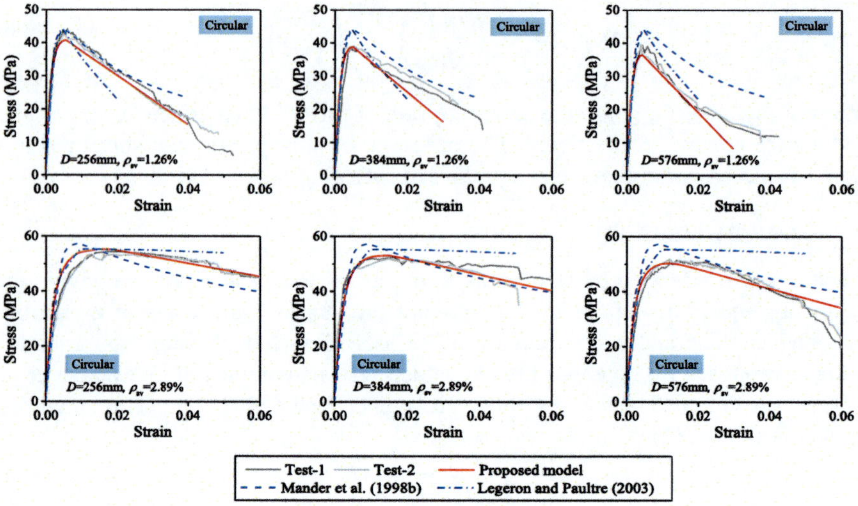

Fig. 7.69 Comparisons of predicted and experimental stress-strain curves for circular columns

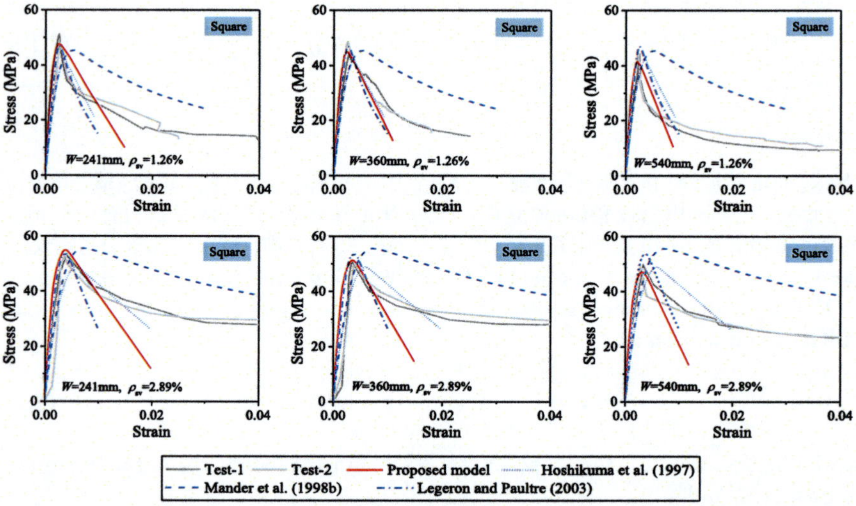

Fig. 7.70 Comparisons of predicted and experimental stress-strain curves for square columns

Hoshikuma et al. 1997; and Legeron and Paultre 2003) generally match well with the stress-strain curves of smaller-sized columns. However, with the increase of the specimen dimension, the difference between the test results and these models (Mander et al. 1988; Hoshikuma et al. 1997; and Legeron and Paultre 2003) is more significant. The stress-strain model without considering size effect has overestimated the peak axial stress and peak axial strain of larger-sized specimens, which decreases the reliability of engineering design.

7.6 Compressive Failure of FRP-Confined Concrete Column

The experiment of 30 geometrically-similar CFRP-wrapped square concrete columns under monotonic and repeated axial compression tested in Jin et al. (2020a) is presented in this section. Moreover, the numerical simulation that was extended to study the axial-compressive failure of FRP-wrapped square and circular concrete columns (Jin et al. 2020e) are summarized.

7.6.1 Experimental Investigations on CFRP-Confined RC Column

1. *Experimental program*

In order to reduce the dispersion of the results, all the specimens were built from the same batch of concrete. The measured average compressive strength of the concrete cube specimens (sized of 150 mm × 150 mm × 150 mm) was 41.65 MPa. The longitudinal rebar was selected as the "HRB400" grade, and the stirrup rebar was selected as the "HPB300" grade. The measured mechanical properties of the CFRP and steel bars used in the test are listed in Tables 7.21 and 7.22, respectively.

A total of 30 CFRP-wrapped square concrete columns with different geometries (in the ratio of 1:2:3) were designed. The slenderness ratio ($\lambda = L/B_c$) was set to 3 for all columns, and the size were designed as Small (200 mm × 200 mm × 600 mm), Medium (400 mm × 400 mm × 1200 mm), and Large (600 mm × 600 mm × 1800 mm). The corners of all the columns were ground into chamfers after the concrete specimens were cured. The chamfer radius was 0.16 times the length of the column sections. The geometric details of the columns designed for the test are detailed in Fig. 7.71 and Table 7.23.

The experimental arrangement and loading system are plotted in Fig. 7.72. The 30 CFRP-wrapped concrete columns were subject to cyclic axial compression tests by the 112,000 kN capacity MESLTS (shown in Fig. 7.72a). The schematic of the loading arrangement and the displacement measurement is shown in Fig. 7.72b. The

Table 7.21 Mechanical parameters of CFRP

Variables	Ultimate tensile strength f_f (MPa)	Single layer thickness t (mm)	Elasticity modulus E_f (GPa)	Ultimate rupture strain ε_f (%)
CFRP	3461	0.167	241	1.61

Table 7.22 Mechanical parameters of steel bars

Grade of steel	Diameter (mm)	Ultimate strength (MPa)	Yield strength (MPa)
HPB300	6	536	381
HPB300	8	452	386
HPB300	10	424	316
HRB400	8	570	419
HRB400	16	694	496
HRB400	25	609	453

Note "HPB" and "HRB" are "Hot-rolled Plain-steel Bar" and "Hot-rolled Ribbed-steel Bar", respectively

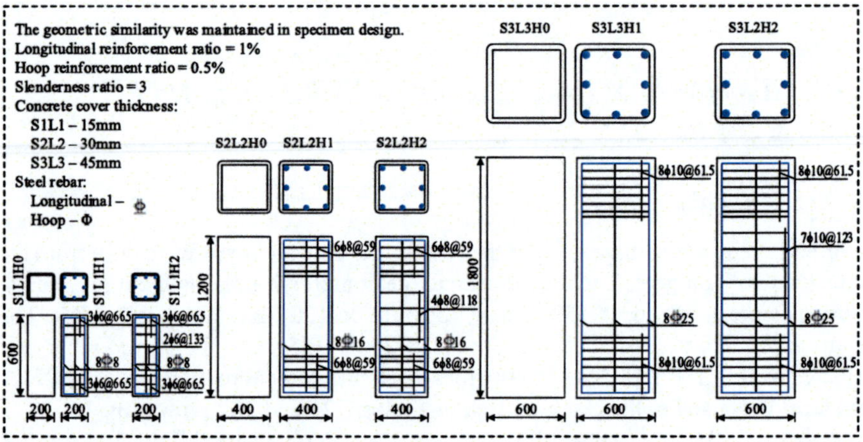

Fig. 7.71 The cross-sectional dimensions and reinforcement details

test specimens were supported on the instrument by both ends and the load was applied axially. Both ends of the test specimens were placed on the spherical seating supports so that the eccentricity of the load could be minimized.

The control was based on a constant increment of averaged forces during the test. Small (S1), medium (S2) and large (S3) sizes of components were loaded in increments of 200 kN, 1000 kN and 2000 kN, respectively. Each stage load was repeated twice, and the test was terminated when the bearing capacity of the specimen decreased to 85% of the peak value. Two loading modes were used including full loading and full unloading (P) and full loading and partial unloading (C). "P" was unloaded to 0 after each full loading, and "C" was unloaded to 50% of each load level after each full loading. The loading modes are shown in Fig. 7.72c.

2. *Failure patterns*

The failure process of specimens with different sizes was basically similar under the same conditions. When the specimen was destroyed, CFRP was pulled apart in

Table 7.23 Specimen design parameters

Label	Size (mm × mm)	Height *H* (mm)	Specimen number	Thickness (mm)	Reinforcement (%)	Stirrup (%)	FRP layer	Radius (mm)
S1L1H0C	200 × 200	600	3	15	–	–	1	32
S1L1H0P			3		–	–	1	32
S1L1H1C	200 × 200	600	1	15	1	–	1	32
S1L1H1P			1		1	–	1	32
S1L1H2C	200 × 200	600	1	15	1	0.5	1	32
S1L1H2P			1		1	0.5	1	32
S2L2H0C	400 × 400	1200	3	30	–	–	2	64
S2L2H0P			3		–	–	2	64
S2L2H1C	400 × 400	1200	1	30	1	–	2	64
S2L2H1P			1		1	–	2	64
S2L2H2C	400 × 400	1200	1	30	1	0.5	2	64
S2L2H2P			1		1	0.5	2	64
S3L3H0C	600 × 600	1800	3	45	–	–	3	96
S3L3H0P			3		–	–	3	96
S3L3H1C	600 × 600	1800	1	45	1	–	3	96
S3L3H1P			1		1	–	3	96
S3L3H2C	600 × 600	1800	1	45	1	0.5	3	96
S3L3H2P			1		1	0.5	3	96

Note "S1", "S2", and "S3" mean "small", "medium", and "large" corresponding to section width of 200 mm, 400 mm and 600 mm; the second letter "L" represents CFRP layers, "L1", "L2" and "L3" mean one, two, and three layers, respectively; the third letter "H" represents the configuration of reinforcing rebar, "H0" expresses the specimen has no rebar, "H1" expresses the specimen only has longitudinal rebar, and "H2" means the specimen has longitudinal and stirrups rebar; in addition, "C" and "P" corresponds to different loading modes, "C" means "full loading and full unloading" and "P" means "full loading and partial unloading"

the middle part of the column, the concrete was crushed locally, and concrete slag splashed around. The CFRP-wrapped plain concrete columns were most seriously damaged, and concrete columns were directly crushed in the middle, as shown in Fig. 7.73. The experimental phenomena of reinforced concrete columns (longitudinal reinforcement only) strengthened by CFRP were similar to those of reinforced plain concrete columns at the initial stage of loading. When the load approached the peak load of concrete columns, CFRP cracked locally and the transverse strain of CFRP increased obviously. When the specimens were destroyed, the longitudinal steel bars configured in the specimens buckled outward, and the concrete around the steel bars spalled in a large area. The CFRP was pulled off at the corner, and the concrete was conical at the damaged part of the specimen. Compared with specimens without stirrup restraint, the buckling degree of longitudinal reinforcement was smaller and

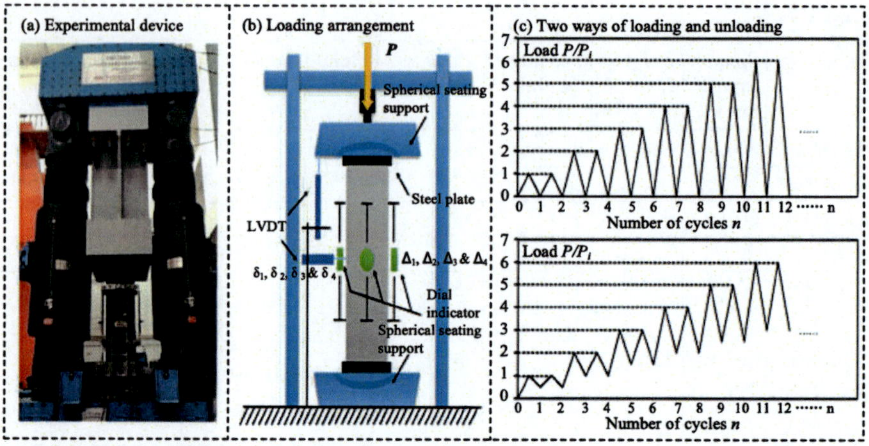

Fig. 7.72 Loading instrumentation and schematic diagram of the test setup: **a** Experimental device; **b** Loading arrangement; and **c** Two ways of loading and unloading

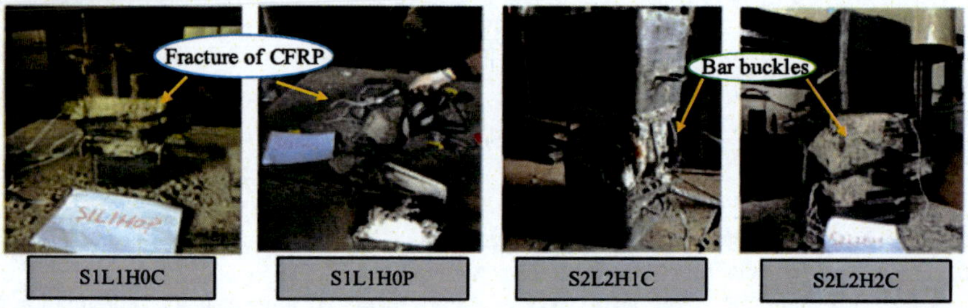

Fig. 7.73 Typical failure modes of tested columns

the spalling of concrete was less. The typical failure phenomena of specimens are shown in Fig. 7.73.

3. *Axial load-displacement characteristic*

The test results of each column are shown in Table 7.24. The nominal compressive strength σ_N (i.e. the nominal stress at peak load) was defined as $\sigma_N = P_{\max}/D^2$. P is the force applied on the top of the test specimens; $P_{\max}$ means the maximum applied loads; and D is the side length of the section before loading. Similarly, the nominal compressive strain was defined as $\varepsilon = \Delta/L$. In order to eliminate the dispersion, the test results of CFRP-wrapped plain concrete columns were the average values of three columns.

From Table 7.24, it can be seen that the failure location of specimens was basically in the middle of the columns, and this result was not affected by the size change. The nominal compressive strength of CFRP-wrapped reinforced concrete columns (H1 or H2) was higher than that of CFRP-wrapped plain concrete columns (H0) under the

Table 7.24 Test results

Specimen label	Peak P_u (kN)	Peak Δ_p (mm)	σ_{Nu} (MPa)	Peak ε_p ($\mu\varepsilon$)	Failure position
S1L1H0P	1515.85	1.45	38.89	2420	Middle
S1L1H0C	1590.56	1.25	40.76	2090	Middle
S1L1H1P	1892.45	1.14	47.31	1900	Middle
S1L1H1C	1861.70	2.23	46.54	3720	Middle
S1L1H2P	1783.52	3.95	44.59	6580	Middle
S1L1H2C	1971.23	1.33	48.58	2220	Middle
S2L2H0P	6833.53	2.52	40.71	2100	Middle
S2L2H0C	7066.02	1.87	41.16	1560	Middle
S2L2H1P	7821.69	2.58	48.89	2150	Middle
S2L2H1C	7246.86	2.07	45.29	1730	Middle
S2L2H2P	7574.04	2.00	47.34	1670	Middle
S2L2H2C	8167.69	1.89	50.05	1580	Middle
S3L3H0P	13,787.76	2.05	38.30	1140	Middle
S3L3H0C	15,064.19	2.22	39.48	1320	Middle
S3L3H1P	16,113.21	1.34	44.76	740	Middle
S3L3H1C	14,545.72	1.64	42.40	910	Middle
S3L3H2P	16,568.00	1.35	46.02	750	Middle
S3L3H2C	16,970.00	1.76	47.14	980	Middle

Note The nominal compressive strength (i.e. the nominal stress at peak load), was defined as $\sigma_{Nu} = P_u/D^2$

same size. With the increase of structural size, this difference turns to be decreased. For "H1" columns (with longitudinal rebar only) and "H0" columns (without rebar), the increased range of peak strength for small (S1), medium (S2) and large (S3) size columns was 20.9%, 8.41% and 6.26%, respectively. For CFRP-wrapped concrete columns, the result shows that the influence of stirrups on the strength enhancement decreases with the increase of structural size. For CFRP-wrapped concrete columns, the influence of reinforcing bars on the peak displacement varies with structural size. Compared with "H0" columns in the small-scale components (200 mm × 200 mm × 600 mm), "H1" columns and "H2" columns have a 24.81% and 56.68% increase in the peak displacement, respectively. However, for large-scale components (600 mm × 600 mm × 1800 mm), the peak displacement does not improve. Compared with "H0" columns, "H1" columns and "H2" columns have a 30.21% and 27.40% decrease in the peak displacement, respectively.

4. *Hysteretic and skeleton curves*

The hysteretic curves of load-displacement in this test under repeated cyclic axial compression loads are plotted in Fig. 7.74. The stiffness, ductility, strength and energy dissipation ability of the specimen can be reflected by the hysteretic curve of

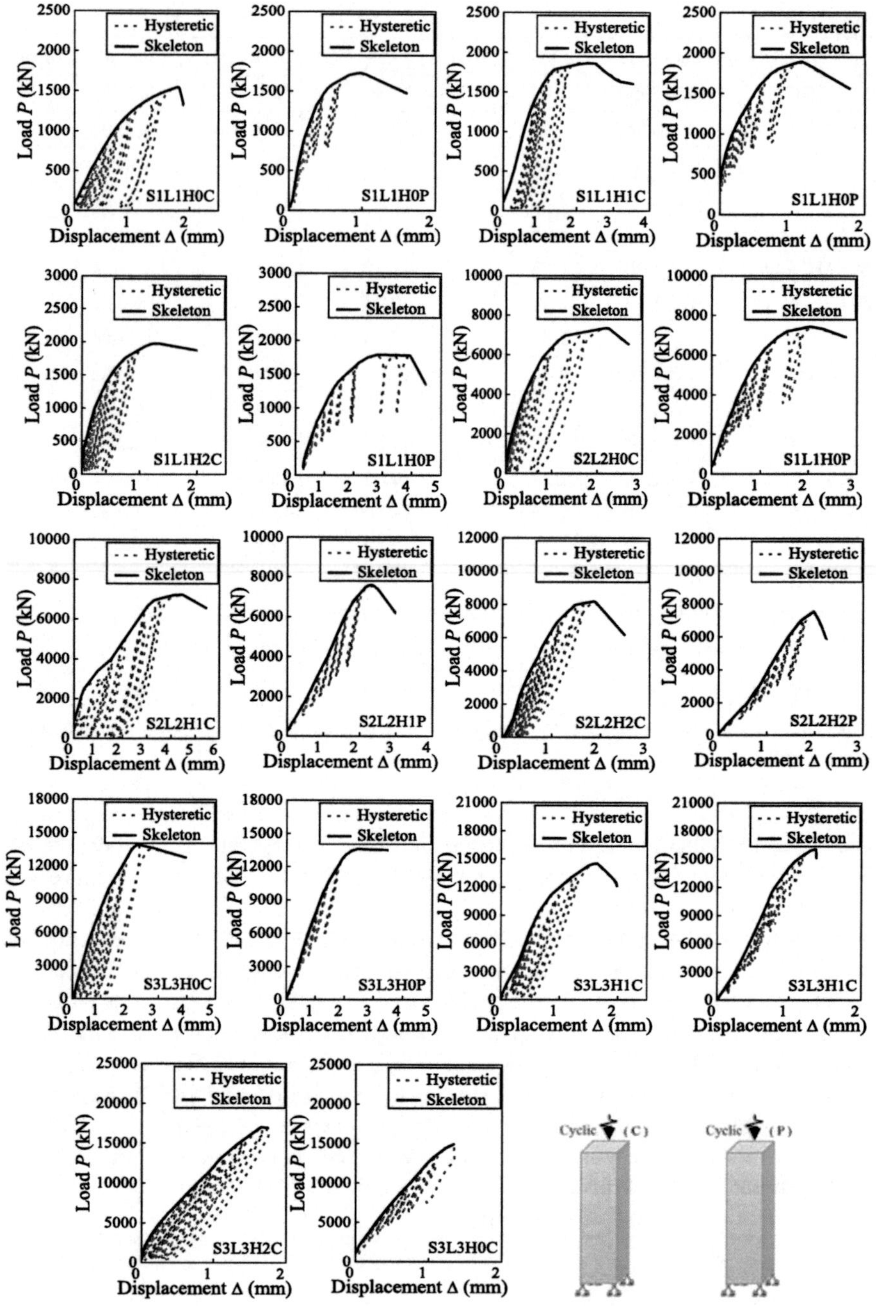

Fig. 7.74 Tested load-displacement (P-Δ) hysteretic and skeleton curves

the specimen under cyclic loading. The skeleton curves (i.e. the link between each maximum points of the hysteresis curve) are also drawn in Fig. 7.74 (the solid curve). It can be seen that the load-displacement curves of "H0" columns, "H1" columns and "H2" columns are similar in shape. Both of them undergo the process of elastic rise, while the decline stage occurs after the fracture of CFRP.

The skeleton curves of specimens under different reinforcement conditions are shown in Fig. 7.75. It can be seen that the stiffness and ultimate bearing capacity of the column increase gradually when the longitudinal reinforcement or stirrups are installed. The skeleton curves of specimens under different loading modes are shown in Fig. 7.76. As can be seen from Fig. 7.76, "C" (full loading and full unloading) and "P" (full loading and partial unloading) have no obvious influence on the axial compression performance.

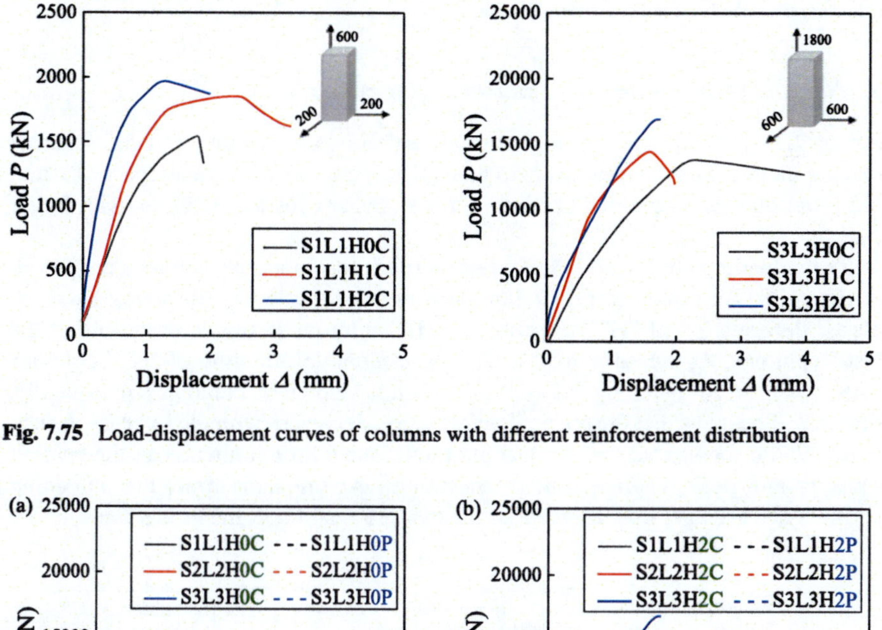

Fig. 7.75 Load-displacement curves of columns with different reinforcement distribution

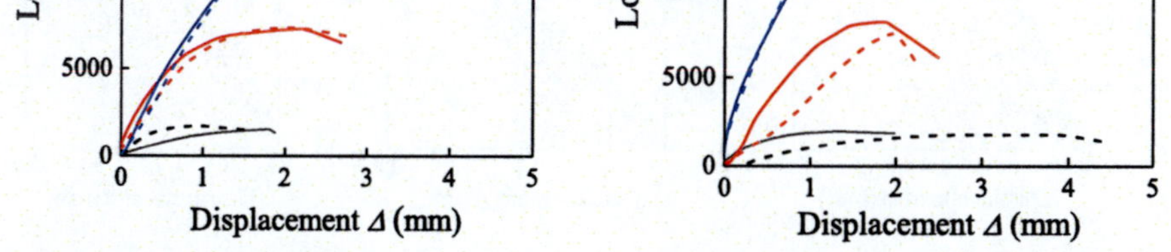

Fig. 7.76 Load-displacement curves of columns under different loading modes: **a** Plain concrete columns; and **b** Reinforced concrete columns

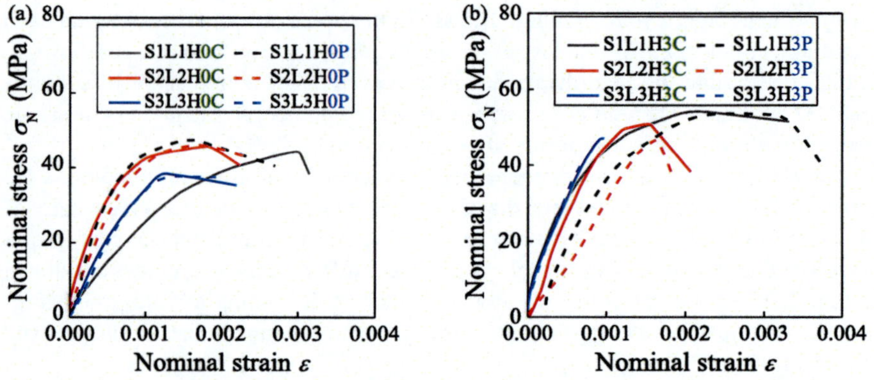

Fig. 7.77 Nominal stress-strain curves of columns under different loading modes: **a** Plain concrete columns; and **b** Reinforced concrete columns

5. *Nominal stress-strain curves and unloading curves*

Figure 7.77 presents the nominal stress-strain curves of columns under different loading modes. The nominal strength gradually decreases with the increase of structure size, and the degree of decline is more obvious for the columns with larger size.

The unloading curves of CFRP-wrapped reinforced concrete columns are shown in Fig. 7.78. As shown in Fig. 7.78a, because of the role of reinforcing bars, the plastic deformation of "H1" columns or "H2" columns is smaller than that of the "H0" columns. At the same load level, the residual deformation of "H2" columns is the smallest, followed by that of "H1" columns and "H0" columns subsequently. Figure 7.78b shows that the residual deformation decreases with the increase in size. Figure 7.78c shows that the two loading modes have little influence on the bearing capacity and peak displacement of specimens. At the same time, the unloading stiffness and residual displacement of the two groups of specimens are close.

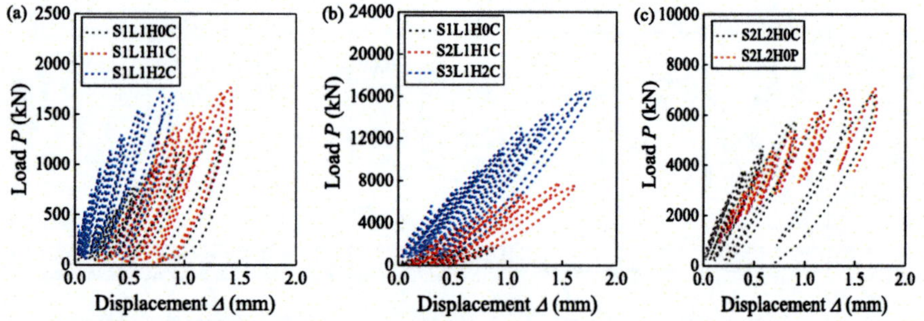

Fig. 7.78 Unloading curves for columns with: **a** Different ways of reinforcement distribution; **b** Different sizes of concrete section; and **c** Different loading modes

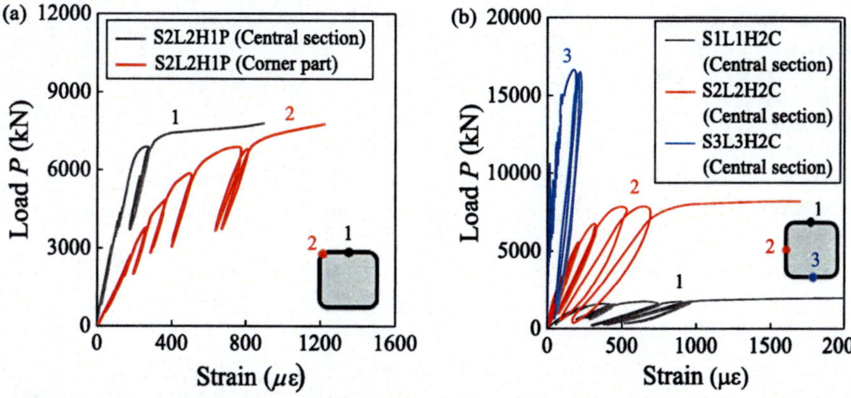

Fig. 7.79 The strain of the wrapped CFRP at: **a** Different locations for the same specimen S2L2H1P; and **b** The central region for the specimens having different cross-section size

6. *Strain curves of CFRP*

As can be seen from Fig. 7.79a that the strain of CFRP at the corner is greater than that of CFRP in the middle of the border under the same load. Stress concentration occurs at the corner, leading to the failure of CFRP, which is consistent with the experimental failure. In Fig. 7.79b, the strain of CFRP is smaller in the early stage of loading. At the later stage of loading, the CFRP strain of larger section columns is much smaller than that of smaller section columns. In order to fully explore this phenomenon, the FRP confining pressure was given by ACI 440.2R (ACI 2017) and expressed as:

$$f_l = \frac{2E_f t_f n \varepsilon_{fe}}{D} \tag{7.12}$$

where f_1 means lateral confining pressure of FRP, E_f means elastic modulus of FRP, t_f means thickness of one FRP wrap, n means the number of CFRP layer, ε_{fe} means lateral strain of FRP, and D is the diameter of column. For noncircular cross-sections, $D = \sqrt{b^2 + h^2}$, where b means the section width and h means the depth of the column section. Therefore, D of the square section is taken as $\sqrt{2}b$.

Based on Eq. (7.12), the restraint of CFRP is affected by the fracture strain of CFRP. Therefore, with the increase of column section size, the decrease of fracture strain will lead to a decreased lateral constraint degree of CFRP. The number of CFRP layers was increased in this experiment based on geometric similarity to ensure that the constraint effect remained unchanged.

7. *Analysis of nominal strength*

Figure 7.80 shows the relationship between nominal compressive strength σ_{Nu} (i.e. nominal stress under peak load) and structural size D (i.e. cross-sectional size) of all CFRP-wrapped reinforced concrete columns. The nominal compressive strength

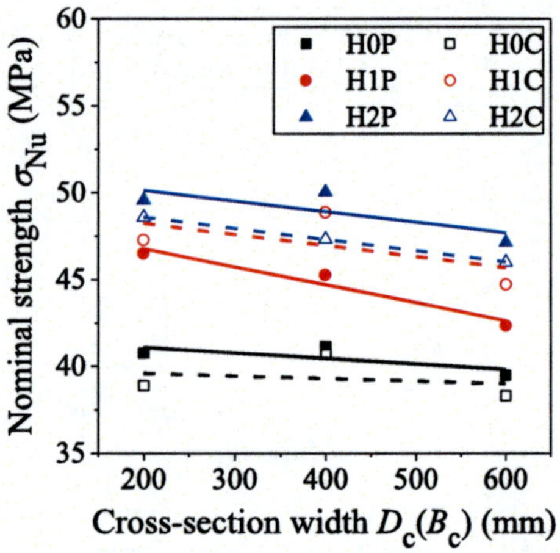

Fig. 7.80 Nominal axial compressive strength versus cross-section width under different ways of reinforcement distribution

σ_{Nu}, is defined as $\sigma_{Nu} = P_u/D^2$. As presented in Fig. 7.80, the nominal compressive strength σ_{Nu} decreases with the increase of structural size D. This diagram clearly demonstrates the nominal strength size-effect of CFRP-wrapped reinforced concrete columns. This conclusion is similar to that of Thériault and Neale (2000). They proposed that for square and rectangular sections, the compressive strength of confined concrete decreased with the increase of column cross-section size. It can be noted from Fig. 7.80 that the decline degrees of the nominal compressive strength (i.e. slope of the fitting curve) for six groups of specimens (i.e. with different reinforcement conditions) are very similar, indicating that the detailed rules of size-effect should be similar. Loading mode has an ignorable influence on the size effect. The influence of stirrups on the structure is very small due to the low(er) stirrup ratios under CFRP reinforcement. In other words, in the case of CFRP reinforcement, low reinforcement ratio has little effect on size effect.

In fact, the size effect of CFRP-wrapped concrete columns is affected by: (1) the properties of steel bars and concrete, (2) the complex interaction between steel bars and concrete, and (3) the restraint degree of CFRP on concrete columns. When the addition of CFRP constraints is ignored, compared with "P" (full loading and partial unloading), the loading mode "C" (full loading and full unloading) further reduces the mechanical properties of the steel bar and concrete, and also reduces the bond performance between the two. Damage in the loading mode "C" is mainly caused by concrete, and theoretically, the dimensional response should be more pronounced. However, as shown in Fig. 7.80, the size effect in the two loading modes is closely similar, and thus it can be inferred that the CFRP constraints have a greater impact on the degree of concrete damage and the size effect.

8. *Analysis of energy dissipation*

Definitions of displacement ductility of CFRP-wrapped reinforced concrete columns are few, but reinforced concrete columns have been studied in this field and formally defined. Therefore, in order to evaluate the displacement ductility of CFRP-wrapped reinforced concrete columns, the definition method mentioned in Sect. 7.2.3 was utilized. According to the definitions, the values of δ_y and $\delta_{0.85}$ were obtained. The displacement ductility coefficient μ_δ of columns under the "C" loading mode are listed in Table 7.25. The relationship between the displacement ductility coefficient μ_δ and section size D_c is plotted in Fig. 7.81, where the ductility coefficient μ_δ decreases with the increase of section size D_c. This result shows that the size effect on displacement ductility coefficient μ_δ exists in CFRP-wrapped reinforced concrete columns. That is, the larger the structural size of the column is, the more brittle failure patterns the column exhibits.

The performance of CFRP-wrapped reinforced concrete columns under cyclic loading was evaluated with the parameters related to the energy dissipation mechanism. The parameters include energy dissipation values (W_{pl} and $\sum W_{pl}$) and hysteretic damping factors (ζ and $\sum \zeta$). The relationship between those two sets of parameters and the number of load cycles "n" are shown in Figs. 7.82 and 7.83,

Table 7.25 Measured displacement response and ductility factors

Dimensions (mm)	Specimen label	δ_y (mm)	$\delta_{0.85}$ (mm)	μ_δ
200 × 200 × 600	S1L1H0C	0.85	2.19	2.58
	S1L1H1C	1.17	3.56	3.04
	S1L1H2C	0.70	2.02	2.88
400 × 400 × 1200	S2L2H0C	0.95	2.69	2.83
	S2L2H1C	3.10	5.50	1.77
	S2L2H2C	1.28	2.24	1.75
600 × 600 × 1800	S3L3H0C	1.84	4.02	2.18
	S3L3H1C	1.13	1.97	1.74
	S3L3H2C	1.44	2.38	1.65

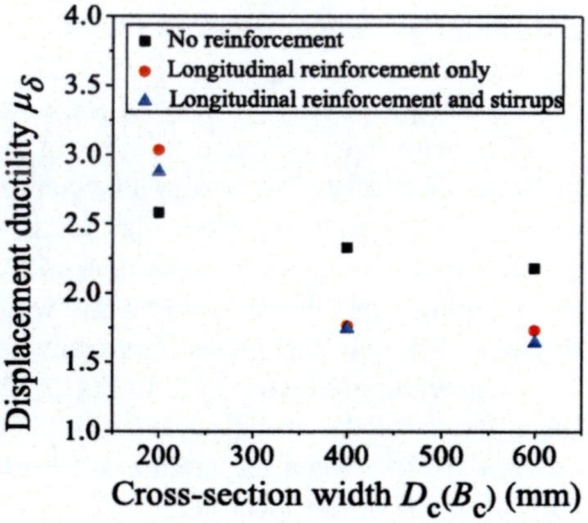

Fig. 7.81 Displacement ductility of the columns with different structural sizes

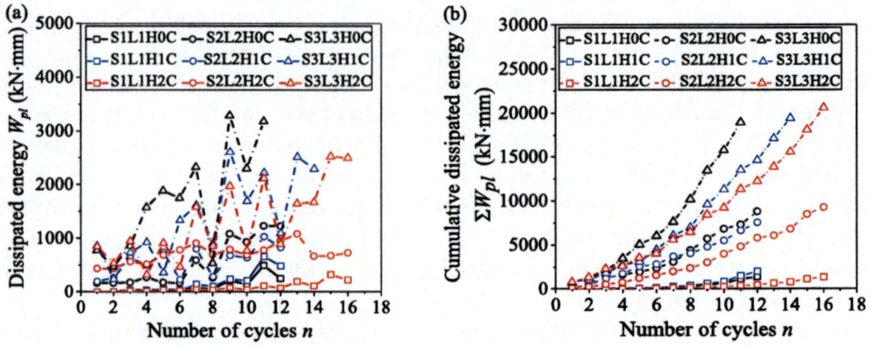

Fig. 7.82 **a** The dissipated energy W_{pl}; and **b** The cumulative dissipated energy $\sum W_{pl}$

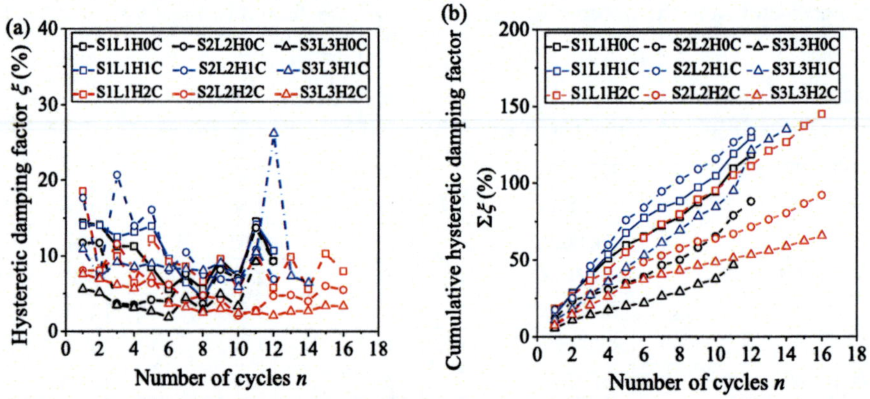

Fig. 7.83 **a** The hysteretic damping factor ξ; and **b** The cumulative hysteretic damping factor $\sum \xi$

respectively. W_{pl} is the corresponding region of each cycle and $\sum W_{pl}$ is all the energy dissipation by the cumulative cycle of the specimen. The definition of hysteretic damping factor ζ was given in Sect. 7.2.2.

The dissipated energy W_{pl} and the corresponding damping factor ζ of CFRP-wrapped reinforced concrete columns in the *i*th cycle are plotted in Figs. 7.82a and 7.83a, respectively. The energy dissipation capacity of the same size has a good consistency. In addition, from Fig. 7.82a, more energy is consumed at each cycle of the larger section specimens than in the smaller section specimens. However, the damping characteristics do not show a similar pattern, as shown in Fig. 7.83a. Figures 7.82b and 7.83b show cumulative dissipated energy $\sum W_{pl}$ and cumulative hysteretic damping factor $\sum \zeta$. In Fig. 7.82b, the larger the section size of the specimen, the more the energy dissipation. However, as shown in Fig. 7.83b, there is no regularity between the cumulative hysteretic damping coefficient $\sum \zeta$ and the section size D of the specimen.

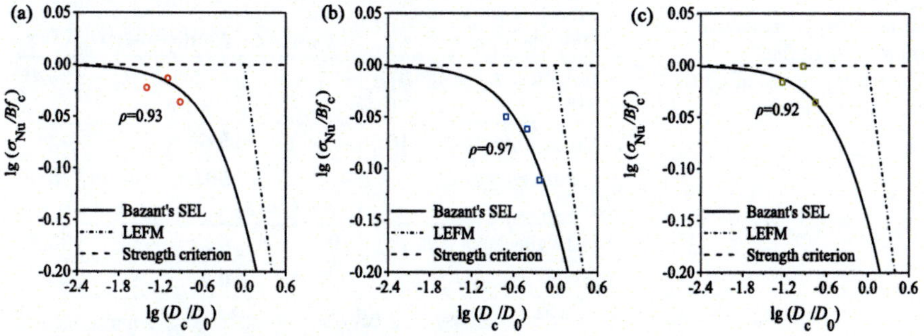

Fig. 7.84 Comparison of the test results and Bažant's SEL for CFRP-wrapped concrete columns: **a** No reinforcement; **b** Longitudinal reinforcement; and **c** Longitudinal reinforcement and stirrups

To sum up, the energy dissipation capacity (W_{pl} and $\sum W_{pl}$) of the CFRP-wrapped RC columns increases with the increase of cross-section size, while the hysteretic damping factors (ζ and $\sum \zeta$) present no obvious relationships with cross-section size.

9. *Analysis of size effect*

The comparison between SEL of Bažant (1984) and the experimental data is presented. Herein, f_c means compressive strength for concrete (i.e. the test data of standard concrete cube specimens which a size of 150 mm × 150 mm × 150 mm), D is the feature size of specimens (i.e. the width of cross-section of square columns). The double logarithmic plots are plotted in Fig. 7.84.

As can be seen from the above analysis, the loading mode has little influence on the size effect. Therefore, experimental data only under the "C" loading mode were used for analysis. It can be seen from Fig. 7.84 that the experimental data are in good agreement with SEL, and the correlation coefficients ρ are 0.93, 0.97 and 0.92 respectively. It shows that the size effect of the CFRP-wrapped reinforced concrete column can be better reacted by SEL. The three sets of data are closer to the ideal elastoplastic line, indicating that the cracking of the concrete is well constrained by CFRP. The size effect on the axial compression failure of the CFRP-wrapped reinforced concrete column can be better represented by the SEL of Bažant (1984).

7.6.2 *Numerical Investigations on CFRP-Confined RC Column*

Herein this section, the numerical results and discussions on size effect of axial failure for CFRP-wrapped columns in Jin et al. (2020a) are presented.

Table 7.26 Mechanical parameters of the meso-components of concrete used in the columns

Materials	Aggregate	Mortar matrix	ITZs
Compressive strength σ_c (MPa)	–	43.5[a]	35.4[b]
Tensile strength σ_t (MPa)	–	3.6[a]	2.9[b]
Fracture energy G_c (J/m^2)	–	50.0[c]	30.0[c]
Elastic modulus E (GPa)	70[a]	42.8[a]	38.7[b]
Poisson's ratio ν	0.2[a]	0.2	0.2

Note [a]Data are obtained from the tests; [b]Data are parameters by repeated trials; and [c]Data refer to the work of Yılmaz and Molinari (2017)

1. *Generation of 3D meso-scale model*

According to the program in Chap. 3, the 3D mesoscale model of concrete was generated. The material properties of the numerical model were identical with that in Chap. 3. The main material parameters for the three meso-components of concrete are illustrated in Table 7.26.

An ideal bonding was utilized to describe the interaction between the CFRP and the inner concrete, i.e. no separation and slip between CFRP and concrete would generate during the loading process. That is a rough assumption since the experimental tests show a pronounced delamination at the interface.

The assembly diagram of the numerical simulation model can be seen in Fig. 7.85a. The loading mode is the same as the "C" mode of the test, as shown in Fig. 7.85b. The prescribed displacement was applied on the top surface of the columns. At the bottom

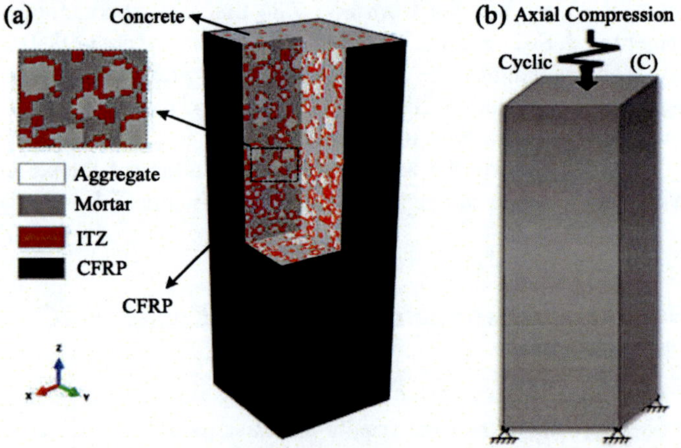

Fig. 7.85 Sketch view of meso-scale simulations: **a** Meso-scale simulation model; and **b** Loading method

Table 7.27 Specimen design parameters of the meso-components in the columns

Specimen label	Section size (mm × mm)	Height h (mm)	Specimen number	Layer of FRP	Constraint ratio ρ_f (%)
S1T1	200 × 200	600	3	1	0.334
S1T2			3	2	0.668
S1T3			3	3	1.002
S2T1	400 × 400	1200	3	2	0.334
S2T2			3	4	0.668
S2T3			3	6	1.002
S3T1	600 × 600	1800	3	3	0.334
S3T2			3	6	0.668
S3T3			3	9	1.002
S4T1	800 × 800	2400	3	4	0.334
S4T2			3	8	0.668
S4T3			3	12	1.002

Note "S" means the section size of specimen (S1-200 × 200, S2-400 × 400, S3-600 × 600, S4-800 × 800); and "T" means constraint type of CFRP (T1- type one, T2- type two, T3- type three)

surface, the fixed constraint was applied in the vertical direction. Other surfaces were left free.

The numerical method was utilized to model the CFRP-wrapped columns with larger size and constrain ratio. The test pieces are designed as shown in Table 7.27. Among these 16 specimens, the S1T1, S2T1 and S3T1 correspond to S1L1H0C, S2L2H0C and S3L3H0C in experiment of Sect. 7.6.1, respectively. The constrain ratio of FRP was defined by

$$\rho_f = \frac{4nt}{D} \text{ or } \rho_f = \frac{4nt}{B} \tag{7.13}$$

in which, ρ_f is the constraint ratio, n is the layer number, t is the single layer thickness of FRP, D is cross-sectional dimeter and B is cross-sectional width.

2. *Influence of the constrain ratio*

Failure modes of the same structural size ($D_c = 200$ mm) are shown in Fig. 7.86a. The typical brittle compression shear failure mode is shown. It can be seen that with the increase of constraint ratio, the angle between the final failure plane and the horizontal plane decreases gradually. The level of restraint has an effect on the failure angle, and the improvement of the level of restraint has a limiting effect on the cracking degree of concrete.

Figure 7.87 shows typical failure modes with different restraint ratios, similar conclusions can be drawn for larger sizes (i.e. $D_c = 400$ mm, $D_c = 600$ mm, and $D_c = 800$ mm). The load-displacement curves of specimens with the same size ($D_c = 200$ mm) and different restraint ratios are plotted in Fig. 7.86b. The peak load

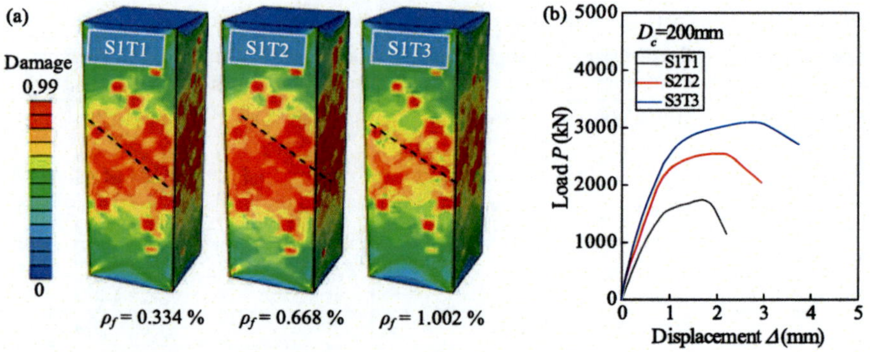

Fig. 7.86 Simulation results for the columns having the cross-sectional width of 200 mm: **a** Final failure modes; and **b** Simulated load-displacement curves

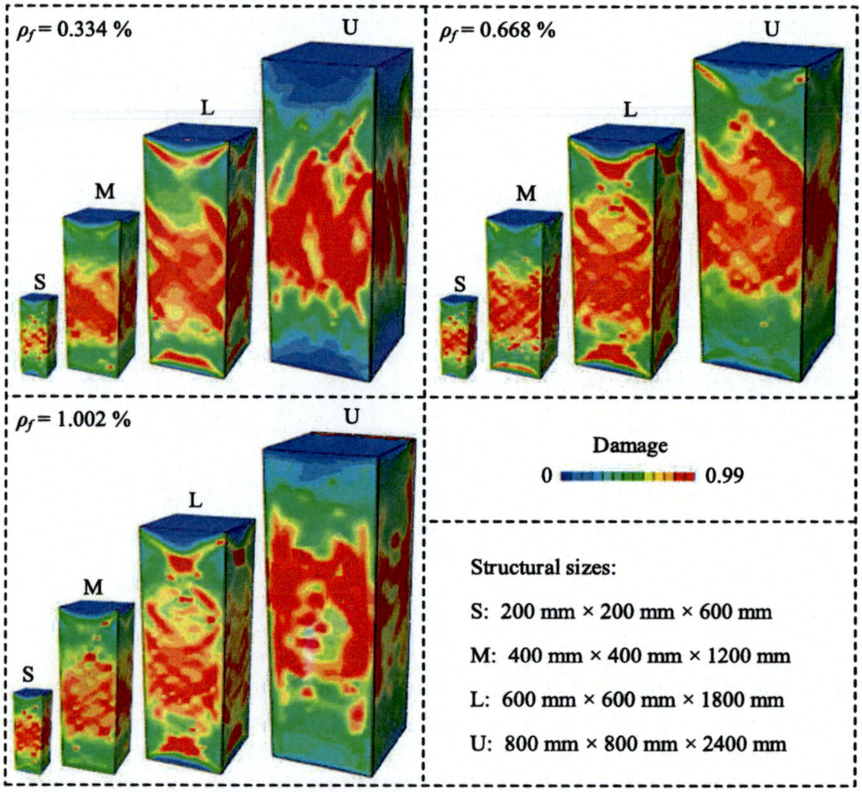

Fig. 7.87 Failure modes of the CFRP columns with different structure sizes and confinement ratios

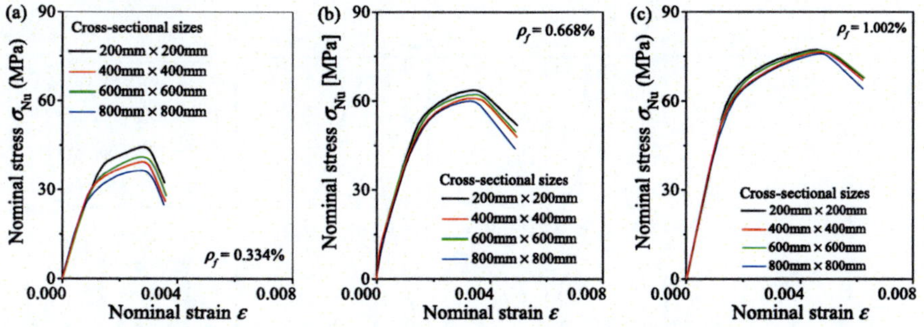

Fig. 7.88 Comparison of the nominal stress-strain curves of the CFRP columns with different structural sizes under the same confinement ratios: **a** ρ_f= 0.334%; **b** ρ_f= 0.668%; and **c** ρ_f= 1.002%

is obviously increased with the increase of restraint ratio, which indicates that the cracking of CFRP confined concrete increases the peak load. The delay in the internal concrete cracking also led to the increasing of the stiffness. From Fig. 7.86b, we can see that the slope of the rising section is increasing. The behavior after peak (softening or yielding) is affected by the constraint ratio. The higher the constraint ratio, the softer the softening curve. With the increase of constraint ratio, the ductility shows the same trend.

From the nominal stress-strain curve, i.e. Figure 7.88, the size effect is more obvious, and the restraining effect of constraint ratios on the size effect is obvious. With the increases of restraint ratio, the degree of nominal strength size effect decreases, which is similar with that in Sect. 7.5.

3. *Hoop strain of CFRP*

The effect of CFRP on core concrete is reflected in the restraint of concrete expansion, which belongs to passive restraint. In order to better reflect the distribution of CFRP strain around the whole column section, Fig. 7.89 was given. Four sizes of members with constraint ratio of 0.668% were selected to measure the strain at eight points of the middle section when the nominal axial strain was 0.5%. In order to avoid the influence of stress concentration on the measurement results, the measurement of corner region points (i.e. Nos. 2, 4, 6, 8) was selected to take the average value near the corner. It can be observed that the strain value in the corner region is significantly smaller than that in the middle region (i.e. points of Nos. 1, 3, 5, 7). According to the test results in the previous chapters, it can be found that the first failure area of CFRP often occurs in the corner, so it can be inferred that the concrete in the middle area is not well restrained. It can be concluded that the effective strain at failure should be considered as the average of the measured results at the corner. With the increase of structure size, the strain value of CFRP at each point decreases. The constraint force of CFRP on core concrete decreases with the increase of structure size, which indirectly affects the size effect of axial compressive strength.

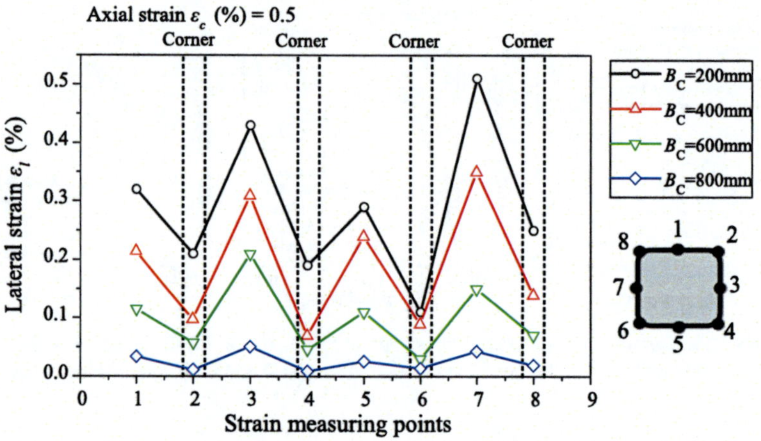

Fig. 7.89 Typical hoop strain distribution on CFRP wrap, when axial strain ε_c as 0.5%

7.6.3 Numerical Investigations on GFRP-Confined Concrete Column

In order to study the failure behavior and size effect of GFRP-confined concrete columns under monotonic axial compression, the numerical results of 30 GFRP-confined concrete columns in Jin et al. (2020e), was presented.

1. *Generation of the numerical model*

The generation of the GFRP-confined circular concrete column model, along with the boundary conditions, were identical with that in Sect. 7.6.2. The main mechanical parameters for the three meso-components of concrete and GFRP jackets are listed in Tables 7.28 and 7.29, respectively.

The circular and square columns have same structural size (shown in Table 7.30). The large-size, medium-size and small-size are defined by cross-sectional widths of 800 mm, 400 mm and 200 mm, respectively.

Table 7.28 Mechanical parameters of the meso-components of concrete used in the columns

Materials	Aggregate	Mortar matrix	ITZ
Compressive strength σ_c (MPa)	–	42.4[a]	35.3[b]
Tensile strength σ_t (MPa)	–	2.6[a]	2.3[b]
Elastic modulus E (GPa)	60[a]	30[a]	24[b]
Poisson's ratio ν	0.2[a]	0.2[a]	0.2[b]

Note [a]Data were obtained from the tests; and [b]Date were parameters by repeated trial

Table 7.29 The mechanical properties of FRP

Variables	Elasticity modulus E_f (GPa)	Ultimate tensile strength f_f (MPa)	Ultimate rupture	Single layer thickness t (mm)
GFRP	87	3271	3.2	0.169

Table 7.30 The geometrical parameters of the larger sized GFRP-concrete columns

Specimen name	Structural size *D or B* (mm)	Height *h* (mm)	Thickness of GFRP *t* (mm)	Constraint ratio ρ_f(%)
C-A-S/S-A-S	200	600	0	0
C-A-M/S-A-M	400	1200	0	0
C-A-L/S-A-L	800	2400	0	0
C-B-S/S-B-S	200	600	0.676	1.352
C-B-M/S-B-M	400	1200	1.352	1.352
C-B-L/S-B-L	800	2400	2.704	1.352
C-C-S/S-C-S	200	600	1.352	2.704
C-C-M/S-C-M	400	1200	2.704	2.704
C-C-L/S-C-L	800	2400	5.408	2.704
C-D-S/S-D-S	200	600	2.535	5.07
C-D-M/S-D-M	400	1200	5.07	5.07
C-D-L/S-D-L	800	2400	10.14	5.07
C-E-S/S-E-S	200	600	5.07	10.14
C-E-M/S-E-M	400	1200	10.14	10.14
C-E-L/S-E-L	800	2400	20.28	10.14

Note C/S = shape of cross section (c = circular, s = square); A-E = group number (divided by constraint ratio); and S/M/L = specimen's size (S-small, M-middle, L-large)

2. *Failure patterns*

The failure patterns of the concrete columns confined with GFRP are presented in Fig. 7.90. As can be seen from Fig. 7.90, for both the circular and square columns, the failure patterns of the small-sized and medium-sized columns are similar under the same constraint ratio ($\rho_f = 1.352\%$). However, the large-sized circular column presents compression-shear failure mode, and there are many inclined cracks generated in the columns. For the large-sized square column, the stress concentration phenomenon occurs at the corner regions of column, as can be observed from Fig. 7.90. The GFRP was pulled off at the corner region due to the stress concentration phenomenon. This indicates that GFRP material has a better restraint effect on small-sized concrete columns than large-sized ones, which makes the failure modes of small-sized columns more uniform.

Moreover, Fig. 7.91a illustrates the effective restraint of GFRP on the core concretes and the stress distribution (the stress S33 in the Z-direction is extracted)

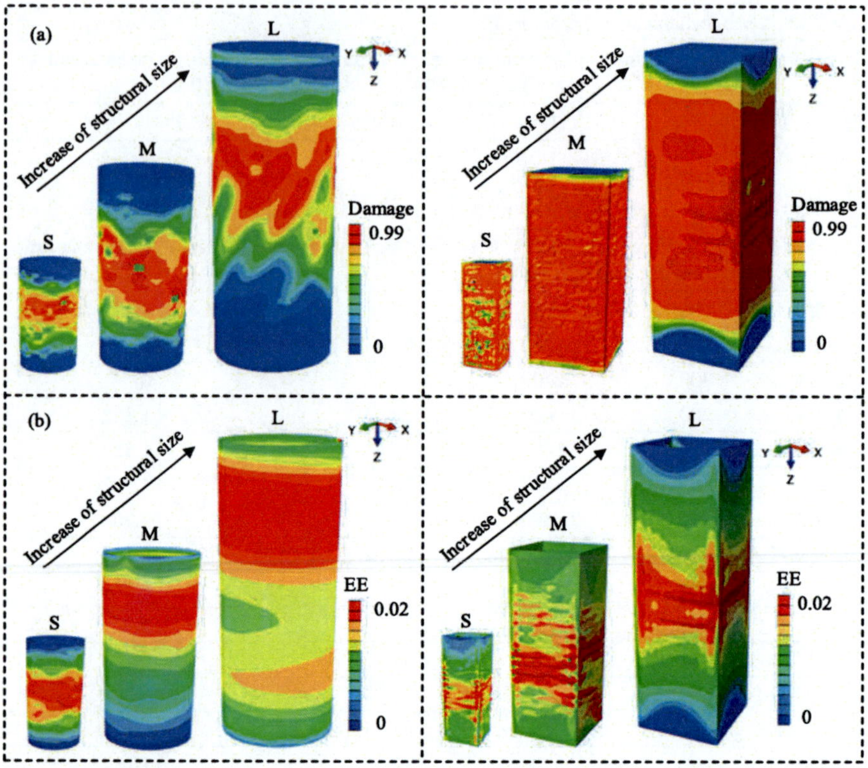

Fig. 7.90 Failure patterns of the columns with different structural sizes (Series CB(SB), ρ_f = 1.352%): **a** Core concrte; and **b** GFRP

of cross-section concrete in the middle region of the columns (i.e. the columns of C-B-S and S-B-S). One can note that the stress distribution of concrete for circular columns is relatively uniform, while that for square columns mainly distributes along two diagonals of square section, and there are weak restraint zones. From Fig. 7.91b, one can find that the fine cracks start to generate at the weaker ITZ phase, and finally develop into macroscopic cracks.

3. *Axial stress-strain curves*

The normalized axial stress-strain curves of all simulated columns are plotted in Fig. 7.92. There are three types of nominal stress-strain curves for the confined concrete columns: (1) when the constraint ratio of GFRP is lower, the stress-strain curves decrease rapidly after reaching the peak point and do not rise anymore due to the weak confinement (corresponding to curve I listed in Fig. 7.92a and b); (2) with the increase of constraint ratio, the stress-strain curves decrease slightly after reaching the peak point and then rise gradually due to the stronger constraint of GFRP (corresponding to curve II listed in Fig. 7.92a and b); (3) with the constraint

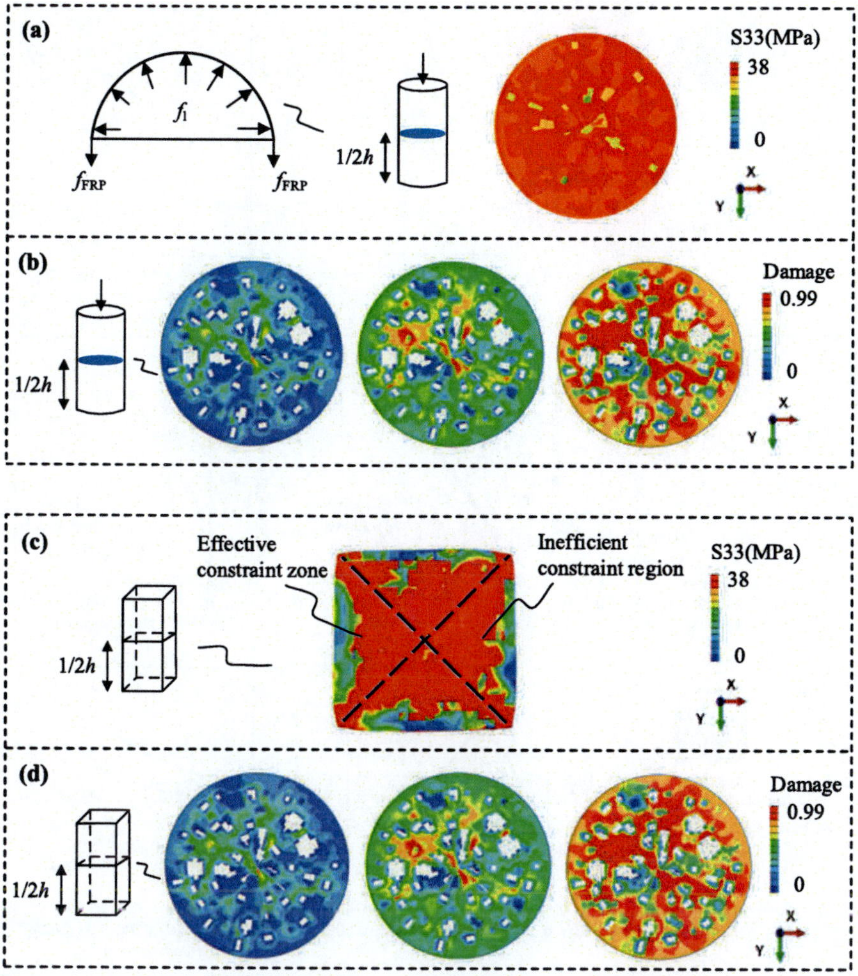

Fig. 7.91 Constraint effective of GFRP at 1/2 column height: **a** Stress distribution of circular column; **b** Damage process in cross section of circular column; **c** Stress distribution of square column; and **d** Damage process in cross section of square column

ratio continues to increase, the curves of circular columns do not appear a descending section, but a bilinear continuous rise for circular columns (corresponding to curve III listed in Fig. 7.92a), while the curves of square columns have a sudden descent at the its end (corresponding to curve III listed in Fig. 7.92b). The case (1) and (2) occurs for the columns having a lower constraint ratio, while the case (3) occurs for the columns having a higher constraint ratio.

When square columns were strongly restraint, the square columns failed at the corner region due to stress concentration and the properties of GFRP material have not been utilized effectively. The results are consistent with those of Yu et al. (2010).

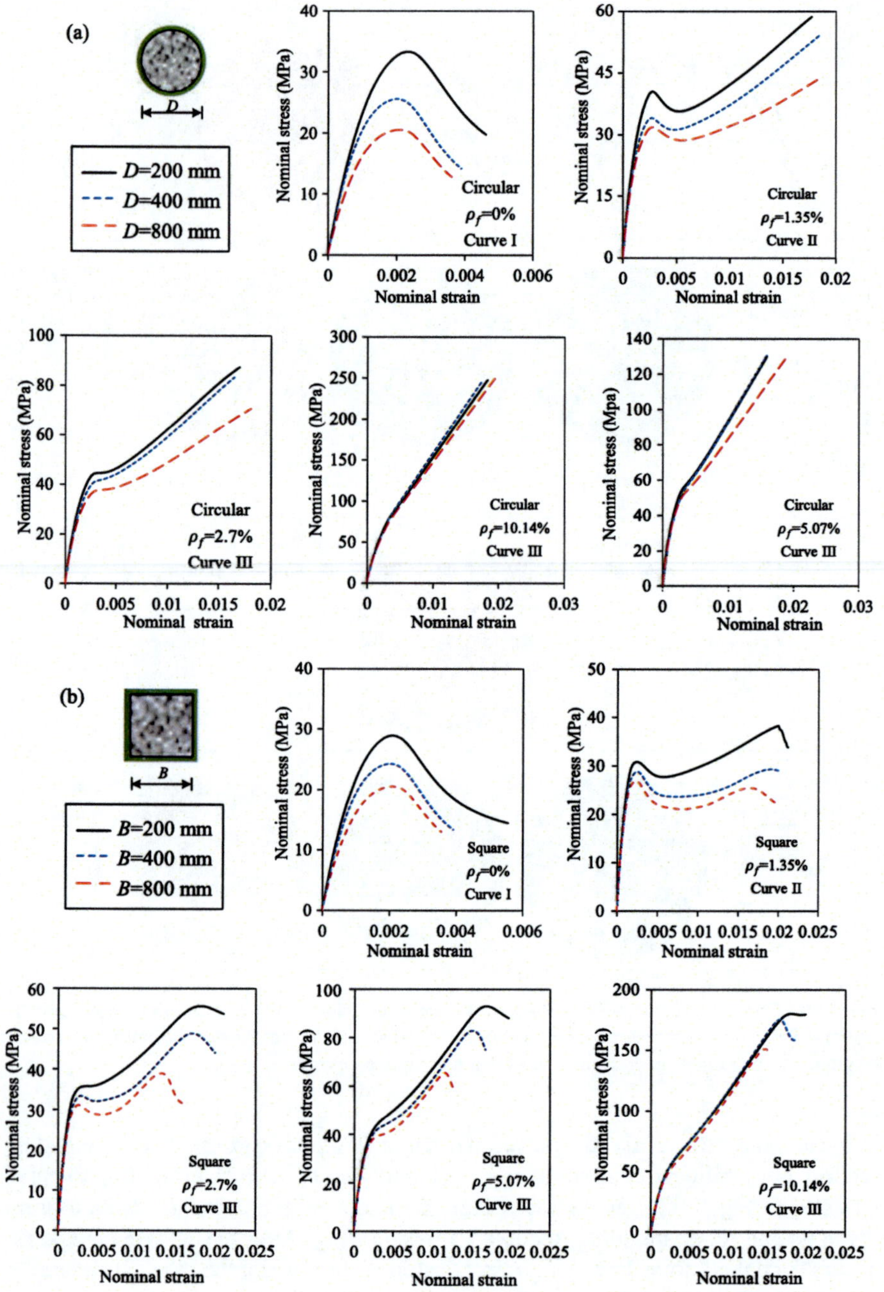

Fig. 7.92 Nominal compressive stress-strain curves: **a** Circular columns; and **b** Square columns

In general, constraint ratio has a significant influence on post-peak softening curves of GFRP-confined concrete columns.

Figure 7.92 shows that the presence of GFRP can significantly enhance the ultimate compressive strength and improve the ductility of concrete columns. Moreover, for the same constraint ratio, the structural size has a light effect on shape of stress-strain curves of the columns.

4. *Effective confinement provided by GFRP*

For concrete columns confined with GFRP, the pressure increases only with the expansion of core concrete thus the lateral confinement provided by the GFRP is regard as a passive-constraint. Furthermore, due to the radial displacement compatibility between the core concrete and GFRP, the circumferential strain of these two parts is considered to be equivalent (Wang et al. 2016). Therefore, the lateral strain of GFRP can be discussed in detail to explore the variation of lateral constraints around the cross-section.

Figure 7.93 shows the difference between the role of GFRP in the middle and bottom regions of columns (i.e. series CB and SB). As mentioned above, the axial stress and strain are defined as positive, while the lateral strain as negative. Herein, the magnitude of the strain denotes the average value of the four points obtained in the circumferential direction of the GFRP. It can be seen from Fig. 7.93 that the lateral strains are larger in the middle regions than the bottom ones when the columns failed. In addition, the GFRP is found to be pulled off in the middle regions of the columns,

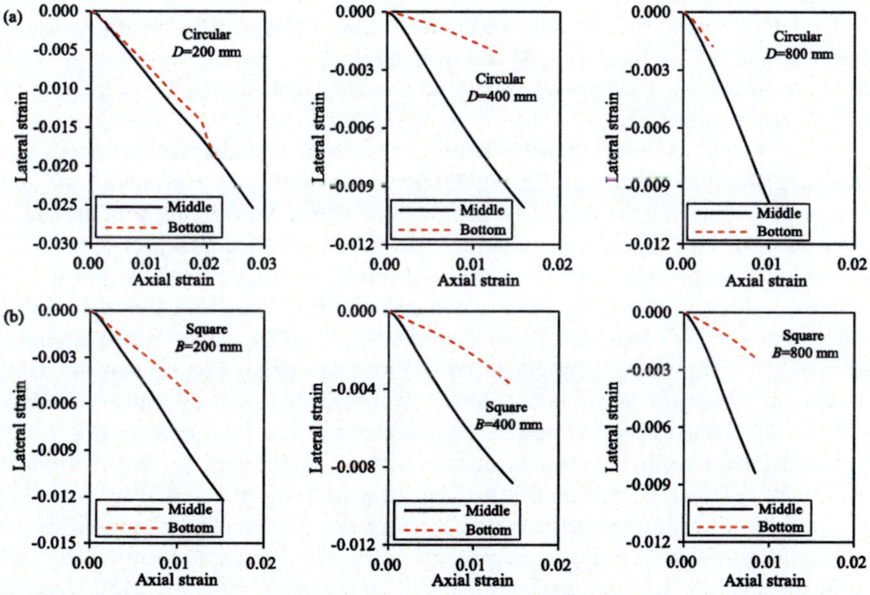

Fig. 7.93 The difference between the role of GFRP in the middle and the bottom of columns have fixed constraint at the end: **a** Circular columns; and **b** Square columns

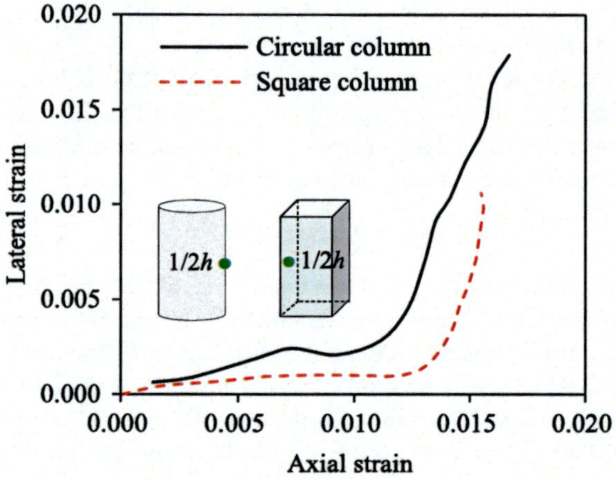

Fig. 7.94 The relationship between lateral strain and axial strain of columns

which implies that GFRP reached its effective strain in these regions, making the fact that the lateral strain of middle regions is larger than those of bottom regions. In addition, structural size has no obvious influence on strain distribution curves in the middle regions but has a slight effect on the strain curves in bottom regions. From Fig. 7.93, it can be seen that the lateral strain decreases with the increase of the structural size of the column. According to the failure patterns, GFRP has a weak constraint on larger sized specimens. This is suitable for both the middle and bottom regions of columns.

The lateral strain versus axial strain curve at one fixed point in the middle region of the column (e.g. C-A-S and S-A-S) is shown in Fig. 7.94. One can see from Fig. 7.94 that initially lateral strain grows slowly with axial strain, subsequently the rate of growth increases sharply.

Figure 7.95 plots the relationship between the nominal compressive strength and the constraint ratio of columns having different structural sizes. For unconfined and confined columns, the nominal axial compressive strength means the nominal stress when the axial load reaches the peak load and the ultimate load, respectively.

One can clearly note that, the simulated nominal compressive strength of the concrete columns having the same structural size increases with the increase of constraint ratio. This means that the restraint effect of GFRP can obviously improve the nominal compressive strength of columns. It can be also found that the axial compressive strength of circular columns is higher than that of square columns with the same constraint ratio and the same structural size. Furthermore, Fig. 7.95b depicts the relationship between the constraint ratio and the increase rate in nominal strength of columns. It can be observed that structural size of columns has a significant influence on the nominal strength increase rate for circular columns with the same constraint ratio. For instance, nominal compressive strength increased 81 and 129% for small and large circular column with the constraint ratio of 1.352%, respectively. When the constraint ratio is 10.14%, the nominal strength increase rate up to

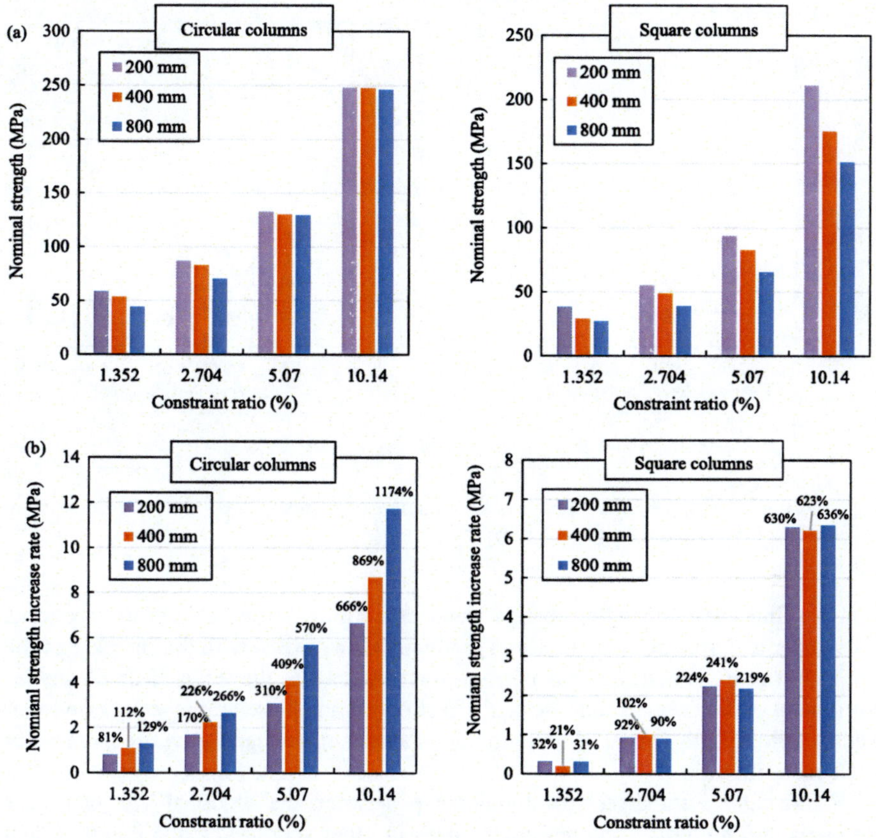

Fig. 7.95 Effect of constraint ratio: a Nominal strength; and b Nominal strength increase rate

666 and 1174% for small and large columns, respectively. Thus the larger the specimen size is, the greater the increase ratio of nominal axial compressive strength is. However, structural size has no obvious influence on the nominal strength increase rate for square columns. It is obvious that the strength and strength increase rate of circular columns are higher than those of the square columns under the same constraint ratio. This should be caused by the reality that the constraint of GFRP on the circular columns is uniform, and weak constraints exist in GFRP constraints on square columns.

5. *Size effect on nominal compressive strength*

The relationship between the simulated nominal strength and cross-sectional width of columns is presented in Fig. 7.96. From Fig. 7.96, it can be found that the nominal strength of the columns decreases with the increase of cross-sectional size, indicating that there exists an obvious size effect on compressive strength of columns. However, the decreasing rate of the nominal strength is slightly different for circular and square columns with the increase of the constraint ratio. For circular columns, the trend of

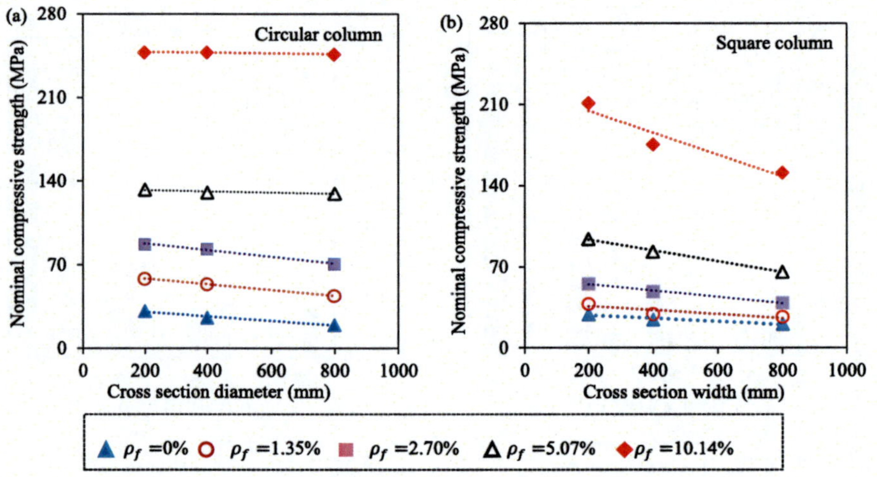

Fig. 7.96 Relationship between nominal axial compressive strength and size of cross section: **a** Circular columns; and **b** Square columns

oblique line (describing the change trend of nominal strength with the sectional size under different constraint ratios) gradually slows down with the increase of the constraint ratio. When the constraint ratio is about 10%, the trend of line becomes horizontal, which means that the size effect on nominal strength is suppressed. For square columns, the trend of oblique line is almost unchanged with the increase of the constraint ratio.

Figure 7.97 summarizes the relationship between the sectional size of square columns and nominal compressive strength in other scholars' experiments (Wang

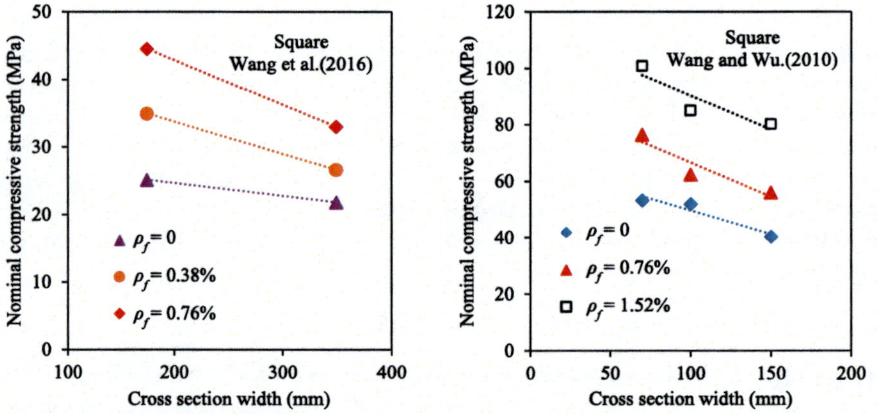

Fig. 7.97 Relationship between nominal compressive strength and cross section width of square columns in available experiments

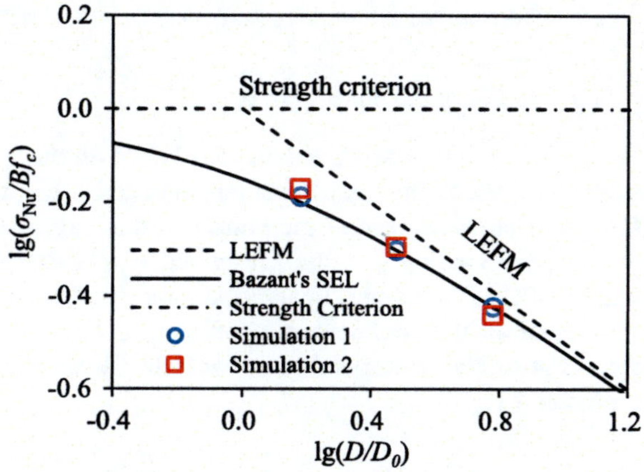

Fig. 7.98 Regression analysis on the simulated data of plain concrete columns

and Wu 2010; Wang et al. 2016). Similar conclusions can be drawn from the analysis of other scholar's experimental data. Therefore, the confinement of GFRP can only improve the axial compressive strength of square columns, but cannot weaken the size effect on compressive strength.

The nominal compressive strength was analyzed according to the SEL of Bažant (1984). Herein, f_c means the compressive strength of a cylinder with 150 mm diameter and 300 mm length ($f_c = 33.31$ MPa). By the linear fitting of nominal compressive strength σ_{Nu} of plain concrete columns having different structural sizes, one can get $B = 1.542$ and $D_0 = 132$.

Figure 7.98 presents the comparison of the simulation data points with the classic SEL, the plastic strength criteria and the linear elastic fracture mechanics (LEFM). It is noted that, simulation 1 and 2 presented in Fig. 7.98 represent two groups of concrete strength, which are calculated by two groups of numerical specimens with different aggregate distributions. One can clearly note from Fig. 7.98 that all simulated data points keep far from the two curves (strength criterion and LEFM), indicating that the failure of the columns should be between ductile and brittle. According to the calculations, it can be found that the obtained fitting correlation coefficient for the simulation results and the SEL curve is 0.98 (i.e. $R^2 = 0.98$), demonstrating the rationality of the SEL proposed by Bažant (1984). Moreover, the presence of size effect on plain concrete columns can also be demonstrated.

7.7 Compressive Failure of CFST Columns

Herein the present section, the simulation results of concrete filled steel tube (CFST) columns with different sizes under the axial compression was presented, in which the influence of the lateral restraint of steel tube, section shape and aggregate type on the axial compression failure behavior and size effect of CFST columns was analyzed (Jin et al. 2019, 2020b).

7.7.1 Numerical Investigations on Ordinary CFST Columns

1. *Generation of the mesoscale model*

The generation of concrete model was based on the verified meso-scale numerical model in Chap. 3. The meso-scale model of CFST columns is illustrated in Fig. 7.99. The columns were fixed at the bottom surface against all degrees of freedom, the compressive loading (i.e. the prescribed displacement) was employed at the top surface of CFST columns, in addition, other surfaces were set free.

In general, the confinement coefficient ζ could be utilized to characterize the constraint effect of steel tube on concrete. According to Han (2000), it could be expressed by:

$$z = \frac{f_y A_s}{f_c A_c} \tag{7.14}$$

in which f_c means axial compressive strength of concrete; f_y is the yield strength of steel tube; and A_c and A_s are respectively the cross-sectional areas of concrete and steel tube. Herein, $\alpha = A_s/A_c$ means the steel ratio of CFST column.

Herein, the cross-section sizes of the simulated columns D_c (B_c) are respectively 200 mm, 400 mm, 800 mm and 1000 mm. B_c denotes the cross-sectional width of square column, and D_c means the cross-sectional diameter of circular column. The confinement coefficients ζ are respectively 0, 0.38, 0.77, 1.12 and 1.53. The axial compressive strength of concrete is 52.7 MPa and the yield strength of steel tube is 350 MPa. The mechanical parameters of the meso-components of concrete and steel tube are listed in Table 7.31. The slenderness ratio λ for all columns is taken as 3, the number of the simulated specimens in each group is 2, and a total of 80 specimens are modelled.

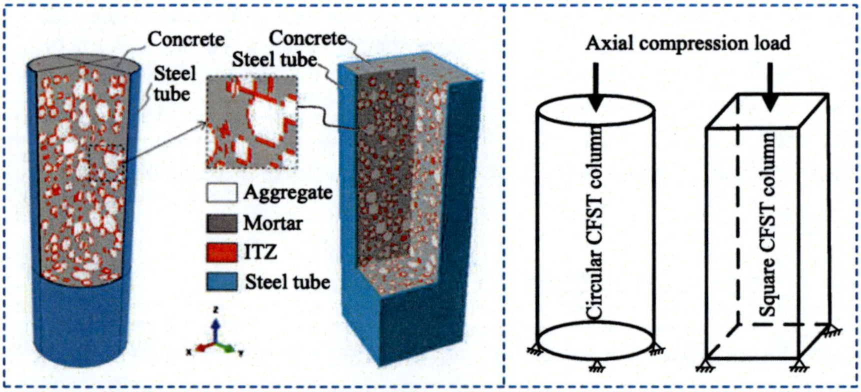

Fig. 7.99 3D mesoscale model and boundary conditions

Table 7.31 Mechanical parameters of the meso-components of concrete and steel tube

Microscopic component	Aggregate	Mortar matrix	ITZ	Steel
Compressive strength σ_c (MPa)	–	53.4[a]	49.50[b]	–
Tensile strength σ_t (MPa)	–	3.14[a]	2.95[b]	–
Elastic modulus E (GPa)	60	35.5[a]	28.40[b]	210[a]
Poisson's ratio ν	0.2	0.2	0.2	0.3[a]
Shear-Dilative angle ψ (°)	–	18	15	–
Yield strength f_y (MPa)	–	–	–	350[a]

Note [a]Data were measured in the Han (2000); and [b]Data were measured by repeated trial

2. *Failure modes*

Here, the influence of size effect on failure patterns for circular/square CFST columns with a low confinement coefficient was discussed. The failure patterns for the circular/square columns with sized between 200 mm to 1000 mm under the confinement coefficient ζ of 0.77 are presented in Fig. 7.100. With the increase of specimen sizes, the oblique angles of shear planes basically did not change, performing a typical compression-shear failure. Furthermore, local failure occurred at both ends of the columns, and may be possibly caused by the loading method. Consequently, one can conclude that the CFST columns having different structural sizes present similar failure patterns.

3. *Stress-strain curves*

Figure 7.101 shows the relationship of the simulated nominal stress and strain of circular and square columns. In Fig. 7.101, the specimens were arranged by the

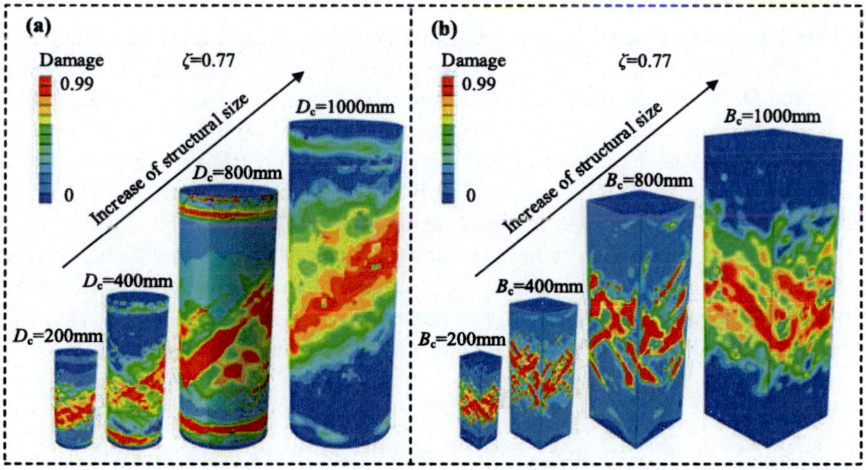

Fig. 7.100 Final failure patterns of the simulation specimens with different dimensions under same confinement coefficient: **a** Circular columns; and **b** Square columns

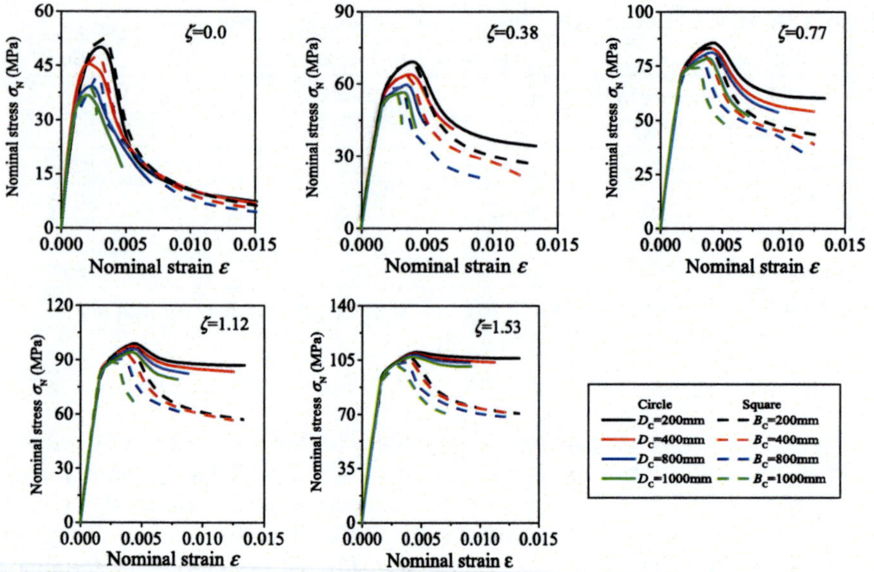

Fig. 7.101 Comparison of the nominal stress-strain curves of the circular and square CFST columns with different structural sizes under same confinement coefficient

confinement coefficients. Herein, the nominal stress σ_N is the ratio of the axial load N to the all cross section area A_{sc} of the columns, and the nominal strain ε is obtained by dividing the overall displacement u of the specimens by the corresponding lengths L. In addition, when the axial load reaches peak load, the nominal compressive strength σ_{Nu} is obtained. The nominal compressive strength σ_{Nu} can be expressed as $\sigma_{Nu} = P_u/A_{sc}$, where P_u is the peak load, adn A_{sc} means the cross-sectional area of columns.

One can clearly note from Fig. 7.101 that, when the cross-sectional diameter D_c and cross-sectional width B_c varied from 200 to 1000 mm, peak strain and compressive strength of the columns under the same confinement coefficient both obviously decrease. In addition, it is worth mentioning that compared to the circular CFST columns with the same size and the confinement coefficient, descent stage is steeper for the square CFST columns. This is because the lateral constraint of circular columns is more uniform than that by square columns.

As plotted in Fig. 7.101, when the confinement coefficient is relatively low(er) (e.g. $\zeta = 0.38$), the size effect on CFST columns can still be clearly observed. However, as the confinement coefficient increases, the failure of the column become less brittle, resulting into weakening of size effect in compressive strength.

4. *Horizontal/vertical strains of steel tube*

The relationships of the nominal stress versus the horizontal/vertical strains of steel tubes are presented in Fig. 7.102. Here, the average strains in the middle part of steel

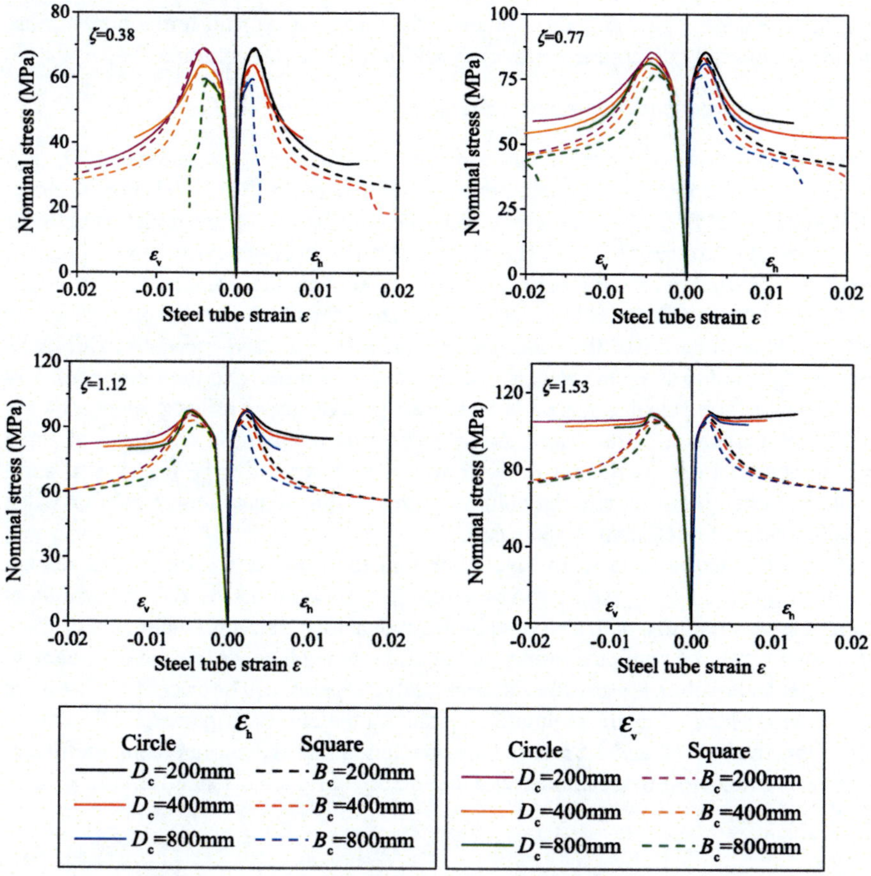

Fig. 7.102 Relationship between the nominal stress and steel tube strain for the circular/square CFST columns with different structural sizes under same confinement coefficient

tubes were calculated and studied. One can note from Fig. 7.102 that, in the elastic phase, the horizontal/vertical strains of steel tubes gradually increase linearly as the loading increased, and the growth speed of the vertical strains is obviously greater than that of the lateral strains. When the loading continued, the horizontal/vertical strains increase nonlinearly. In addition, when the peak stress i.e. axial compressive strength is reached, the steel tube in the middle reaches yield. While after the peak stress, the stress decreases, and the horizontal/vertical strains increase greatly.

In Fig. 7.102, when the confinement coefficient is 0.38, the smaller size specimens have obviously the larger horizontal/vertical strains of steel tubes when the peak stress is reached, especially for the square columns. Compared with the large size specimens, the smaller size specimens have faster growth in the horizontal/vertical strains after the peak stress. This illustrates that the constraint of steel tube for small-sized specimens has been fully played. However, when the confinement coefficient is larger (such as $\zeta = 0.77, 1.12, 1.53$), the influence of specimen size on the horizontal/vertical strains is gradually disappear, more obvious for the circular columns.

5. *Size effect on axial compression strength*

The relationships between the axial compressive strength σ_{Nu} and the size of D_c (B_c) under different confinement coefficients are all plotted in Fig. 7.103. For instance, "C-ζ-0.38" and "S-ζ-0.38" are respectively the circular and square columns with the confinement coefficient of 0.38. The size effect on compressive strength of the plain concrete columns was obviously observed. Moreover, the compressive strengths of square plain concrete column are greater than those of circular plain concrete column. It is to be noted that, the nominal axial compressive strength exhibits an increasing trend as the confinement coefficient ζ increases, and the trend gradually slows down. The possible reason is that when the constraint effect provided by the steel tube reaches a certain limit, the strength improvement is no longer affected by the confinement coefficient ζ. Moreover, the increase degree in nominal axial compressive strength for circular column is greater than that of square column when the confinement coefficient ζ increases.

It can be also noticed from Fig. 7.103 that there exists a critical confinement coefficient (ζ_{cr}). For a given concrete strength and column size, the nominal axial compressive strength would no longer be dependent on the structural dimension of the specimen when the applied confinement coefficient reaches the critical confinement coefficient. Namely, the size effect is completely suppressed. The critical confinement coefficients obtained by the simulation results for the circular/square CFST columns are respectively 1.12 and 1.53. It is thus concluded that the confinement coefficient plays a dominant role in the influence of the size effect for axial-loaded circular/square CFST columns.

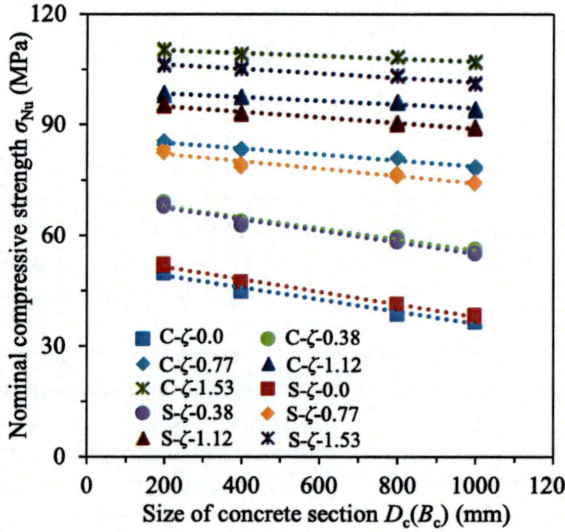

Fig. 7.103 Relationship between nominal axial compressive strength and size of concrete section under different confinement coefficients

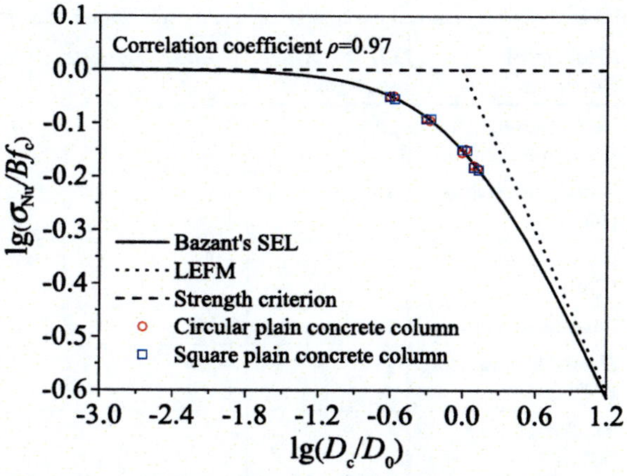

Fig. 7.104 Comparison of the simulation results of plain concrete columns with Bažant's SEL

Figure 7.104 presents the fitting relationship between the numerical data points of circular/square plain concrete columns and the Bažant's SEL. The corresponding correlation coefficient ρ is 0.97, indicating that Bažant's SEL could preferably represent size effect on the axial compressive failure of concrete columns.

7.7.2 Numerical Investigations on LWACFST Columns

1. *Generation of the LWACFST column at meso-scale*

The numerical model and boundary conditions of lightweight-aggregate concrete filled steel tube (LWACFST) column are similar with that in Sect. 7.7.1. However, unlike normal aggregates, which have high tensile and compressive strength, lightweight aggregates were set to be destroyed. Therefore, for the simulation of the LWACFST column, the mesoscale components (light-weight aggregate, mortar matrix, and ITZ) were all described by the plastic damage model. The mechanical parameters of the meso-components of concrete and steel tube are listed in Table 7.32.

Herein, the cross-sectional size and constraint coefficient ζ (defined in Sect. 7.7.1) were chosen as the key simulation parameters. For comparison, the same mechanical parameters (listed in Table 7.32) and geometric size for the CFST and the LWACFST columns are utilized. Herein, three different cross-sectional widths were selected, i.e., 200 , 400 and 800 mm. The steel tubular thickness t for the LWACFST (CFST) columns sized of 200 mm are 0 mm, 3 mm, 6 mm, 9 mm and 12 mm, respectively. In this way, the confinement coefficients ζ are evaluated as 0.00, 0.37, 0.78, 1.13 and 1.54. Therefore, for other LWACFST (CFST) columns having different structural dimensions, the same confinement coefficients would be utilized (Table 7.33).

Moreover, square and circular section types were employed to explore the section effect on size-effect for LWACFST columns under uniaxial compression. Similar to

Table 7.32 Mechanical parameters of the meso-components of concrete and steel tube

Mechanical parameter	Lightweight aggregate	Normal aggregate	Mortar matrix	ITZ	Steel
Compressive strength σ_c (MPa)	18[b]	–	42.6[d]	35.7[d]	–
Tensile strength σ_t (MPa)	5.1[b]	–	2.5[d]	2.2[d]	–
Elastic modulus E (GPa)	16[b]	73[a]	38[a]	39.3[d]	210[c]
Poisson's ratio ν	0.22[b]	0.16[a]	0.2[a]	0.2[d]	0.3[c]
Shear-Dilative angle ψ (°)	15	–	18	15	–
Yield strength f_y (MPa)	–	–	–	–	350[c]

Note [a]Data are quoted from Dilger et al. (1984); [b]Data are based on Wang and Chen (2008); [c]Data are quoted from Han (2000); and [d]Data are measured by repeated trial value

the circular/square CFST under axial compression (Han 2000), the slenderness ratio ($\lambda = L/B_c$, L denoted the height of the columns) for all columns was chosen as 3.0.

2. *Failure modes*

The failure modes of the inner concretes for CFST and LWACFST columns having the same cross-sectional width of 200 mm are compared in Fig. 7.105. One can note that, the failure modes of LWAC columns and NC columns have obvious similarities and differences. The cracks of both inner concretes pass through the specimens at a 45-degree angle, as shown in Fig. 7.106. Figure 7.106 presents the damage distribution at the cross-section in the middle height of the columns having two different aggregate types. Different from that of normal concrete columns, damage would generate through the weaker lightweight aggregates within LWACFST columns, as observed in Fig. 7.106. The distinct crack paths would lead to the difference in the ductility of the columns.

The typical final failure modes of squared CFST/LWACFST components with the confinement coefficient of ζ 1.13 with different structural sizes are respectively shown in Fig. 7.107a and b. Under the same confinement coefficient ζ, with the increase of the cross-section size, both squared CFST and LWACFST columns exhibit more brittle failure. The larger the structural size of the column, the more brittle failure pattern the columns exhibit. The oblique crack and the failure angle of the column (i.e., 800 mm) with larger structural sizes is larger than those with smaller sizes, indicating the failure turns out to be more brittle. Generally, the failure angle of LWACFST column is slightly larger than that of the NC ones. Furthermore, it is noted that the aggregate type has a negligible effect on the failure mode, which is similar to the experimental results of Ghannam et al. (2010).

The influence of confinement effect and cross-sectional shape were also discussed in Jin et al. (2020b). In summary, the cross-sectional type, the confinement of steel

Table 7.33 Parameters of the LWACFST and CFST columns

Specimen designation	Structural size B (mm)	Height h (mm)	f_y (MPa)	f_{ck} (MPa)	Confinement coefficients ζ
N-AS-S-1(2)	200	600	350	53.4	0
LW-AS-S(C)-1(2)				52.1	
N-AM-S-1(2)	400	1200		53.4	
LW-AM-S(C)-1(2)				52.1	
N-AL-S-1(2)	800	2400		53.4	
LW-AL-S(C)-1(2)				52.1	
N-BS-S-1(2)	200	600	350	53.4	0.37
LW-BS-S(C)-1(2)				52.1	
N-BM-S-1(2)	400	1200		53.4	
LW-BM-S(C)-1(2)				52.1	
N-BL-S-1(2)	800	2400		53.4	
LW-BL-S(C)-1(2)				52.1	
N-CS-S-1(2)	200	600	350	53.4	0.78
LW-CS-S(C)-1(2)				52.1	
N-CM-S-1(2)	400	1200		53.4	
LW-CM-S(C)-1(2)				52.1	
N-CL-S-1(2)	800	2400		53.4	
LW-CL-S(C)-1(2)				52.1	
N-DS-S-1(2)	200	600	350	53.4	1.13
LW-DS-S(C)-1(2)				52.1	
N-DM-S-1(2)	400	1200		53.4	
LW-DM-S(C)-1(2)				52.1	
N-DL-S-1(2)	800	2400		53.4	
LW-DL-S(C)-1(2)				52.1	
N-ES-S-1(2)	200	600	350	53.4	1.54
LW-ES-S(C)-1(2)				52.1	
N-EM-S-1(2)	400	1200		53.4	
LW-EM-S(C)-1(2)				52.1	
N-EL-S-1(2)	800	2400		53.4	
LW-EL-S(C)-1(2)				52.1	

Note N/LW = aggregate type (N = normal aggregate concrete, LW = lightweight aggregate concrete); A-E = group number (divided by confinement coefficients); S/M/L = specimen size (S-small, M-middle, L-large); and S/C = specimen shape (S-square, C-circle)

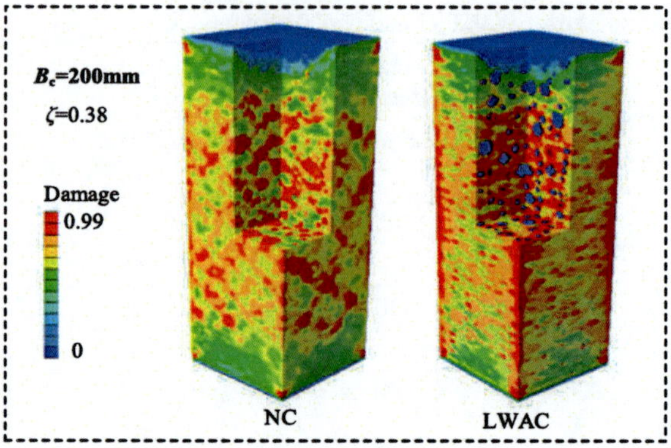

Fig. 7.105 Final failure modes of the simulation specimens under different aggregate

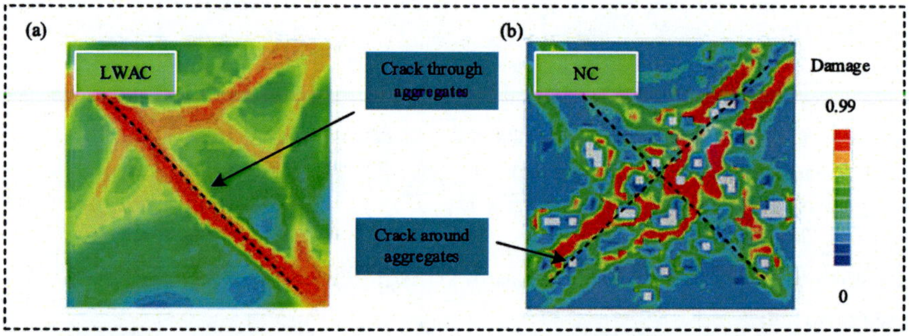

Fig. 7.106 Comparison of failure modes between: **a** Lightweight aggregate concrete; and **b** Normal concrete

tube, and the size of the specimen will affect the failure mode, but the type of aggregate has little effect on the failure mode.

3. *Load-deformation curves*

Figure 7.108 plots the connection between the load and the prescribed displacement of LWACFST/CFST columns sized of 200 mm having different confinement coefficients. For the ascending section of the curve, both the initial stiffness and the peak loads of lightweight aggregate concrete columns are slightly smaller than those of normal-weight one. Moreover, the more brittle characteristics of lightweight aggregates can be seen from the descending curves. The simulation results are consistent with the experimental results of Yu et al. (2016). Unlike that of plain concrete columns, the point of turning occurred after the elastic phase of core concrete for both CFST and LWACFST columns. At this point, the steel tube began to provide constraints for the internal concrete, thereby increase the axial compressive strength. The location of the breakpoint is related to the aggregate types.

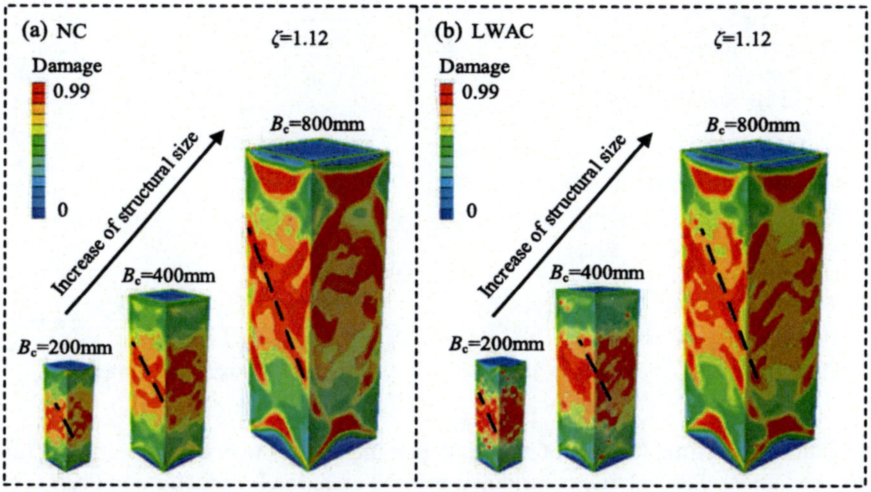

Fig. 7.107 Final failure modes of the simulation specimens with different dimensions under same confinement coefficient: **a** NC columns; and **b** LWAC columns

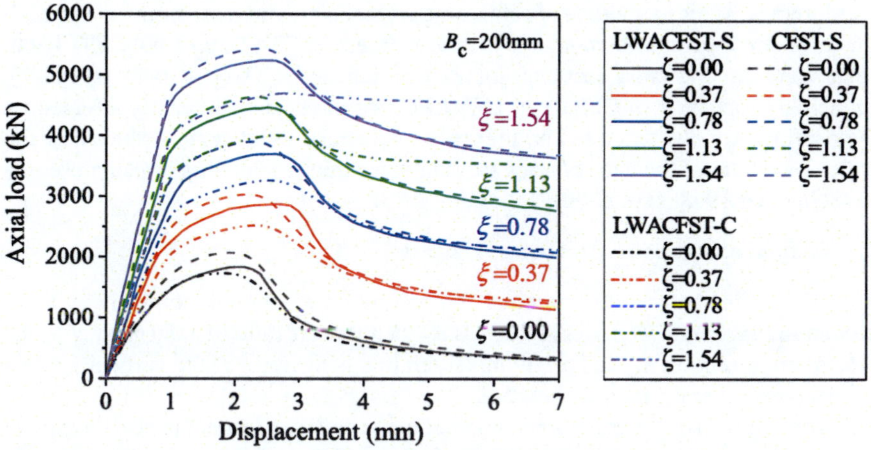

Fig. 7.108 Comparison of the axial load-displacement curves of the LWACFST and CFST columns with same dimension under different confinement coefficients

In Fig. 7.109, the lateral-to-axial strain ratios for steel tube are used to reflect the degree of lateral constraint. At the initial stage, one can see that the ratio is about 0.3, which is identical to the Poisson's ratio for low carbon steel (Liew and Xiong 2012). For LWACFST columns, when the axial load reaches about $0.2N_u$, the steel tubes have an early restraint on concrete, but it has little effect on the strength enhancement. Because the elastic modulus of LWC is much lower than that of NC, the early volume expansion of LWC is caused. At this time, since there are no large-scale cracks, only cracks which occurs on the surface of concrete are constrained by

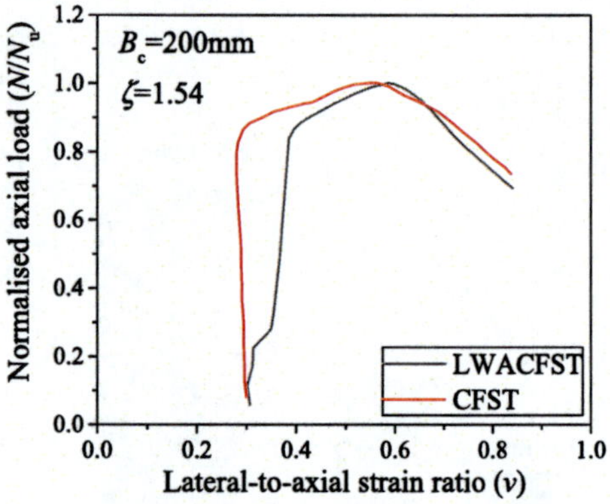

Fig. 7.109 Development of lateral-to-axial strain ratios for steel tube of the LWACFST and CFST columns

steel tube. Therefore, it does not have a greater impact on the bearing capacity of the columns under uniaxial compression. When the axial load reaches about $0.9N_u$ for both LWAC and NC columns, large cracks generated within the concretes, and the stiffness of the columns decreases obviously. At the moment, the restraints of steel tube increased rapidly, making the increase in the loading capacity.

Moreover, as the magnitude of confinement coefficients increase from zero to 1.54, the axial strength enhancement and elastic stiffness of CFST and LWACFST (both square and circular cross sections) increase significantly. The post-peak (softening or yielding) behavior of CFST and LWACFST columns is significantly affected by the confinement coefficients. The higher the constraint coefficient is, the softer the softening curves of the two columns are. The same trend in ductility is also presented in (Schneider 1998; Han 2000).

4. *Nominal compressive stress-strain curves*

The relationship between the nominal compressive stress σ_N and nominal strain ε is plotted in Fig. 7.110. The nominal stress σ_N denotes the ratio of axial-load N to the area of cross section A_{sc} and the nominal strain ε is the ratio of the vertical applied displacement u to the column length L.

From Fig. 7.110, under the same confinement coefficient, both the peak compressive stress and the corresponding nominal strain of the column decrease obviously with the increase of structural size. In addition, it is worth notice that the descent stage (i.e. softening curve) of LWACFST columns is steeper than that of CFST columns in the same case. In order to show the trend of decreasing section quantitatively, the softening rate is defined. The softening rates of CFST and LWAFST columns are plotted in Fig. 7.111a, the softening rate k (unit: MPa) is defined as the absolute value of the slope from the peak nominal strength σ_0 to 85% σ_0 in the nominal compressive stress-strain curve. From the results of Fig. 7.111b, it can be seen that the softening rate of LWACFST columns is generally higher than that of CFST columns, and the change of softening rate of LWACFST columns is more obvious with the increase of size. This is because the failure of lightweight aggregate concrete is much more brittle than that of the normal one. Meanwhile, the restraining effect of constraint

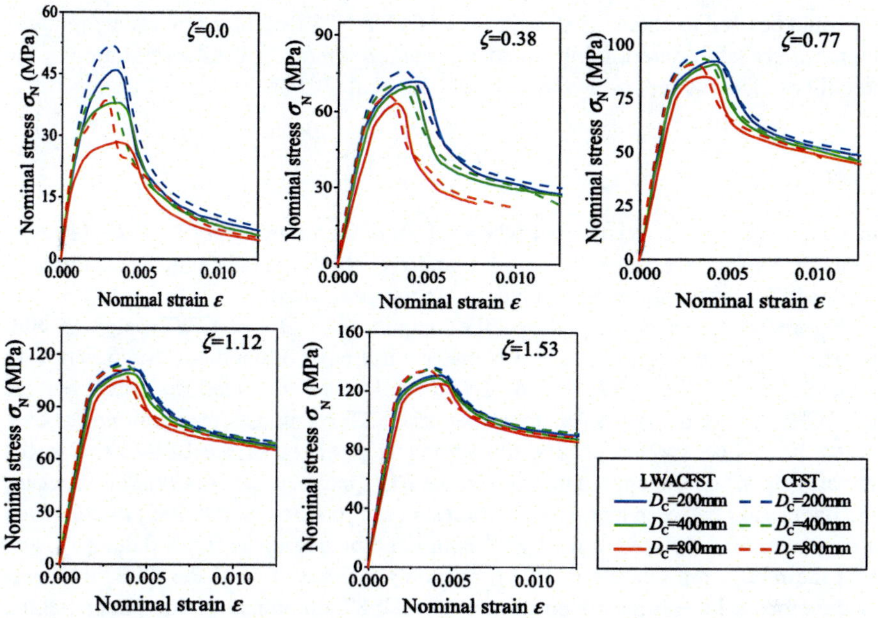

Fig. 7.110 Comparison of the nominal stress-strain curves of the LWACFST and CFST columns with different structural sizes under the same confinement coefficient

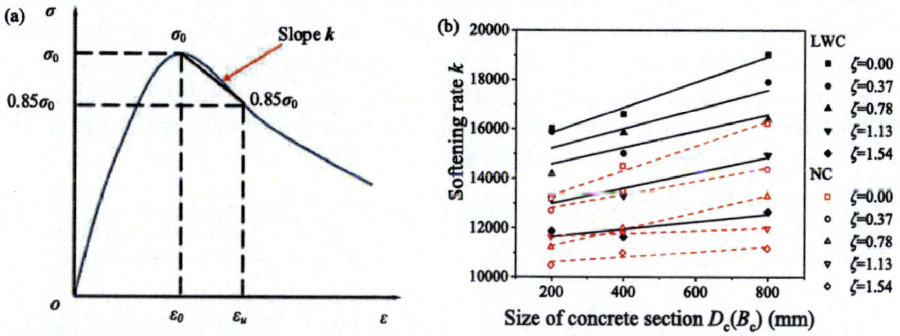

Fig. 7.111 Softening rate of the LWACFST and CFST columns: **a** Definition of softening rate *k*; and **b** Comparison of softening rate

coefficient on the softening rate is obvious. With the increase of constraint degree, the softening rates decrease. This result is similar with the experimental results observed in Yu et al. (2016).

5. *Ductility*

The ductility index (DI) is often used to evaluate whether the plastic deformations of LWACFST/CFST short columns have obvious strength degradation characteristics.

DI could be used to check whether the LWACFST/CFST columns meets the requirements or not, after determining the ductility requirements. A similar definition of the ductility index has the following form as (Tao et al. 2011)

$$DI = \frac{\varepsilon_{85\%}}{\varepsilon_y} \tag{7.15}$$

in which $\varepsilon_{85\%}$ is the axial strain when the load decreased to 85% of the ultimate strength, and ε_y is equal to $\varepsilon_{75\%}/0.75$, and $\varepsilon_{75\%}$ is the axial strain when the load attained 75% of the ultimate strength in the pre-peak stage.

Figure 7.112 respectively shows the magnitude of DI for CFST columns with respect to the two aggregate types under different confinement coefficients. In general, due to the brittleness of LWAC columns, the ductility coefficient of LWACFST columns is weaker than that of CFST columns. As mentioned above, when the failure surface occurs, the aggregate particles of the LWAC specimens are broken, while the aggregate particles of NC specimens are not broken. Because the interaction between aggregate particles would produce an interlock force, which would only be generated in the CFST columns (the aggregate is not damaged), and will indirectly improve the ductility. The obtained ductility of the CFST columns is therefore a little higher than that of LWACFST columns. In this phenomenon, constraint coefficient also plays a role. A higher degree of restraint will enhance the aggregate internal locking force of CFST columns and the failure surface friction of LWACFST columns, indirectly improving the ductility, and increasing the *DI* values. That is to say, the columns having larger confinement index present a better ductility.

Furthermore, one can note from Fig. 7.112 that the magnitude of *DI* is also associated with the structural sizes, i.e., the DI value decreases with the increase of structural size. Larger size leads to more internal defects, and the effect of internal force on failure surface is relatively reduced, which means that the larger the structure size, the worse the ductility of the columns.

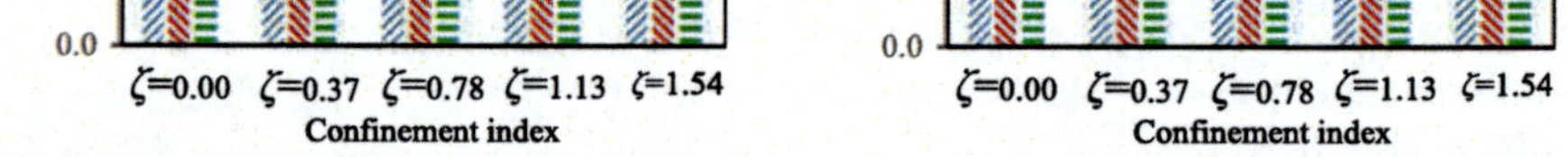

Fig. 7.112 Comparison of ductility indexes (*DI*) under different confinement coefficients: **a** LWACFST; and **b** CFST

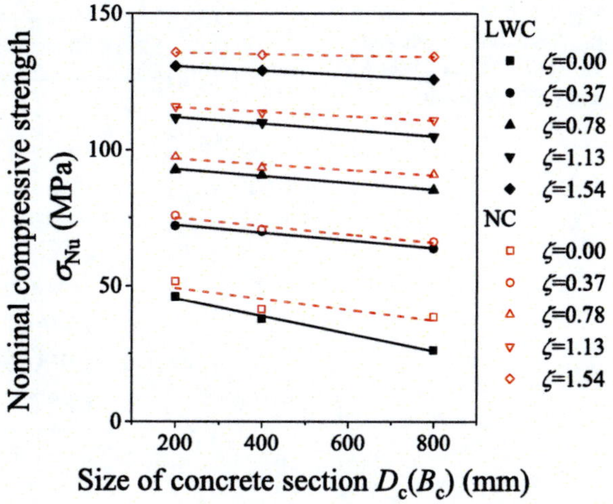

Fig. 7.113 Relationship between the obtained nominal axial compressive strength and size of concrete section under different confinement coefficients

6. *Nominal compressive strength*

Figure 7.113 shows the variation trend of nominal compressive strength with size under different confinement coefficients for square cross-section columns. The nominal strength of both LWACFST and CFST columns show obvious size-effect, and the size-effect is tended to decrease with the increase of confinement coefficient. Similar conclusions can be drawn from Samoel (2017). Nominal compressive strength improves with the increase of confinement coefficient, but this trend slows down gradually. This may be due to the weakening of the effect on strength improvement when the restraint degree reaches some level. In addition, the nominal compressive strength size-effect for LWACFST columns is more obvious than that for CFST columns, under different confinement coefficients. Under high confinement coefficient (e.g., $\zeta = 1.54$), this effect on nominal strength could be neglected for CFST columns, but this could not be neglected for LWACFST columns.

According to the SEL of Bažant et al. (1984), three sets of data were extracted and compared. LWAC columns, NC columns and LWACFST columns with confinement coefficient of 1.53 are extracted respectively, as shown in Fig. 7.114. Herein, f_c means uniaxial compressive strength for LWAC/NC materials (i.e. the simulated compressive strength for plain LWAC/NC specimens sized of 200 mm × 200 mm × 600 mm). The fitting parameters of plain LWAC/NC columns and LWACFST columns ($\zeta = 1.53$) are shown in Table 7.34. In Fig. 7.114, the data for plain LWAC columns was closer to LEFM curve than that for the NC. When the steel tube is utilized, the ductility coefficient increases obviously, making the failure of columns ductile. On the whole, it can be concluded that Bažant's SEL can well describe the effect of structural size on the axial compression failure of LWACFST columns. Nevertheless, for different constraint coefficients, the obtained fitting parameters of B and D_0 are significantly distinct. Therefore, Bažant's SEL cannot quantitatively describe the influence of constraint coefficient on the size effect in axial-loaded CFST columns.

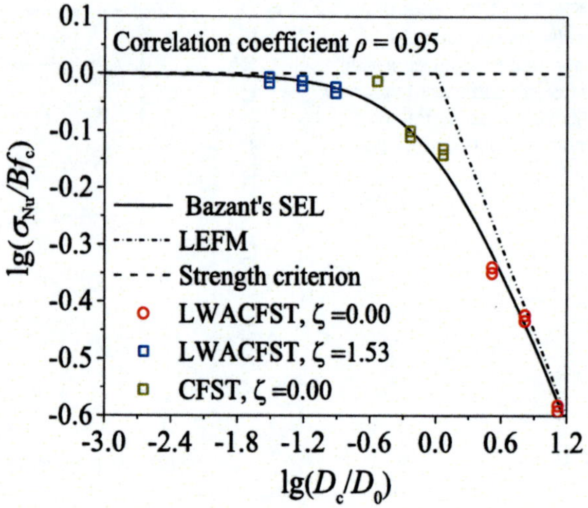

Fig. 7.114 Comparison of the simulation results and Bažant's SEL for LWACFST and CFST columns

Table 7.34 Parameters of Bažant size effect law for LWAC columns

	A	C	B	D_0
LWAC columns	0.0038	0.2290	2.0897	60.261
NC columns	0.0011	0.7544	1.1513	685.818
LWACFST columns	0.00002	0.1309	2.7639	6545

7.8 Size Effect Law for Axial-Loaded Confined Concrete Columns

In this section, the size effect law of nominal compressive strength for the confined concrete columns tested in Sect. 7.5–7.7 was presented (Jin et al. 2019, 2020a, b, c, e).

7.8.1 Basic Assumptions

It can be known from the experimental and numerical investigation that, for concrete columns, the size effect behavior is much more complicated than that of concrete materials. Therefore, the rationality that the available SELs for concrete materials were utilized to describe the size effect in RC components should be furtherly explored and discussed. In addition, the available SELs for concrete materials can only present the unique factor—"structural size" on the nominal strength, and they cannot describe the influence of other important factors. Accordingly, the size effect laws suitable for RC components are presented, in which the influences of the other factors on the "size effect" can be explicitly taken into account.

In reality, for the concrete columns in Sects. 7.1–7.6, the axial load-carrying capacity N contains two aspects, involving (1) load carrying capacity of longitudinal reinforcement, N_s, and (2) the load carrying capacity of concrete columns without longitudinal reinforcement, N_{sc}. The total load carrying capacity N of concrete columns (confined with stirrups, FRP or steel tube) can be described as

$$N = N_{sc} + N_s \tag{7.16a}$$

with

$$N_s = f_y' A_s \tag{7.16b}$$

$$N_{sc} = \sigma_{Nu} A_c \tag{7.16c}$$

in which, f_y' is the yield strength of the longitudinal reinforcement; A_s and A_c is respectively the cross-sectional area of longitudinal reinforcement and the concrete column; and σ_{Nu} is the nominal strength of the concrete columns (confined with stirrups, FRP or steel tube) without longitudinal reinforcement.

Actually, the mechanical properties of steel reinforcement can be treated as size independent, and then the loading capacity N_s should be size independent. One can note from Eqs. (7.16a, b, c) that only the load carrying capacity N_{sc} is closely related to dimensions of the column. That is to say, the loading capacity N_{sc} (or the nominal strength σ_{Nu}) should be size dependent. In other words, the size effect in concrete columns (confined with stirrups, FRP or steel tube) under axial compression is merely contributed by the failure of confined concrete columns, but not the longitudinal reinforcements.

In Sects. 7.1–7.7, the influence of various parameters on size effect of axially loaded concrete columns were studied, including concrete strength, loading arrangement, cross-sectional shape, and confinement effect. Among these parameters, the confinement effect has significant influence on the size effect of axially loaded concrete columns. Therefore, the confinement effect was taken as the main parameter to establish the size effect law.

7.8.2 *Size Effect Law for Nominal Axial Compressive Strength*

1. *Establishment of size effect law*

According to the experimental and numerical results in Sects. 7.1–7.7, it is not hard to conclude that, the confinement effect on the failure behavior of axially-loaded concrete columns contain the following two aspects.

(1) Enhancement on the nominal compressive strength. Due to the confinement effect generated by stirrups (or FRP, steel tube), the core concrete is under complicated tri-axial compressive stress status, making the nominal compressive strength of the columns improves Moreover, the larger the constrain ratio is, the larger nominal compressive strength the columns possess. These can be obviously noticed from the obtained test and simulation results. This behavior can be characterized by a strength improvement coefficient, φ.
(2) Weakening the size effect. Due to the confinement effect, the core concrete is under a tri-axial compression stress state. The corresponding deformation of the core concrete is restricted, making that the confined concrete column exhibits less-brittle failure. This finally weaken the size effect of concrete columns under axial compression. In addition, for the case that the confinement effect is extremely strong, the axial compressive failure of the columns is almost plastic, and the size effect would disappear completely (Şener et al. 2004; Bažant 1984). This influence can be described by the size effect weakening coefficient β.

According to the above two influence mechanisms of the confinement effect and based on the classic SEL (Bažant 1984) for concrete materials, the nominal compressive strength σ_{Nu} of confined concrete columns can be expressed as:

$$\sigma_{Nu} = \frac{Bf_c}{\sqrt{1+D/D_0}} \cdot \varphi \cdot \beta \tag{7.17}$$

in which, the parameters of φ and β are both functions of constrain ratio.

It is to be noted that, the size effect law can quantitatively describe the confinement effect on the size effect in nominal strength of confined RC columns under uniaxial compression. The parameters of φ and β would be determined and discussed below.

2. *Determination of φ*

(1) Stirrup-confined circular and square concrete columns

The calculated expressions proposed by Mander et al. (1988), which has gained world-wide acceptance in the research of confined concrete, were utilized to determine the function of φ of stirrup-confined concrete columns.

For the circular hoops, the strength enhancement coefficient φ is expressed as follows:

$$\varphi_c = \frac{f_{cc}}{f_c} = -1.254 + 2.254\sqrt{1 + \frac{7.94 f_l}{f_c}} - 2\frac{f_l}{f_c} \tag{7.18a}$$

$$f_l = \frac{1}{2} k_e \rho_{su} f_{yh} \tag{7.18b}$$

$$k_e = \left(1 - \frac{s'}{2d_s}\right)^2 / (1 - \rho_{cc}). \tag{7.18c}$$

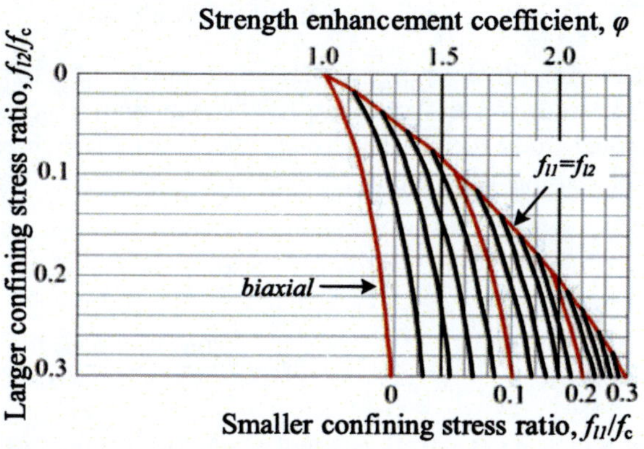

Fig. 7.115 Relationship between strength enhancement coefficient and effective lateral confining stresses for square or rectangular hoops

in which, f_l = effective lateral confining stress generated by stirrups, k_e = the effective confinement coefficient, f_{yh} = the yield strength of the transverse reinforcement, s' = the clear vertical spacing between transverse reinforcements, ρ_{cc} = the longitudinal reinforcement ratio, and d_s = the diameter of circle between bar centers.

As for the square or rectangular hoops, the strength enhancement coefficient φ is determined by two lateral confining stresses, f_{l1} and f_{l2} (i.e., lateral confining stresses in the x and y directions, and $f_{l1} \le f_{l2} \le 0.3f_c$), which is shown in Fig. 7.115.

El-Tawil and Deierlein (1996) described the curves (Fig. 7.115) by a series of equations:

1. When the confined concrete is under f_{l1} and $f_{l2} = 0.3f_c$, the upper limit of strength enhancement coefficient, φ_u, can be expressed as:

$$\varphi_{s,u} = \frac{f_{cc}}{f_c} = 1.3 + 5.55\frac{f_{l1}}{f_c} - 7.5\left(\frac{f_{l1}}{f_c}\right)^2 \tag{7.19a}$$

2. When the confined concrete is under $f_{l1} = f_{l2} \le 0.3f_c$, the lower limit of strength enhancement coefficient, φ_l, can be expressed as:

$$\varphi_{s,l} = \frac{f_{cc}}{f_c} = -1.254 + 2.254\sqrt{1 + \frac{7.94f_l}{f_c}} - 2\frac{f_l}{f_c} \tag{7.19b}$$

Therefore, for the square or rectangular hoops, the required strength enhancement coefficient φ can be obtained through interpolating between the upper and lower limits:

$$\varphi_s = \frac{f_{cc}}{f_c} = \varphi_l + (\varphi_u - \varphi_l)\sqrt{\frac{f_{l2}/f_c - f_{l1}/f_c}{0.3 - f_{l1}/f_c}} \tag{7.19c}$$

(2). GFRP-confined circular concrete columns

The simplified strength calculation model of FRP-confined concrete cylinders proposed by Lam and Teng (2002) was widely accepted. According to the theoretical study of Lam and Teng (2002), the strength enhancement coefficient φ for columns having circular cross section can be described as:

$$\varphi = \frac{f_{cc}}{f_c} = 1 + 2\frac{f_l}{f_c} \tag{7.20a}$$

$$f_l = \frac{2f_{FRP}t}{D} = \frac{E_{FRP}t\varepsilon_h}{R} \tag{7.20b}$$

in which, f_l is lateral restraint strength, f_{FRP} is tensile strength of FRP, t is total thickness of FRP, D is column cross-sectional dimeter, E_{FRP} is elastic modulus of FRP, ε_h is fracture strain of FRP, and R is sectional radius.

It can be seen from Eqs. (7.20a, b) that φ is mainly related to strength grade of concrete, tensile strength and thicknesses of FRP.

(3). CFRP-confined square RC columns

For the CFRP-confined reinforced concrete columns in Sect. 7.6.1 and 7.6.2, both the stirrup and CFRP have influence on the axial compressive strength. Therefore, the strength improvement coefficient φ was defined as:

$$\varphi = \varphi_1 \cdot \varphi_2 \tag{7.21a}$$

where φ_1 means the strength enhancement factor generated by CFRP, and φ_1 can be given by Eq. (7.20a, b). In addition, φ_2 presents the strength enhancement effect of stirrups, which can be described according to Park et al. (1982):

$$\varphi_2 = 1 + \rho_{sv}\frac{f_y}{f_c} \tag{7.21b}$$

where ρ_{sv} means volumetric stirrup ratio, and f_y is strength of stirrup.

(4). CFST (concrete filled steel tube) columns

There are two methods can be used to describe the function of φ. **Method 1** was according to the Chinese design codes named DL/T5085-1999 (NETC 1999) and GJB4142-2000 (NMSC 2001). **Method 2** was based on the CECS 159-2004 (CECS 2004) and European technical codes ECS (2004).

Method 1:

According to the Chinese design codes named DL/T5085-1999 (NETC 1999) and GJB4142-2000 (NMSC 2001), the axial compression strengths of circular/square CFST (i.e. f_{scy}^c and f_{scy}^s) are respectively described as

$$f_{scy}^{c} = \left(1.212 + B_1\zeta + C_1\zeta^2\right)f_{ck} \tag{7.22a}$$

$$f_{scy}^{s} = \left(1.212 + B_2\zeta + C_2\zeta^2\right)f_{ck} \tag{7.22b}$$

where f_{ck} is the standard value of axial compressive strength of concrete; $B_1 = 0.1759f_{\mathrm{y}}/235 + 0.974$; $C_1 = -0.1038f_{\mathrm{ck}}/20 + 0.0309$; $B_2 = 0.1381f_{\mathrm{y}}/235 + 0.7646$; and $C_2 = 0.0727f_{\mathrm{ck}}/20 + 0.0216$ (NETC 1999; NMSC 2001).

Based on the axial compressive strength of concrete, and considering the restraint effect of steel tube, the confinement coefficient is introduced to calculate the axial compressive strength of CFST members. In this section, utilizing this calculation method, and taking the axial compressive strength for concrete material as the reference strength, the strength improvement coefficients for circular/square CFST columns i.e. φ_{c} and φ_{s} can respectively be defined by

$$\varphi_c = \frac{f_{cc}}{f_c} = \gamma_c \times \left(1.212 + B_1\zeta + C_1\zeta^2\right) \tag{7.23a}$$

$$\varphi_s = \frac{f_{cc}}{f_c} = \gamma_s \times \left(1.212 + B_2\zeta + C_2\zeta^2\right) \tag{7.23b}$$

where, γ_{c} and γ_{s} are respectively the conversion coefficients of the reference strengths for circular/square CFST columns, $\gamma = f_{ck}/f_c$. Herein, according to the fitting analysis in Fig. 7.116, the two parameters of γ_{c} and γ_{s} are respectively calculated as 0.83 and 0.87.

Method 2:
On the basis of CECS 159-2004 (CECS 2004) and Eurocode 4 (CEN 2004b), the axial compression strengths of CFST columns is similarly described as

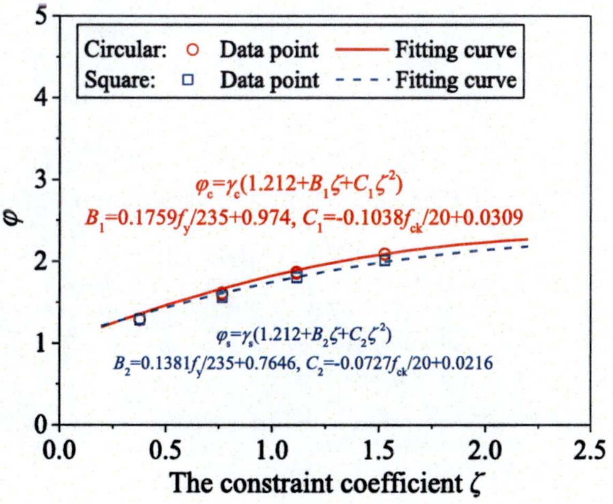

Fig. 7.116 Determination of the strength improvement coefficient φ

$$N_u = f_c A_c (1 + \zeta) \tag{7.24}$$

The simulated compressive strength of concrete specimen (sized by 150 mm × 150 mm × 150 mm) under uniaxial compression is taken for reference. Therefore, the strength improvement coefficient φ can be evaluated as:

$$\varphi = \frac{f_{cc}}{f_c} = \gamma \times (1 + \zeta) \tag{7.25}$$

where γ is the adjustment coefficient; and ζ means confinement coefficient, $\zeta = f_y A_s / f_c A_c$.

Similar to CECS 159-2004 (CECS 2004), an adjustment function $(\mu \cdot \sqrt{\zeta})$ is proposed, Eq. (7.25) is transformed into

$$\varphi_L = \frac{f_{ccl}}{f_c} = \gamma_L \times \left(1 + \zeta + \mu_L \cdot \sqrt{\zeta}\right) \tag{7.26a}$$

$$\varphi_N = \frac{f_{ccn}}{f_c} = \gamma_N \times \left(1 + \zeta + \mu_N \cdot \sqrt{\zeta}\right) \tag{7.26b}$$

where f_{ccl} (f_{ccn}) is the strength of LWCFST (CFST) columns under uniaxial compression; γ_L and γ_N are the adjustment coefficients for LWCFST and CFST columns. By comparing with the simulated data, μ_L and μ_N are respectively taken as 2.0 and 1.0.

In Fig. 7.117, the magnitude of the nominal strength under different confinement coefficients is given, and the curve (φ/γ) is drawn with black lines. Based on the fitting results given in Fig. 7.117, γ_L and γ_N are calculated as 0.60 and 0.70 respectively. Figure 7.118 shows the comparison between the design codes and the proposed strength enhancement coefficient. It can be seen from Fig. 7.118 that the size effect law is more compatible with the simulation values.

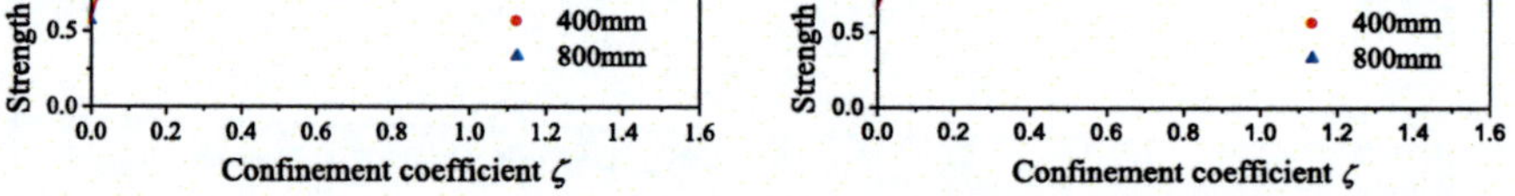

Fig. 7.117 Determination of the strength improvement coefficient φ: **a** LWACFST; and **b** CFST

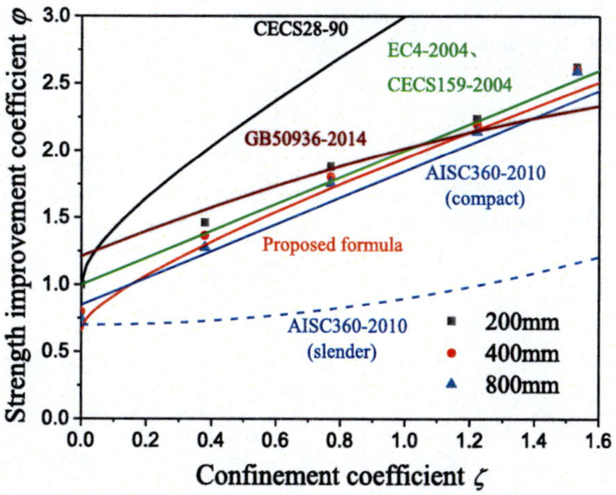

Fig. 7.118 Comparison with design codes of the strength improvement coefficient φ for CFST column

Considering the influence of the shape-effect on the nominal compressive strength, the section coefficient ϕ is added to describe the strength improvement coefficient, and it has the form as

$$\phi = \sigma_C / \sigma_S \tag{7.27}$$

where σ_C is the nominal strength of Circular cross-section members, and σ_S is the nominal strength of Square cross-section ones. That is, the strength of the CFST columns having the square cross-sectional shape is set as the reference strength. Therefore, the nominal strength of the columns having the circular cross-section shape can be directly obtained through

$$\varphi = \phi \times \varphi_L \tag{7.28a}$$

$$\varphi = \phi \times \varphi_N \tag{7.28b}$$

As shown in Fig. 7.119, for circular cross-section, the magnitude of the section coefficient ϕ could be recommended 1.10. Simply to say, for the columns having the same structural sizes, the nominal strength of circular columns would be 1.10 times as large as that of square columns.

3. *Determination of β*

(1). Stirrup-confined circular and square concrete columns

According to the analyses above, the size effect of peak axial stress decreases with the increase in stirrup ratio. Besides, the stronger the constraint effect is, the more obviously the size effect weakens. Figure 7.120 shows the relationship between the

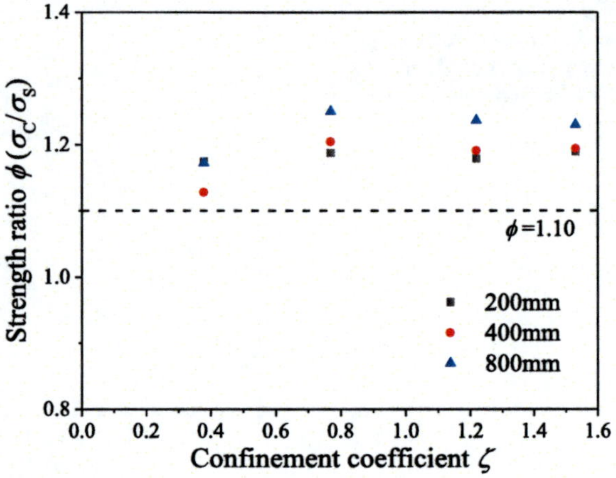

Fig. 7.119 Determination of the strength ratio ϕ for LWACFST

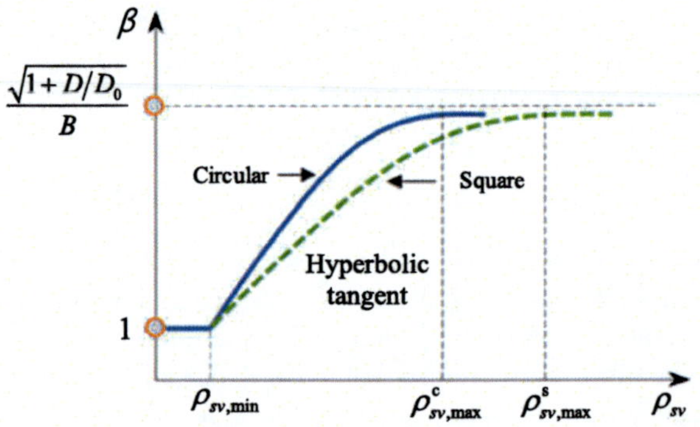

Fig. 7.120 The size effect weakening coefficient versus volumetric stirrup ratio

weakening coefficient β and the stirrup ratio ρ_{sv}. This will be specifically described below.

1. For $\rho_{sv} \leq \rho_{sv,min}$ (the minimum stirrup ratio), the confinement effect is quiet weak, the existence of stirrups has almost no influence on the size effect of peak axial stress of stirrup-confined RC column, thus $\beta = 1$. Herein, the minimum stirrup ratio $\rho_{sv,min}$ is taken as 0.25% according to the code (Ministry of Construction of the People's Republic of China 2011).
2. For $\rho_{sv} \rightarrow \rho_{sv,max}$ ($\rho_{sv,max}$ is the critical stirrup ratio that can provide extremely strong confinement), the confinement effect generated by stirrups turns out to be very strong, and the corresponding failure behavior of RC column tends to be plastic. The size effect on the nominal compressive strength would be completely suppressed, as also described in (Şener et al. 2004; Bažant 1984). In this case, the nominal axial compressive strength is size independent, and the corresponding nominal strength of confined concrete would be:

$$\sigma_{Nu}|_{\rho_{sv}\to\rho_{sv,max}} = f_{cc} \tag{7.29}$$

By substituting Eq. (7.29) into Eq. (7.17) and rearranging, one can get the corresponding weakening coefficient as:

$$\beta|_{\rho_{sv}\to\rho_{sv},\max} = \frac{\sqrt{1+D/D_0}}{B} \tag{7.30}$$

3. When the stirrup ratio ρ_{sv} is between $\rho_{sv,min}$ and $\rho_{sv,max}$, the size effect in nominal strength of confined concrete columns would exist. Moreover, the corresponding size effect would be gradually weakened or suppressed with the increase of stirrup ratio. The relationship between weakening coefficient β and stirrup ration ρ_{sv} can be assumed to be a non-linear curve asymptotically to $\left(\sqrt{1+D/D_0}\right)/B$.

According to the above analyses and assumptions, the hyperbolic tangent curve was employed to characterize the influence of stirrup ratio on the size effect. The coefficient β can be expressed as:

$$\beta = \begin{cases} 1, & \rho_{sv} \le \rho_{sv,\min} \\ \left(\frac{\sqrt{1+D/D_0}}{B} - 1\right)\tanh[\alpha(\rho_{sv} - 0.0025)] + 1, & \rho_{sv} > \rho_{sv,\min} \end{cases} \tag{7.31}$$

in which, α is an adjustment coefficient that reflects the change rate of weakening coefficient along with stirrup ratio, and it would be determined by fitting with the test and simulation results. Certainly, the coefficient α is closely related with constraint effect.

(2). GFRP-confined circular concrete columns

For GFRP-confined circular concrete columns, the minimum constrain ratio was set as zero. Therefore, the size effect weakening factor β can be expressed as follows:

$$\beta = \frac{\sqrt{1+D/D_0}}{B} - 1\tanh\left(\alpha\rho_f\right) + 1, \rho_f \ge 0 \tag{7.32}$$

in which, the constraint ratio ρ_f is described by Eq. (7.13).

(3). CFRP-confined square RC columns

As mentioned in Sect. 7.6.1, the size effect would be weakened by the confinement of CFRP and stirrups. For the CFRP-confined RC columns in Sect. 7.6.1, the stirrups ratio is relative smaller, and the weakening effect of stirrups can be ignored. The weakening effect generated by CFRP was considered only. Similar with that of GFRP-confined circular column, size effect weakened parameter, the weakened coefficient β of CFRP-confined RC columns has the same form as Eq. (7.32).

Therefore, the size effect law of size effect of axial-loaded CFRP-wrapped reinforced square columns can be finally determined by Eqs. (7.17, 7.20a, b, 7.32). As mentioned previously, the parameter fc is the uniaxial compressive strength of standard concrete specimens.

(4). CFST (concrete filled steel tube) columns

Similar to that in FRP confined concrete columns, the weakening coefficient β_ζ for LWACFST and CFST columns can be fixed respectively as:

$$\beta_{L\zeta} = \left(\frac{\sqrt{1 + D/D_0}}{B} - 1\right) tanh(\varepsilon \alpha_L \zeta) + 1, \zeta \geq 0 \tag{7.33a}$$

$$\beta_{N\zeta} = \left(\frac{\sqrt{1 + D/D_0}}{B} - 1\right) tanh(\varepsilon \alpha_N \zeta) + 1, \zeta \geq 0 \tag{7.33b}$$

where α_L and α_N are the adjustment coefficient for LWACFST and CFST columns respectively, reflecting the speed of β_ζ varies with confinement coefficient, similar to the strength enhancement coefficient, ε is the shape influence factor.

7.8.3 Validation of Size Effect Law

1. *Stirrup-confined concrete column*

For stirrup-confined concrete columns, there are three parameters leaving to be determined, i.e., B, D_0 and α. To determine the parameters, the standard prismatic concrete sample sized of 150 mm × 150 mm × 300 mm is taken as reference specimen for square column, and the cylindrical concrete specimen sized of Φ150 mm × 300 mm is for circular column. Therefore, the axial compressive strength f_c is 39 MPa for square column, and 34.76 MPa for circular column (numerical results in Sect. 7.5.3). The parameters B and D_0 obtained from regression analysis of simulation results of plain concrete columns (Sect. 7.5.3) were selected as basis parameters. For circular column, $B = 1.017$ and $D_0 = 800$ and for square column, $B = 1.176$ and $D_0 = 714$.

It is noteworthy that Eq. (7.31) is a transcendental equation and the analytical solution of α is difficult to obtain. Therefore, the inversion method (repetitive experimental algorithm) was applied to approximate the numerical solution. When the predicted values closely approximate to the simulation results, the value of α was selected. For stirrup-confined circular column, good agreement can be found when the coefficient α is between 6 and 10. For stirrup-confined square column, good agreement was obtained when α changes from 0.8 to 1.5. Therefore, α is recommended to be 8 for stirrup-confined circular concrete column, and 1 for stirrup-confined square concrete column. The details about the determination of α was reported in Jin et al. (2020c).

With the parameters fixed, Fig. 7.121 presents comparisons between the predictive value of nominal compressive strength and the extended simulation results (including larger structure size),. From Fig. 7.121, one can know that the predictive results are in good agreement with the simulation and test results (Sect. 7.7), indicating the accuracy of the size effect law for the nominal compressive strength of stirrup - confined concrete columns.

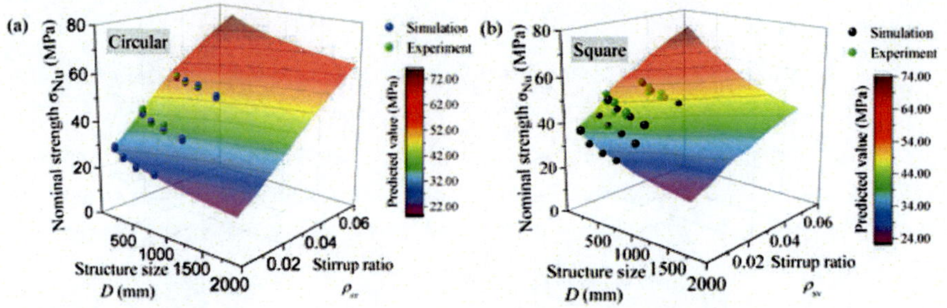

Fig. 7.121 Comparison of the predicted results with the experimental and simulation results: **a** Circular columns; and **b** Square columns

Furthermore, Fig. 7.122 compares the predictions of peak axial stress with the collected experimental data. $\sigma_{\mathrm{Nu,exp}}$ and $\sigma_{\mathrm{Nu,pre}}$ presents the experimental and predicted values, respectively. The ratio of predicted value to experimental data varies from 0.76 to 1.15 for all specimens. The average value and variation coefficient were 0.99 and 0.07 for circular column, and 1.01 and 0.07 for square column. It is noteworthy that the predictions of peak axial stress distribute closely around the $y = x$ line. In addition, most of the experimental results are within a $\pm$ 15% margin of error, indicating that the predictions of peak axial stress closely agree with the experimental results. In general, the size effect law can predict the peak axial stress of stirrup-confined concrete column reasonably and correctly.

2. *GFRP-confined circular concrete column*

For validation of size effect law of GFRP-confined concrete columns, f_{c} is selected as the numerical result of compressive strength of concrete cylinder with 150 mm diameter and 300 mm length in Sect. 7.6.3, i.e. $f_{\mathrm{c}} = 33.31$ MPa. The parameters B and D_0 obtained from the regression analysis of the simulation results in Sect. 7.6.3

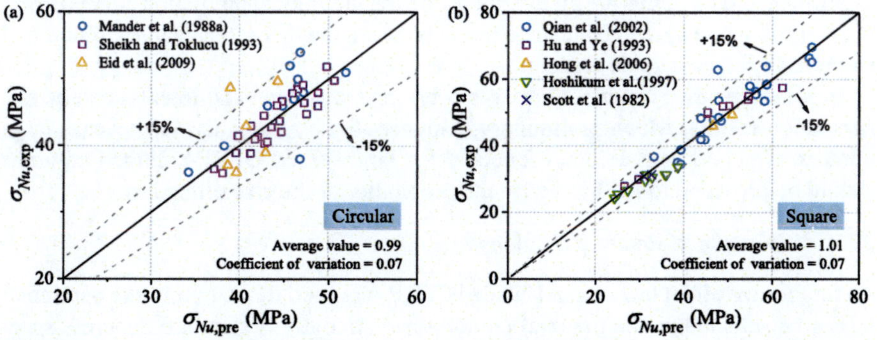

Fig. 7.122 Comparison between the predictions and experimental data: **a** Stirrup-confined circular columns; and **b** Stirrup-confined square columns

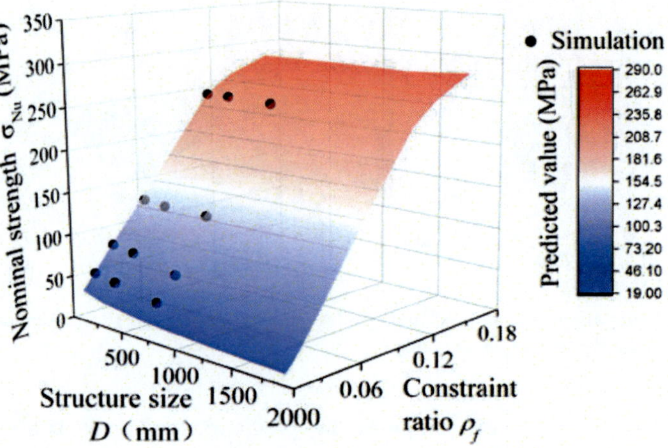

Fig. 7.123 Comparisons between the calculated values and numerical results for α = 26

are taken as the reference parameters, i.e. $B = 1.542$, $D_0 = 132$. It is found that the predicted values are in good agreement with the simulation results when the magnitude of the adjustment coefficient α is between 24 and 28. Therefore, α is recommended to be 26. For more details about the determination of α, please see Jin et al. (2020e). It is worth noting that the value of α is variable, it is related to the strength grade of concrete, but independent of GFRP.

The predicted values of nominal compressive strength are compared with the simulation results, as presented in Fig. 7.123. It can be seen from Fig. 7.123, the predicted results (theoretical value) obtained are in good agreement with the simulation results in Sect. 7.6.3, which preliminary demonstrates the accuracy and rationality of the size effect law on nominal strength of circular concrete columns confined with GFRP.

In addition, the test results of Manuel and Rodrigues (2006) and Bu (2015) were selected for the comparisons, as shown in Fig. 7.124a. It can be found that the experimental data points are all distributed on the theoretical prediction curves. From Fig. 7.124b, good agreement can be found between the theoretical values of nominal strength and the available test data.

In summary, the size effect law was consistent with the experimental results and simulation results, which preliminary demonstrate its rationality. However, a large number of experimental data, especially for larger-sized GFRP-confined columns should be provided to further verify the rationality of the size effect law.

3. *CFRP-confined square RC columns*

As for the size effect law of axial-loaded CFRP-wrapped RC columns, the parameter f_c can be selected as the uniaxial compressive strength of standard concrete specimens, which is 41.65 MPa. Three plain concrete specimens having different sizes were produced and tested numerically. The compressive strength of the three specimens was obtained by regression analysis, and the obtained parameters B and D_0

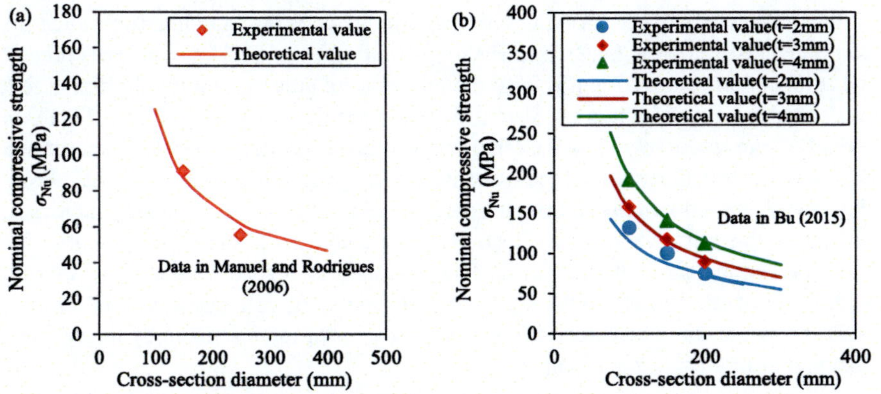

Fig. 7.124 Comparisons between the theoretical results and the available experimental data: **a** Manuel and Rodrigues (2006); and **b** Bu (2015)

were used to the reference factors, i.e. $B = 1.112$ and $D_0 = 103$. Similar with that of stirrup-confined concrete columns, the same repeated trial method was employed to find the proper value of parameter α. A better fit can be achieved when the adjustment coefficient α is between 130.0 and 140.0. The coefficient α is suggested to be taken as 135.0, which can better show the tendency of the size effect to be suppressed. The determination of α was detailed in Jin et al. (2020a).

It can be observed from Fig. 7.125, the tested data is consistent with the calculate value of the size effect law, and the trend of strength decline for the two groups is also consistent. However, the theoretical results are slightly larger than the real testes. This should be caused by reality that the utilized model (Lam and Teng 2002) may slightly overestimate the strength enhancement for present cases. Nevertheless, the reliability of the theoretical formula can be proved by the experimental data.

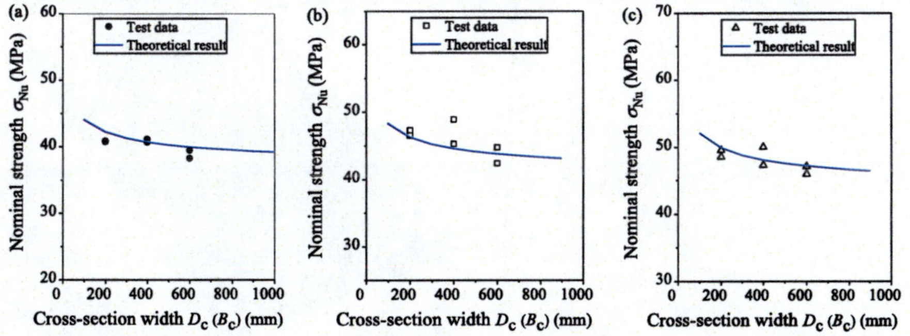

Fig. 7.125 Comparison of the test results and proposed size effect law for CFRP-wrapped concrete columns: **a** No reinforcement; **b** Longitudinal reinforcement; and **c** Longitudinal reinforcement and stirrups

In addition, Fig. 7.126 shows the comparison between the calculated value and the numerical results (Sect. 7.6.2). It can be found that the size effect law can better predict the axial compression strength. With the increases of constraint ratio, the size-effect tends to decrease, which is consistent with the previous argument (Sect. 7.7.1). In order to more intuitively view the relationship among test data, numerical simulation data and theoretical result, a three-dimensional diagram is drawn in Fig. 7.127. From Fig. 7.127, the numerical simulation data is reduced as the section size increases, indicating the existence of size effect. With the increase of constraint ratio, the size effect of nominal strength is reduced. The results of numerical simulation agree well with those of the formula, which proves that the use of constraint ratio in the new SEL is reliable. In conclusion, the new SEL is proved to be reliable by the real test data and numerical simulation data.

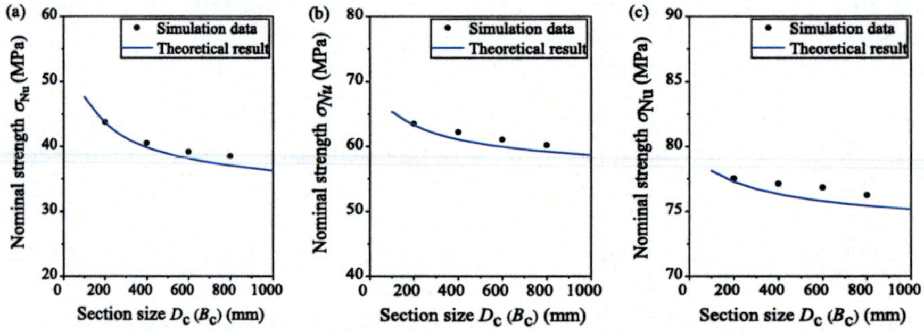

Fig. 7.126 Comparison of the simulation results and pro posed size effect law for CFRP-wrapped concrete columns: **a** $\rho_f = 0.334\%$; (b) $\rho_f = 0.668\%$; and **c** $\rho_f = 1.002\%$

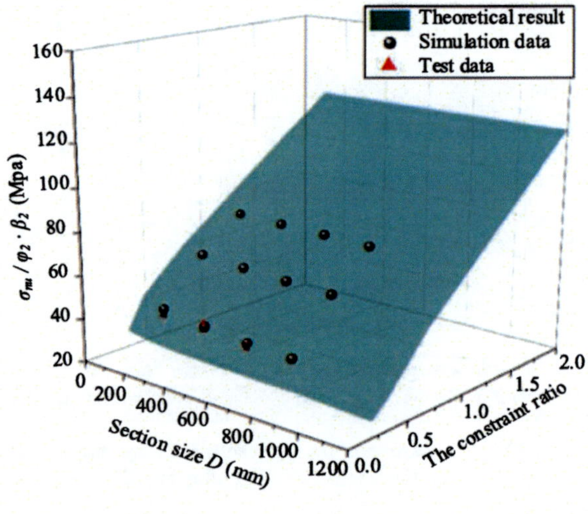

Fig. 7.127 Comparisons of the simulations results, test data and the theoretical results

4. *CFST (concrete filled steel tube) columns*

The validation of the **Method 1** and **Method 2** were conducted using the numerical results of Sect. 7.7.1 and Sect. 7.7.2, respectively.

Validation of **Method 1**:

The standard cubic concrete sample sized of 150 mm × 150 mm × 150 mm are taken as examples, and the corresponding axial compressive strength f_c (in Sect. 7.7.1) was 52.7 MPa. The size effect parameters B and D_0 gained by regression analysis in simulation results of circular/square plain concrete columns are used as benchmark parameters. According to Sect. 7.7.1, for circular CFST columns, $B = 1.06$ $D_0 = 803.09$; for square CFST columns, $B = 1.11$ $D_0 = 806.4$. When the adjustment coefficients for circular and square CFST columns are respectively taken as 12 and 8, good agreement can be found between the theoretical values and the simulated results of circular/square CFST columns (see Jin et al. (2019) for details).

Figure 7.128 plots the comparisons between experimental results in Han (2000), the simulation results and the present theoretical results of the nominal axial compressive strength of CFST columns. As presented in Fig. 7.128, almost all the experimental data are close to the theoretical surface, which demonstrates the accuracy and the rationality of the size effect law. In addition, the experimental results in Wang et al. (2017) test are selected as comparative analysis, as shown in Fig. 7.129. One

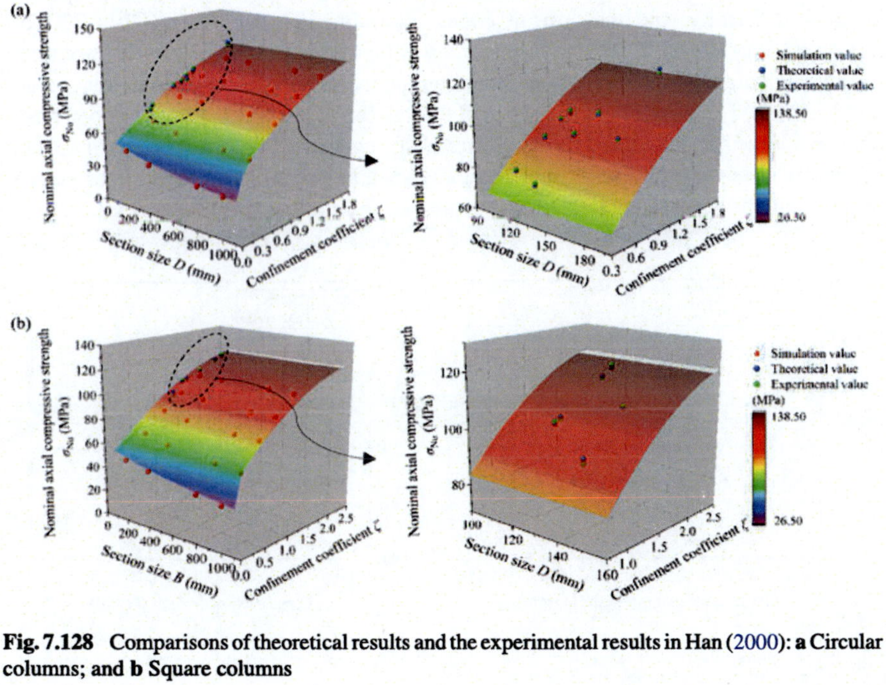

Fig. 7.128 Comparisons of theoretical results and the experimental results in Han (2000): **a** Circular columns; and **b** Square columns

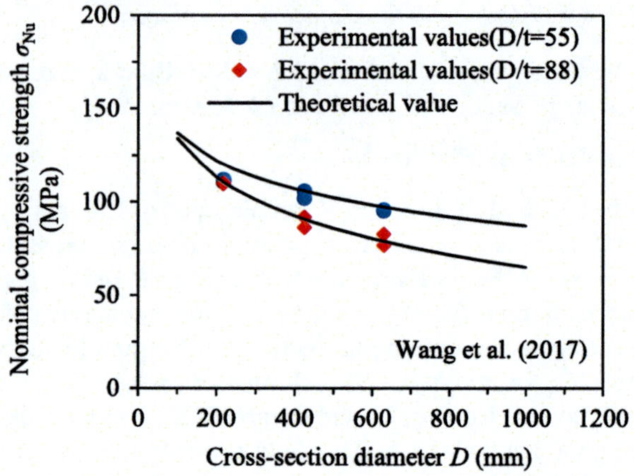

Fig. 7.129 Comparison between the theoretical and available experimental data in Wang et al. (2017)

can again note that the experimental axial compressive strengths are all distributed on the theoretical prediction curves.

Furthermore, the size effect formula and current codes are respectively adopted for a comparison with the available test data collected by Sakino et al. (2004). Figure 7.130 shows the influence of the cross-section dimension on the ratio between theoretical and experimental values for the nominal axial compression strength. The theoretical results (i.e. predicted results) are respectively calculated by the proposed theoretical formula and the current codes, including EN 1994-1-1 (CEN 2004b) and AISC 360-10 (AISC 2010). Figure 7.130 illustrates that the ratio of the nominal strength calculated by the codes and that obtained from the tests presents an increasing trend with the increase of cross-sectional dimension. This is because the codes have not considered the size effect on the nominal strength of CFST short columns. It can be clearly seen that the increasing trend of the ratio between the theoretical values of nominal strength calculated by the proposed theoretical formula and test results

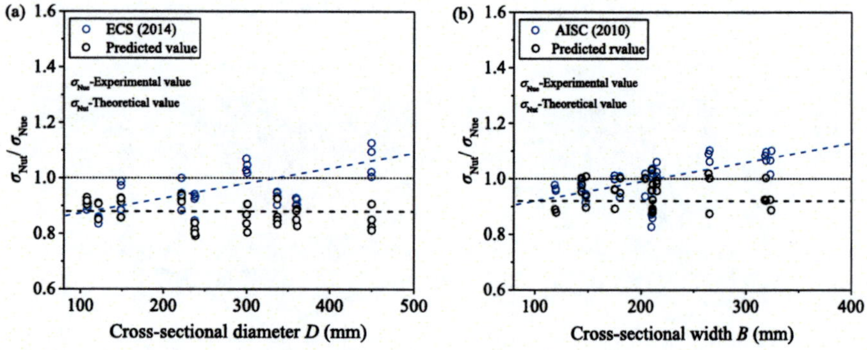

Fig. 7.130 Comparisons of the predicted results with experimental result in Sakino et al. (2004): **a** Circular columns; and **b** Square columns

almost disappears. This shows enough reliability in predicting the nominal axial compressive strength of CFST columns.

Validation of **Method 2**:

For the CFRP-confined square RC columns, the standard concrete cube specimens sized of 150 mm × 150 mm × 150 mm was set as reference. The parameters $f_c = 52.3$, $B = 2.0897$, and $D_0 = 60.261$ were utilized. The adjustment coefficient (α_L and α_N are taken as 10 and 12, respectively) and the shape influence factor ε (1 for square section, 1.5 for circular section) are obtained by calculation.

In Wang et al.'s (2017) work, the failure of CFST columns was tested under axial compression. The confinement coefficients in the test were different, and the cross-section width ranged from 220 mm to 630 mm. Here, a comparison between the present theoretical results and Wang et al.'s (2017) test data is plotted in Fig. 7.131. On the whole, the predicted results match well with Wang et al.'s (2017) test data.

In 2004, Sakino et al. (2004) carried out some tests on the failure of CFST columns. The test data measured by Sakino et al. (2004) were also adopted for the comparison to verify the accuracy of the proposed theoretical formula, as presented in Fig. 7.132.

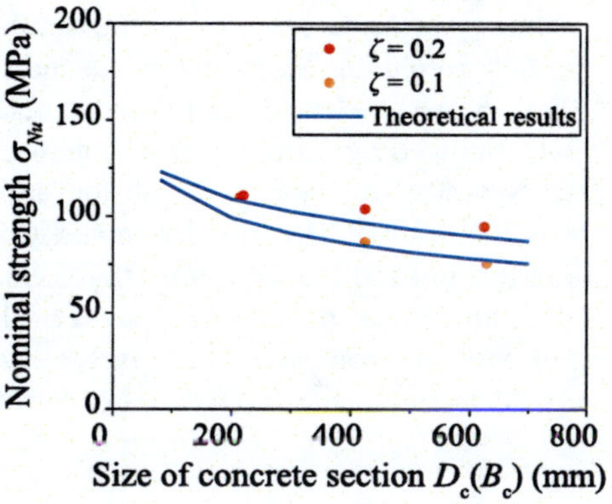

Fig. 7.131 Comparison of the predictions and Wang et al.'s (2017) experimental results for CFST columns

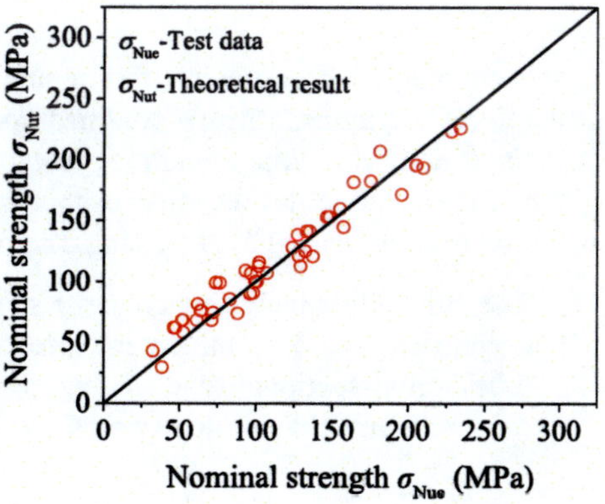

Fig. 7.132 Comparison of the predictions with Sakino et al.'s (2004) experimental data for CFST columns

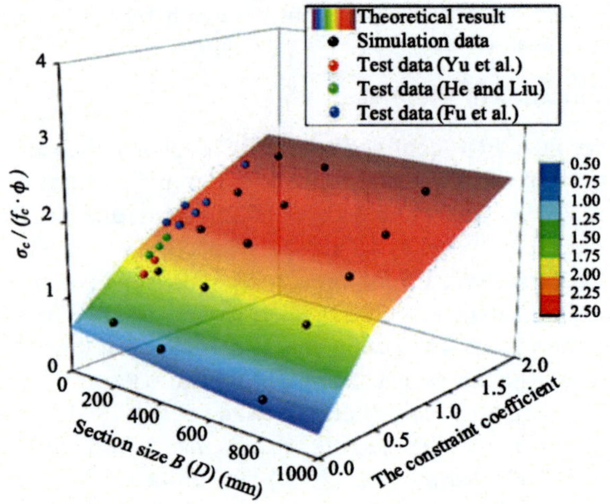

Fig. 7.133 Comparison of experimental results and predicted value for LWACFST columns

Again, one can find that the present theoretical value σ_{Nut} are consistent with the test value σ_{Nue}.

Moreover, the test results on the nominal compressive strength σ_{Nu} of square LWACFST columns measured by He and Liu (2008), Yu et al. (2016) and Fu et al. (2011) are also selected for the comparisons, as described in Fig. 7.133. On the whole, the well agreement can be obtained between theoretical values and the test data. Nevertheless, more test data for large-scale LWACFST columns under axial compression should be provided to verify the theoretical formula for describing the size effect in axial-loaded LWACFST columns.

It is noted that, the structural sizes of the above selected columns are still very small. More tests on the failure of larger-sized LWACFST and CFST columns should be performed to verify the theoretical formula.

7.9 Summary

In this chapter, the compressive failure and size effect behavior of normal and high strength RC columns, stirrups-confined concrete columns, FRP-confined concrete columns, and normal /lightweight aggregate concrete filled steel tube columns were given. The size effect law considering various parameters was presented, which was based on the classical SEL of Bažant. The following main conclusions are obtained:

(1) The parameters including concrete strength, slenderness ratio, loading pattern, eccentricity ratio, confinement effect (generated by stirrups, FRP, and steel tube), cross-sectional type, and aggregate type, present great influences on the compressive behavior and size effect of concrete columns.

(2) The confinement effect was the main parameters that influence the size effect behavior of concrete columns, which including two aspects: enhancement on nominal compressive strength and weakening of size effect.
(3) The size effect laws can well describe the size effect on nominal compressive strength of stirrup/FRP confined concrete columns and concrete filled steel tube columns.
(4) A size-dependent stress-strain model of stirrup-confined concrete columns was presented, which considers the size effect of peak axial stress, the corresponding strain and softening rate.

References

Abbasnia R, Ziaadiny H (2010) Behavior of concrete prisms confined with FRP composites under axial cyclic compression. Eng Struct 32:648–655

ACI (American Concrete Institute) (2014) Building code requirements for structural concrete and commentary. ACI 318-141, ACI, Farmington Hills, MI

ACI (American Concrete Institute) (2017) Guide for the design and construction of externally bonded FRP systems for strengthening concrete structures. ACI 440.2R-17, ACI, Farmington Hills, MI

AISC (American Institute of Steel Construction) (2010) Specification for structural steel buildings. AISC 360-10, Chicago, USA

Bahn BY, Hsu CTT (1998) Stress-strain behavior of concrete under cyclic loading. ACI Mater J 95(2):178–193

Barbhuiya S, Choudhury AM (2015) A study on the size effect of RC beam-column connections under cyclic loading. Eng Struct 95:1–7

Bažant ZP (1984) Size effect in blunt fracture: concrete, rock, metal. Journal of engineering mechanics 110:518–535

Bažant ZP, Kwon YW (1994) Failure of slender and stocky reinforced concrete columns: tests of size effect. Mater Struct 27(2):79–90

Bažant ZP, Xiang YY (1997) Size effect in compression fracture: splitting crack band propagation. Journal of engineering mechanics 123:162–172

Bu FT (2015) Experimental study on the size effect of axial compression performance of concrete columns confined with GFRP. Guangdong University of Technology, Guangdong (in Chinese)

CECS (China Association for Engineering Construction Standardization) (2004) Technical specification for structures with concrete-filled rectangular steel tubular members: CECS 159-2004. China Planning Press, Beijing. (in Chinese)

CEN (Comité Européen de Normalisation), (2004a) Design of concrete structures. Part 1–1: General rules and rules for buildings: EN-1992-1-1, Eurocode 2. CEN, Brussels, Belgium

CEN (Comité Européen de Normalisation) (2004b) Design of composite steel and concrete structures, Part 1.1 General rules and rules for buildings: EN 1994-1-1, Eurocode 4. CEN, Brussels

Chen P, Liu CY, Wang YY (2018) Size effect on peak axial strain and stress-strain behavior of concrete subjected to axial compression. Constr Build Mater 188:645–655

Dilger WH, Koch R, Kowalczyk R (1984) Ductility of plain and confined concrete under different strain rates. ACI journal 81(1):73–81

Du XL, Jin L, Ma GW (2013) Meso-element equivalent model for macro-scopic mechanical properties analysis of concrete materials. Int J Damage Mech 29(5):654–661

Du M, Jin L, Du X, Li D (2017) Size effect tests of stocky reinforced concrete columns confined by stirrups. Struct Concr 18(3):454–465

El-Tawil SM, Deierlein GG (1996) Fiber element analysis of composite beam-column cross-sections. Cornell University, New York, p 1996

Fu ZQ, Ji BH, Zhou Y, Wang XL (2011) An experimental behavior of lightweight aggregate concrete filled steel tubular stub under axial compression. Geo Hunan Int Conf 2011:24–32

Ghannam S, Al-Ani HR, Al-Rawi O (2010) Comparative study of load carrying capacity of steel tubular columns filled with lightweight concrete and normal concrete. Jordan J Civil Eng 4:109–112

Guo Z, Bažant ZP (2006) Size effect on buckling strength of eccentrically compressed column with fixed or propagating transverse crack. Int J Fract 142(1–2):151–162

Han LH (2000) Concrete-filled steel tubular structures. China Science Press, Beijing (in Chinese)

He MS, Liu XY (2008) Study on behavior of lightweight aggregate concrete filled square thin-walled steel tubulars under axial load. Sichuan Build Sci 32(2):18–21 (in Chinese)

Hoover CG, Bažant ZP (2014) Universal size-shape effect law based on comprehensive concrete fracture tests. J Eng Mech 140(3):473–479

Hoshikuma J, Kawashima K, Nagaya K, Taylor AW (1997) Stress-strain model for confined reinforced concrete in bridge piers. J Struct Eng 123(5):624–633

Ichinose T, Kanayama Y, Inoue Y, Bolander JE Jr (2004) Size effect on bond strength of deformed bars. Constr Build Mater 18(7):549–558

Jiang T, Teng JG (2007) Analysis-oriented stress–strain models for FRP–confined concrete. Eng Struct 29(11):2968–2986

Jin L, Li D, Du XL (2016) Mechanical behavior and size effect of moderate high-strength RC columns under monotonic and cyclic axial compression. Eng Struct 124:269–285

Jin L, Du M, Li D, Du X, Xu H (2017a) Effects of cross section size and transverse rebar on the behavior of short squared RC columns under axial compression. Eng Struct 142:223–239

Jin L, Li D, Du X L, Lu AZ, Ding Z X (2017b) Experimental and numerical study on size effect in eccentrically loaded stocky RC columns. J Struct Eng 143(2):04016170

Jin L, Ding Z, Li D, Du X (2018) Experimental and numerical investigations on the size effect of moderate high-strength reinforced concrete columns under small-eccentric compression. Int J Damage Mech 27(5):657–685

Jin L, Fan LL, Li P, Du XL (2019) Size effect of axial-loaded concrete-filled steel tubular columns with different confinement coefficients. Eng Struct 198:109503

Jin L, Chen H, Wang Z, Du X, (2020a) Size effect on axial compressive failure of CFRP-wrapped square concrete columns: tests and simulations. Compos Struct 112843

Jin L, Chen H, Fan L, Li P, Du X (2020b) Size effect on nominal strength of lightweight and normal concrete-filled steel tube columns under axial compression: meso-scale simulation. J Struct Eng 146(12):04020265

Jin L, Li P, Du X, Fan L (2020c) Size effect on nominal strength of circular stirrup-confined RC columns under axial compression: mesoscale study. J Struct Eng 146(3):04019213

Jin L, Li P, Du X (2020d) Compressive behavior of stirrup-confined concrete columns: size effect and a size-dependent stress-strain model. Eng Struct (under review)

Jin L, Li X, Fan L, Du X (2020e) Size Effect on compressive strength of GFRP-confined concrete columns: numerical simulation. J Compos Constr 24(5):04020038

Karsan ID, Jirsa JO (1969) Behavior of concrete under compressive loadings. J Struct Div 95(12):2543–2564

Kim JK, Kim YY (1996) Experimental study of the fatigue behavior of high strength concrete. Cem Concr Res 26(10):1513–1523

Kim JK, Yang JK (1995) Buckling behaviour of slender high-strength concrete columns. Eng Struct 17(95):39–51

Lam L, Teng JG (2002) Strength models for fiber-reinforced plastic-confined concrete. J Struct Eng 128(5):612–623

Lam L, Teng JG (2009) Stress-strain model for FRP-confined concrete under cyclic axial compression. Eng Struct 31:308–321

Lam L, Teng JG, Cheung CH, Xiao Y (2006) FRP-confined concrete under axial cyclic compression. Cement Concr Compos 28(10):949–958

Legeron F, Paultre P (2003) Uniaxial confinement model for normal-and high-strength concrete columns. J Struct Eng 129(2):241–252

Liew JYR, Xiong DX (2012) Ultra-high strength concrete filled composite columns for multi-storey building construction. Adv Struct Eng 15(9):1487–1503

Li D, Jin L, Du XL, Fu J, Lu AZ (2016) Size effect tests of normal-strength and high-strength RC columns subjected to axial compressive loading. Eng Struct 109:43–60

Lu Y, Vintzileou E, Zhang GF, Tassios TP (1999) Reinforced concrete scaled columns under cyclic actions. Soil Dyn Earthq Eng 18(2):151–167

Mander JB, Priestley MJN, Park R (1988) Theoretical stress-strain model for confined concrete. J Struct Eng 114(8):1804–1826

Manuel AG, Rodrigues CC (2006) Size and relative stiffness effects on compressive failure of concrete columns wrapped with glass FRP. J Mater Civ Eng 18(3):334–342

Marzec I, Bobiñski J, Tejchman J (2007) Simulations of spacing of localized zones in reinforced concrete beams using elasto-plasticity and damage mechanics with non-local softening. Comput Concr 4(5):377–402

Ministry of Construction of the People's Republic of China (2003) Standard for test method of mechanical properties on ordinary concrete: GB/T 50081-2002. China Architecture & Building Press, Beijing (in Chinese)

Ministry of Construction of the People's Republic of China (2011) Code for design of concrete structures: GB 50010-2010 [S]. China Architecture & Building Press, Beijing (in Chinese)

Němeček J, Bittnar Z (2004) Experimental investigation and numerical simulation of post-peak behavior and size effect of reinforced concrete columns. Mater Struct 37(3):161–169

NETC (National Economic and Trade Commission of the People's Republic of China) (1999) Code for design of steel-concrete composite structure: DL/T5085-1999. China Electric Power Press, Beijing (in Chinese)

NMSC (National Military Standard of the People's Republic of China) (2001) Technical specifications for early-strength model composite structure used for navy port emergency repair in wartime: GJB4142-2000. General Logistics Department of the Chinese People's Liberation Arm. (in Chinese)

Ouyang CS, Shah SP (1994) Fracture energy approach for predicting cracking of reinforced concrete tensile members. ACI Struct J 91(1):69–78

Park R, Rriestley MJN, Gill WD (1982) Ductility of square confined concrete columns. J Struct Div 108(4):929–951

Popovics S (1973) A numerical approach to the complete stress–strain curve of concrete. Cem Concr Res 3(5):583–599

Porras R, Carmona JR, Yu RC, Ruiz G (2016) Experimental study on the fracture of lightly reinforced concrete elements subjected to eccentric compression. Mater Struct 49:87–100

Razvi S, Saatcioglu M (1999) Confinement model for high-strength concrete. J Struct Eng 125(3):281–289

Ruiz G, Elices M, Planas J (1999) Size effect and bond-slip dependence of lightly reinforced concrete beams. Eur Struct Integrity Soc 24:67–97

Sakino K, Nakahara H, Morino S, Nishiyama I (2004) Behavior of centrally loaded concrete-filled steel- tube short columns. J Struct Eng 130(2):180–188

Samoel MS (2017) Size effect on the load carrying capacity of normal and lightweight concrete filled square steel tubular composite columns. Int J Appl Eng Res 12(15):5261–5266

Saroushian P, Choi KB (1989) Local bond of deformed bars with different diameters in confined concrete. ACI Struct J 86(2):217–222

Schneider SP (1998) Axially loaded concrete-filled steel tubulars. J Struct Eng 124(10):1125–1138

Şener S, Barr B, Abusiaf H (2004) Size effect in axially loaded reinforced concrete columns. J Struct Eng 130:662–670

Syroka-Korol E, Tejchman J (2014) Experimental investigations of size effect in reinforced concrete beams failing by shear. Eng Struct 58(7):63–78

Syroka-Korol E, Tejchman J, Mróz Z (2014) FE analysis of size effects in reinforced concrete beams without shear reinforcement based on stochastic elasto-plasticity with non-local softening. Finite Elem Anal Des 88(10):25–41

Tao Z, Uy B, Liao FY, Han LH (2011) Nonlinear analysis of concrete-filled square stainless steel stub columns under axial compression. J Constr Steel Res 67(11):1719–1732

Thériault M, Neale KW (2000) Design equations for axially loaded rein forced concrete columns strengthened with fibre rein-forced polymer wraps. Can J Civ Eng 27:1011–1020

Verderame GM, Fabbrocino G, Manfredi G, (2008) Seismic response of RC columns with smooth reinforcement. Part I: Monotonic tests. Eng Struct 30(9):2277–2288

Wang LC, Chen GB (2008) Mesoscopic simulation for mechanical properties of lightweight aggregate concrete by rigid body spring model. J Hydraul Eng 39(5):80–87

Wang Y, Wu H (2010) Size effect of concrete short columns confined with aramid FRP jackets. J Compos Constr 15(4):535–544

Wang Z, Wang D, Smith ST, Lu D (2012) Experimental testing and analytical modeling of CFRP-confined large circular RC columns subjected to cyclic axial compression. Eng Struct 40:64–74

Wang DY, Wang ZY, Smith ST, Yu T (2016) Size effect on axial stress-strain behavior of CFRP-confined square concrete columns. Constr Build Mater 118:116–126

Wang WJ, Ma H, Li ZB, Tang ZY (2017) Size effect in circular concrete-filled steel tubulars with different diameter-to-thickness ratios under axial compression. Eng Struct 151:554–567

Xu C, Jin L, Ding Z, Li D, Du X (2016) Size effect tests of high-strength RC columns under eccentric loading. Eng Struct 126:78–91

Yerlici VA, Ozturan T (2000) Factors affecting anchorage bond strength in high-performance concrete. ACI Struct J 97(3):499–507

Yılmaz O, Molinari JF (2017) A mesoscale fracture model for concrete. Cem Concr Res 97:84–94

Yu Q, Bažant ZP (2011) Can stirrups suppress size effect on shear strength of RC beams. J Struct Eng 137(5):607–617

Yu T, Teng JG, Wong YL, Dong SL (2010) Finite element modeling of confined concrete-II: Plastic-damage model. Eng Struct 32(3):680–691

Yu X, Tao Z, Song YY (2016) Effect of different types of aggregates on the performance of concrete-filled steel tubular stub columns. Mater Struct 49:3591–3605

Chapter 8
Seismic Performances and Size Effect in Columns

The concrete columns under the action of wind and seismic load usually subject to the combined action of compression, shearing and bending. The columns often show characteristics of brittle shear failure and less-ductile bending failure. Therefore, the seismic performance of concrete columns under combined lateral and axial loads may present size effect.

Extensive experimental and numerical researches on the failure behavior of concrete columns under combined lateral and axial loads have been carried out (Tomii and Sakino 1979; Ye et al. 2016; Shamim and Grace 2002). As can be concluded from the above literature, the shear behavior of concrete columns is affected by many factors, mainly including shear-span ratio, axial compression ratio and constraint effect of stirrup, steel tube and FRP. The different failure modes are mainly caused by shear-span ratio that has great influences on the shear capacity and plastic deformation capacity of concrete columns. Moreover, both the axial compression ratio and constraint effect of stirrup, steel tube and FRP affect the shear capacity and ductility of concrete columns as well. At present, little experimental studies on size effect in shear behavior of RC columns and FRP confined concrete columns have been conducted (Xie et al. 2013; Cui 2018), and it is found that size effect does exist in the shear performance of concrete columns. In view of this, it is necessary to further study the shear behavior and size effect of concrete columns. In order to investigate the shear behavior and size effect of concrete columns, a series of systematically designed experimental and numerical tests of RC columns with different structure sizes are carried out in the work (Jin et al. 2018, 2019). The influential factors including the loading pattern, shear-span ratio, axial compression ratio, and the confinement effect of steel tube and FRP would be reported in this Chapter. Moreover, the size effect law of concrete columns considering the influence of shear-span ratio and confinement effect is also presented.

X. Du and L. Jin, *Size Effect in Concrete Materials and Structures*,
https://doi.org/10.1007/978-981-33-4943-8_8

8.1 Compression-Shear Failure in Stocky RC Columns

Based on a combined experimental and numerical analysis, Jin et al. (2018) investigated the failure mechanism of monotonically and cyclically loaded short RC columns with different structural sizes and axial compression ratios, and revealed the impact of size effect on the seismic behavior of short RC columns.

8.1.1 Seismic Tests on Failure of Stocky RC Columns

1. *Experimental description*

A total of 8 RC columns with different structural sizes (in the ratio of 3:5:7) were designed and tested. A schematic view of the loading protocol and displacement measurements used in the test is illustrated in Fig. 8.1, respectively for different sized specimens. The top ends of the RC columns were located against spherical seating supports. In the tests, the collected data involves the horizontal displacement at the loading point measured by the linear variable differential transformer (i.e. LVDT), and the reaction force at the loading point. At the same time, the bending deflection, shear deformation and bond-slip deformation which are in the plastic hinge in the bottom region of the columns, are measured via dial indicators.

The geometric details of the RC columns employed are respectively listed in Table 8.1 by two series (i.e. CC and CM). It has to be noted here, the effective length L_{eff} of the columns of CM-3-0.6 and CC-3-0.6 is the distance from the loading point

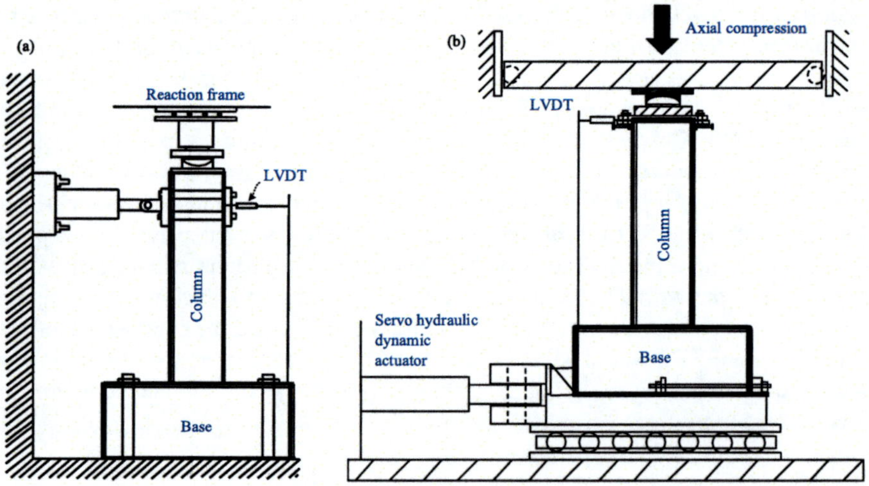

Fig. 8.1 Schematic diagram of loading apparatus: **a** Loading protocol and **b** Displacement measurements

Table 8.1 Physical parameters of the RC columns

Specimen	$b \times h$ (mm × mm)	L_{eff} (mm)	c (mm)	λ	n	N(kN)	ρ_l (%)	ρ_{sv} (%)
CM-3-0.6	300 × 300	546	15	2.02	0.6	1560	1.51	0.338
CC-3-0.6	300 × 300	546	15	2.02	0.6	1560	1.51	0.338
CM-5-0.6	500 × 500	910 + 250	25	2.58	0.6	4500	1.51	0.338
CC-5-0.6	500 × 500	910 + 250	25	2.58	0.6	4500	1.51	0.338
CM-7-0.4	700 × 700	1274 + 250	35	2.42	0.4	5880	1.51	0.338
CC-7-0.4	700 × 700	1274 + 250	35	2.42	0.4	5880	1.51	0.338
CM-7-0.6	700 × 700	1274 + 250	35	2.42	0.6	8900	1.51	0.338
CC-7-0.6	700 × 700	1274 + 250	35	2.42	0.6	8900	1.51	0.338

Note The first letter "C" in the specimen name denotes "Column", and the second letters "C" and "M" denote "Cyclic load" and "Monotonic load", respectively. The three numbers of "3", "5" and "7" mean the cross-sectional sizes, i.e. 300 mm × 300 mm, 500 mm × 500 mm and 700 mm × 700 mm. "0.6" and "0.4" denote the applied axial compression ratio. In addition, L_{eff} = effective length, c = cover thickness, λ = shear-span ratio, n = axial compression ratio, N = axial force, ρ_l = longitudinal reinforcement ratio, ρ_{sv} = transversal reinforcement ratio

to the bottom surface, i.e. 546 mm, by using the loading device in Fig. 8.1a. While for the other 6 larger-sized columns, i.e. CM-5-0.6, CC-5-0.6, CM-7-0.4, CC-7-0.4, CM-7-0.6 and CC-70.6, the loading device is provided with a scooter under the base because of the large dimension of the specimens (see Fig. 8.1b), the corresponding effective length L_{eff} is therefore the sum of the actual column length (L) and the spherical hinge length (250 mm). Shear-span ratio is defined as $\lambda = M/Vh_0$, where $M = PL_{eff}$, $V = P$ are respectively the moment and the shear force acting at the end cross-section, and $h_0 = h - 2c$ is the effective depth of the element. Therefore, the shear-span ratios of the test columns are respectively 2.02, 2.58 and 2.42.

All specimens were cast from the same batch of continually mixed concrete to minimize statistical scatter of the results. Table 8.2 gives the proportions of the concrete mixture in the tests. The corresponding average values of compressive and tensile strengths of standard concrete cube samples, sized of 150 mm × 150 mm × 150 mm, were 30.2 MPa and 2.81 MPa, respectively. The detailed information about the mechanical parameters of the reinforcing bars utilized in the tests are presented in Table 8.3.

Generally, all geometric dimensions, including the diameter and spacing of the longitudinal steel bars as well as the stirrups, and the concrete cover thickness, were scaled to obtain geometrically proportional specimens with the ratio of 3:5:7. The cross-sectional dimensions and the reinforcement details for the eight RC columns are depicted in Fig. 8.2.

The columns were subjected to constant axial compressive load and monotonic/cyclic horizontal displacements. Herein, the force–displacement hybrid control was utilized for all RC columns. The columns of CM-3-0.6, CM-5-0.6, CM-7-0.4 and CM-7-0.6 were subjected to monotonic loading, and the other four columns of

Table 8.2 Concrete mixture proportions utilized

Aggregate (kg/m^3)				Cement (kg/m^3)	Water (kg/m^3)	*w*/*c* ratio	Sum (kg/m^3)	Compressive Strength f_c (MPa)	Tensile strength f_t (MPa)
0–5 mm	5–10 mm	10–20 mm	20–30 mm						
1008	165	412	247	380	228	0.6	2440	30.2	2.81

Table 8.3 Mechanical properties of the reinforcing bars utilized

Diameter of rebar	Grade of rebar	Elastic modulus (GPa)	Elongation (%)	Yield strength (MPa)	Ultimate strength (MPa)
6	HPB300 ϕ	209	16.11	460	545
10	HPB300 ϕ	207	15.00	382	435
14	HPB300 ϕ	206	29.05	327	462
12	HRB400 ⏀	203	20.56	433	606
20	HRB400 ⏀	201	28.50	424	589
28	HRB400 ⏀	202	26.43	422	616

Note "HRB" means "Hot-rolled Ribbed-steel Bar", and "HPB" means "Hot-rolled Plain-steel Bar"

CC-3-0.6, CC-5-0.6, CC-7-0.4 and CC-7-0.6 were loaded under low cyclic lateral loading. The detailed loading protocol was recorded in Jin et al. (2018).

It has to be noted that, for experimental studies, at least three groups of specimens for each series should be required to narrow the discreteness of the test data. However, due to the consideration of experimental expense, only one test solely performed for each case. Moreover, due to operational errors, it is a pity that the tests on the two cases of CM-7-0.4 and CM-7-0.6 were not successful. Herein, all available experimental results which could preliminarily reveal the seismic behavior of short RC column were remained.

2. *Failure patterns*

Figure 8.3 presents the final failure patterns of the six successfully tested specimens with different structural sizes or axial compression ratios. Figure 8.3a shows the photos of the two smallest (i.e. the cross-sectional size is 300 mm × 300 mm) RC columns (i.e. CM-3-0.6 and CC-3-0.6) with a shear-span ratio of 2.02. It can be clearly seen that, CM-3-0.6 is destroyed by a single narrow diagonal shear crack and the other inclined cracks are distributed at the bottom left area of the specimen, as the load was monotonically applied from the left side. CC-3-0.6 is also destroyed by a single narrow diagonal shear crack in a similar way with CM-3-0.6. However, compared with CM-3-0.6, the main crack causing the failure of CC-3-0.6 generates from the opposite direction, and the inclined cracks distribute at both sides, as the load was cyclically applied from both directions. Moreover, it can be clearly seen from Fig. 8.3a that, the damage area of CC-3-0.6 is larger than that of CM-3-0.6, which is also caused by the different loading modes.

Figure 8.3b and c respectively show the photos of the two groups of larger sized RC columns (i.e. CM-5-0.6, CC-5-0.6, CC-7-0.4 and CC-7-0.6) with 500 mm × 500 mm and 700 mm × 700 mm cross-sectional sizes. It has to be noted here, although the RC columns were all designed in the structural ratio of 3:5:7, their shear-span ratios did not remain the same during the test because of the use of different loading devices as presented in Fig. 8.1b and c. The shear-span ratios (i.e. 2.58 and 2.42) of the four larger sized specimens are apparently much larger than that of the two smallest

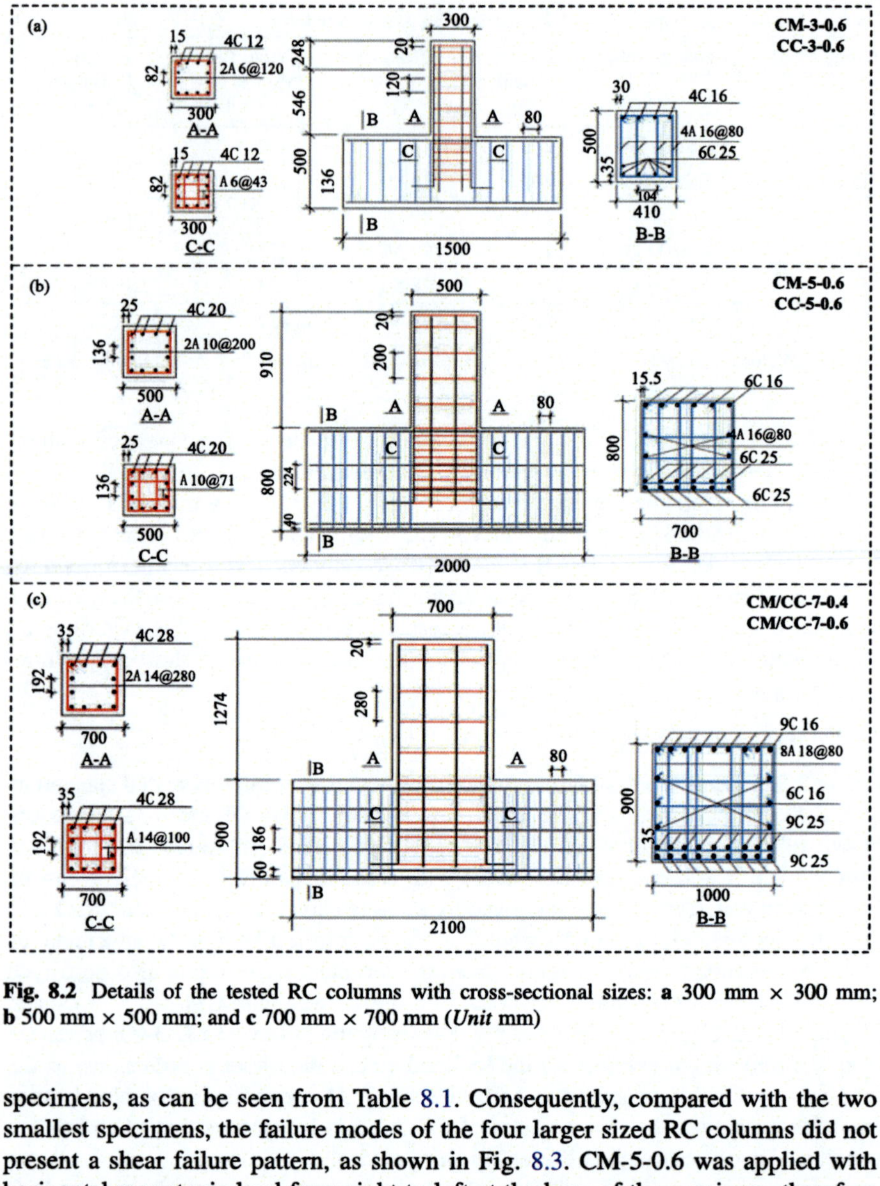

Fig. 8.2 Details of the tested RC columns with cross-sectional sizes: **a** 300 mm × 300 mm; **b** 500 mm × 500 mm; and **c** 700 mm × 700 mm (*Unit* mm)

specimens, as can be seen from Table 8.1. Consequently, compared with the two smallest specimens, the failure modes of the four larger sized RC columns did not present a shear failure pattern, as shown in Fig. 8.3. CM-5-0.6 was applied with horizontal monotonic load from right to left at the base of the specimen, therefore the cracks generated from the bottom right area of the specimen. However, the main failure crack appeared at the left side of the column, which was apparently not a shear crack but a splitting crack causing by the axial load. The specimen was loaded with both axial compression and horizontal loading with a shear-span ratio of 2.58, the failure was consequently not controlled by shear but moment. Therefore, the specimen presented a failure pattern similar with an eccentrically loaded RC column

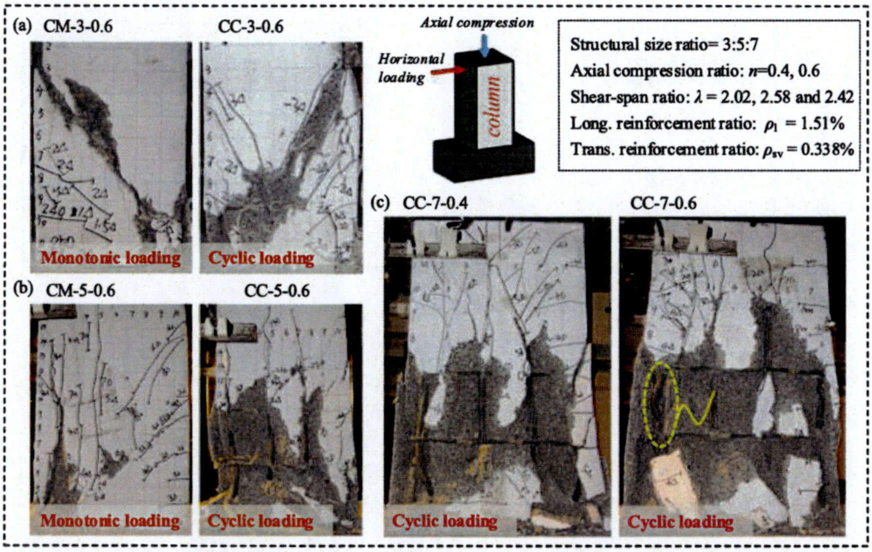

Fig. 8.3 Final failure patterns of the six RC columns having different structural sizes or axial compression ratios: **a** 300 mm × 300 mm; **b** 500 mm × 500 mm; and **c** 700 mm × 700 mm

(Majewski et al. 2008; Xu et al. 2016). CC-5-0.6, CC-7-0.4 and CC-7-0.6 were all under cyclic horizontal loads at the base of the specimen, and they exhibited similar failure patterns with large damage area at bottom region of the column. Because of the large shear-span ratios of 2.58 and 2.42, the failures of the aforementioned three columns were not controlled by shear, either.

Additionally, it can be seen from Fig. 8.3c that, the axial compression ratio could also influence the failure behavior of the RC columns. Both of these two columns were subjected to cyclic loading at the horizontal direction, one was under an axial compression ratio of 0.4, while the other was under an axial compression ratio of 0.6. They have nearly the same failure patterns (similar with that of CC-5-0.6) at the end of the loading history. However, CC-7-0.6 has a larger damage area than CC-7-0.4, which is definitely caused by the different ratio of axial compression.

3. *Load–deflection curve*

Figure 8.4 depicts the entire collection of load P versus horizontal deflection Δ curves obtained from the test for the six successfully tested RC columns. It can be found clearly from the curves that, for the columns under both monotonic and low cyclic loadings, there is an initial elastic response for all specimens and the load-horizontal deflection relationship can be considered roughly to be linear. Figure 8.4a and b show the P-Δ curves of the two smallest specimens. Corresponding to the obvious shear failure pattern demonstrated in Fig. 8.3a, both the columns of CM-3-0.6 and CC-3-0.6 failed in a sudden with short or no plateau after the peak point of the curve. The hysteretic hoop for CC-3-0.6 is apparently shriveled, indicating weak ductility and

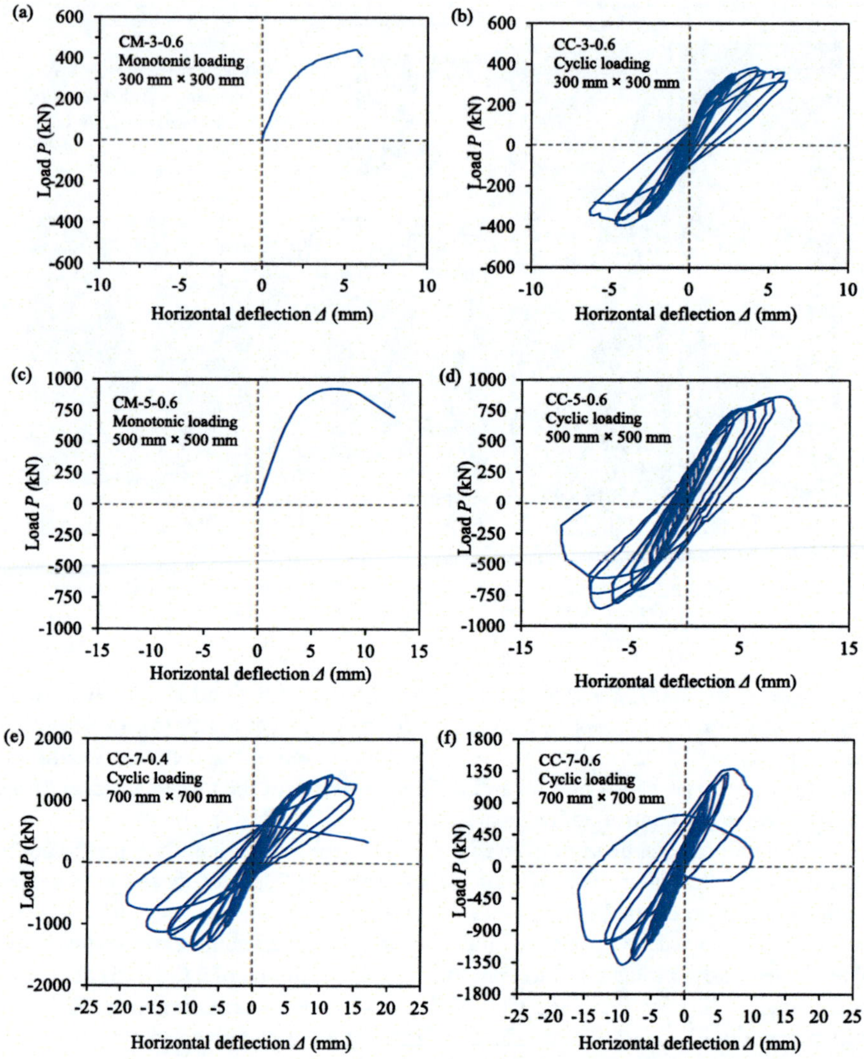

Fig. 8.4 Load–deflection curves for the six RC columns having different structural sizes and axial compression ratios: **a** CM-3-0.6; **b** CC-3-0.6; **c** CM-5-0.6; **d** CC-5-0.6; **e** CC-7-0.4; and **f** CC-7-0.6

low energy dissipation capacity. Figure 8.4c and f show the *P*-*Δ* curves of the four larger sized specimens. As illustrated in the Sect. 3.1, although the RC columns were all designed in the structural ratio of 3:5:7, their shear-span ratios did not remain the same during the test because of the use of different loading devices. The shear-span ratios (i.e. 2.58 and 2.42) of the four larger sized specimens are apparently much larger than that of the two smallest specimens with a shear-span ratio of 2.02, and consequently causing the difference in the final failure patterns. The *P*-*Δ* curve for

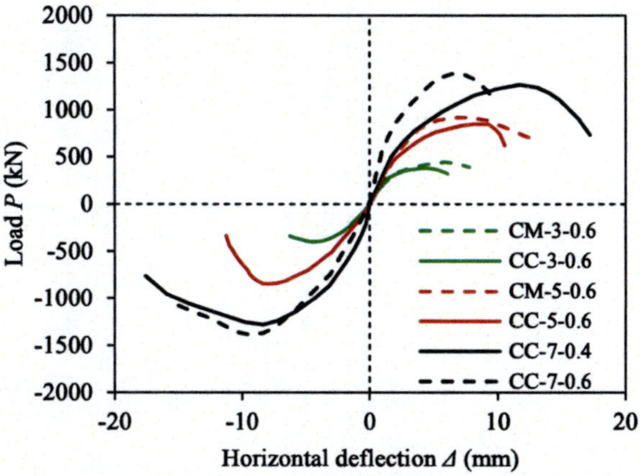

Fig. 8.5 Tested envelop curves of the six RC columns

CM-5-0.6 is much ductile than that of CM-3-0.6 with a longer plateau after the peak point and a gentler descending branch before the final failure of the specimen. The hysteretic hoops for the three larger sized columns (i.e. CC-5-0.6, CC-7-0.4 and CC-7-0.6) under cyclic loading are shriveled either, but they are still a little bit plumper than that of CC-3-0.6.

A comparison of the envelope curves (the link of the maximum points of the load–deflection hysteretic curve) is provided in Fig. 8.5. The load bearing capacity and ultimate deformation improved dramatically as the structural size of the columns increases. The load capacity of cyclically loaded specimens is lower than that of monotonically loaded ones, causing by the fatigue damage generated in the cyclic actions. One can also note from Fig. 8.5 that, the peak load of CC-7-0.6 is larger than that of CC-7-0.4, indicating that high axial compression ratio could improve the load capacity of the column. However, the ductility or the deformation capacity become weaker with the increase of axial compression ratio. Moretti and Tassios (2007) also investigated the influence of the axial compression ratio on the seismic behavior of short RC columns (shear-span ratio $\lambda = 2$) by comparing two different axial compression ratios of 0.3 and 0.6, and they presented similar conclusions. Moreover, it is to be noted that the load carrying capacities for the RC columns are different in the positive and negative loading cycles. This should be caused by (1) the discreteness of the mechanical properties of materials, especially for concrete, and (2) the strengthening of steel. These two factors cause the differences in crack distribution and crack width, finally result in different load carrying capacity in the two loading directions.

In addition, the measured yield/peak lateral loads P_y/P_{max}, the yield/ultimate displacements $\Delta y/\Delta u$, the nominal strength σ_{max}, the displacement ductility coefficient $\mu = \Delta u/\Delta y$ and the ultimate displacement angle $\theta_u = \Delta u/L_{eff}$, are provided in Table 8.4. Herein, the ultimate displacement Δu is defined as the displacement corresponding to $0.85p_{max}$ after the peak load.

Table 8.4 Test results of the RC columns

Specimen	Loading direction	P_y (kN)	Δy (mm)	P_{max} (kN)	Δu (mm)	σ_{max} (MPa)	$\mu = \Delta u / \Delta y$	Mean μ	θ_u
CM-3-0.6	Positive	312.1	2.79	442.8	5.45	5.19	1.95	1.95	0.01
CC-3-0.6	Positive	307.9	2.42	373.2	4.7	4.5	1.94	1.85	0.0086
	Negative	314.4	2.68	388.9	4.73	4.59	1.76		
CM-5-0.6	Positive	693.2	4.22	926.8	7.6	3.9	1.8	1.8	0.0088
CC-5-0.6	Positive	744.7	4.33	866.2	7.81	3.58	1.8	1.71	0.0082
	Negative	744.3	4.75	859	7.71	3.54	1.62		
CM-7-0.4	Positive								
CC-7-0.4	Positive	1200.1	6.97	1250.3	11.78	2.69	1.69	1.67	0.008
	Negative	1201.9	5.17	1274.9	8.53	2.73	1.65		
CM-7-0.6	Positive								
CC-7-0.6	Positive	1178.9	4.66	1389.1	6.92	2.98	1.48	1.49	0.0064
	Negative	1205.1	6.24	1380	9.33	2.96	1.49		

4. *Displacement ductility and load capacity*

Figure 8.6 present the data points of displacement ductility coefficients μ, the ultimate displacement angle θ_u and the nominal strength σ_{max} for the test RC columns with different structural size D (i.e. the cross-sectional width). It can be clearly seen from Fig. 8.6 that, the μ, θ_u and σ_{max} for monotonically loaded specimens are all larger than that of cyclically loaded ones, indicating that the mechanical properties of RC column would be deteriorated under seismic actions. The RC member subjected to cyclic loading could present a more brittle failure mode and the size effect behavior therefore could be expected (Jin et al. 2016a, b). Herein Fig. 8.6, a preliminary trend of decrease of displacement ductility and load capacity with the increase of structural size could be sensed from the two larger sized cyclically loaded RC columns (i.e. CC-5-0.6 and CC-7-0.6) with approximate shear-span ratios of 2.58 and 2.42. A more detailed size effect study will be conducted by numerical trials in the subsequent section.

Moreover, it can be clearly seen from Fig. 8.6a and b that, μ and θ_u for CC-7-0.4 are apparently larger than that of CC-7-0.6, indicating that the energy dissipation capacity of the column with a smaller axial compression ratio of 0.4 is better than the one of 0.6. However, the load capacity of the column with larger axial compression

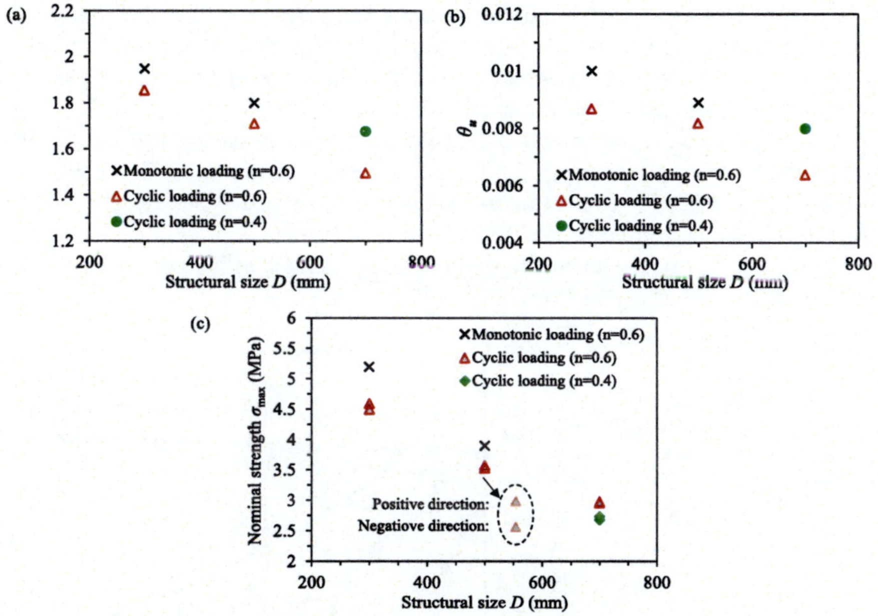

Fig. 8.6 **a** Displacement ductility coefficients $\Delta u/\Delta y$; **b** Ultimate displacement angle θ_u of the six RC columns; and **c** Nominal strength σ_{max} for the RC columns with different size D

ratio of 0.6 is larger than the one of 0.4, as can be seen from Fig. 8.6c. To sum up, increasing the axial compression ratio could improve the load capacity of the RC column, but the deformation capacity would be restricted.

5. *Stiffness degradation*

The relationship between the secant stiffness K and the rotation angle θ of the four cyclically loaded RC columns at the horizontal direction is demonstrated in Fig. 8.7. It can be clearly seen that, as the rotation angle increases, the stiffness of all specimens drops rapidly. A larger specimen presents a larger initial stiffness and the stiffness of all specimens evolve in a similar way. It could also be preliminarily noted that larger column presents a more brittle failure, noting that the stiffness degradation of CC-7-0.6 is sharper than that of CC-5-0.6.

Moreover, the columns of CC-7-0.4 and CC-7-0.6 present the close initial stiffness, however, the stiffness degradation of CC-7-0.6 evolves in a more brittle way with a sharper slope and a much smaller rotation angle at the end of the test process. Consequently, a larger axial compression ratio could make the failure of specimen more brittle.

6. *Energy dissipation*

Figure 8.8a presents the relationship between the equivalent viscous damping coefficient h_e and rotation angle θ. It can be seen from Fig. 8.8a that the equivalent viscous damping coefficient of all specimens are getting larger as the rotation angle increases. The equivalent viscous damping coefficient is almost the same for all specimens as the magnitude of rotation angle is less than 0.2%. Moreover, by comparing the two larger sized cyclically loaded RC columns (i.e. CC-5-0.6 and CC-7-0.6) with approximate shear-span ratios of 2.58 and 2.42, it can be preliminarily noted that, as the rotation angle increases, larger-sized columns dissipated more energy as the equivalent viscous damping coefficients increases. A more appropriate concept for evaluating the energy dissipation capacity is also presented as below

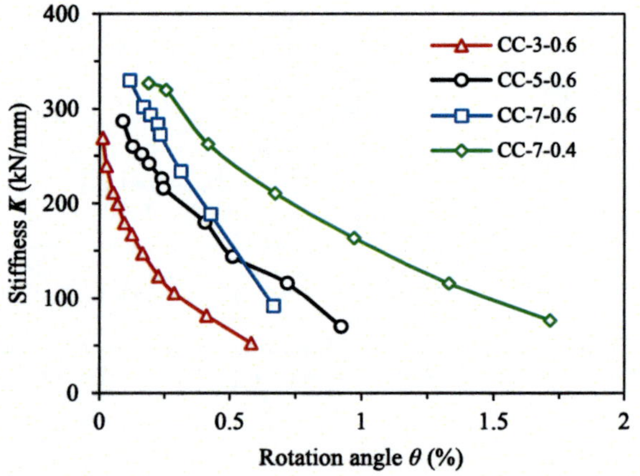

Fig. 8.7 Relationship between the secant stiffness K and the rotation angle θ of the four cyclically loaded RC columns at the horizontal direction

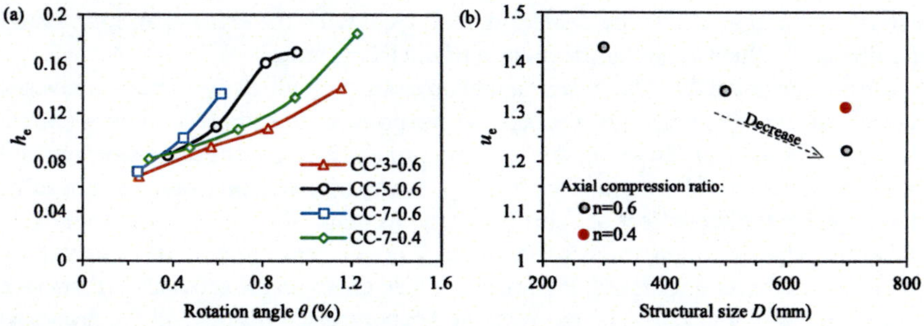

Fig. 8.8 **a** Relationship between the equivalent viscous damping coefficient h_e and rotation angle θ and **b** Average energy dissipation coefficient u_e of the RC columns

$$u_e = \frac{E}{mE_y} \tag{8.1}$$

in which, u_e is the average energy dissipation coefficient, E is the total area of the hysteresis loops between the yield and the failure, E_y is the nominal elastic energy, and m is the cycle index after the yield. Figure 8.8b presents the relationship between the average energy dissipation coefficient and the structural size. It can be seen from Fig. 8.8b that, the average energy dissipation coefficient decreases with the increase of structural size, indicating that the energy dissipation capacity exhibits size effect (preliminarily concluded by comparing CC-5-0.6 and CC-7-0.6).

In addition, the column with a smaller axial compression ratio (i.e. CC-7-0.4) possesses a higher energy dissipation coefficient than the one with a larger axial compression ratio (i.e. CC-7-0.6), indicating that the energy dissipation capacity could be deteriorated with the increase of axial compression ratio.

7. *Discussion on the failure of RC columns with different shear-span ratios*

Herein the experimental program, columns with shear-span ratio of 2.02 fail in obvious shear failure pattern with a more brittle manner, while columns with shear-span ratios of 2.58 and 2.42 exhibit a less brittle failure mode. This phenomenon is similar with that in the work of Moretti and Tassios (2007), corresponding to the conclusion that columns with shear ratio values less than 2 behave in a particularly brittle manner when subjected to cyclic displacements. Moreover, for the columns with small shear-span ratios (Fig. 8.3a), the mechanism of the diagonal concrete strut is more activated compared to the truss mechanism of force resistance, a fact which leads to increased diagonal cracking of concrete and enhanced brittleness. However, for the columns with larger shear-span ratios, the lateral force should be carried by a direct compression strut, as can be observed by the final failure patterns in Fig. 8.3b and c. The columns failed in crack-splitting failure and the load transfer mechanism represents rather a structural behavior than a shear behavior (Moretti and Tassios 2007). Therefore, the load bearing capacity should be more exactly determined by a strut-and-tie model than by a shear model. Generally, the shear-span ratio could significantly influence the failure pattern of the RC member.

It has to be noted that, the experimental program was initially designed to investigate the impact of size effect on the seismic performance of short RC columns (with a shear-span ratio of 2). However, due to the improper design of the test procedure and the unexpected error during the test process, the experimental program was not sufficiently reveal the size effect of short RC columns under seismic actions. Therefore, taking the experimental tests as calibrations, a good deal of numerical trials will be conducted to thoroughly reveal the impact of size effect on the seismic performance of short RC columns based on the restricted experimental results and the proposed mesoscopic simulation method in the subsequent section.

8.1.2 Simulations on Failure of Stocky RC Columns

Due to operational errors, it is a pity that the tests on the two cases of CM-7-0.4 and CM-7-0.6 were not successful. Therefore, the numerical analysis was conducted to systematically study the influence of axial compression ratio and size effect behavior of short RC columns under cyclic loading (Jin et al. 2018).

1. *Generation of the mesoscale model*

According to the simulation method in Chap. 3, the 3D mesoscale model of RC columns was generated, as shown in Fig. 8.9. The interaction between concrete and rebar was considered as mentioned in Chap. 3. The material properties of concrete and steel rebar were listed in Table 8.5.

As mentioned in Sect. 8.1.1, the shear-span ratios of tested columns did not remain the same during the test because of the use of different loading devices. Herein, in order to investigate the impact of size effect on the seismic performance of short

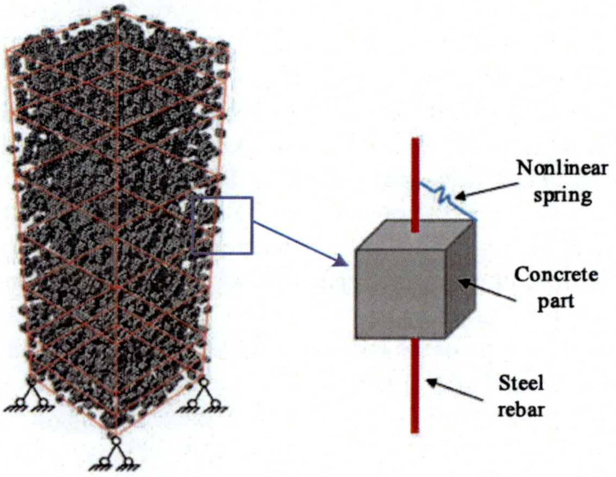

Fig. 8.9 Typical meso-scale mechanical model for the RC column

Table 8.5 Mechanical parameters of the meso components and steel rebar

	Mortar matrix	ITZ	Aggregate particle	Steel rebar HRB400/HPB300
Young's modulus E (GPa)	32.5[a]	26[b]	70[a]	210[a]
Poisson ratio ν	0.20[a]	0.22[b]	0.16[a]	0.3[a]
Dilatant angle ψ (°)	18	15	–	–
K_c	0.667	0.667	–	–
Eccentricity η (%)	0.1	0.1	–	–
Stress ratio σ_{b0}/σ_{c0}	1.16	1.16	–	–
Compressive strength f_c (MPa)	31.6[a]	23.9[b]	80[a]	–
Tensile strength f_t (MPa)	2.92[a]	2.4[b]	8[a]	–
Yield strength f_y (MPa)	–	–	–	400/300[a]

Note [a]Data are obtained from the tests; [b]Parameters are given with reasonable assumption; and the other parameters are default values or determined by referring to Marzec et al. (2007) and Syroka-Korol et al. (2014)

RC columns with a shear-span ratio of 2 as initially designed. Three groups of numerical specimens with different characteristic dimensions, i.e. 300 mm × 300 mm × 546 mm, 500 mm × 500 mm × 910 mm, and 700 mm × 700 mm × 1274 mm, are established for supplementing and extending the restricted experimental research.

2. *Influence of axial compression ratio*

As demonstrated in the analysis of the experimental results, the axial compression ratio could have great influence on the seismic behavior of RC columns, based on the two largest specimens of CC-7-0.4 and CC-7-0.6. However, the values of the shear-span ratios of the two experimentally tested columns are both 2.42, which is not the initial purpose (the shear-span ratios of the columns should be 2.0 and the columns should fail in shear) of this experiment. Furthermore, test results obtained by only two specimens could not substantially summarize any adequate conclusions, either. In light of this consideration, some numerical tests are conducted to thoroughly investigate the influence of axial compression ratio on the seismic behavior of short RC columns.

Figure 8.10 presents the results of a short RC column sized of 300 mm × 300 mm × 546 mm with different axial compression ratios of 0, 0.2, 0.4 and 0.6. The final failure patterns of the short RC column exhibit a very clear evolution along with the axial compression ratios, as shown in Fig. 8.10a. The specimen with no vertical load presents a pure shear failure mode at the end of the test, which is similar with that in a cyclically loaded cantilever beam (Jin et al. 2016a). As the axial compression ratio increases (i.e. $n = 0.2, 0.4$ and 0.6), the vertical load causes some initial damages (e.g. splitting cracks) before the horizontal load is applied, consequently the orientation of the distributed cracks on the surface of the specimen also varies. Generally, the slope

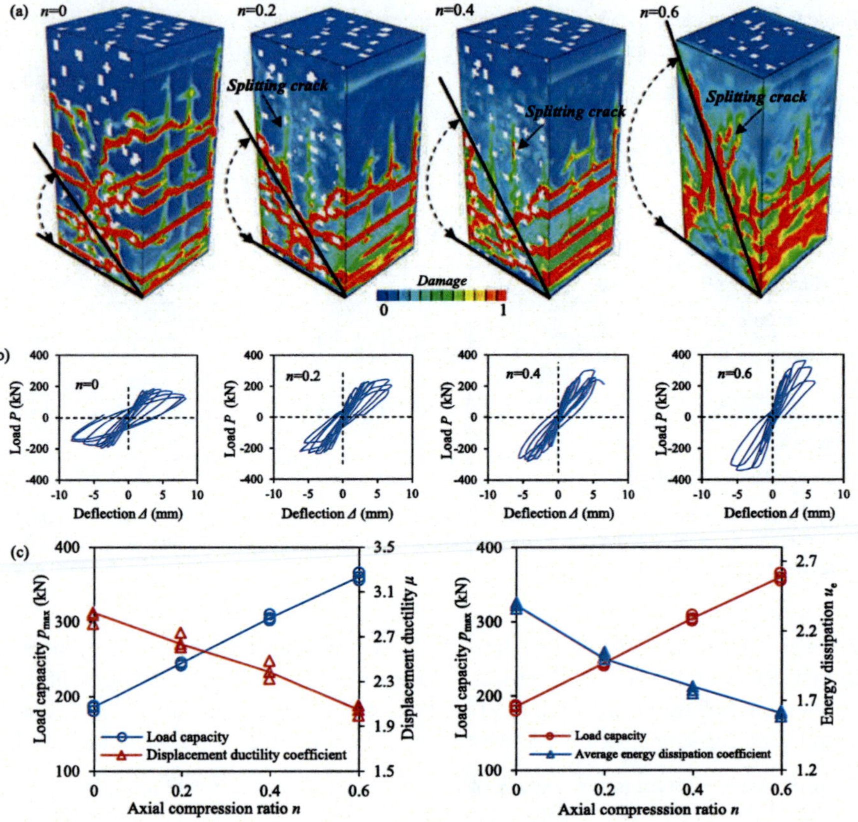

Fig. 8.10 Simulation results of a short RC column with different axial compression ratios of 0, 0.2, 0.4 and 0.6: **a** Final failure pattern; **b** Load–deflection curves; and **c** Trends of load capacity, displacement ductility and energy dissipation capacity with axial compression ratio

of the main crack causing the final failure of the column is getting steeper with the increase of the axial compression ratio, this phenomenon could also be illustrated by a strut-and-tie model, corresponding to the conclusion in Moretti and Tassios (2007). The load–deflection curves for the cyclically loaded short RC column with different axial compression ratios are respectively placed under the failure patterns in Fig. 8.10b, and trends of load capacity, displacement ductility and energy dissipation capacity with axial compression ratio are plotted in Fig. 8.10c. It can be clearly seen from Fig. 8.10b and c that, the load capacity is getting larger with the increase of axial compression ratio, however the displacement ductility and the energy dissipation capacity are getting weaker.

Correspondingly, the numerical results coincide with the initial conclusion obtained by the experimental program, which, increasing the axial compression ratio could improve the load capacity of the RC column, but the deformation capacity would be restricted and the failure mode would be more brittle, consequently the energy dissipation capacity could be deteriorated.

3. *Failure pattern and seismic behavior*

Figure 8.11 gives the final failure patterns of different sized short RC columns with a unified shear-span ratio of 2 and a unified axial compression ratio of 0.6. It can be clearly seen from the figure that all short RC columns fail in a same pattern with a clear diagonal shear crack on the surface of the specimen. Furthermore, the angles between the main crack and the bottom edge are approximate, which is definitely caused by the influence of compression ratio.

Figure 8.12 shows the relationships between displacement ductility/energy dissipation capacity and structural size. It can be noted from the figure that, these two parameters both decrease with the increase of structural size. Namely, the size effect could significantly affect the seismic behavior of the short RC column.

4. *Size effect on nominal strength*

The current used code GB 50010-2010 for the design of concrete structures in GB 50010 (2010) has defined the calculation formula for the shear strength of RC columns as below.

$$V = \frac{1.75}{\lambda + 1} f_t b h_0 + f_{yv} \frac{A_{sv}}{s} h_0 + 0.07N \tag{8.2}$$

in which, λ is the shear-span ratio, f_t is the tensile strength of the utilized concrete, f_{yv} is the yield strength of transversal reinforcement, s is the distance of stirrup spacing,

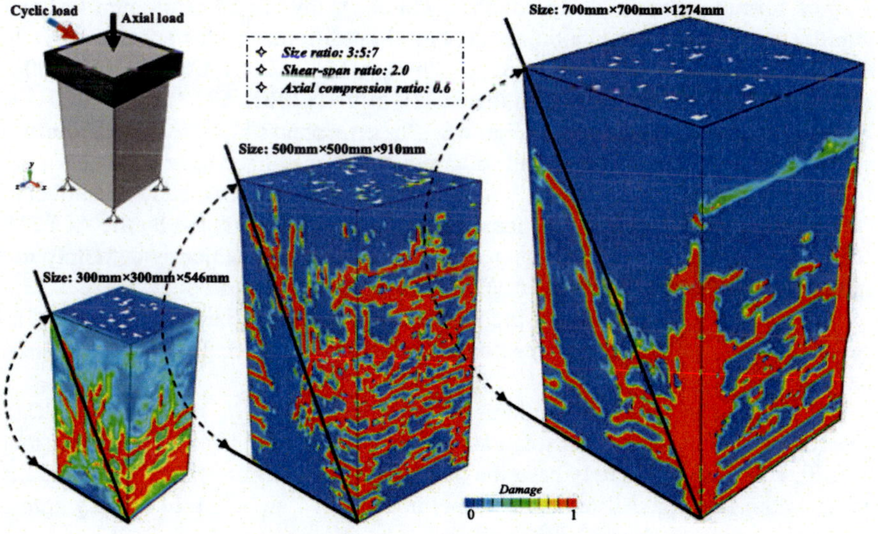

Fig. 8.11 Final failure patterns of different sized short RC columns

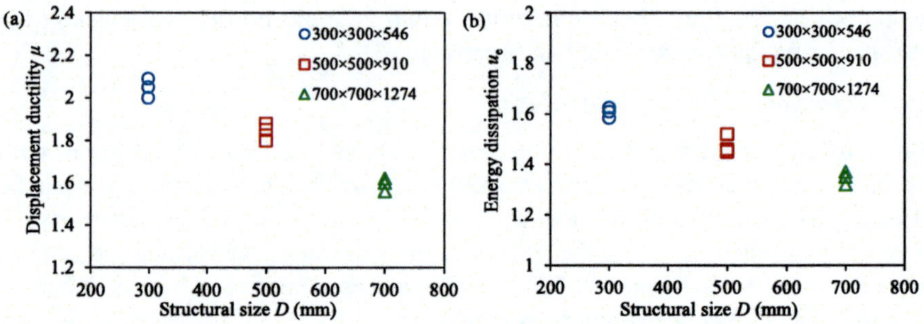

Fig. 8.12 Relationships between **a** displacement ductility and structural size and **b** energy dissipation capacity and structural size

and N is the design value of the axial compression. When N is larger than $0.3f_cA_0$, N is taken as $0.3f_cA_0$. Here, A_0 is the cross-sectional areas. The shear reinforcement ratio is defined as $\rho_{sv} = A_{sv}/bs$. One can note from Eq. (8.2) that, the current design code GB 50010 (2010) of China could not reveal the fact that the nominal strength of concrete structures decreases with the increase of the structural size.

Herein, some calculation formulae for the design of shear load capacity from other codes are collected, e.g. EN 1992-1-1 (2004a), ACI 318 (2011), CSA A 23.3 (2004), and BS 8110 (1997). Combing with the test data obtained by the numerical trials, a relationship chart between the shear load capacity/nominal shear strength and the structural size was plotted in Fig. 8.13. It can be clearly seen from the Fig. 8.13a that, all design values are lower than the test values and the shear load capacity increases along with the increase of structural size. However, due to the fact that the adopted formulae for the design of shear load capacity are all semi-empirical and semi-theoretical, the calculation results based on different codes are dramatically different. The design values of BS 8110 (1997), AS 3600-2001, CSA-A 23.3 (2004) and ACI 318 (2011) are safer than that of GB 50010 (2010). EN 192-1-1 (2004) is apparently the least conservative. Figure 8.13b correspondingly give the relationship between nominal shear strength and structural size. The utilized formulae in EN 192-1-1 (2004), GB 50010 (2010) and BS 8110 (1997) could not reveal any size effect in the design of the specimen. The nominal shear strength calculated by the ACI 318 (2011) even get larger with the increase of structural size, which is definitely not agree with the natural law. Nevertheless, among these design codes, CSA-A 23.3 (2004) and AS 3600-2001 could preliminarily consider the size effect behavior, as the calculated nominal shear strength decreases with the increase of the structural size.

Figure 8.14 plots the bi-logarithmic diagram of log $(\sigma_{Nu}/Bf_t{'})$ versus log (D/D_0). One can clearly see that, the correlation coefficient R^2 equals up to 0.9, indicating that the SEL could approximately describe the size effect behavior of short RC columns under cyclic loading. All test data points in Fig. 8.14 are close to the asymptote

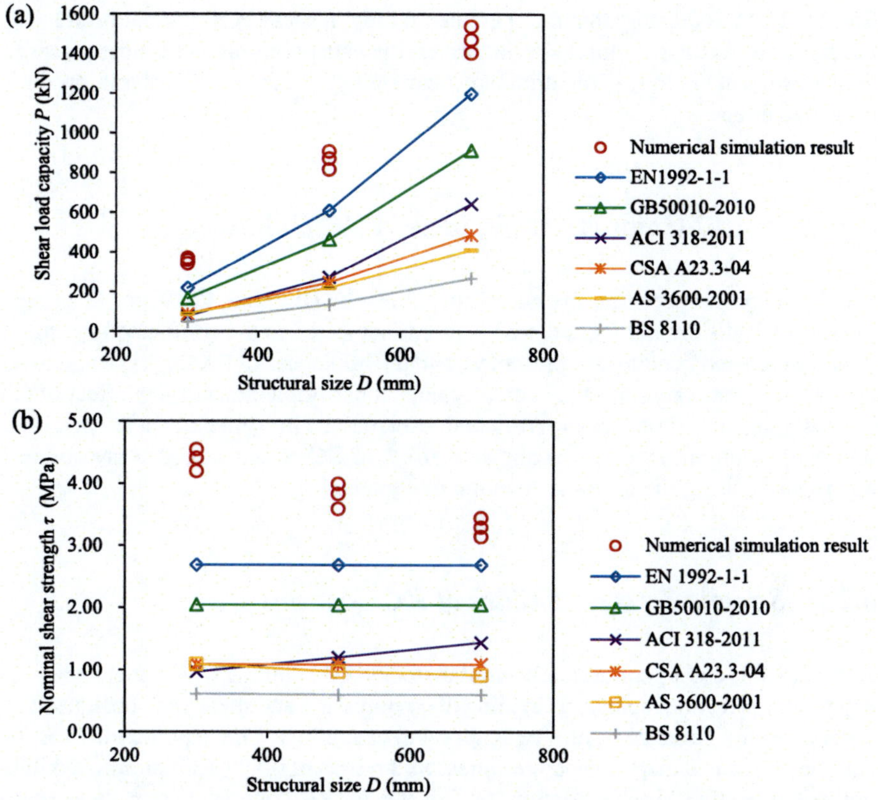

Fig. 8.13 Relationship chart between **a** shear load capacity and structural size and **b** nominal shear strength and structural size

Fig. 8.14 Bi-logarithmic diagram for the tested short RC columns

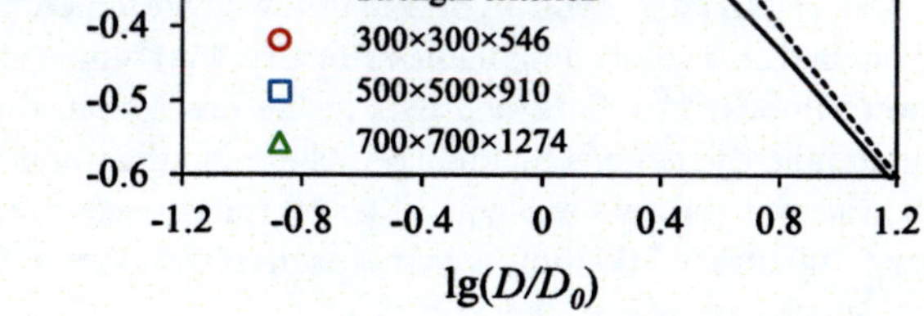

line of LEFM, indicating that the specimen failing in shear exhibit great brittleness. Furthermore, as the structural size increases, the data points are even more closer to the LEFM. That is to say, the nominal shear strength of short RC column presents clear size effect.

8.2 Flexural-Compressive Failure of RC Columns

Herein the present section, the experimental results (Jin et al. 2019) on 12 groups of geometrically similar RC squared columns subjected to the combined loading of compression and bending are presented, and the influences of loading type and axial compression ratio are introduced on mechanical performances and size effect of the RC columns. Based on this experimental research, the influence of shear-span ratio on flexural-compressive failure and size effect of RC columns is also investigated and reported using the meso-scale numerical method.

8.2.1 Seismic Tests on Failure of RC Columns

In Sect. 8.1.1, eight geometrically similar stocky RC columns with a shear span-to-depth ratio of 2 under combined loading of compression and shear were designed and tested to study shear-compressive failure and size effect. The experimental results have demonstrated that the shear-compressive strength of stocky RC columns exhibits obvious size effect. In this Section, twelve slender RC columns with a shear span-to-depth ratio of 4.0 have been tested by Jin et al. (2019) to study the seismic performances and examine the size effect in term of flexural compressive failure of RC columns under constant axial loading and monotonic/cyclic lateral loading.

1. *Experimental program*

12 RC columns having different cross-sectional sizes (i.e. 300 mm × 300 mm, 500 mm × 500 mm and 700 mm × 700 mm) and different axial load ratios were designed and tested. The test program was carried out in the Key Laboratory of Urban Security and Disaster Engineering of Ministry of Education of P.R. China at Beijing University of Technology. A representative loading photo of the RC column with cross-sectional dimension of 300 mm was presented in Fig. 8.15. The specimen was installed on a solid component vertically. Monotonic and cyclic lateral displacements were imposed to the specimens in the horizontal direction by means of the servo hydraulic dynamic actuator. The schematic diagrams for columns sized of 300 mm × 300 mm and columns with other two cross-sectional sizes (i.e. 500 mm × 500 mm and 700 mm × 700 mm) were respectively plotted in Fig. 8.1a and b, and the details of which can refer to the Sect. 8.1.1.

Three groups of geometrically similar RC columns are listed in Table 8.6. There were four specimens designed for each group of tested specimens having the same

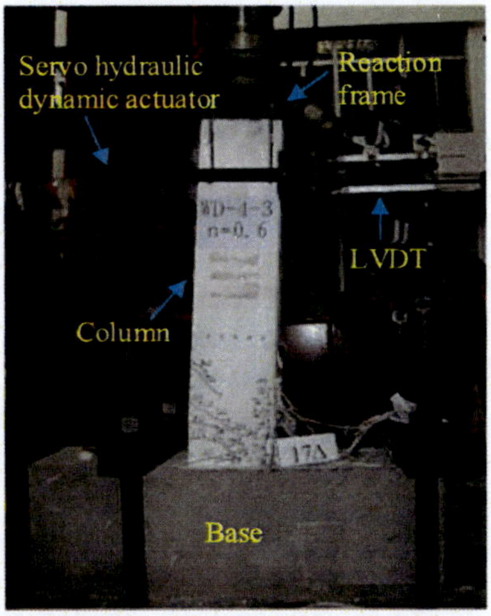

Fig. 8.15 Loading apparatus for testing the columns with cross-sectional width of 300 mm

Table 8.6 Physical parameters of the tested RC columns

Column name	$b \times h$ (mm × mm)	H (mm)	c (mm)	n	N (kN)	ρ (%)	ρ_{sv} (%)
CC-3-0.4/CM-3-0.4	300 × 300	1092	15	0.4	1044	1.51	1.89
CC-3-0.6/CM-3-0.6	300 × 300	1092	15	0.6	1566	1.51	1.89
CC-5-0.4/CM-5-0.4	500 × 500	1820 + 250	25	0.4	3000	1.51	1.89
CC-5-0.6/CM-5-0.6	500 × 500	1820 + 250	25	0.6	4500	1.51	1.89
CC-7-0.4/CM-7-0.4	700 × 700	2548 + 250	35	0.4	5880	1.51	1.89
CC-7-0.6/CM-7-0.6	700 × 700	2548 + 250	35	0.6	8820	1.51	1.89

Note The first letter "C" in the column name denotes "Column", and the second letter "C" or "M" denote "Cyclic" or "Monotonic" loading. The number of "3", "5" and "7" represent the structural dimension of the specimens, i.e. cross-sectional widths of 300, 500 and 700 mm. Furthermore, "0.4" and "0.6" respectively denote the axial load ratio n applied vertically on the columns. H is the total length of columns and N is the axial force applied to the top of the columns

structural sizes. In each group, two of specimens named as CM were applied to monotonic load, and the other two named as CC were designed to bear cyclic load. Table 8.6 presents the detailed information about the tested columns. The lengths (L) of the columns are respectively 1092 mm, 1820 mm and 2548 mm. What should be noticed is that the effective length L_{eff} of the columns with cross-sectional sizes of 300 mm × 300 mm is 1092 mm, namely the actual column height, as plotted in Fig. 8.1a. However, for the columns with cross-sectional sizes of 500 mm × 500 mm and 700 mm × 700 mm, the effective length L_{eff} is the sum of the spherical hinge length (250 mm) and the actual column height owing to different loading apparatus (see Fig. 8.1b). A shear-span ratio λ (effective length to depth) of 4.0 is utilized for all columns. The cover thicknesses c are 15, 25 and 35 mm over the transverse reinforcement for the columns having different structural sizes. In addition, the selected stirrup ratio ρ_{sv} and the reinforcement ratio ρ are 1.89% and 1.51% respectively.

Herein, the compressive and tensile strengths of concrete were measured by uniaxial compressive and splitting tensile tests of the standard cube specimens (sized 150 mm × 150 mm × 150 mm) based on the currently utilized code of "Standard for test method of mechanical properties on ordinary concrete" of China (GB/T 50081 2003). The compressive and tensile strengths of the specimens were 32.6 MPa and 2.4 MPa, respectively. HRB400 (i.e. the Hot-rolled Ribbed-steel Bar with the scheming average yield strength of 400 MPa) is used as the longitudinal reinforcement and HPB300 (i.e. the Hot-rolled Plain-steel Bar with the scheming average yield strength of 300 MPa) is used as transverse reinforcement. Figure 8.16 shows the reinforcement details and the cross-sectional sizes about the twelve tested RC columns. In order to maintain geometrically proportional columns (in the ratio of 3:5:7), all geometric dimensions were scaled strictly, including the diameter of steel rebar, their covers and spacing, as well as the column sizes.

The specimens were applied to combined effects of constant axial loading and monotonic/cyclic lateral displacements. The test adopted a force–displacement hybrid control. The detailed loading scheme was listed as follows:

(1) Axial compressive loading: the axial force was imposed by means of a hydraulic jack until reached the specified load at the beginning of the test, and it was kept at all times during the tests.
(2) Monotonic loading: the specimens were applied to the averaged horizontal forces controlled by constant increment before the peak load. In order to get the post-peak softening curve of the load–displacement curve, it was switched to constant increment control of averaged horizontal deflection. After that, it was converted to horizontal displacement control until the load carrying capacity of the columns dropped to 85% peak load or serious damage occurred in the columns. To guarantee the pseudo-static loading, the loading process cost about 5–10 min for each column.
(3) Cyclic loading: Fig. 8.17 given the typical force and displacement history applied to specimens. Firstly, the force control at the horizontal direction (i.e. positive and negative directions, respectively) was imposed on the columns and all subsequent cycles repeated once. The applied force in horizontal direction was loaded on the columns with 20, 40, 60, 80 and 100% of the yield load, and only one cycle was utilized at each level of load. The displacement control was adopted after the yielding of the columns. The yielding displacement Δ_y that was defined as the displacement when the RC columns yielded, was then employed as the level difference for the loading. Herein, two cycles were applied at each level of the lateral displacement history. Similarly, it was utilized until the load descended to 85% of the peak bearing capacity or when significant damage occurred in the columns. It should pay attention to is that the process cost about 10 min for each cyclic to guarantee the pseudo-static loading.

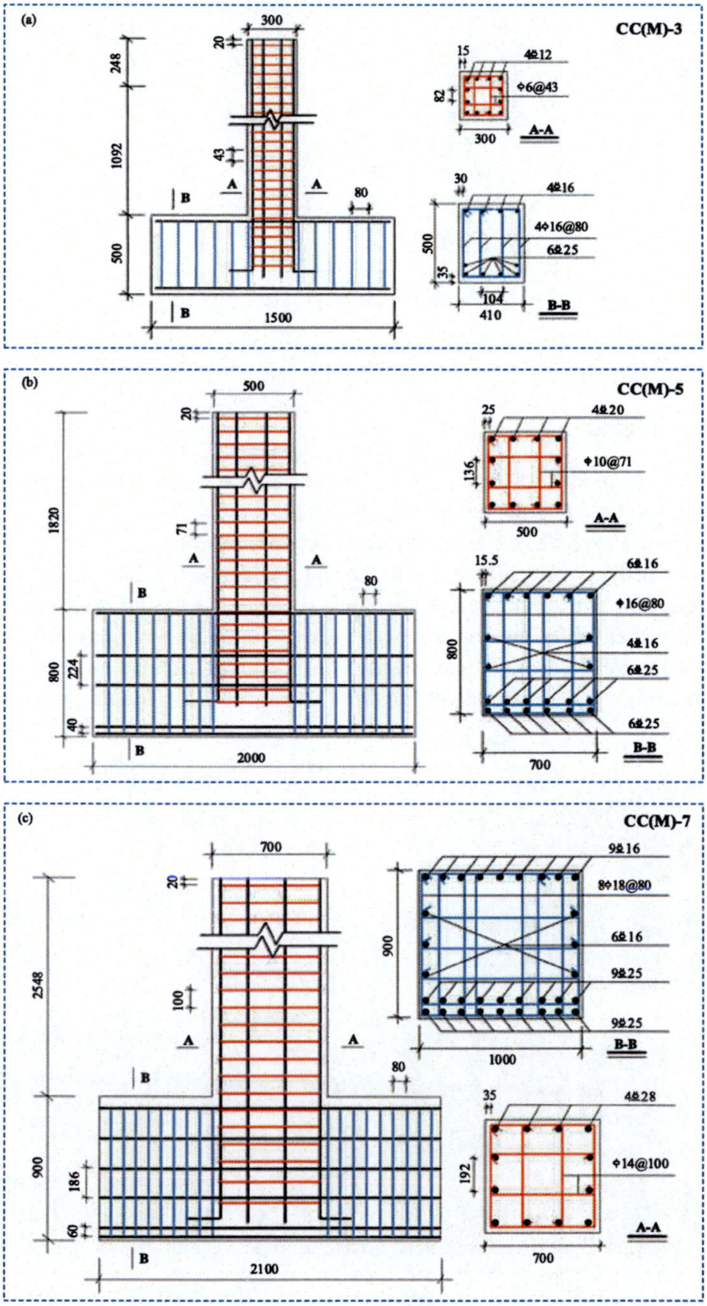

Fig. 8.16 Geometry and reinforcement details of 12 tested RC columns: **a** CC(M)-4-3; **b** CC(M)-4-5; and **c** CC(M)-4-7 (*Unit* mm)

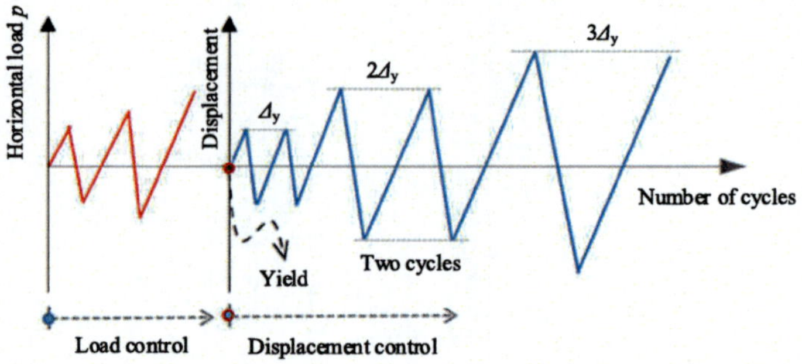

Fig. 8.17 Loading scheme for the cyclic lateral loading

2. *Failure patterns*

The final failure modes of the twelve columns under monotonic and reverse lateral loads are respectively presented in Figs. 8.18 and 8.19.

As shown in Fig. 8.18, for the columns under monotonic load, the flexural-tensile cracks generated on front surface due to the bending of the columns, and the concrete crushing occurred at the bottom of the back surface due to compression. It can be noted clearly that the failure modes of the six columns under monotonic loads are closely similar. When the columns reached their bearing capacities, about 5–7 main flexural tensile cracks generated at the front surface. As the loading continued, the

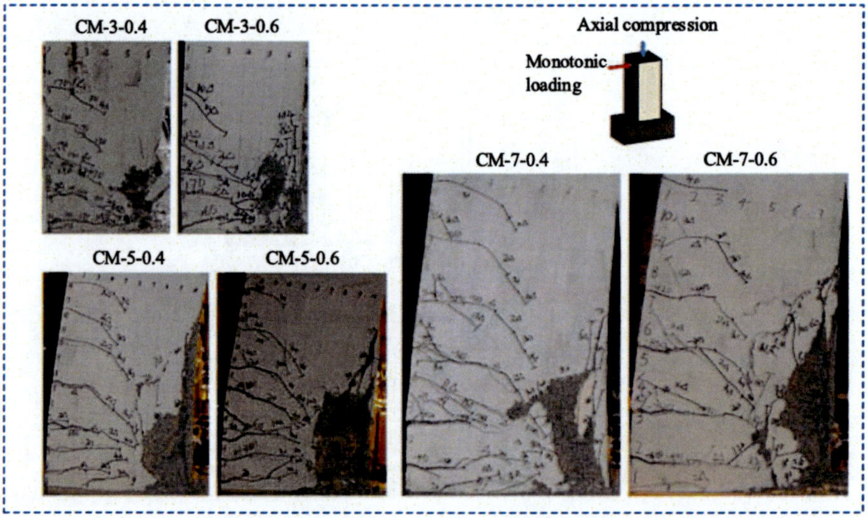

Fig. 8.18 Final failure patterns of the RC columns subjected to monotonic lateral loading

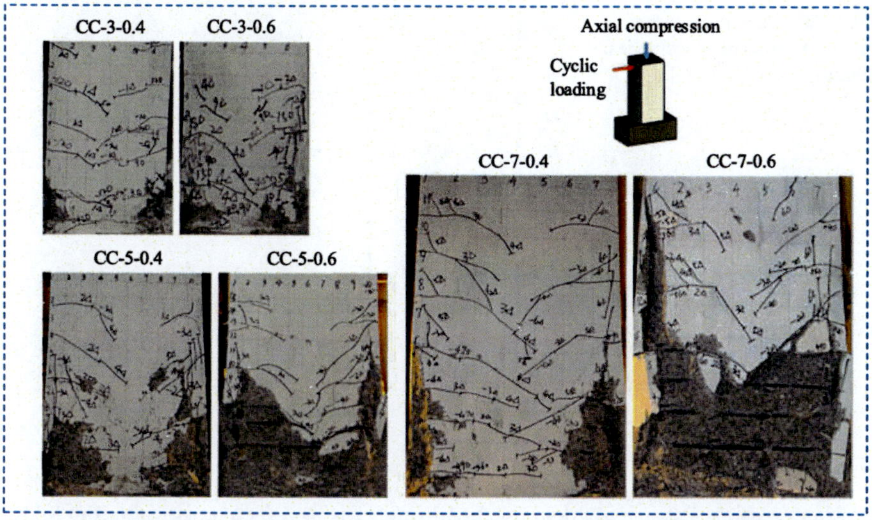

Fig. 8.19 Final failure patterns of the RC columns subjected to cyclic lateral loading

number of cracks no longer increased, while the crack width increased significantly. In addition, the cracks propagated obliquely with 45^0. Finally, the concrete crushing region at the back surface increased, and the typical compression-flexure failure occurred.

The failure modes of the six cyclically loaded RC columns are presented in Fig. 8.19. It can be seen from Fig. 8.19 that the cracks of the geometrically similar columns have different structural sizes developed in the similar way in the whole test process. On the whole, the damage concentrated at the root regions of the columns. There are about five to seven major cracks generated at the front and back surfaces when the columns reached their bearing capacities. This is similar with that under monotonic loading. However, under low cyclic fatigue loading, the columns failed more brittle. Moreover, one can clearly note that the concrete spalling areas and the number of vertical cracks are much larger under cyclic load, as can be observed from the comparison of Figs. 8.18 and 8.19. The spalling of concrete finally leads to the buckling of the longitudinal reinforcement, which would be illustrated in the next section.

3. *Hysteretic curves and Skelton curves*

For the seismic design of composite structures, there are some important parameters including load carrying capacity, ductility, stiffness degradation capacity and energy dissipation capacity. Herein, some parameters would be simply illustrated. A typical load P versus displacement Δ curve is plotted in Fig. 8.20. In Fig. 8.20, Δ_y denotes the yielding drift when the force reaches the yield strength. In other words, the yielding displacement corresponding to the applied load P_y. In the manner of described above, Δ_m characterizes the peak displacement when the applied force gets to the maximum

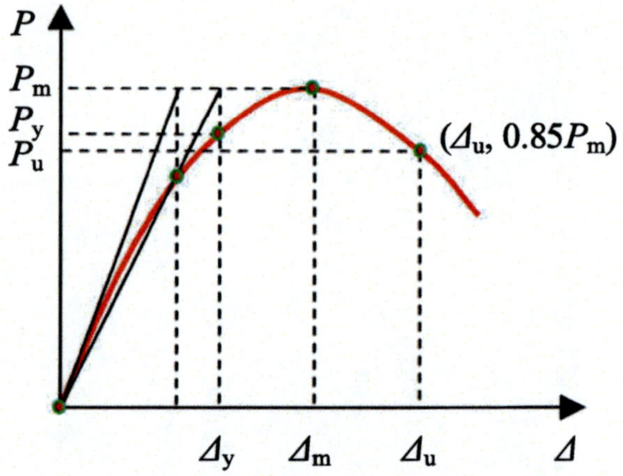

Fig. 8.20 Definition of the imposed cyclic displacement ductility factor

bearing capacity P_m, and Δ_u denotes the ultimate displacement corresponding to the ultimate bearing capacity, i.e. P_u.

The method proposed by Rodrigues et al. (2013) was employed to determine the yielding displacement and ductility. Referring to the work of Rodrigues et al. (2013), the ductility capacity of the RC columns is quantitatively analyzed by the displacement ductility factor, which is set as $\mu = \Delta_u/\Delta_y$. The ultimate displacement is regarded to the displacement when the bearing capacity fell to 85% peak load carrying capacity in the decline period (Jin et al. 2016a; Lu et al. 2015), as plotted in Fig. 8.20. In addition, the ultimate elastic–plastic displacement rotation angle θ_u was defined as $\theta_u = \Delta_u/H$ (H is the height of the column).

Figure 8.21 plots the recorded relationships of the load P and the lateral displacement Δ of the six geometrically similar RC columns under monotonic lateral load. It can be found that the columns with a larger axial compressive force (i.e. $n = 0.6$) have larger bearing capacity. However, the larger the axial compression ratio of the columns is, the more brittle failure the columns exhibits. This phenomenon can be concluded from the peak-post softening curves in Fig. 8.21. The columns with a larger axial compression ratio have steeper softening curves, and thus they have weaker deformation capacity. Verderame et al. (2008a) have also reported similar observations.

The recorded horizontal force versus displacement hysteretic relationships and the corresponding skeleton curves of the six RC columns subjected to cyclic lateral loads are plotted in Fig. 8.22. The hysteretic curves provide important information for investigating the seismic performances of RC components under combined actions of axial loading and cyclic lateral loading, including energy dissipation capacity, ductility as well as degradation of stiffness of RC columns. It can be concluded from Fig. 8.22 that similar with those under monotonic load, under cyclic load the columns with a larger axial load ratio have larger bearing capacities and steeper softening curves. Moreover, it can be found clearly from the skeleton curves in Fig. 8.22 that the specimens were in the elastic phase at the beginning of the test. However, the

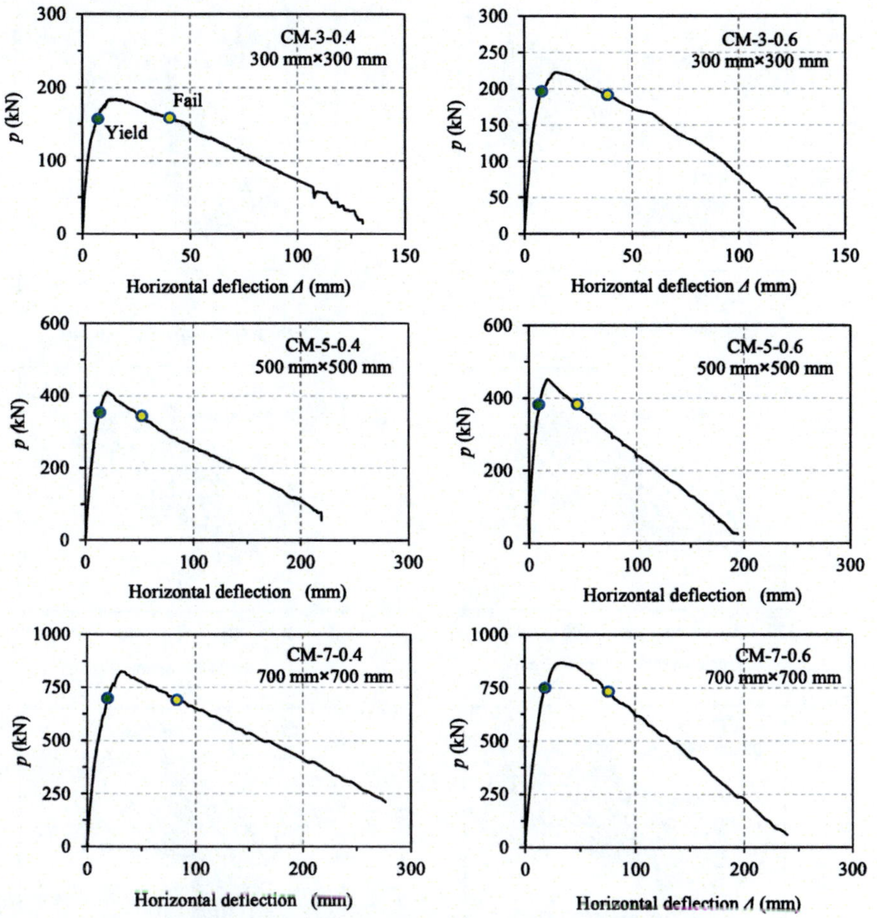

Fig. 8.21 Tested hysteretic curves of the RC columns under monotonic loading

specimens entered the plastic stage with increasing the applied horizontal loading. At the same time, the plastic deformation emerged in the columns as well as the stiffness of the columns degraded gradually. Moreover, the more times of the loading cycles are, the faster the degree of the stiffness degradation is.

Comparing Figs. 8.21 with 8.22, one can note that the envelop curves (i.e. the skeleton curve) under monotonic lateral load are all greater than those under cyclic lateral load. This should be caused by the low cyclic fatigue effect. The similar results have also been reported in the researches (Zhang et al. 2013; Pujol et al 2006; Kawashima and Koyama 1988).

4. *Load carrying capacity*

To explore the influences of structural size on the flexural capacity of the RC columns, the dimensionless nominal bending M' ($M' = M/f_c bh_0^2$) was utilized and calculated.

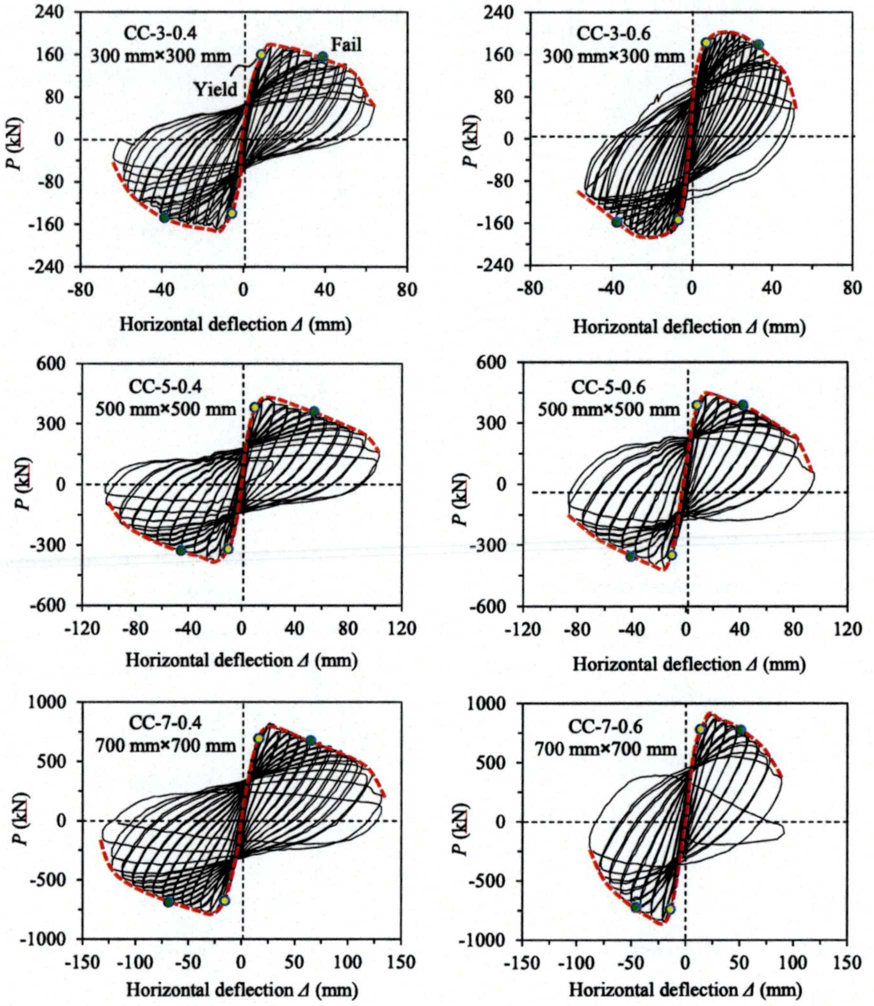

Fig. 8.22 Tested hysteretic curves of the RC columns under cyclic loading

Here, f_c denotes the concrete uniaxial compressive strength, b and $h0$ mean the cross sectional width and the effective cross-sectional height of the columns respectively. M characterizes the moment at the bottom of the columns, and it can be evaluated as $M = PH + N\Delta$, in which H denotes the height of columns and N means the applied compressive loads at the top surface of the columns. It is worth noting that the formula $M = PH + N\Delta$ considers the so-called $P - \Delta$ effect.

The nominal flexural moments M'_y, M'_m and M'_u corresponding to the yielding load, peak load and ultimate load can be calculated as $M'_y = M_y/f_c bh_0^2$, $M'_m = M_m/f_c bh_0^2$, and $M'_u = M_u/f_c bh_0^2$.

Figure 8.23a and b presents the recorded relationships between the nominal moments and structural size (i.e. the cross-section width) of the columns having the axial load ratios of 0.4 and 0.6, respectively. It can be seen that the dimensionless nominal moments decrease with increasing the structural size, indicating an obvious size effect. From Fig. 8.23, one can clearly note that the obtained nominal moments, including the dimensionless nominal yielding moment, peak moment and ultimate-displacement moment for the columns under monotonic load are basically larger than those under cyclic load.

Comparing Fig. 8.23a with b, one can notice that a larger axial load ratio leads to a larger nominal moment. Furthermore, one can clearly see that the degradation of the bearing capacity is getting faster with increasing the axial load ratio. This should be caused by the fact that, the columns with larger axial load ratio exhibit more brittle failure.

The theoretical results calculated from ACI 318 (2011), EN 1992-1-1 (2004a) as well as GB50010-2010 (2010) were employed for the comparison with the present test data on the flexural capacity of RC columns. The comparison of the tested data and design values calculated by these codes is made and plotted in Fig. 8.24. One can clearly see from the Fig. 8.24a and b that the experimental values are greater than all calculated values. The flexural capacity M_m increases as cross sectional size D increases. Since the formulae in the above codes are all semi-empirical and semi-theoretical, there is a great difference in the design values according to different codes.

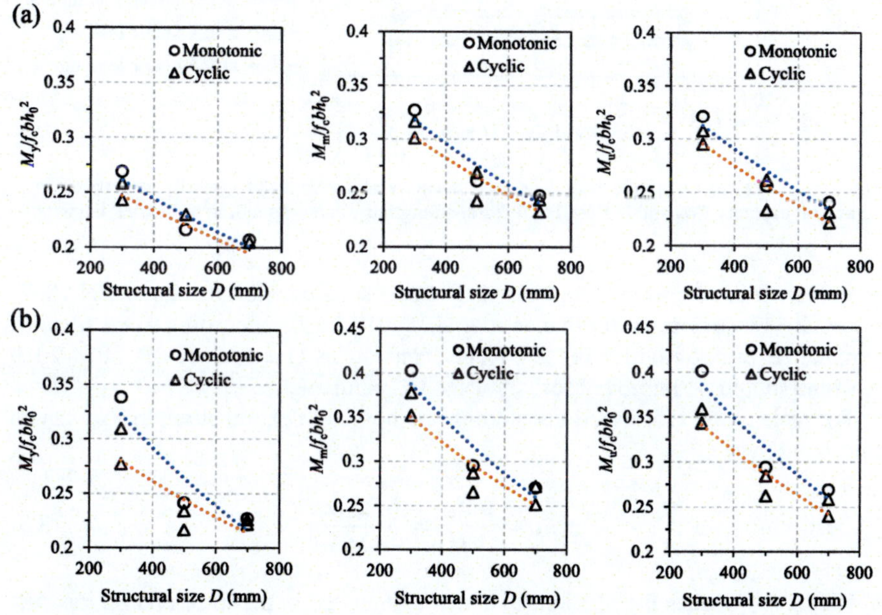

Fig. 8.23 Relationship of dimensionless moments and structural size for the tested RC columns: **a** $n = 0.4$ and **b** $n = 0.6$

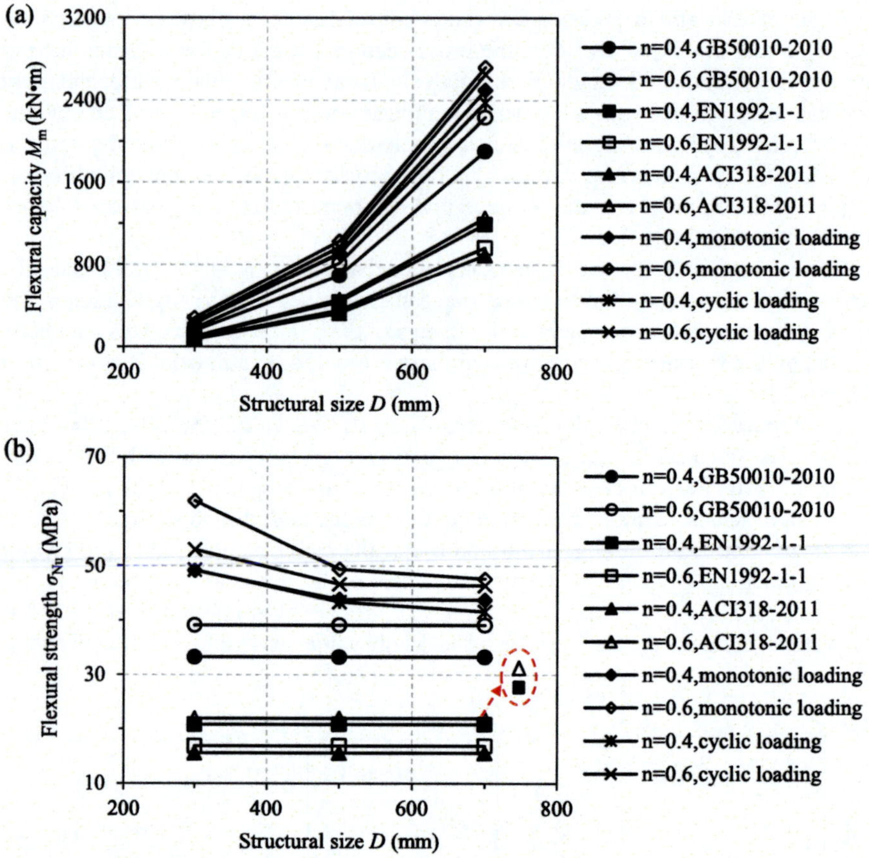

Fig. 8.24 Comparison of the tested flexural capacity and calculated ones by different codes: **a** Flexural capacity M_m and **b** Nominal flexural strength σ_{Nu} corresponding to flexural capacity

Moreover, it can be found from Fig. 8.24 that the calculation results of EN 1992-1-1, ACI 318-2011 are safer than that of GB50010-2010. According to the trend of nominal flexural strength in Fig. 8.24b, one can note that the code of GB50010-2010 of China should be improved for large-size RC columns.

For rectangular cross-section columns, the nominal flexural strength σ_{Nu} can be calculated by

$$\sigma_{Nu} = \frac{6M_m}{bh_0^2} = \frac{6P_{max}l}{bh_0^2} \tag{8.3}$$

in which, M_m means the maximum moment generated in the fixed end region, $M_m = P_{max}\, l$. The length l is the distance from the loading point to the fixed end, P_{max} is the maximum applied loading, *b* is the column width, and *h0* is the effective cross-sectional height of the column. Figure 8.24b presents the relationship between the nominal flexural strength and structural size. It can be seen from Fig. 8.24b that the obtained flexural strength of the columns under cyclic load is smaller than that under monotonic load due to the fatigue damage. Moreover, the obtained flexural strength under the axial load ratio of 0.6 is larger than that under the axial load ratio of 0.4 except the design values calculated by EN 1992-1-1. Nevertheless, the adopted formulae based on GB 50010-2010, EN 1992-1-1and ACI 318-2011 do not consider the effect of structural size, as can be observed in Fig. 8.24b.

5. *Ductility capacity and Degradation of stiffness*

For RC columns, the displacement ductility factor μ and the ultimate drift ratio θ_u were often utilized to describe the ductility capacity. The relationships between mentioned two parameters and cross sectional width are respectively plotted in Figs. 8.25 and 8.26. It is to be noted that, for cyclic-loading, there are two groups of test data for μ and θ_u, corresponding to the positive and negative directions, respectively. One can find that both μ and θ_u decrease with increasing the cross-sectional size. This means that increasing the cross-sectional size would weaken deformation capacity. For all cases, the displacement ductility would be restricted as the axial compression ratio increases. Similar results have also been reported in Mo and Wang's (2000) work. In addition, as the axial load ratio increases, both μ and θ_u decrease remarkably. Just as mentioned previously, the larger axial load ratio leads to larger bearing capacity and weaker deformation capacity.

Compared Fig. 8.25a and b, one can note that the displacement ductility coefficients obtained under cyclic loading are smaller than those under monotonic loading. Furthermore, it is obvious that the size effect also exists in the displacement ductility coefficient. Namely, as the structural size increases, the displacement ductility coefficient μ decreases obviously for the two different axial compressive ratios. One can also note from Fig. 8.25 that, the tested displacement ductility coefficients for axial

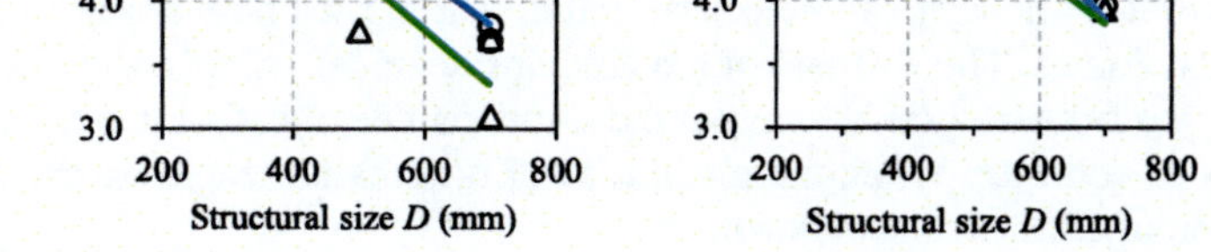

Fig. 8.25 Displacement ductility coefficients $\mu = \Delta_u/\Delta_y$ of the columns: **a** Under cyclic load and **b** Under monotonic load

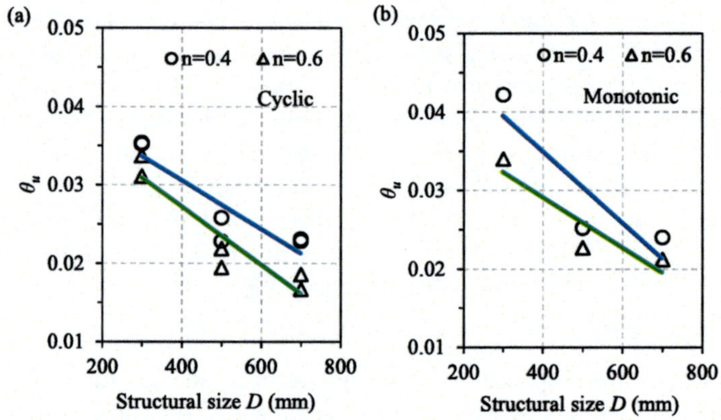

Fig. 8.26 Ultimate elastic–plastic displacement rotation angle $\theta_u = \Delta_u/H$ of the RC columns: **a** Under cyclic load and **b** Under monotonic load

ratio of $n = 0.4$ are obviously larger than those for $n = 0.6$. Similar conclusions can also be achieved for the ultimate drift ratio θ_u, as can be concluded from the comparison of Fig. 8.26a and b. One can see from Fig. 8.26 that, on the whole, the values of chord rotations of cyclically loaded columns are lower than that of monotonically loaded ones, reducing approximate 27%. It is noted that such a difference seems to have no relation with the level of axial loading. The results coincide with the conclusions obtained by Verderame et al.'s (2008a; b). Moreover, there is an obvious size effect in the ultimate elastic–plastic displacement rotation angle $\theta_u = \Delta_u/H$ of the RC columns. As the structural size of the columns increases from 300 to 700 mm, the ultimate elastic–plastic displacement rotation angle decreases clearly.

According to the above analyses, the size effect in both displacement ductility coefficient and ultimate displacement ration angle should be taken into account in the design of concrete structures.

In reality, under low cyclic loading, the tensile/compressive strengths of concrete materials and the yield strength of steel reinforcements, as well as the bonding strength at the concrete/steel interface degrade obviously. This finally makes the failure of the columns under cyclic loading more brittle than that under monotonic loading, as also observed in (Verderame et al. 2008a; b; Li et al. 2016; Zhu et al. 2016).

Figure 8.27 presents the relationship between the secant stiffness K and the rotation angle θ (= Δ/H) level at the lower ends of the RC columns under cyclic loading. On the whole, it can be clearly found that the stiffness develops similarly for all RC columns. The stiffness of the columns degrades sharply with the rotation angle varying from 0 to 2% because of the significant degradation in the mechanical properties of concrete. When the magnitude of the rotation angle exceeds 2%, the degradation of stiffness slows down.

One can note from Fig. 8.27 that, the larger the cross-sectional size is, the more rapid the stiffness degradation of the columns is. Moreover, it can also be noted from

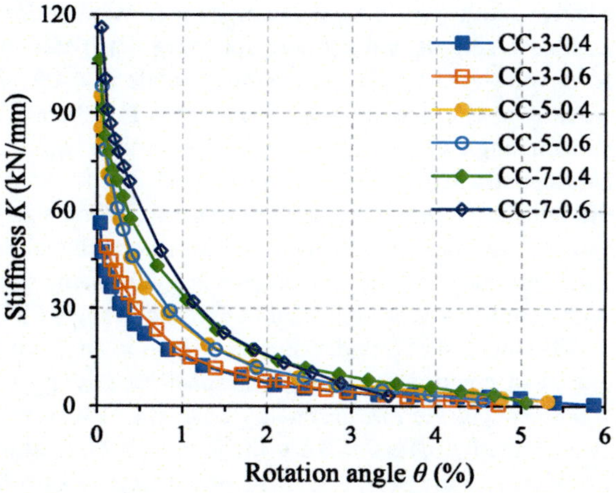

Fig. 8.27 Relationship between the secant stiffness K and the rotation angle θ of the six RC columns subjected to combined action of axial loading and cyclic lateral loading

Fig. 8.27 that, when the rotation angle is less than 2%, the greater the axial load ratio is, the faster the columns stiffness degrades. As the rotation angle is larger than 2%, the flexural stiffness is dominated by the longitudinal bars. At this time, the effect of axial compression ratio on the stiffness degradation could be ignored.

6. *Energy dissipation capacity*

The equivalent viscous damping coefficient ζ is usually employed to assess the energy dissipation capacity of the RC columns, and the details of which can refer to Chap. 6. The variation with rotation angle θ (= Δ/H) at the ends of the columns subjected to axial compression levels of $n = 0.4$ and 0.6 is depicted in Fig. 8.28a and b respectively. For all columns, the equivalent viscous damping coefficient ζ has a trend of increase as the rotation angle θ increases. This indicates that as the

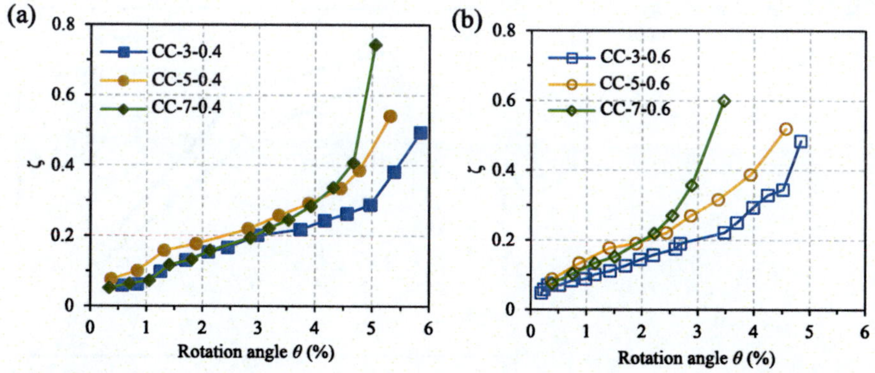

Fig. 8.28 Relationship between the equivalent viscous damping coefficient ζ and rotation angle θ of the RC columns under the different axial compressive ratios: **a** $n = 0.4$ and **b** $n = 0.6$

rotation angle increases, the damage generated within the columns gets to be more and more serious and the corresponding dissipated energy increases. For the axial force ratio of 0.4, when the rotation angle $\theta < 4\%$, the ζ–θ relation is linear, and when $\theta > 4\%$, ζ increases significantly. For the axial load ratio of 0.6, when the rotation angle $\theta < 2.5\%$, the ζ–θ relation is linear, and when $\theta > 2.5\%$, ζ increases significantly. Comparing Fig. 8.28a with b, one can note that, for the columns with the same structural sizes, the larger the axial force ratio is, the larger the equivalent viscous damping coefficient is. It can be concluded that columns with higher axial load ratio have greater dissipated energy for same rotation angle. This is consistent with the observations reported in Mo and Wang's (2000) work.

The coefficient ζ under the same rotation angle could not completely explain the energy dissipation capacity of the specimens owing to the difference of the ultimate rotation angle. Herein, the energy dissipation capacity after yielding of the specimens is reflected by means of the average energy dissipation coefficient μ_e that has been introduced in the Sect. 8.1.1. The obtained μ_e versus D (i.e. the cross-section width) is plotted in Fig. 8.29. It can be clearly found from in Fig. 8.29 that the capacity of dissipated energy gets weaker with the increase of axial load ratio. What's more, increasing the cross-sectional size D results in remarkable decline of the average energy consumption.

7. *Rotation capacity of plastic hinge*

The rotation capacity of plastic hinge is often utilized to describe the plastic deformation capacity of the columns. In general, the total deformation involves elastic deformation and plastic deformation for the members with the flexural failure mode. Therefore, the parameter of θ_{pl} is utilized to describe the rotation capacity of plastic hinge, and it has the form as

$$\theta_{pl} = \arctan\left(\Delta_{pl}/H\right) \tag{8.4}$$

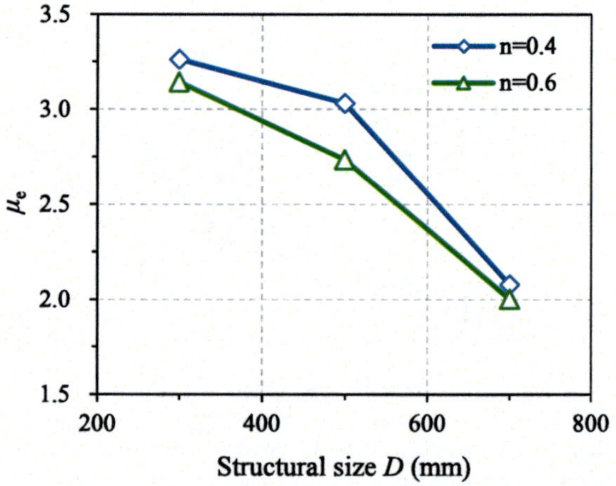

Fig. 8.29 Average energy dissipation coefficient μ_e of RC columns

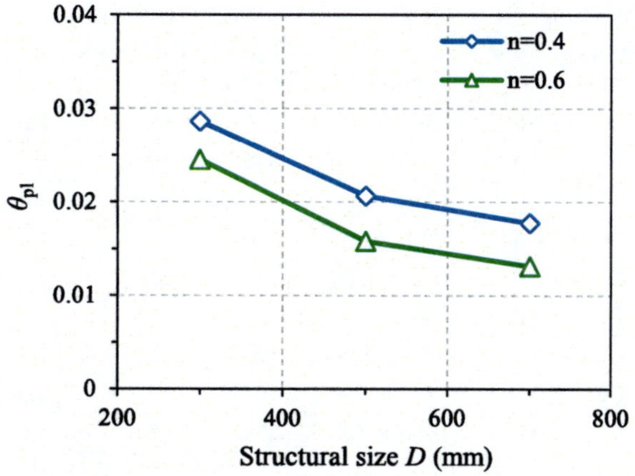

Fig. 8.30 Rotation capacity of plastic hinge of the tested columns

where Δ_{pl} means the plastic deformation of the columns, and it can be evaluated by $\Delta_{pl} = \Delta_u - \Delta_y$.

Figure 8.30 plots the relationship between the rotation angle θ_{pl} of plastic hinge and structural size. It can be stated that the magnitude of the rotation angle of plastic hinge decreases significantly as the structural size increases, as shown in Fig. 8.30. In addition, the smaller the axial compression ratio is, the larger the rotation angle of plastic hinge of the columns is. It should pay attention to is that, the effect of structural size on the plastic hinge length would be explored in details in another work.

8.2.2 Simulations on Failure of RC Columns

In Sects. 8.1.1 and 8.2.1, the shear-span ratios of the tested RC columns are respectively designed as 2.0 and 4.0. Based on the above experiments, the RC columns sized of 300 mm × 300 mm, 500 mm × 500 mm and 700 mm × 700 mm with shear-span ratio of 3.0 are supplemented in this section by means of numerical simulation method. The establishment of meso-scale model of RC columns and mechanical parameters of the meso components and steel rebar can refer to the Sect. 8.1.2. There are 9 groups of RC columns with different sizes and shear-span ratios, and the relevant geometric parameters are shown in Table 8.7. The values "2, 3 and 4" in the column name represent the shear-span ratio; "0.6" is the axial compression ratio of the tested specimens; and other letters and values are the same as those in Table 8.6.

1. *Failure pattern*

The ultimate failures mode of RC columns with different structure sizes and shear-span ratios under horizontal cyclic loading are shown in Fig. 8.31. It can be observed that the columns with shear-span ratio of 2.0 exhibit the shear-diagonal compression

Table 8.7 Geometrical parameters of the RC columns

Column name	$b \times h$ (mm)	L_{eff} (mm)	c (mm)	λ	n	ρ (%)	ρ_{sv} (%)
CC-2-3-0.6	300 × 300	546	15	2.0	0.6	1.51	1.890
CC-3-3-0.6	300 × 300	819	15	3.0	0.6	1.51	1.890
CC-4-3-0.6	300 × 300	1092	15	4.0	0.6	1.51	1.890
CC-2-5-0.6	500 × 500	910	25	2.0	0.6	1.51	1.890
CC-3-5-0.6	500 × 500	1365	25	3.0	0.6	1.51	1.890
CC-4-5-0.6	500 × 500	1820	25	4.0	0.6	1.51	1.890
CC-2-7-0.6	700 × 700	1274	35	2.0	0.6	1.51	1.890
CC-3-7-0.6	700 × 700	1911	35	3.0	0.6	1.51	1.890
CC-4-7-0.6	700 × 700	2548	35	4.0	0.6	1.51	1.890

failure while the columns with shear-span ratio of 3.0 and 4.0 show the bending failure.

For the specimens with shear-span ratio of 2.0, there are horizontal bending cracks on the root of the specimens at the initial stage of loading. With the increase of horizontal load, the number of horizontal bending cracks is decreasing, while the number of inclined cracks is increasing. Finally, the concrete fails to crush, the horizontal load drops sharply, and the specimen is damaged. In this process, the stirrups yield but the longitudinal bars do not, indicating that the shear capacity of the RC columns is mainly provided by stirrups and concrete. For the specimens with shear-span ratio of 3.0 and 4.0, horizontal cracks appear on the root of the specimens at the beginning of loading. With the increase of horizontal load, the horizontal bending cracks develop along the direction of column height, and the number of which is increasing. The concrete cover is crushed, the longitudinal bars buckle, the horizontal load drops sharply, and the specimens are damaged. In this process, the longitudinal bars yield but the stirrups do not yield, which indicates that the influence of the longitudinal bars and concrete on the shear capacity of the RC columns has replaced that of the stirrups.

2. *Hysteretic curves and Skelton curves*

The recorded horizontal force versus displacement hysteretic relationships of the simulated RC columns subjected to cyclic lateral loads are plotted in Fig. 8.32. It can be seen that under the same shear-span ratio, the shape of the hysteretic curves of the specimens with different structure sizes is approximately the same. For the specimens with the same structure size, the larger the shear-span ratio is, the more full the hysteretic curve becomes, the weakened the pinch effect is, and the better deformation capacity and energy consumption capacity become, which shows that the shear-span ratio increases from 2.0 to 4.0, the failure mode of the specimens gradually transforms from brittle shear failure into the ductile bending failure.

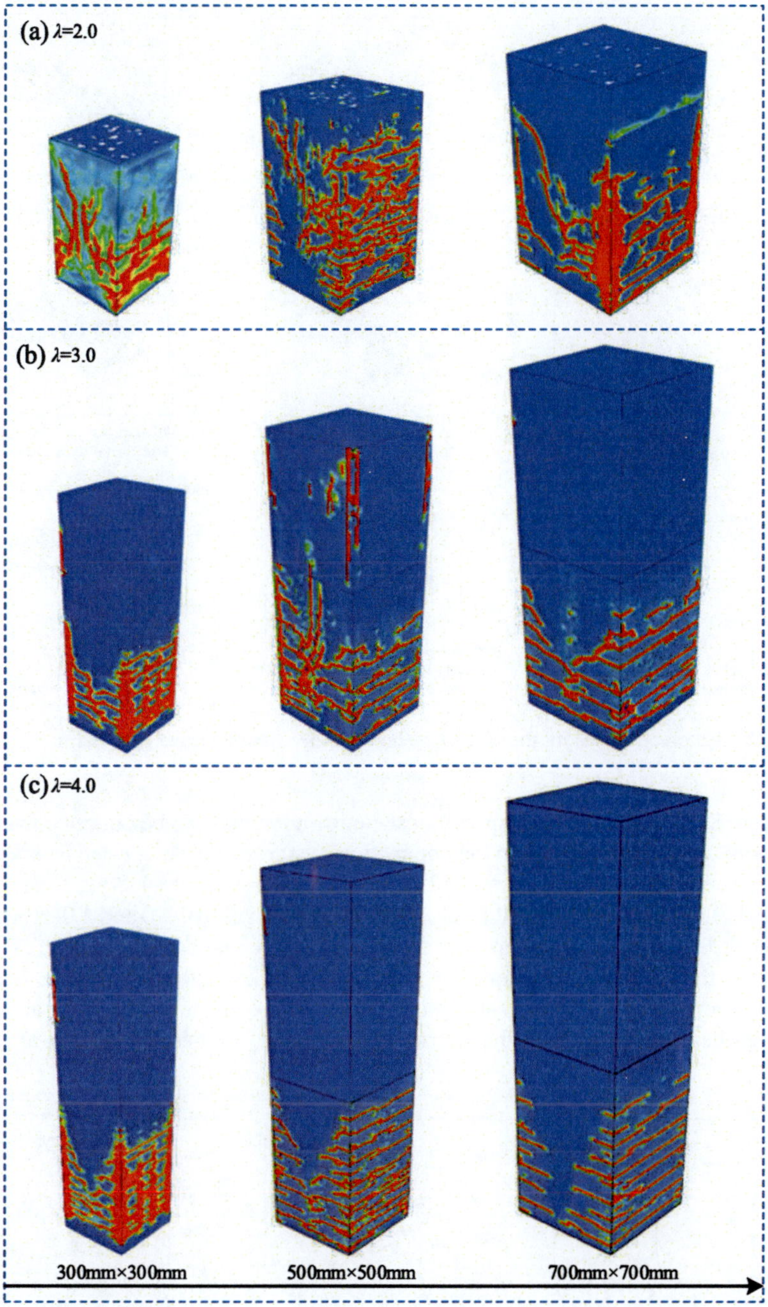

Fig. 8.31 Failure pattern of the RC columns having different structural sizes and shear span ratios: **a** $\lambda = 2.0$; **b** $\lambda = 3.0$; and **c** $\lambda = 4.0$

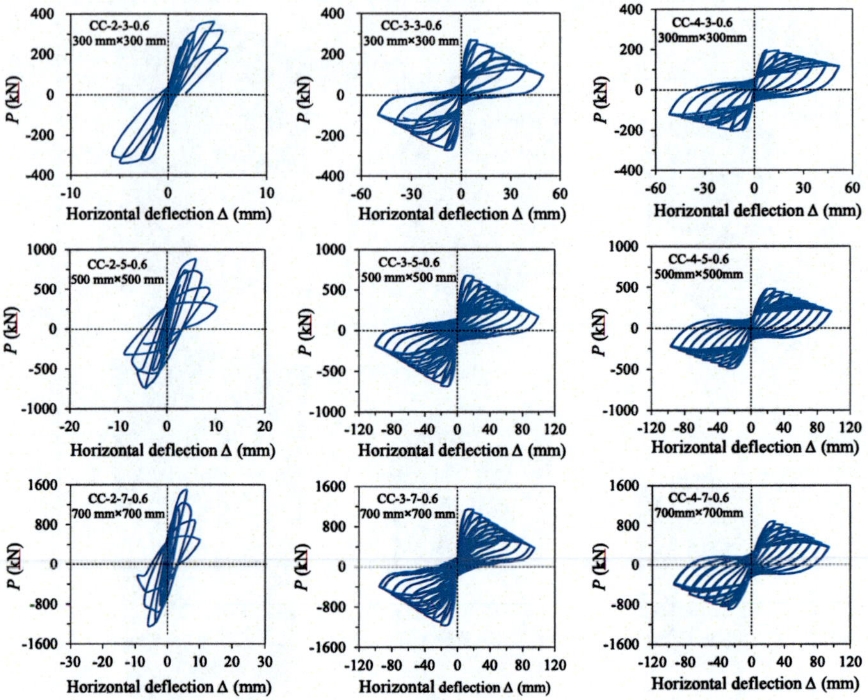

Fig. 8.32 Hysteretic curves for the nine RC columns under different shear span ratios

Figure 8.33 shows the corresponding skeleton curve that can be obtained from the peak point of each stage load in the connection hysteresis curve. As can be seen, for the specimen with shear span ratio of 2.0, its initial stiffness is very large, and its peak load is also significantly larger than that with shear span ratio of 3.0 and 4.0. However, after peak load, the bearing capacity decreases rapidly, the descent section is steep and short, the stiffness degradation is serious, and its seismic performance is poor. With the increase of shear span ratio (shear span ratio is equal to 3.0 and 4.0), although the shear bearing capacity of the specimen is reduced, the descent section after the

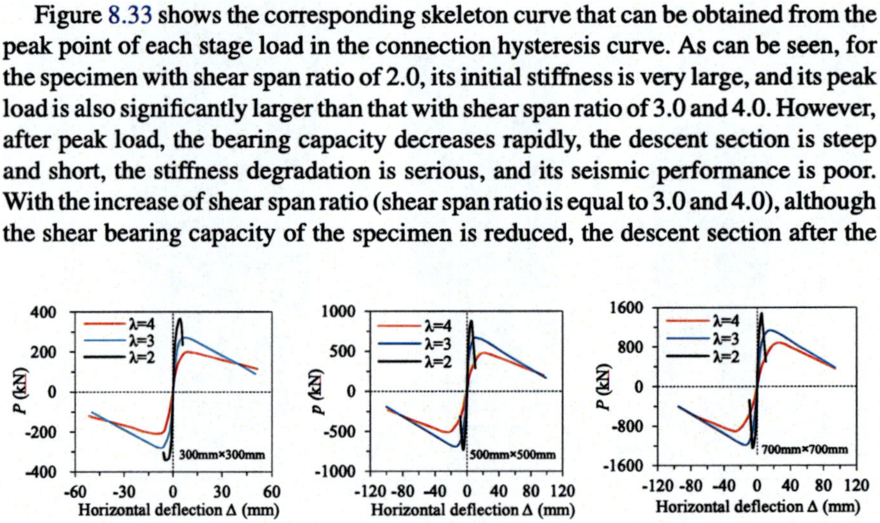

Fig. 8.33 Skeleton curves for the nine RC columns under different shear span ratios

peak load is relatively gentle, and the ultimate displacement is greatly increased, the rigidity degradation is relatively moderate, and the seismic performance is good.

3. *Ductility capacity and Energy dissipation capacity*

The relationship between the displacement ductility factor μ and structure size of RC columns under different shear-span ratios is plotted in Fig. 8.34. For the RC columns with the same shear-span ratio, the displacement ductility coefficient decreases with the increase of the structure size, showing the obvious size effect, but this decreasing trend decreases with the increase of the shear-span ratio. As mentioned before, the failure mode of columns with shear-span ratio of 2.0 is brittle compression-shear failure with poor ductility, while that of columns with shear-span ratio of 3.0 and 4.0 is ductile compression-bending failure with good ductility. Therefore, under the same structure size, the displacement ductility coefficient of columns increases with the increase of shear-span ratio, indicating that the shear-span ratio is one of the influencing factors affecting the deformation capacity of RC columns. It is to be noted that the increasing trend decreases with the increase of the shear-span ratio because the columns may occur buckling failure that is more brittle when the shear-span ratio is too large.

Figure 8.35 shows the relationship between the average energy dissipation coefficient μ_e and the structure size of RC columns under different shear-span ratios. The average energy dissipation coefficient decreases with the increase of the structure size of RC columns with the same shear-span ratio, but the decreasing trend increases with the increase of the shear-span ratio, which is different from the displacement ductility coefficient. Under the same structure size, the average energy dissipation coefficient increases and this increasing trend decreases with the increase of shear-span ratio.

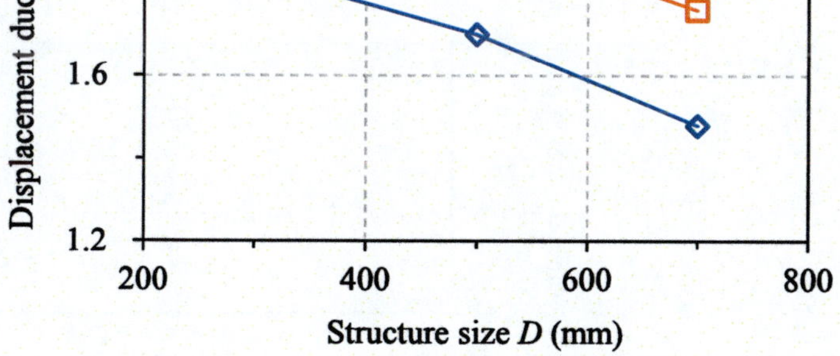

Fig. 8.34 Relationship between structure size and displacement ductility coefficient

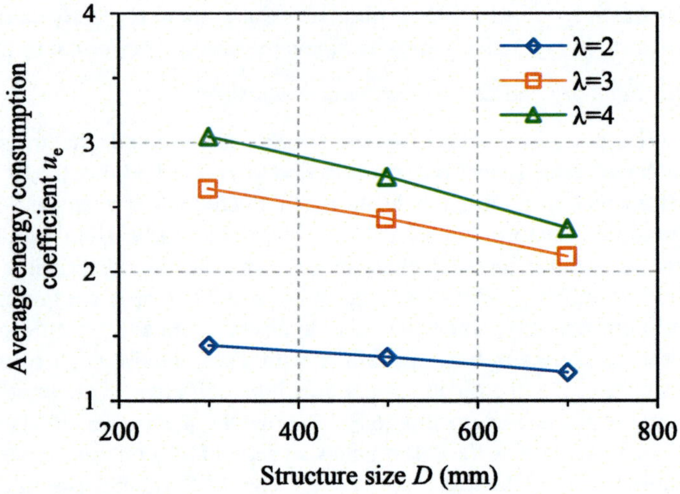

Fig. 8.35 Relationship between structure size and average energy consumption coefficient

4. ***Influence of structural size on the nominal strength***

The relationship between the nominal shear strength ($\sigma_{Nu} = V/bh_0$) and structure size for simulated RC columns under different shear-span ratios is presented in Fig. 8.36. It can be obviously seen that the nominal shear strength of RC columns under combined compression, bending, and shearing does exist size effect, that is, it decreases with the increase of structure size. However, the decreasing trend decreases with the increase of shear-span ratio. In addition, the nominal shear strength of RC columns decreases with the increase of shear-span ratio. This is because when the shear-span ratio is smaller, the columns exhibit brittle shear failure, and the shear

Fig. 8.36 Nominal strength vs structure size

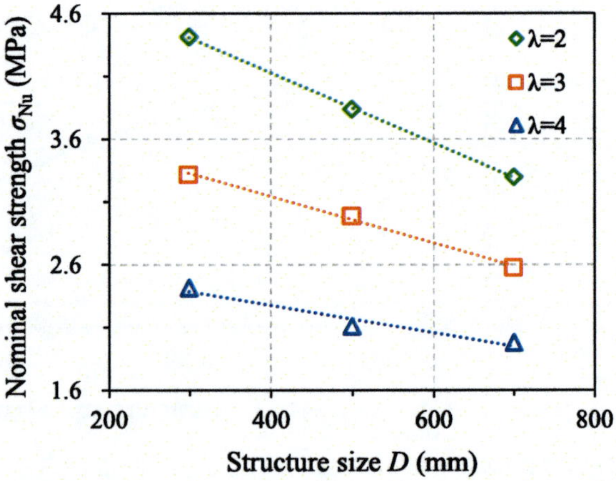

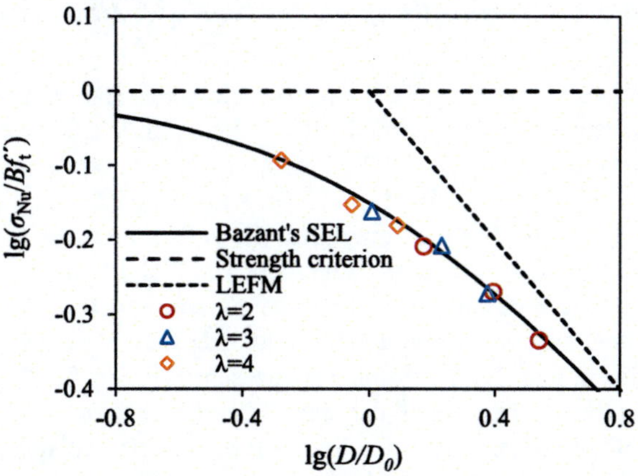

Fig. 8.37 Logarithmic curve of nominal strength versus structure size

capacity provided by stirrups having yielded dominates in the total shear capacity of RC columns. When the shear span-ratio increases, the stirrup fails to yield and the column shows ductile bending failure. The bending stress replacing the shear stress gradually becomes the main factor affecting the nominal strength of RC columns, resulting in the decrease of the nominal shear strength.

In order to compare with the Bažant's SEL, the double logarithm curve of the nominal strength of RC columns with different structure size can be obtained by fitting the simulated data, as shown in Fig. 8.37. The results show that the size effect exists in the nominal strength of RC columns with different shear-span ratios (λ = 2.0, 3.0, and 4.0). However, the data points with small shear-span ratio are closer to the curve with a finite slope of −1/2 (i.e. linear elastic fracture mechanics, LEFM) than those with large shear-span ratio, which shows that the size effect on the columns with small shear-span ratio is more obvious.

8.3 Compression-Shear Failure in CFST Columns

Here in this Section, based on the meso-scale numerical model established in Chap. 3 and the verification of simulation approach, and the simulation results of square CFST columns with different sizes under combined lateral and axial loads are presented (Jin et al. 2020). Moreover, the influences of the shear-span ratio, axial compression ratio and steel ratio on the shear failure and size effect of square CFST columns are also provided.

8.3.1 Simulations on Failure of CFST Columns

Based on the established meso-scale numerical method in Chap. 3, a series of simulation tests were carried out by Jin et al. 2020 to investigate shear behavior and size effect of square CFST columns under combined lateral and axial loads. A 3D meso-scale numerical model for square CFST column is established, as depicted in Fig. 8.38a. The dark grey area, the yellow area, the light grey area and the blue area respectively denote the aggregate, the ITZ, the mortar matrix and the steel tube. All the meso-components of concrete were modelled by eight-node solid elements and steel tube was simulated by four-node shell elements. The boundary conditions and loading of CFST column were presented in Fig. 8.38b. The bottom surface of column is fixed to restrain all degrees of freedom and the top surface of column is free. Both of the axial compression load N (i.e. prescribed displacement) and the lateral loading P are respectively applied in the Y and X directions of the top surface of column. Table 8.8 gives the detailed mechanical parameters of meso-components of concrete and steel tube. The concrete cube compressive strength f_{cu} is 33.9 MPa and the yield strength f_y of steel is 297.9 MPa according to the test data in the test (Tomii and Sakino 1979).

The shear-span ratio *m*, axial compression ratio *n* and steel ratio *a* are the main parameters that affects the mechanical performances of CFST columns, and they can be respectively expressed as

$$m = \frac{L}{B} \tag{8.5}$$

$$n = \frac{N}{N_u} \tag{8.6}$$

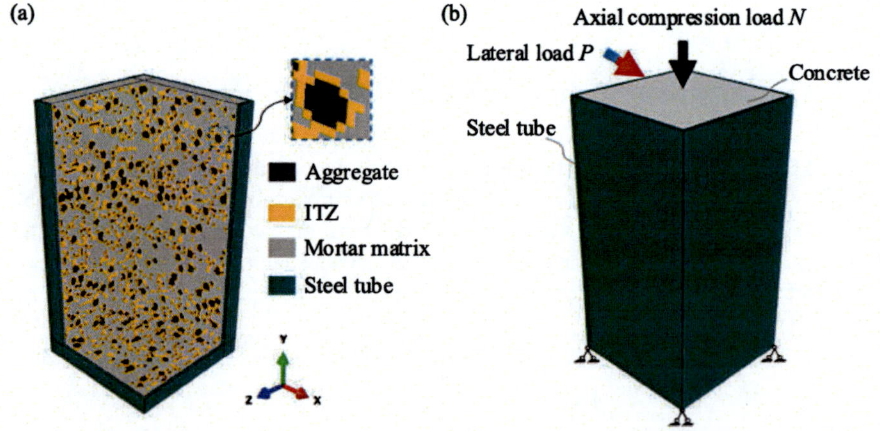

Fig. 8.38 **a** 3D meso-scale simulation model of square CFST column and **b** Loading schematic diagram

Table 8.8 Mechanical parameters of the meso-components of concrete and steel tube

	Aggregate	Mortar matrix	ITZ	Steel
Compressive strength σ_c/MPa	–	33.9[a]	30.8[b]	–
Tensile strength σ_t/MPa	–	3.39[a]	3.08[b]	–
Elastic modulus E/GPa	60	30[a]	24[b]	209[a]
Poisson's ratio ν	0.2	0.2	0.2	0.3[a]
Shear-Dilative angle ψ(°)	–	18	15	–
Yield strength f_y/MPa	–	–	–	297.9[a]

Note [a]Data are measured in the tests (Tomii and Sakino 1979); [b]Data are measured by repeated trial; and the other parameters are default value

$$a = \frac{A_s}{A_c} \tag{8.7}$$

where A_c and A_s are the cross-sectional areas of concrete and steel tube, respectively. Here, the shear-span ratio is taken as 1.0, 2.0 and 3.0 and the axial compression ratio is set as 0, 0.2, 0.4, 0.6 and 0.7. According to the design code for CFST columns in China (GB 50936 2014), the steel ratio should not be less than 0.036, thus here the steel ratio is selected as 0.05, 0.10, and 0.15. The plain concrete columns with the steel ratio of 0.0 is also designed for the comparison with CFST columns. The cross-sectional width B_c of concrete for the square CFST columns is expanded to 200, 400 and 800 mm on the basis of Tomii and Sakino's (1979) test. The scheme of parametric studies is given in Table 8.9, where the axial bearing capacity N_u of the CFST column is determined by the EC4 code (EN 1994-1-1 2004b). In Table 8.9, taking "S2-10-1-0" as an example, the detailed meaning of the specimen label is explained as following: "S2" refers to square column with the shear-span ratio of 2.0; the number "10" stands for the steel ratio of 10%; the number "1" means that the cross-sectional width of concrete is 100 mm; and the number "0" stands for the axial compression ratio of zero.

8.3.2 Simulated Results and Size Effect Analysis

1. *Failure mode*

Taking the specimen named "S2-10-4-4" as an example, the shear force P versus lateral deflection Δ relationships and the expanding process of cracks for the specimen are both presented in Fig. 8.39. It can be observed that, the P–Δ curve was close to the straight line at the elastic stage, and a few bending cracks first appeared on the bottom of specimen, as shown in point A. As lateral deformation increased, the specimen began to enter the elastic–plastic stage at point B. The steel tube in the compression area gradually developed plasticity as shown in Fig. 8.39b, and the

Table 8.9 Scheme of parametric design

Parameters	Specimen label	$B_c \times t \times L$ (mm)	m	n	a	N_u (kN)	N (kN)
Axial compression ratio	S2-10-1-0	100 × 2.5 × 200	2	0.0	0.10	860	0
	S2-10-1-2	100 × 2.5 × 200	2	0.2	0.10	860	172
	S2-10-1-4	100 × 2.5 × 200	2	0.4	0.10	860	344
	S2-10-1-6	100 × 2.5 × 200	2	0.6	0.10	860	516
	S2-10-1-7	100 × 2.5 × 200	2	0.7	0.10	860	688
	S2-10-2-0	200 × 4.9 × 400	2	0.0	0.10	3216	0
	S2-10-2-2	200 × 4.9 × 400	2	0.2	0.10	3216	643
	S2-10-2-4	200 × 4.9 × 400	2	0.4	0.10	3216	1286
	S2-10-2-6	200 × 4.9 × 400	2	0.6	0.10	3216	1929
	S2-10-2-7	200 × 4.9 × 400	2	0.7	0.10	3216	2251
	S2-10-4-0	400 × 9.8 × 800	2	0.0	0.10	12,864	0
	S2-10-4-2	400 × 9.8 × 800	2	0.2	0.10	12,864	2572
	S2-10-4-4	400 × 9.8 × 800	2	0.4	0.10	12,864	5145
	S2-10-4-6	400 × 9.8 × 800	2	0.6	0.10	12,864	7718
	S2-10-4-7	400 × 9.8 × 800	2	0.7	0.10	12,864	9004
	S2-10-8-0	800 × 19.5 × 1600	2	0.0	0.10	51,457	0
	S2-10-8-2	800 × 19.5 × 1600	2	0.2	0.10	51,457	10,291
	S2-10-8-4	800 × 19.5 × 1600	2	0.4	0.10	51,457	20,582
	S2-10-8-6	800 × 19.5 × 1600	2	0.6	0.10	51,457	30,874
	S2-10-8-7	800 × 19.5 × 1600	2	0.7	0.10	51,457	36,019
Steel ratio	S2-0-1-4	100 × 0 × 200	2	0.4	0.0	268	107
	S2-0-2-4	200 × 0 × 400	2	0.4	0.0	1071	428
	S2-0-4-4	400 × 0 × 800	2	0.4	0.0	4285	1714
	S2-0-8-4	800 × 0 × 1600	2	0.4	0.0	23,539	9416
	S2-5-1-4	100 × 1.3 × 200	2	0.4	0.05	535	214
	S2-5-2-4	200 × 2.5 × 400	2	0.4	0.05	2143	857
	S2-5-4-4	400 × 4.9 × 800	2	0.4	0.05	8572	3429
	S2-5-8-4	800 × 9.8 × 1600	2	0.4	0.05	34,297	13,719
	S2-15-1-4	100 × 3.6 × 200	2	0.4	0.15	1070	428
	S2-15-2-4	200 × 7.2 × 400	2	0.4	0.15	4287	1715
	S2-15-4-4	400 × 14.5 × 800	2	0.4	0.15	17,152	6861
	S2-15-8-4	800 × 28.9 × 1600	2	0.4	0.15	68,615	27,446

(continued)

Table 8.9 (continued)

Parameters	Specimen label	$B_c \times t \times L$ (mm)	m	n	a	N_u (kN)	N (kN)
Shear-span ratio	S1-10-1-4	100 × 2.5 × 100	1	0.4	0.10	860	344
	S1-10-2-4	200 × 4.9 × 200	1	0.4	0.10	3216	1286
	S1-10-4-4	400 × 9.8 × 400	1	0.4	0.10	12,864	5145
	S1-10-8-4	800 × 19.5 × 800	1	0.4	0.10	51,457	20,582
	S3-10-1-4	100 × 2.5 × 300	3	0.4	0.10	860	344
	S3-10-2-4	200 × 4.9 × 600	3	0.4	0.10	3216	1286
	S3-10-4-4	400 × 9.8 × 1200	3	0.4	0.10	12,864	5145
	S3-10-8-4	800 × 19.5 × 2400	3	0.4	0.10	51,457	20,582

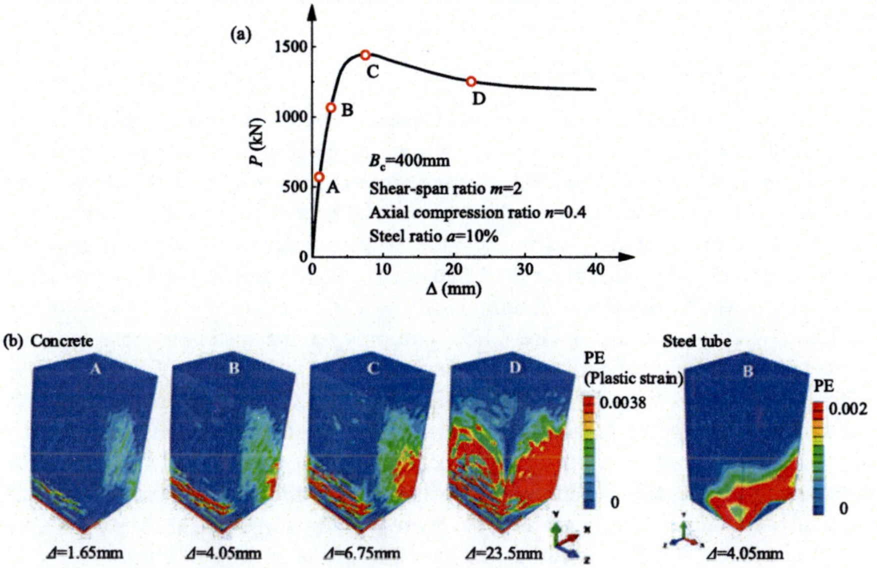

Fig. 8.39 Failure process of the square CFST column (S2-10-4-4): **a** Shear force versus lateral deflection relationships and **b** Crack propagation process

rigidity of specimen tended to decline. With the load further increasing, both the number and width of bending cracks increased accompanied by the expansion of tension area of concrete. At point C, the shear force reached the peak value. After point C, the P–Δ curve appeared declining segment as shown in point D, and last the concrete in the tensile zone occurred tensile failure and concrete in the compression zone happened local buckling.

The failure pattern of CFST columns usually is determined by the shear-span ratio, which has been proved by lots of researches (Tomii and Sakino 1979; Xiao et al. 2005; Qian et al. 2007; Ye et al. 2016; Han et al. 2008). The ultimate failure modes of square CFST columns with different shear-span ratios ($m = 1.0$, 2.0 and 3.0) under combined lateral and axial loads are shown in Fig. 8.40. As can be observed that the CFST columns with the shear-span ratio of 1.0 were dominated by shear failure. The concrete in compression area was crushed and failing in diagonal compression failure, which belongs to brittle failure. When the shear-span ratio increased to 2.0 and 3.0, the path of force transfer was changed, the horizontal flexural cracks of concrete in tensile area increased, indicating that the failure mode of CFST columns was from brittle shear failure to ductile bending failure. The above simulated results have a good agreement with the experimental results (Tomii and Sakino 1979). Furthermore, under the same shear-span ratio, the CFST columns having different structural sizes exhibited the similar final failure mode, besides the more obvious local bucking for the large-size members.

2. *Nominal shear stress-rotation curves*

In order to analyze the influence of each parameter investigated in this paper on size effect of square CFST columns under lateral shear and axial compression, the nominal shear stress τ-rotation R curves are given in Figs. 8.41, 8.42 and 8.43. Here, the nominal shear stress τ is defined by the ratio of lateral load P and cross-sectional area A of column, and the maximum value of the nominal shear stress is regarded as the nominal shear strength τ_u. Moreover, the rotation R is the ratio of lateral deflection Δ and the length of column L.

The τ-R curves of the square CFST columns having different cross-sectional widths under different shear-span ratios are presented in Fig. 8.41. When the shear-span ratio increases from 1.0 to 3.0, the failure mode of CFST columns changes from shear failure to bending failure, the nominal shear strength decreases significantly, but the ductility improves. Under same shear-span ratio, the nominal shear strengths decreases with the cross-sectional width increasing, showing obvious size effect, especially for smaller shear-span ratios. Moreover, the larger the sectional size, the less the ductility of columns and the steeper the softening stage of curves.

The τ-R curves of CFST columns under different steel ratios are plotted in Fig. 8.42. Without the restraint of steel tube, the nominal shear strength and ductility performance of plain concrete columns are obviously weaker than that of CFST columns. With the increase of steel ratio, the restraint effect of steel tube is enhanced, so that the nominal shear strength and ductility of CFST columns are improved. Under the same steel ratio, both of the nominal shear strength and ductility of columns decrease as the cross-sectional size increases, showing the size effect behavior.

Figure 8.43 gives the τ-R curves of CFST columns under five kinds of axial compression ratios. When $n = 0.0$, i.e. the column is subjected to pure shearing, the τ-R curves have no falling stage and the columns exhibit good ductile property. With the increase of axial compression ratio, the ductility of columns is obviously reduced and the falling stage of curves becomes steeper. Moreover, size effect on the

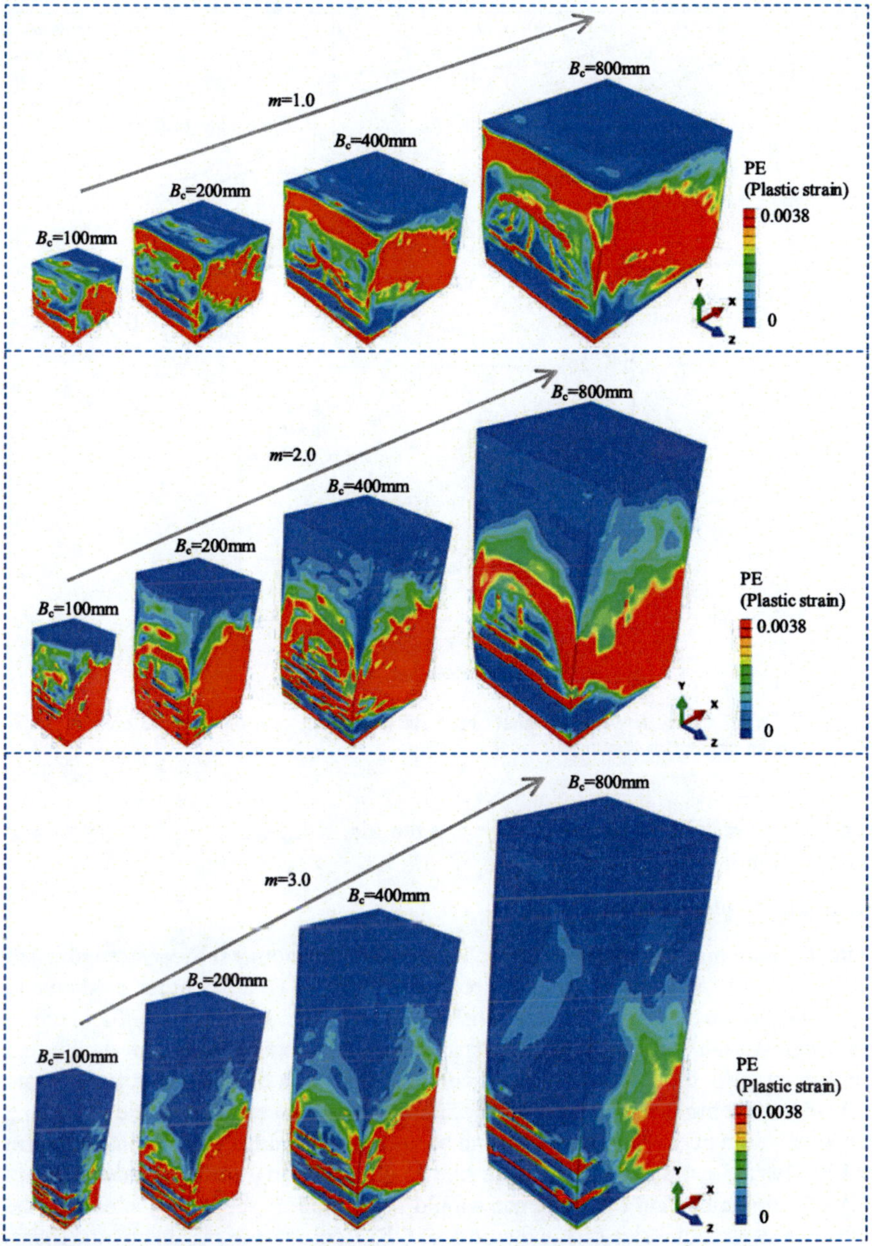

Fig. 8.40 Final failure modes of the square CFST columns with different shear-span ratios and structural sizes

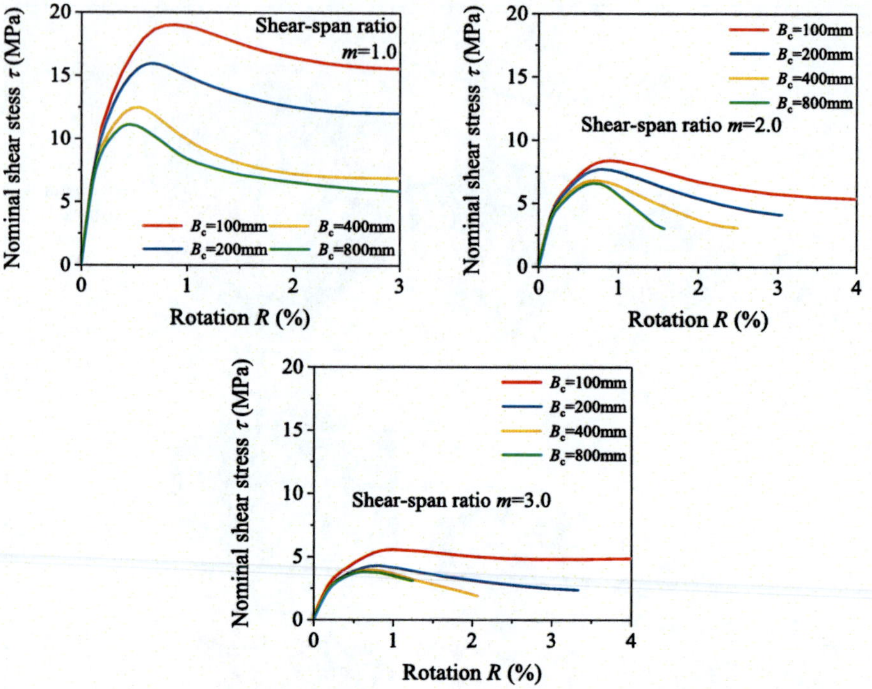

Fig. 8.41 Relationship between nominal shear stress and rotation of concrete column under different shear-span ratios

nominal shear strength can be observed in the Fig. 8.43, especially for larger axial compression ratios.

3. *Ductility analysis*

Here, the softening rate K (absolute value) between the nominal shear strength τ_u to $0.85\tau_u$ in the τ–R curve is used to reflect the ductility of CFST column, as shown in Fig. 8.44. Figure 8.44 gives the relationships between the softening rate K and cross-sectional width under different shear-span ratios. As it can be seen, the smaller the shear-span ratio, the less the brittleness of columns and the larger the softening rate K. Under the same shear-span ratio, the softening rate K increases with the increasing sectional width of columns and this trend becomes slow gradually, especially when $m = 1.0$. This indicates that the structure size has an apparent influence on the ductility of CFST columns, and the influence should be limited. It should be noted that the influence of structure size on the ductility of CFST columns is affected by the shear-span ratio, i.e. this influence is no longer obvious with the increase of shear-span ratio.

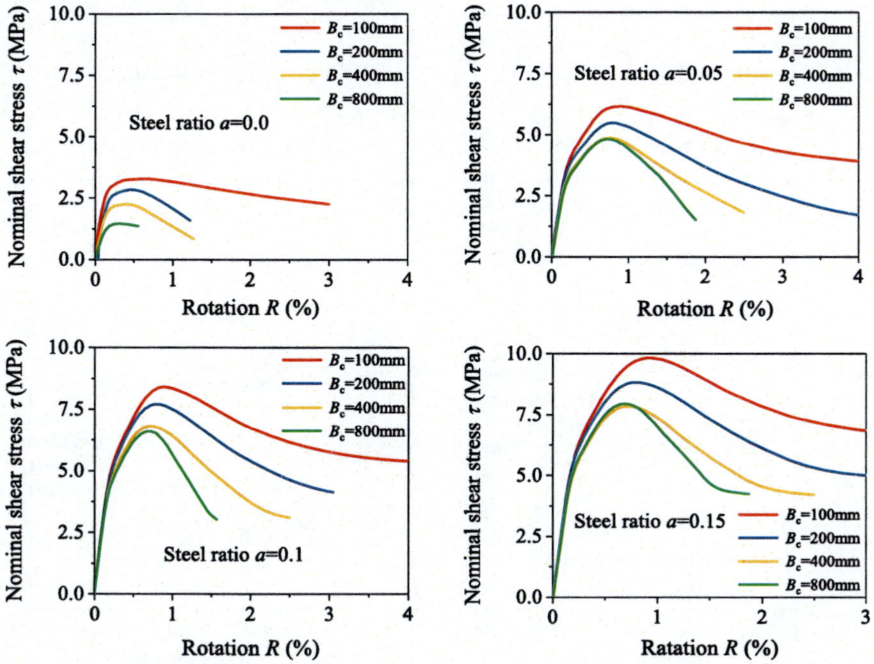

Fig. 8.42 Relationship between nominal shear stress and rotation of concrete column under different steel ratios

4. *Size effect on nominal shear strength*

The main aim of this section is to explain the influence of the parameters investigated in the study on size effect in the nominal shear strength of CFST columns. Figure 8.45 gives the nominal shear strength versus cross-sectional width relationships and corresponding strength reduction percentage for the square CFST columns under different shear-span ratios, axial compression ratios and steel ratios. The strength reduction percentage p is expressed by

$$p = \frac{|\tau_o - \tau_{100}|}{\tau_{100}} \times 100\% \tag{8.8}$$

in which, τ_{100} and τ_o are the nominal shear strength of columns with cross-sectional width of 100 mm and with other cross-sectional width under the same condition.

As shown in Fig. 8.45a, the nominal shear strength obviously decreases with the increase of shear-span ratio. When the shear-span ratio is smaller, the CFST columns occur brittle failure, size effect on the nominal shear strength is obvious. Specifically, when $m = 1.0$, the nominal shear strength of columns with size of 800 mm is about 41% lower than that of columns with size of 100 mm. While when the shear-span ratio increases to 2.0 and 3.0, the failure mode of CFST columns gradually changes

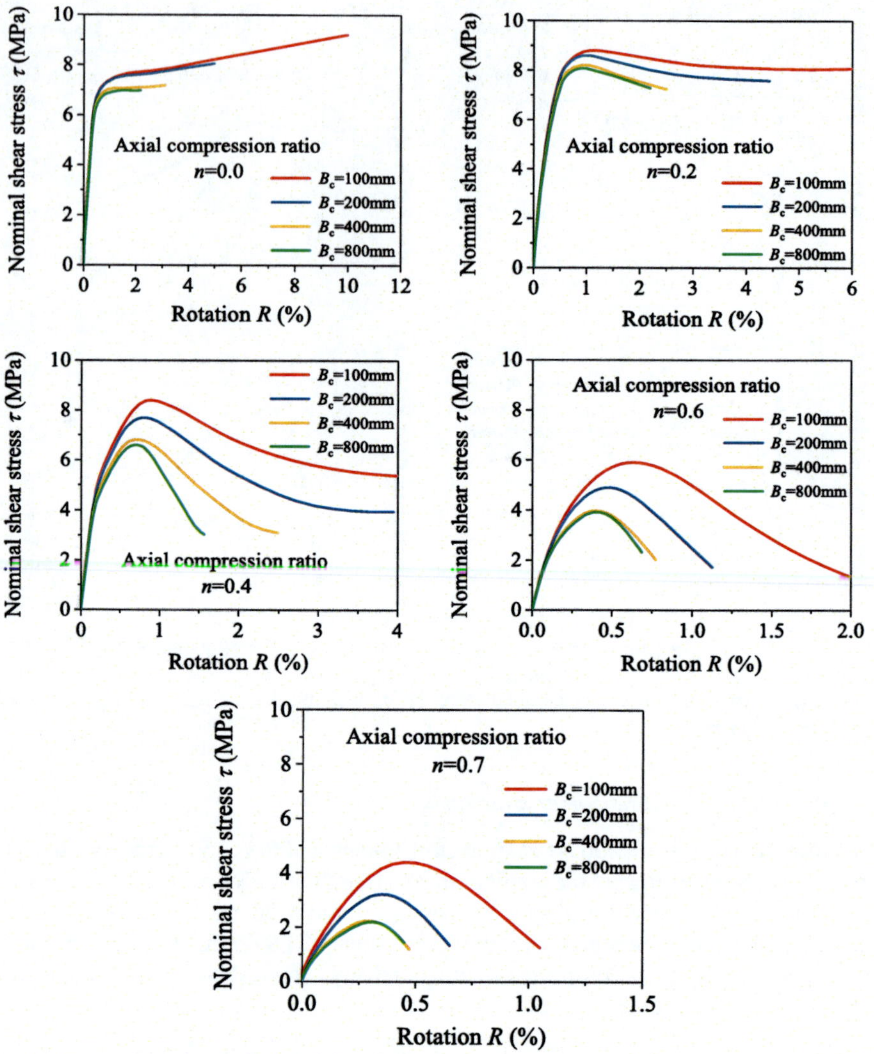

Fig. 8.43 Relationship between nominal shear stress and rotation of concrete column under different axial compression ratios

to ductile failure, the size effect behavior becomes not significant. This indicates that size effect on the nominal shear strength is obviously weakened with the increase of shear-span ratio. This phenomenon can also be observed from Fig. 8.46a. In addition, there is little difference in the strength of columns with size of 400 and 800 mm, illustrating the effect of structure size is limited.

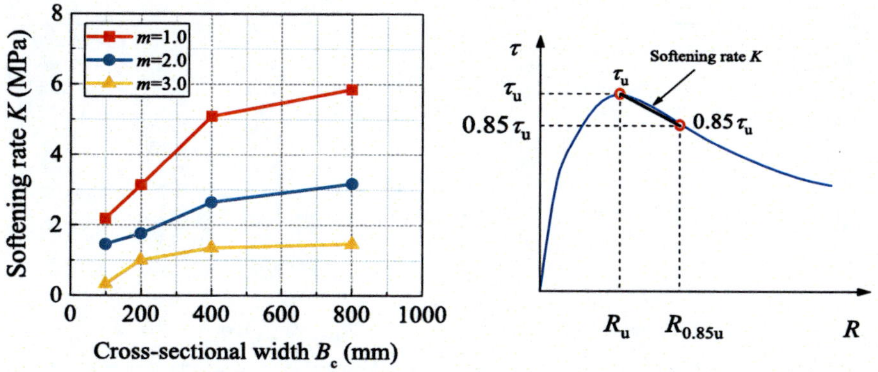

Fig. 8.44 Relationship between softening rate and cross-sectional width of square CFST columns

In Fig. 8.45b, the nominal shear strength increases with the increasing axial load level of less than 0.4, while that gradually decreases with the increasing axial compression ratio of larger than 0.4. This is consistent with the results described in (Tomii and Sakino 1979; Han et al. 2008). Additionally, for $n \le 0.4$, the CFST columns perform good ductility, the influence of size effect on nominal shear strength is not obvious and can be ignored. However, for $n > 0.4$, the CFST columns show stronger brittleness, the size effect becomes to be more obvious with the increase of axial compression ratio. For instance, when $n = 0.7$, the nominal shear strength of columns with size of 800 mm is approximately 50% less than that of columns with size of 100 mm.

The influence of steel ratio on size effect on the nominal shear strength is plotted in Fig. 8.45c. Compared with the plain concrete columns, the nominal shear strength of CFST columns improves obviously due to the restraint effect provided by steel tube. However, as the steel ratio increases, the trend that the nominal shear strength increases with the increase of steel ratio is gradually slowing down. Moreover, compared with the plain concrete columns, the ductile performance of CFST columns improves, size effect on the nominal shear strength is weakened significantly, but this weakening effect is not obvious when the steel ratio increases from 0.05 to 0.15. That is to say, the size effect behavior always exists at any steel ratio, and even for a greater steel ratio (e.g. 0.15) it cannot be suppressed completely, as presented in Fig. 8.45c. That is consistent with the conclusion proposed by Yu and Bažant (2011).

For size effect in shear failure, a size effect formula for the shear strength of concrete components has been developed by Bažant and Kim (1984), and the details of which can refer to Chap. 6. The two empirical factors i.e. V_0 and D_0 are determined by regression analysis based on the simulation results, as listed in Table 8.10. The comparison between Bažant's SEL (Bažant and Kim 1984) and simulation results of columns under different shear-span ratios, axial compression ratios and steel ratios is shown in Fig. 8.46. For comparisons, both of the plastic strength criteria (horizontal

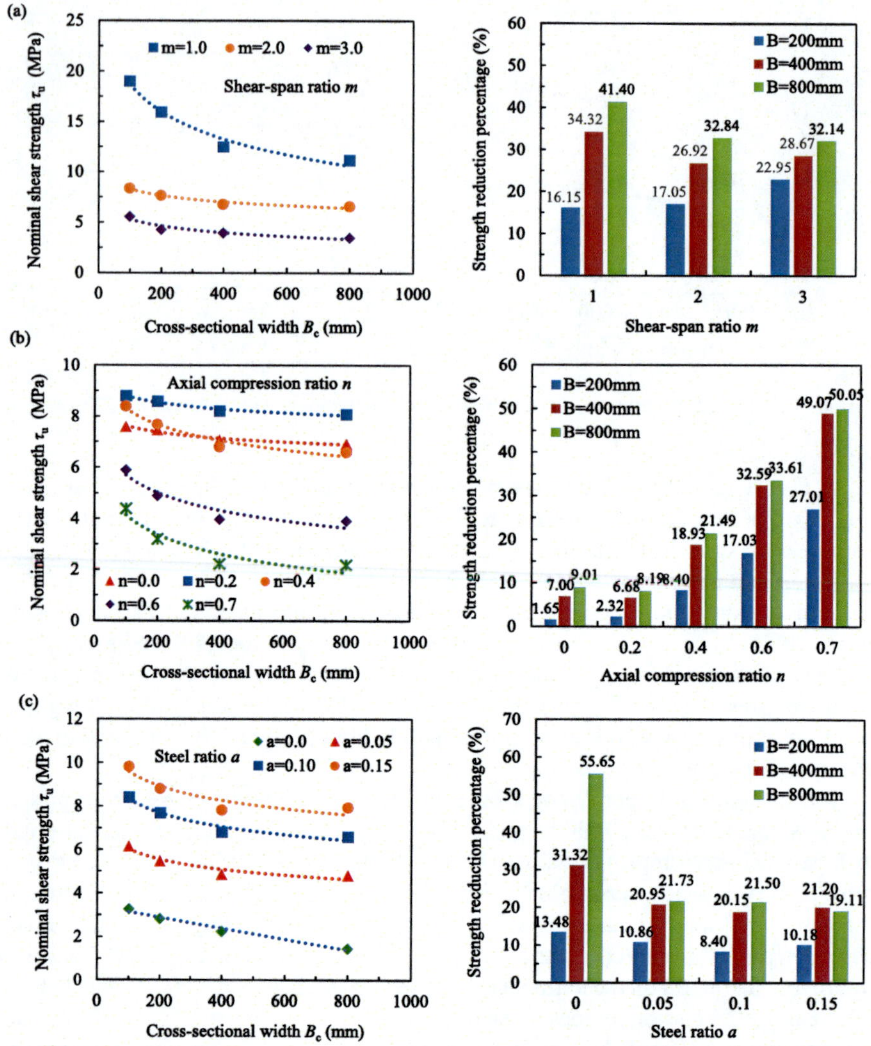

Fig. 8.45 Nominal shear strength versus cross-sectional width, and strength reduction percentage of the square CFST columns: **a** Under different shear-span ratios; **b** Under different axial compression ratios; and **c** Under different steel ratios

line, for plastic materials) and the linear elastic fracture mechanics theory (LEFM, for pure brittle materials) with a finite slope of −1/2 were also plotted in Fig. 8.46. For the cases having different shear-span ratios, axial compression ratios and steel ratios, the magnitudes of the correlation coefficients R^2 between the simulated data and Bažant's SEL are respectively 0.89, 0.87 and 0.87, indicating that size effect on nominal shear strength can be well described by Bažant's SEL.

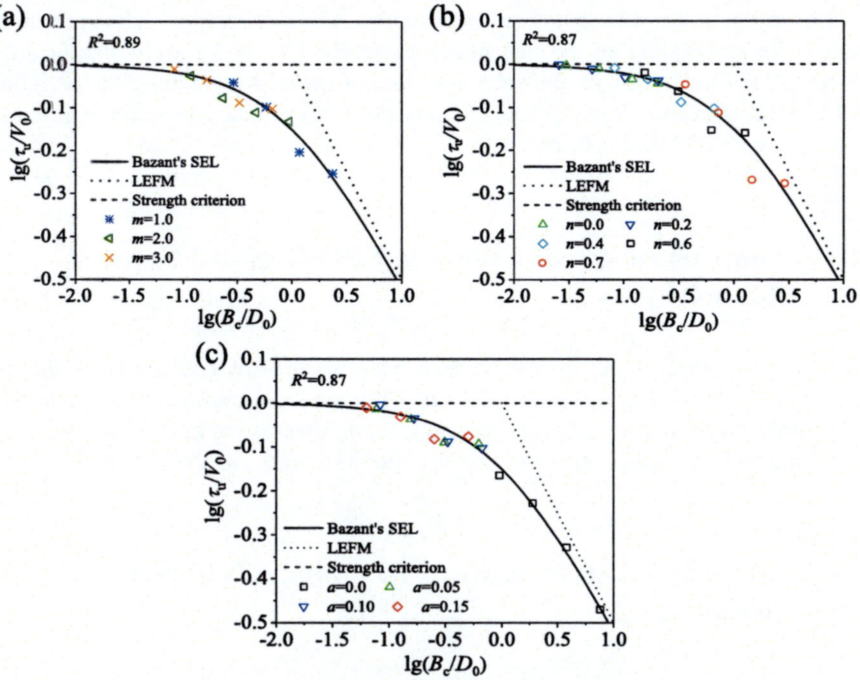

Fig. 8.46 Size effect fitting of nominal shear strength of square CFST columns: **a** Under different shear-span ratios; **b** Under different axial compression ratios; and **c** Under different steel ratios

Table 8.10 Parameters of Bažant size effect law for CFST columns with different shear-span ratios

Specimen label	m	τ_u (MPa)	A	C	V_0	D_0
S1-10-1-4	1.0	19.02	9e-5	0.0288	19.98	320
S1-10-2-4		15.95				
S1-10-4-4		12.50				
S1-10-8-4		11.15				
S2-10-1-4	2.0	8.42	0.0001	0.1647	8.35	1647
S2-10-2-4		7.71				
S2-10-4-4		6.83				
S2-10-8-4		6.61				
S3-10-1-4	3.0	5.59	0.0005	0.4341	5.15	1754
S3-10-2-4		4.31				
S3-10-4-4		3.99				
S3-10-8-4		3.79				

Nevertheless, for different shear-span ratios, axial compression ratios and steel ratios, the magnitudes of the two fitting parameters V_0 and D_0 are significantly distinct. This indicates that Bažant's SEL cannot quantitatively describe the influences of shear-span ratio, axial load ratio and steel ratio on the size effect in nominal shear strength of CFST columns.

8.4 Compression-Shear Failure in FRP-Confined Concrete Columns

This section presents the simulation results on a series of geometrically similar concrete columns confined with CFRP with different cross-sectional sizes under the horizontal cyclic load. The influences of axial compression ratio and the constraint ratio of CFRP on the size effect of nominal shear strength are also reported.

8.4.1 Simulations on Failure of FRP-Confined Concrete Columns

According to the approach recoded in Chap. 3, a 3D meso-scale geometrical model of FRP-confined concrete columns was established, as presented in Fig. 8.47. Fixed constraint was set at the bottom surface of the columns. The predetermined axial load was firstly applied to the top surface of columns, then the horizontal cyclic load was applied to the columns. The horizontal load is controlled by displacement as shown in Fig. 8.48, where Δy is the yield displacement of square RC columns confined with CFRP. The main mechanical parameters for the three meso-components of concrete and CFRP materials are listed in Tables 8.11 and 8.12, respectively.

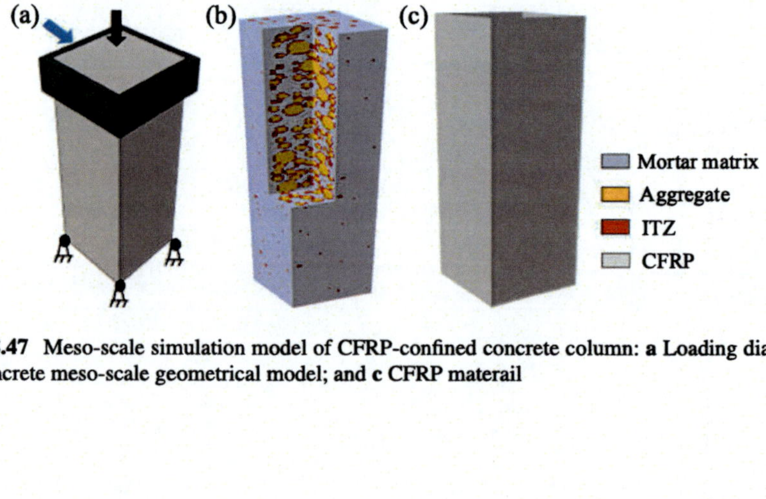

Fig. 8.47 Meso-scale simulation model of CFRP-confined concrete column: **a** Loading diagram; **b** Concrete meso-scale geometrical model; and **c** CFRP materail

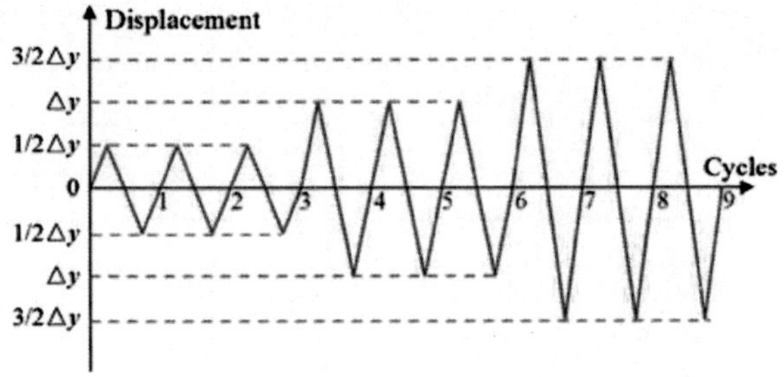

Fig. 8.48 Scheme of loading

Table 8.11 Mechanical parameters of the meso-components of concrete and reinforcing bars utilized

Materials	Aggregate	Mortar matrix	ITZ	Longitudinal steel bars	Stirrup
Compressive strength σ_c (MPa)	–	53.1[a]	48.2[b]	–	–
Tensile strength σ_t (MPa)	–	5.2[a]	4.7[b]	–	–
Elastic modulus E (GPa)	60[a]	42.5[a]	38.4[b]	210[a]	200[a]
Poisson's ratio ν	0.2[a]	0.2[a]	0.2[b]	0.3[a]	0.3[a]
Yield strength (MPa)	–	–	–	380.71[a]	419.09[a]
Reinforcement ratio (%)	–	–	–	1[a]	0.5[a]

Note [a]Data are obtained from the tests and [b]data are parameters by repeated trial

Table 8.12 Mechanical properties of CFRP

Variables	Elasticity modulus E_f (GPa)	Ultimate tensile strength f_f (MPa)	Ultimate rupture	Single layer thickness t (mm)
CFRP	240	3461	1.61%	0.167

To study the influence of constraint ratio and the axial compression ratio on the shear strength and size effect of CFRP-confined concrete columns, the cyclic loading tests of CFRP-confined concrete columns were simulated. The geometrical parameters of the CFRP-confined concrete columns with different structure size as shown in Table 8.13 were utilized.

Table 8.13 Geometrical parameters of the larger sized CFRP-concrete columns

Specimen name	Structural size B (mm)	Height h (mm)	Axial compression ratio n	Constraint ratio ρ_f
SC-200-0.4-1	200	600	0.4	0
SC-200-0.4-2	200	600	0.4	0.334%
SC-200-0.4-3	200	600	0.4	1.336%
SC-200-0.6-1	200	600	0.6	0
SC-200-0.6-2	200	600	0.6	0.334%
SC-200-0.6-3	200	600	0.6	1.336%
SC-400-0.4-1	400	1200	0.4	0
SC-400-0.4-2	400	1200	0.4	0.334%
SC-400-0.4-3	400	1200	0.4	1.336%
SC-400-0.6-1	400	1200	0.6	0
SC-400-0.6-2	400	1200	0.6	0.334%
SC-400-0.6-3	400	1200	0.6	1.336%
SC-800-0.4-1	800	2400	0.4	0
SC-800-0.4-2	800	2400	0.4	0.334%
SC-800-0.4-3	800	2400	0.4	1.336%
SC-800-0.6-1	800	2400	0.6	0
SC-800-0.6-2	800	2400	0.6	0.334%
SC-800-0.6-3	800	2400	0.6	1.336%

Note The large-size, medium-size and small-size of columns are defined by cross-sectional widths of 800 mm, 400 mm and 200 mm, respectively. As shown in Table 8.13, take "SC-200-0.4-1" for example, the first letter "SC" represent the square concrete column; the first number "200" represents the columns with the cross-sectional width of 200 mm; and the second number "0.4" describes that the axial compression ratio is 0.4. The last number from "1" to "3" represent three different constraint ratios

8.4.2 Simulated Results and Size Effect Analysis

1. *Failure pattern*

The failure patterns of the concrete columns confined with CFRP are presented in Fig. 8.49. As can be seen from Fig. 8.49, first cracks appeared in the plastic hinge area of the columns and gradually developed with the increase of the load. Finally, the bottom of the columns presents an "X" type crack. The concrete at the root of the columns crushed and the longitudinal reinforcements yielded. The structure size had no significant effect on the failure mode of the concrete columns confined with CFRP.

2. *Load-displacement curves*

Figure 8.50 shows the hysteretic curves of concrete columns confined with CFRP. In the initial stage of loading, the hysteretic curve of the column was basically in

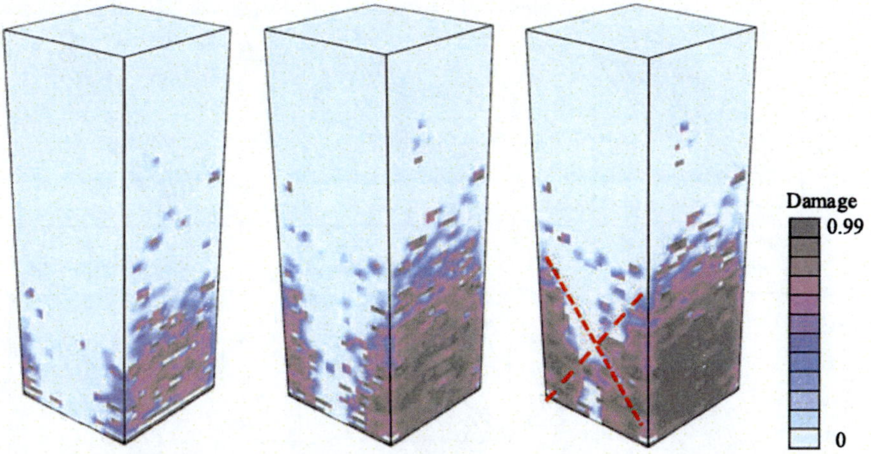

Fig. 8.49 Failure modes of concrte columns confined with CFRP

the form of a straight line and the area surrounded by the hysteretic loop was very small. The rigidity of the column was unchanged basically and a linear relationship was presented between the displacement and load. When the load achieved the yield load, it no longer increased linearly with the increase of displacement. In addition, the hysteresis curve presented significant pinch phenomenon. The bearing capacity and ultimate displacement of concrete columns confined with CFRP having three different structure sizes were improved. In addition, the area surrounded by the hysteretic curve and the energy absorbed increased, which shows that the seismic performance of the confined concrete column was improved compared with unconfined column.

3. *Skeleton curve*

Figure 8.51 shows the skeleton curves of concrete columns confined by CFRP. The skeleton curve dropped rapidly after reaching the peak load for the unconfined concrete columns, which indicates that the columns presented obvious brittle failure. The axial compression ratio had little influence on the ultimate bearing capacity and ductility of the unconfined concrete columns with the structure size of 200 mm. For the unconfined columns with the structure sizes of 400 or 800 mm, the smaller the axial compression ratio was, the larger the ultimate bearing capacity was. The form of skeleton curve for columns confined with CFRP was similar to that of unconfined columns. Compared with the unconfined columns, the bearing capacity and deformation capacity of confined columns had been improved.

4. *Ductility analysis*

In general, displacement ductility coefficient μ is used to reflect the ductility capacity of members and can be expressed as $\mu = \Delta_\mu/\Delta_y$, where Δ_μ is the displacement

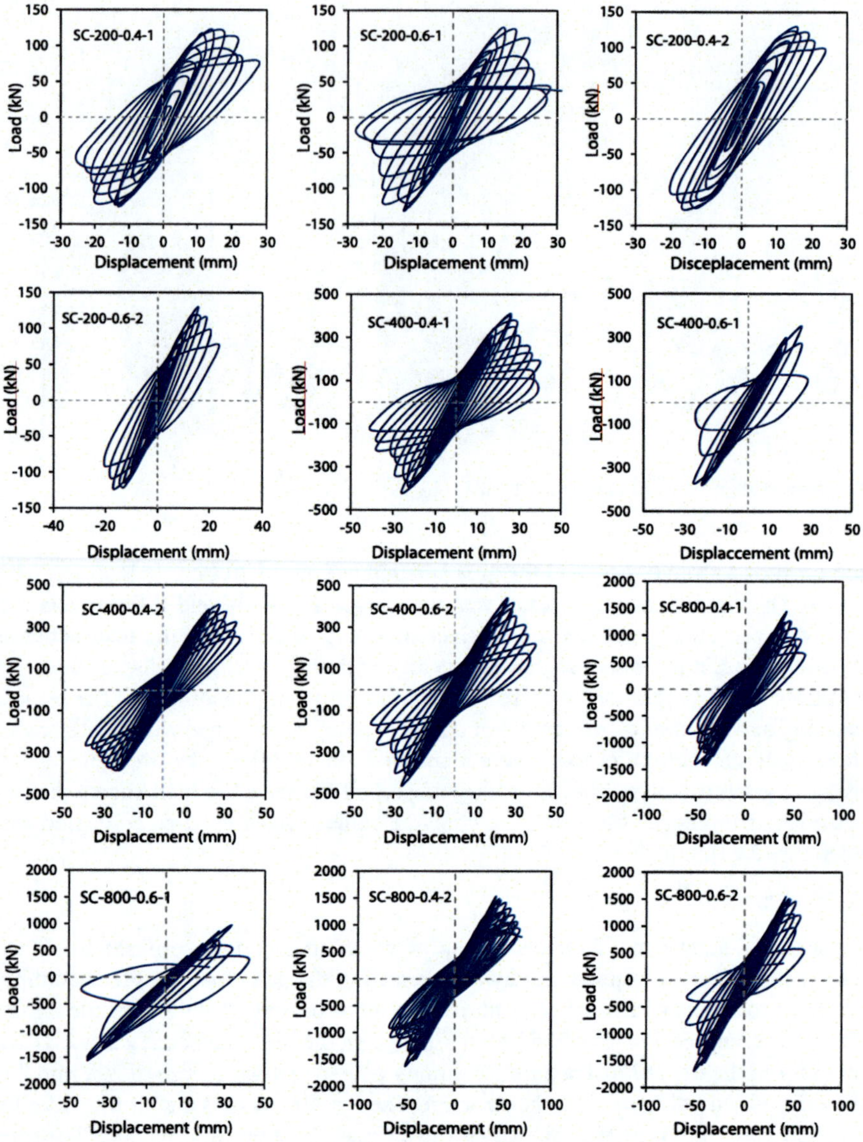

Fig. 8.50 Hysteretic curves of concrete columns confined with CFRP

corresponding to the component at $P_u = 0.85\ P_{max}$; Δ_y is the displacement corresponding to the yield load of the member, and can be calculated by the energy method.

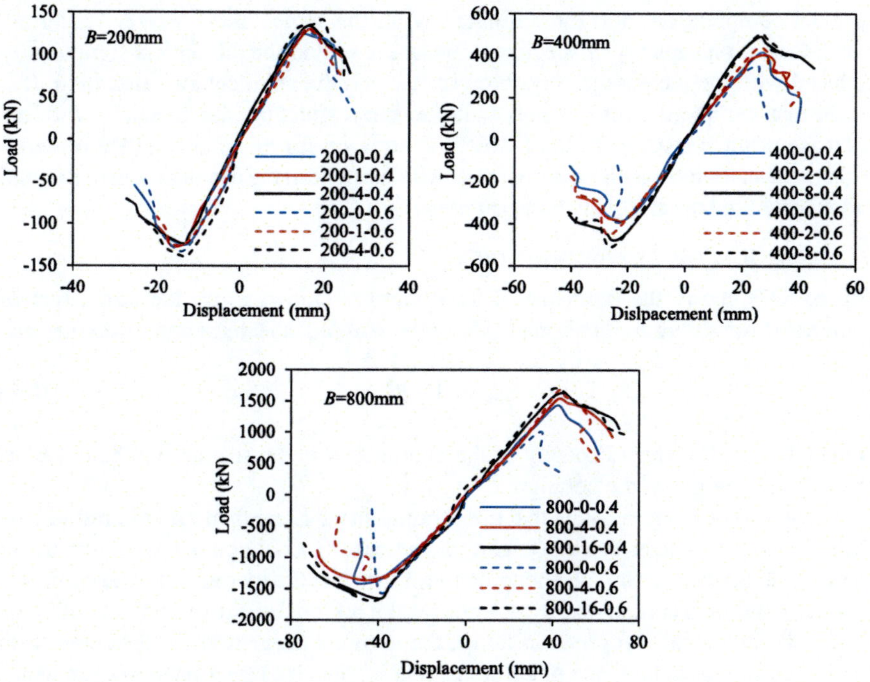

Fig. 8.51 Skeleton curves of concrete columns confined with CFRP

Figure 8.52 shows the relationship between the displacement ductility coefficient and the structure size of columns. It can be seen that the displacement ductility coefficient decreased and this decreasing trend decreased with the increase of structure size. In addition, the displacement ductility coefficient decreases with the increase

Fig. 8.52 The relationship between displacement ductility coefficient and structure size of columns

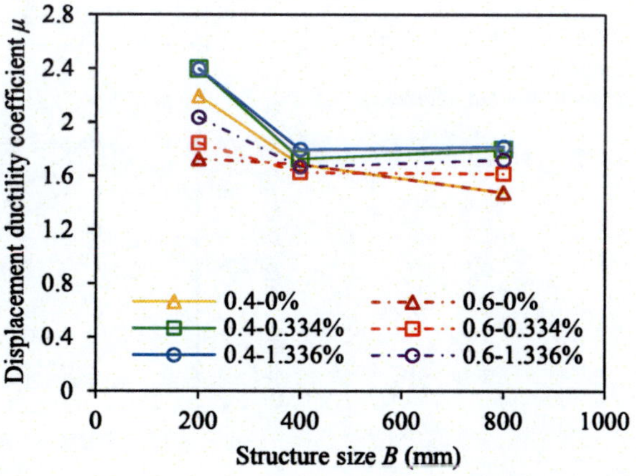

of axial compression ratio for columns having the same structure size. Therefore, the deformation capacity of specimens is weakened and the ductility is significantly reduced with the increase of the structure size and the axial compression ratio. It is noted that the ductility of columns with the same structure size increases with the increase of the constraint ratio of CFRP. The reason for this is that CFRP provides effective constraints on the core concrete, which can limit the development of internal cracks and the lateral expansion of columns.

5. *Influence of axial compression ratio*

Figure 8.53 shows the influence of axial compression ratio on the size effect of nominal shear strength of columns, where the nominal shear strength is defined as

$$\tau_u = V/Bh_0 \tag{8.9}$$

in which, V is the shear capacity of the columns; B is the section width; and $h0$ is the effective height of the columns.

It can be seen from Fig. 8.54 that the nominal shear strength of the columns having different structure sizes increases with the increase of axial compression ratio under the same constraint ratio. The phenomenon shows that the increase of axial compression ratio within a certain range can improve the nominal shear strength of columns. Under the two axial compression ratios, the nominal shear strength decreases with the increase of structure size of columns, that is, the size effect phenomenon exists in the nominal shear strength of columns confined with CFRP.

Herein, the unconfined concrete columns with axial compression ratio of 0.2, 0.4 and 0.6 are used to analyze the influence of axial compression ratio on the size effect of shear strength. Figure 8.54a presents the comparison of the simulation data points (i.e. the shear strength of circular plain concrete columns with different sizes), along with the classic SEL, the plastic strength criteria (horizontal line, for plastic materials) and the linear elastic fracture mechanics (LEFM) having a finite slope of −1/2 (for pure brittle materials). In Fig. 8.54a, as the axial compression ratio is 0.2

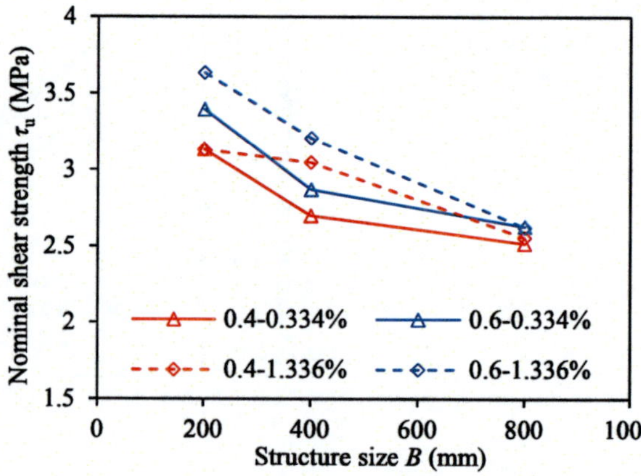

Fig. 8.53 Influence of axial compression ratio on nominal strength of columns

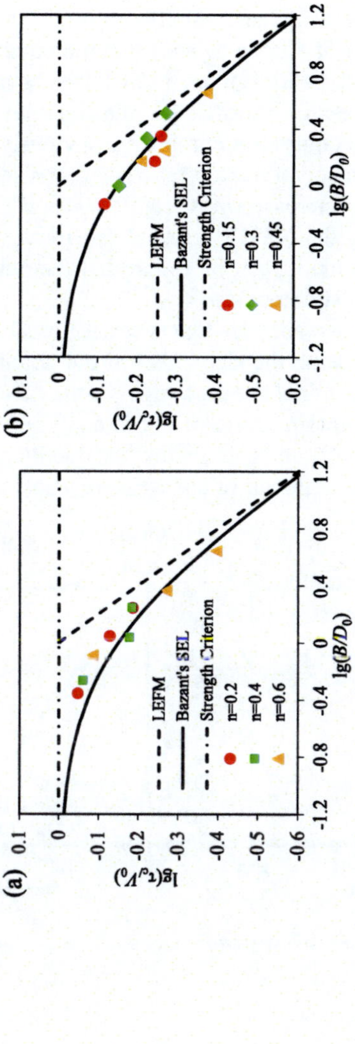

Fig. 8.54 Size effect fitting of nominal shear strength of plain concrete columns: **a** Simulation results and **b** Experimental results in Cui (2018)

or 0.4, the data points of columns with the same structure size are basically coincide. The data points of columns with the axial compression ratio of 0.6 are closed to the curve with a finite slope of −1/2. When the axial compression ratio is less than 0.4, the influence of the axial compression ratio on the size effect of shear strength is small and can be ignored. In addition, the same conclusion can be drawn by analyzing the experimental data in Cui (2018) as shown in Fig. 8.54b.

6. *Influence of constraint ratio*

The relationship between the simulated nominal strength and structure size of columns is presented in Fig. 8.55. It can be seen that the nominal shear strength of concrete columns confined with CFRP is improved with the increase of the constraint ratio. When the axial compression ratio is 0.4, the nominal shear strength with medium-sized and large-sized columns are improves obviously with the increase of constraint ratio. When the axial compression ratio is 0.6, the nominal shear strength of the medium-sized columns increases most significantly with the increase of the constraint ratio, which shows that CFRP material can improve the shear strength of concrete columns. In addition, it can be found that the nominal shear strength decreases with the increase of structure size.

It can be seen from the above analysis that when the axial compression ratio is 0.4, it has no obvious influence on the size effect of the nominal shear strength of the columns. Therefore, the nominal shear strength of the CFRP confined concrete columns with the axial compression ratio of 0.4 is analyzed using the regression method, as shown in Fig. 8.56. It can be found that the ductility of concrete columns is enhanced and the size effect behavior of concrete are weakened with the increase of constraint ratio.

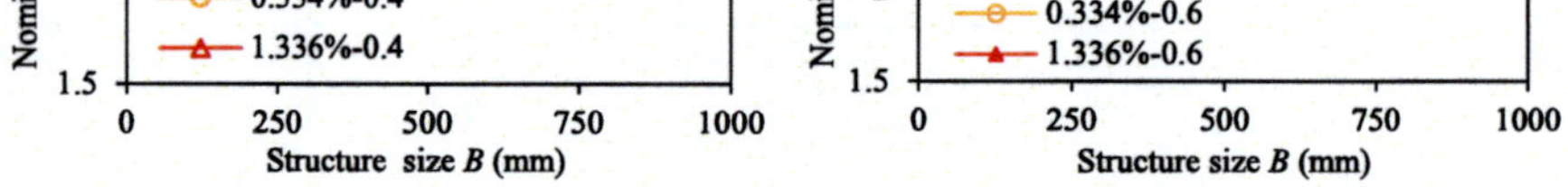

Fig. 8.55 Nominal shear strength vs structure size

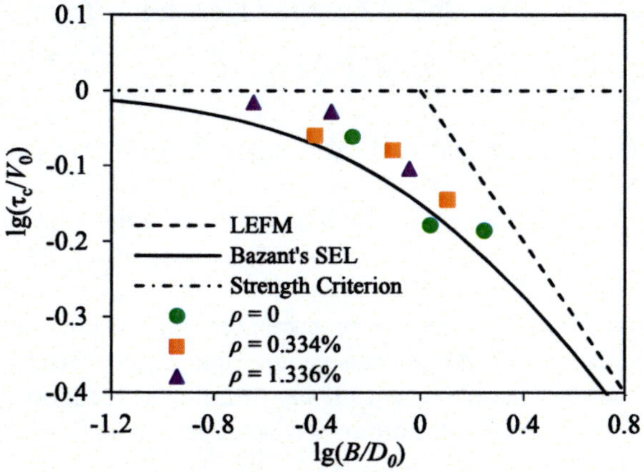

Fig. 8.56 Size effect fitting of nominal shear strength of CFRP-confined concrete columns ($n = 0.4$)

8.5 Size Effect Law for Compression-Shear Failure of Concrete Columns

In this section, the size effect law for describing the nominal shear strength of concrete columns would be presented.

8.5.1 Basic Assumptions

The shear capacity V_u of concrete columns consists of three aspects, i.e. (1) shear capacity V_c of concrete; (2) shear capacity V_s of stirrups, steel tube or FRP; and (3) shear capacity V_n provided by axial compression load. The total shear capacity V_u of concrete columns can be expressed as

$$V_u = V_c + V_s + V_n \tag{8.10}$$

Obviously, V_c provided by the plain concrete should be affected by the size effect, while V_s and V_n can be regfarded to be independent of size effect. According to the above experimental and simulated results in Sects. 8.1–8.4, it can be concluded that the shear-span ratio, axial compression ratio and constraint coefficient present significant influences on the shear failure and size effect of concrete columns. In detail, as shear-span ratio increases, the nominal shear strength of concrete columns decreases and ductility increases. The size effect in shear would be weakened as the shear-span ratio increases since the corresponding failure turns from brittle to ductile. The axial compression ratio also has an influence on size effect on nominal shear strength i.e. as axial compression ratio increases, the size effect is more obvious. In addition, the size effect on nominal shear strength can be weakened but cannot be suppressed as the constraint coefficient increases.

8.5.2 Size Effect Law for Nominal Shear Strength

1. *Shear capacity*

Based on the analysis in Sect. 8.5.1, for the CFST columns, the total shear capacity V_u can be defined as

$$V_u = V_c + V_{st} + V_n \tag{8.11}$$

in which, V_{st} is the shear capacity provided by the steel tube. The contribution by the steel tube, axial compressive load and plain concrete on the shear capacity of CFST columns would be respectively determined as following.

For the CFRP confined RC columns can be described as

$$V_u = V_{Rc} + V_f \tag{8.12a}$$

$$V_{Rc} = V_c + f_{yv}\frac{A_{sv}}{s}h_0 + 0.07N \tag{8.12b}$$

$$\mathrm{V_f} = \rho_f \varepsilon_{fe} E_f bhcot\theta \tag{8.12c}$$

in which, the calculation method of shear capacity of RC columns V_{Rc} is taken from the code for design of concrete structures in China (GB50010-2010), V_f is the shear capacity provided by FRP and is calculated by the formula in the reference (Ye et al. 2000); f_{yv} is the design value of stirrup tensile strength; A_{sv} is the cross-sectional area of stirrups; s is the spacing of stirrups; N is the design value of axial pressure; $\rho_f$is the constraint ratio of CFRP; ε_{fe} is the effective tensile strain of CFRP; E_f is the elastic modulus of CFRP; θ is the angle between the diagonal crack and the column axis, which is generally taken as 45°. Considering the influence of size effect, the contribution by the plain concrete on the shear capacity of CFRP confined RC columns would be also determined as following.

2. *Contribution by steel tube*

According to the strength theory of material mechanics (Fan 2000), the shear capacity V_{st} provided by steel tube would be discussed considering the fact that CFST columns exhibit different failure modes under different shear-span ratios.

For the shear-span ratio $m = 0$, the steel tube is in a pure shear state, so that according to the moment balance principal and the Mises equivalent stress following the fourth strength theory of material mechanics, the shear capacity V_{st} provided by steel tube can be calculated as (Fan 2000)

$$V_{st} = \frac{\tau A_s}{2} = \frac{f_{yA_s}}{2\sqrt{3}} \approx 0.3 f_{yA_s} \tag{8.13}$$

where τ is shear stress; f_y is the yield strength of steel; and A_s is the cross-sectional area of steel tube.

For the shear-span ratio $m \geq 3.0$, the failure mode of CFST columns is mainly bending failure (Tomii and Sakino 1979; Ye et al. 2016). For this case, V_{st} can be obtained by the following expression (Fan 2000)

$$V_{st} = \frac{1 + a_s^2}{6m} f_y A_s \tag{8.14}$$

where a_s is a ratio between the width of concrete to outer width of steel tube, i.e. $a_s = B_c/B$.

For the shear-span ratio of $0 < m < 3.0$, V_{st} can be calculated by the method of linear interpolation based on Eqs. (8.13) and (8.14) (Qian et al. 2007). One can clearly note that from Eqs. (8.13) and (8.14), V_{st} represents the shear capacity directly generated by steel tube, which should be independent of size effect.

3. *Contribution by axial compressive load*

The available experimental observations (Tomii and Sakino 1979; Han et al. 2008; Guo 2013) have shown that, when the value of axial compression load is less than a certain value, the axial force could delay the appearance of cracks and reduce the crack width, which can improve the shear capacity of compressed members. One can note from the test data (Tomii and Sakino 1979) shown in Fig. 8.57, when the value of the axial compression ratio is less than 0.4, the non-dimensionless shear capacity V_u/f_cD^2 of columns increases as the axial load ratio n increases. While, when the magnitude of the axial load ratio is larger than 0.4, the shear capacity decreases with increasing the axial load ratio. Therefore, only the condition that the axial load is beneficial to the shear capacity of CFST columns i.e. $n \leq 0.4$ is discussed.

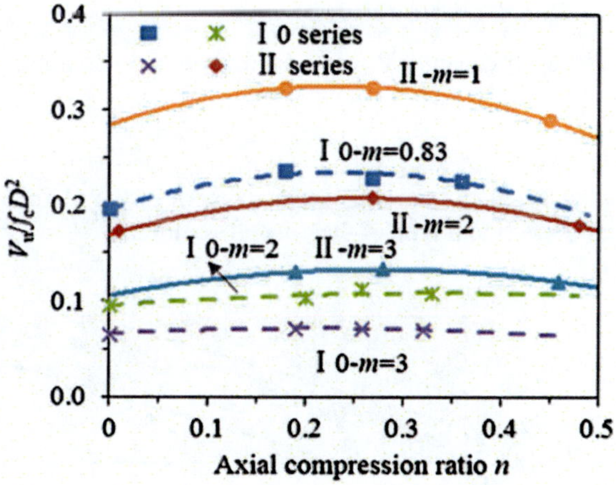

Fig. 8.57 Relationship of shear capacity of tested specimens (Tomii and Sakino 1979) and axial compression ratio

For $n \leq 0.4$, the shear capacity of CFST columns increases with the increase of axial load, so that V_n can be regarded as the improved shear capacity due to the existing of axial load. Here, an influence coefficient γ is introduced to described the effect of axial load (Qian et al. 2007), and V_n can be expressed as

$$V_n = \gamma N \tag{8.15}$$

in which, N is the axial load applied to the CFST columns.

Here, the typical truss-arch model (Guo 2013) would be used in order to determine the detailed form of the shear capacity V_n provided by steel tube. As shown in Fig. 8.58, the axial force N is mainly transmitted through the arch action in the truss-arch model, and the lateral force V_n that is the shear capacity of arch can be defined as (Qian et al. 2007)

$$V_n = \alpha N \tan\theta = \alpha N \frac{B}{L} = \frac{\alpha}{m} N \tag{8.16}$$

where α is the influence coefficient of axial force acting on the arch. Combined Eqs. (8.15) and (8.16), the influence coefficient γ can be evaluated as

$$\gamma = \frac{\alpha}{m} \tag{8.17}$$

Equation (8.17) indicates that the influence coefficient γ is an inverse function of the shear-span ratio. Based on the fitting of large amount of test data (Tomii and Sakino 1979; Ye et al. 2016), the relationship between the influential coefficient γ (average value) and shear-span ratio m can be obtained, as shown in Fig. 8.59. According to the fitting results in Fig. 8.59, the influence coefficient α should be taken as 0.05.

4. *Contribution by plain concrete*

The shear capacity V_c provided by the plain concrete part can be expressed as

$$V_c = \tau_c A_c \tag{8.18}$$

where τ_c is the nominal shear strength of the plain concrete column, and A_c is the cross-sectional area of the concrete column.

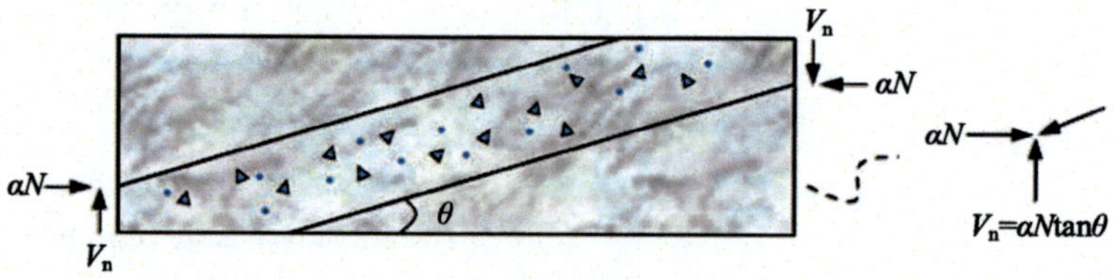

Fig. 8.58 Sketch of arch model

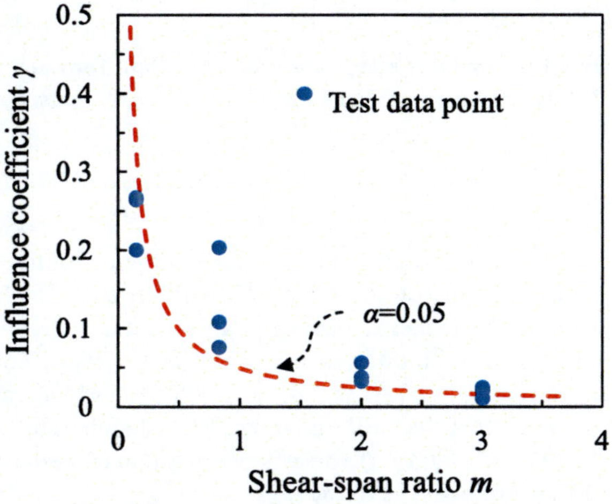

Fig. 8.59 Relationship between the influential coefficient γ and shear-span ratio m (Tomii and Sakino 1979; Ye et al. 2016)

The nominal shear strength τ_c contributed by concrete part should be related to the factors including concrete strength, structure size and shear-span ratio, which means that the nominal shear strength τ_c should exhibit obvious size effect. According to the simulated results in Sects. 8.3 and 8.4, size effect on the nominal shear strength of concrete columns is significantly affected by shear-span ratio, axial compression ratio and constraint coefficient.

The influence of shear-span ratio is firstly discussed. As mentioned in Sect. 8.3, the shear-span ratio has an obvious influence on the nominal shear strength and corresponding size effect. The nominal shear strength of CFST columns decreases with increase of shear-span ratio, which can be described by an influential coefficient φ. Moreover, it can be concluded that from above simulated results, size effect on the nominal shear strength can be weaken with increase of shear-span ratio. This behavior can be characterized by a size effect coefficient β_s.

As discussed previously, when the axial compression ratio is less than 0.4, the CFST columns show good ductile performance, the influence of axial compression ratio on size effect on nominal shear strength can be ignored. Based on the statistics for CFST columns applied in practical civil engineering, the steel ratio of CFST columns is predominantly distributed in the range of 4–10% (Wang et al. 2017). Therefore, the effect of steel ratio less than 0.04 on size effect is not considered. According to the simulation results, the weakening effect of steel ratio on size effect on the nominal shear strength is not obvious with the increase of steel ratio larger than 0.05.

For CFRP confined RC columns, the axial compression ratio have little influence on the size effect behavior of nominal shear strength of columns when the axial compression ratio of columns is less than 0.4. In addition, the existence of CFRP material limits the development of concrete internal cracks, which enhances the ductility of concrete columns and weakens the size effect behavior.

5. *Size effect law*

Based on analysis above and the size effect formula proposed by Bažant and Kim (1984), the nominal shear strength τ_c of plain concrete can be expressed as

$$\tau_c = \frac{V_0}{\sqrt{1 + B/D_0}} \cdot \varphi \cdot \beta \tag{8.19}$$

in which, φ is the strength influence coefficient of the shear-span ratio; and β is the size effect influence coefficient. The parameters V_0 and D_0 are obtained by regression analysis of simulation data of plain concrete columns (i.e. the columns with steel ratio of 0.0) with different cross-sectional widths. For CFST columns, V_0 and D_0 are respectively taken as 1.42 and 105.51, and they are respectively taken as 3.67 and 364.00 for the CFRP confined RC columns. The determination of the influence coefficients φ and β (i.e. coefficient β_s due to steel tube, and coefficient β_f due to FRP) is discussed as following.

(1) Strength influential coefficient φ

The influential coefficient φ should be in inverse proportion to m according to the analysis before, and the determination of form can refer to the work in the literature (Qian et al. 2007). The collected test results including 54 experimental data (Tomii and Sakino 1979; Ye et al. 2016) and simulated results are fitted, as presented in Fig. 8.60. According to the fitting results, φ can be defined as

$$\varphi = \frac{2}{m + 0.2} \tag{8.20}$$

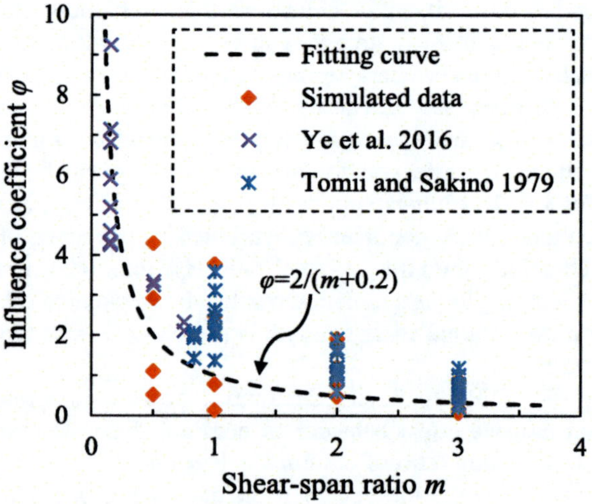

Fig. 8.60 Determination of the influential coefficient of strength φ

It should be noticed that the further verification of reasonability of the influence coefficient φ here is needed using extensive experimental data.

(2) Size effect influential coefficient β_s

The size effect on the nominal shear strength of column would be weakened by the confinement effect generated by steel tube. The determination of the size effect influence coefficient β_s would be explained in details as following.

For $m \to 0$, the nominal shear strength τ_c is greatly affected by the structure size, thus the size effect influential coefficient β_s is taken as 1.0. At this case, τ_c can be expressed as

$$\tau_c = \frac{V_0}{\sqrt{1 + B/D_0}} \cdot \varphi \tag{8.21}$$

For $m \to m_{cr}$ (m_{cr} is the critical value of shear-span ratio), the CFST columns show the ductile bending failure, the nominal shear strength τ_c is independent of structure size and can be expressed as

$$\tau_c = \varphi \cdot f_t \tag{8.22}$$

in which, f_t is the tensile strength of concrete material. According to Eqs. (8.19) and (8.22), one can get the coefficient β_s as

$$\beta_{s|m \to m_{cr}} = \frac{\sqrt{1 + B/D_0}}{V_0} \cdot f_t \tag{8.23}$$

For $0 < m < m_{cr}$, the coefficient β_s is a non-linear curve asymptotic to $\frac{\sqrt{1+B/D_0}}{V_0} \cdot f_t$, and the change trend of which is described by the hyperbolic tangent curve here, as shown in Fig. 8.61. Combined Eq. (8.23), the coefficient β_s can be expressed as

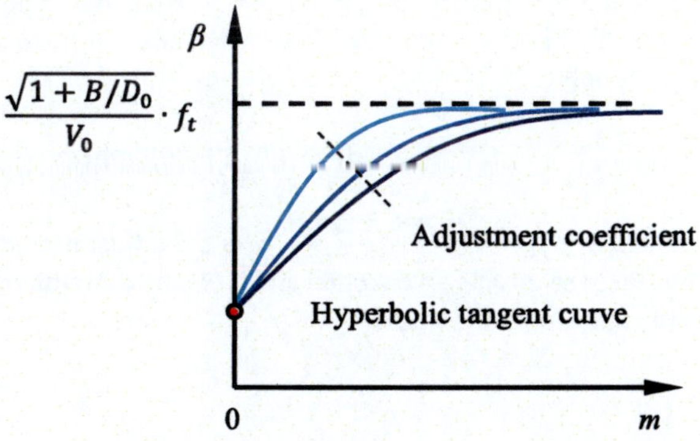

Fig. 8.61 Determination of the influential coefficient of size effect β

$$\beta_s = \begin{cases} 1 m \to 0 \\ (\beta_{scr} - 1)\tanh(\omega m) + 1 0 < m < m_{cr} \end{cases} \tag{8.24}$$

where $\beta_{scr} = \frac{\sqrt{1+B/D_0}}{V_0} \cdot f_t$; ω is the adjustment coefficient reflecting the rate of β_s varying with the shear-span ratio, and can be determined by comparing with the simulation results.

(3) Influence coefficient of size effect β_f

The size effect on the nominal shear strength of column would be weakened by the confinement effect generated by CFRP. The relationship between the size effect influence coefficient β_f and the constraint ratio ρ_f would be explained in details as following.

For $\rho_f = 0$ (i.e. unconfined concrete column), the size effect influence coefficient β_f has no influence on the size effect on the nominal strength of plain concrete columns, thus $\beta_f|_{\rho_f=0} = 1.0$.

For the $\rho_f \to \infty$, the CFRP material has a strong constraint effect on core concrete. Then the size effect of concrete shear strength is greatly weakened and the shear strength of concrete columns gradually tends to the shear strength of concrete materials. The relationship between the shear strength and tensile strength of concrete materials proposed in reference (Zhang and Guo 1992).

$$\tau_c = 1.2 f_t^{0.89} \tag{8.25}$$

The size effect weakening factor β_f in this case can be obtained as:

$$\beta_f = \frac{1.2 f_t^{0.89}\sqrt{1 + B/D_0}}{V_0} \tag{8.26}$$

For $\rho_f > 0$, the shear strength of concrete columns is gradually weakened as the increase of CFRP constraint, then the weakening coefficient β_f is gradually approaching from 1.0 to $\frac{1.2 f_t^{0.89}\sqrt{1+B/D_0}}{V_0}$. Here, the hyperbolic tangent function shown in Fig. 8.61 is also employed to describe the change trend of the weakening coefficient β_f, namely:

$$\beta_f = (\beta_{fcr} - 1)\tanh(\omega \rho_f) + 1 \tag{8.27}$$

in which, $\beta_{fcr} = \frac{1.2 f_t^{0.89}\sqrt{1+B/D_0}}{V_0}$, ω is the adjustment coefficient that represents how fast the size effect is weakened and it can be determined by fitting with the simulated and experimental results.

8.5.3 *Validation of Size Effect Law*

In summary, the theoretical formula predicting the shear capacity of concrete columns is established, and its accuracy and rationality should be verified through numerical simulation results and a large number of reliable test data.

1. *CFST columns*

Figure 8.62 shows the comparison between nominal shear strengths calculated by the proposed theoretical formula and simulated values gained by meso-scale numerical method for CFST columns under different shear-span ratios and steel ratios. In Fig. 8.62, the red data points are the simulation results and the colored surface refers to the predicted values, i.e. theoretical values. It is found that when the value of the adjustment coefficient ω is between 5.0 and 8.0, the predicting theoretical results would have a good agreement with the simulated results. Herein, for simplicity, it is recommended that ω is taken as 6.0.

Additionally, the applicability of the theoretical prediction formula is further illustrated by using the existing test data from the tests of Ye et al. (2016) and Tomii and Sakino (1979). The predicted results are compared with the experimental data (Ye et al. 2016), as shown in Fig. 8.63, where V_{ut} and V_{ue} are the predicted and experimental shear capacities of specimens, respectively. As can be seen that, V_{ut} and V_{ue} are in good agreement. Figure 8.64 gives the comparison of shear capacity between theoretical results and test data (Tomii and Sakino 1979) involving three series in total of 21 specimens under different axial compression ratios. It can be found that the theoretical values have a good agreement with the test values, so the theoretical formula built in this paper could give a reasonable prediction of shear capacity for square CFST columns.

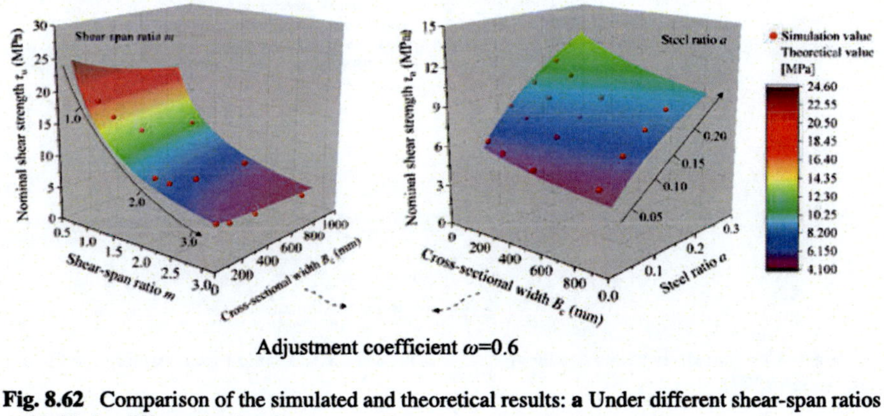

Fig. 8.62 Comparison of the simulated and theoretical results: **a** Under different shear-span ratios and **b** Under different steel ratios

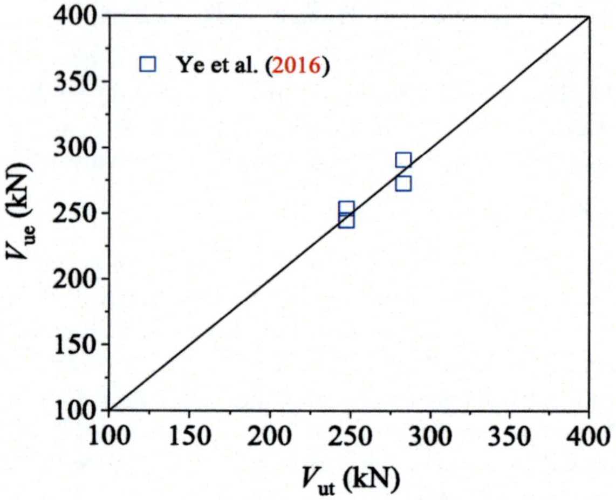

Fig. 8.63 Comparison of the theoretical and experimental results (Ye et al. 2016)

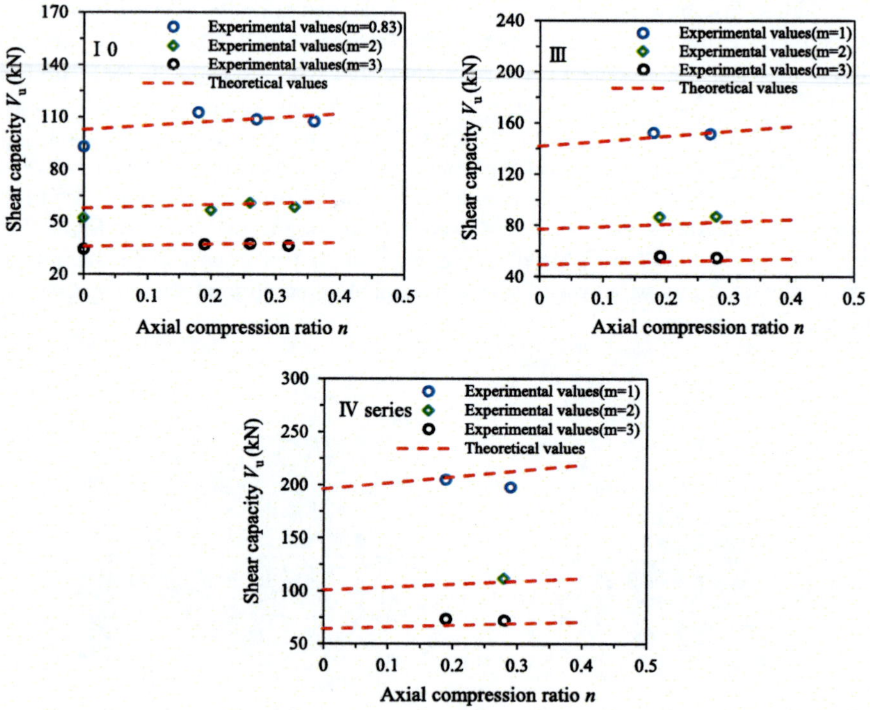

Fig. 8.64 Comparison between the theoretical results and the available experimental data (Tomii and Sakino 1979)

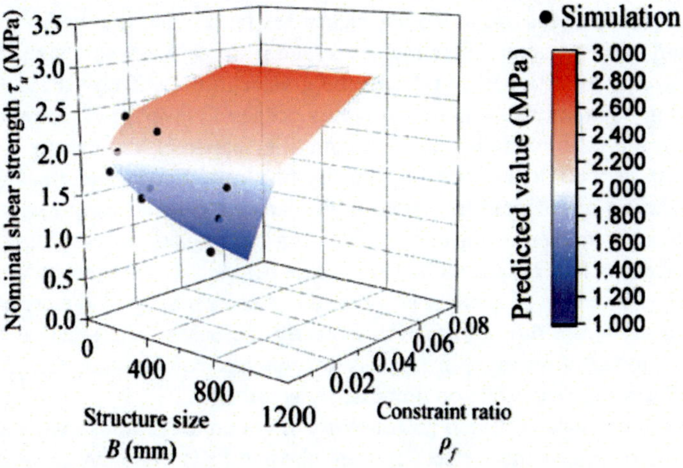

Fig. 8.65 Comparison between simulation results and theoretical calculation values

2. *CFRP-confined columns*

Equation (8.27) is a transcendental equation, and it is difficult to get the analytical solution of the coefficient ω. It's found that the value of the adjustment coefficient ω is 42 or 43, the theoretical calculation value is in good agreement with the simulation result after a lot of trial calculation, Herein, ω is recommended to be 43.

In summary, the size effect law is consistent with the simulation results as shown in the Fig. 8.65, which preliminary demonstrate the rationality of the size effect law. However, a large number of experimental data should be provided to further verify the rationality of the size effect law.

8.6 Summary

In this chapter, the experimental results on the seismic behavior of RC columns are introduced. The influences of loading pattern, shear-span ratio, axial compression ratio, and the confinement effect of steel tube and FRP on the seismic performance and size effect of concrete columns are also presented. Moreover, the theoretical formula (i.e. size effect law) for predicting the shear capacity of concrete columns is introduced. Several conclusions can be drawn as follows:

(1) As the cross-sectional dimension of the RC columns increases, the nominal shear strength, the dimensionless nominal moment, the ductility capacity, the energy dissipation capacity and the rotation capacity of plastic hinge declines obviously, showing obvious size effect.
(2) The RC columns under the horizontal cyclic load show less ductility because of the low cyclic fatigue effect. The dimensionless nominal moment of the

RC columns under cyclic loading decreases faster than that under monotonic loading as the structural size increases, and it shows a stronger size effect.

(3) The greater the axial force ratio is, the greater initial stiffness and lateral load carrying capacity the concrete columns possess. Nevertheless, improving the axial force ratio tends to lower the ductility capacity, the energy dissipation capacity as well as the rotation capacity of plastic hinge, while accelerate the degradation of stiffness. In addition, the larger the axial compression ratio is, the stronger brittleness the concrete column shows, and the more significant the size effect on the compression-shear failure behavior of concrete column shows.

(4) With the increase of shear-span ratio, the shear capacity of concrete columns decreases, while the deformation capacity increases. The smaller the shear-span ratio is, the stronger brittleness the concrete column shows, and the more significant the size effect on nominal shear strength is.

(5) The nominal shear strength and ductility of concrete columns increase with the increase of confinement effect of steel tube and FRP. The increase of constraint coefficient within a certain range can weaken but cannot suppress the size effect on nominal shear strength of concrete columns.

(6) The size effect law for describing the size effect in shear strength of concrete columns can quantitatively show the influence of shear-span ratio and constraint coefficient.

References

American Concrete Institute (2011) Building code requirements for structural concrete (ACI 318–11) and commentary (ACI 318R–11). Farmington Hills, MI

Bažant ZP, Kim JK (1984) Size effect in shear failure of longitudinally reinforced beams. ACI J 81:456–468

British Standards Institution (1997) Structural use of concrete. BS 8110-1997, London, UK

British Standards Institution (2004a) Design of concrete structures-Part 1-1: general rules and rules for buildings. EN 1992-1-1, CEN, Brussels

British Standards Institution (2004b) Design of composite steel and concrete structures, Part 1-1: general rules and rules for building. EN 1994-1-1, CEN, Brussels

Cui YQ (2018) Research of size effect on compressive and seismic behavior of CFRP reinforced concrete column. M.Sc. thesis, Beijing: University of Civil Engineering and Architecture. (in Chinese)

Canadian Standard Association (2004) Design of concrete structures. CSA-A23.3-04, Mississauga, Ont.

Fan QS (2000) Material Mechanics. High Education Press, Beijing (in Chinese)

Guo ZH (2013) Principle for reinforced concrete structure. Tsinghua University Press, Beijing (in Chinese)

Han LH, Tao Z, Yao GH (2008) Behaviour of concrete-filled steel tubular members subjected to shear and constant axial compression. Thin-Walled Struct 46(7–9):765–780

Jin L, Du X, Li D, Su X (2016) Seismic behavior of RC cantilever beams under low cyclic loading and size effect on shear strength: an experimental characterization. Eng Struct 122:93–107

Jin L, Li D, Du X (2016) Mechanical behavior and size effect of moderate high-strength RC columns under monotonic and cyclic axial compression. Eng Struct 124:269–285

Jin L, Zhang S, Han JY, Li D, Du XL (2019) Effect of cross-section size on flexural compressive failure of RC columns: monotonic and cyclic tests. Eng Struct 186:456–470

Jin L, Zhang S, Li D, Xu H, Du X, Li Z (2018) A combined experimental and numerical analysis on the seismic behavior of short reinforced concrete columns with different structural sizes and axial compression ratios. Int J Damage Mech 27(9):1416–1447

Kawashima K, Koyama T (1988) Effect of number of loading cycles on dynamic characteristics of reinforced concrete bridge pier columns. Doboku Gakkai Ronbunshu 392:205–213

Li D, Jin L, Du XL, Fu J, Lu A (2016) Size effect tests of normal-strength and high-strength RC columns subjected to axial compressive loading. Eng Struct 109:43–60

Lu Y, Li N, Li S, Liang H (2015) Behavior of steel fiber reinforced concrete-filled steel tube columns under axial compression. Constr Build Mater 95:74–85

Majewski T, Bobinski J, Tejchman J (2008) FE analysis of failure behaviour of reinforced concrete columns under eccentric compression. Eng Struct 30:300–317

Marzec I, Bobiñski J, Tejchman J (2007) Simulations of spacing of localized zones in reinforced concrete beams using elasto-plasticity and damage mechanics with non-local softening. Comput Concr 4(5):377–402

Ministry of Construction of the People's Republic of China (2003) Standard for test method of mechanical properties on ordinary concrete GB/T 50081–2002. China Architecture & Building Press, Beijing

Ministry of Construction of the People's Republic of China (2010) Code for design of concrete structures GB 50010–2010. China Architecture & Building Press, Beijing

Ministry of Housing and Urban-Rural Development of the People's Republic of China (2014) Code for design of concrete filled steel tubular structures GB 50936-2014. China Architecture & Building Press, Beijing

Moretti M, Tassios TP (2007) Behavior of short columns subjected to cyclic shear displacements: experimental results. Eng Struct 29:2018–2029

Mo YL, Wang SJ (2000) Seismic behavior of RC columns with various tie configurations. ASCE J. Struct Eng 126(10):1122–1130

Pujol S, Sozen MA, Ramirez JA (2006) Displacement history effects on drift capacity of reinforced concrete columns. ACI Mater J 103(2):253–262

Qian JR, Cui Y, Fang XD (2007) Shear strength tests of concrete filled steel tube columns. China Civil Eng J 40(5):1–9 (in Chinese)

Rodrigues H, Arêde A, Varum H, Costa AG (2013) Experimental evaluation of rectangular reinforced concrete column behaviour under biaxial cyclic loading. Earthquake Eng Struct Dynam 42(2):239–259

Shamim AS, Grace Y (2002) Seismic behavior of concrete columns confined with steel and fiber reinforced polymers. ACI Struct J 99(1):72–80

Syroka-Korol E, Tejchman J, Mróz Z (2014) FE analysis of size effects in reinforced concrete beams without shear reinforcement based on stochastic elasto-plasticity with non-local softening. Finite Elem Anal Des 88(10):25–41

Tomii M, Sakino K (1979) Experimental studies on concrete filled square steel tubular beam-columns subjected to monotonic shearing force and constant axial force. Trans Architectural Inst Japan 281:81–92

Verderame GM, Fabbrocino G, Manfredi G (2008a) Seismic response of RC columns with smooth reinforcement. Part I: monotonic tests. Eng Struct 30(9):2277–2288

Verderame GM, Fabbrocino G, Manfredi G (2008b) Seismic response of RC columns with smooth reinforcement. Part II: cyclic tests. Eng Struct 30(9):2289–2300

Wang YY, Chen P, Liu CY, Zhang Y (2017) Size effect of circular concrete-filled steel tubular short columns subjected to axial compression. Thin-Walled Struct 120:397–407

Xiao CZ, Cai SH, Xu CL (2005) Experimental study on shear resistance performance of concrete filled steel tube columns. China Civil Eng J 38(4):5–11 (in Chinese)

Xie YP, Zhen ZB, Du XL, Ma H, Guo EW, Chen LM (2013) Experimental study on size effect of seismic behavior for reinforced concrete columns under low cycle reversed loading. J Build Struct 34(12):86–93 (in Chinese)

Xu C, Jin L, Ding Z, Li D, Du X (2016) Size effect tests of high-strength RC columns under eccentric loading. Eng Struct 126:78–91

Ye LP, Zhao SH, Yue QR, Zhang K (2000) Calculation of shear strength of concrete column strengthened with carbon fiber reinforced plastic sheet. J Build Struct 2:59–67 (in Chinese)

Ye Y, Han LH, Tao Z, Guo SL (2016) Experimental behaviour of concrete-filled steel tubular members under lateral shear loads. J Constr Steel Res 122:226–237

Yu Q, Bažant ZP (2011) Can stirrups suppress size effect on shear strength of RC beams? ASCE J Struct Eng 137(5):607–617

Zhang Q, Gong JX, Zhang YQ (2013) Lateral-load behavior prediction and pushover analysis of reinforced concrete columns including shear effects. Adv Struct Eng 16(4):741–758

Zhang Q, Guo ZH (1992) Investigation on shear strength and shear strain of concrete. J Build Struct 13(5):17–24 (in Chinese)

Zhu W, Jia J, Gao J, Zhang F (2016) Experimental study on steel reinforced high-strength concrete columns under cyclic lateral force and constant axial load. Eng Struct 125:191–204

Chapter 9
Size Effect on Shear Failure of RC Beam-to-Column Joints

Check for updates

In the seismic-resistant design of building structures, the structural system and members should have enough strength to resist the earthquake action. To prevent catastrophic collapse of the structures during a severe earthquake, the stable and reliable capacity for energy dissipation should be provided through proper design and detailing of the system, members, and joints (Chen et al. 1996). Reinforced concrete (RC) beam-to-column joints are considered vulnerable structural elements during earthquakes. The failure of a joint or a group of joints can result in at least partial collapse of the structure (Tsonos 1999). The seismic failures of RC beam-to-column joints have been paid much attention in recent years. Typical brittle shear failure often occurs in the core region of joint, and shear failure of joints may result in non-ductile performance of RC moment-resisting frames. Moreover, the brittle shear failure may cause a strong size effect on the nominal shear strength of the RC joints.

This chapter mainly introduces the seismic behavior and size effect of RC beam-to-column joints from three aspects: experiment, numerical simulation and theory.

9.1 Shear Failure of Interior RC Beam-to-Column Joints

9.1.1 Seismic Tests on Interior RC Beam-to-Column Joints

1. *Introduction of experiment*

There were 14 geometrically similar RC beam-to-column joints (designed by Chinese code (Ministry of Construction of the People's Republic of China 2011)). To present the effect of loading type on the failure behavior of the RC joints, both the monotonic and cyclic loadings were applied on joints. The axial force was applied to the top of the column, and the monotonic and cyclic vertical loadings were applied at the beam end.

X. Du and L. Jin, *Size Effect in Concrete Materials and Structures*,
https://doi.org/10.1007/978-981-33-4943-8_9

The structural sizes of joints were 300 mm × 300 mm, 500 mm × 500 mm, 700 mm × 700 mm and 900 mm × 900 mm. The cross-sectional size (i.e. width b_b and depth h_b) of the beams ranged from 225 mm × 450 mm to 675 mm × 1350 mm, and the effective length (L_{eff}) of the beams ranged from 850 mm to 2550 mm. In addition, the cross-sectional size (including cross-sectional width b_c and height h_c) of columns ranged from 300 mm × 300 mm to 900 mm × 900 mm, and the length of columns ranged from 1500 to 4500 mm.

Furthermore, to introduce the effect of stirrups on the failure behavior and the size effect of RC joints, two different stirrup ratios were utilized, i.e. $\rho_{sv} = 0$ and 0.84%. In summary, there were 14 specimens and they were BCJ-300-M-i, BCJ-300-M-ii, BCJ-300-C-i, BCJ-300-C-ii, BCJ-500-M-i, BCJ-500-M-ii, BCJ-500-C-i, BCJ-500-C-ii, BCJ-700-M-i, BCJ-700-M-ii, BCJ-700-C-i, BCJ-700-C-ii, BCJ-900-C-i, and BCJ-900-C-ii, respectively. The detailed information of all RC joints are given in Table 9.1.

The mixture proportions of the concrete are given in Table 9.2. The grade of concrete was C30. The detailed data of concrete are shown in Table 9.2. The steel bars in the specimen are generally divided into two kinds, namely the longitudinal reinforcement (HRB400) and the stirrup (HPB300).

The steel bars of grade HPB300 are also used as the tie bars (in order to improve the integrity of the reinforcement skeleton). The detailed information of the longitudinal reinforcement bars and stirrups of the specimens are listed in Table 9.3 and Fig. 9.1.

The installation of specimens and the test setup are given in Fig. 9.2. The loading scheme is plotted in Fig. 9.3. The detailed introduction of the experiment is in article of Jin et al. (2018).

Table 9.1 Physical parameters of the RC beam-to-column joint

Specimen name	Sizes $b \times h$ (mm)		Effective length L_{eff} (mm)	Cover thickness c (mm)	Concrete strength grade	Axial compression ratio n	Longitudinal reinforcement ratio ρ (%)	Stirrup ratio ρ_{sv} (%)
BCJ-300	Beam	225 × 450	850	12	C30		0.78	0.67
	Column	300 × 300	1500	15	C30	0.3	1.00	0.63
	Joint	300 × 300		15	C30	0.3		0(0.84)
BCJ-500	Beam	375 × 750	1417	20	C30		0.78	0.67
	Column	500 × 500	2500	25	C30	0.3	1.00	0.63
	Joint	500 × 500		25	C30	0.3		0(0.84)
BCJ-700	Beam	525 × 1050	2170	28	C30		0.78	0.67
	Column	700 × 700	3500	35	C30	0.3	1.00	0.63

(continued)

Table 9.1 (continued)

Specimen name	Sizes $b \times h$ (mm)		Effective length L_{eff} (mm)	Cover thickness c (mm)	Concrete strength grade	Axial compression ratio n	Longitudinal reinforcement ratio ρ (%)	Stirrup ratio ρ_{sv} (%)
	Joint	700 × 700		35	C30	0.3		0(0.84)
BCJ-900	Beam	675 × 1350	2550	36	C30		0.78	0.67
	Column	900 × 900	4500	45	C30	0.3	1.00	0.63
	Joint	900 × 900		45	C30	0.3		0(0.84)

Note The first letter "B" in the specimen's name denotes "Beam"; the second letter "C" in the specimen's name denotes "Column"; the third letter "J" denotes "Joint"; and the numbers of "300", "500", "700" and "900" mean the cross-sectional sizes, i.e. 300 mm × 300 mm, 500 mm × 500 mm, 700 mm × 700 mm and 900 mm × 900 mm

Table 9.2 Concrete mixture proportions utilized

Fine aggregate 0–5 mm (kg/m^3)	Coarse aggregate 5–30 mm (kg/m^3)	Cement (kg/m^3)	Water (kg/m^3)	w/c ratio	Sum (kg/m^3)	Compressive strength f_{cu} (MPa)	Splitting tensile strength f_t' (MPa)	Elastic modulus E_c($\times 10^4$ N/mm^2)
1008	824	380	228	0.6	2440	34.55	3.06	3.18

Table 9.3 Material properties of reinforcement bars

Steel types	Diameter (mm)	Grade	Yield strength (MPa)	Ultimate strength (MPa)	Elastic modulus (× 10^2 GPa)	Elongation (%)
Longitudinal reinforcement	36	HRB400 Φ	463.34	644.23	1.76	20.86
	28	HRB400 Φ	406.98	561.91	1.77	21.12
	20	HRB400 Φ	443.51	599.48	1.64	19.40
	12	HRB400 Φ	492.20	643.99	2.04	20.89
Stirrup	18	HPB300 ϕ	313.59	458.21	2.05	24.57
	14	HPB300 ϕ	385.44	540.91	2.02	22.92
	10	HPB300 ϕ	420.17	447.33	2.07	18.47
	6	HPB300 ϕ	496.33	571.78	2.03	27.06
Tie bar	14	HPB300 ϕ	385.44	540.91	2.02	22.92
	10	HPB300 ϕ	420.17	447.33	2.07	18.47
	8	HPB300 ϕ	405.18	473.49	2.10	29.55
	5	HPB300 ϕ	542.57	638.32	2.04	24.32

Note "HRB" means "hot-rolled ribbed-steel bar"; and "HPB" means "hot-rolled plain-steel bar"

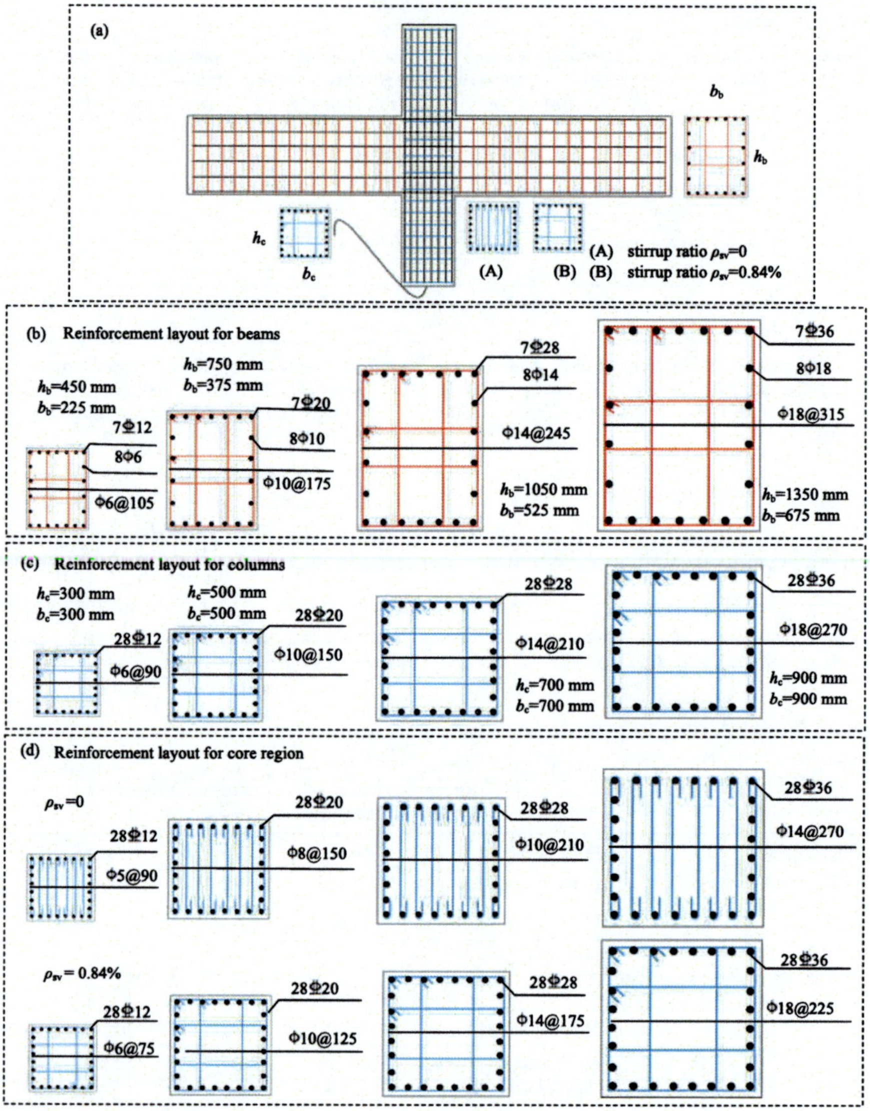

Fig. 9.1 Reinforcement details of the tested RC joints

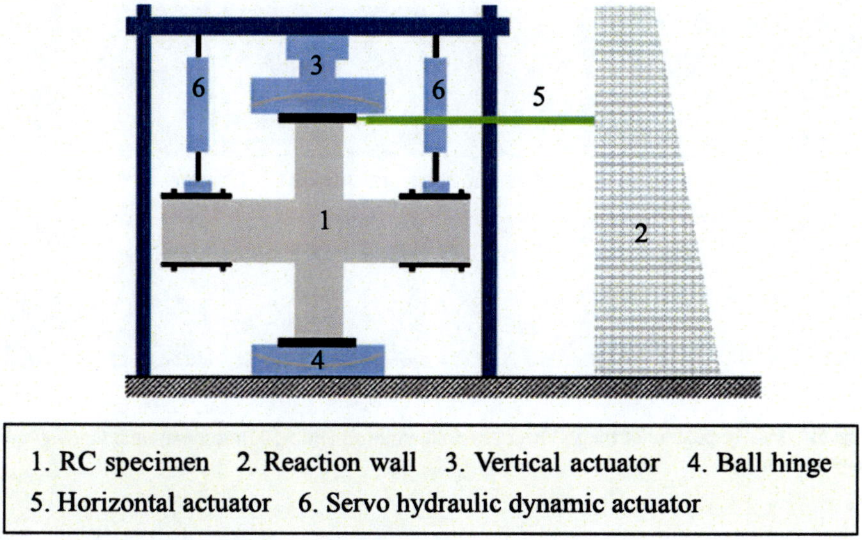

Fig. 9.2 Test setup of RC beam-to-column joints

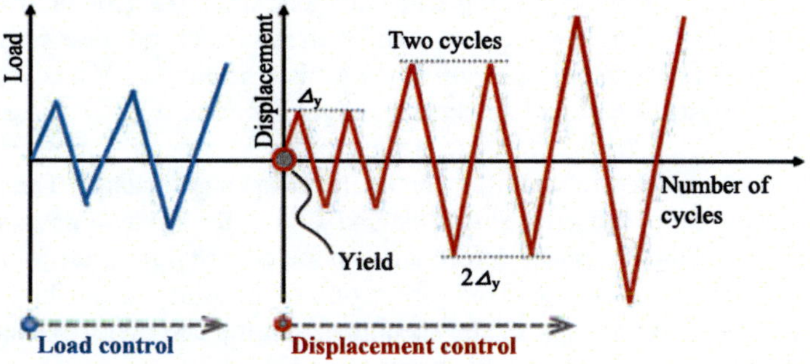

Fig. 9.3 Loading scheme

2. *Failure mode of RC beam-to-column joints*

Taking the failure process of the specimens of BCJ-500 as an example, the development of cracks in the core region of the beam-to-column joints subjected to monotonic and cyclic loadings are presented respectively in Fig. 9.4a and b. On the whole, it can be clearly found from Fig. 9.4a and b that the typical shear failure modes generated in the core region of the two RC beam-to-column joints. For RC beam-to-column joints, the core region of the joint is in a complex stress state caused by the combined effects of axial force, bending moment and shear force.

One can see from Fig. 9.4a that the initial first diagonal crack appeared in the center of core region of the joint when the load P applied to the beam end reached 170 kN. As the load increased, the diagonal cracks extended towards the ends of the

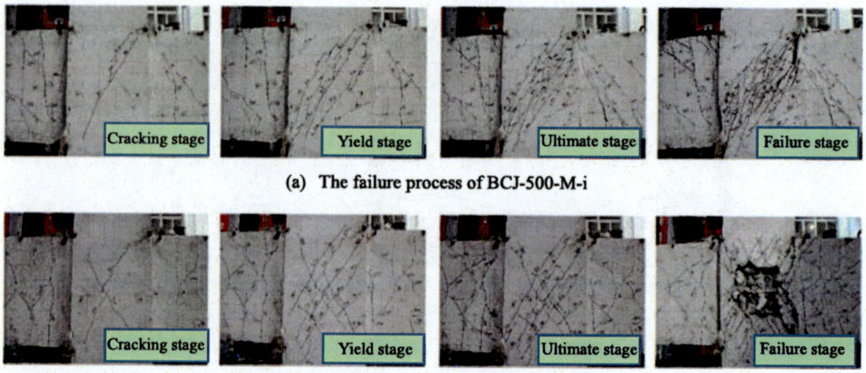

(a) The failure process of BCJ-500-M-i

(b) The failure process of BCJ-500-C-i

Fig. 9.4 Failure process of the RC beam-to-column joints (BCJ-500): **a** monotonic loading; and **b** cyclic loading

joint, and then some new diagonal cracks generated parallel to the initial diagonal crack. In the loading process, the number of cracks increased obviously, the tensile strains of stirrups in the core region increased gradually. As the load continued to be applied, the joint entered the yielding stage, and then when the applied load reached the bearing capacity of the specimen, the specimen entered the ultimate stage. During the loading process, it can be seen from Fig. 9.4a that the number of cracks increases continuously, while the width of cracks has a slight increase. With the increase of the load, the joint entered the failure stage. As the load continued, the number of cracks did not increase, while the width of cracks increased significantly. The widening of the diagonal cracks caused the rapid degradation of the strength and stiffness of the joint. At this time, the shear deformation in the core region of the joint became dominant. Finally, the concrete in the core region of the joint was crushed, resulting in the loss of bearing capacity of the specimen, and the compressive shear failure occurred in the core region of the RC beam-to-column joint.

As shown in Fig. 9.4b, the cracks' development of the specimen under cyclic loading is similar to that under monotonic loading. After the occurrence of the initial crack, cracks along the diagonal direction occurred in the RC specimen under reverse loading. When the loading P got to 180 kN, some vertical cracks appeared at the beam end of the specimen. At the cracking stage of the specimen, two intersect cracks through the core region appeared within the specimen. With the number of loading cycles increasing, the core region of the RC joint entered the yielding stage and then the ultimate stage. Similarly, in the loading process the number of cracks increased, while the crack width did not have an obvious change. After the ultimate stage, the bearing capacity decreased gradually, and the diagonal cracks widened greatly. The diagonal cracks were no longer closed in when the specimen was subjected to negative loading. At this time, the spalling of concrete occurred in the core region of the specimen, it can be clearly noticed from Fig. 9.4b. Finally, the shear failure occurred in the core region of RC joint, which was similar with that of the specimen under monotonic loading. According to the above analysis, one can conclude that the existence of stirrups can effectively control the development of diagonal cracks, and increases the friction between aggregates in concrete.

The final failure modes of all tested RC joints under both monotonic and cyclic loadings are shown in Fig. 9.5. It can be observed from Fig. 9.5 that the failure modes of six beam-to-column joints subjected to monotonic loading are basically similar. The first cracks (marked with blue lines) appeared at the end of the beam. The concrete in the core region of the joint was crushed, and obvious large diagonal cracks (marked with red lines) generated in the core region, leading to the loss of bearing capacity and shear failure of the joints. However, compared with the failure of the smaller one, the larger specimens' failure was much more serious. The larger the specimen was, the wider the crack was, and the larger the damaged region was. Moreover, compared with the RC beam-to-column joints with the stirrup ratio $\rho_{sv} = 0.84\%$ (i.e. the specimens numbered with "i"), the shear failure of joints without stirrups (i.e. the specimens numbered as "ii") was more serious. For instance, the damage of BCJ-700-M-ii was obviously more serious than that of BCJ-700-M-i.

Similarly, one can observe from Fig. 9.5 that the final failure modes of the eight RC joints under cyclic loading are also closely similar. The first crack also formed at the end of the beam (marked with blue lines). Concrete spalling occurred on the surface of core region of all joints under cyclic loading. On the whole, the spalling areas (marked with red lines) of the joints with stirrups were smaller than that without stirrups. That is to say, the failure of the core region of the joints without stirrups was more serious. Moreover, one can note from the failure modes of joints under cyclic loading that, the larger the structural size of the specimen was, the more serious the failure of the specimen was.

3. *Strains of steel bars*

Some typical measured data are given. Typically, the strains of steel bars embedded in the three RC beam-to-column joints (i.e. BCJ-300-M-ii, BCJ-500-M-ii, and BCJ-700-M-ii) having different structural sizes under monotonic loading are shown in Fig. 9.6. One can see clearly from Fig. 9.6 that, when the load applied at the beam end of the three RC joints reached the peak load (i.e. the bearing capacity), the measured strains were smaller than yielding strain of the steels. That is to say, the steel bars were in elastic stage and did not reach their yield strength. This leads to the fact that the failure of the core region of RC beam-to-column joints is mainly caused by the destruction of concrete. This would be discussed in the following analysis.

4. *Hysteretic and skeleton curves*

In order to study the shear bearing capacity of RC beam-to-column joints, the hysteretic and skeleton curves of the RC joints were recorded. The cyclic hysteretic load-displacement angle curves (i.e. the solid grey lines) and the corresponding skeleton curves (i.e. the red solid lines) of the RC specimens subjected to cyclic loading are shown in Fig. 9.7. Herein, the displacement angle at the beam end is the vertical displacement of the loading point at the beam end divided by the calculated length of the beam (i.e. the length from the loading point to the junction of the beam

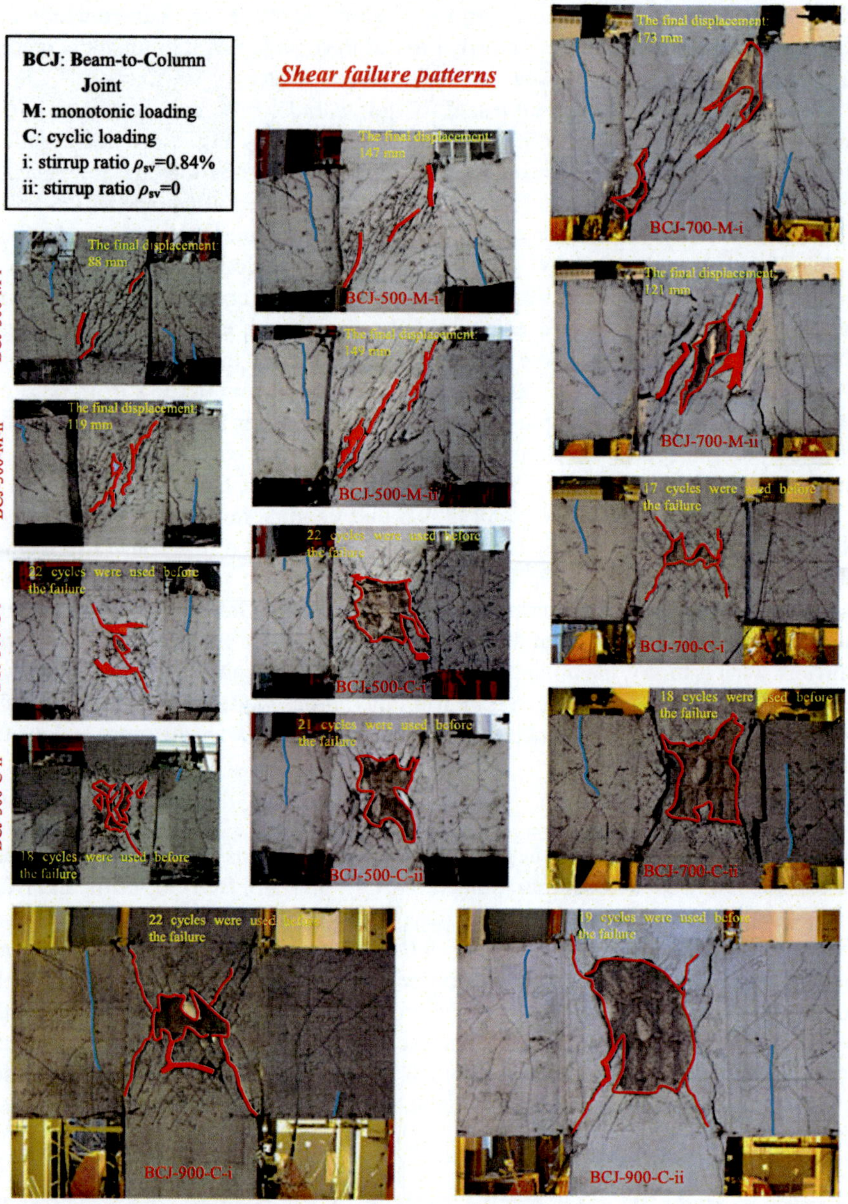

Fig. 9.5 Shear failure modes at the core regions of all RC beam-to-column connections (the first cracks of all the specimens appeared at beam end)

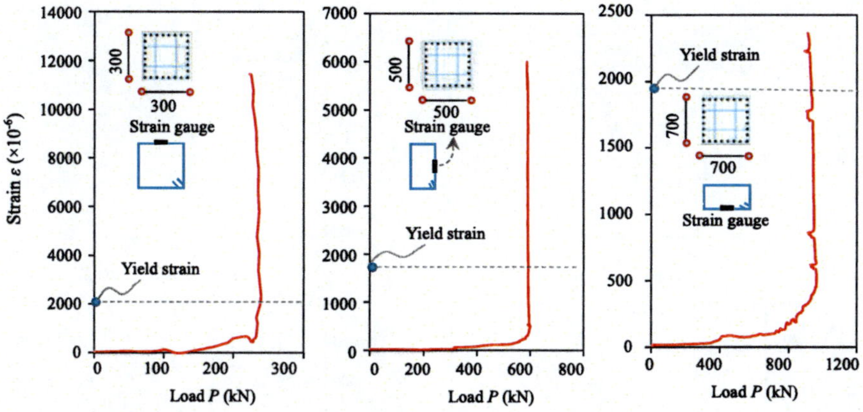

Fig. 9.6 The strain curves of steel bars at the joints

and column). The displacement angle represents the relative rotation angle of the beam to the column. Moreover, the "Yield" and "Fail" points are also marked in Fig. 9.7. The "Yield" points denote that the RC specimens yielded, and the "Fail" ones mean the point corresponding to 85% bearing capacity in the decline period. Furthermore, the test results are listed in Table 9.4, including the yield load P_y, the corresponding yield displacement Δy, the peak load (i.e. the bearing capacity) P_{max} and the displacement Δu corresponding to ultimate loading defined as 85% bearing capacity during the descent stage.

In general, for the eight RC beam-to-column joints subjected to cyclic loading, the shapes of the recorded hysteretic curves are all plump, as can be observed in Fig. 9.7. At the first stage, all RC specimens were in elastic stage. As the load continued, the RC joints under cyclic loading gradually entered the elastic-plastic stage, and unrecoverable plastic deformation occurred. At this moment, the residual deformation of specimens became larger and larger after loading, and the cracks developed quickly, making that the stiffness degraded quickly. In the last stage, the diagonal cracks widened, the concrete of the core region spalled off, and the shear failure of joints occurred. The corresponding skeleton curves exhibited the 'S-shape', as observed from Fig. 9.7.

During the cyclic loading process, the cracks of the RC specimens opened and closed ceaselessly, resulting in the decrease of shear capacity of RC joints. The larger the shear deformation is, the more seriously the shear stiffness degrades. Furthermore, one can note from Fig. 9.7 that, the softening curves after peak loading become steeper with increasing the size of RC joints, indicating that the brittleness of the specimen increases with the increase of the structural dimension.

Figure 9.8a and b plot the skeleton curves of the RC joints under monotonic and cyclic loadings respectively. One can note from Fig. 9.8 and Table 9.4 that, the bearing capacity of specimens under cyclic loading is obviously smaller than that under monotonic loading. In addition, the deformation capacity of specimens under

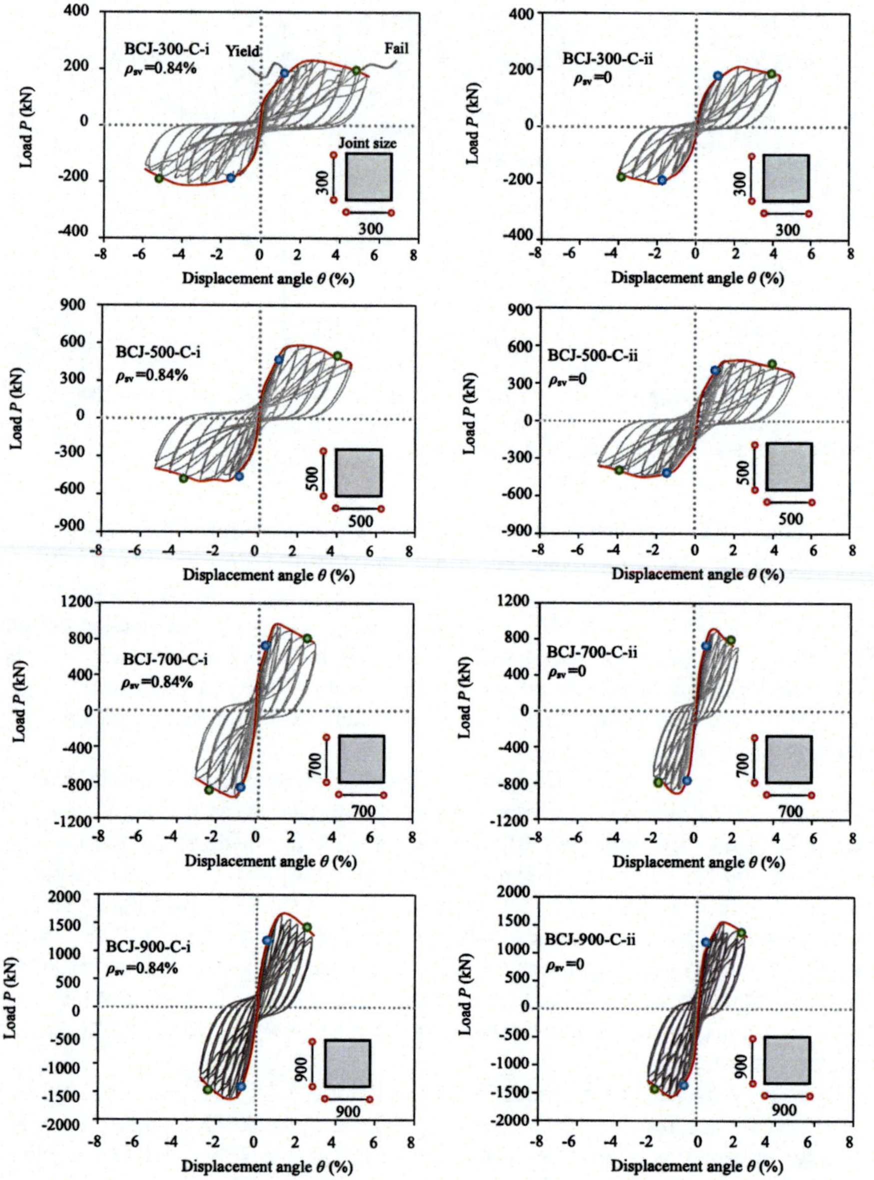

Fig. 9.7 The hysteretic curves and skeleton curves of the RC beam-to-columns joints under cyclic loading

Table 9.4 The test results of the beam-to-column joint

Specimen name	P_y (kN)	Δy (mm)	P_{max} (kN)	Δu (mm)	$\mu = \Delta u/\Delta y$
BCJ-300-M-i	198.8	11.40	238.9	105.75	9.27
BCJ-300-M-ii	150.3	10.44	169.6	118.64	11.37
BCJ-300-C-i	182.4	10.64	218.1	46.32	4.35
BCJ-300-C-ii	169.6	9.58	207.0	34.41	3.59
BCJ-500-M-i	537.0	17.69	598.2	148.26	8.38
BCJ-500-M-ii	457.4	14.62	520.5	148.54	10.16
BCJ-500-C-i	463.1	14.09	537.0	64.99	4.61
BCJ-500-Cii	385.2	15.23	468.0	65.92	4.33
BCJ-700-M-i	841.26	16.69	977.0	172.66	10.35
BCJ-700-M-ii	822.25	16.59	945.1	120.86	7.29
BCJ-700-C-i	828.8	14.75	965.7	65.20	4.42
BCJ-700-Cii	757.6	13.61	892.3	46.06	3.38
BCJ-900-C-i	1382.5	19.26	1640.9	65.40	3.40
BCJ-900-Cii	1266.6	17.54	1563.8	61.28	3.49

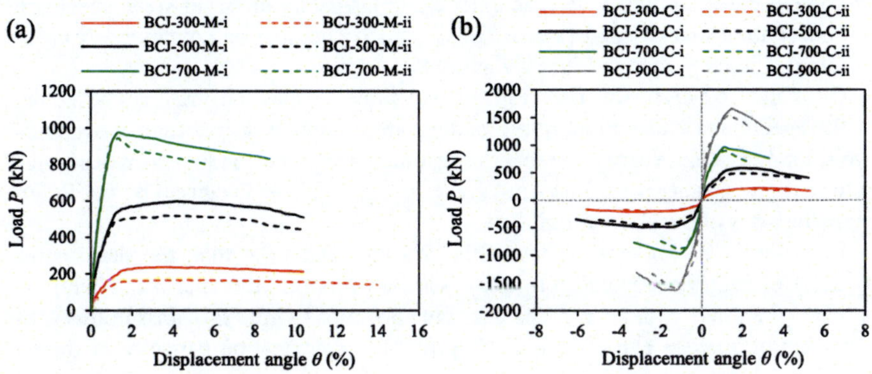

Fig. 9.8 Skeleton curves of the RC beam-to-column joints: **a** monotonic load; and **b** cyclic load

cyclic loading is greatly weaker than that under monotonic loading, and the failure of specimens under cyclic loading is much more brittle. This means the deformation capacity of the RC joints under cyclic loading is significantly weaker than that under monotonic loading.

5. *Ductility capacity*

The relationships between the ductility coefficients and structure size D (herein the width of cross section of the column was utilized) are plotted in Fig. 9.9. Table 9.4 listed the test results of RC joints. It can be seen from Table 9.4 that, the yield load P_y and the peak load P_{max} of the tested RC joints under both monotonic and

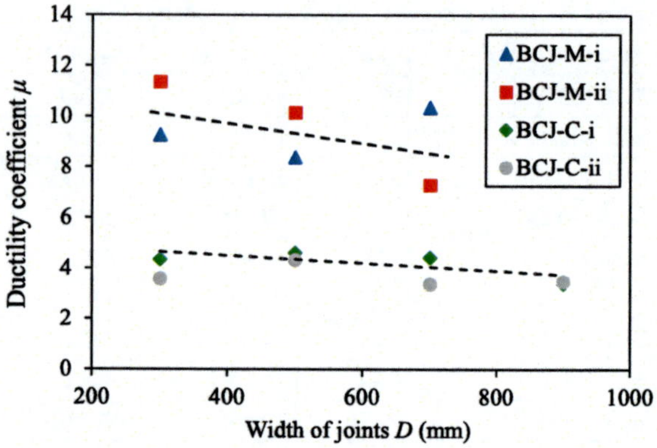

Fig. 9.9 The ductility coefficient curve of all specimens

cyclic loadings increase with the increase of specimens' size. However, the increase extent decreases gradually, illustrating the presence of size effect. Moreover, one can clearly see from Fig. 9.9 that with the increase of the specimen size, the ductility coefficient of the specimens under both monotonic and cyclic loadings generally exhibit a downward trend. That is, the larger the structure size is, the lower ductility coefficient the RC joints have. The deformation capacity of larger-sized specimens is weaker than that of smaller-sized ones, which can also be concluded from the softening curves plotted in Figs. 9.7 and 9.8.

One can also clearly note from Fig. 9.9 and Table 9.4 that, the ductility coefficients of the specimens under monotonic loading are obviously larger than those under cyclic loading. That is to say, the specimens under monotonic loading present stronger deformation capacity than those under cyclic loading. This is caused by the fatigue characteristics of cyclic loading.

In addition, it can be seen from Fig. 9.9 and Table 9.4 that, the deformation capacity of RC beam-to-column joints with stirrups (i.e. the specimens with "i") is better than that without stirrups (i.e. the specimens with "ii"). In summary, the existence of stirrups can effectively improve the deformation capacity of the RC beam-to-column joints.

6. *Stiffness degradation*

For the eight RC joints of different structural sizes, the corresponding stiffness degradation curves are plotted in Fig. 9.10.

One can clearly find from Fig. 9.10a and b that, when the value of the displacement angle θ is less than 4%, the stiffness of larger-sized specimens degrades obviously faster than that of small(er)-sized specimens. However, the stiffness degradation curves of the same-sized specimens with and without stirrups are also overlapped especially for the ones under cyclic loading. One can also conclude from Fig. 9.10a and b that, from the cracking stage to the yielding stage, the global stiffness of specimens degrades rapidly, while the stiffness degradation turns out to be slower in the later stage. Furthermore, one can note from the comparison of Fig. 9.10a and b

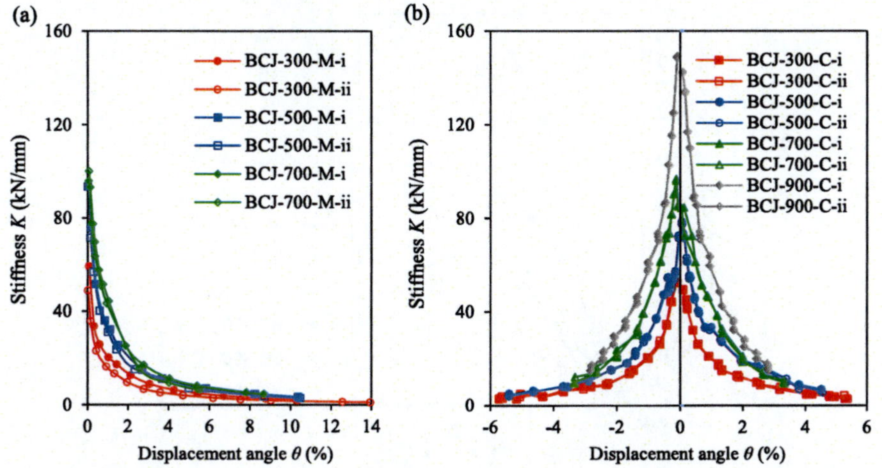

Fig. 9.10 The stiffness degradation curves of specimens: **a** monotonic loading; and **b** cyclic loading

that, the stiffness of the specimens under cyclic loading degrades much faster than that under monotonic loading.

7. *Energy dissipation capacity*

The relationship between the energy dissipation coefficient E_j and the displacement angle θ of the RC joints under cyclic loading is shown in Fig. 9.11. One can find that, when the magnitude of the displacement angle θ is less than 0.5%, the corresponding energy dissipation coefficient increases slowly. However, as the displacement angle θ is larger than 1%, the energy dissipation coefficient per unit area develops rapidly and nonlinearly.

8. *Nominal shear strength*

As discussed earlier, the typical brittle shear failure occurred in the core region of all RC beam-to-column joints in the test. The brittle shear failure leads to the fact that the nominal shear strength exhibits a strong size effect. Herein this part, the existence of size effect of the nominal shear strength would be checked. According to Chinese code (Ministry of Construction of the People's Republic of China 2011), the actual shear force V_{jt} in the core area of RC beam-to-column joint is

$$V_{jt} = \left(\frac{\sum M_b}{h_{b0} - a_s'}\right)\left(1 - \frac{h_{b0} - a_s'}{H_c - h_b}\right) \tag{9.1}$$

in which, M_b is the sum of the design values of counterclockwise or clockwise combined bending moments at the beam ends on both sides of the joint; h_{b0} is the effective height of the beam section; a_s' is the distance from resultant point of compression reinforcement of the beam to the near side of the beam cross section;

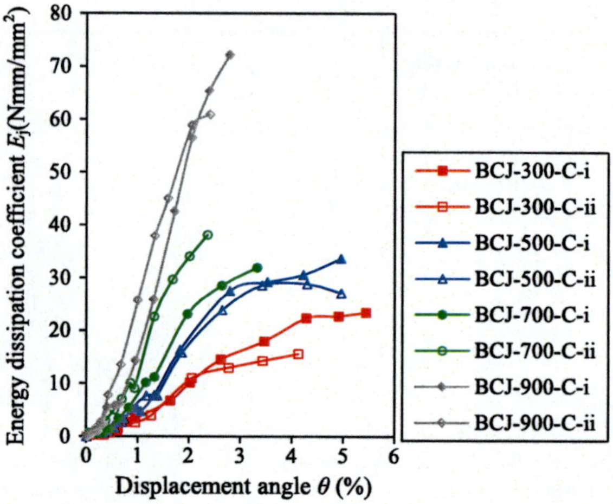

Fig. 9.11 Energy consumption coefficient per unit area of the specimens under cyclic loading

H_c is the distance between the reverse bending points of the upper column and the lower column of the joint; and h_b is the height of the beam section.

The nominal shear strength τ_{max} is calculated by

$$\tau_{max} = V_{max}/b_j h_j \tag{9.2}$$

in which, V_{max} denotes the bearing capacity of RC joints; and b_j and h_j respectively mean the width and the height of the RC beam-to-column joints.

Figure 9.12 shows the nominal shear strength of all tested RC beam-to-column joints having different structural sizes. It can be seen from Fig. 9.12a and b that with increasing the specimen's size, the nominal shear strengths of the RC joint with and without stirrups decrease obviously. In a word, the nominal shear strength exhibits an obvious size effect.

What's more, it can also be found that, the tested nominal shear strengths of the RC specimens under cyclic loading are obviously smaller than those under monotonic loading. In reality, due to the damage accumulation of the RC joints under cyclic loading, not only the tensile/compressive strength of concrete and yield strength of the steel bars degrade, but also the bonding capacity of the interface of concrete and steel bars decreases.

It can be found from the comparison of Fig. 9.12a and b that, the nominal shear strengths of the specimens with the stirrup ratio ρ_{sv} of 0.84% are larger than those without stirrups (i.e. $\rho_{sv} = 0$). However, the difference on the shear strength is not large. This should be attributed to the fact that, when the RC specimens with stirrups failed, almost all stirrups were in the elastic stage and they did not reach their yield strengths, as discussed previously in Sec. 9.1.1 (3. Strains of steel bars). This, therefore, indicates that in the present tests, the shear bearing capacity of the

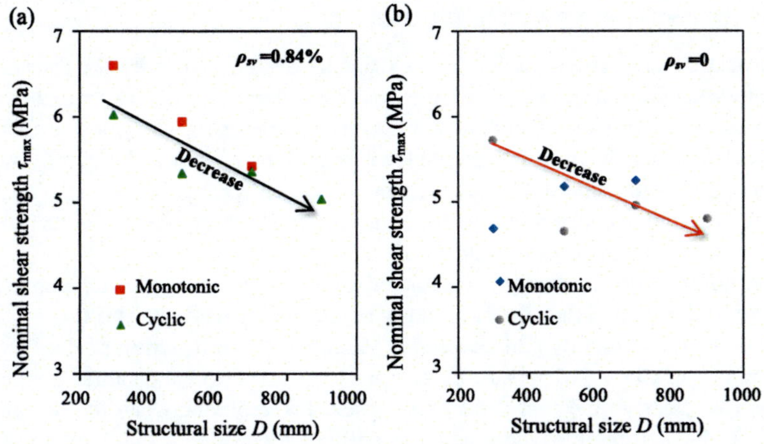

Fig. 9.12 Tested nominal shear strengths for all RC beam-to-column joints: **a** stirrup ratio $\rho_{sv} = 0.84\%$; and **b** stirrup ratio $\rho_{sv} = 0$

RC beam-to-column joints with stirrups is mostly contributed by the concrete part rather than the stirrups. Of course, this should be the main reason that the nominal shear strength of the RC specimen with stirrups shows obvious size effect.

9. *Comparison with some codes*

There are some formulas for calculating the shear bearing capacity of RC beam-to-column joints from four standard codes, e.g. NZ S3 101 (2006), ACI 318-05 (2005), AIJ (1998) and Chinese code (Ministry of Construction of the People's Republic of China 2011). The comparison of the tested values and the design values of RC joints under cyclic loading is plotted in Fig. 9.13.

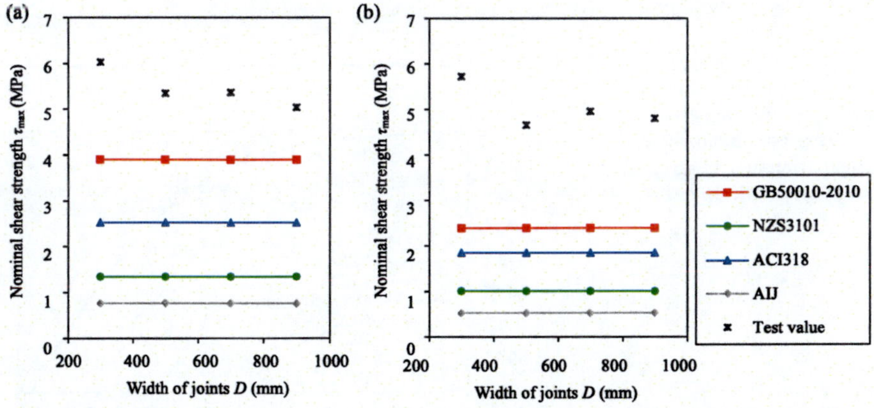

Fig. 9.13 Comparison of the test data (under cyclic loading) and the design value calculated by different codes: **a** stirrup ratio $\rho_{sv} = 0.84\%$; and **b** stirrup ratio $\rho_{sv} = 0$

It can be seen clearly from Fig. 9.13 that:

1. All tested shear strengths are larger than the design values by the four codes. However, as the structural size increases, the difference narrows gradually.
2. Because of the lack of understanding on the mechanism of shear effect, the calculated shear bearing capacity of the core region of joints varies greatly among the four different codes. Obviously, there's a big difference between the nominal shear strength calculated by the four codes. Among them, the AIJ code is found to be the most conservative one.
3. The nominal shear strengths calculated by the four codes are independent on structural size, and thus they could not reveal any size effect. The shear bearing capacity calculated by Chinese code (Ministry of Construction of the People's Republic of China 2011) is the closest one to the experimental value. In addition, one can clearly note that there exists great difference between the design and tested shear strength of the specimens without stirrups.

10. *Size effect on the nominal shear strength*

According to the test data, the double logarithmic curves of the nominal shear strength of the RC beam-to-column joints under different loading types and different stirrup ratios are plotted in Fig. 9.14. Furthermore, the size effect law (SEL) for shear strength of RC beams without stirrups (Bažant and Kim 1984), the strength criterion without considering size effect (for plastic materials) and the classic linear-elastic fracture mechanics (i.e., the LEFM) for pure brittle materials are also plotted in Fig. 9.14 for comparisons.

Furthermore, the test data on the nominal shear strength are in good agreement on Bažant size effect law (SEL) of RC beam. However, the size effect law does not quantitatively describe the effect of stirrup ratios on size effect of specimens. The shear mechanism of RC joints with stirrups and RC beams without stirrups is quite different. Therefore, size effect law which can quantitatively describe the

Fig. 9.14 The double logarithmic curves of the nominal shear strength of the RC beam-to-column joints under different loading types and different stirrup ratios

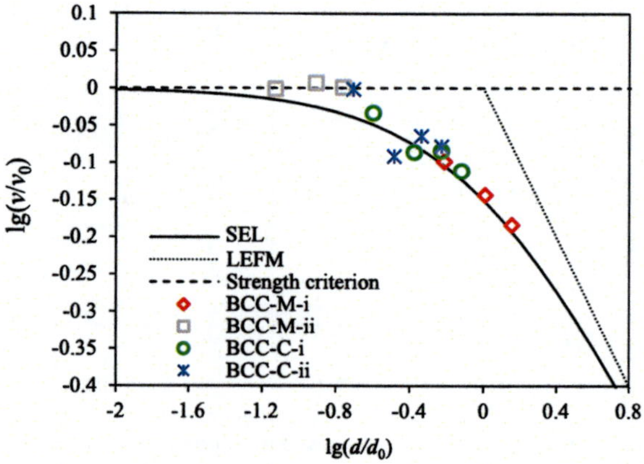

effect of stirrups on size effect of shear strength of RC joints should be developed in subsequent studies.

It can be seen from Fig. 9.14 that, the shear strength of the RC joints under both monotonic and cyclic loadings presents an obvious size effect, namely, the magnitude of the strength decreases with increasing the structural sizes. Compared with the test data of the RC beam-to-column joints with the stirrup ratio $\rho_{sv} = 0.84\%$, that of joints with $\rho_{sv} = 0$ lies closer to the inclined asymptote of slope $-1/2$ (LEFM), indicating a more obvious size effect.

Moreover, one can also note from Fig. 9.14 that, the test data for the RC specimens under cyclic loading lies much closer to the inclined asymptote of slope $-1/2$ compared with that under monotonic loading. This should be attributed to the fact that, the shear failure of RC joints under cyclic loading is much more brittle, which results in a stronger size effect in the nominal shear strength.

11. *Discussion on the source of size effect*

Generally speaking, due to the existence of transverse reinforcement (e.g. stirrups), the size effect of shear strength is supposed to be alleviated, weakened or even suppressed, as observed in Angelakos et al. (2001) and Lubell et al. (2004). In reality, stirrups or other shear resistant steel-bars can prominently improve the shear strength due to the strengthening of compression strut. Nevertheless, a series of test results, e.g. Bažant (2005), Bažant and Xiang (1997), Walraven and Lehwalter (1994) as well as Jin et al. (2016) have reported that there still exists a significant size effect on geometrically similar deep beams having stirrups. Moreover, the results of slender beams simulated by Yu and Bažant (2011) have indicated that no matter how much the stirrups ratio is, the stirrups cannot suppress the size effect in the shear strength absolutely.

Similarly, as the test results have shown that, the shear failure of RC beam-to-column joints with and without stirrups exhibits a strong size effect. The source of such strong size effect may be as follows:

1. Though having transverse reinforcements, the shear failure at core region of RC beam-to-column joints under cyclic loads is still brittle. Simply to say, it attributes to the fatigue nature of cyclic loading.
2. From the standpoint of mechanism, before the yielding of stirrup, the compression-shear brittle failure has occurred in the concrete of the core region of the joint having stirrups. The tested results on bearing capacity have demonstrated this fact. That is to say, the failure of the RC beam-to-column joints having stirrups is governed by concrete failure, but not stirrups yielding. In the present tests, most of the shear force was found to be supported by concrete part, but not the stirrups.
3. The test observations have shown that, the larger the specimen's size is, the greater the crack width is. The crack with larger width results in a decrease in biting force of aggregates and friction of shear forces. Consequently, the greater the crack width is, the smaller the bite force and friction force are, so also is the ultimate capacity.

9.1.2 Simulations on Interior RC Beam-to-Column Joints

It can be seen from the experimental analysis of Sect. 9.1.1 that the stirrup ratio in the core area of the joint has a significant influence on the seismic performance and size effect of the joint. On the other hand, Parker and Bullman (1997) showed that axial load of column influences the behavior of joints by changing their failure modes. Scott et al. (1999) found an enhancement in the ultimate shear capacity in specimens with higher axial stresses. Fu et al. (2000), Haach et al. (2014) and Masi et al. (2013) found axial load applied to columns influenced the failure modes and mechanical performance of RC joints. Hayashi et al. (1983) and Wakabayashi et al. (1983) have reported that the mechanical properties of stirrups are closely related to the anchorage length and the axial compression ratio of joints. Therefore, based on the experiment in Sect. 9.1.1, more parameters (section size of joints: 300 mm × 300 mm, 500 mm × 500 mm, 700 mm × 700 mm, 900 mm × 900 mm, 5 groups of axial compression ratio: 0, 0.15, 0.3, 0.45, 0.6, and 4 groups of stirrup ratios: 0, 0.42%, 0.84% and 1.26%) were considered to study the influence mechanism and law of these parameters on the size effect of beam-to-column joint in this section.

1. *Establishment and verification of meso-scale numerical model*

The establishment and verification of the mesoscopic numerical model of RC joints were shown in Sec. 3.6.4. The detail information of mesoscopic numerical model of interior joint can be found in the article of Jin et al. (2020a).

2. *Failure mode*

Figure 9.15 shows the final failure mode of joints with section size of 500 mm × 500 mm under different stirrup ratios and axial compression ratios. As can be seen from Fig. 9.15a, the failure of specimens was shear failure occurred in the core area. Nevertheless, as stirrup ratio increased, the failure range and degree of joints decreased. When the stirrup ratio of joints is zero, the longitudinal bars of beams yield. When the stirrup ratio is 0.42%, 0.84% and 1.26% respectively, the number of the yielding longitudinal bars is reduced. When the stirrups bear part of the shear force, part of the stirrups yield. Figure 9.15b shows the final failure mode of joints under axial compression ratio of 0.3 and different stirrup ratios. Like the specimens with axial compression ratio of zero, the shear failure of joints having axial compression ratio of 0.3 also occurred in core area, and the failure degree of joints decreased with the increase of stirrup ratio. However, due to the increase of axial compression ratio, more shear force is transferred to the core area of the joint, which increases the shear force in core area of joints, and joints are destroyed more seriously.

Figure 9.16 shows the shear capacity provided by concrete and stirrups of joints with width of 500 mm under different stirrups ratios and axial compression ratios. As can be seen from Fig. 9.16, the concrete bears most of the shearing force in the simulation results, and the failure of the joints is mainly caused by the failure of concrete. The shear capacity of joints provided by stirrups in the core area increases

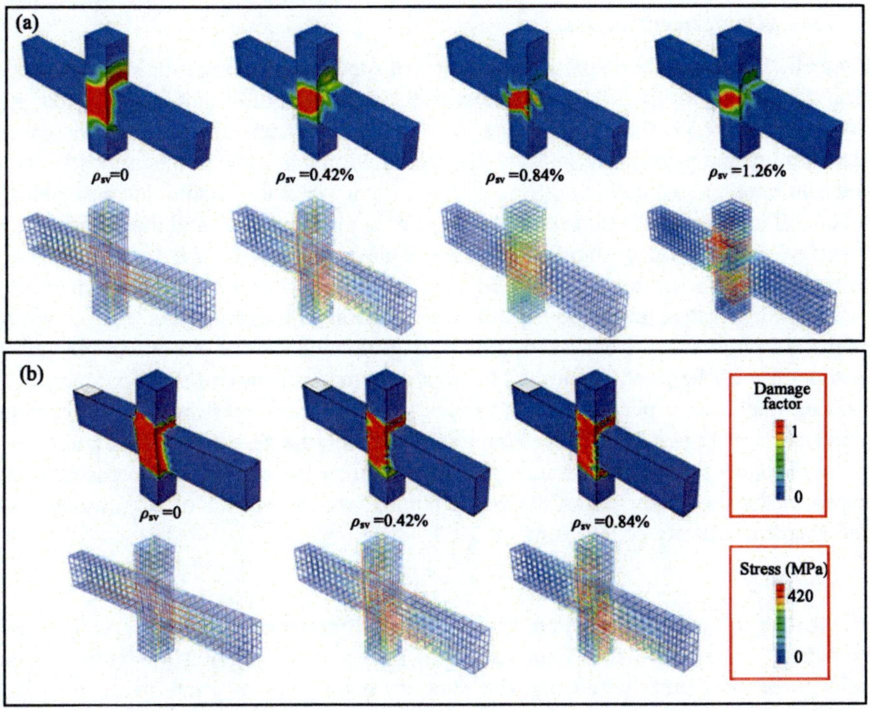

Fig. 9.15 Failure mode of the RC beam-to-column joints with different stirrup ratios ($D = 500$ mm): **a** axial compression ratio $n = 0$; and **b** $n = 0.3$

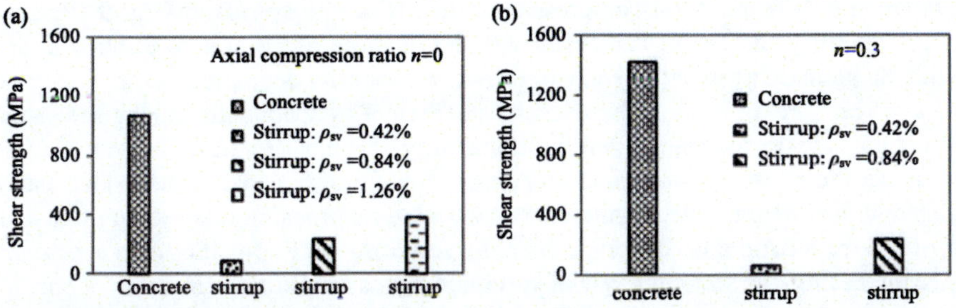

Fig. 9.16 Shear capacity provided by concrete and stirrups of joints: **a** axial compression ratio $n = 0$; and **b** $n = 0.3$

with the increase of stirrup ratio. This is because more stirrups bear the shear force transferred to the core area of the joint. As the axial compression ratio increasing, aggregate bite force and the friction force of concrete at the crack increase, as a result, the shear capacity of joints contributed by concrete increases.

3. *load-displacement curves*

Figure 9.17 shows load-displacement curves of specimens having four kinds of structural sizes, four kinds of stirrup ratios, five kinds of axial compression ratios. As shown in Fig. 9.17, the curve of the joint with the stirrup ratio of zero decreases rapidly after the peak point, that is, the deformation capacity of specimens is poor and brittleness is strong. In addition, with the stirrup ratio increasing, the descending section of load-displacement curve of RC joint becomes gentle, and the deformation capacity of RC joints gradually increases. This is actually because stirrups in the core area of joints inhibited the development of shear cracks, increased the bite force between aggregates, and enhanced the deformation capacity and ductility of specimens. The higher the stirrup ratio, the stronger the constraint effect and the better the ductility of the joint. Figure 9.17 also shows the influence of axial compression ratio on the seismic performance of joints. As axial compression ratio increasing, the value of peak points on curves increase, however, the descending segment of the curve becomes steeper, which indicates that the increase of axial compression ratio improves the shear capacity of the joints, but leads to the decline of ductility and the deformation capacity of RC joint.

4. *Ductility*

Figure 9.18 shows the ductility coefficient μ and softening rate k of joints. As obtained from Fig. 9.18a and b that with structural size of joints having the same stirrup ratio in the core area increasing, the ductility coefficient of joints decreases, that is, the deformation capacity decreases. With the same size of the specimen, the ductility coefficient of the joints increases with the increase of stirrup ratio, and the deformation capacity and ductility increase. Figure 9.18c and d shows the relationship between the ductility coefficient and the structural size of joints with the stirrup ratio of zero and different axial compression ratios. It can be seen from Fig. 9.18c and d that ductility coefficient decreases with the increase of the size of joints, that is, with the increase of structural size of joints, the ductility decreases. As the increase of axial compression ratio, the ductility coefficient of specimens having the same structural size decreases, the deformation ability of joints decreases.

In addition, size effect behavior of shear strength of the specimens is also associated with softening rate. Figure 9.18c shows the relationship of softening rate and size and the stirrup ratio of joints. As observed from the Fig. 9.18c, with specimen size increasing, the softening rate of specimen increases. With the increase of stirrup ratio, the softening rate is generally decline, and the deformation ability of joints is enhanced. It can be noted from Fig. 9.18d that with the increase of structural size, softening rate increases, and the deformation capacity of the joint is reduced. The softening rate of the joints having the same size increases with the increase of the axial compression ratio, indicating that the increase of the axial compression ratio reduces the deformation capacity of the joint.

5. *Nominal shear strength*

Figure 9.19 shows the relationship of nominal shear strength and width of joints with different sizes, stirrup ratios and axial compression ratios. It can be seen

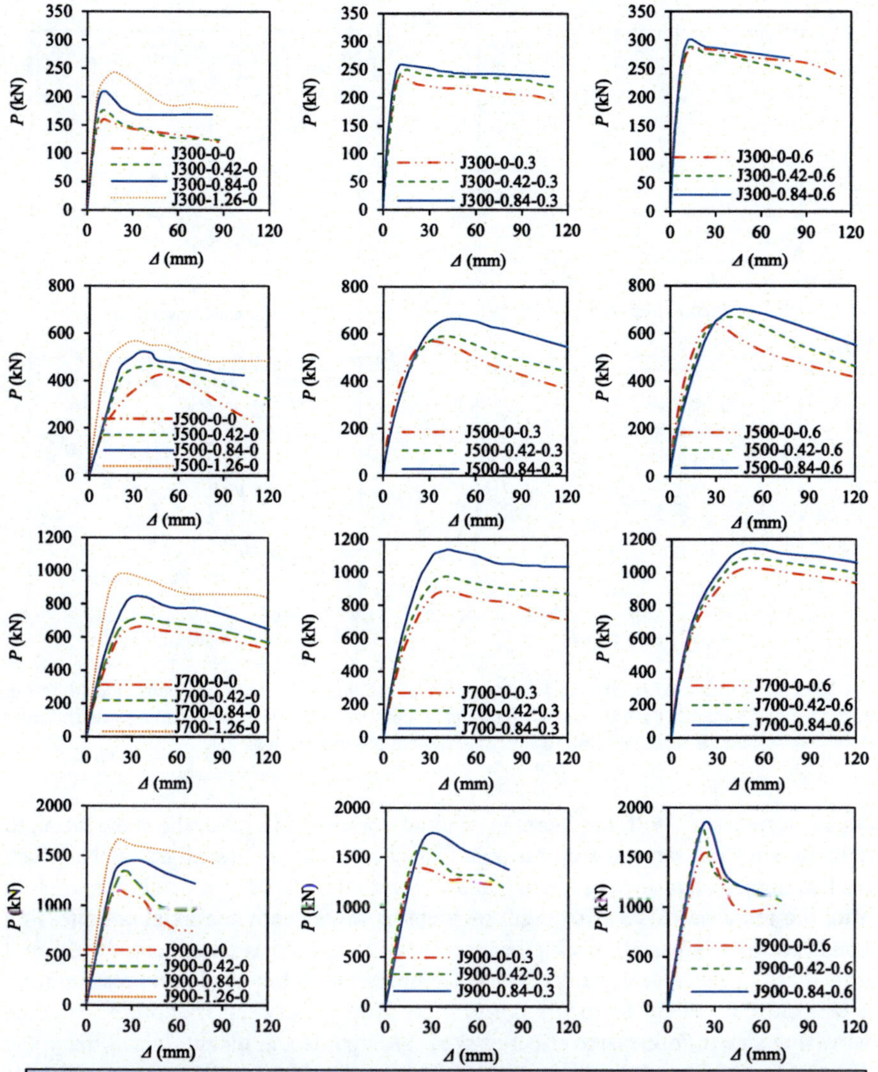

P: load of the beam end. *Δ*: displacement of the beam end. **J300-0.42-0.3**: The letter "J" denotes joint; 300 means the width of joints is 300 mm; 0.42 means stirrup ratio is 0.42%; and 0.3 means axial compression ratio is 0.3.

Fig. 9.17 *P-Δ* curves of joints with different sizes, stirrup ratios and axial compression ratios

from Fig. 9.19a that nominal shear strength of specimens having the same stirrup ratio decreases with the increase of specimen size, that is, size effect is obvious. However, as stirrup ratio of joints increasing, the decreasing trend of nominal shear strength is alleviated, that is, the stirrup inhibits the size effect. As can be noted from Fig. 9.19b that, with the increase of stirrup ratio, the size effect behavior of joints

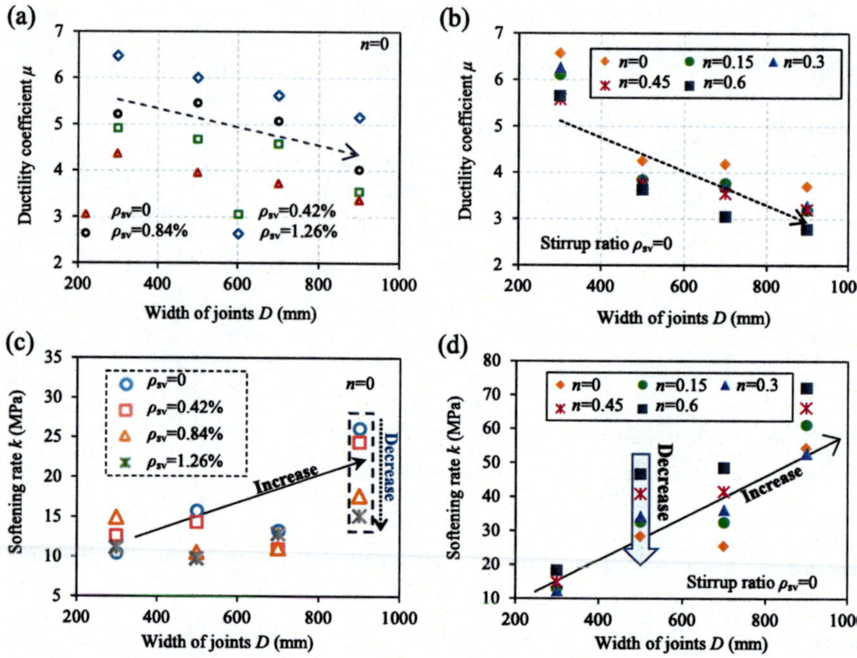

Fig. 9.18 Ductility coefficient and softening rate of RC joints: **a** ductility coefficient (different stirrup ratios); **b** ductility coefficient (different axial compression ratios); **c** softening rate(different stirrup ratios); and **d** softening rate (different axial compression ratios)

can be suppressed. With the increase of axial compression ratio, the shear strength of joints with the same stirrup ratio and structural size increases gradually. When axial compression ratio is between 0 and 0.3, with the increase of axial compression ratio, the increase of shear strength of joints is obvious. When axial compression ratio is between 0.3 and 0.6, with the increase of axial compression ratio, the amount of increase in shear strength decreases. Compared with the axial compression ratio of 0.3 and 0.6, when the axial compression ratio is between 0 and 0.3, with the increasing size of joints, size effect of shear strength is significant. In summary, the stirrup ratios and axial compression ratios have a significant effect on size effect of RC joints.

In order to intuitively reflect the weakening degree of stirrup ratios of size effect on shear strength, there defines size effect degree value D_{SE}.

$$D_{\mathrm{SE}} = \frac{\tau_{\min} - \tau_{\max}}{\tau_{\min}} \times 100\% \tag{9.3}$$

where $\tau_{\min}$ is the nominal shear strength of joints having the smallest width (300 mm) under the same working condition (the same stirrup ratio and axial compression ratio);

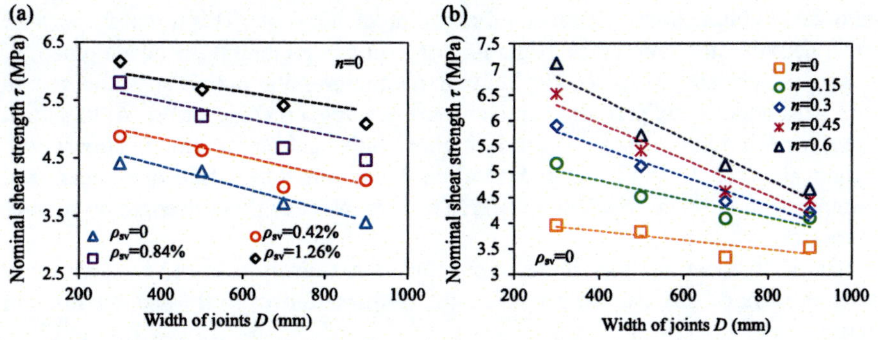

Fig. 9.19 Nominal shear strength of RC joints under different stirrup ratios and axial compression ratios: **a** axial compression ratio $n = 0$; and **b** stirrup ratio $\rho_{sv} = 0$

and τ_{max} is nominal shear strength of the joint with the largest width (900 mm) under the same working condition.

The size effect degree value represents the degree to which the shear strength of the joint decreases with the increase of its size. The bigger size effect degree value indicates that size effect on shear strength is significant. The comparison of the size effect degree to shear strength of joints with different axial compression ratios and stirrup ratios is plotted in Fig. 9.20a. It can be observed that under the same axial compression ratio, degree of size effect on shear strength of joints decreased with increment of stirrup ratio, which shows that the increasing stirrup ratio weakens size effect on shear strength. Because of stirrup ratio increasing, the mechanical properties of joints are improved, including increasing shear strength and ductility of joints. Especially for large-sized joints, their concrete strength is affected by size effect to a certain extent, which makes large-sized joints be more dependent on the contribution of stirrups in the middle and later stage of loading, so the increase of stirrup ratio can weaken size effect on shear strength of joints. At the same time, with the increase of axial compression ratio, the effect of stirrups in the core area of the

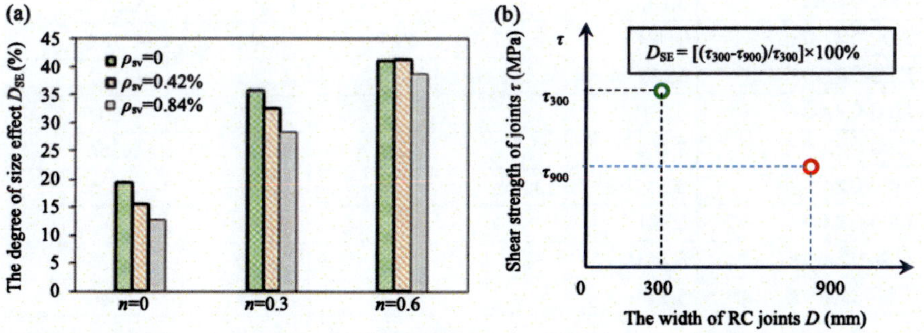

Fig. 9.20 Comparison of the degree of size effect: **a** value of D_{SE}; and **b** definition of D_{SE}

joint on inhibiting the size effect on shear strength decreases. This is mainly because axial compression ratio at the beginning of loading can inhibit the development of cracks in core area of joints, but larger axial compression ratio makes concrete in core area under more pressure, which makes concrete crushing earlier. At this time, the shear failure of joints is obviously higher brittle, and the increase of stirrup ratio cannot significantly inhibit its brittle failure. Therefore, for joints having larger axial compression ratio, the weakening degree on size effect of shear strength by stirrups is lower.

Based on above analysis, it can be concluded that stirrup ratio and axial compression ratio have obvious influence on the deformation capacity and ductility of joints.

6. *Size effect on shear strength*

Size effect parameters are obtained by the regression analysis of the simulation results of joints having different stirrup ratios and axial compression ratios presented in Tables 9.5 and 9.6 (the naming of the specimens can be seen in Fig. 9.17). The plotting of the double logarithmic curve of nominal shear strength is made.

Figure 9.21 shows the regression analysis curve of nominal shear strength of joints having different stirrup ratios and axial compression ratios. The shear strength of joints obtained by simulated was compared and fitted with the size effect law (SEL) of shear strength of RC beam without stirrups, the theory of linear elastic fracture mechanics (LEFM), and strength criterion. It can be noted from Fig. 9.21 that the simulation data points are between elastic-plastic curve and linear elastic theory curve, namely the shear failure behavior is between the plastic and brittle failure.

Table 9.5 Fitting parameters for the RC beams having different stirrup ratios

Specimen name	v_u	v_0	d_0	d/d_0	$\lg(v_u/v_0)$
J300-0-0.6	7.800	18.898	70	4.286	−0.384
J500-0-0.6	6.353			7.143	−0.473
J700-0-0.6	5.709			10.000	−0.520
J900-0-0.6	4.800			12.857	−0.595
J300-0.42-0.6	7.984	12.217	223	1.343	−0.185
J500-0.42-0.6	6.683			2.239	−0.262
J700-0.42-0.6	6.030			3.134	−0.307
J900-0.42-0.6	5.404			4.030	−0.354
J300-0.84-0.6	8.344	13.363	187	1.607	−0.205
J500-0.84-0.6	6.981			2.679	−0.282
J700-0.84-0.6	6.353			3.750	−0.323
J900-0.84-0.6	5.570			4.821	−0.380

Note The naming of the specimens: such as J300-0.42-0.6, the letter "J" represents joints; 300 represents the height *d* of joints is 300 mm; 0.42 represents the stirrup ratio of joints is 0.42%; and 0.6 represents axial compression ratio of joints is 0.6

Table 9.6 Fitting parameters for the RC beams having different axial ratios

Specimen name	v_u	v_0	d_0	d/d_0	$\lg(v_u/v_0)$
J300-0-0	4.411	5.832	490	0.612	−0.121
J500-0-0	4.270			1.020	−0.135
J700-0-0	3.719			1.428	−0.195
J900-0-0	3.400			1.836	−0.234
J300-0-0.15	5.740	7.930	318	0.943	−0.140
J500-0-0.15	5.021			1.572	−0.198
J700-0-0.15	4.555			2.201	−0.240
J900-0-0.15	4.111			2.830	−0.285
J300-0-0.3	6.567	13.867	104	2.884	−0.324
J500-0-0.3	5.678			4.807	−0.387
J700-0-0.3	4.921			6.730	−0.449
J900-0-0.3	4.220			8.653	−0.516
J300-0-0.45	7.254	20.851	46	6.521	−0.458
J500-0-0.45	6.024			10.869	−0.539
J700-0-0.45	5.140			15.217	−0.608
J900-0-0.45	4.450			19.565	−0.670
J300-0-0.6	7.800	18.898	70	4.286	−0.384
J500-0-0.6	6.353			7.143	−0.473
J700-0-0.6	5.709			10.000	−0.520
J900-0-0.6	4.800			12.857	−0.595

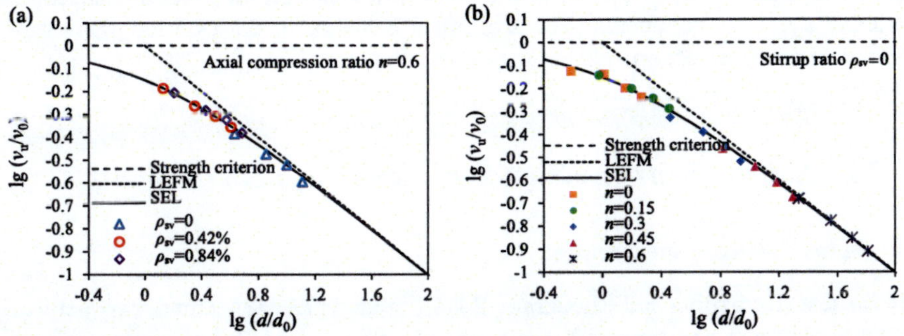

Fig. 9.21 Double logarithmic curve of nominal shear strength of joints of different stirrup ratios and axial compression ratios: **a** different stirrup ratios; and **b** different axial compression ratios

The larger the stirrup ratio of the joint is, the closer the simulation results are to strength criterion curve. The lager the axial compression ratio is, the closer the simulation results are to LEFM. It indicts that size effect is inhibited by stirrups, and the increasing axial compression ratio makes size effect more obvious.

In spite of this, it can be found from Tables 9.5 and 9.6 (the values of v_0 and d_0 obtained by fitting under different working conditions), under the different stirrup ratios, the fitting value of v_0 and d_0 are obviously different. Equation (6.2) is the size effect law (SEL) for the shear strength of RC beam without stirrups, and the size effect law does not quantitatively describe the effect of stirrup ratios and axial compression ratios on size effect of specimens. At the same time, it is different from the shear mechanism of specimens without stirrups, the shear force in joints with stirrups is mainly borne by concrete when inclined cracks are not formed. When inclined cracks occurs, the shear force is mainly borne by stirrups in core area of joints. The existence of stirrups limits the development of cracks and enhances the ductility of joints. Therefore, the size effect on shear strength of joints having stirrups is weaker than that of joints without stirrups. In other words, Eq. (6.2) cannot fully reflect the size effect on shear strength of joints. In Sect. 9.3, an equation for size effect on joints is introduced, which can quantitatively reflect the influence of stirrup and axial force on size effect of shear strength of joints.

9.2 Shear Failure of Exterior RC Beam-to-Column Joints

In order to explore the size effect behavior and law of exterior joints, experimental (Barbhuiya and Choudhury 2015) and numerical study on the exterior joint in the middle layer was conducted. The size effect behavior of the exterior joints were introduced in this section.

9.2.1 Seismic Tests on Exterior RC Beam-to-Column Joints

1. *Introduction of experiment*

In the test (Barbhuiya and Choudhury 2015), three types of specimens, namely, (a) Beam-column connection with beam weak in flexure, (b) Beam-column connection with beam weak in shear and (c) Beam-column connection with column weak in shear were considered. In each type, three geometrically similar specimens: full scaled, two-third scaled and one-third scaled sizes were considered. Two specimens were considered under each case and, hence, there were eighteen specimens. All the three dimensions of two-third and one-third scaled specimens were arrived at by geometrically scaling down the dimensions of full size specimen. Similarly, reinforcement and coarse aggregates were also geometrically scaled down for satisfying the similitude requirement.

The aggregate size effect was nullified by using different mix designs for different aggregates and it was ensured that all gives the same target strength. Two grades of concretes with target strength 30 and 25 MPa were used. The former was used for beam-column joint specimens with beam weak in flexure and for beam-column joint specimens with beam weak in shear, while the later was used for beam-column joint specimens with column weak in shear.

The naming of the specimens was done with five alphabets, the first three alphabets cover the deficiency type, the fourth for size and the fifth for material type of specimens, which is concrete in this case. For example, BWFLC stands for beam weak in flexure large concrete specimen; similarly CWSSC stands for column weak in shear small concrete specimens.

Reinforcing steel of diameters 20, 12 and 8 mm were of High Yield Steel Deformed (HYSD, Fe 500) bar, while reinforcing steel of diameters 6, 4 and 2 mm were of mild steel (Fe 250) type. The details of these specimens are shown in Fig. 9.22.

The installation of specimens and the test setup are given in Fig. 9.23. The loading scheme is plotted in Fig. 9.24. The detail introduction of specimens, loading device and the loading process are shown in (Barbhuiya and Choudhury 2015).

2. *Analyses of experimental results*

The experiment was stopped at a stage when the load came down in the range of 60–70% of the ultimate load carrying capacity. Figure 9.25 shows a typical photograph of BWFMC specimen at the completion of the experiment. The development of deep crack along the beam-column interface can be observed from the figure and the beam was observed to rotate about this plane during the final stage of the experiment. The hysteretic responses for typical three large specimens belonging to the three deficient categories are shown in Fig. 9.26a–c. The average of peak loads in the push and pull direction in the first cycle of each amplitude was calculated. The maximum load as obtained by such averaging was considered as ultimate load carrying capacity of the specimens. Ultimate load carrying capacity of all the tested specimens is presented in Table 9.7. The beam weak in shear was made shear deficient and it was expected to fail at a lower load than beam weak in flexure specimens. However, both showed the similar failure load and this may be due experimental uncertainties.

3. *Size effect on the variation of stress versus relative deflection*

The relative deflections for different scaled models were calculated as the ratio of actual deflection to the scale of the model. Peak value of the load of each displacement cycle was used to calculate the stresses. The evaluated relative deflection and average stresses (considering two specimens) were plotted for different sizes and types of specimens in Fig. 9.27. The plot corresponding to the smallest specimen shows highest stresses at every level of relative deflection, while it is least for largest specimen in most of the cases. Therefore, it can be inferred that the principle of size effect is valid for stress and relative deflection relationship as well.

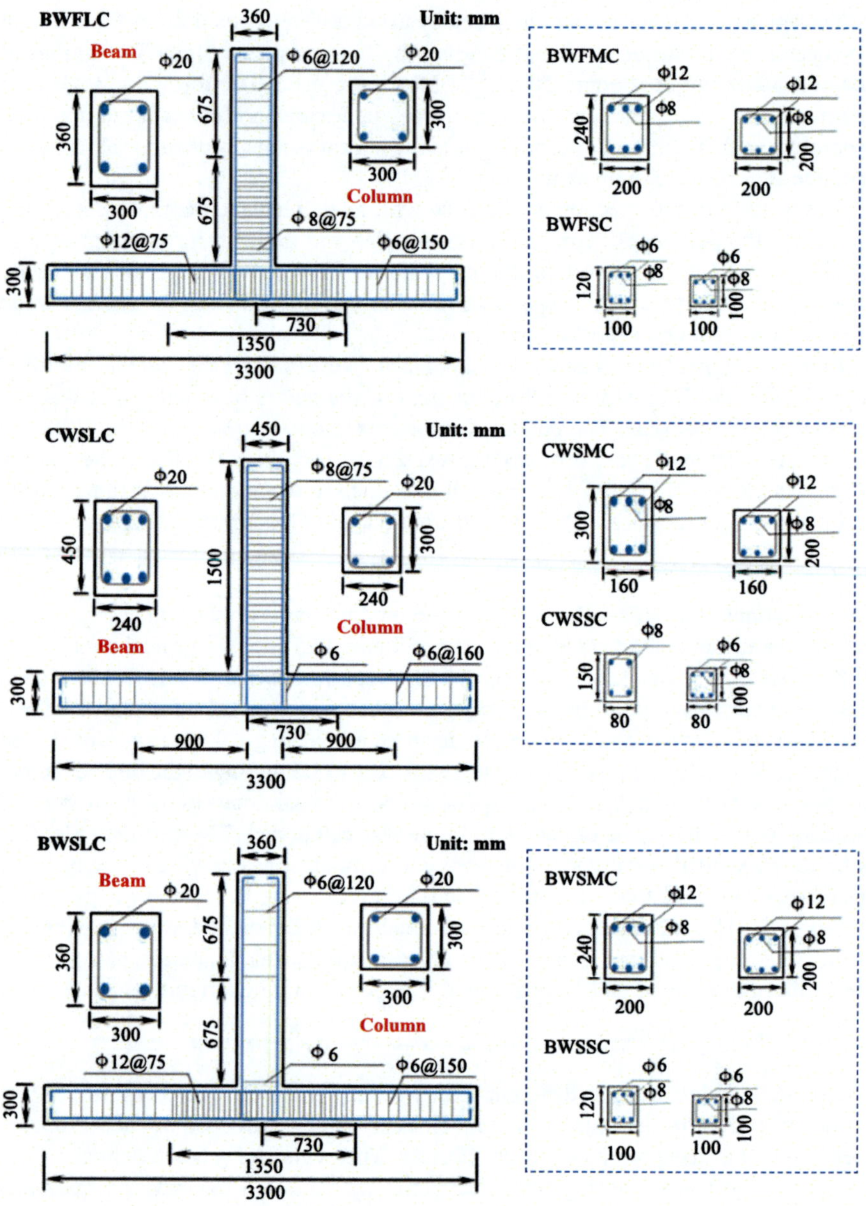

Fig. 9.22 Detail of specimens (Barbhuiya and Choudhury 2015)

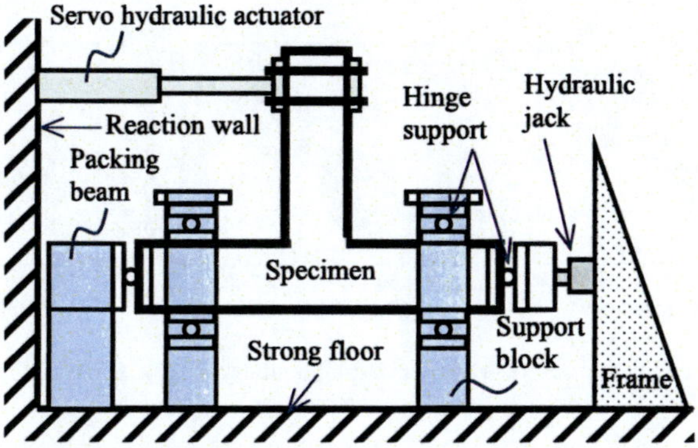

Fig. 9.23 Typical schematic diagram of test set-up

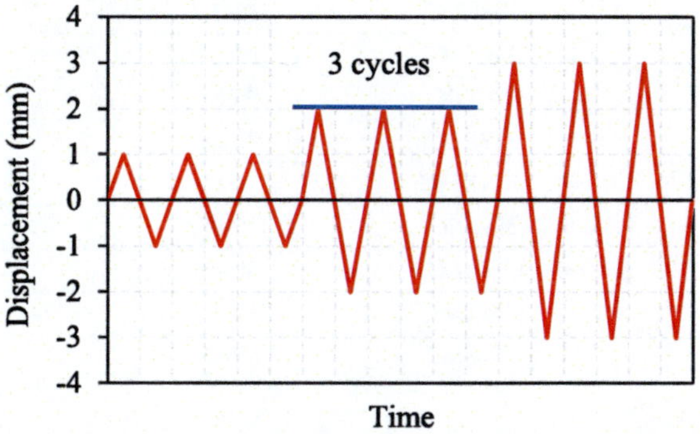

Fig. 9.24 Typical displacement history applied to large specimen

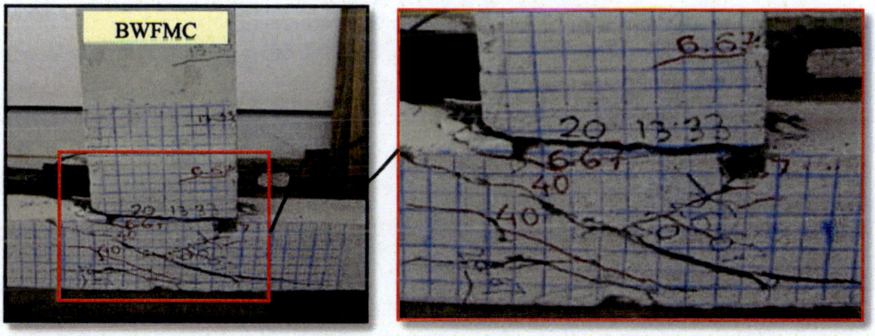

Fig. 9.25 Shear failure modes (Barbhuiya and Choudhury 2015)

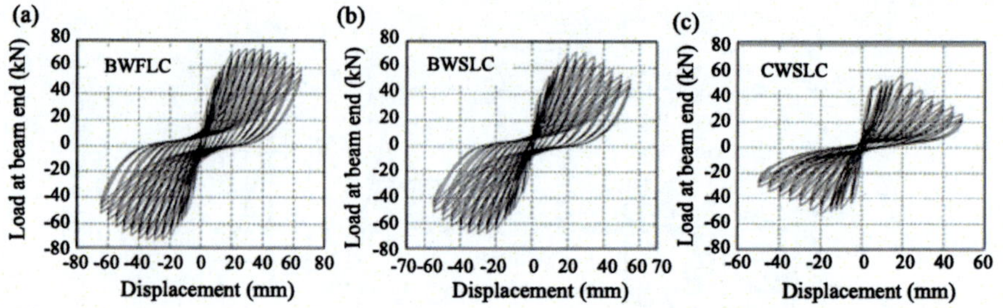

Fig. 9.26 Hysteretic responses (Barbhuiya and Choudhury 2015): **a** BWFLC; **b** BWSLC; and **c** CWSLC

Table 9.7 Ultimate load carrying capacity of specimens

Name of specimen	Ultimate load carrying capacity (kN)
BWFLC	73.285
BWFMC	34.845
BWFSC	9.885
BWSLC	72.085
BWSMC	35.1
BWSSC	10.053
CWSLC	58.01
CWSMC	30
CWSSC	8.635

4. *Size effect on energy dissipated per unit volume*

The ability of a structural element to resist an earthquake depends to a large extent on its capacity to dissipate the energy. The area of hysteresis loop is a measure of the energy dissipated. The cumulative energy dissipated at a particular amplitude was calculated by summing up the energy dissipated in all the preceding cycles including that amplitude. To compare the energy dissipation of specimens having different sizes of beam-column joints, a parameter energy dissipation per unit volume of joint (E_N)

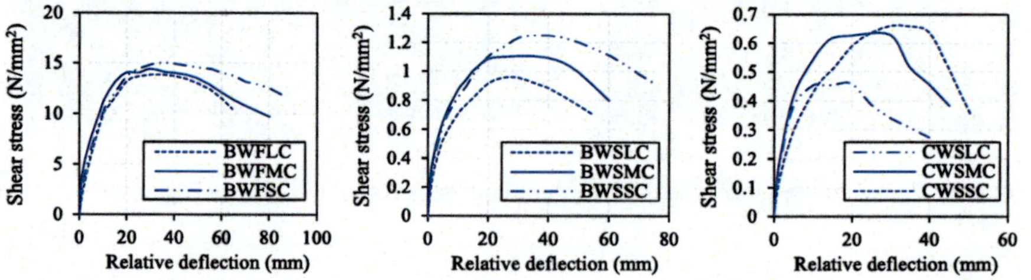

Fig. 9.27 Relationship between relative deflection and average stress (Barbhuiya and Choudhury 2015)

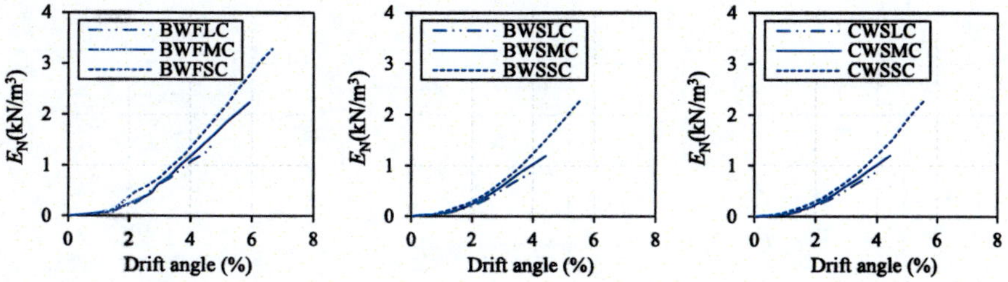

Fig. 9.28 Variation of average cumulative energy dissipated per unit volume (Barbhuiya and Choudhury 2015)

was introduced. The volume which is common to both beam and column in a sub-assemblage is considered as the volume of a joint. Cumulative energy dissipated up to a particular amplitude was divided by the joint volume gives the energy dissipation per unit volume of joint (E_N). The variation of average cumulative energy (considering two specimens) dissipated per unit volume (E_N) for all the specimen sizes and types were plotted with respect to drift angle. The plots for different types of specimens are shown in Fig. 9.28. It can be clearly observed that the uppermost curve is for the smallest specimen and lowermost curve for the largest specimen in all the cases. Thus, the energy dissipation per unit volume increases as the specimen size decreases. This establishes the fact that the energy dissipated per unit volume is dependent on specimen size.

5. *Bi-logarithmic plot*

The bi-logarithmic plot was drawn with $\lg(D/D_0)$ in the X-axis and $\lg(\tau_{Nu}/Bf_t)$ in the Y-axis as shown in Fig. 9.29. It is observed from Fig. 9.29a that it follows a horizontal line at the initial part, indicating no size effect. The curve approaches a straight line with slope of $-1/2$ towards the end (LEFM). In intermediate zone there is a smooth curved transitional part. Thus, it can be concluded that the plot shows presence of size effect in accordance with size effect law. Similarly, the bi-logarithmic plots were drawn for other types of specimens. The bi-logarithmic plots for beams and columns weak in shear are presented in Figs. 9.29b and c respectively. These plots also show the presence of significant size effect and follows size effect law. Moreover, the slope of the bi-logarithmic plot towards the end of the curve for beam weak in flexure specimens is found to be the least, while it is highest for column weak in shear specimens. Column weak in shear specimens are the most brittle specimens out of three types of specimens. Thus, it can be concluded that as the specimen exhibits more and more brittle mode till failure, the size effect also become more pronounced.

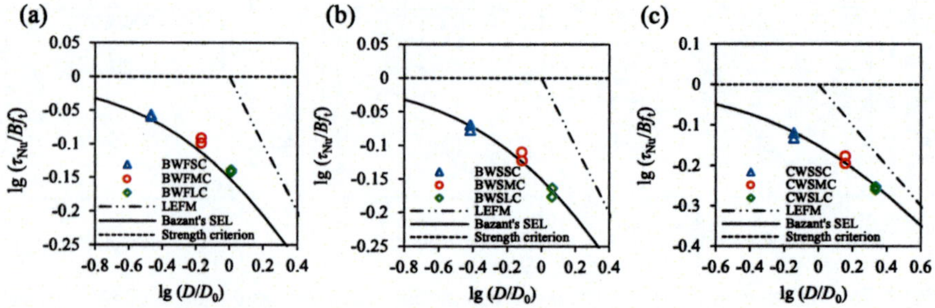

Fig. 9.29 The bi-logarithmic plot: **a** BWF; **b** BWS; and **c** CWS

9.2.2 *Simulations on Exterior RC Beam-to-Column Joints*

1. *Mesoscopic numerical model of RC joints*

The mesoscopic models of RC exterior joints with different structural sizes (300 mm × 300 mm 500 mm × 500 mm, 700 mm × 700 mm, 900 mm × 900 mm) were established. The axial compression ratios were 0, 0.3 and 0.6, and stirrup ratios were 0, 0.42 and 0.84%. The aggregate volume fraction of all specimens is 35%. The aggregate is only randomly distributed in mesoscopic model. The process of modeling is similar with Sect. 3.6.4. The material mechanics parameters of the mesoscopic composition of concrete and the steel were listed in Table 9.8.

To verify the reliability of the simulation method, the result of specimen named JD4 (Wang 2010) was compared with the simulation results. The section size of column is 300 mm × 300 mm, the length is 2800 mm, the section size of beam is 300 mm × 400 mm, the length of beam is 1500 mm. The failure mode of joints is shown in Fig. 9.30a. The failure mode obtained by simulation is in conformity with that obtained by experiment. Figure 9.30b shows the relationship between load at beam end P and displacement Δ obtained by simulation and experiment. It can be seen from the Fig. 9.30b that the simulation value is in good agreement with

Table 9.8 Mechanical parameters of the meso-components of concrete and reinforcing bars utilized

Meso-scale composition	Aggregate	Mortar matrix	ITZ	Reinforcement
Compressive strength f_c (MPa)	–	36.17[a]	28.94[b]	–
Tensile strength f_t (MPa)	–	3.41[a]	2.72[b]	–
Young's modulus E (GPa)	70[a]	32.1[a]	25.68[b]	203[a]
Fracture energy Gc (J/m^2)		45	27	
Poisson ratio ν	0.2[a]	0.2[a]	0.2[b]	0.3[a]
Yield strength f_y (MPa)	–	–	–	496[a]

Note [a]Data were obtained from the tests of Jin et al. (2018); and [b]data were got by the inversion method

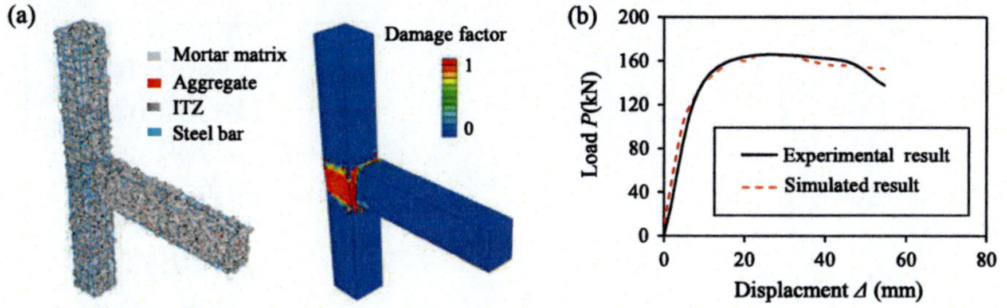

Fig. 9.30 Comparison of simulation and experimental result: **a** RC exterior beam-column joints model and failure process; and **b** load-displacement curves

the experimental value, which proves the reliability of the mesoscopic numerical simulation method.

2. *Simulation results*

Figure 9.31 shows the relationship curves between load and displacement angle of joints having width of 300, 500, 700 and 900 mm. The axial compression ratio is 0.3, and stirrup ratio is 0.42%. It can be seen from Fig. 9.31 that as the structural size increases, the descending section changes from gentle to rapid after the peak load at the beam end, which shows that the ductility of the exterior joint decreased with increasing size, and the displacement angle at beam end decreases with increasing size, the deformation of joint decreases.

Figure 9.32 shows the failure mode of exterior joints with different structure sizes. The axial compression ratio is 0.3, and stirrup ratio is 0.42%. The damage degree of the joint is expressed by the damage factor. It can be noted from Fig. 9.32 that failure modes of joints are shear failure in the core area. There are several oblique fractures in the core area, it indicated that the failure mode of joints is typical shear failure. The degree of shear failure increases with the increase of structural size. In addition, not only the core area of joints with large size is seriously damaged, but also the beam ends and column ends are damaged more and more seriously with the size increasing. The above phenomenon indicates that the size effect of the joint is significant.

The curve of nominal shear strength of exterior joints and width of joint was shown in Fig. 9.33. Different legends represent different working conditions, and the working condition named "0.3–0.42%" represents "axial compression ratio is 0.3 and stirrup ratio is 0.42%". As can be seen from Fig. 9.33, the nominal shear strength of the exterior joint decreases obviously with the increase of the size, which indicates that the size effect on shear failure of joints is obvious. However, axial compression ratio and stirrup ratio have certain influence on nominal shear strength of exterior joints, and the details are described in the following section.

3. *Influence of axial compression ratio*

Figure 9.34 shows the curve of load-displacement angle of the exterior joint under different sizes and different axial compression ratios. It can be noted from the change

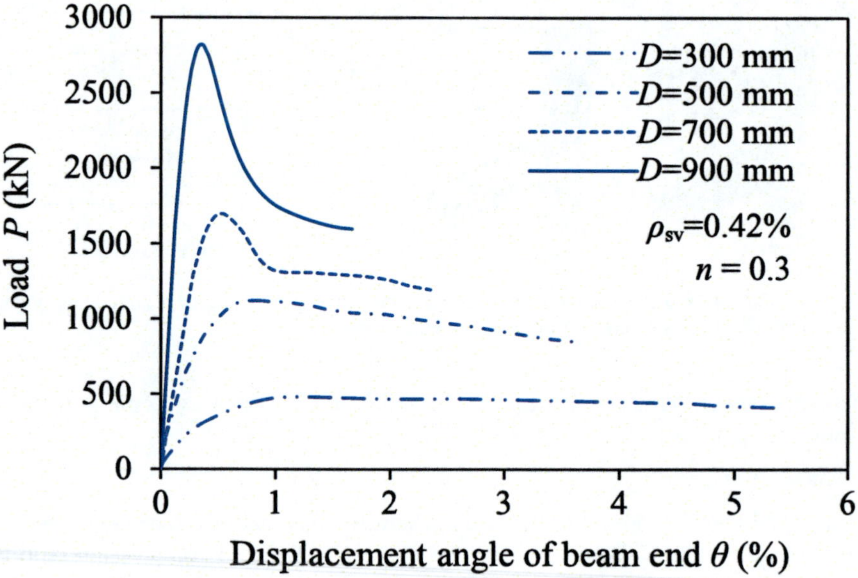

Fig. 9.31 Load-displacement curves of exterior joints with $\rho_{sv} = 0.42\%$ and $n = 0.3$

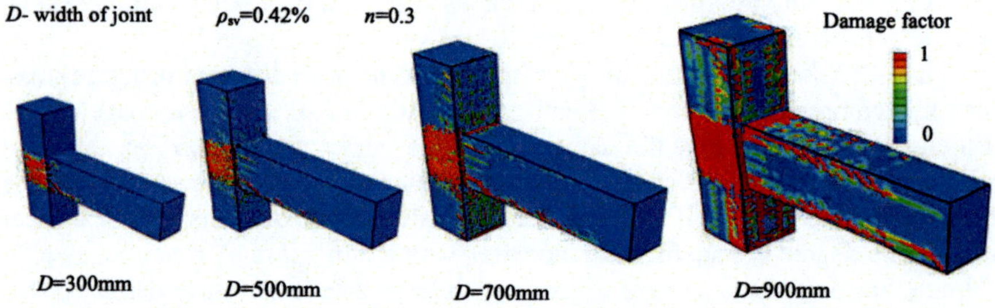

Fig. 9.32 Failure modes of the RC beam-column exterior joints having different structural sizes

of load-displacement angle curves that increasing the axial compression ratio will increase the peak load at the beam end, but for joints with width of 300 mm it has little effect on descending segment of load-displacement angle. For joints with cross-section widths of 500, 700 and 900 mm, increasing the axial compression ratio will slightly increase the peak load at the beam end, and the descent section of load-displacement angle curves becomes significantly steeper, that is the ductility of the joint decreases.

The stress mechanism of the exterior joint is different from that of the interior joint. Therefore, there is a certain difference in the influence of axial compression ratio on the failure modes of exterior and interior joints. Figure 9.35 shows the failure modes of exterior joints with different axial compression ratios (0, 0.3 and 0.6). The

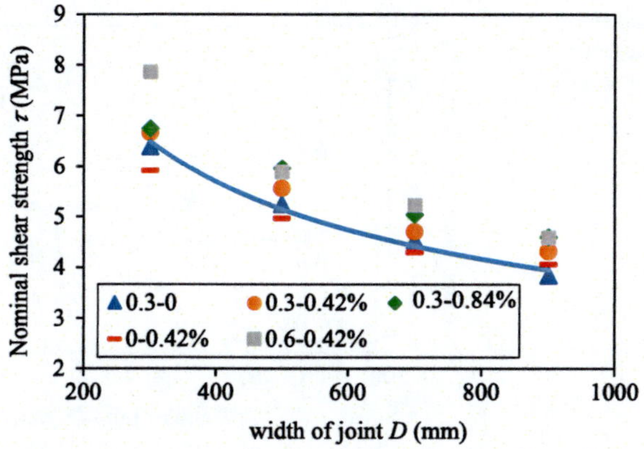

Fig. 9.33 Nominal shear strength of RC joints under different stirrup ratios and axial compression ratios

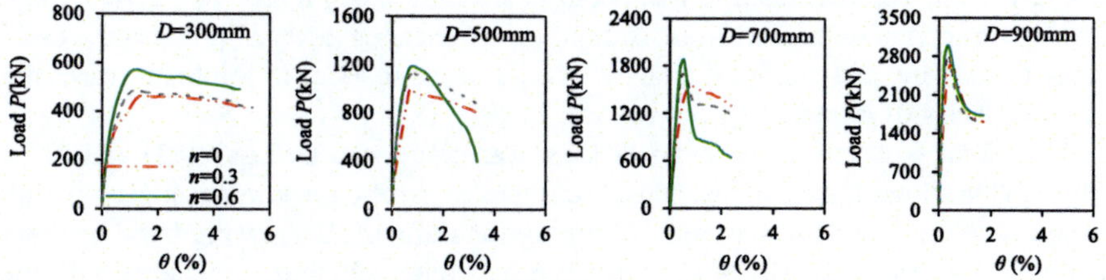

Fig. 9.34 The curves of load P-displacement angle θ

Fig. 9.35 Failure modes of the RC beam-column joints of different axial compression ratios

section size of joint is 300 mm × 300 mm, and stirrup ratio is 0.42%. The shear failure is the main failure of the joint with axial compression ratio of zero, and the bending failure of column end is the auxiliary failure mode. The failure mode of joint with axial compression ratio of 0.3 is shear failure in the core region. The failure mode of the joint with the axial compression ratio of 0.6 is the shear failure in the

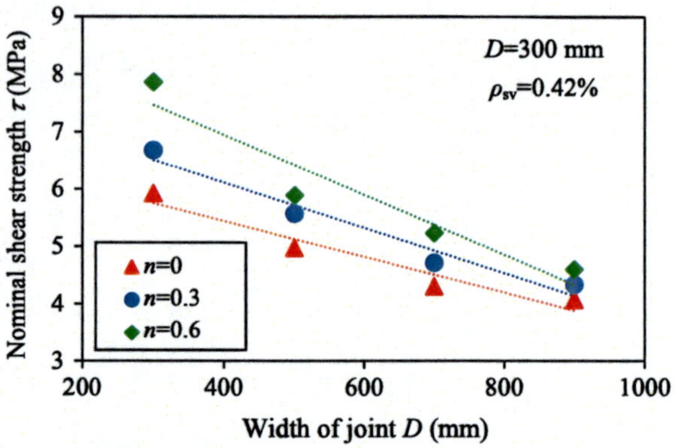

Fig. 9.36 Nominal shear strength of the joint with different axial compression ratios

core area. At the same time, bending failure occurred at the beam end and column end. With the increase of axial compression ratio, the damage degree of the core area is aggravated. The existence of axial compression ratio can inhibit the development speed of fractures in the core area, strengthen the shear capacity of joints at the early stage of loading, and a certain amount of axial force to exterior joints increases the shear capacity of joints.

The shear strength of the exterior joint was calculated by Eqs. (9.1) and (9.2). Figure 9.36 shows the influence of axial compression ratio on nominal shear strength of joints. With the increase of axial compression ratio, the decreasing trend of shear strength with the increase of size is more and more obvious, indicating that the increase of axial compression ratio will enhance the size effect of exterior joints. Although nominal shear strength of joints with larger axial compression ratio and same section size is improved, this positive effect can only play from the early stage of loading to the peak load stage. When the concrete in the core area cracks and the diagonal compression strut mechanism occur in joints, larger axial compression ratio will accelerate the crush of concrete, resulting in a rapid decline of ductility.

4. *Influence of stirrup ratio*

The stirrup ratio is one of the important factors affecting the shear capacity of joints. It can be seen from Fig. 9.37 that with the increase of stirrup ratio, the peak load at the beam end increases slightly, and the decline trend of load at beam end after the peak load is alleviated. The ductility is also improved.

The stirrup ratio also affects the failure mode of joints. Figure 9.38 shows the failure mode of the exterior joints with different stirrup ratios (0, 0.42 and 0.84%) and the yield phenomenon of steel bar. In the Fig. 9.38a, the section size of joints is 500 mm × 500 mm, and the axial compression ratio is 0.3. It can be seen from the failure mode that with the increase of the stirrup ratio, the damage degree of the core area of the joint decreases slightly, indicating that the increase of the stirrup ratio has a certain mitigation effect on the damage of the core area of the joint.

The yield condition of steel bar was shown in Fig. 9.38b. The increase of stirrup ratio has little effect on the stress of steel bar. The longitudinal bars of beam and

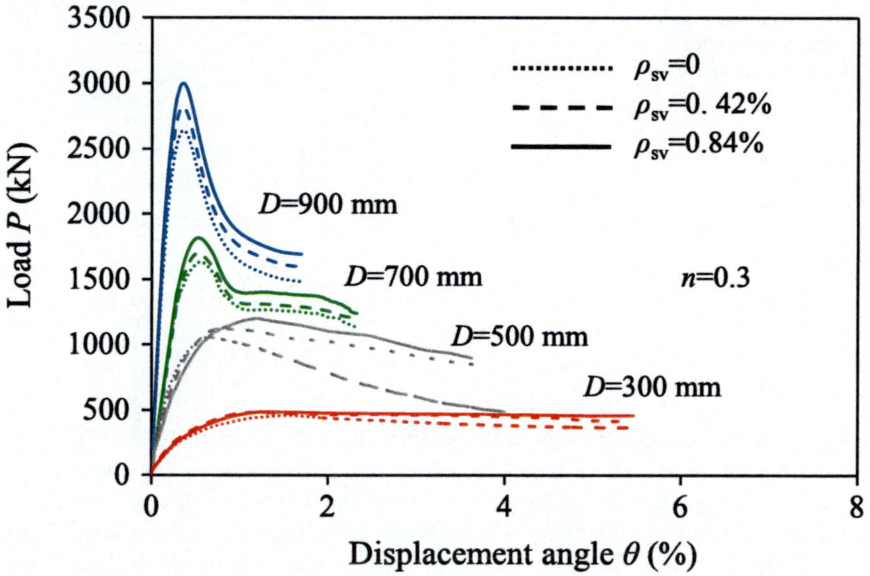

Fig. 9.37 The relationship curves of load and displacement angle

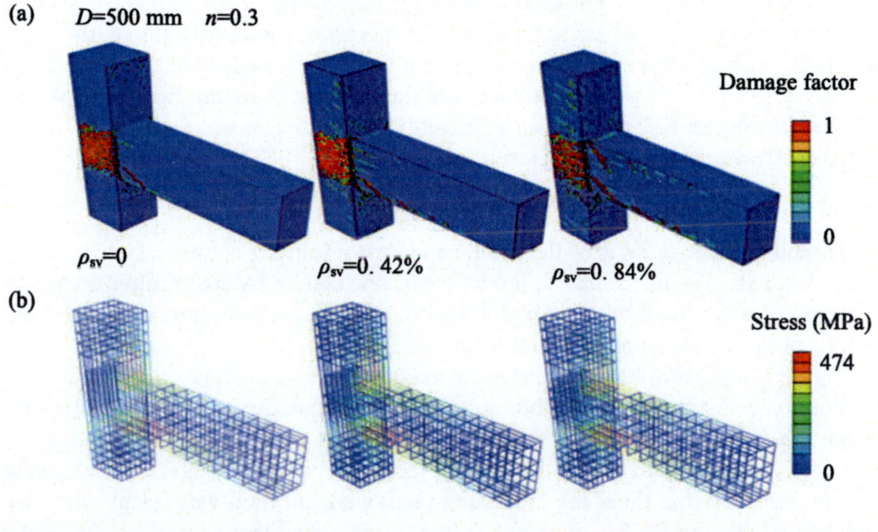

Fig. 9.38 Failure modes of the RC beam-column joints of different stirrup ratios: **a** failure mode; and **b** yield condition of steel bar

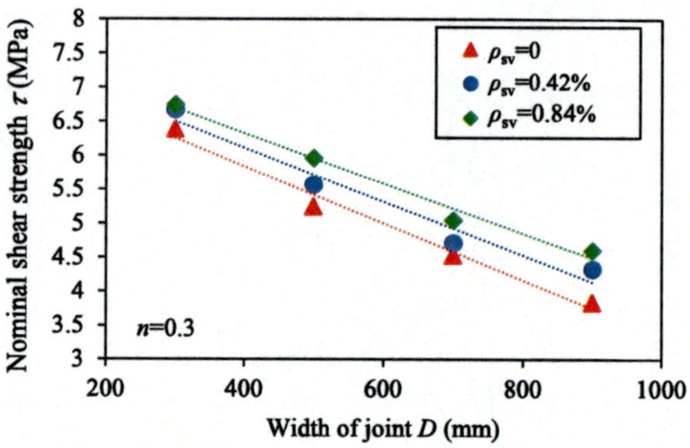

Fig. 9.39 Nominal shear strength of joint with different stirrup ratios

column and stirrups in core area undertake the force acting on the joint. The longitudinal bar at lower part of the beam yield first, and cracks can be found at the corresponding position in the lower part of the beam. The stirrups in core area of joints do not yield and their contribution to the shear capacity of the exterior joints is small. The failure of joints in different structural positions under the same stress condition has some differences. Compared with the interior joint, the enhancement of stirrup ratio has less effect on the shear capacity of the exterior joint.

Figure 9.39 shows the relationship between nominal shear strength and the width of joints under different stirrup ratios. The nominal shear strength of the joint increases with the increase of the stirrup ratio. As the width of the exterior joint increases, the shear strength decreases, and the increase of stirrup ratio is helpful to alleviate the decreasing trend of shear strength. This indicates that the stirrup inhibits the size effect behavior of the exterior joint.

5. *The analysis and discussion of size effect*

The nominal shear strength of the simulated exterior joints is compared with the size effect law (SEL) of the beam without stirrups, and the double logarithmic curves of nominal shear strength of exterior joints with different axial compression ratios and stirrup ratios are obtained, as can be seen in Fig. 9.40.

Figure 9.40a shows the influence of axial compression ratio on size effect of nominal shear strength of exterior joints. The simulated results are between the linear elastic theoretical curve (LEFM) and strength criterion, indicating that the shear failure of the core area of the exterior joint is between brittle failure and plastic failure. At the same time, the simulated results are in good agreement with size effect law (SEL) of RC beam without stirrups, indicating that size effect law (SEL) of RC beam without stirrups can well describe the size effect law of shear failure of exterior joints. The increase of axial compression ratio makes the data point close to the LEFM, which means that the shear failure of the exterior joint is more and more brittle, indicating that the axial compression ratio enhances the size effect behavior. This is consistent with the interior joints.

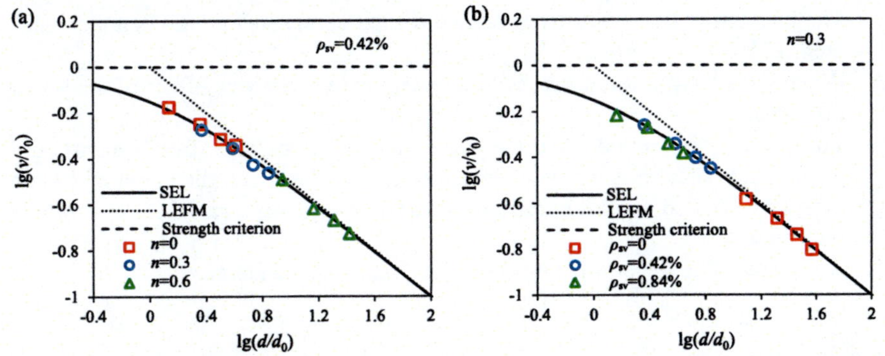

Fig. 9.40 Double logarithmic curve of nominal shear strength of joints: **a** different axial compression ratios; and **b** different stirrup ratios

Figure 9.40b shows the influence of stirrup ratio on size effect of nominal shear strength of exterior joints. The result of exterior joints with smaller stirrup ratio is closer to the LEFM, which means that the failure of joint is more brittle and the size effect is more significant. It can be seen from Fig. 9.40b that the stirrup ratio has slight influence on the distribution of data points, indicating that the stirrup ratio has a weaker extinction effect on the size effect of shear strength of exterior joints.

9.3 Size Effect Law for Shear Failure of Beam-to-Column Joints

This section mainly introduces a semi-empirical and semi-theoretical formula which can quantitatively reflect the influence of stirrup ratio and axial compression ratio on the size effect of shear strength of RC joints.

9.3.1 Basic Assumptions

Through the analysis of test (Jin et al. 2018) and numerical simulation results (Jin et al. 2020a, b), it can be seen that the stirrup ratio and axial compression ratio are two important factors affecting the size effect of shear strength of RC joints. Therefore, the formula introduced in this section mainly reflects the influence of stirrup ratio and axial compression ratio.

Since the factors affecting the size effect of shear strength of joints are relatively complex, the following assumptions should be made in order to facilitate the analysis.

1. Only consider the influence of the size effect of stirrup ratio and axial compression ratio on the shear strength of RC joints.

2. The interaction between stirrup ratio and axial compression ratio is not considered.
3. When the stirrup ratio increases to a certain extent, the size effect of the joints disappears.
4. The size only has an effect on the shear strength of concrete (the shear strength of the joints decreases when the size increases). Therefore, the effect of stirrup ratio and axial compression ratio on size effect of shear strength of concrete is only considered.
5. The effect of aggregate size on size effect was not considered.

9.3.2 Size Effect Law for Shear Strength

According to Chinese code (Ministry of Construction of the People's Republic of China 2011), the calculation formula of the shear capacity of joints is

$$v_j = \frac{1}{\gamma_{RE}}\left(0.9\eta_j f_t b_j h_j + 0.05\eta_j N\frac{b_j}{b_c} + f_{yv}A_{svj}\frac{h_{b0} - a_s'}{s}\right) \tag{9.4}$$

where γ_{RE} is the seismic adjustment coefficient of bearing capacity; η_j is the influence coefficient of orthogonal beams; f_t is the design value of axial tensile strength of concrete; b_j is the sectional width of the core area of joints; h_j is sectional height of core area of joints; N is the axial force on columns; b_c is the sectional width of columns; f_{yv} is the design value of tensile strength of stirrups; A_{svj} is the total cross section area of stirrups in the same cross section check direction within the effective checking width range of the core area of joints;h_{b0} is the effective sectional height of beams; a_s' is the distance from resultant force point on reinforcement of the beam to the compression edge of the beam; and s is the spacing of stirrups.

According to Eq. (9.4), the shear bearing capacity of joints (v) consists of three parts: (1) the shear bearing capacity v_c provided by concrete, (2) shear capacity v_N provided by axial force, (3) the shear bearing capacity v_s provided by stirrups in core area of joints. Therefore, the shear capacity of the joints can be expressed as

$$v = v_c + v_N + v_s \tag{9.5a}$$

$$\tau = v/A \tag{9.5b}$$

$$\tau = \tau_{Nu} + \tau_s + \tau_N \tag{9.5c}$$

in which, τ is shear strength of joints, A is sectional area of joints; τ_{Nu} is shear strength of concrete in joints; τ_s is the shear strength provided by stirrups; and τ_N is the shear strength provided by axial force.

According to the basic assumption above, the specimen is considered to be composed of steel reinforcement and concrete. Only the influence of stirrups and axial compressive force on size effect of shear strength of concrete are considered in the following section.

Based on the understanding of the influence of stirrup ratio on size effect and theoretical formula for size effect on shear strength of RC beam without stirrups, a size effect equation which can reflect the effect of stirrup ratio and axial compression ratio is

$$v_u = \frac{v_0}{\sqrt{1 + d/d_0}} \gamma \varphi \tag{9.6}$$

in which, γ is the influence coefficient of stirrup ratio on size effect of shear strength of RC joint; and φ is the influence coefficient of axial compression ratio on size effect of shear strength of RC joint.

1. *Determination of* γ

Based on previous understanding of the effect of stirrups on shear strength of joints, it is discussed on inhibiting effect of stirrups on size effect of shear strength of RC joints.

When $\rho_{sv} \leq \rho_{sv,min}$ ($\rho_{sv,min}$ is the minimum stirrup ratio of core area of joints that specified by Chinese code (Ministry of Construction of the People's Republic of China 2011)), the inhibiting effect of the stirrup on size effect of shear strength of joints is weak, the influence of the stirrup can be negligible, at this point $\gamma = 1$.

When $\rho_{sv} \to \rho_{sv,max}$ ($\rho_{sv,max}$ is the upper limit for the stirrup ratio), the restriction of the stirrup is strong, and greatly weakened the influence of size effect on shear strength of the joint. At this moment, the shear strength τ_{Nu} of concrete of joints tends towards shear strength of concrete τ_c. According to experimental research (Shisheng 1999), the relationship of compressive strength f_c and shear strength τ_c of concrete materials can be expressed as follows

$$\tau_c = 0.121 f_c \tag{9.7}$$

In combination with Eqs. (9.6) and (9.7), γ can be expressed as follows

$$\gamma = 0.121 f_c \sqrt{1 + d/d_0} / v_0 \tag{9.8}$$

When $\rho_{sv} \geq \rho_{sv,min}$, the function γ becomes a non-linear curve which tends continuously from 1.0 to $0.121 f_c \sqrt{1 + d/d_0} / v_0$, as shown in Fig. 9.41, and $A = 0.121 f_c \sqrt{1 + d/d_0} / v_0$.

The hyperbolic tangent function is adopted here to describe the relationship of size effect weakening coefficient γ and stirrup ratio.

Based on the previous analysis, the specific function of γ is obtained as follows

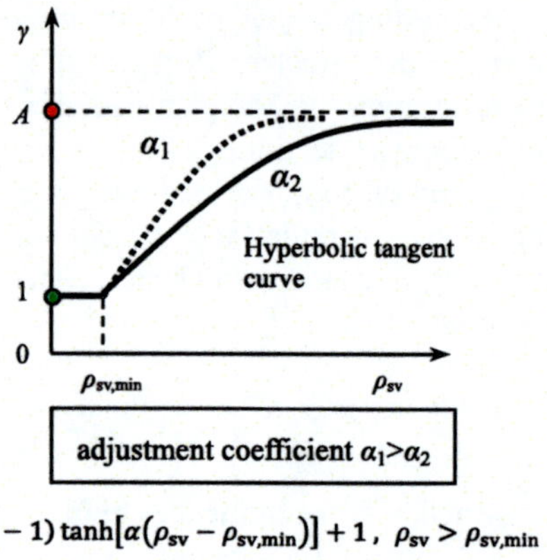

Fig. 9.41 Determination of the weakening coefficient γ

$$\gamma = \begin{cases} 1 & \rho_{sv} \leq \rho_{sv,min} \\ (A-1)tanh\left[\alpha\left(\rho_{sv} - \rho_{sv,min}\right)\right] + 1 & \rho_{sv} > \rho_{sv,min} \end{cases} \tag{9.9}$$

in which, α is an adjustment coefficient, reflecting how fast the weakening coefficient γ changes with stirrup ratio, which can be determined by fitting with simulated results.

The adjustment coefficient α, which can reflect how fast a function curve approaches its asymptote, is determined by contrasting and fitting with simulated results. According to Chinese code (Ministry of Construction of the People's Republic of China 2011), the minimal stirrup ratio in core area of RC joints is given as 0.4%. Combined with v_0 and d_0 obtained by regression analysis above, through data fitting and comparison, it was found from Fig. 9.42 that when adjustment coefficient α was between 50 and 60, the theoretical and simulation results are in good agreement. Here α is set as 55.

For the convenience of engineering application, Fig. 9.43 shows the basic determination method of the upper limit of stirrup ratio. It can be seen from Fig. 9.43 that when the stirrup ratio of joints reaches 4 to 6%, the value of γ starts to be stable and almost no longer increases with the increase of stirrup ratios, that is, the inhibition effect of stirrup on the size effect does not change, and it is suggested that the upper limit of stirrup ratio (that is, the critical stirrup rate) is 5%.

2. *The determination of* φ

Since only the influence of axial compression ratio on size effect of shear strength of concrete in the core area of RC joint is considered, the simulation results in Fig. 9.44 are all obtained by the joint with the stirrup ratio of zero. The shear strength τ_0 of joints with stirrup ratio of zero and axial compression ratio of zero is standard shear strength, and the changing law of shear strength of the concrete with the change of axial compression ratio is investigated. The influence of axial compression ratio on the shear strength of concrete of joints is defined as the influence coefficient of axial compression ratio on size effect.

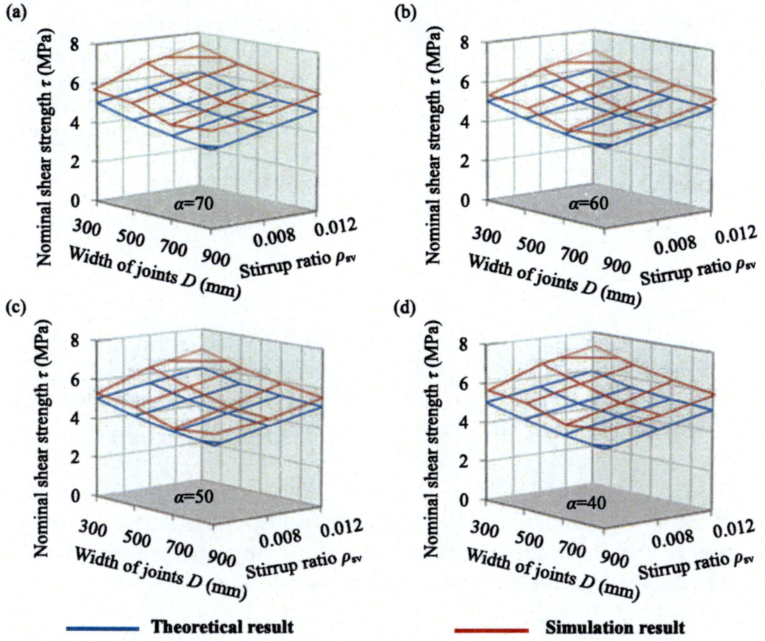

Fig. 9.42 Definition of adjustment coefficient α: **a** $\alpha = 70$; **b** $\alpha = 60$; **c** $\alpha = 50$; and **d** $\alpha = 40$

Fig. 9.43 Determination of upper limit value of stirrup ratio

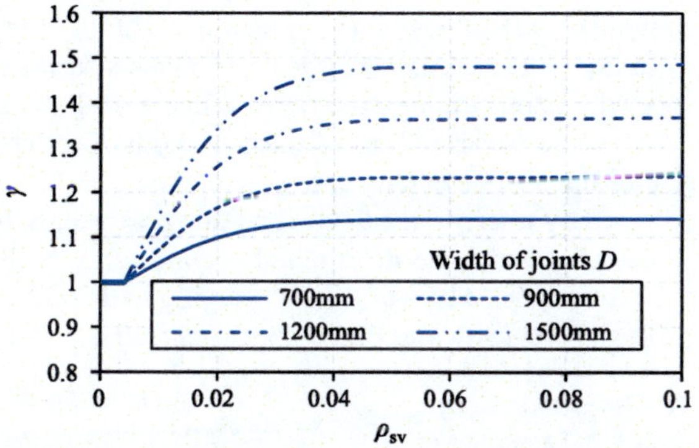

$$\varphi_{\text{sim}} = (\tau - \tau_0)/\tau_0 \tag{9.10}$$

in which, φ_{sim} is the effect coefficient of axial compression ratio on size effect obtained by simulated result; τ_0 is the shear strength of joints with stirrup ratio of zero and axial compression ratio of zero; and τ is the shear strength of concrete of joints.

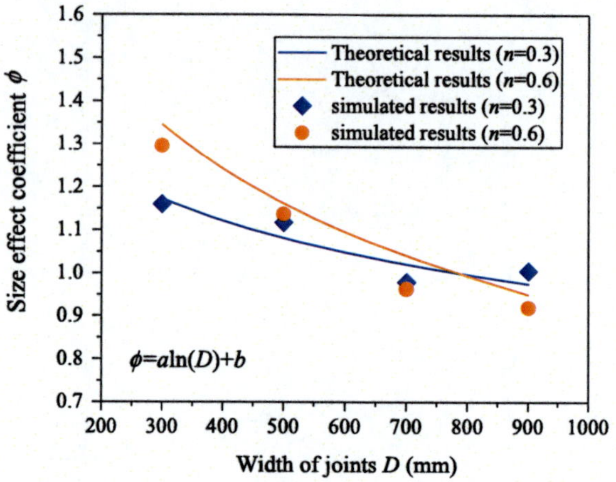

Fig. 9.44 Size effect coefficient of joints of different axial compression ratios

According to the simulation results shown in Fig. 9.44, it is assumed that the change rule of the influence coefficient of axial compression ratio of size effect conforms to the logarithmic curve, then φ has the form as

$$\varphi = a\ln(D) + b \tag{9.11}$$

in which, φ is theoretical coefficient of influence of axial compression ratio on size effect obtained by calculation; a and b are the coefficient affected by axial compression ratio; and D is the section width of joints.

In Eq. (9.11), a and b are the coefficients affected by axial compression ratio and vary with axial compression ratio. When $n = 0$, the axial compression ratio has no influence on the size effect of shear strength of concrete in joints, no matter how the size of the specimen changes, $\varphi = 1$, then $a = 0$, $b = 1$. When $n = 0.6$, it can be obtained by many trial calculations that $a = -0.36$, $b = 3.4$. When $0 < n < 0.6$, the values of a and b are determined according to the linear interpolation method. In this section, the values of a and b varying with axial compression ratio are shown in Table 9.9. As shown in Fig. 9.45, the correlation between the theoretical value and

Table 9.9 Parameters of formula for size effect coefficient of axial compression ratio of joints

Axial compression ratio n	a	b
0	0	1
0.15	−0.09	1.6
0.3	−0.18	2.2
0.45	−0.27	2.8
0.6($n > 0.6$)	−0.36	3.4

Note When $0 \leq n \leq 0.6$, a and b are determined by linear interpolation determines

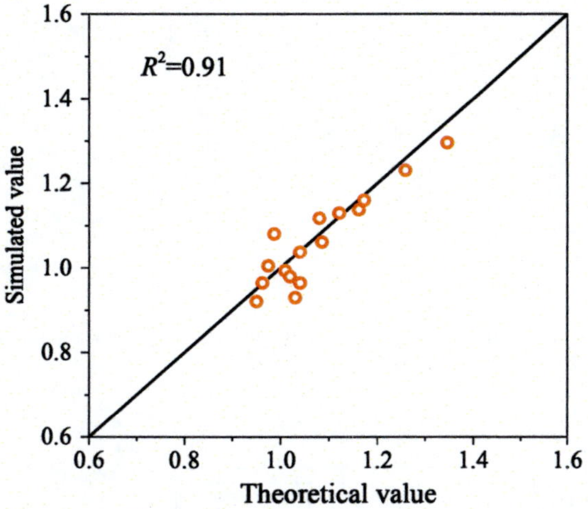

Fig. 9.45 The comparison of correlation between the theoretical value and the simulated value of the size effect coefficient

the simulated value is analyzed, and the correlation coefficient $R^2 = 0.91$. That is to say, the theoretical value and the simulated value have a high correlation. In other words, Eq. (9.11) can better describe the variation rule of the size effect coefficient of axial compression ratio varying with axial compression ratio.

Here, RC joints with axial compression ratio of zero and stirrup ratio of zero are used as reference specimens to determine v_0 and d_0, and the specific circumstance of reference specimens is shown in Table 9.9. Regression analysis was performed on the simulated value of shear strength of reference specimens, then v_0 and d_0 were obtained ($v_0 = 5.83$, $d_0 = 490$). When the strength grade of concrete is fixed, the parameters of v_0 and d_0 are fixed to a certain magnitude which will not change with axial compression ratio and stirrup ratio.

9.3.3 Validation of the Theoretical Formula

Based on the above analysis, the size effect equation of shear strength of RC joint, which can reflect the influence of stirrup ratio and axial compression ratio, is obtained. In order to verify the rationality and accuracy of Eq. (9.6), the theoretical value of shear strength of RC joints calculated by Eq. (9.6) is compared with the simulated value of shear strength, as shown in Fig. 9.46. It can be found that the theoretical value is in good agreement with the simulated value. For further verification of the accuracy and rationality of Eq. (9.6), the theoretical value was compared with the experimental value of Jin et al. (2018), and the comparison results were shown in Fig. 9.47a. The experimental value of shear strength in Fig. 9.47a is obtained by the shear strength value of RC joints with stirrup ratio of 0.84% and axial compression

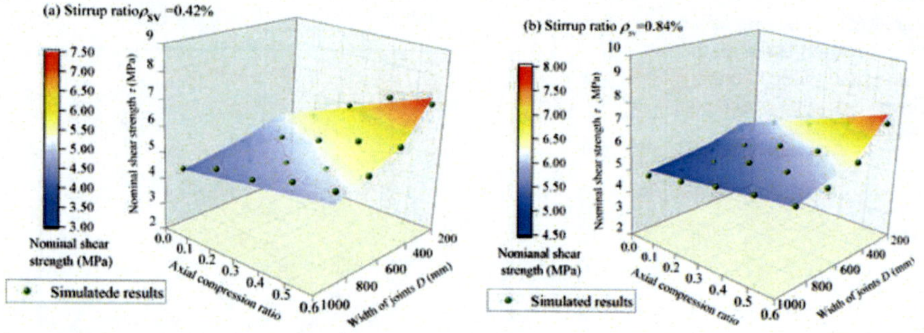

Fig. 9.46 Comparison between the theoretical values and simulated results: **a** stirrup ratio ρ_{sv} is 0.42%; and **b** stirrup ratio ρ_{sv} is 0.84%

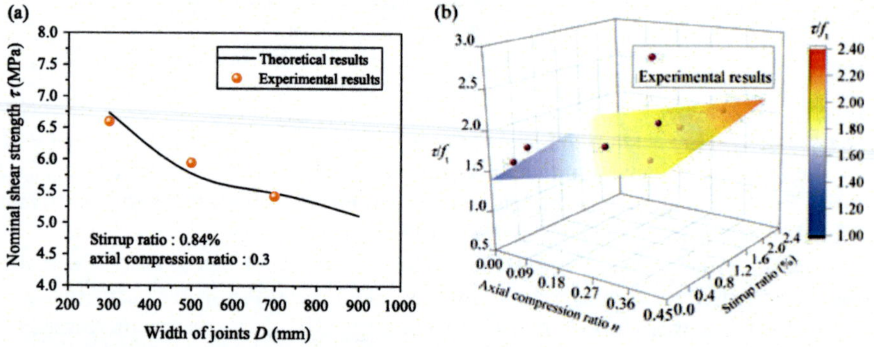

Fig. 9.47 Comparison between the theoretical values and test results: **a** Different section sizes of joints (Jin et al. 2018); and **b** width of joints $D = 400$ mm (Gao et al. 2014; Wu 2007; You et al. 1997)

ratio of 0.3. It can be seen from Fig. 9.47a that there is little difference between shear strength of RC joints calculated by Eq. (9.6) and that obtained by experiment.

In order to verify the universality of the formula, the theoretical results are compared with other experimental results (Gao et al. 2014; Wu 2008; You et al. 1997), and the comparison results are shown in Fig. 9.47b. In Fig. 9.47b, f_t is the design value of axis tensile strength of concrete. The test results shown in Fig. 9.47b are all obtained from RC joints with cross section width of 400 mm. Among them, the stirrup ratios of joints (Gao et al. 2014) are 1.03 and 1.202%, and the axial compression ratio is 0.28. In the paper (Wu 2008), the stirrup ratios of joints are 1.48 and 1.97%, and the axial compression ratios are 0.3 and 0.35. In the paper (You et al. 1997), the stirrup ratios of joints are 0.25 and 0.49%, and the axial compression ratios are 0.024 and 0.24. As can be seen from Fig. 9.47b, the theoretical results are in good agreement with the experimental results.

Through the previous analysis and discussion, the rationality and accuracy of the Eq. (9.6) are verified. However, due to the limited quantity of experiments, there are few data available to verify the formula. Therefore, more experiments are needed to further verify the accuracy and rationality of the formula.

9.4 Summary

In this chapter, the shear failure behavior of RC beam-to-column joints with different stirrup ratios and axial compression ratios is presented by means of experiment and mesoscopic numerical simulation. The semi-empirical and semi-theoretical formula is introduced, and it can quantitatively reflect the influence of stirrup ratio and axial compression ratio on size effect of shear strength of joints. In addition, the main conclusions are given as follows:

1. The existence of stirrups in core area of joints increases the shear strength of RC joints, and limits the development of cracks in the core area of joints. At the same time, stirrups would improve the ductility and weaken the size effect on shear strength of RC joints.
2. The increase of axial compression ratio improves the shear strength of joints, limits the development of fractures in core area of RC joints, and increases the bite force between the aggregates. However, the increase of axial compression ratio reduces the ductility of joints, thus enhancing the size effect on shear strength of RC joints.
3. The semi-empirical and semi-theoretical formula can well reflect the influence of the stirrup ratio and axial compression ratio on the size effect of shear strength of RC joint.

It is worth noting that the semi-empirical and semi-theoretical formula introduced in this chapter still need to be verified by a large number of experimental results. Moreover, this formula only considers the effect of the stirrup ratio and axial compression ratio, and concrete strength, stirrup spacing, stirrup strength and other factors still have an impact on size effect of shear strength of RC joints.

References

ACI (American Concrete Institute) (2005) Building code requirements for structural concrete. ACI 318-05, Farmington Hills, Mich: ACI

AIJ (Architectural Institute of Japan) (1998) Calculation standard and explanation of steel reinforced concrete structure. Feng Naiqian, et al. Trans. Beijing: Atomic Energy Press (in Chinese)

Angelakos D, Bentz EC, Collions MP (2001) Effect of concrete strength and minimum stirrups on shear strength of large members. ACI Struct J 98(3):290–300

Barbhuiya S, Choudhury AM (2015) A study on the size effect of RC beam-column connections under cyclic loading. Eng Struct 95:1–7

Bažant ZP (2005) Scaling of structural strength (second edition). Scaling of structural strength, pp 269–312

Bažant ZP, Kim JK (1984) Size effect in shear failure of longitudinally reinforced beams. ACI J 81(5):456–468
Bažant ZP, Xiang Y (1997) Size effect in compression fracture: splitting crack band propagation. ASCE J Struct Eng 123(2):167–172
Chen SJ, Yeh CH, Chu JM (1996) Ductile steel beam-to-column connections for seismic resistance. J Struct Eng 122(11):1292–1299
Fu J, Chen T, Wang Z, Bai S (2000) Effect of axial load ratio on seismic behavior of interior beam-column joints. In: Proceedings of 12th world conference on earthquake eng
Gao F, Huang S, Liu B, Zhou D (2014) Experimental research in influence of shear-compression ratio on the interior beam-column joint with HRB500E reinforcements. J Civil Eng Manage 4:1–6 (in Chinese)
Haach VG, El ALHDC, El Debs MK (2014) Influence of high column axial loads in exterior R/C beam-column joints. KSCE J Civil Eng 18(2):558–565
Hayashi S, Morimoto T, Kokusho S (1983) The bond behavior of reinforcing bars through the interior beam column joints. Trans JCI 5(11):325
Jin L, Du XL, Li D, Su X (2016) Seismic behavior of RC cantilever beams under low cycle loading and size effect on shear strength: an experimental characterization. Eng Struct 122:93–107
Jin L, Miao LY, Han J, Du X, Wei N, Li D (2018) Size effect tests on shear failure of interior RC beam-to-column joints under monotonic and cyclic loadings. Eng Struct 175:591–604
Jin L, Miao L Y, Du XL, (2020a) Size effect on shear failure of interior RC beam-to-column joints with different stirrup ratios: formulation and validation. Eng Struct (under review)
Jin L, Miao L Y, Du XL, Li D (2020b) Size effect on shear failure of interior RC beam-to-column joints with different axial compression ratios: formulation and validation (under review)
Lubell A, Sherwood T, Bentz EC, Collins MP (2004) Safe shear design of large, wide beams. Concr Int 26(1):67–78
Masi A, Santarsiero G, Lignola GP, Verderame GM (2013) Study of the seismic behavior of external RC beam-column joints through experimental tests and numerical simulations. Eng Struct 52:207–219
Ministry of Construction of the People's Republic of China (2011) Code for design of concrete structures: GB 50010-2010. China Architecture & Building Press, Beijing (in Chinese)
Parker DE, Bullman PJM (1997) Shear strength within reinforced concrete beam-column joints. Struct Eng 75(4):53–57
Scott RH, Hamil SJ, Baglin PS (1999) Behavior of high-strength concrete beam-column connections. ACI Spec Publicat 186:699–714
Shisheng S (1999) Shear strength, modulus of rigidity and young's modulus of concrete. China Civil Eng J 32(2):47–52 (in Chinese)
Standard Association of New Zealand (2006) New Zealand standard code of practice for the design of concrete structures. Standards New Zealand (SNZ), Wellington
Tsonos AG (1999) Lateral load response of strengthened reinforced concrete beam-to-column joints. ACI Struct J 96:46–56
Wakabayashi M, Minami K, Nishimura Y, Imanaka N (1983) Anchorage of bent bar in reinforced concrete exterior joints. Trans Japan Concrete Instit 5(11):317–324
Walravena J, Lehwalter N (1994) Size effect in short beams loaded in shear. ACI Struct J 91(5):585–593
Wang Y (2010) Experimental study on the behaviors of exterior joints of RC frame under low cyclic reversed loading. Doctoral thesis, Harbin Institute of Technology, Harbin (in Chinese)
Wu X (2008) Experimental and design methodological research for the seismic behavior of interior joint with 500 MPa longitudinal bar. Doctoral thesis, Chongqing University, Chongqing (in Chinese)
You Y, Fu J, Bai S, Tang H (1997) Re-recognition of shear capacities in joints of RC frames. J Chongqing Jianzhu University 19(1):14–25 (in Chinese)
Yu Q, Bažant ZP (2011) Can stirrups suppress size effect on shear strength of RC beams? J Struct Eng 137(5):607–617